37 Springer Series in Solid-State Sciences

Edited by Hans-Joachim Queisser

Springer Series in Solid-State Sciences

Editors: M. Cardona P. Fulde K. von Klitzing H.-J. Queisser

40 **Semiconductor Physics** An Introduction
3rd Edition By K. Seeger

41 **The LMTO Method**
Muffin-Tin Orbitals and Electronic Structure
By H. L. Skriver

42 **Crystal Optics with Spatial Dispersion, and Excitons**
By V. M. Agranovich and V. L. Ginzburg

43 **Resonant Nonlinear Interactions of Light with Matter**
By V. S. Butylkin, A. E. Kaplan, Yu. G. Khronopulo, and E. I. Yakubovich

44 **Elastic Media with Microstructure II**
Three-Dimensional Models
By I. A. Kunin

45 **Electronic Properties of Doped Semiconductors**
By B. I. Shklovskii and A. L. Efros

46 **Topological Disorder in Condensed Matter**
Editors: F. Yonezawa and T. Ninomiya

47 **Statics and Dynamics of Nonlinear Systems**
Editors: G. Benedek, H. Bilz, and R. Zeyher

48 **Magnetic Phase Transitions**
Editors: M. Ausloos and R. J. Elliott

49 **Organic Molecular Aggregates,** Electronic Excitation and Interaction Processes
Editors: P. Reineker, H. Haken, and H. C. Wolf

50 **Multiple Diffraction of X-Rays in Crystals**
By Shih-Lin Chang

51 **Phonon Scattering in Condensed Matter**
Editors: W. Eisenmenger, K. Laßmann, and S. Döttinger

52 **Superconductivity in Magnetic and Exotic Materials**
Editors: T. Matsubara and A. Kotani

53 **Two-Dimensional Systems, Heterostructures, and Superlattices**
Editors: G. Bauer, F. Kuchar, and H. Heinrich

54 **Magnetic Excitations and Fluctuations**
Editors: S. Lovesey, U. Balucani, F. Borsa, and V. Tognetti

55 **The Theory of Magnetism II**
Thermodynamics and Statistical Mechanics
By D. C. Mattis

56 **Spin Fluctuations in Itinerant Electron Magnetism** By T. Moriya

57 **Polycrystalline Semiconductors,**
Physical Properties and Applications
Editor: G. Harbeke

58 **The Recursion Method and Its Applications**
Editors: D. Pettifor and D. L. Weaire

59 **Dynamical Processes and Ordering on Solid Surfaces**
Editors: A. Yoshimori and M. Tsukada

60 **Excitonic Processes in Solids**
By M. Ueta, H. Kanzaki, K. Kobayashi, Y. Toyozawa, and E. Hanamura

61 **Localization, Interaction, and Transport Phenomena**
Editors: B. Kramer, G. Bergmann, and Y. Bruynseraede

62 **Theory of Heavy Fermions and Valence Fluctuations**
Editors: T. Kasuya and T. Saso

63 **Electronic Properties of Polymers and Related Compounds**
Editors: H. Kuzmany, M. Mehring, and S. Roth

64 **Symmetries in Physics:** Group Theory Applied to Physical Problems
By W. Ludwig and C. Falter

65 **Phonons: Theory and Experiments II**
Experiments and Interpretation of Experimental Results By P. Brüesch

66 **Phonons: Theory and Experiments III**
Phenomena Related to Phonons
By P. Brüesch

67 **Two-Dimensional Systems: Physics and New Devices**
Editors: G. Bauer, F. Kuchar, and H. Heinrich

68 **Phonon Scattering in Condensed Matter V**
Editors: A. C. Anderson and J. P. Wolfe

69 **Nonlinearity in Condensed Matter**
Editors: A. R. Bishop, D. K. Campbell, P. Kumar and S. E. Trullinger

70 **From Hamiltonians to Phase Diagrams**
The Electronic and Statistical-Mechanical Theory of sp-Bonded Metals and Alloys
By J. Hafner

71 **High Magnetic Fields in Semiconductor Physics**
Editor: G. Landwehr

72 **One-Dimensional Conductors**
By S. Kagoshima, T. Sambongi, and H. Nagasawa

73 **Quantum Solid-State Physics**
Editors: S. V. Vonsovsky and M. I. Katsnelson

74 **Quantum Monte Carlo Methods** in Equilibrium and Nonequilibrium Systems
Editor: M. Suzuki

Volumes 1–39 are listed on the back inside cover

Modern Crystallography IV

Physical Properties of Crystals

Editor: L. A. Shuvalov

With Contributions by
L. A. Shuvalov, A. A. Urusovskaya, I. S. Zheludev,
A. V. Zalessky, S. A. Semiletov, B. N. Grechushnikov,
I. G. Chistyakov and S. A. Pikin

With 270 Figures and 65 Tables

Springer-Verlag Berlin Heidelberg New York
London Paris Tokyo

Professor Dr. *L. A. Shuvalov*
Professor Dr. *I. S. Zheludev*
Professor Dr. *S. A. Semiletov*
Dr. *I. G. Chistyakov*
Dr. *A. A. Urusovskaya*
Dr. *A. V. Zalessky*
Professor Dr. *B. N. Grechushnikov*
Dr. *S. A. Pikin*

Institute of Crystallography, Academy of Sciences of the USSR, 59 Leninsky prospect, SU-117333 Moscow, USSR

Series Editors:

Professor Dr. Manuel Cardona
Professor Dr. Peter Fulde
Professor Dr. Klaus von Klitzing
Professor Dr. Hans-Joachim Queisser

Max-Planck-Institut für Festkörperforschung, Heisenbergstrasse 1
D-7000 Stuttgart 80, Fed. Rep. of Germany

Title of the original Russian edition:
Sovremennaja kristallografija

Library of Congress Cataloging-in-Publication Data. (Revised for volume 4)
Modern crystallography. (Springer series in solid-state sciences; 15, 21, 36–37)
Translation of Sovremennaja kristallografija. Includes bibliographies and indexes.
Contents: 1. Vaĭnshteĭn, B. K. Symmetry of crystals. Methods of structural crystallography. – 2. Vaĭnshteĭn, B. K., Fridkin, V. M., Indenbom, V. L., Structure of crystals. – [etc.] – 4. Physical properties of crystals / L. A. Shuvalov.
1. Crystallography. I. Vaĭnshteĭn, B. K. (Boris Konstantinovich), 1921– . II. Series: Springer series in solid-state sciences; 15, etc.
QD905.2.S6813 548 80-17797
ISBN-13: 978-3-642-81840-0 e-ISBN-13: 978-3-642-81838-7
DOI: 10.1007/ 978-3-642-81838-7

Softcover reprint of the hardcover 1st edition 1988

Typesetting: K+V Fotosatz GmbH, 6124/Beerfelden
2154/3020-543210

Modern Crystallography

in Four Volumes *

I Symmetry of Crystals. Methods of Structural Crystallography

II Structure of Crystals

III Crystal Growth

IV Physical Properties of Crystals

Foreword

Crystallography – the science of crystals – has undergone many changes in the course of its development. Although crystals have intrigued mankind since ancient times, crystallography as an independent branch of science began to take shape only in the 17th – 18th centuries, when the principal laws governing crystal habits were found, and the birefringence of light in crystals was discovered. From its very origin crystallography was intimately connected with mineralogy, whose most perfect objects of investigation were crystals. Later, crystallography became associated more closely with chemistry, because it was apparent that the habit depends directly on the composition of crystals and can only be explained on the basis of atomic-molecular concepts. In the 20th century crystallography also became more oriented towards physics, which found an ever-increasing number of new optical, electrical, and mechanical phenomena inherent in crystals. Mathematical methods began to be used in crystallography, particularly the theory of symmetry (which achieved its classical completion in space-group theory at the end of the 19th century) and the calculus of tensors (for crystal physics).

Early in this century, the newly discovered x-ray diffraction by crystals made a complete change in crystallography and in the whole science of the

---

* Published in *Springer Series in Solid-State Sciences*, I: Vol. 15; II: Vol. 21; III: Vol. 36; IV: Vol. 37

atomic structure of matter, thus giving a new impetus to the development of solid-state physics. Crystallographic methods, primarily x-ray diffraction analysis, penetrated into materials sciences, molecular physics, and chemistry, and also into many other branches of science. Later, electron and neutron diffraction structure analyses became important since they not only complement x-ray data, but also supply new information on the atomic and the real structure of crystals. Electron microscopy and other modern methods of investigating matter – optical, electronic paramagnetic, nuclear magnetic, and other resonance techniques – yield a large amount of information on the atomic, electronic, and real crystal structures.

Crystal physics has also undergone vigorous development. Many remarkable phenomena have been discovered in crystals and then found various practical applications.

Other important factors promoting the development of crystallography were the elaboration of the theory of crystal growth (which brought crystallography closer to thermodynamics and physical chemistry) and the development of the various methods of growing synthetic crystals dictated by practical needs. Man-made crystals became increasingly important for physical investigations, and they rapidly invaded technology. The production of synthetic crystals made a tremendous impact on the traditional branches: the mechanical treatment of materials, precision instrument making, and the jewelry industry. Later it considerably influenced the development of such vital branches of science and industry as radio engineering and electronics, semiconductor and quantum electronics, optics, including nonlinear optics, acoustics, etc. The search for crystals with valuable physical properties, study of their structure, and development of new techniques for their synthesis constitute one of the basic lines of contemporary science and are important factors of progress in technology.

The investigation of the structure, growth, and properties of crystals should be regarded as a single problem. These three intimately connected aspects of modern crystallography complement each other. The study, not only of the ideal atomic structure, but also of the real defect structure of crystals makes it possible to conduct a purposeful search for new crystals with valuable properties and to improve the technology of their synthesis by using various techniques for controlling their composition and real structure. The theory of real crystals and the physics of crystals are based on their atomic structure as well as on the theoretical and experimental investigations of elementary and macroscopic processes of crystal growth. This approach to the problem of the structure, growth, and properties of crystals has an enormous number of aspects, and determines the features of modern crystallography.

The branches of crystallography and their relation to adjacent fields can be represented as a diagram showing a system of interpenetrating branches which have no strict boundaries. The arrows show the relationship between the branches, indicating which branch influences the activity of the other, although, in fact, they are usually interdependent.

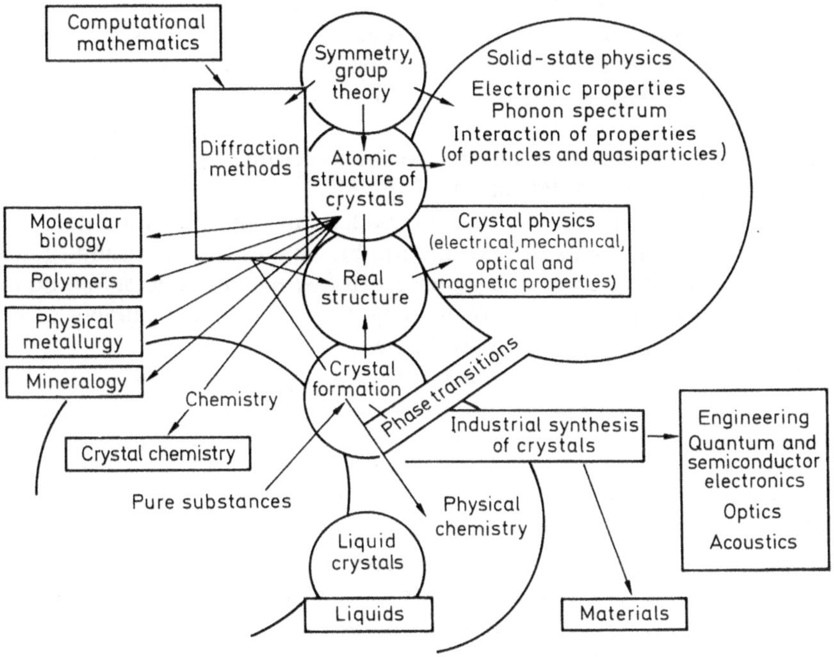

Branches of crystallography and its relation to other sciences

Crystallography proper occupies the central part of the diagram. It includes the theory of symmetry, the investigation of the structure of crystals (together with diffraction methods and crystal chemistry), and the study of the real structure of crystals, their growth and synthesis, and crystal physics.

The theoretical basis of crystallography is the theory of symmetry, which has been intensively developed in recent years.

The study of the atomic structure has been extended to extremely complicated crystals containing hundreds and thousands of atoms in the unit cell. The investigation of the real structure of crystals with various disturbances of the ideal crystal lattices has been gaining in importance. At the same time, the general approach to the atomic structure of matter and the similarity of the various diffraction techniques make crystallography a science not only of the structure of crystals themselves, but also of the condensed state in general.

The specific applications of crystallographic theories and methods allow the utilization of structural crystallography in physical metallurgy, materials science, mineralogy, organic chemistry, polymer chemistry, molecular biology, and the investigation of amorphous solids, liquids, and gases. Experimental and theoretical investigations of crystal growth and nucleation processes and their development draw on advances in chemistry and physical chemistry and, in turn, contribute to these areas of science.

Crystal physics deals mainly with the electrical, optical, and mechanical properties of crystals closely related to their stucture and symmetry, and adjoins solid-state physics, which concentrates its attention on the analysis of laws defining the general physical properties of crystals and the energy spectra of crystal lattices.

The first two volumes are devoted to the structure of crystals, and the last two, to the growth of crystals and their physical properties. The authors present the material in such a way that the reader can find the basic information on all important problems of crystallography. Due to the limitation of space the exposition of some sections is concise, otherwise many chapters would have become separate monographs. Fortunately, such books on a number of crystallographic subjects are already available.

The purpose of such an approach is to describe all the branches of crystallography in their interrelation, thus presenting crystallography as a unified science to elucidate the physical meaning of the unity and variety of crystal structures. The physico-chemical processes and the phenomena taking place in the course of crystal growth and in the crystals themselves are described, from a crystallographic point of view, and the relationship of properties of crystals with their structure and conditions of growth is elucidated.

This four-volume edition is intended for researchers working in the fields of crystallography, physics, chemistry, and mineralogy, for scientists studying the structure, properties, and formation of various materials, for engineers and those engaged in materials science technology, particularly in the synthesis of crystals and their use in various technical devices. We hope that this work will also be useful for undergraduate and graduate students at universities and higher technical colleges studying crystallography, solid-state physics, and related subjects.

Modern Crystallography is written by a large group of authors from the Institute of Crystallography of the USSR Academy of Sciences, who benefited from the assistance and advice of many other colleagues. The English edition of all four volumes of *Modern Crystallography* is being published almost simultaneously with the Russian edition. The authors have included in the English edition some of the most recent data. In several instances some additions and improvements have been made.

B. K. Vainshtein

Preface

This fourth volume of *Modern Crystallography* deals with the physical properties of crystals. It completes the presentation of the science of crystallography, the other components of which – the theory of symmetry of crystals, their structure, and crystallization – are elucidated in the preceding three volumes. The present volume contains a systematic exposé of the modern conceptions of the mechanical, electrical, magnetic, and optical properties of crystals, and of transport phenomena in them.

In distinction to textbooks and monographs on solid-state physics, the authors attempted to render the material in the crystallographic key. In particular, they took into account the symmetry (including the space symmetry) in analyzing the properties of crystals and the associated anisotropy of the crystalline substance, and also paid considerable attention to the interrelation of the properties with the atomic and real structure of particular crystals and with the conditions of their growth. The basic notions of the electron and phonon spectra of crystals are covered by the second volume; therefore, the present book treats the relevant aspects within a relatively narrow scope.

The introductory chapter describes the tensor apparatus of crystal physics and considers the general problems of symmetry of the physical properties of crystals. The second chapter is devoted to the mechanical properties of crystals. Along with the usual description of the elastic properties of crystals, it provides detailed data on their plastic deformation, mechanical twinning, and failure.

The electrical and electromechanical properties of crystals are considered in the third chapter. Special attention is given to ferroelectrics – the most important class of dielectric crystals – and to their properties. The fourth chapter studies the magnetic properties of crystals. There, magnetically ordered crystals are investigated comprehensively, mainly from the crystallographical viewpoint. The physical properties of semiconductor crystals, as well as the operating principles of semiconductor diodes, transistors, and lasers, are described in the fifth chapter.

The sixth chapter is dedicated to transport phenomena; in addition to electrical and thermal conductivity, it discusses various thermoelectrical, galvano and thermomagnetic properties of crystals. The optical properties of crystals are described in the seventh chapter. Together with a nonconventional presentation of the traditional and new branches of crystal optics, it elucidates the problems of optical and radiospectroscopy and the generation of coherent radiation. The last chapter deals with the properties of liquid crystals, a rapidly developing branch of crystal physics.

Despite the considerable effort taken in editing, the great number of contributors to this volume proved an obstacle in avoiding some divergency in style, depth of presentation, and even symbols in the different chapters. We hope, nevertheless, that the book will provide a sufficiently comprehensive picture of the present-day state of the art in crystal physics and, taken together with the other volumes, of the entire crystallographical science.

In the first three volumes the general bibliography (monographs, surveys, and the most important works) is given for the volume as a whole, while the special literature (original papers) is referred to the separate chapters. In view of the numerous general sources concerning the properties of crystals cited in the present volume, we refer to them also in the respective chapters.

The book has been written by a group of research workers at the Institute of Crystallography of the U.S.S.R. Academy of Sciences: L.A. Shuvalov, A.A. Urusovskaya, I.S. Zheludev, A.V. Zalessky, S.A. Semiletov, B.N. Grechushnikov, I.G. Chistyakov, S.A. Pikin. The authors deeply appreciate the kind help of V.A. Koptsik, I.M. Silvestrova, M.V. Klassen-Nekhlyudova, V.L. Indenbom, D.G. Sannikov, A.P. Levanyuk, A.F. Konstantinova, T.F. Veremeichik, I.N. Kalinkina, O.V. Kachalov, L. Li, and others, who took part in the writing and discussion of various chapters.

Moscow, September 1987 *L.A. Shuvalov*

Contents

1. Fundamentals of Tensor and Symmetry Description of the Physical Properties of Crystals

1.1 Introduction

1.1.1 Crystal as a Continuous Homogeneous Anisotropic Medium

As noted in [Ref. 1.1, Chap. 1], in considering the macroscopic physical properties of crystals, one can abstract oneself from their discrete microperiodic structure. The crystal is then treated as a continuous homogeneous anisotropic medium.

Indeed, when investigating the macroscopic physical properties of crystals, we deal with distances considerably larger than the largest spacing of the crystal lattice and with volumes greatly exceeding the volume of the unit cell. Therefore, a crystal can be regarded as a *continuous* medium.

It can further be assumed that the properties of a crystal are identical at all its points. In other words, the unit volume to be tested can be taken from any place in the crystal (except in special cases when we are interested in the specific properties of the surface and the adjoining layers, or when they should be taken into account). Hence, a crystal can be considered not only a continuous, but also a *homogeneous* medium. Here we abstract ourselves not only from the discrete structure of the crystal, but also from the fact that a real crystal contains various impurities and imperfections with different volume distribution (sectorial, zonal, etc.). Therefore, it should be borne in mind that a crystal can be regarded as a continuous homogeneous medium only to a certain degree of accuracy, which is different, generally speaking, for different properties, at different levels of consideration, and in different particular cases.

Finally, at least some physical properties of a crystal are anisotropic, i.e., they depend on the direction, or, to put it differently, their description depends on the orientation of the coordinate system. Hence, the crystal medium is *anisotropic*.

Thus, in considering the macroscopic physical properties of a crystal, it can be regarded as a continuous homogeneous anisotropic medium. This means that the dependence of the properties on the direction is the same at all points of a crystal, while the properties in a given direction are independent of the choice of the unit volume to be tested; in other words, the symmetry group of the crystal contains, as its subgroup, a group of three-dimensional infinitely small translations $T_{\tau_1\tau_2\tau_3}$, and hence the properties of the crystal are invariant to any parallel translation of the coordinate system in the crystal medium.

The symmetry group of any physical property is a direct product of $T_{\tau_1\tau_2\tau_3}$ and some definite point group of symmetry G_0^3, which may be a crystallographic or limiting group. Since group $T_{\tau_1\tau_2\tau_3}$ is equally inherent in all crystals in their macroscopic description, in studying the symmetry of a particular property one restricts oneself to indicating its group G_0^3, denoted as G_{pr} (Sect. 1.5).

1.1.2 Cartesian Coordinate System and Its Transformations

The physical properties of crystals are usually described in *Cartesian (orthogonal normalized) systems of coordinates* X_1, X_2, X_3 (or X, Y, Z). The right-handed coordinate system is generally used. In such a system (Fig. 1.1 a) the shortest rotation from the X_1 to the X_2 axis is performed anticlockwise; if we move, simultaneously with this rotation, along the positive direction of the X_3 axis, it will be a motion according to the right-handed screw rule. Only in special cases (when considering left-hand enantiomorphous forms of crystals), the left-handed coordinate system is sometimes used (Fig. 1.1 b).

To eliminate ambiguity in the description of the physical properties of crystals, it is customary to use a Cartesian coordinate system oriented in a definite way with respect to the crystallographic axes [Ref. 1.1, Sect. 3.3], which is called the *crystallophysical* system of coordinates. The rules for selecting it are given below in Sect. 1.5. But in solving many crystallophysic problems it is more convenient to use, not the crystallophysical, but a different, special Cartesian coordinate system.

Transition from the X_1, X_2, X_3 system to X_1', X_2', X_3' with the same origin and metrics (*orthogonal transformation*) is described by a set of equations, which is expressed, with the use of the summation indices, as follows:

$$e_i' = \alpha_{ij} e_j \, , \tag{1.1}$$

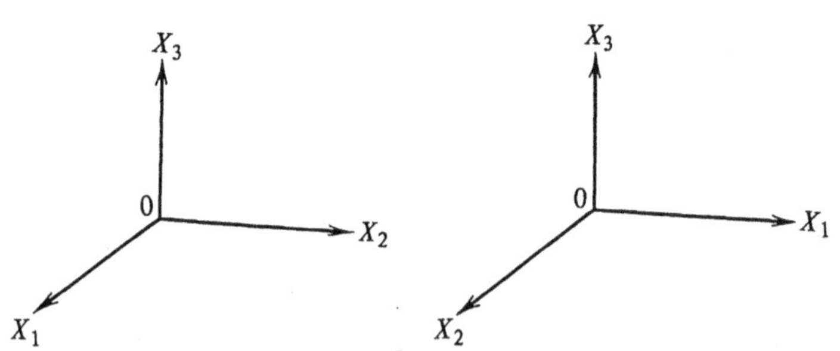

a b

Fig. 1.1. Right-handed (**a**) and left-handed (**b**) orthogonal systems of coordinates

where e_i' and e_j are the unit vectors of the new and old coordinate system, respectively, and α_{ij} are the direction cosines of the angles between the new X_i' and old X_j axes. The nine coefficients α_{ij} defining the transformation of the coordinate system can be written in the form of a matrix, which is called the orthogonal transformation matrix:

$$\|\alpha_{ij}\| = \begin{Vmatrix} \alpha_{11} & \alpha_{12} & \alpha_{13} \\ \alpha_{21} & \alpha_{22} & \alpha_{23} \\ \alpha_{31} & \alpha_{32} & \alpha_{33} \end{Vmatrix} . \tag{1.2}$$

There are always six relationships (orthogonality relations) among the nine cosines α_{ij} [1.2],

$$\alpha_{ik}\alpha_{jk} = \begin{cases} 1 & \text{for} \quad i = j \\ 0 & \text{for} \quad i \neq j \end{cases} , \tag{1.3}$$

so that only three cosines are mutually independent. This is in line with the fact that transition from one coordinate system to another can always be fully assigned by three independent parameters (for instance, three Euler angles).

It is obvious that the reverse transformation from the new X_i' to the old X_j axes is described by the set of equations

$$e_j = \alpha_{ij}' e_i' , \tag{1.4}$$

and that the matrix of this orthogonal transformation will be transposed with respect to matrix $\|\alpha_{ij}\|$:

$$\|\alpha_{ij}'\| = \|\alpha_{ji}\| = \begin{Vmatrix} \alpha_{11} & \alpha_{21} & \alpha_{31} \\ \alpha_{12} & \alpha_{22} & \alpha_{32} \\ \alpha_{13} & \alpha_{23} & \alpha_{33} \end{Vmatrix} . \tag{1.5}$$

As has been discussed in [Ref. 1.1, Sect. 2.2], the determinant of the matrix of any orthogonal transformation

$$|\alpha_{ij}| = \pm 1$$

for *first-kind* transformations (proper, i.e., simple rotations) is

$$|\alpha_{ij}| = 1 ,$$

and for *second-kind* transformations (improper rotations: reflection in a plane, inversion, mirror or inversion rotation) is

$$|\alpha_{ij}| = -1 .$$

It follows that on first-kind transformation a right-handed coordinate system remains right-handed, and a left-handed system remains left-handed, whereas on second-kind transformation a right-handed system changes to a left-handed one, and vice versa.

1.2 Tensors and Their Transformations

1.2.1 Scalars, Pseudoscalars, Vectors, and Tensors

If some quantity, say one describing some physical property of a crystal, is simply numerical, i.e., is not associated with the concept of direction in space and remains unaltered on transformation of the coordinates, it is called a *scalar*[1]. Among the scalars are such physical quantities as the temperature, entropy, and heat capacity.

Some quantities, although retaining their numerical value on any transformations of the coordinate system, change their sign on second-kind transformations. Such quantities are called *pseudoscalars*. An example of a pseudoscalar physical quantity is the specific optical rotation. Thus, the scalar and the modulus of the pseudoscalar are invariant to any transformation of the coordinates.

In distinction to the scalars and pseudoscalars, other quantities – vectors and tensors – are anisotropic and generally change their numerical value on transformation of the system of coordinates. The simplest anisotropic quantity is the vector. *Vector* $\boldsymbol{a}$ is completely defined by assignment of its length and direction, or by its *components*, i.e., its projections on coordinate axes X_1, X_2, X_3: $\boldsymbol{a} = [a_1, a_2, a_3]$, and the vector length $\boldsymbol{a}$:

$$|\boldsymbol{a}| = \sqrt{a_1^2 + a_2^2 + a_3^2} \; . \tag{1.6}$$

One vector quantity can be a function of another vector quantity: $\boldsymbol{b} = f(\boldsymbol{a})$; or, as is said sometimes, one vector can be induced by another. The relationship between vectors is expressed, in the simplest case, by a scalar, i.e., $\boldsymbol{b} = s\boldsymbol{a}$.

Generally, however (this is precisely true for crystals and other anisotropic solids), the relationship between a pair of vectors $\boldsymbol{b}$ and $\boldsymbol{a}$ depends on the direction. If each component of $\boldsymbol{b}$ is a linear function of each component of $\boldsymbol{a}$, then

---

[1] We emphasize once more that henceforward in this chapter we mean transformations leaving the origin unmoved. If we lift this restriction and consider scalar field $a(X, Y, Z)$ with a translation symmetry $T_{\tau_1 \tau_2 \tau_3}$, then a transformation of the coordinate system including $T_{\tau_1' \tau_2' \tau_3'} \neq T_{\tau_1 \tau_2 \tau_3}$ will evidently change the value of a.

$$b_1 = T_{11}a_1 + T_{12}a_2 + T_{13}a_3 ,$$
$$b_2 = T_{21}a_1 + T_{22}a_2 + T_{23}a_3 , \quad (1.7)$$
$$b_3 = T_{31}a_1 + T_{32}a_2 + T_{33}a_3 .$$

The quantity relating two vectors $\boldsymbol{b} = [b_1, b_2, b_3]$ and $\boldsymbol{a} = [a_1, a_2, a_3]$ by (1.7) is tabulated as

$$\begin{Vmatrix} T_{11} & T_{12} & T_{13} \\ T_{21} & T_{22} & T_{23} \\ T_{31} & T_{32} & T_{33} \end{Vmatrix} = T_{ij} \quad (1.8)$$

and called a *second-rank tensor*. Each of the nine coefficients T_{11}, $T_{12}, \ldots,$ called *tensor components*, has a definite physical and geometric meaning. Thus, T_{11}, T_{21}, and T_{31} define the components of vector $\boldsymbol{b}$ along axes X_1, X_2, and X_3, respectively, if vector $\boldsymbol{a}$ is parallel to the X_1 axis. Components T_{11}, T_{22}, and T_{33}, which stand on the main diagonal, are the *longitudinal* components of the tensor, because they relate parallel components of $\boldsymbol{b}$ and $\boldsymbol{a}$, while the other tensor components are *transverse*, i.e., they relate mutually perpendicular components of $\boldsymbol{b}$ and $\boldsymbol{a}$.

Using the summation indices, we can rewrite (1.7) in a concise form:

$$b_i = T_{ij}a_j . \quad (1.9)$$

We shall now give examples of physical quantities and properties of crystals described by vectors and second-rank tensors.

Assigned vector	Induced vector	Tensor property
Electric field	Dielectric polarization	Dielectric susceptibility
Electric Field	Electrical induction	Dielectric permittivity
Electric Field	Electrical current density	Specific electrical conductivity
Temperature gradient	Heat flux density	Thermal conductivity coefficients

1.2.2 Transformation of the Components of Vectors and Second-Rank Tensors

If vector $\boldsymbol{a}$ has components a_1, a_2, a_3 (Fig. 1.2) in the old system of coordinates X_1, X_2, X_3 and components a'_1, a'_2, a'_3 in the new system of axes X'_1, X'_2, X'_3 assigned by (1.1), the new component a'_1 will be the sum of the projections of all the old components of the vector onto the new axis X'_1:

$$a'_1 = a_1 \cos \widehat{X'_1 X_1} + a_2 \cos \widehat{X'_1 X_2} + a_3 \cos \widehat{X'_1 X_3} = \alpha_{11}a_1 + \alpha_{12}a_2 + \alpha_{13}a_3 . \quad (1.10a)$$

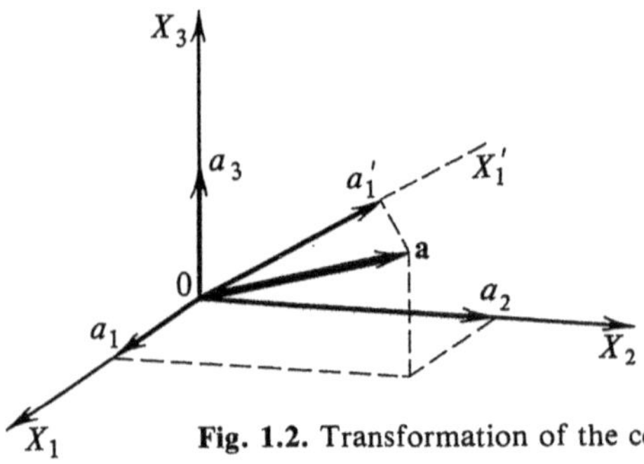

Fig. 1.2. Transformation of the components of vector $\boldsymbol{a}$

We similarly find

$$a_2' = \alpha_{21}a_1 + \alpha_{22}a_2 + \alpha_{23}a_3 \ , \qquad a_3' = \alpha_{31}a_1 + \alpha_{32}a_2 + \alpha_{33}a_3 \ . \tag{1.10b}$$

Finally, using the abbreviations, we write, in place of (1.10), the vector transformation equation in the general form:

$$a_i' = \alpha_{ij}\, a_j \ . \tag{1.11}$$

By reasoning in like manner it is easy to obtain an analogous equation for the reverse transformation, i.e., the transition from the components in the new system of coordinates to the components in the old system:

$$a_i = \alpha_{ji}\, a_j' \ . \tag{1.12}$$

The matrix of the direction cosines of the reverse transformation α_{ji} is, as always (1.5), the transposed matrix α_{ij} of the direct transformation. Attention should be drawn to the order of the subscripts in (1.11, 12): in the direct transformation the summation subscripts stand together, while in the reverse transformation they stand apart.

Note that with (1.11) it is easy to show that

$$a_j' a_j' = a_i a_i \ ,$$

i.e., the sum of the squares of the vector components, which determines the vector length (1.6), is invariant to orthogonal transformation.

Suppose now that in coordinate system X_i the two vectos $\boldsymbol{b}$ and $\boldsymbol{a}$ are related by the expression

$$b_k = T_{kl}\, a_l \ , \tag{1.13}$$

which is evidently similar to (1.7) because we can use any letters for the subscripts. Hence, T_{kl} is a second-rank tensor.

On transition to the new system of axes X_i' we have, by (1.11, 12),

$$b_i' = \alpha_{ik} b_k , \quad a_l = \alpha_{jl} a_j' . \tag{1.14}$$

Combining (1.13, 14), we get

$$b_i' = \alpha_{ik} b_k = \alpha_{ik} T_{kl} a_l = \alpha_{ik} T_{kl} \alpha_{jl} a_j' = T_{ij}' a_j' , \tag{1.15}$$

where

$$T_{ij}' = \alpha_{ik} \alpha_{jl} T_{kl} . \tag{1.16}$$

Analogously to (1.13), (1.15) relates the new components of vectors $\boldsymbol{b}$ and $\boldsymbol{a}$. Consequently, nine coefficients T_{ij}' are components of second-rank tensor T_{kl} in the new system of coordinates, and (1.16) is the *law of transformation of a second-rank tensor.* Recall that (1.16) is a concise notation of the set of nine equations, each of which contains nine terms in the right-hand side.

It is easy to show that the reverse transformation expressing the old tensor components in terms of the new ones has a similar form, but the sequence of the subscripts at α is reversed (matrices $\|\alpha\|$ have been transposed):

$$T_{kl} = \alpha_{ik} \alpha_{jl} T_{ij}' .$$

We emphasize that on transformation of the tensor it is not, of course, the physical quantity described by the tensor that changes (it is independent of the choice of the particular coordinate system), but only the method for representing this quantity.

1.2.3 Tensors of Different Ranks

In the same way as we have just done with the second-rank vectors and tensors, we can, by writing the transformation equations for the quantities

$$T_{nop}' = \alpha_{ni} \alpha_{oj} \alpha_{pk} T_{ijk} , \tag{1.17}$$

$$T_{nopq}' = \alpha_{ni} \alpha_{oj} \alpha_{pk} \alpha_{ql} T_{ijkl} , \tag{1.18}$$

$$T_{nopqr}' = \alpha_{ni} \alpha_{oj} \alpha_{pk} \alpha_{ql} \alpha_{rm} T_{ijklm} , \tag{1.19}$$

. . . .

use these expressions as definitions. Obviously, (1.17) defines a third-rank tensor, and (1.18) and (1.19), fourth- and fifth-rank tensors, respectively. It would be logical to supplement this equation series, for the sake of completeness not only with (1.16), but also with (1.11). Then it will be natural to call the vector a *first-rank tensor* as well. Extending the extrapolation, it is logical to call the scalar a *zero-rank tensor.*

Thus, in the general case an Nth-rank tensor has N subscripts, each of which (in three-dimensional space) can take values from 1 to 3. Therefore an Nth-rank tensor has 3^N components.

Just as a second-rank tensor relates two vectors (1.7), a third-rank tensor (for instance, the tensor of piezoelectric coefficients) relates a second-rank vector and tensor, i.e.,

$$a_i = T_{ijk} Q_{jk} \ . \tag{1.20}$$

A fourth-rank tensor (for instance, the tensor of elastic coefficients) can relate two second-rank tensors, i.e.,

$$R_{ij} = T_{ijkl} Q_{kl} \ , \tag{1.21}$$

or a third-rank vector and tensor, i.e.,

$$a_i = T_{ijkl} R_{jkl} \ .$$

In general, if an Nth-rank tensor expresses the relationship between tensors of ranks L and M, then $L+M=N$.

Tensors can also be regarded as the result of differentiation of tensors with respect to tensors. Thus, second-rank tensors include a derivative of a vector with respect to a vector or the second derivative of a scalar with respect to vector arguments, i.e., as is easy to show, nine quantities of the form

$$T_{ij} = \frac{\partial a_i}{\partial b_j} \quad \text{or} \quad R_{ij} = \frac{\partial a}{\partial b_i \partial c_j} \tag{1.22}$$

are transformed according to (1.15) and hence are components of a second-rank tensor.

In general, partial derivatives of the components of a Kth-rank tensor with respect to tensors of ranks L and M are themselves components of a tensor of rank $N = K + L + M$.

1.2.4 Pseudotensors (Axial Tensors)

By analogy with the pseudoscalar concept introduced above, we can introduce the concept of a pseudotensor. A *pseudotensor* differs from a tensor only in that the transformation of its components also includes multiplication by the determinant of transformation $|\alpha_{ij}|$. Consequently, for an Nth-rank pseudotensor the transformation law, which can be used as its definition, has the form:

$$P_{ijkl} = |\alpha_{ij}| \, \alpha_{ip} \alpha_{jq} \alpha_{kr} \alpha_{ls} \dots P_{pqrs\dots} \ . \tag{1.23}$$

It is clear that under first-kind transformations ($|\alpha_{ij}| = +1$) a pseudotensor behaves as an ordinary tensor, and under second-kind transformations ($|\alpha_{ij}| = -1$) the pseudotensor components additionally change sign with respect to the components of an ordinary tensor.

To emphasize the difference between pseudotensors (which are also called *axial tensors*) and ordinary tensors, the latter are sometimes called *polar tensors*. (In the following text we continue to call ordinary, polar tensors simply tensors, except where this may cause confusion.)

As an example of a *zero-rank pseudotensor* (*pseudoscalar*) we have already mentioned specific optical rotation. Examples of *first-rank pseudotensors* (axial vectors) are the magnetic field and the magnetic induction. A second-rank pseudotensor is exemplified by the gyration tensor, which describes the optical activity of crystals.

It is obvious that if there exists a relationship between a polar and an axial vector $\boldsymbol{a}$ and $\boldsymbol{q}$ it is expressed by a second-rank pseudotensor; the relationship between two axial vectors is expressed by a second-rank polar tensor; the relationship between a polar vector (axial vector) and a second-rank pseudotensor (second-rank polar tensor) is expressed by a third-rank pseudotensor. In general, the product of a polar tensor and a pseudotensor is a pseudotensor, the product of two pseudotensors being a polar tensor.

We have already noted in Sect. 1.1 that not all the anisotropic physical properties of crystals are necessarily described by tensors (polar or axial). Indeed, it is easy to show that $\sqrt{T_{ij}}$ is not a tensor, because it is not transformed by law (1.15). Hence, for instance, refractive indices $n_i = \sqrt{\varepsilon_i}$ describe anisotropic but nevertheless not tensor properties of a crystal.

1.2.5 Symmetric and Antisymmetric Tensors. Interior Symmetry of Tensors

Both polar and axial tensors can possess properties of symmetry with respect to their subscripts. If the tensor components remain unchanged on permutation of two or more subscripts, the tensor is called *symmetric* with respect to these subscripts. Thus, for example, if

$$T_{ij} = T_{ji} , \tag{1.24}$$

$$T_{ijk} = T_{ikj} , \tag{1.25}$$

$$T_{ijkl} = T_{klij} , \tag{1.26}$$

then tensor T_{ij} is symmetric, tensor T_{ijk} is symmetric with respect to the last two subscripts, and tensor T_{ijkl} is symmetric with respect to permutation of the first and second pairs of subscripts.

Naturally, the number of independent tensor components decreases in accordance with the number of relations of the type (1.24–26). Thus, a symmetric second-rank tensor generally has only six independent components among the nine:

$$\begin{Vmatrix} T_{11} & T_{12} & T_{13} \\ T_{21} & T_{22} & T_{23} \\ T_{31} & T_{32} & T_{33} \end{Vmatrix} = \begin{Vmatrix} T_{11} & T_{12} & T_{13} \\ T_{12} & T_{22} & T_{23} \\ T_{13} & T_{23} & T_{33} \end{Vmatrix} ;$$

in a third-rank tensor symmetric with respect to two subscripts only 18 components out of $3^3 = 27$ are independent, etc. In general, the symmetricity of an Nth-rank tensor with respect to a pair of subscripts yields 3^{N-1} relations among its components and reduces the number of independent components to

$$3^N - 3^{N-1} = 2 \cdot 3^{N-1} ; \quad (1.27)$$

the symmetricity of an Nth-rank tensor with respect to two pairs of subscripts yields $5 \cdot 3^{N-2}$ relations among its components and reduces the number of independent components to

$$3^N - 5 \cdot 3^{N-2} = 4 \cdot 3^{N-2} ; \quad (1.28)$$

etc.

If the tensor components remain unaltered after an even number of permutations of subscripts and change their sign after an odd number of permutations, the tensor is said to be *antisymmetric* with respect to these subscripts. Thus, if

$$T_{ij} = -T_{ji} , \quad (1.29)$$

$$T_{ijk} = -T_{ikj} , \quad (1.30)$$

then tensor T_{ij} is antisymmetric, and tensor T_{ijk} antisymmetric with respect to the second and the third subscript.

Owing to relations of the type (1.29, 30), the number of independent components of the antisymmetric tensor not only decreases, but some of the components reduce identically to zero. Thus, for a second-rank tensor we have, from condition (1.29),

$$T_{ii} = -T_{ii} = 0 ,$$

and the tensor takes the form

$$\begin{Vmatrix} 0 & -T_{21} & T_{13} \\ T_{21} & 0 & -T_{32} \\ -T_{13} & T_{32} & 0 \end{Vmatrix} , \quad (1.31)$$

having only three independent components.

It is not hard to prove that the tensor remains symmetric or antisymmetric on changing to a new coordinate system, i.e., the property of symmetricity and antisymmetricity of the tensor with respect to some subscripts or a group of subscripts is invariant to orthogonal transformation. This property characterizes the so-called *interior symmetry* of tensors.

We shall now show that any second-rank tensor can be represented as the sum of a symmetric one and an antisymmetric one. Indeed, an arbitrary second-rank polar tensor b_{ij} can be written in the form

$$b_{ij} = \beta_{ij} + \omega_{ij} \ , \quad \text{where} \tag{1.32}$$

$$\beta_{ij} = \tfrac{1}{2}(b_{ij} + b_{ji}) \quad \text{and} \quad \omega_{ij} = \tfrac{1}{2}(b_{ij} - b_{ji}) \ . \tag{1.33}$$

It is easy to ascertain that tensor β_{ij} thus assigned is symmetric, since $\beta_{ij} = \beta_{ji}$, and tensor ω_{ij} is antisymmetric, since $\omega_{ij} = -\omega_{ji}$. Note that in order to prove the symmetricity of a tensor describing a particular physical property of a crystal it is necessary, as a rule, to resort to thermodynamic consideration.

The following symbols are generally used for describing the interior symmetry of tensors [1.3]. If an Nth-rank polar tensor is symmetric with respect to L subscripts, its interior symmetry is denoted by $[V^L]V^{N-L}$ or $V^{N-L}[V^L]$. Thus, the interior symmetry of a tensor, characterized by (1.24), is denoted by $[V^2]$ and the one characterized by (1.25), $V[V^2]$. In all cases the total power of V is equal to rank N of the tensor. Thus, if a tensor of an even rank N is symmetric with respect to all the pairs of subscripts, its interior symmetry is denoted by $[V^2]^{N/2}$, and if, besides, it is symmetric with respect to permutations of all the pairs of subscripts, its symmetry is denoted by $[[V^2]^{N/2}]$. If, however, a tensor of an even rank N is symmetric only with respect to permutations of the pairs of subscripts, its interior symmetry is denoted by $[(V^2)^{N/2}]$. Thus, the interior symmetry of a tensor, which is characterized by (1.26), is denoted by $[(V^2)^2]$.

Similar symbols are used for describing the interior symmetry of antisymmetric polar tensors, except that the square brackets are replaced by braces. Thus, the interior symmetry of a tensor, which is characterized by (1.29), is denoted by $\{V^2\}$, and the one characterized by (1.30), $V\{V^2\}$.

Finally, in describing the interior symmetry of pseudotensors, one uses the same symbols, but symbol ε (denoting the interior symmetry of a pseudoscalar) is placed also in the front.

It has been shown above (1.31) that an antisymmetric second-rank polar tensor has only three independent components and, as is easy to see, on second-kind transformations the tensor components will change sign. As can easily be proved, this reflects the fact that an antisymmetric second-rank polar tensor is dual to an axial vector, i.e., either of them can equally be used for describing one and the same geometric (physical) object. At the same time an antisymmetric second-rank pseudotensor is dual to a polar vector. Using the symbols describing the interior symmetry of tensors it is not hard to obtain other duality relations.

1.2.6 Reciprocal Tensors

In describing the physical properties of crystals represented by tensors of different ranks and defined by dependences of the type (1.15 – 19) it is often found expedient to consider the reciprocal dependences, and hence it becomes necessary to introduce the concepts of reciprocal properties and reciprocal tensors.

Suppose there is a tensor relationship

$$R_{ijk\ldots} = T_{ijk\ldots pqr\ldots} Q_{pqr\ldots} , \tag{1.34}$$

where $R_{ijk\ldots}$ and $Q_{pqr\ldots}$ are tensors of the same rank N, and hence $T_{ijk\ldots pqr\ldots}$ is a tensor of even rank $2N$. Then there exists a tensor relationship

$$Q_{ijk\ldots} = T^{-1}_{ijk\ldots pqr\ldots} R_{pqr\ldots} , \tag{1.35}$$

where $T^{-1}_{ijk\ldots pqr\ldots}$ is a component of a tensor of rank $2N$ reciprocal to tensor $T_{ijk\ldots pqr\ldots}$. It should be emphasized that dependence (1.35) exists only when the determinant compiled from the components of tensor $T_{ijk\ldots pqr\ldots}$, Δ_T does not equal zero.

The components of the reciprocal tensor are not generally reciprocal to the respective components of the initial tensor, but are related to them in a more complicated way. Since (1.34) is a concise notation for a set of 3^N inhomogeneous equations with 3^N terms in the right-hand side, it can be resolved for $Q_{pqr\ldots}$. Using Kramer's equations, we obtain

$$Q_{ijk\ldots} = \frac{A^T_{pqr\ldots ijk}}{\Delta_T} R_{pqr\ldots} , \tag{1.36}$$

where Δ_T is a determinant compiled of the components of tensor $T_{ijk\ldots pqr\ldots}$, and $A^T_{pqr\ldots ijk}$ is an algebraic complement for its element $T_{pqr\ldots ijk}$.

Comparing (1.35, 36), we get the equation for finding the components of the reciprocal tensor in terms of the components of the initial tensor:

$$T^{-1}_{ijk\ldots pqr\ldots} = A^T_{pqr\ldots ijk\ldots}/\Delta_T . \tag{1.37}$$

It is obvious that by regarding (1.35) as the initial relationship, we can obtain the equation for finding the components of the initial tensor via those of the reciprocal one:

$$T_{ijk\ldots pqr\ldots} = A^{T-1}_{pqr\ldots ijk}/\Delta_{T-1} , \tag{1.38}$$

where $\Delta_{T^{-1}}$ is a determinant compiled from the components of tensor $T^{-1}_{ijk\ldots pqr\ldots}$, and $A^{T-1}_{pqr\ldots ijk\ldots}$ is the algebraic complement of this determinant for its element $T^{-1}_{pqr\ldots ijk}$. Attention should be drawn to the reverse order of the subscripts in the algebraic complement and the sought-for component of the tensor in (1.37, 38).

We emphasize that transition to the reciprocal tensor is possible only for even-rank tensors. It is not hard to prove that the polarity and symmetricity of the tensor are retained after the transition. Thus, the symmetric second-rank tensor of dielectric "nonpermittivity" is reciprocal to the symmetric second-rank tensor of the dielectric permittivity. Fourth-rank tensors of elastic stiffnesses and elastic compliances c_{ijkl} and s_{ijkl}, which are symmetric with respect to both pairs of subscripts and to their permutation, are mutually reciprocal and, therefore, according to (1.37, 38),

$$s_{ijkl} = A^c_{klij}/\Delta_c \ , \quad c_{ijkl} = A^s_{klij}/\Delta_s \ . \tag{1.39}$$

1.2.7 Matrix Notation

For symmetric (fully or with respect to some subscripts) tensors, one could use, instead of tensor notation, more concise *matrix notation* with a reduced number of subscripts, owing to the smaller number of their independent components, and that would make the equations more compact.

Thus, two subscripts of a symmetric second-rank tensor with six independent components, which run through the values from 1 to 3, can be replaced by a single subscript running through the values from 1 to 6. The relationship among the subscripts is then as follows:

$$\begin{array}{lcccccc} \text{Tensor subscripts:} & 11 & 22 & 33 & 23=32 & 31=13 & 12=21 \\ \text{Matrix subscripts:} & 1 & 2 & 3 & 4 & 5 & 6 \end{array} \tag{1.40}$$

To summarize:

$$\left\| \begin{matrix} T_{11} & T_{12} & T_{13} \\ T_{21} & T_{22} & T_{23} \\ T_{31} & T_{32} & T_{33} \end{matrix} \right\| \rightarrow \left\| \begin{matrix} T_1 & T_6 & T_5 \\ T_6 & T_2 & T_4 \\ T_5 & T_4 & T_3 \end{matrix} \right\| . \tag{1.41}$$

The same rule (1.40) is also followed by transition from tensor to matrix notation for higher-rank tensors symmetric with respect to one or more pairs of subscripts.

Thus, for a third-rank tensor symmetric with respect to the first two subscripts and hence having 18 independent components according to (1.27), the first and second subscripts can, by rule (1.40), be replaced by one, running through values 1,...,6. As a result, the unfolded tensor is, in matrix form, a table of six lines and three columns (a 6×3 table), whereas in tensor form it was a 9×3 table. An example of such a tensor is the tensor of the coefficients of the linear electro-optical effect.

Similarly, if a fourth-rank tensor is symmetric with respect to both pairs of subscripts, it has, by (1.28), $4 \cdot 3^{4-2} = 36$ independent components. In tensor notation it is represented in unfolded form by a 9×9 table, and on passing to matrix form according to rule (1.40), for each pair of subscripts it is represented by a 6×6 table, i.e., it contains only 36 components instead of 81. Examples of such a tensor are tensors of electrostriction and piezo-optical coefficients.

It must be borne in mind that for higher-than-second-rank tensors, in transition from the tensor to the matrix form (and vice versa) it is sometimes necessary, when equating the corresponding components according to rule (1.40), to introduce additional numerical factors (2, 4, etc.). The need for introducing such factors becomes clear from inspection of particular equations of the type (1.20 – 22).

The use of matrix notation for symmetric tensors considerably shortens the expressions and is therefore convenient in solving many particular problems and also in using the matrix computation techniques in crystal physics. It should always be remembered, however, that in order to transform the tensor components according to rules (1.16 – 19), when passing from one coordinate system to another, one should revert from the matrix to the tensor notation. Thus, although the components of symmetric third- and fourth-rank tensors have two subscripts each in matrix notation, they are by no means components of a second-rank tensor and do not transform similarly to them.

Rules (16 – 19), although simple, are unwieldy, and therefore more complicated but handier rules for transforming tensor components in matrix form have been developed, (sec, for instance, [1.4]).

1.3 Geometric Interpretation of Tensors. Indicatrice Surfaces

1.3.1 Characteristic Surface for a Symmetric Second-Rank Tensor

One of the most widespread tensors in description of the physical properties of crystals is the symmetric second-rank polar tensor. Therefore, we shall dwell on it in more detail. We begin with its geometric interpretation.

It is known from analytical geometry that the general equation of the central second-order surface with center at the origin can be written

$$T_{ij} x_i x_j = 1 \quad . \tag{1.42}$$

We transform this equation to a new system of coordinates. To do this we take into account that the coordinates of the moving point of the surface are simultaneously the components of the radius vector and hence are transformed according to law (1.11), i.e.,

$$x_i = \alpha_{ki} x'_k \ , \quad x_j = \alpha_{lj} x'_l \ . \tag{1.43}$$

Substituting (1.43) into (1.41), we get

$$T_{ij}\, \alpha_{ki}\, \alpha_{lj}\, x'_k x'_l = 1 \quad \text{or} \quad T'_{kl} x'_k x'_l = 1 \ ,$$

where

$$T'_{kl} = \alpha_{ki}\, \alpha_{lj} T_{ij} \ . \tag{1.44}$$

Comparing (1.44) and (1.16), we see that they are identical.

Consequently, the law for transforming the equation of a second-order surface at transition to the new coordinate system coincides with the law for transforming a symmetric second-rank tensor. Thus, to find how the components of a symmetric second-rank tensor are transformed, it is sufficient to consider the transformation of the equation of a central second-order surface with center at the origin and with the coefficients equal to the tensor components. Therefore, such a surface is called the *characteristic surface for the symmetric second-rank tensor* and can be used for describing any property of the crystal represented by such a tensor.

1.3.2 Principal Axes of a Symmetric Second-Rank Tensor

Any central surface of the second order possesses *principal axes,* i.e., three mutually perpendicular directions, on selection of which as coordinate axes the general equation of the surface (1.42) is reduced to the simplified form

$$T_1 x_1^2 + T_2 x_2^2 + T_3 x_3^2 = 1 \ . \tag{1.45}$$

Hence, a symmetric second-rank tensor can also always be reduced to its *principal axes.* Then the tensor

$$T_{ij} = \begin{Vmatrix} T_{11} & T_{12} & T_{13} \\ T_{12} & T_{22} & T_{23} \\ T_{13} & T_{23} & T_{33} \end{Vmatrix}$$

takes the form

$$\begin{Vmatrix} T_1 & 0 & 0 \\ 0 & T_2 & 0 \\ 0 & 0 & T_3 \end{Vmatrix} , \tag{1.46}$$

and numbers T_1, T_2, and T_3 are called the *principal values of the components of tensor* T_{ij} (or the property described by it) and coincide with the coefficients of the respective equation (1.45), while the system of coordinate axes parallel to the principal axes of the tensor is called the *principal system of coordinates of the tensor*. Obviously, the principal axes coincide with the symmetry elements (axes) of the characteristic surface.

In a tensor related to the principal axes (or, as is customary to say, to the *diagonal form*) the number of independent components decreases to three. The number of the "degrees of freedom," however, generally remains equal to six, since the three independent values are now needed for determining the direction of the fixed principal coordinate axes (the principal axes of the tensor).

If a tensor relating two vectors by the equation

$$a_i = T_{ij}\, b_j$$

is symmetric, then on passing from arbitrary coordinate axes to the tensor this equation simplifies and takes the form

$$a_i = T_{ii}\, b_i = T_i\, b_i \ , \quad \text{i.e.,} \quad a_1 = T_1 b_1 \ , \quad a_2 = T_2 b_2 \ , \quad a_3 = T_3 b_3 \ . \qquad (1.47)$$

From (1.47) it follows at once that when vector $\boldsymbol{b}$ is directed along any one of the principal axes of the tensor, vector $\boldsymbol{a}$ is parallel to it. However, the proportionality coefficients of the vectors for all the three axes are different (the relationship is tensorial, not scalar!). On the other hand, if vector $\boldsymbol{b}$ is not collinear to any one of the principal axes of tensor T_{ij}, then vector $\boldsymbol{a}$ is not parallel to vector $\boldsymbol{b}$ because of the same difference in the coefficients of proportionality of the various components of $\boldsymbol{a}$ and $\boldsymbol{b}$.

1.3.3 Property in a Given Direction

The concept of a property in a given direction, which is often used when considering the properties of crystals, requires no explanation if the property is described by a polar or an axial vector. The length and sign of the vector will describe the magnitude and sign of the property in a given direction. But as we pass to the properties represented by second-rank tensors, this concept already becomes more complicated, since a second-rank tensor is already characterized by a second-order surface (and then only in the case of symmetry), and the vectors related by it are generally nonparallel to one another (1.3.2).

Let $a_i = T_{ij}\, b_j$. Then the quantity T characterizing the tensor property in a given direction will be the ratio of the projection of vector $\boldsymbol{a}$ onto this direction to the length of vector $\boldsymbol{b}$ laid off in the same direction.

From the definition of T follows the method for its analytical expression. Suppose a given direction in an arbitrary coordinate system has direction cosines e_1, e_2, and e_3. Then vector $\boldsymbol{b}$ laid off in this direction can be written $\boldsymbol{b} = [e_1 b, e_2 b, e_3 b]$, i.e., $b_i = e_i b$. The projection of $\boldsymbol{a}$ onto $\boldsymbol{b}$, if the angle between them is θ, is equal to

$$a \cos\theta = ab \cos\theta / b = (\boldsymbol{ab})/b = a_i b_i / b \ ,$$

in accordance with the properties of a scalar product of vectors. Consequently, by definition,

$$T = a_i b_i / b^2 = T_{ij} b_j b_i / b^2 \ ,$$

or, finally,

$$T = T_{ij} e_i e_j \ . \tag{1.48}$$

If tensor T_{ij} is symmetric and if its main coordinate system, rather than an arbitrary one, is used, (1.48) simplifies to

$$T = T_{ii} e_i^2 = e_1^2 T_1 + e_2^2 T_2 + e_3^2 T_3 \ . \tag{1.49}$$

In considering a property in a given direction the coordinate system is often chosen so that one of the axes, usually X_1', coincides with this direction. Then, proceeding from the definition of T and using (1.47), we have

$$T = a_1'/b_1 = T_{11}' = \alpha_{1i} \alpha_{1j} T_{ij} \ . \tag{1.50}$$

Taking into account that by (1.4) $e_i = \alpha_{1i}$, it is easy to see that expressions (1.50, 48) are equivalent.

In accordance with (1.50), quantity T is often also called the *"longitudinal" component of the tensor in a given direction.*

By analogy with (1.50) we can also introduce the concept of the "longitudinal" component of the tensor in a given direction for the properties described by tensors of any ranks. Quantity T – the "longitudinal" component of tensor $T_{ijk\ldots p}$ in a given direction taken as the X_1' axis – is defined by the equation

$$T = T_{111\ldots 1}' = \alpha_{1i} \alpha_{1j} \alpha_{1k} \ldots \alpha_{1p} T_{ijk\ldots p} \ . \tag{1.51}$$

While for second-rank tensors only the "longitudinal" component of the tensor in a given direction is usually considered, for higher-rank tensors one sometimes introduces, besides the "longitudinal" component in a given direction defined by (1.51), the concepts of various other components in a given direction. Thus, if we have a tensor relationship

$$R_{ijk\ldots p} = T_{ijk\ldots pqrs\ldots z} Q_{qrs\ldots z} \ ,$$

then the quantity

$$T = T_{222\ldots 2111\ldots 1} = \alpha_{2i}\alpha_{2j}\alpha_{2k}\ldots\alpha_{2p}\alpha_{1q}\alpha_{1r}\alpha_{1s}\ldots\alpha_{1z}T_{ijk\ldots pqrs\ldots z} \tag{1.52}$$

is called the *"transverse" component of tensor $T_{ijk\ldots pqrs\ldots z}$ in a given direction* taken as the X_1' axis. In the general case, the quantity

$$T = T'_{ijk\ldots} = \alpha_{is}\alpha_{jt}\alpha_{ku}\ldots T_{stu\ldots} \tag{1.53}$$

can be called the $ijk\ldots$th component of tensor $T_{stu\ldots}$ in a given direction, taken as the X_1' axis.

1.3.4 Geometric Properties of the Characteristic Surface of a Symmetric Second-Rank Tensor

Suppose $\boldsymbol{r}$ is the radius vector of the characteristic surface of a symmetric second-rank tensor given by (1.42);

$$T_{ij}x_ix_j = 1$$

and e_i are its direction cosines in an arbitrary coordinate system. Then the components of the radius vector or, what is the same, the coordinates of its terminus describing surface (1.42), can be written in the form

$$x_i = re_i \ .$$

Substituting this expression into (1.42), we have

$$r^2T_{ij}e_ie_j = 1 \ .$$

Finally, using (1.48), we obtain

$$T = 1/r^2 \ , \quad r = 1/\sqrt{T} \ . \tag{1.54}$$

Thus, the length of the arbitrarily oriented radius vector of the characteristic surface of a symmetric second-rank tensor T_{ij} is equal numerically to the reciprocal value of the square root of quantity T characterizing the given tensor property in the direction of the radius vector.

In particular, since the values of T in the direction of the principal axes are equal to T_1, T_2, and T_3, respectively (1.47, 49), from (1.37) it follows that the lengths of the semiaxes of the characteristic surface are

$$1/\sqrt{T_1} \ , \quad 1/\sqrt{T_2} \ , \quad 1/\sqrt{T_3} \ . \tag{1.55}$$

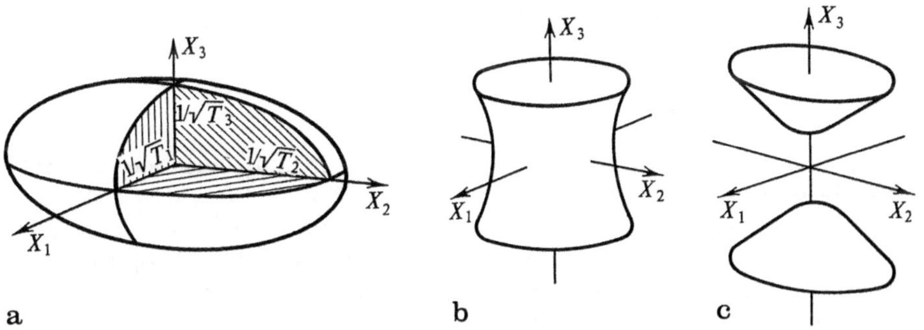

Fig. 1.3. Characteristic surfaces of a symmetric second-rank tensor: (**a**) ellipsoid; (**b**) hyperboloid of one sheet; (**c**) hyperboloid of two sheets

For most of the physical properties of crystals described by symmetric second-rank tensors all the principal components of tensor T_{ij} are essentially positive (for instance, for tensors of the dielectric constant, electrical conductivity, etc.). But there are also such tensor properties for which one, two, or all the three principal components of T_{ij} can be negative (for instance, the thermal expansion tensor). Since the semiaxes of the characteristic surface are, by (1.55), equal to $1/\sqrt{T_1}$, it is obvious that the characteristic surface for T_{ij} is an ellipsoid if all the $T_i>0$ (Fig. 1.3). If one principal component of T_i is less than zero, the characteristic surface is a hyperboloid of one sheet, if two, a hyperboloid of two sheets; and if all the $T_i<0$, the characteristic surface is an imaginary ellipsoid.

If there exist $T_i<0$, one considers the real branches of surface $T_{ij}\,x_i\,x_j = \pm 1$ to avoid encountering imaginary radii vectors in construction. It is then taken into account that for $T_{ij}\,x_i\,x_j = -1$, $r = 1/\sqrt{-T}$ and $T = -1/r^2$.

We shall now consider the property of the radius vector and the normal. Suppose the relationship between vectors $\boldsymbol{a}$ and $\boldsymbol{b}$ is described by symmetric tensor T_{ij}. Let us choose its principal axes as the coordinate axes. Then, by (1.47),

$$a_i = T_i\,b_i \ ,$$

and according to (1.45) the equation of the characteristic surface of tensor T_{ij} has the form

$$T_1 x_1^2 + T_2 x_2^2 + T_3 x_3^2 = 1 \ .$$

We assume, for simplicity, that $T_i>0$.

In an arbitrary direction, defined by direction cosines e_1, e_2, and e_3, we assign $\boldsymbol{b} = [e_1 b,\, e_2 b,\, e_3 b]$. Then

$$\boldsymbol{a} = [T_1 e_1 a,\, T_2 e_2 a,\, T_3 e_3 a] \ .$$

Hence, the direction cosines of vector $\boldsymbol{a}$ are proportional to $T_1 e_1$, $T_2 e_2$, and $T_3 e_3$. We now direct radius vector $\boldsymbol{r}$ parallel to $\boldsymbol{b}$ (Fig. 1.4). Clearly,

$$r = [re_1, re_2, re_3] \ .$$

It is known from analytic geometry that the direction cosines of a normal to surface $F(x_i) = 0$ at a given point are proportional to $\partial E(x_i)/\partial x_i$ calculated at this point. Consequently, the direction cosines of a normal erected to surface $F = T_i x_i^2 - 1 = 0$ at a point with coordinates $\boldsymbol{r}e_i$ will be proportional to $2T_1 re_1, 2T_2 re_2, 2T_3 re_3$, or $T_1 e_1$, $T_2 e_2$, $T_3 e_3$, i.e., will be proportional to the direction cosines of vector $\boldsymbol{a}$.

Thus, if $a_i = T_{ij} b_j$ (T_{ij} being equal to T_{ji}), direction $\boldsymbol{a}$ for the given $\boldsymbol{b}$ coincides with that of the normal to the characteristic surface of tensor T_{ij} erected at the terminus of the radius vector parallel to $\boldsymbol{b}$.

So far we have assumed that all $T_i > 0$. If, however, there exist $T_i < 0$, the real branches of surface $T_i x_i^2 = \pm 1$ are constructed, and for surface $T_i x_i^2 = -1$ the direction of $\boldsymbol{a}$ is taken parallel to the outward, rather than inward, normal.

1.3.5 Finding the Principal Axes of a Symmetric Second-Rank Tensor

As we shall see in Sect. 1.4, in monoclinic and triclinic crystals the principal axes of symmetric second-rank tensors describing their physical properties have an arbitrary (in a plane or in space, respectively) orientation. Therefore, for such crystals we encounter the problem of determining the orientations of the principal axes of a symmetric second-rank tensor with respect to a given orthogonal system of coordinates fixed in a crystal. Let us consider briefly the solution of this problem.

The equation of the characteristic surface of some symmetric second-rank tensor, not reduced to the principal axes, can be written as (1.42)

$$F(x_i) = T_{ij} x_i x_j - 1 = 0 \ , \quad \text{where} \quad T_{ij} = T_{ji} \ . \tag{1.56}$$

Denote by x_i the components of an arbitrary radius vector of this characteristic surface. As we have already mentioned in the preceding section, the direction cosines of the normal to surface (1.56) erected at the terminus of radius vector $\boldsymbol{x}$ are proportional to $\partial F(x_i)/\partial x_i$, i.e., vector $T_{ij} x_j$ is parallel to this normal.

The radius vector laid off along one of the principal axes of characteristic surface (1.56), which we denote by X_i, is parallel to the normal to the surface, erected at its terminus. Hence, vector $T_{ij} x_j$ is parallel to vector $\boldsymbol{x}$, and their components are proportional, i.e.,

$$T_{ij} x_j = \lambda x_i \ , \tag{1.57}$$

where λ is a constant.

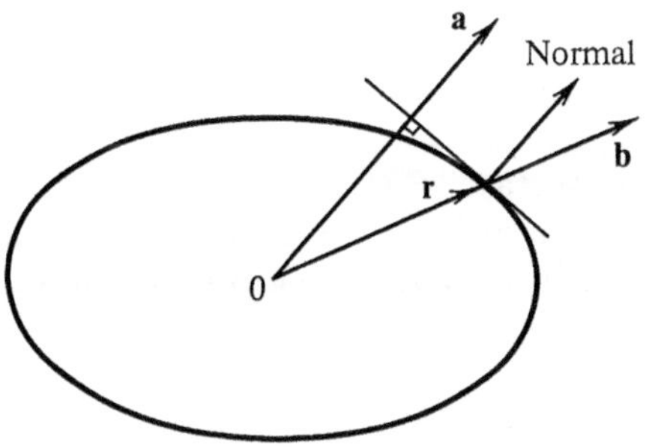

Fig. 1.4. Diagram explaining the properties of a radius vector and normal of the characteristic surface of a symmetric second-rank tensor

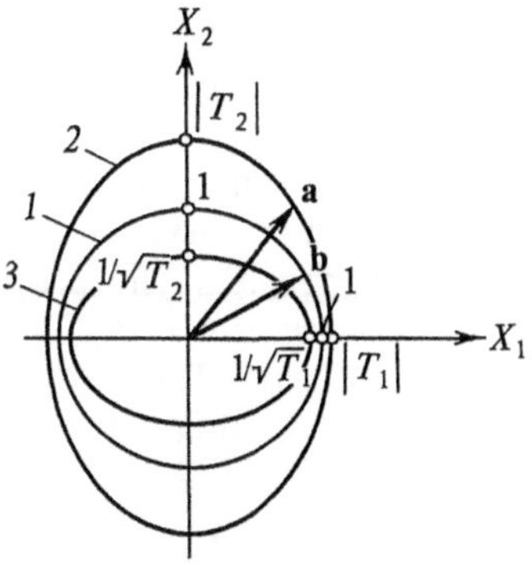

Fig. 1.5. Central sections of a sphere (*1*) described by a unit vector $\boldsymbol{b}$, the ellipsoid of the values of a symmetric tensor (*2*) given by the equation $a_i = T_{ij} b_j$ and of the characteristic surface of this tensor (*3*)

Relation (1.57) is a system of three linear homogeneous equations in variables x_i. This system has a nonzero solution if the determinant consisting of the coefficients of the equations is equal to zero, i.e.,

$$\begin{vmatrix} T_{11}-\lambda & T_{12} & T_{13} \\ T_{12} & T_{22}-\lambda & T_{23} \\ T_{13} & T_{23} & T_{33}-\lambda \end{vmatrix} = 0 \ . \qquad (1.58)$$

Expression (1.58) is a cubic equation with respect to λ and is called the *secular equation*. Each of its three roots defines the direction in which the radius vector of surface (1.56) is parallel to its normal, i.e., it defines the direction of one of the principal axes of the characteristic surface and the respective tensor.

The direct method for finding the principal axes of the tensor, which involves solving cubic equation (1.58), is not very convenient for numerical calculations. Therefore, in practice one generally uses approximate methods: the least squares method for monoclinic crystals and the method of successive approximations for triclinic crystals (see, for instance, [1.5]).

1.3.6 Other Indicatrice Surfaces for a Second-Rank Tensor

For the property described by symmetric second-rank tensor T_{ij} we may introduce, besides the characteristic surface, another useful indicatrice surface, whose equation, reduced to the main axes T_{ij}, has the form

$$\frac{x_1^2}{T_1^2}+\frac{x_2^2}{T_2^2}+\frac{x_3^2}{T_3^2}=1 \ . \tag{1.59}$$

This surface, with any signs of the main components of the tensor, is an ellipsoid with semiaxes $|T_1|$, $|T_2|$, and $|T_3|$.

Let us consider the properties of this surface. Suppose $a_i = T_{ij} b_j$. We assign unit vector $\boldsymbol{b} = [b_1, b_2, b_3]$ in the main coordinate system of tensor T_{ij}. Then $\boldsymbol{a} = [T_1 b_1,\ T_2 b_2,\ T_3 b_3]$. Since according to the condition we have $b_1^2 + b_2^2 + b_3^2 = 1$, then

$$\frac{a_1^2}{T_1^2}+\frac{a_2^2}{T_2^2}+\frac{a_3^2}{T_3^2}=1 \ .$$

Comparing the last expression with (1.59), we can draw the following conclusion (Fig. 1.5). While unit vector $\boldsymbol{b}$, in rotating, describes a circle with its terminus, the terminus of vector $\boldsymbol{a}$ will describe an ellipsoid (1.59). The orientations of vectors $\boldsymbol{b}$ and $\boldsymbol{a}$ will generally be different and will coincide only in the direction of the axes of ellipsoid (1.59).

Thus, surface (1.59) characterizes the length of vector $\boldsymbol{a}$ (for a unit $\boldsymbol{b}$) if the direction of $\boldsymbol{a}$ has already been defined, for instance, with the aid of the radius vector and the normal to the characteristic surface of tensor T_{ij}. Therefore, surface (1.59) is called the *ellipsoid of the values of symmetric second-rank tensor* T_{ij}.

For tensor T_{ij} (not necessarily symmetric) it is possible to construct one more indicatrice surface whose radius vector will be equal to T, the value characterizing the property in the given direction. This surface, which is given by (1.50),

$$T = T'_{11} = \alpha_{1i}\,\alpha_{1j}\,T_{ij} \ ,$$

can naturally be called the *surface of the longitudinal component of tensor* T_{ij}.

For symmetric second-rank tensor T_{ij} with essentially positive principal components T_i, the equation of this surface, reduced to the principal axes of the tensor, takes the form (1.49)

$$x_1 = e_1 T \ , \quad x_2 = e_2 T \ , \quad x_3 = e_3 T \ .$$

Substituting these expressions into (1.49) and bearing in mind that $T^2 = x_1^2 + x_2^2 + x_3^2$, we get

$$(T_1 x_1^2 + T_2 x_2^2 + T_3 x_3^2)^2 = (x_1^2 + x_2^2 + x_3^2)^3 \ , \tag{1.60}$$

which is the ovaloid equation.

Thus, the surface of the longitudinal component of symmetric tensor T_{ij} (for $T_i > 0$) is a sixth-order surface, i.e., an ovaloid.

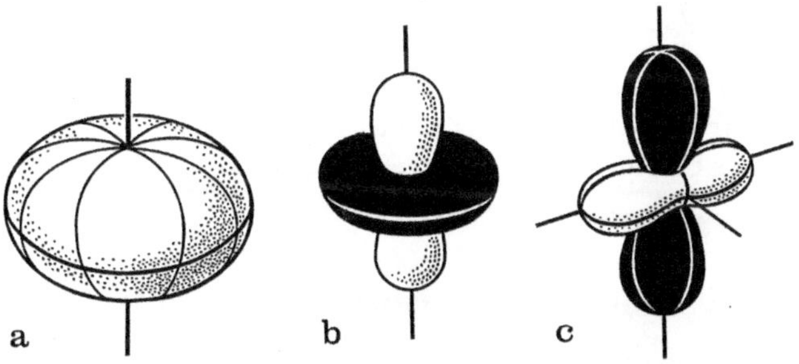

Fig. 1.6. Examples of indicatrice surfaces of the tensor of thermal expansion α_{ij} at various relationships among the signs of the main components of the tensor α_1, α_2, and α_3: **(a)** uniaxial crystals with $\alpha_3 > 0$ and $\alpha_1 > 0$, surface symmetry ∞/mmm; **(b)** uniaxial crystals with $\alpha_3 > 0$ and $\alpha_1 < 0$, surface symmetry ∞/mmm; **(c)** biaxial crystals with $\alpha_1 \neq \alpha_2 \neq \alpha_3$; α_1, $\alpha_2 > 0$, while $\alpha_3 < 0$; surface symmetry *mmm*. The regions corresponding to $\alpha > 0$ are white, and those corresponding to $\alpha < 0$ are black

It is easy to see that if one or two values T_i of symmetric tensor T_i are negative, the surface of the longitudinal component of this tensor will not be a surface of one sheet. Such a case is encountered, for instance, for the tensor of the coefficients of linear expansion of some crystals (Fig. 1.6).

1.3.7 Indicatrice Surfaces for Higher-Than-Second-Rank Tensors

A tensor of a rank higher than the second rank cannot be interpreted geometrically by any single indicatrice surface. For each tensor of any rank, however, it is possible to construct various indicatrice surfaces, whose radius vectors are equal to value T characterizing some tensor component or other in a given direction.

Thus, the surface described by the radius vector assigned by (1.51),

$$T = T'_{111\ldots1} = \alpha_{1i}\alpha_{1j}\alpha_{1k}\ldots\alpha_{1p}T_{ijk\ldots p} \,,$$

will represent the surface of the "longitudinal" component of tensor $T_{ijk\ldots p}$. In general, the surface whose equation is written in the form of (1.53),

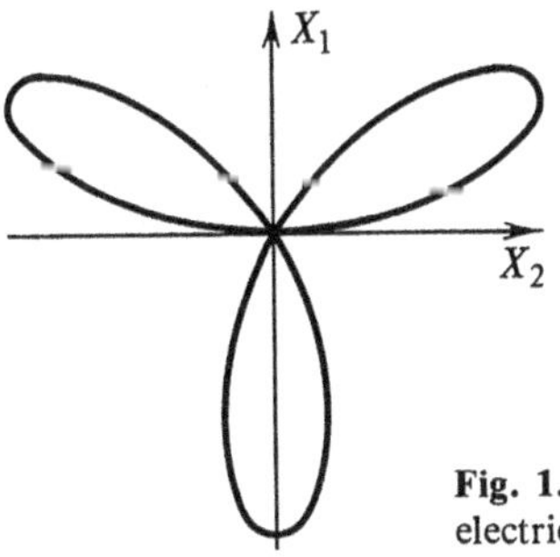

Fig. 1.7. Sections of the indicatrice surface of the longitudinal piezoelectric coefficient d_{33} for a quartz crystal

$$T = T'_{ijk\ldots} = \alpha_{is}\,\alpha_{jt}\,\alpha_{ku}\ldots T_{stu\ldots}\ ,$$

will be the surface of the $ijk\ldots$th component of tensor $T_{stu\ldots}$. The construction of such surfaces is often used for pictorial description of piezoelectric, elastic, and other properties of particular crystals (Fig. 1.7).

1.4 Proper (Exterior) Symmetry of Tensors

1.4.1 General Notions of the Proper Symmetry of Tensors

In the preceding volumes, when considering the concept of symmetry, we dealt with the symmetry of geometric or material objects: crystal individuals, crystal structure, reciprocal lattice, etc. As already noted in Sect. 1.1, symmetry is naturally also inherent in the physical properties of crystals. Since most of these properties are described by tensors, we first consider the proper (or, as it is often called, exterior) symmetry of tensors [1.6].

In accordance with the general definition of point symmetry, whose operations are assigned by some matrix of orthogonal transformations $\| \alpha_{ij} \|$, we assume that some Nth-rank tensor possesses a given symmetry operation if this operation transforms all the components of this tensor into themselves, i.e.,

$$\alpha_{pi}\,\alpha_{rj}\,\alpha_{sk}\ldots T_{ijk\ldots} = T'_{prs\ldots} = T_{ijk\ldots}\ . \tag{1.61}$$

Thus, the group of the *proper symmetry of a tensor* is point group G_0^3 of orthogonal transformations with respect to which all the tensor components are invariant. These components are naturally also invariant with respect to any other subgroup of group G_0^3.

We emphasize that the proper symmetry of a tensor, and also the number of its independent components (Sect. 1.5) are invariant properties of the tensor, i.e., are independent of its notation in a particular coordinate system.

For pictorialness, whenever possible tensors are supplemented with their geometric images, which are definite geometric figures whose symmetry completely coincides with that of the respective tensors.

We shall now dwell on the general symmetry properties of tensors. Let us consider the inversion operation, which is assigned by the matrix

$$\left\| \begin{matrix} -1 & 0 & 0 \\ 0 & -1 & 0 \\ 0 & 0 & -1 \end{matrix} \right\| .$$

In accordance with such a matrix, the equations for the inversion transformation of the components of an Nth-rank polar tensor are written

$$T'_{ijk\ldots} = \alpha_{ip}\,\alpha_{jq}\,\alpha_{kr}\ldots T_{pqr\ldots}$$
$$= \alpha_{ii}\,\alpha_{jj}\,\alpha_{kk}\ldots T_{ijk\ldots}\ldots = (-1)^N T_{ijk\ldots} \; . \quad (1.62)$$

From (1.62) it follows that under inversion of an even-rank tensor, $T'_{ijk\ldots} = T_{ijk\ldots}$, and under the same transformation of an odd-rank tensor, $T'_{ijk\ldots} = -T_{ijk\ldots}$. Consequently, any even-rank polar tensor has a center of symmetry, and any odd-rank polar tensor has no center of symmetry.

Since an inversion transformation is a second-kind transformation, then, in accordance with (1.23), the equation for the inversion transformation of the components of an Nth-rank pseudotensor (axial tensor) will differ from (1.62) by an additional factor (-1) in the right-hand side, i.e.,

$$P'_{ijk\ldots} = (-1)^{N+1} P_{ijk\ldots} \; . \quad (1.63)$$

Hence, any even-rank pseudotensor has no center of symmetry, and any odd-rank pseudotensor is centrosymmetric. Thus, a scalar, axial vector, or second-rank polar tensor has a center of symmetry, whereas a pseudoscalar, polar vector, or second-rank pseudotensor has none.

1.4.2 Limiting Groups of Symmetry

The proper symmetry groups of tensors are not exhausted by finite point groups, but may also belong to so-called limiting groups of symmetry, which we shall consider here (see also [Ref. 1.1, Chap. 2]).

Limiting (infinite, continuous) point symmetry groups are point groups containing operations of infinitely small rotations, i.e., symmetry axes of infinite order ∞. P. Curie showed that there are seven such groups:

$$\infty, \quad \infty mm, \quad \infty 22, \quad \infty/m, \quad \infty/mmm, \quad \infty/\infty, \quad \infty/\infty mm \; .$$

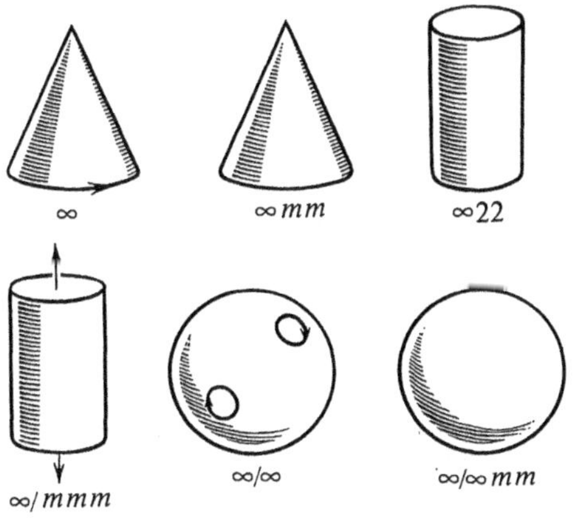

Fig. 1.8. Specimen figures for limiting point symmetry groups. One arrow on a figure indicates its rotation, two arrows indicate curling; a closed arrow on the surface of the sphere indicates that all its radii are curled in one direction, either to the right or to the left

They can easily be memorized from the simplest model figures possessing the respective symmetry (Fig. 1.8).

Of the seven limiting groups five describe anisotropic media and have the unique axis (∞ axis): ∞, ∞mm, $\infty 22$, ∞/m, ∞/mmm. Two groups describe isotropic media: ∞/∞, $\infty/\infty mm$. Of the seven groups, the following have a center of symmetry: ∞/m, ∞/mmm, $\infty/\infty mm$. The remaining four – ∞, ∞mm, $\infty 22$, ∞/∞ – are noncentrosymmetric. Among the latter, three permit of enantiomorphism; these are ∞, $\infty 22$, ∞/∞. One of the three groups (group ∞), together with the fourth noncentrosymmetric one (∞mm), is polar, i.e., has a unique polar axis (∞ axis). Finally, the same group ∞ and group ∞/m are axial, i.e., they have a unique axial axis (∞ axis). The scheme of subordination of limiting groups is given in [Ref. 1.1, Sect. 2.6.4], to which we refer the reader.

1.4.3 Proper Symmetry of Scalars, Pseudoscalars, and Vectors

It is easy to see that the symmetry groups of a scalar and a pseudoscalar are limiting groups $\infty/\infty mm$ and ∞/∞, respectively. Indeed, from Fig. 1.8 it is clear that as a scalar coincides with itself on any transformation of a full orthogonal group $\infty/\infty mm$, a pseudoscalar, too, coincides with itself on any rotations or first-kind transformations, i.e., it is described by group ∞/∞. On any second-kind transformations a pseudoscalar changes the "sign" of curling of the radii (it transfers to the enantiomorphous form).

Let us now find the symmetry elements of polar vector $\boldsymbol{a}$. We choose a coordinate system so that the positive direction of one of the axes, for instance, X_3, coincides with direction of vector $\boldsymbol{a}$ ($\boldsymbol{a} = [0, 0, a_3]$). Then on any rotation about the X_3 axis assigned by the matrix

$$\left\| \begin{matrix} \alpha_{11} & -\alpha_{12} & 0 \\ \alpha_{12} & \alpha_{11} & 0 \\ 0 & 0 & 1 \end{matrix} \right\| ,$$

we have $a_3' = a_3 = a_i$. Hence, the polar vector has a symmetry axis ∞. In the same way, a reflection in a plane longitudinal with respect to the vector, for instance X_1X_3, leads to $a_i' = a_3 = a_i$. At the same time a reflection in transverse plane X_1X_2 leads to $a_i' = -a_3 = -a_i$. Consequently, longitudinal plane X_1X_3 is the symmetry plane of vector $\boldsymbol{a}$, and the number of such symmetry planes is infinite because the vector has a symmetry axis ∞. On the other hand, vector $\boldsymbol{a}$ does not have a transverse symmetry plane (or a center of symmetry, as already noted in Sect. 1.4.1).

Thus, the polar vector possesses a symmetry ∞mm. It is customary to use a straight arrow

as the geometric image of a polar vector.

Proceeding to find the symmetry of an axial vector, of which we so far know only that, being a first-rank pseudotensor, it is centrosymmetric (see the beginning of this section), we recall that under first-kind transformations an axial vector behaves as a polar one (Sect. 1.2). The ∞ axis is a first-kind symmetry element, and hence an axial vector, as well as a polar one, possesses an ∞ axis. A mirror reflection is a second-kind transformation. This means that on reflection the components of the axial vector will change sign additionally as compared with those of the polar vector. Therefore, from the consideration carried out for a polar vector, it is clear that an axial vector has no longitudinal symmetry plane, whereas it does have a transverse one.

Thus, the symmetry of an axial vector is ∞/m. Obviously a straight arrow can no longer be the geometric image of an axial vector (although it is often erroneously assumed that it can), because it does not reflect its symmetry. *Shubnikov* suggested that an axial vector be represented by a figure of the respective symmetry, a line segment piercing the center of a circular arrow:

As with a polar vector, we can distinguish two ends (south and north poles) in an axial vector. But in a polar vector the ends are not equal, either congruently, or specularly, whereas the ends of an axial vector are equal specularly, although not congruently. This vividly demonstrates the difference between electrical and magnetic vectors.

1.4.4 Proper Symmetry of Second-Rank Polar Tensors

We now pass on to the symmetry of a symmetric second-rank polar tensor. Instead of seeking its symmetry analytically, we shall consider the symmetry of its characteristic surface, recalling that this symmetry, being its geometric image, possesses all its properties and hence the identical symmetry (Sect. 1.3).

The characteristic surface of a symmetric second-rank tensor T_{ij}, reduce to the diagonal form, is described by (1.45). We assume, for definiteness, that all the principal coefficients of surface (1.45) (the principal components of tensor T_{ij}) $T_i>0$, and find the symmetry of surface (1.45) at different relationships between T_i, taking into account the known centrosymmetricity of tensor T_{ij}.[2]

It is easy to see that at $T_1 = T_2 = T_3$ the characteristic surface degenerates into a sphere, and hence the symmetry of tensor T_{ij} is then $\infty/\infty mm$, i.e., tensor T_{ij} degenerates into a scalar. At $T_1 = T_2 \neq T_3$ surface (1.45) is an ellipsoid of rotation with the ∞ axis along X_3. Consequently, the symmetry of tensor T_{ij} is then ∞/mm. Finally, at $T_1 \neq T_2 = T_3$, surface (1.45) turns into a triaxial ellipsoid with a symmetry *mmm*. The same symmetry is exhibited by tensor T_{ij}. Here the twofold axes and the normals to the symmetry planes coincide with the principal axes of the characteristic surface and of tensor T_{ij}.

Thus, a symmetric second-rank polar tensor has a proper symmetry *mmm*, ∞/mmm, or $\infty/\infty mm$, depending on the relationship between its principal components. Since only a symmetric tensor has a characteristic surface describing it unambiguously and can be reduce to the diagonal form, the reverse is also true, namely, any second-rank polar tensor which has such a proper symmetry is a symmetric tensor.

By using the general approach in accordance with (1.61), it is easy to show that an antisymmetric polar tensor has a symmetry ∞/m, the same as an axial vector. This is understandable considering that an antisymmetric second-rank polar tensor is dual with respect to an axial vector (Sect. 1.2.5).

It is now easy to find the types of symmetry of an nonsymmetric second-rank polar tensor. We take into account (Sect. 1.2.5) that it can always be represented as a sum of a symmetric and an antisymmetric second-rank polar tensor, and the mutual orientation of the principal axes of these parts may be different; besides, a symmetric tensor may have different relationships between the principal (diagonal) components. Then, in accordance with the Curie principle (Sect. 1.5), a nonsymmetric second-rank polar tensor has a symmetry defined by the common symmetry elements of its symmetric and antisymmetric parts at a given mutual arrangement of these elements. If the symmetry group of the symmetric part contains an ∞ axis and if this axis is parallel to the ∞ axis of the antisymmetric part, the total symmetry of their set and hence the symmetry of the nonsymmetric second-rank polar tensor is ∞/m. If the ∞ axes are mutually perpendicular or if the symmetry axes and planes coincide, when the symmetric part has a symmetry *mmm*, the total symmetry of the set is $2/m$. Finally, at an arbitrary mutual orientation of the two parts, their set retains only center of symmetry $\bar{1}$.

Thus, a nonsymmetric second-rank polar tensor possesses the following types of proper symmetry: ∞/m, $2/m$, or $\bar{1}$.

---

[2] The case of $T_i<0$ does not affect the consideration, since the uniaxial and biaxial hyperboloids have the same symmetry as the respective ellipsoid. The presence of $T_{ij}=0$ does not affect the consideration either because then we deal with a degenerate figure retaining its symmetry.

1.4.5 Proper Symmetry of Second-Rank Pseudotensors

We now proceed to an analysis of the symmetry of second-rank pseudotensors. A symmetric second-rank pseudotensor P_{ij}, as well as a symmetric second-rank polar tensor T_{ij}, can be reduced to the diagonal form. The characteristic surface for tensor P_{ij} will (for $P_i > 0$) have the same form and symmetry as for tensor T_{ij} (for $T_i > 0$), except that it cannot possess a center of symmetry in accordance with the noncentrosymmetricity of P_{ij} (1.62). All the radius vectors of characteristic surface P_{ij} must be visualized as wound to the left or to the right. Therefore, excluding the center of symmetry from the number of symmetry elements of tensor T_{ij}, we get, at $P_1 = P_2 = P_3$, symmetry group ∞/∞ for tensor P_{ij} and for its characteristic surface (the pseudotensor degenerates into a pseudoscalar); at $P_1 = P_2 \neq P_3$, group $\infty 22$; and at $P_1 \neq P_2 \neq P_3$, group 222. By adding to the consideration an incomplete symmetric pseudotensor with $P_1 = -P_2$, $P_3 = 0$, it is possible to show that such a tensor has a symmetry $\bar{4}2m$ (axis $\bar{4}$ coincides with X_3, and axes 2 coincide with X_1 and X_2). For a symmetric second-rank polar tensor, such a case does not lead to a new symmetry group.

Thus, a symmetric second-rank pseudotensor possesses a symmetry ∞/∞, $\infty 22$, 222 or $\bar{4}2m$, depending on the relationships among its principal components.

It is easy to show by direct checkup that an antisymmetric second-rank pseudotensor has a symmetry ∞mm, the same as a polar vector. This is understandable, since an antisymmetric second-rank pseudotensor is dual with respect to a polar vector (Sect. 1.2.5).

For a total nonsymmetric second-rank pseudotensor we obtain, at different relationships among its diagonal components and different mutual orientation of the principal axes of its symmetric and antisymmetric parts, groups ∞, 2, and 1 by excluding the center of symmetry from the number of symmetry elements of the respective nonsymmetric second-rank polar tensor. By adding to the consideration an incomplete nonsymmetric pseudotensor

Table 1.1. Proper symmetry of scalars, pseudoscalars, vectors, and second-rank tensors

Tensor quantity	Symmetry group
Scalar (zero-rank polar tensor)	$\infty/\infty mm$
Pseudoscalar (zero-rank pseudotensor)	∞/∞
Polar vector (first-rank polar tensor)	∞mm
Axial vector (first-rank pseudotensor)	∞/m
Symmetric second-rank polar tensor	$\infty/\infty mm$, ∞/mmm, mmm
Nonsymmetric second-rank polar tensor	∞/m, $2/m$, $\bar{1}$
Antisymmetric second-rank polar tensor	∞/m
Symmetric second-rank axial tensor (pseudotensor)	∞/∞, $\infty 22$, 222, $\bar{4}2m$
Nonsymmetric second-rank axial tensor	∞, 2, 1, $mm2$, m
Antisymmetric second-rank axial tensor	∞mm

with $P_3 = 0$, it is easy to show that the latter will have a symmetry *mm*2, *m*, or 1, depending on the mutual orientation of the symmetry elements of groups $\bar{4}2m$ and ∞mm.

The results of the above consideration are summarized in Table 1.1 (see also [1.7]).

1.4.6 Proper Symmetry of Higher-Rank Tensors

For tensors of ranks above the second rank it is more difficult to find the proper symmetry because one cannot construct characteristic surfaces for them; neither can one, incidentally, for nonsymmetric second-rank tensors. Although these tensors cannot be reduced to the diagonal form, there does exist for them an orthogonal coordinate system as well, in which the matrix of tensor components is written in the simplest form. It would be natural to call such a coordinate system for a tensor of any rank, as well as for a symmetric second-rank tensor (Sect. 1.3), the *principal coordinate system of the tensor.* Obviously, the principal coordinate axes coincide with the symmetry elements of the tensor (if they exist). If the tensor symmetry is spherical, then any coordinate system is the principal one. If the symmetry group of the tensor contains a unique axis ∞, then in the plane perpendicular to the ∞ axis any axes are the principal ones.

The problem of finding the proper symmetry of a tensor of any rank reduces to writing the tensor in the principal coordinate system and verifying the fulfillment of equalities (1.61) $T'_{ijk\ldots} = T_{ijk\ldots}$ (or $P'_{ijk\ldots} = P'_{ijk}\ldots$) under symmetry transformations.

1.5 Symmetry of Physical Properties

1.5.1 Material and Field Tensors

So far, in considering the general properties and the proper symmetry of tensors we could ignore their particular physical content. Now, however, it is necessary to mention that physical tensors may be of two different types, depending on their relation to the object, namely, tensors describing the properties of the crystal (i.e., the relationships of the measurable values), so-called *material tensors*, and tensors describing the effects on the crystal and its response, so-called *field tensors*.

The symmetry of material tensors must correspond to that of the crystal in accordance with the Neumann principle (Sect. 1.5.4), and the symmetry elements of these tensors and of their characteristic and indicatrice surfaces must coincide with the existing respective symmetry elements of the crystal.

In contrast to the material tensors, the symmetry of field tensors is not associated with that of the crystal, and tensors may have any orientation with respect to the symmetry elements of the crystal. For instance, one can apply to a crystal of any symmetry an arbitrarily oriented electric field (polar vector) or mechanical stress (a symmetric second-rank polar tensor); one can assign, with equal right, an arbitrarily directed polarization or any deformation component in a crystal of any symmetry. We should remember, however, that when we assign, for example, an arbitrarily oriented mechanical stress (deformation), the reaction in the form of deformation (stress) obtained will already depend on the symmetry of the crystal, because the reaction is defined by the elastic properties of the crystal.

In different situations the same physical tensors may play the role of either material or field tensors. Thus, the vector of polarization $\boldsymbol{P}$, as just noted, is usually a field tensor, but in pyroelectrics (and ferroelectrics) vector $\boldsymbol{P}_s$ describes the property of spontaneous polarization and must correspond to the symmetry of the crystal [see (1.66) below]. The deformation tensor, which is usually a field tensor, becomes material and must correspond to the symmetry of the crystal when we describe the spontaneous deformation of ferroelastics (Sect. 1.5.7).

As far as field tensors are concerned, there is no difference between isotropic and anisotropic media. This difference exists with reference to material tensors describing the physical properties. Therefore, we shall consider only the latter type.

1.5.2 Crystallophysical System of Coordinates

As has just been noted, the elements of the proper symmetry of any tensors describing the physical properties of crystals must coincide with the respective elements of the point symmetry of the crystal (if they exist). But in a number of crystal classes symmetry is insufficient for rigid fixation of the principal system of coordinates of a tensor of some rank or other with respect to the crystal. Understandably, for classes m, 2, $2/m$, 4, $\bar{4}$, $4/m$, 3, $\bar{3}$, $\bar{6}$, 6 and $6/m$ only one axis of the principal coordinate system is fixed, generally speaking, and for classes 1 and $\bar{1}$, none at all. As a result, the principal coordinate system of the tensor in crystals of these classes may change its orientation, for instance, because of the temperature or frequency dispersion of the property described or owing to a change in other scalar parameters. These changes in orientation will generally be different for tensors describing different properties of crystals. In most cases it is inconvenient to use such a mobile coordinate system. Considering this and also the fact that the notation of the tensor in a particular coordinate system depends on the orientation of this system with respect to the principal coordinate system of the tensor, it is customary to use, for describing tensor and, in general, anisotropic physical properties of crystals, a system of orthogonal axes chosen according to definite rules and called the *crystallophysical system of coordinates*.

Clearly in the general case it is impossible to use the crystallographic system of coordinates as such a system [Ref. 1.1, Chap. 3], since the former is

Table 1.2. Rules for selecting a crystallophysical coordinate sytem

Syngony	Orientation with respect to crystallographic axes	Orientation with respect to symmetry elements of crystal
Triclinic	$X_3 \parallel c$, or $X_2 \parallel b$, or $X_1 \parallel a$	
Monoclinic	$X_2 \parallel b$ and $X_1 \parallel a$ (or $X_3 \parallel c$) Sometimes $X_3 \parallel b$ and $X_1 \parallel a$ (or $X_2 \parallel c$)	$X_2 \parallel$ to axis 2 or $\perp$ to plane m. Sometimes $X_3 \parallel$ to axis 2 or $\perp$ to plane m
Orthorhombic	$X_3 \parallel c$, $X_2 \parallel b$, $X_1 \parallel a$	$X_3 \parallel$ to axis 2; X_1, $X_2 \parallel$ to other axes 2 or $\perp$ to planes m
Tetragonal	$X_3 \parallel c$, $X_2 \parallel b$, $X_1 \parallel a$	$X_3 \parallel$ to axis 4 (or $\bar{4}$); X_1, $X_2 \parallel$ to axes 2 or $\perp$ to planes m (if they exist); for class $\bar{4}2m$ usually X_1, $X_2 \parallel$ to axes 2
Trigonal and hexagonal	$X_3 \parallel c$, $X_1 \parallel a$. For classes $3m$ and $\bar{6}m2$ sometimes $X_2 \parallel b$	$X_3 \parallel$ to axis 3, $\bar{3}$, 6 or $\bar{6}$; $X_1 \parallel$ to axis 2 (if they exist, except class $\bar{6}m2$); for classes $3m$ and $\bar{6}m2$ usually $X_1 \perp$ to plane m, but sometimes $X_2 \perp m$
Cubic	$X_3 \parallel c$, $X_2 \parallel b$, $X_1 \parallel a$	X_1, X_2, $X_3 \parallel$ to three mutually perpendicular axes 4 (or $\bar{4}$), and if they do not exist, then to similar axes 2

not orthogonal in crystals of low-symmetry syngonies. But the crystallophysical axes are bound with the crystallographic axes as rigidly as possible: with all of them (for cubic, tetragonal, and orthorhombic crystals), with two (for hexagonal, trigonal, and monoclinic crystals) or, at least, with one (for triclinic crystals).

The generally accepted rules for selection of crystallophysical axes of coordinates and their versions are given in Table 1.2 (we denote them also by X_1, X_2, and X_3). As well as the crystallographic system, the crystallophysical one is usually chosen to be right-handed, and the angles between the positive direction of the respective axes of both systems are taken to be less than 90°.

Attention should be given to the ambiguity of the choice of the crystallophysical axes for a number of classes. The ambiguity of the choice of the crystallographic axes increases the arbitrariness still further. This must be taken into consideration when comparing the results of different authors' measurements of the physical properties of particular crystals. The same circumstance obliges the researcher to ensure unambiguous setting of the crystal in experiment and describe it in clear terms when presenting the results.

Note that sometimes, in measurements carried out with monoclinic and triclinic crystals, it is convenient, in order to ensure a close relationship between the "crystallophysical" and the laboratory system of coordinates, to take, as the "crystallophysical" system, the principal system of the tensor of the property being measured under definite conditions (for instance, in optical measurements, the main axes of the optic indicatrix with fixed values of wavelength λ and temperature T).

1.5.3 Relationship Between the Proper Symmetry of Tensors and the Symmetry of the Physical Properties Described by Them

As mentioned in the preceding section, in crystals of some classes (they are in the majority), the principal system of coordinates of any tensor describing a physical property always retains its orientation in the crystal and coincides with the crystallophysical coordinate system fixed in the crystal. In crystals of other classes the principal coordinate systems of some, or even any, tensors describing physical properties do not retain their orientation in the crystal upon a change in temperature, the frequency of the measuring field, and other scalar parameters. In this case, if the crystallophysical system of coordinates is fully fixed in the crystal, such a property will be described in the crystallophysical system by a set of differently oriented tensors, whose principal axes rotate about one of the crystallophysical axes which coincides with the fixed main axis of the tensors. If, however, the crystallophysical system is not fixed completely with respect to the crystallographic axes, the property will be described by a set of differently oriented tensors whose principal axes rotate about the fixed origin. Thus, the principal coordinate system of the tensor describing the physical properties may have in the crystal additional degrees of freedom as compared with the crystallophysical system of coordinates.

Obviously, in this case the general symmetry of the set of tensors describing the given physical property, as viewed in the crystallophysical system of coordinates, will be below the proper symmetry of the tensor, which manifests itself only in its mobile main system of coordinates. Therefore, it is expedient to introduce, along with the concept of the proper symmetry of a tensor, the concept of the symmetry of the physical property described by it. The *symmetry of a physical property* is the general symmetry of a set of tensors describing the given property in the crystallophysical coordinate system at different values of the scalar parameters (temperature, frequency, hydrostatic pressure, etc.).

When the principal coordinate system of the tensor retains its orientation in the crystal, the symmetry of the physical property and the proper symmetry of the tensor describing it are equal; in the opposite case, the symmetry group G_{pr} of the property is a subgroup of the group of the proper symmetry of the respective tensor G_{T}, i.e.,

$$G_{\text{pr}} \subseteq G_{\text{T}} \tag{1.64}$$

(the symbol $\subset$ means inclusion). Then in the case $G_{pr} \neq G_T$, to the given G_{pr} corresponds group G_T, which is its nearest supergroup (if there are several groups G_T).

It should be emphasized that, as well as the proper symmetry of the tensor, the symmetry of the physical property described by it is invariant to transformations of the coordinates, i.e., is independent of the choice of particular orthogonal crystallophysical axes. We also note that if we do not take into account the possible dependence of the tensor orientation on the scalar parameters (for instance, if we consider the optical properties of monoclinic and triclinic crystals at a constant temperature and wavelength), then always $G_{pr} = G_T$.

Let us consider some examples. In a crystal of class m, vector $\boldsymbol{a}$ describing some physical property lies necessarily in the mirror plane, but its direction in this plane is not fixed and in a crystal of class 1 the direction of vector $\boldsymbol{a}$ is not fixed in space. It is easy to see that although the proper symmetry of vector $\boldsymbol{a}$ is ∞mm, the polar-vector property in the crystals of these crystal classes will have a symmetry m and 1, respectively. Similarly, for crystals of Classes $\bar{1}$

Table 1.3. Symmetry of physical properties of crystals

Tensor quantity describing property	Symmetry groups of property, G_{pr}
Scalar (zero-rank polar tensor)	$\infty/\infty mm$
Pseudoscalar (zero-rank pseudotensor)	∞/∞
Polar vector (first-rank polar tensor)	$1, m, \infty mm$
Axial vector (first-rank pseudotensor)	$\bar{1}, \infty m$
Symmetric second-rank polar tensor	$\bar{1}, 2/m, mmm, \infty/mmm, \infty/\infty mm$
Nonsymmetric second-rank tensor	$\bar{1}, 2/m, \infty/m$
Antisymmetric second-rank tensor	$\bar{1}, \infty/m$
Symmetric second-rank pseudotensor	$1, 2, 222, \infty 22, \infty/\infty, m, \bar{4}, \bar{4}2m$
Nonsymmetric second-rank pseudotensor	$1, 2, \infty, m, mm2$
Antisymmetric second-rank pseudotensor	$1, m, \infty mm$
Third-rank polar tensor symmetric with respect to two subscripts	$1, 2, 222, 3, 32, \infty, \infty 22, m, mm2, 3m, \infty mm, \bar{4}, \bar{2}m, \bar{6}, \bar{6}m2, \bar{4}3m$
Third-rank axial tensor symmetric with respect to two subscripts	$\bar{1}, 2/m, mmm, \infty/m, \infty/mmm, \bar{3}, \bar{3}m, m3m$
Fourth-rank polar tensor symmetric with respect to two pairs of subscripts and to their permutation	$\bar{1}, 2/m, mmm, 4/m, 4mmm, \bar{3}, \bar{3}m, \infty/mmm, m3m$
Fourth-rank polar tensor symmetric with respect to two pairs of subscripts	$\bar{1}, 2/m, mmm, 4/m, 4/mmm, \bar{3}, \bar{3}m, \infty/m, \infty/mmm, m3, m3m$

and 1, the axial vector describing the physical property may have any orientation in space. The axial vector fan retains only the center of symmetry. Hence, the axial-vector property in triclinic crystals has a symmetry $\bar{1}$.

For crystals of the monoclinic system, the orientation of only one axis of the characteristic surface of the symmetric second-rank polar tensor (the triaxial ellipsoid with a symmetry *mmm*) is fixed, and, for crystals of the triclinic system, none at all. Therefore, the symmetry of sets of differently oriented characteristic surfaces of such a tensor for monoclinic and triclinic crystals will clearly be $2/m$ and $\bar{1}$, respectively. It is obvious that the physical properties described by symmetric second-rank polar tensors with a proper symmetry *mmm* in monoclinic and triclinic crystals will possess a symmetry $2/m$ and $\bar{1}$, respectively.

Similarly, the physical properties described by symmetric second-rank pseudotensors will have a symmetry 2 and 1, respectively, in monoclinic and triclinic crystals, while the proper symmetry of the tensors will be described by group 222. For a crystal of class 4, only one principal axis of the tensor is fixed, and therefore the symmetry of the property described by an incomplete symmetric second-rank pseudotensor with a proper symmetry $\bar{4}/2m$ will be $\bar{4}$.

The summary of the kinds of symmetry for the physical properties described by tensors of different ranks is given in Table 1.3 (cf. Table 1.2). The symmetry groups of properties which are not simultaneously groups of proper symmetry of the tensors describing them are underlined in the table. The closest group standing in the line on the right-hand side and not underlined (i.e., the nearest higher symmetry group of the property, coinciding with the symmetry group of the tensor) will be the group of the proper symmetry of the tensor for such cases.

1.5.4 Relationship Between the Point Symmetry of a Crystal and the Symmetry of Its Physical Properties

We have considered the proper symmetry of tensors and the symmetry of the physical properties described by them. Let us now see how the point symmetry of a crystal (the symmetry of its ideal habit) is related to the symmetry of its physical properties. This question can be answered by using the fundamental postulate of crystal physics known as the *Neumann principle*. It can be formulated as follows.

The symmetry group of any physical property, G_{pr}, must include all the elements of the point group (crystal class) of the crystal K; in other words, group K either coincides with group G_{pr} or its subgroup. Consequently, allowing for (1.64), we can write

$$K \subseteq G_{\mathrm{pr}} \subseteq G_{\mathrm{T}} \ . \tag{1.65}$$

The Neumann principle is obvious enough. Indeed, if, in violation of the principle, the crystal had a physical property whose symmetry group G_{pr}

did not contain one of the symmetry operations of group K, then, under the action of this operation, the crystal would, on the one hand, completely coincide with itself and, on the other, change the physical properties. If the Neumann principle is fulfilled, there will be no such contradiction.

Note that the Neumann principle only asserts the possibility of the existence in crystals of the properties satisfying condition (1.65), but does not require their mandatory presence, i.e., the fulfillment of (1.65) is a necessary, but not a sufficient condition for the existence of particular physical properties in a crystal. At the same time, the presence of a property not obeying condition (1.65) is a priori impossible, i.e., nonfulfillment of (1.65) is a necessary and a sufficient condition for the absence of the given property in the crystal: "Symmetry definitely forbids the existence of a property, but if it admits a property, it does it with a probability only" (A. V. Shubnikov). If, however, a particular crystal does show a physical property not obeying condition (1.65), this will only mean that the point group of the crystal has been determined incorrectly.

Let us consider some consequences of (1.65) (see also Sects. 1.5.1, 2).

In accordance with (1.62) centrosymmetric crystals cannot possess properties described by odd-rank polar tensors and even-rank pseudotensors. Hence, centrosymmetric crystals cannot have pyro- or piezoelectric, linear electro-optical properties, optical activity, etc. This circumstance is utilized, for instance, in rapid checking of a substance for noncentrosymmetricity by testing for the presence of piezoelectric properties or the effect of second-harmonic generation (SHG) under laser beam radiation; both methods allow one to use a fine crystalline material. Obviously, the detection of the piezoelectric and SHG effects suggests unambiguously that the substance is noncentrosymmetric.[3] On the other hand, the nondetection of the effect does not allow one to draw an equally unambiguous opposite conclusion, because the effects may simply be weak.

Polar-vector properties can be found only in crystals whose groups K are subgroups of polar vector symmetry group ∞mm, i.e., crystals of ten crystal classes:

$$1, 2, 3, 4, 6, m, 2mm, 3m, 4mm, 6mm \ . \tag{1.66}$$

These crystal classes are called *polar*. Clearly, these and only these classes may include pyroelectric crystals and ferroelectric crystals in ferroelectric phases (Chap. 3).

---

[3] Strictly speaking, the presence of inhomogeneous internal stresses and other distortions in crystals may lower their symmetry to such an extent that morphic (Sect. 1.5.7) piezoelectric and second-harmonic generation effects may be observed even when the fine-crystalline substance is centrosymmetric.

Axial-vector properties may be found only in crystals whose groups K are subgroups of the axial-vector symmetry group ∞/m, i.e., crystals of 13 classes:

$$1, 2, 3, 4, 6, m, 2/m, 4/m, 6/m, \bar{1}, \bar{3}, \bar{4}, \bar{6} \ . \tag{1.67a}$$

Obviously, ferromagnetic crystals can belong to these and only these crystal classes below the Curie point (Chap. 4). Considering that the Neumann principle is valid within not only the ordinary, but also the magnetic symmetry, we can obtain a list of magnetic classes similar to (1.67a). To do this we must use the limiting groups of magnetic symmetry (antisymmetry) [Ref. 1.1, Chap. 2]. Then, since the magnetic symmetry of the axial magnetization vector is $\infty/mm'm'$, we can easily single out the following 31 subgroups of group $\infty/mm'm'$ from among the 90 point groups of magnetic symmetry K':

$$\begin{gathered}1, \bar{1}, m', 2', 2'/m', m, mm2', 2, m'm'2, 22'2', 2/m, m'm'm,\\ \bar{4}, \bar{4}2'm', 4, 4m'm', 42'2', 4/m, 4/mm'm', 3, 3m', 32', \bar{6}, \bar{6}m'2',\\ \bar{3}, \bar{3}m', 6, 62'2', 6m'm', 6/m, 6/mm'm' \ .\end{gathered} \tag{1.67b}$$

Obviously, the ferromagnetic phases of any ferromagnetic crystals can belong to these and only these classes of magnetic symmetry [1.8].

Since the highest proper symmetry of any symmetric second-rank polar tensors $\infty/\infty mm$ is the symmetry of a scalar, the properties described by these tensors can evidently be inherent in all crystals. Hence, we draw a rather trivial conclusion that all crystals can possess dielectric susceptibility, and electrostrictive, piezo-optical, etc. properties.

Similar consideration can be carried out for any other tensor properties. It should only be remembered that not for all tensors all the G_T (and the respective G_{pr}) are subgroups of one group. Therefore, the symmetry groups of crystals which can possess such properties will also be subgroups, not of one, but of two or several groups (see below, example on piezoelectric properties).

The Neumann principle (1.65) can be formulated in a stronger form: the symmetry groups of the tensor, G_T, and of the corresponding property, G_{pr}, are not simply supergroups of the point group K of the crystal, but are its closest supergroups among all the possible symmetry groups of a tensor of this rank with a given interior symmetry and the symmetry groups of the corresponding property.

Consequently, through Tables 1.2, 3 and knowledge of the group K of the crystal, we can not only predict the tensor physical properties it can have, but also indicate immediately the symmetries of the corresponding tensors. Further, using the same tables we can not only establish the class of crystals which can have such a property, but also indicate at once to which group K each group G_{pr} (and G_T) corresponds.

Table 1.4. Symmetry of piezoelectric properties G_{pr} and of the tensors G_T describing them for crystal classes K permitting of these properties

G_T	G_{pr}	K	G_T	G_{pr}	K	G_T	G_{pr}	K
222	1	1	$\infty 22$	$\infty 22$	422	$\bar{4}2m$	$\bar{4}$	$\bar{4}$
	2	2		$\infty 22$	622		$\bar{4}2m$	$42m$
	222	222	$mm2$	m	m	$\bar{6}m2$	$\bar{6}$	$\bar{6}$
32	3	3		$mm2$	$mm2$		$\bar{6}m2$	$\bar{6}m2$
	32	32	$3m$	$3m$	$3m$	$\bar{4}3m$	$\bar{4}3m$	23
∞	∞	4	∞mm	∞mm	$4mm$		$\bar{4}3m$	$\bar{4}3m$
	∞	6		∞mm	$6mm$			

For example, a crystal of class $\bar{4}$ cannot possess properties described by a pseudoscalar and a polar vector. At the same time, crystals of class $\bar{4}$ can have properties described by (1) a scalar, (2) a symmetric second-rank polar tensor, (3) a third-rank polar tensor symmetric with respect to two subscripts, (4) a fourth-rank polar tensor symmetric with respect to two pairs of subscripts, (5) a fourth-rank polar tensor symmetric with respect to two pairs of subscripts and to their permutation, (6) an axial vector, (7) a symmetric second-rank pseudotensor, (8) a third-rank pseudotensor symmetric with respect to two subscripts, etc. The symmetry of these properties will be as follows (wherever G_T is different, it is indicated in parentheses): (1) ∞/mm, (2) ∞/mmm, (3) $\bar{4}$ ($\bar{4}2m$), (4) $4/m$ ($4/mmm$), (5) $4/m$ ($4/mmm$), (6) ∞/m, (7) $\bar{4}$ ($\bar{4}2m$), (8) ∞/m (∞/mmm), etc. As an example, let us now dwell on the properties described by third-rank polar tensors symmetric with respect to two subscripts (i.e. piezoelectric and linear electrooptical properties). From Table 1.3 it follows that, in accordance with (1.65), such properties can be exhibited by crystals of 20 classes out of the total number of 21 of the acentric crystal class distributed over the 16 groups G_{pr} and the 10 groups of G_T, according to Table 1.4. In crystals of the twenty-first class (432), no piezoelectric or symmetry-similar properties are possible, since their group K is not a subgroup of any G_{pr} for the piezoelectric tensor (Table 1.4).

Table 1.5 summarizes the results of such a consideration of the symmetry of the physical properties described by tensors of different ranks, for all the crystal classes (a dash in a line means that this property is impossible in crystals of the given class).

1.5.5 Form in Different Coordinate Systems of the Matrices of Tensors Describing the Physical Properties of Crystals of Different Classes

As we have mentioned more than once, a physical property, its symmetry, and a symmetry of the tensor describing it remain unchanged on transformation of the coordinate system. But the form of the matrix of a tensor of any rank

Table 1.5. Relationships between the symmetry of crystals and the symmetry of their physical properties

Group K	Symmetry of tensor properties[a]									
	1	2	3	4	5	6	7	8	9	10
1	1	$\bar{1}$	$\bar{1}$	$\bar{1}$	1	1	1	$\bar{1}$	$\bar{1}$	$\bar{1}$
$\bar{1}$	–	$\bar{1}$	$\bar{1}$	$\bar{1}$	–	–	–	$\bar{1}$	$\bar{1}$	$\bar{1}$
m	m	∞/m	$2/m$	$2/m$	m	m	m	$2/m$	$2/m$	$2/m$
2	∞mm	∞/m	$2/m$	$2/m$	2	2	2	$2/m$	$2/m$	$2/m$
$2/m$	–	∞/m	$2/m$	$2/m$	–	–	–	$2/m$	$2/m$	$2/m$
222	–	–	mmm	–	222	–	222	mmm	mmm	mmm
$mm2$	∞mm	–	mmm	–	–	$mm2$	$mm2$	mmm	mmm	mmm
mmm	–	–	mmm	–	–	–	–	mmm	mmm	mmm
4	∞mm	∞/m	∞/mmm	∞/m	$\infty 22$	∞	∞	∞/m	$4/m$	$4/m$
$\bar{4}$	–	∞/m	∞/mmm	∞/m	$\bar{4}$	–	$\bar{4}$	∞/m	$4/m$	$4/m$
$4/m$	–	∞/m	∞/mmm	∞/m	–	–	–	∞/m	$4/m$	$4/m$
422	–	–	∞/mmm	–	$\infty 22$	–	$\infty 22$	∞/mmm	$4/mmm$	$4/mmm$
$4mm$	∞mm	–	∞/mmm	–	–	–	∞mm	∞/mmm	$4/mmm$	$4/mmm$
$\bar{4}2m$	–	–	∞/mmm	–	$\bar{4}2m$	–	$\bar{4}2m$	∞/mmm	$4/mmm$	$4/mmm$
$4/mmm$	–	–	∞/mmm	–	–	–	–	∞/mmm	$4/mmm$	$4/mmm$
3	∞mm	∞/m	∞/mmm	∞/m	$\infty 22$	∞	3	$\bar{3}$	$\bar{3}$	$\bar{3}$
$\bar{3}$	–	∞/m	∞/mmm	∞/m	–	–	–	$\bar{3}$	$\bar{3}$	$\bar{3}$
32	–	–	∞/mmm	–	$\infty 22$	–	32	$\bar{3}m$	$\bar{3}m$	$\bar{3}m$

(Continued)

[a] The numerals in the column headings stand for the properties described by the following tensors: (1) a polar vector, an antisymmetric second-rank pseudotensor V, and $\varepsilon\{V^2\}$; (2) an axial vector, an antisymmetric second-rank polar tensor εV, and $\{V^2\}$; (3) a symmetric second-rank polar tensor $[V^2]$; (4) a nonsymmetric second-rank polar tensor V^2; (5) a symmetric second-rank pseudotensor $\varepsilon[V^2]$; (6) a nonsymmetric second-rank pseudotensor εV^2; (7) a third-rank polar tensor symmetric with respect to a pair of subscripts $V[V^2]$; (8) an axial third-rank tensor symmetric with respect to a pair of subscripts $\varepsilon V[V^2]$; (9) a fourth-rank polar tensor symmetric with respect to two pairs of subscripts and to their permutation $[[V^2]^2]$; (10) a fourth-rank polar tensor symmetric with respect to two pairs of subscripts $[V^2]^2$.

Table 1.5 (continued)

Group K	Symmetry of tensor properties[a]									
	1	2	3	4	5	6	7	8	9	10
$3m$	∞mm	–	∞/mmm	–	–	–	$3m$	$\bar{3}m$	$\bar{3}m$	$\bar{3}m$
$\bar{3}m$	–	–	∞/mmm	–	–	–	–	$\bar{3}m$	$\bar{3}m$	$\bar{3}m$
6	∞mm	∞/m	∞/mmm	∞/m	$\infty 22$	∞	∞	∞/m	∞/mmm	∞/m
$\bar{6}$	–	∞/m	∞/mmm	∞/m	–	–	$\bar{6}$	∞/m	∞/mmm	∞/m
$6/m$	–	∞/m	∞/mmm	∞/m	–	–	–	∞/m	∞/mmm	∞/m
622	–	–	∞/mmm	–	$\infty 22$	–	$\infty 22$	∞/mmm	∞/mmm	∞/mmm
$6mm$	∞mm	–	∞/mmm	–	–	–	∞mm	∞/mmm	∞/mmm	∞/mmm
$\bar{6}m2$	–	–	∞/mmm	–	–	–	$\bar{6}m2$	∞/mmm	∞/mmm	∞/mmm
$6mmm$	–	–	∞/mmm	–	–	–	–	∞/mmm	∞/mmm	∞/mmm
23	–	–	$\infty/\infty mm$	–	∞/∞	–	$\bar{4}3m$	$m3m$	$m3m$	$m3$
$m3$	–	–	$\infty/\infty mm$	–	–	–	–	$m3m$	$m3m$	$m3$
432	–	–	$\infty/\infty mm$	–	∞/∞	–	–	–	$m3m$	$m3m$
$\bar{4}3m$	–	–	$\infty/\infty mm$	–	–	–	$\bar{4}3m$	–	$m3m$	$m3m$
$m3m$	–	–	$\infty/\infty mm$	–	–	–	–	–	$m3m$	$m3m$

(not degenerated into a scalar or a pseudoscalar) essentially depends on the choice of the coordinate axes.

The matrix of a tensor has the simplest form in its principal coordinate system. For a tensor of a given rank with a given internal symmetry, to each G_T corresponds a particular form of matrix with definite relationships among the components and a definite number of independent components. Hence, the total number of different forms of the matrix of a tensor in its principal coordinate system is equal to the number of kinds of its G_T.

Since it is customary to describe the physical properties of crystals in the crystallophysical coordinate system, the tabulated values of the components of the tensor describing a given physical property are the values of its components in this system. When the crystallophysical system coincides with the principal coordinate system of the tensor, these values are evidently the principal values of the tensor components. When two or all the three crystallophysical axes do not coincide with the principal axes of the tensor, i.e., when the orientation of the principal tensor axes is not fully or not at all fixed in the crystal, the form of its matrix in the crystallophysical system becomes more complicated than in the principal system (the number of nonzero components increases). Obviously, in this case the symmetry transformations for the tensor components will consist only of those transformations of the crystallophysical axes which are contained in the respective group G_{pr}. Consequently, the total number of distinct types of matrix of a tensor of a given rank (and with a given interior symmetry) in the crystallophysical coordinate system is equal to the number of the respective distinct G_{pr}. Moreover, for crystal classes in which G_{pr} is identical, the form of the tensor matrix is the same.

In general, when a coordinate system deviates from the principal system for a given tensor G_T, its form becomes more complicated because the transformations of the coordinate axes cease to be symmetry transformations of the tensor components, i.e., they fail to satisfy (1.61). The form of the tensor matrix then becomes the same as in the crystallophysical coordinate system, when G_{pr} contains only those symmetry elements which coincide with (correspond to) the axes of a given coordinate system.

Let us consider some examples. For orthorhombic crystals a symmetric second-rank polar tensor has $G_T = mmm$, and $G_{pr} = G_T$ (Table 1.3). On changing from the principal system of coordinates of the tensor, X_1, X_2, X_3 (coinciding with the crystallophysical one) to system X_1', X_2', X_3' rotated about X_3 through an angle not equal to 90° or 180°, the form of the tensor matrix will be the same as in the crystallophysical coordinate system, when $G_{pr} = 2/m$, i.e., the same as for monoclinic crystals. On passing to arbitrary axes X_1', X_2', X_3' the form of the matrix will be the same as in the crystallophysical system, when $G_{pr} = \bar{1}$, i.e., the same as for triclinic crystals.

For crystals of class $\bar{4}2m$ the principal axes of a tensor of any rank (coinciding with the crystallophysical ones) are oriented as follows (Table 1.2): the X_3 axis is parallel to axis $\bar{4}$, and the X_1 and X_2 axes, to axes 2. If we now change to the system of axes X_1', X_2', X_3' rotated about the X_3 axis through an angle not equal to 90° or 45° [4], the form of the matrices of a symmetric second-rank polar tensor and a third-rank axial tensor symmetric with respect to a pair of subscripts will remain unaltered, because in such cases $G_{pr} = G_T = \infty/mmm$, and any X_1' and X_2' axes are the main axes. As for a symmetric second-rank pseudotensor and a third-rank polar tensor symmetric with respect to a pair of subscripts for which $G_{pr} = G_T = \bar{4}2m$, and fourth-rank polar tensors with $ijkl = jikl = ijlk = klij$ (interior symmetry $[[V^2]]^2)$ and with $ijkl = jikl = ijlk$ (interior symmetry $[[V^2]]^2$), for which $G_{pr} = G_T = 4/mmm$, – in these cases in system X_1', X_2', X_3' the form of their matrices will be the same as in the crystallophysical system of coordinates, when $G_{pr} = \bar{4}$ and $G_{pr} = 4/m$, respectively, i.e., the same as for crystals of class $\bar{4}$.

For crystals of classes 222 and $mm2$, a third-rank polar tensor symmetric with respect to a pair of subscripts, with $G_{pr} = G_T = 222$ and $G_{pr} = G_T = mm2$, respectively, will, on passing to the system of axes X_1', X_2', X_3', rotated about the X_3 axis through an angle not equal to 90° or 180°, have the identical form, the same as in the crystallophysical coordinate system, when $G_{pr} = 2$ (with the X_3 axis parallel to axis 2), i.e., the same form as for crystals of class 2 (with the X_3 axis parallel to axis 2). Finally, for a crystal of any class the matrix of any tensor (not degenerated into a scalar or a pseudoscalar) will have, in an arbitrary coordinate system, the general form (all the components are nonzero), i.e., the same as for triclinic crystals. Thus, even without finding the actual form of the matrices of tensors and knowing only their symmetry, it is possible to draw definite conclusions about their form in some coordinate system or other.

The general method for finding the type of the matrix of tensor $T_{ijk...}$ for crystals of a given class in some chosen system of coordinates is as follows: having established how the symmetry elements of the crystal (and also of the tensor) correlate with the axes of this system, we perform the appropriate symmetry transformations of the tensor components and equate $T'_{ijk...}$ to $T_{ijk...}$ by (1.61). Solving the equations obtained for $T_{ijk...}$, we ascertain which $T_{ijk...}$ are equal to zero and find the relationships (if they exist) among the nonzero components. There are also a number of other less general but simpler methods for finding the form of matrix of the tensors in the chosen system of coordinates [1.4, 5, 9, 10].

The form of matrices of tensors of various ranks in the crystallophysical coordinate system for crystals of different classes is given in the following chapters, when considering the relevant physical properties. We refer the reader to those chapters.

Note that although the form of a tensor matrix becomes more complicated on changing from the principal coordinate system to a rotated one, the number of independent tensor components remains unchanged, i.e., all the components in the rotated system can be expressed in terms of the principal

---

[4] On rotation through 45°, the X_1 and X_2 axes will coincide with the symmetry planes, and we shall simply pass to another variant of the principal coordinate system of the tensor.

ones. For crystal classes with $G_{pr} \neq G_T$, instead of assigning independent angular coefficients for the principal axes with one or three degrees of freedom, it is customary to write (as mentioned above) the tensor matrix in the crystallophysical system of coordinates and to regard the components appearing additionally as independent ones (precisely this principle was used below in compiling Table 1.7). On passing from the crystallophysical to a more general coordinate system the same rule applies: the number of independent tensor components remains unchanged, i.e., all the tensor components in this new system can be expressed via the tabulated components in the crystallophysical system.

1.5.6 Determining the Number of Independent Components of Tensors Describing the Physical Properties of Crystals

Direct determination of the form of tensors and independent components of tensors of a rank higher than the third rank is rather cumbersome. Using the methods of group theory it is easy enough to find the number of independent components of tensors of any rank describing the physical properties of crystals.

Since on transformation of components of rth-rank tensors under operations of point symmetry they transform as products of r components of the vectors, by analogy with vector representations of point groups we can introduce the concept of *tensor representations* of point groups [Ref. 1.1, Chap. 2]. Any tensor representation of power 3^r of a point group is a vector representation $D(G)$ of this group raised to the rth power.

The number of nonzero independent components of a tensor with a given interior symmetry for the some crystallographic point group K, is equal to number n_i, which shows how many times a fully symmetric irreducible representation τ_i is encountered in the relevant tensor representation of this point group.

As shown in group theory (see, for instance, [1.11]) number n_i is defined by the equation

$$n_i = \frac{1}{N} \sum_j \eta_j \chi_j'(R) \chi_i(R) \ . \tag{1.68}$$

In this formula, N is the total number of elements of group K, η_j is the order of the jth class, χ_j' is the character of the matrix of transformation of tensor components corresponding to operation K, and $\chi_i(R)$ is the character of the symmetry operation R belonging to the jth class in representation τ_i. Since here we consider fully symmetric representations, character $\chi_i(R)$ is equal to unity for all R.

Thus, number n_i can be found easily if character $\chi_j'(R)$ is known. In simple cases, expression $\chi_j'(R)$ can easily be represented in the form of a function depending on the angle of rotation of the coordinate system.

For instance, in a simple or mirror rotation about the X_3 axis through angle φ, the components of a polar vector transform according to equations

$$P_1' = P_1 \cos\varphi + P_2 \cos\varphi \ , \quad P_2' = -P_1 \sin\varphi + P_2 \cos\varphi \ , \quad P_3' = \pm P_3 \qquad (1.69)$$

(the upper sign is taken for simple rotations, and the lower, for mirror rotations).

The character of the matrix of this transformation is equal to $2\cos\varphi \pm 1$. It is easy to establish that under transformations of components of a symmetric second-rank polar tensor the matrix character will be equal to $4\cos^2\varphi \pm 2\cos\varphi$.

For axial tensors, all the coefficients in equations of the type (1.69) are multiplied by (-1) for the operations of mirror rotations. Thus, the character

Table 1.6. Characters of representations $\chi_j'(R)$ and the maximum number of components of the tensors describing various physical properties

Tensor quantity describing property	Character $\chi_Q(r)$ ($C = \cos\varphi$)	Maxium number of tensor components
Polar vector V and antisymmetric second-rank axial tensor $\varepsilon\{V^2\}$	$2C \pm 1$	3
Axial vector εV and antisymmetric second-rank polar tensor $\{V^2\}$	$\pm 2C + 1$	3
Symmetric second-rank polar tensor $[V^2]$	$4C^2 \pm 2C$	6
Symmetric second-rank axial tensor $\varepsilon[V^2]$	$\pm 4C^2 + 2C$	6
Nonsymmetric second-rank polar tensor V^2	$(2C \pm 1)^2$	9
Nonsymmetric second-rank axial tensor εV^2	$(2C \pm 1)(\pm 2C + 1)$	9
Third-rank polar tensor symmetric with respect to two subscripts, $V[V^2]$	$(2C \pm 1)(4C^2 \pm 2C)$	18
Third-rank axial tensor symmetric with respect to two subscripts, $\varepsilon V[V^2]$	$(\pm 2C + 1)(4C^2 \pm 2C)$	18
Fourth-rank polar tensor symmetric with respect to two pairs of subscripts and to their permutation, $[[V^2]^2]$	$16C^4 \pm 8C^3 - 4C^2 + 1$	21
Fourth-rank polar tensor symmetric with respect to two pairs of subscripts $[[V^2]^2]$	$(4C^2 \pm 2C)^2$	36

of the matrix for the transformation of an axial vector under simple and mirror rotations about the X_3 axis is equal to $\pm 2\cos\varphi + 1$.

Table 1.6 lists the characters of representations $\chi_j'(R)$ of various material tensors and the maximum number of components of tensors describing different physical properties in the crystallophysical coordinate system.

As an example, let us calculate the number of independent components of a tensor with the aid of Table 1.6. Thus, from the table it follows that number n_i for the tensor of elastic coefficients is defined by the character

$$\chi_j'(R) = 16\cos^4\varphi \pm 8\cos^3\varphi - 4\cos^2\varphi + 1 \ . \tag{1.70}$$

Table 1.7. Number of independent components of various material tensors in the crystallophysical coordinate system for all crystal classes

Class K	Tensor type									
	V	εV	$[V^2]$	$\varepsilon[V^2]$	V^2	εV^2	$V[V^2]$	$\varepsilon V[V^2]$	$[[V^2]^2]$	$[V^2]^2$
1	3	3	6	6	9	9	18	18	21	36
$\bar{1}$	0	3	6	0	9	0	0	18	21	36
m	2	1	4	2	5	4	10	8	13	20
2	1	1	4	4	5	5	8	8	13	20
$2/m$	0	1	4	0	5	0	0	8	13	20
$mm2$	1	0	3	1	3	2	5	3	9	12
222	0	0	3	3	3	3	3	3	9	12
mmm	0	0	3	0	3	0	0	3	9	12
4	1	1	2	2	3	3	4	4	7	10
$\bar{4}$	0	1	2	2	3	2	4	4	7	10
$4/m$	0	1	2	0	3	0	0	4	7	10
$4mm$	1	0	2	0	2	1	3	1	6	7
$\bar{4}2m$	0	0	2	1	2	1	2	1	6	7
422	0	0	2	2	2	2	1	1	6	7
$4/mmm$	0	0	2	0	2	0	0	1	6	7
3	1	1	2	2	3	3	6	6	7	12
$\bar{3}$	0	1	2	0	3	0	0	6	7	12
$3m$	1	0	2	0	2	1	4	2	6	8
3/2	0	0	2	2	2	2	2	2	6	8
$\bar{3}m$	0	0	2	0	2	0	0	2	6	8
$\bar{6}$	0	1	2	0	3	0	2	4	5	8
6	1	1	2	2	3	3	4	4	5	8
$6/m$	0	1	2	0	3	0	0	4	5	8
$\bar{6}m2$	0	0	2	0	2	0	1	1	5	6
$6mm$	1	0	2	0	2	1	3	1	5	6
622	0	0	2	2	2	2	1	1	5	6
$6/mmm$	0	0	2	0	2	0	0	1	5	6
23	0	0	1	1	1	1	1	1	3	4
$m3$	0	0	1	0	1	0	0	1	3	4
$\bar{4}3m$	0	0	1	0	1	0	1	0	3	3
432	0	0	1	1	1	0	0	0	3	3
$m3m$	0	0	1	0	1	0	0	0	3	3

For class 1, number n_i is evidently maximal, and is determined, in accordance with (1.68, 70), by expression ($\cos\varphi = 1$)

$$n_i = \tfrac{1}{1}\,[1\,(16+8-4+1)] = 21 \ .$$

For class *mmm* we get $n_i = \frac{1}{8}\,(21+5+5+5+21+5+5+5) - 9$, where each term corresponds to one of the eight operations of the group.

The number of independent coefficients of various tensors in the crystallophysical coordinate system for all the crystal classes is given in Table 1.7.

1.5.7 The Curie Principle and Its Application

Along with the Neumann principle, one more symmetry postulate, which is usually called the *Curie principle,* is very important for crystal physics [1.12]. While the Neumann principle relates the symmetry of a property to the initial, unperturbed symmetry of the crystal, i.e., the symmetry of the crystal before the external influence, the Curie principle enables one to determine the symmetry of the crystal after the external influence. It can be formulated as follows.

A crystal subjected to the influence of an external force changes its point symmetry so as to preserve only the symmetry elements common with the symmetry elements of the influencing force. In other words, the point symmetry of the crystal, K, changes, as a result of the external influence, to $\tilde{K}$, i.e., the highest common subgroup of the symmetry groups of the crystal, K, and of the influence, G_{inf}, at a given mutual orientation of the symmetry elements of these groups, which means that $\tilde{K}$ is the intersection of groups K and G_{inf}:

$$\tilde{K} = K \cap G_{\text{inf}} \ . \tag{1.71}$$

The Curie principle becomes almost obvious if we imagine that a figure with the symmetry of the influence in the given orientation is superimposed on a geometric figure with the symmetry of the crystal. It is easy to see that the new figure retains only the common (in fact, *all* the common) symmetry elements of the initial figures.

From (1.71) it follows at once that if $K \subset G_{\text{inf}}$, then $\tilde{K} = K$, i.e., if the initial symmetry group of the crystal is a subgroup of the symmetry group of the influence, then the symmetry of the crystal remains unchanged after this external influence. Thus, after scalar effect ($G_{\text{inf}} = \infty/\infty mm$ and hence always $G_{\text{inf}} \supset K$), the symmetry of the crystal remains unchanged. Hence, for instance, the thermal expansion of the crystal may lead to a change in the angles between the faces of the crystal, but it cannot lead to a change in its symmetry (unless there are phase transitions).

When the symmetry of the crystal changes after an anisotropic influence, the Curie principle enables one to find this change immediately and hence the

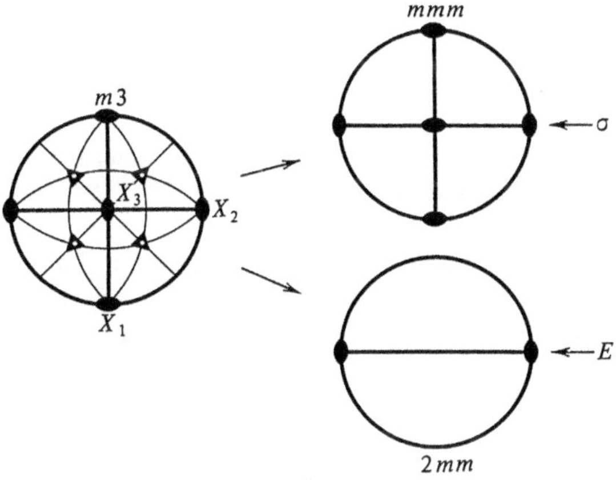

Fig. 1.9. Stereographic projections demonstrating the change in the symmetry of an $m3$-class crystal on application of external uniaxial extension stress σ and electric field E along the [100] axis

corresponding change in the symmetry of the physical properties as well (the Neumann principle must be fulfilled as before). This latter circumstance will make it possible to take into account at once the role of the so-called *morphic effects,* i.e., effects caused by the lowering in the symmetry of the crystal. Morphic effects are due to the appearance, in the crystal, of *morphic properties* inherent in group $\tilde{K}$ and described by either morphic tensors or morphic tensor components forbidden in group K, or else by morphic deviations from the relationships between the tensor components existing in group K. The morphic properties are proportional to some degree of influence.

Let us consider, by way of illustration, a cubic crystal with $K = m3$ (Fig. 1.9). A uniaxial extension stress ($G_{\mathrm{inf}} = \infty/mm$) applied to a crystal along [100] will transfer it to orthorhombic class $\tilde{K} = mmm$ according to (1.71). As a result of this change in symmetry the crystal will become optically biaxial; it will acquire morphic (additional) components of the tensor of the elastic coefficients proportional to the effect, etc. If we apply an electric field ($G_{\mathrm{inf}} = \infty mm$) to the same crystal with $K = m3$ along the same axis [100], then, according to (1.71), $\tilde{K} = mm2$. Now, in addition to the previous changes, the crystal loses the center of symmetry and accordingly acquires a morphic tensor of piezoelectric moduli, etc. If the same actions are applied to a crystal with $K = m3$ in an arbitrary direction $[hkl]$, its symmetry will decrease to triclinic in accordance with (1.71), but the centrosymmetric mechanical stress will retain the center of symmetry ($\tilde{K} = \bar{1}$), while the polar electric field will eliminate it ($\tilde{K} = 1$); the corresponding morphic properties will then appear.

The Curie principle is applicable for determining the changes, under an external influence, not only in the point symmetry of the crystal, but also in its space symmetry. The external influence is here regarded as a homogeneous spatial field, i.e., its space group can be written as $G_{\mathrm{inf}} \otimes T_{\tau_1 \tau_2 \tau_3}$ (Sects. 1.1.1, 1.3.7 and [Ref. 1.1, Chap. 2]). Then we get

$$\tilde{\Phi} = \Phi \cap G_{\text{inf}} \otimes T_{\tau_1\tau_2\tau_3} \ . \tag{1.72}$$

Under ordinary measuring influences the morphic effects are very small, and the corresponding lowering of the symmetry can be neglected. But under very strong influences and, sometimes, in crystallooptical measurements, morphic effects may be sufficiently large and must be taken into account (Chap. 8).

In the case of structural phase transitions in a crystal, accompanied by changes in its point symmetry and described by some macroscopic anisotropic parameter whose spontaneous value appears in the low-symmetry phase of the crystal, we can regard this parameter as an external influence and apply the Curie principle to define the change in the point symmetry of the crystal at phase transition [1.13, 14, 15]. Thus, if we know the symmetry of transition parameter G_{tr}, then, by (1.71), the symmetry of the low-symmetry phase, K_{F}, is the highest common subgroup of symmetry group K of the initial phase of the crystal and of group G_{tr} at a given orientation of the elements of symmetry G_{tr} in the axes of group K, i.e.,

$$K_{\text{F}} = K \cap G_{\text{tr}} \ . \tag{1.73}$$

This approach facilitates the determination of the change in the point symmetry of crystals on ferroelectric, ferromagnetic, ferroelastic, and other parametric structural transitions, and, for cases of so-called proper phase transitions, also the change in space symmetry in accordance with relation

$$\Phi_{\text{F}} = \Phi \cap G_{\text{tr}} \otimes T_{\tau_1\tau_2\tau_3} \ . \tag{1.74}$$

Let us consider, for instance, the change in symmetry on ferroelectric phase transitions in barium titanate ($BaTiO_3$). We take into account that in the initial (paraelectric) phase $K = m3m$, $G_{\text{tr}} = \infty mm$ (the symmetry of spontaneous polarization vector $\boldsymbol{P}_{\text{s}}$). In the first ferroelectric phase, vector $\boldsymbol{P}_{\text{s}}$ is oriented along $\langle 001 \rangle$ of the initial phase. Obviously, in this case $K_{\text{F}} = (K \cap G_{\text{tr}}) = 4mm$. In the second ferroelectric phase, $\boldsymbol{P}_{\text{s}}$ is oriented along $\langle 110 \rangle$ of the initial phase; hence, $K_{\text{F}} = (K \cap G_{\text{tr}}) = mm2$. In the third ferroelectric phase $\boldsymbol{P}_{\text{s}} \parallel \langle 111 \rangle$, and therefore $K_{\text{F}} = 3m$. (For more details see Chap. 3.)

In the case of a ferroelastic phase transition [1.16], the change in the point symmetry of the crystal is fully described by the appearance of spontaneous deformation, which may consist in a shear or a uniaxial extension/contraction. Considering that a shear can be replaced by a uniaxial extension/contraction by rotating the crystal through 45° about the shear axis, we can always take the deformation of a uniaxial extension/contraction as the spontaneous deformation, i.e., assume $G_{\text{tr}} = \infty/mmm$. Let us consider, as an example, a ferroelastic transition in $KH_3(SeO_3)_2$ with $K = mmm$. By orienting G_{tr} along $\langle 101 \rangle$ and using (1.73), we obtain $K_{\text{F}} = 2/m$. For $NH_4H_2PO_4$ with $K = \bar{4}2m$ we orient G_{tr} along $\langle 100 \rangle$, thus obtaining transition to $K_{\text{F}} = 222$.

It should be noted that the ferroelastic phase in $NH_4H_2PO_4$ is also an antiferroelectric one. This is not surprising, since a change in the point symmetry of a crystal on transition to the antiferroelectric phase, which accompanies the appearance of antipolarization, can be described with the aid of $G_{tr} = \infty/mmm$. In other words, at the point symmetry level, ferroelastic phase transitions are indistinguishable from antiferroelectric transitions, and all the latter (except transitions without a change in point symmetry) will simultaneously be ferroelastic transitions. (Note that ferroelastic phase transitions may occur not only in dielectrics, examples being transitions $m3m \leftrightarrow 4/mmm$ in Vi_3Si, and $4/mmm \leftrightarrow 2/m$ in VO_2.)

The Curie principle, as well as that of Neumann, also holds true within the framework of magnetic symmetry, and therefore by using (1.73) we can determine at once the change in the point magnetic symmetry of a crystal, K', on transition to ferromagnetic phase K_F', considering that $G_{tr}' = \infty/mm'm'$ (magnetic symmetry of the spontaneous magnetization vector $\boldsymbol{M}_s$) [1.14]. Thus, for Fe and Ni, $K' = m3m1'$; therefore, when $\boldsymbol{M}_s$ is oriented along $\langle 001 \rangle$ we have $K_F' = 4/mm'm'$ in the case of Fe, and we have $K_F' = \bar{3}m'$ when $\boldsymbol{M}_s$ is oriented along $\langle 111 \rangle$ in the case of Ni.

It should be emphasized that in all the examples considered above, as well as under structural phase transitions of this type in general, during transition to the low-symmetry phase the crystal is divided into transformation twins (domains), and group K_F (K_F') describes the symmetry of a single domain, while a polydomain crystal as a whole may have a higher (pseudo)symmetry, up to $K(K')$.

2. Mechanical Properties of Crystals

The mechanical properties of solids depend on their response to the mechanical load applied. Three basic characteristics are used in describing these properties of materials.

The first characteristic is the *elasticity,* which reflects the ability of a solid to resume its shape when a force is applied to it for some time and then removed. This property is manifested in the initial stage of deformation, which is called the *elastic* (reversible) deformation.

The second characteristic is the *plasticity,* which indicates how rapidly the material changes its shape under the effect of prolonged loading or how large the force must be for this change to occur at a definite rate. The plasticity characterizes the behavior of solids in the second stage of deformation, which is irreversible and called the *plastic deformation.*

The third mechanical characteristic is the strength, or the resistance to failure. The failure occurs in the final stage of the deformation.

All these characteristics differ greatly in value from one crystal to another. For instance, the elasticity, which is measured by Young's modulus, varies within about two orders of magnitude (from 10^{10} to 10^{12} dyn/cm^2). The plasticity and strength, which are measured by the force which must be applied to a unit surface of a body to cause plastic deformation or failure, respectively, vary still more, between 10^5 and 10^{12} dyn/cm^2.

In crystals, the elastic properties depend on the behavior of their constituent particles (atoms, ions, and molecules); the plastic ones, mainly on the behavior of chains of such particles (dislocations); and the strength, on the properties of the surfaces built up of these particles. The methods for describing the elastic, plastic, and strength properties have been developed accordingly.

2.1 Elastic Properties of Crystals

2.1.1 Stress

The mechanical properties of crystals, which are sets of many particles (atoms, ions, or molecules) forming a regular lattice, depend on the forces of interaction of these particles. In any type of crystal – atomic, ionic, metallic,

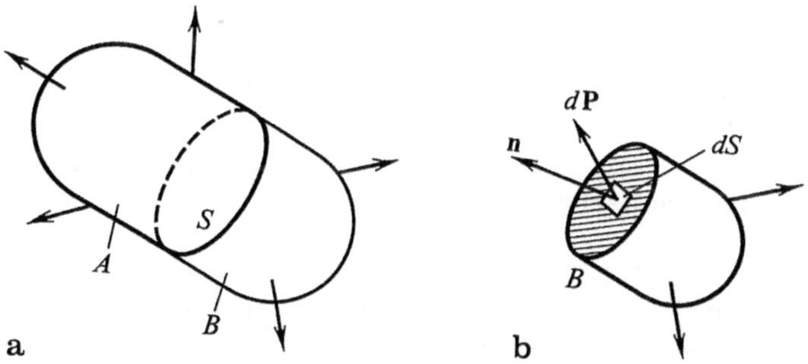

Fig. 2.1. Distribution of balanced (**a**) and unbalanced (**b**) interaction forces in a solid

and molecular – the particle interaction forces decrease with the distance, the repulsion forces diminishing more rapidly than those of attraction. The equilibrium distance between particles corresponds to the equality of the attraction and repulsion forces. If the crystal is subjected to mechanical effects, the balance of these forces is disturbed, the particles are displaced, and the lattice spacing changes. The resulting forces tend to bring the body back to the equilibrium state. Macroscopic changes in lattice spacing manifest themselves as elastic strains, and changes in particle interaction forces, as stress.

Let us consider a solid in which the forces of interaction between its different parts in the absence of external forces offset each other (Fig. 2.1 a). After an external load is applied, the resultant of the internal forces (indicated by arrows in Fig. 2.1) differs from zero. Let us mentally dissect a body subjected to external forces on a surface S into two parts A and B. The effect of part A on part B consists in the fact that surface S is subjected to forces which are internal for the body as a whole. Assume that the internal forces are dis-

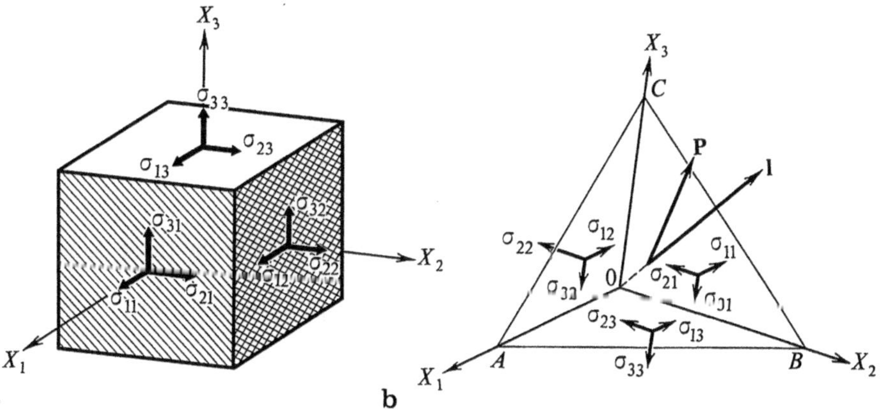

Fig. 2.2. Forces acting on (**a**) the faces of a unit cube in a uniformly stressed body and (**b**) the faces of a tetrahedron formed by three coordinate planes and the ABC face

tributed uniformly over the interface. If elementary area dS is acted upon by a force $d\boldsymbol{P}$ (Fig. 2.1 b), the ratio $\boldsymbol{P}_n = d\boldsymbol{P}/dS$ is called the *stress vector* on area dS. Subscript n means that the external normal to the area has direction $\boldsymbol{n}$.

A stress is called *homogeneous* if the forces acting on the surface of the part of a solid of definite shape and orientation are independent of the position of this part in the solid. The state of a solid with uniform stresses is called a *homogeneous stressed state.*

If we pick out a cube with edges along the X_1, X_2, and X_3 axes (Fig. 2.2a), the interior part of the cube is subjected to stresses transmitted through all its boundaries from the surrounding material. The stresses acting on each face can be resolved into three components. We denote by σ_{ij} the stress component acting in direction X_i on a cube edge perpendicular to X_j axis. The stress components σ_{ij}

$$\begin{matrix} \sigma_{11} & \sigma_{12} & \sigma_{13} \\ \sigma_{21} & \sigma_{22} & \sigma_{23} \\ \sigma_{31} & \sigma_{32} & \sigma_{33} \end{matrix} \tag{2.1}$$

form a second-rank polar tensor.

To prove this we consider a volume element in the shape of a tetrahedron (Fig. 2.2b), which is in equilibrium with the surrounding material. Suppose an area of tetrahedron surface ABC, perpendicular to unit vector $\boldsymbol{l}$, is acted upon by stress $\boldsymbol{P}[P_1, P_2, P_3]$. The force transmitted through area ABC is equal to the product of stress vector $\boldsymbol{P}$ by the area of ABC. The force component acting on face ABC along the X_1 axis can be written as

$$P_1 S_{ABC} = \sigma_{11} S_{BOC} + \sigma_{12} S_{AOC} + \sigma_{13} S_{AOB} \ ,$$

where S_{ABC}, S_{BOC}, S_{AOC}, and S_{AOB} are the areas of the tetrahedron faces. Dividing both sides of the equation by the area of triangle ABC, we obtain

$$P_1 = \sigma_{11} l_1 + \sigma_{12} l_2 + \sigma_{13} l_3 \ ;$$

similarly,

$$P_2 = \sigma_{21} l_1 + \sigma_{22} l_2 + \sigma_{23} l_3 \ , \quad P_3 = \sigma_{31} l_1 + \sigma_{32} l_2 + \sigma_{33} l_3 \ ,$$

where l_1, l_2, and l_3 are the components of vector $\boldsymbol{l}$ along the three coordinate axes. To summarize,

$$P_i = \sigma_{ij} l_j \ . \tag{2.2}$$

As shown in Chap. 1, the coefficients relating the components of the polar vector form a second-rank polar tensor, and hence the stress components σ_{ij} also form a second-rank polar tensor.

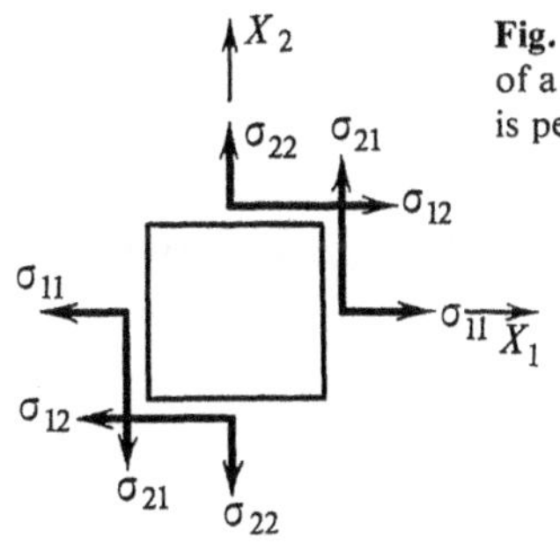

Fig. 2.3. Forces acting on faces (perpendicular to axes X_2 and X_1) of a unit cube picked out in a homogeneously stressed body (X_3 axis is perpendicular to the plane of the drawing)

Stress components σ_{11}, σ_{22}, and σ_{33} are called *normal stresses,* because they act perpendicular to the respective areas. The other components act along the areas and are called *tangential stresses.* As can be seen from Fig. 2.3, the tangential stresses applied to the surface always form a force couple, which is counteracted by another force couple (cf., for instance, the force couples formed by σ_{12} and σ_{21}). For the equilibrium to be maintained, these force couples must be equal to one another; thus

$$\sigma_{ij} = \sigma_{ji} \ . \tag{2.3}$$

It follows that tensor (2.1) is symmetric and can be reduced to the principal axes (Chap. 1). Then the shear components of the stresses vanish, and the stress tensor (2.1) is written

$$\left\| \begin{matrix} \sigma_1 & 0 & 0 \\ 0 & \sigma_2 & 0 \\ 0 & 0 & \sigma_3 \end{matrix} \right\| , \tag{2.4}$$

where σ_1, σ_2, and σ_3 are the *principal compression* or *tensile stresses.* This form of stress tensor corresponds to the three-dimensionally stressed state (triaxial contraction or extension). Under the effect of a uniaxial or biaxial compression or extension stress, tensor σ_{ij} is transformed, respectively, to

$$\left\| \begin{matrix} \sigma & 0 & 0 \\ 0 & 0 & 0 \\ 0 & 0 & 0 \end{matrix} \right\| \quad \text{and} \quad \left\| \begin{matrix} \sigma_1 & 0 & 0 \\ 0 & \sigma_2 & 0 \\ 0 & 0 & 0 \end{matrix} \right\| .$$

Under the effect of tangential stresses *alone,* when, for instance, $\sigma_{12} \neq 0$, while the other $\sigma_{ij} = 0$, we get

$$\left\| \begin{matrix} 0 & \sigma_{12} & 0 \\ \sigma_{12} & 0 & 0 \\ 0 & 0 & 0 \end{matrix} \right\| .$$

2.1.2 Characteristic Stress Surface

Like any symmetric second-rank polar tensor (Chap. 1), tensor σ_{ij} can be interpreted geometrically as a second-order characteristic surface with center at the origin of coordinates $x_1 = x_2 = x_3 = 0$. This surface is generally described by an equation of the type

$$\sigma_{ij} x_i x_j = 1 \ . \tag{2.5}$$

On passing to the principal axes the stress-surface equation will be written

$$\sigma_1 x_1^2 + \sigma_2 x_2^2 + \sigma_3 x_3^2 = 1 \ . \tag{2.6}$$

Under triaxial tension the principal stresses are positive, and the characteristic surface is a triaxial ellipsoid with axes $1/\sqrt{\sigma_1}$, $1/\sqrt{\sigma_2}$, and $1/\sqrt{\sigma_3}$. An imaginary ellipsoid corresponds to triaxial compression (all the σ_i are negative). If two of the principal stresses are positive and one is negative, (2.6) describes a hyperboloid of one sheet. If one of the principal stresses is positive and two are negative, it describes a hyperboloid of two sheets. If one of the principal stresses is zero, the characteristic surface is a cylinder (elliptical or hyperbolic, depending on the signs of the other principal stresses). Finally, if two of the principal stresses are zero, the characteristic surface degenerates into a pair of parallel planes perpendicular to the only principal stress.

2.1.3 Strain

The main types of strain are longitudinal (elongation and contraction) and shear deformations. *Elongation (contraction)* is defined as the ratio of the change in the length of a body (segment) to its initial length (Fig. 2.4a):

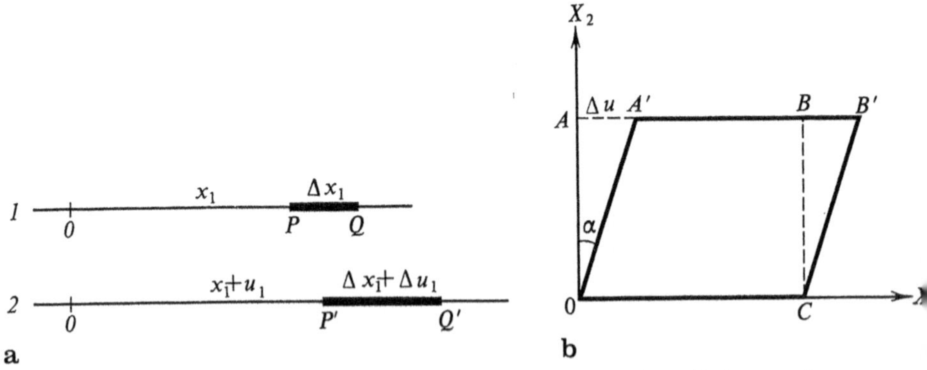

Fig. 2.4. (**a**) Elongation strain: (*1*) before extension, (*2*) after extension; (**b**) shear strain

$$\frac{P'Q'-PQ}{PQ}=\frac{\Delta u_1}{\Delta x_1}=e_{11}\ . \tag{2.7}$$

Shear strain is the relative displacement of one part of a body with respect to another along a certain plane. The magnitude of shear strain is defined as the ratio of the displacement to the distance from this plane. According to Fig. 2.4b the shear strain

$$e_{12}=\Delta u_1/\Delta x_2=\tan\alpha\ .$$

In other words, shear can be regarded as the measure of a change in the angle between two straight lines arbitrarily oriented in the body under strain. The strain at a point is defined by the expression

$$e=\lim_{\Delta x\to 0}\frac{\Delta u}{\Delta x}=\frac{du}{dx}\ , \tag{2.8}$$

whence

$$du=e\,dx\ .$$

Let us consider the strain of a segment in a plane. Suppose segment PQ, lying in plane (X_2X_1) transforms to segment $P'Q'$ after strain (Fig. 2.5). The coordinates of point P are (x_1, x_2), and those of point P' are (x_1+u_1, x_2+u_2). The components of the displacement vector of point P are $\boldsymbol{u}=\boldsymbol{PP'}=(u_1, u_2)$. The coordinates of point Q are $(x_1+\Delta x_1, x_2+\Delta x_2)$. The components of the displacement vector of point Q are $\boldsymbol{QQ'}=(u_1+\Delta u_1, u_2+\Delta u_2)$. It is easy to see that

$$\begin{aligned}\Delta u_1&=\frac{\partial u_1}{\partial x_1}\Delta x_1+\frac{\partial u_1}{\partial x_2}\Delta x_2\ ,\\ \Delta u_2&=\frac{\partial u_2}{\partial x_1}\Delta x_1+\frac{\partial u_2}{\partial x_2}\Delta x_2\ .\end{aligned} \tag{2.9}$$

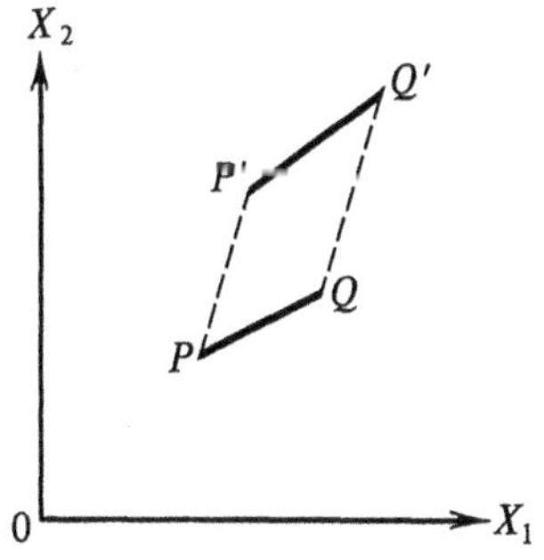

Fig. 2.5. Schematic representation of the strain of a segment

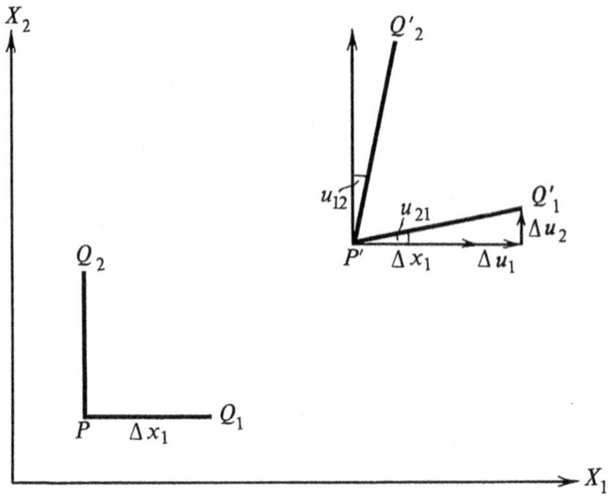

Fig. 2.6. Illustration of the physical meaning of coefficients u_{11} and u_{21}

Introducing notation

$$\frac{\partial u_1}{\partial x_1} = u_{11}, \frac{\partial u_1}{\partial x_2} = u_{12}, \frac{\partial u_2}{\partial x_1} = u_{21}, \frac{\partial u_2}{\partial x_2} = u_{22} , \tag{2.10}$$

we can write (2.9, 10) in the general form:

$$\Delta u_i = u_{ij} \Delta x_j \quad (j = 1, 2) . \tag{2.11}$$

Since Δu_i and Δx_j are vectors, the coefficients u_{ij} relating them form a second-rank polar tensor, which is called the *elastic-distortion tensor.* Let us analyze the physical significance of these coefficients. Assume that line Q_2PQ_1, made up of segments parallel to the coordinate axes (Fig. 2.6), transforms to line $Q_2'P'Q_1'$ as a result of strain. Putting $dx_2 = 0$ for segment $PQ_1 dx_2$ and taking (2.9, 10) into account, we have

$$\Delta u_1 = \frac{\partial u_1}{\partial x_1} \Delta x_1 = u_{11} \Delta x_1 , \tag{2.12}$$

$$\Delta u_2 = \frac{\partial u_2}{\partial x_1} \Delta x_1 = u_{21} \Delta x_1 . \tag{2.13}$$

It is seen from Fig. 2.6 that u_{11} measures the elongation of segment PQ_1, and u_{21} measures the anticlockwise rotation of this segment (if u_{11} and u_{21} are small). Similarly, u_{22} corresponds to the elongation of segment PQ_2, and u_{12} to its clockwise rotation.

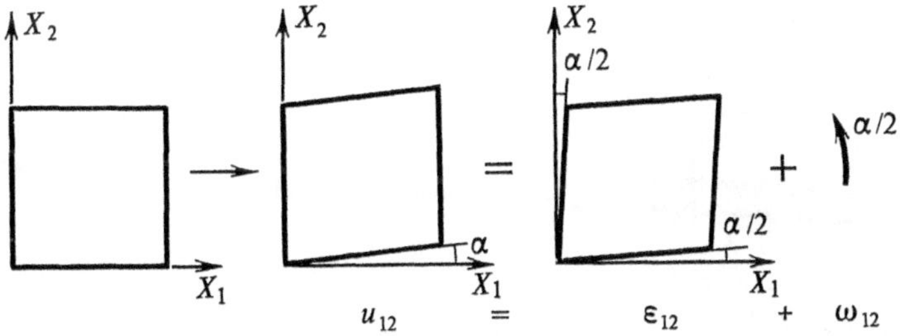

Fig. 2.7. Geometric interpretation of (2.14)

Tensor u_{ij} characterizes not only the deformation of a specimen, but also the lattice rotation, because the body may be undistorted at nonzero components u_{ij} (on rotations).

If u_{ij} are constant throughout the body, the corresponding strain is said to be *homogeneous*. Under uniform strain a straight line remains straight; parallel lines remain parallel; all the mutually parallel lines are extended or contracted to the same extent; an ellipse transforms to an ellipse; and, in particular, a circle becomes an ellipse.

Let us resolve elastic-distortion tensor u_{ij} into a deformation tensor and a lattice-rotation tensor. To do this, we represent it as a sum of the antisymmetric and symmetric tensor (Chap. 1):

$$u_{ij} = \tfrac{1}{2}(u_{ij} - u_{ji}) + \tfrac{1}{2}(u_{ij} + u_{ji}) \ . \tag{2.14}$$

Then $\omega_{ij} = (u_{ij} - u_{ji})/2$ measures the lattice rotation, and $\varepsilon_{ij} = (u_{ij} + u_{ji})/2$ determines the pure elastic strain.

Figure 2.7 gives the geometric interpretation of (2.14). Tensor ω_{ij} is called the *rotation tensor* and has the form

$$\omega_{ij} = \begin{Vmatrix} 0 & -\omega_{21} & \omega_{31} \\ \omega_{12} & 0 & -\omega_{32} \\ -\omega_{13} & \omega_{23} & 0 \end{Vmatrix} \ . \tag{2.15}$$

This tensor helps us to find the components of the axial vector of rotations:

$$\omega_i = \omega_{ij} x_j \ .$$

Polar tensor ε_{ij} is called the *strain tensor*. Since this tensor is symmetric, it can be reduced to the principal axes:

$$\begin{Vmatrix} \varepsilon_{11} & \varepsilon_{12} & \varepsilon_{13} \\ \varepsilon_{12} & \varepsilon_{22} & \varepsilon_{23} \\ \varepsilon_{13} & \varepsilon_{23} & \varepsilon_{33} \end{Vmatrix} \rightarrow \begin{Vmatrix} \varepsilon_1 & 0 & 0 \\ 0 & \varepsilon_2 & 0 \\ 0 & 0 & \varepsilon_3 \end{Vmatrix} , \tag{2.16}$$

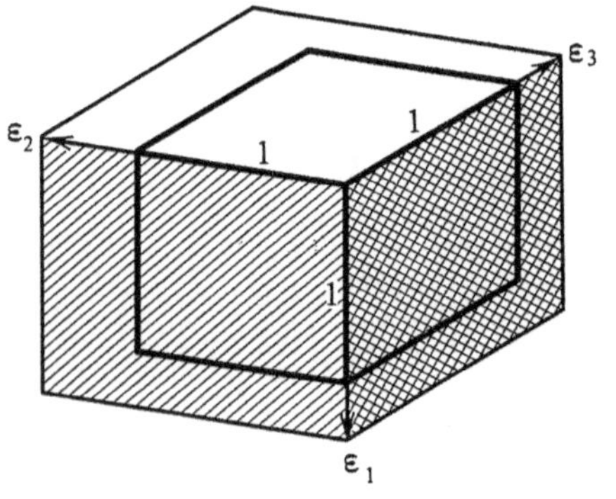

Fig. 2.8. Deformation of a unit cube with edges parallel to the three principal strain axes

where ε_{11}, ε_{22}, and ε_{33} are components of the contraction or extention strain, while the other ε_{ij} are components of the shear strain; ε_1, ε_2, and ε_3 are the *principal strains,* whose meaning is clear from Fig. 2.8 (the figure only shows the displacement u_i caused by deformation proper, but does not show the displacement and rotation of the body as a whole).

Let the edges of a cube picked out of a body be parallel to the principal axes. The main property of the principal axes is their mutual perpendicularity before and after the strain; therefore, ε_1, ε_2, and ε_3 are changes in the length of the edges of the unit cube picked out of the body. After transition to the principal axes, the shear components of the strain vanish, and the faces of the unit cube carved out of the body, which are parallel to the axes, will be acted upon by forces perpendicular to these faces. It would be wrong to think that the principal strain axes lie in directions which remain unaltered after the strain. This is true only when rotation ω_{ij} is equal to zero, i.e., in pure shear. In the general case, the principal axes rotate on strain.

2.1.4 Characteristic Surface and the Strain Ellipsoid

The equation of the characteristic surface of elastic strain has the form

$$\varepsilon_{ij} x_i x_j = 1 \quad . \qquad (2.17)$$

After transition to the principal axes this equation is written as

$$\varepsilon_1 x_1^2 + \varepsilon_2 x_2^2 + \varepsilon_3 x_3^2 = 1 \quad . \qquad (2.18)$$

Principal strains ε_1, ε_2, and ε_3 may be either positive or negative. Like the characteristic surface of stresses, the strain surface may be either a real or an imaginary ellipsoid or hyperboloid, a cylinder or a pair of parallel planes.

In describing the homogeneous three-dimensional elastic strain of a body by means of the strain of a unit sphere, one introduces the strain ellipsoid concept. The sphere equation is:

$$x_1^2+x_2^2+x_3^2=1 \ .$$

After strain, the lengths of the edges of the picked-out cube, which are parallel to the principal axes, will be

$$x_1'=x_1(1+\varepsilon_1) \ , \quad x_2'=x_2(1+\varepsilon_2) \ , \quad x_3'=x_3(1+\varepsilon_3) \ . \tag{2.19}$$

Substituting these values into the sphere equation, we get

$$\frac{x_1'^2}{(1+\varepsilon_1)^2}+\frac{x_2'^2}{(1+\varepsilon_2)^2}+\frac{x_3'^2}{(1+\varepsilon_3)^2}=1 \ . \tag{2.20}$$

Surface (2.20) is always an ellipsoid and is called a strain ellipsoid. From (2.20) it is seen that under uniaxial tension a strain ellipsoid is uniaxial. Under a plane strain, when one of the principal strains is zero, and under a pure shear strain, which is a particular case of uniaxial strain, a strain ellipsoid is biaxial. The circular section of this ellipsoid is parallel to the shear plane. A triaxial strain ellipsoid corresponds to the three-dimensionally stressed state.

A strain ellipsoid should not be confused with the characteristic strain surface (2.18), which may be either a real or an imaginary ellipsoid or hyperboloid, since main strains ε_1, ε_2, and ε_3 may be either positive or negative.

The difference between them is illustrated in Fig 2.9. The figure, which corresponds to a homogeneous three-dimensional strain of a body, displays strain-induced displacement $\boldsymbol{u}$ of the terminus of unit vector $\boldsymbol{l}$ taken in an unstrained body (Fig. 2.9a), the plane cross section of the characteristic strain surface (Fig. 2.9b), and the strain ellipsoid (outer curve), corresponding to the strain of a unit sphere whose cross section is denoted by the inner curve (Fig. 2.9c). In Fig. 2.9b the direction of $\boldsymbol{u}$ coincides with that of the normal to

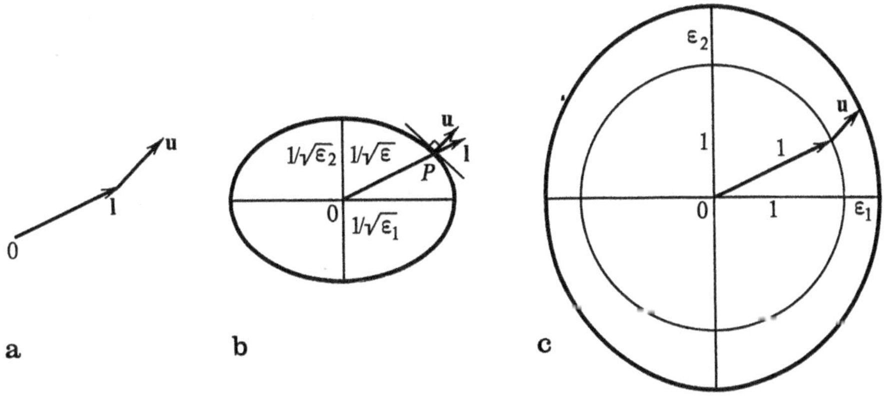

Fig. 2.9. Homogeneous three-dimensional strain. (**a**) strain shift u of the terminus of unit vector l taken in an unstrained body; (**b**) section of the characteristic deformation surface; (**c**) strain ellipsoid (outer curve, cross section of unit sphere/inner curve)

the characteristic surface at point P, and extention ε in the direction of $\boldsymbol{l}$ is defined by relation $OP = 1/\sqrt{\varepsilon}$.

In choosing the signs of strains and stresses we assume that stresses σ_{11}, σ_{22}, and σ_{33} are positive when they cause contraction and negative when they cause elongation, while strains ε_{11}, ε_{22}, and ε_{33} are positive for elongation and negative for contraction. It should be borne in mind that some authors, for instance *Nye* [2.1] and *Wooster* [2.2], use the inverted rules of signs for strains.

Thus, a positive longitudinal strain corresponds to a negative stress, and vice versa. A shear strain will be assumed positive if a rectangle deforms so that the acute angle of the parallelogram formed lies between the positive directions of two axes. The shear stress will be considered positive when it causes a negative shear strain.

2.1.5 Hooke's Law for Crystals

For the simplest case of uniaxial strain of a solid the stress-strain relationship was found as far back as 300 years ago by R. Hooke, who established proportionality between strain ε and stress σ (Hooke's Law):

$$\varepsilon = s\sigma \ , \tag{2.21}$$

where s is the elastic compliance coefficient, or simply compliance. Hooke's Law can also be written differently:

$$\sigma = c\varepsilon \ , \tag{2.22}$$

where c is the elastic stiffness coefficient, or simply stiffness. In a crystal to which a homogeneous arbitrary stress σ_{ij} is applied, the resulting uniform strain will be such that each of its components ε_{ij} is linearly related to all the stress components, for instance:

$$\begin{aligned}\varepsilon_{11} = {} & s_{1111}\sigma_{11} + s_{1112}\sigma_{12} + s_{1113}\sigma_{13} + s_{1121}\sigma_{21} + s_{1122}\sigma_{22} + s_{1123}\sigma_{23} + s_{1131}\sigma_{31} \\ & + s_{1132}\sigma_{32} + s_{1133}\sigma_{33} \ .\end{aligned} \tag{2.23}$$

For the other eight components ε_{ij} there are eight similar equations. Therefore, Hooke's Law in the generalized form can be written

$$\varepsilon_{ij} = s_{ijkl}\sigma_{kl} \ . \tag{2.24}$$

Here s_{ijkl} are the elastic compliance coefficients of the crystal. Expression (2.24) is a set of nine equations, whose right-hand side generally consists of nine terms. Consequently, there are 81 coefficients s_{ijkl} altogether. And, conversely, we can express the stresses via strains as in (2.22); then the general form of the equation will be

$$\sigma_{kl} = c_{klmn}\varepsilon_{mn} , \tag{2.25}$$

where c_{klmn} are the elastic stiffness coefficients, whose number is also 81. According to Chap. 1 the coefficients relating the components of symmetric second-rank polar tensors form a fourth-rank polar tensor. Thus, 81 coefficients s_{ijkl} and 81 coefficients c_{klmn} form fourth-rank polar tensors. Since $s_{ij} = s_{ji}$ and $c_{ij} = c_{ji}$,

$$s_{ijkl} = s_{jikl} = s_{ijlk} = s_{klij}$$

and

$$c_{klmn} = c_{lkmn} = c_{klnm} = c_{mnkl} .$$

Generally, only 21 coefficients of elasticity remain instead of the 81.

2.1.6 Matrix Notation of Coefficients of Elasticity

The coefficients of elasticity s_{ijkl} and c_{klmn} of (2.24, 25) can be written more conveniently by expressing the four subscripts by means of two. This results in the following system of subscripts:

to tensor symbols	11	22	33	23	32	31	13	12	21
correspond the matrix symbols	1	2	3	4		5		6	

Besides, we introduce factors 2 and 4 as follows:

$s_{ijkl} = s_{mn}$ when m and n are equal to 1, 2, or 3;
$2s_{ijkl} = s_{mn}$ when only m or only n is equal to 4, 5, or 6;
$4s_{ijkl} = s_{mn}$ when m and n are simultaneously equal to 4, 5, or 6.

In matrix notation, (2.23) is written

$$\varepsilon_1 = s_{11}\sigma_1 + s_{12}\sigma_2 + s_{13}\sigma_3 + s_{14}\sigma_4 + s_{15}\sigma_5 + s_{16}\sigma_6 , \quad \text{or}$$

$$\varepsilon_1 = s_{1j}\sigma_j . \tag{2.26}$$

Consequently, all the nine equations (2.24) can be written briefly:

$$\varepsilon_i = s_{ij}\sigma_j \quad (i,j = 1,2,\ldots,6) . \tag{2.27}$$

By analogy, (2.25) takes the form

$$\sigma_j = c_{jk}\varepsilon_k \quad (j,k = 1,2,\ldots,6) . \tag{2.28}$$

Table 2.1. Relationship between coefficients of elasticity s_{ij} and c_{ji} for high symmetry

Syngony	Equation
Cubic	$s_{11} = \dfrac{c_{11}+c_{12}}{(c_{11}-c_{12})(c_{11}+2c_{12})}$, $s_{12} = \dfrac{-c_{12}}{(c_{11}-c_{12})(c_{11}+2c_{12})}$, $s_{44} = 1/c_{44}$
Hexagonal	$s_{11}+s_{12} = c_{33}/c$, $s_{11}-s_{12} = 1/(c_{11}-c_{12})$, $s_{13} = -c_{13}/c$, $s_{33} = (c_{11}+c_{12})/c$, $s_{44} = 1/c_{44}$, $c = c_{33}(c_{11}+c_{12})-2c_{13}^2$
Tetragonal	$s_{11}+s_{12} = c_{33}/c$, $s_{11}-s_{12} = 1/(c_{11}-c_{12})$, $s_{13} = -c_{13}/c$, $s_{33} = (c_{11}+c_{12})/c$, $s_{14} = 1/c_{44}$, $s_{66} = 1/c_{66}$, $c = c_{33}(c_{11}+c_{12})-2c_{13}^2$
Triclinic	$s_{11}+s_{12} = c_{33}/c$, $s_{11}-s_{12} = c_{44}/c$, $s_{13} = -c_{13}/c$, $s_{14} = -c_{14}/c'$, $s_{33} = (c_{11}+c_{12})/c$, $s_{44} = (c_{11}-c_{12})/c'$, $c = c_{33}(c_{11}+c_{12})-2c_{13}^2$, $c' = c_{44}(c_{11}-c_{12})-2c_{14}^2$

In matrix notation the number of elastic compliance and elastic stiffness coefficients is 36 and, as $s_{ij}=s_{ji}$, and $c_{ij}=c_{ji}$, the number of independent coefficients generally becomes 21.

Matrices s_{ij} and c_{jk} are mutually inverted, i.e., $s_{ij}c_{jk}=c_{ij}s_{jk}=\delta_{ik}$, where δ_{ik} is a unit matrix.

In the general form, the relationships between s_{ij} and c_{jk} are rather complicated. Table 2.1 gives the relations between s_{ij} and c_{jk} for four high-symmetry syngonies. The stiffnesses are expressed via compliances through identical equations by replacing c_{jk} with s_{ij}.

2.1.7 Effect of the Crystal Symmetry on the Tensor of Coefficients of Elasticity

Because of the symmetry of a crystal, some of the components s_{ij} and c_{ij} become equal to zero or to each other, and the number of independent coefficients among the nonzero ones reduces.

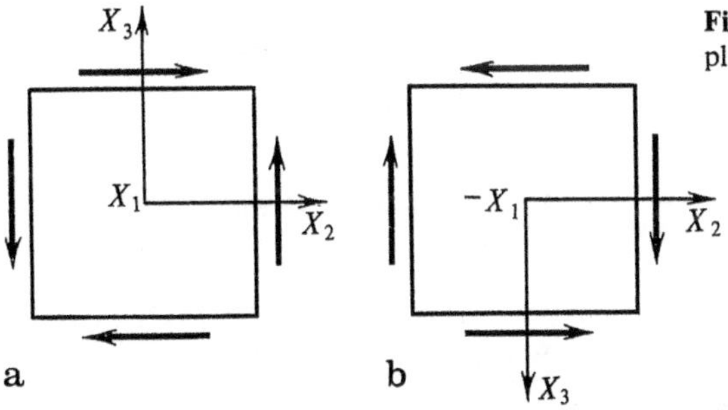

Fig. 2.10. Diagram explaining why compliance s_{34} is zero for class 222

Let us consider an orthorhombic crystal of class 222. We analyze the effect of the symmetry on compliance s_{34}, which relates strain ε_{33} to stress σ_{23}. Strain ε_{33} is elongation along the X_3 axis (Fig. 2.10a). We rotate the whole system (crystal plus shear forces) about a twofold axis parallel to the X_2 axis. The crystal and its elongation along X_3 remain unaltered, while the forces applied reverse their direction (Fig. 2.10b). This is possible only at $s_{34}=0$. By studying in this manner the effect of the symmetry on all the s_{ij} and c_{jk} for crystals of different crystallographic classes, we obtain the type of the corresponding matrices (Table 2.2).

Table 2.2 shows that the number of symmetry groups of the tensors of the elasticity coefficients of crystals and uniform media is ten. These symmetry groups contain two limiting ones: ∞/mmm and $\infty/\infty mm$.

The first one includes the tensors of the elasticity coefficients of hexagonal and trigonal crystals, which have, in addition to a sixfold or threefold axis, a symmetry plane perpendicular to it. All the directions in the plane perpendicular to the principal axis are equivalent with respect to the elastic properties in such crystals (transversely isotropic medium) (Tables 2.3, 4 list the numerical values of elasticity coefficients for a number of cubic and hexagonal crystals).

Class $\infty/\infty mm$ describes the symmetry of the elastic properties of an isotropic body. It is seen from Table 2.2 that the elastic properties of an isotropic body are characterized by only two coefficients of elasticity, s_{11} and s_{12}, or c_{11} and c_{12}. These coefficients can be expressed through Lamé coefficients λ and μ known from theoretical mechanics,

$$\lambda=c_{12}\ , \quad \mu=c_{44}=1/s_{44}\ , \quad \lambda+2\mu=c_{11}\ , \tag{2.29}$$

or via Young's modulus $E=\sigma/\varepsilon$, shear modulus G, and Poisson coefficient $\nu=-\varepsilon'/\varepsilon$ (ε and ε', respectively, are the longitudinal and transverse strains with respect to the straining force):

$$s_{11}=1/E\ , \quad s_{12}=\nu/E\ , \quad 2(s_{11}-s_{12})=1/G\ , \quad G=E/2(1+\nu)\ , \tag{2.30}$$

and $\lambda=2G\nu/(1-2\nu)$, $\mu=G$.

Table 2.2. Matrices of coefficients of elasticity for crystals of different symmetry

Group I. Triclinic syngony, classes 1,$\bar{1}$

$$\begin{array}{cccccc} c_{11} & c_{12} & c_{13} & c_{14} & c_{15} & c_{16} \\ c_{12} & c_{22} & c_{23} & c_{24} & c_{25} & c_{26} \\ c_{13} & c_{23} & c_{33} & c_{34} & c_{35} & c_{36} \\ c_{14} & c_{24} & c_{34} & c_{44} & c_{45} & c_{46} \\ c_{15} & c_{25} & c_{35} & c_{45} & c_{55} & c_{56} \\ c_{16} & c_{26} & c_{36} & c_{46} & c_{56} & c_{66} \end{array} \qquad \begin{array}{cccccc} s_{11} & s_{12} & s_{13} & s_{14} & s_{15} & s_{16} \\ s_{12} & s_{22} & s_{23} & s_{24} & s_{25} & s_{26} \\ s_{13} & s_{23} & s_{33} & s_{34} & s_{35} & s_{36} \\ s_{14} & s_{24} & s_{34} & s_{44} & s_{45} & s_{46} \\ s_{15} & s_{25} & s_{35} & s_{45} & s_{55} & s_{56} \\ s_{16} & s_{26} & s_{36} & s_{46} & s_{56} & s_{66} \end{array}$$

Group II. Monoclinic syngony, classes 2,*m*,2/*m*

$$\begin{array}{cccccc} c_{11} & c_{12} & c_{13} & 0 & c_{15} & 0 \\ c_{12} & c_{22} & c_{23} & 0 & c_{25} & 0 \\ c_{13} & c_{23} & c_{33} & 0 & c_{35} & 0 \\ 0 & 0 & 0 & c_{44} & 0 & c_{46} \\ c_{15} & c_{25} & c_{35} & 0 & c_{55} & 0 \\ 0 & 0 & 0 & c_{46} & 0 & c_{66} \end{array} \qquad \begin{array}{cccccc} s_{11} & s_{12} & s_{13} & 0 & s_{15} & 0 \\ s_{12} & s_{22} & s_{23} & 0 & s_{25} & 0 \\ s_{13} & s_{23} & s_{33} & 0 & s_{35} & 0 \\ 0 & 0 & 0 & s_{44} & 0 & s_{46} \\ s_{15} & s_{25} & s_{35} & 0 & s_{55} & 0 \\ 0 & 0 & 0 & s_{46} & 0 & s_{66} \end{array}$$

Group III. Orthorhombic syngony, classes 222, *mm*2, *mmm*

$$\begin{array}{cccccc} c_{11} & c_{12} & c_{13} & 0 & 0 & 0 \\ c_{12} & c_{22} & c_{23} & 0 & 0 & 0 \\ c_{13} & c_{23} & c_{33} & 0 & 0 & 0 \\ 0 & 0 & 0 & c_{44} & 0 & 0 \\ 0 & 0 & 0 & 0 & c_{55} & 0 \\ 0 & 0 & 0 & 0 & 0 & c_{66} \end{array} \qquad \begin{array}{cccccc} s_{11} & s_{12} & s_{13} & 0 & 0 & 0 \\ s_{12} & s_{22} & s_{23} & 0 & 0 & 0 \\ s_{13} & s_{23} & s_{33} & 0 & 0 & 0 \\ 0 & 0 & 0 & s_{44} & 0 & 0 \\ 0 & 0 & 0 & 0 & s_{55} & 0 \\ 0 & 0 & 0 & 0 & 0 & s_{66} \end{array}$$

Group IV. Tetragonal syngony, classes 4, 4/*m*, 422, 4/*mmm*

$$\begin{array}{cccccc} c_{11} & c_{12} & c_{13} & 0 & 0 & 0 \\ c_{12} & c_{11} & c_{13} & 0 & 0 & 0 \\ c_{13} & c_{13} & c_{33} & 0 & 0 & 0 \\ 0 & 0 & 0 & c_{44} & 0 & 0 \\ 0 & 0 & 0 & 0 & c_{44} & 0 \\ 0 & 0 & 0 & 0 & 0 & c_{66} \end{array} \qquad \begin{array}{cccccc} s_{11} & s_{12} & s_{13} & 0 & 0 & 0 \\ s_{12} & s_{11} & s_{13} & 0 & 0 & 0 \\ s_{13} & s_{13} & s_{33} & 0 & 0 & 0 \\ 0 & 0 & 0 & s_{44} & 0 & 0 \\ 0 & 0 & 0 & 0 & s_{41} & 0 \\ 0 & 0 & 0 & 0 & 0 & s_{66} \end{array}$$

Group V. Tetragonal syngony, classes $\bar{4}$, 4*mm*, $\bar{4}2m$

$$\begin{array}{cccccc} c_{11} & c_{12} & c_{13} & 0 & 0 & c_{16} \\ c_{12} & c_{11} & c_{13} & 0 & 0 & -c_{16} \\ c_{13} & c_{13} & c_{33} & 0 & 0 & 0 \\ 0 & 0 & 0 & c_{44} & 0 & 0 \\ 0 & 0 & 0 & 0 & c_{44} & 0 \\ c_{16} & -c_{16} & 0 & 0 & 0 & c_{66} \end{array} \qquad \begin{array}{cccccc} s_{11} & s_{12} & s_{13} & 0 & 0 & s_{16} \\ s_{12} & s_{11} & s_{13} & 0 & 0 & -s_{16} \\ s_{13} & s_{13} & s_{33} & 0 & 0 & 0 \\ 0 & 0 & 0 & s_{44} & 0 & 0 \\ 0 & 0 & 0 & 0 & s_{44} & 0 \\ s_{16} & -s_{16} & 0 & 0 & 0 & s_{66} \end{array}$$

Group VI. Trigonal syngony, classes $3, \bar{3}$

$$
\begin{array}{cccccc}
c_{11} & c_{12} & c_{13} & c_{14} & -c_{25} & 0 \\
c_{12} & c_{11} & c_{13} & -c_{14} & c_{25} & 0 \\
c_{13} & c_{13} & c_{33} & 0 & 0 & 0 \\
c_{14} & -c_{14} & 0 & c_{44} & 0 & c_{25} \\
-c_{25} & c_{25} & 0 & 0 & c_{44} & c_{14} \\
0 & 0 & 0 & c_{25} & c_{14} & \frac{1}{2}(c_{11}-c_{12})
\end{array}
\qquad
\begin{array}{cccccc}
s_{11} & s_{12} & s_{13} & s_{14} & -s_{25} & 0 \\
s_{12} & s_{11} & s_{13} & -s_{14} & s_{25} & 0 \\
s_{13} & s_{13} & s_{33} & 0 & 0 & 0 \\
s_{14} & -s_{14} & 0 & s_{44} & 0 & s_{25} \\
-s_{25} & c_{25} & 0 & 0 & s_{44} & s_{14} \\
0 & 0 & 0 & s_{25} & s_{14} & \frac{1}{2}(s_{11}-s_{12})
\end{array}
$$

Group VII. Trigonal syngony, classes 32, $3m$, $\bar{3}m$

$$
\begin{array}{cccccc}
c_{11} & c_{12} & c_{13} & c_{14} & 0 & 0 \\
c_{12} & c_{11} & c_{13} & -c_{14} & 0 & 0 \\
c_{13} & c_{13} & c_{33} & 0 & 0 & 0 \\
c_{14} & -c_{14} & 0 & c_{44} & 0 & 0 \\
0 & 0 & 0 & 0 & c_{44} & c_{44} \\
0 & 0 & 0 & 0 & c_{14} & \frac{1}{2}(c_{11}-c_{12})
\end{array}
\qquad
\begin{array}{cccccc}
s_{11} & s_{12} & s_{13} & s_{14} & 0 & 0 \\
s_{12} & s_{11} & s_{13} & -s_{14} & 0 & 0 \\
s_{13} & s_{13} & s_{33} & 0 & 0 & 0 \\
s_{14} & -s_{14} & 0 & s_{44} & 0 & 0 \\
0 & 0 & 0 & 0 & s_{44} & s_{44} \\
0 & 0 & 0 & 0 & s_{14} & \frac{1}{2}(s_{11}-s_{12})
\end{array}
$$

Group VIII. Hexagonal syngony, classes 6, $\bar{6}$, $6/m$, 622, $6mm$, $\bar{6}m2$, $6/mmm$

$$
\begin{array}{cccccc}
c_{11} & c_{12} & c_{13} & 0 & 0 & 0 \\
c_{12} & c_{11} & c_{13} & 0 & 0 & 0 \\
c_{13} & c_{13} & c_{33} & 0 & 0 & 0 \\
0 & 0 & 0 & c_{44} & 0 & 0 \\
0 & 0 & 0 & 0 & c_{44} & 0 \\
0 & 0 & 0 & 0 & 0 & \frac{1}{2}(c_{11}-c_{12})
\end{array}
\qquad
\begin{array}{cccccc}
s_{11} & s_{12} & s_{13} & 0 & 0 & 0 \\
s_{12} & s_{11} & s_{13} & 0 & 0 & 0 \\
s_{13} & s_{13} & s_{33} & 0 & 0 & 0 \\
0 & 0 & 0 & s_{44} & 0 & 0 \\
0 & 0 & 0 & 0 & s_{44} & 0 \\
0 & 0 & 0 & 0 & 0 & \frac{1}{2}(s_{11}-s_{12})
\end{array}
$$

Group IX. Cubic syngony, classes 23, $m3$, 432, $\bar{4}32$, $m3m$

$$
\begin{array}{cccccc}
c_{11} & c_{12} & c_{12} & 0 & 0 & 0 \\
c_{12} & c_{11} & c_{12} & 0 & 0 & 0 \\
c_{12} & c_{12} & c_{11} & 0 & 0 & 0 \\
0 & 0 & 0 & c_{44} & 0 & 0 \\
0 & 0 & 0 & 0 & c_{44} & 0 \\
0 & 0 & 0 & 0 & 0 & c_{44}
\end{array}
\qquad
\begin{array}{cccccc}
s_{11} & s_{12} & s_{12} & 0 & 0 & 0 \\
s_{12} & s_{11} & s_{12} & 0 & 0 & 0 \\
s_{12} & s_{12} & s_{11} & 0 & 0 & 0 \\
0 & 0 & 0 & s_{44} & 0 & 0 \\
0 & 0 & 0 & 0 & s_{44} & 0 \\
0 & 0 & 0 & 0 & 0 & s_{44}
\end{array}
$$

Group X. Isotropic medium

$$
\begin{array}{cccccc}
c_{11} & c_{12} & c_{12} & 0 & 0 & 0 \\
c_{12} & c_{11} & c_{12} & 0 & 0 & 0 \\
c_{12} & c_{12} & c_{11} & 0 & 0 & 0 \\
0 & 0 & 0 & \frac{1}{2}(c_{11}-c_{12}) & 0 & 0 \\
0 & 0 & 0 & 0 & \frac{1}{2}(c_{11}-c_{12}) & 0 \\
0 & 0 & 0 & 0 & 0 & \frac{1}{2}(c_{11}-c_{12})
\end{array}
$$

(Continued)

Table 2.2 (continued)

Group X. (contd.)	s_{11}	s_{12}	s_{13}	0	0	0
	s_{12}	s_{11}	s_{12}	0	0	0
	s_{12}	s_{12}	s_{11}	0	0	0
	0	0	0	$\frac{1}{2}(s_{11}-s_{12})$	0	0
	0	0	0	0	$\frac{1}{2}(s_{11}-s_{12})$	0
	0	0	0	0	0	$\frac{1}{2}(s_{11}-s_{12})$
	$\lambda+2\mu$	λ	λ	0	0	0
	λ	$\lambda+2\mu$	λ	0	0	0
or	λ	λ	$\lambda+2\mu$	0	0	0
	0	0	0	μ	0	0
	0	0	0	0	μ	0
	0	0	0	0	0	μ

In an isotropic medium, $c_{44}=\frac{1}{2}(c_{11}-c_{12})$ and $s_{44}=2(s_{11}-s_{12})$.
For cubic crystals, relation $A=2c_{44}/(c_{11}-c_{12})$, which is called the elastic-anisotropy parameter, is the measure of relative resistance of a crystal to two types of shear strain, since c_{44} characterizes the shear resistance along the (010) plane in the [001] direction, while $(c_{11}-c_{12})/2$ measures the shear resistance along the (110) plane in the [$\bar{1}$10] direction.

Compliance (s_{ij}) and stiffness (c_{ij}) coefficients of crystals can be compared with technical characteristics such as Young's modulus, the shear modulus, and the Poisson coefficient

$$E=1/s_{11}\ ,\quad G=\tfrac{1}{2}(c_{11}-c_{12})\ ,\quad \nu=s_{12}/s_{11}\ .$$

2.1.8 Hooke's Law for an Isotropic Body

Taking into account the relationship between the elastic constants and moduli E, G, and ν, Hooke's Law for an isotropic medium can be written

$$\varepsilon_{11}=\frac{1}{E}[\sigma_{11}-\nu(\sigma_{22}+\sigma_{33})]\ . \tag{2.31}$$

The expressions for ε_{22} and ε_{33} are obtained from (2.31) by cyclical permutation of the subscripts, while components ε_{12}, ε_{23}, and ε_{31} are expressed as

$$\varepsilon_{12}=\frac{1}{2G}\sigma_{12}\ ,\quad \varepsilon_{23}=\frac{1}{2G}\sigma_{23}\ ,\quad \varepsilon_{31}=\frac{1}{2G}\sigma_{31}\ . \tag{2.32}$$

Table 2.3. Coefficients of elasticity c_{ij} (10^{11} dyn/cm^2) and s_{ij} (10^{-12} cm^2/dyn), and the elastic anisotropy parameter A of some cubic crystals at room temperature

Crystal	c_{11}	c_{44}	c_{12}	s_{11}	s_{44}	s_{12}	A
Ag	12.40	4.61	9.34	2.29	2.17	−0.983	3
Al	10.82	2.85	6.13	1.57	3.51	−0.568	1.2
Au	18.6	4.20	15.7	2.33	6.38	−1.065	2.9
Cu	16.84	7.54	12.14	1.498	1.326	−0.629	3.21
Ni	24.65	12.47	14.73	0.734	0.802	−0.274	2.52
Pb	4.95	1.49	4.23	9.51	6.72	−4.38	4.14
Fe	22.8	11.65	13.2	0.762	0.858	−0.279	2.43
Mo	46.0	11.0	17.6	0.28	0.91	−0.078	7.75
Na	0.732	0.419	0.625	64.0	23.9	−29.5	7.85
Nb	24.55	2.93	13.90	0.690	3.42	−0.249	5.48
Ta	26.7	8.25	16.1	0.685	1.21	−0.258	1.5
V	22.8	4.26	11.9	0.683	2.35	−0.234	0.8
W	50.1	15.14	19.8	0.257	0.660	−0.073	1
C (diamond)	107.7	57.8	12.50	0.0953	0.174	−0.0099	1.21
Ge	12.89	6.71	4.83	0.978	1.490	−0.266	4.25
Si	16.57	7.96	6.39	0.768	1.256	−0.214	1.56
NaCl	4.87	1.26	1.24	2.29	7.94	−0.465	0.7
LiF	11.12	6.28	4.20	1.135	1.59	−0.31	1.82
MgO	28.92	15.46	8.80	0.403	0.647	−0.094	1.54
TiC	50.0	17.5	11.3	0.218	0.572	−0.040	0.9

Table 2.4. Coefficients of elasticity c_{ij} (10^{11} dyn/cm^2) and s_{ij} (10^{-12} cm^2/dyn) of some hexagonal crystals at room temperature

Crystal	c_{11}	c_{33}	c_{44}	c_{12}	c_{13}	s_{11}	s_{33}	s_{44}	s_{12}	s_{13}
Be	29.23	33.64	16.25	2.67	1.4	0.348	0.298	0.616	−0.030	−0.0131
C (graphite)	116	4.66	0.23	29	10.9	0.111	3.32	43.5	−0.0046	−0.249
Cd	11.58	5.14	2.04	3.98	4.06	1.24	3.52	4.98	−0.076	−0.920
Co	30.7	35.81	7.83	16.5	10.3	0.472	0.319	1.324	−0.231	−0.069
Hf	18.11	19.69	5.57	7.72	6.61	0.715	0.613	1.80	−0.247	−0.157
Mg	5.97	6.17	1.64	2.62	2.17	2.20	1.97	6.1	−0.785	−0.50
Re	61.25	68.27	16.25	27.0	20.6	0.212	0.170	0.616	−0.080	−0.40
Ti	16.24	18.07	4.67	2.20	6.90	0.958	0.698	2.14	−0.462	−0.189
Zn	16.1	6.10	3.83	3.42	5.01	0.838	2.838	2.61	−0.053	−0.731
ZnO	20.97	21.09	4.25	12.11	10.51	0.787	0.694	2.35	−0.334	−0.221
Zr	14.34	16.48	3.20	7.28	6.53	1.013	0.799	3.13	−0.404	−0.241

Hooke's Law for an isotropic medium can be written as follows, using the Lamé coefficients as well:

$$\sigma_{ij} = \lambda \varepsilon_{ii} \delta_{ij} + 2\mu \varepsilon_{ij} \ , \tag{2.33}$$

where $\varepsilon_{ii} = \varepsilon_{11} + \varepsilon_{22} + \varepsilon_{33}$, $\delta_{ii} = 1$ for $i = j$ and $\delta_{ij} = 0$ for $i \neq j$.

2.1.9 Young's Modulus in an Arbitrary Direction and the Characteristic Surfaces of Elastic Properties

To determine Young's modulus in an arbitrary direction in a crystal, the coordinate system must be transformed so that the X_3' axis is disposed along the selected direction. Obviously, always $\varepsilon_3 = -s_{33}'\sigma_3$, and $1/s_{33}'$ is Young's modulus in the direction of the X_3' axis.

For an arbitrary direction with guiding cosines l_1, l_2, and l_3 for any crystallographic classes (in transforming, one must switch to four-digit tensor symbols), we have the following expressions for s_{ij}':

$$s_{ij}' = l_i l_j l_m l_n s_{ijmn} \ . \tag{2.34}$$

For high-symmetry classes, the right-hand side of (2.34) in the developed form looks thus:

tetragonal syngony, classes 4, $\bar{4}$, $4/m$:

$$(l_1^4 + l_2^4)s_{11} + l_3^4 s_{33} + l_1^2 l_2^2 (2s_{12} + s_{66}) + l_3^2 (1 - l_3^2)(2s_{13} + s_{44}) + 2l_1 l_2 (l_1^2 - l_2^2) s_{16} \ ;$$

tetragonal syngony, classes $4mm$, 4, 2, 422, $4/mmm$:

$$(l_1^4 + l_2^4)s_{11} + l_3^4 s_{33} + l_1^2 l_2^2 (2s_{12} + s_{66}) + l_3^2 (1 - l_3^2)(2s_{13} + s_{44}) \ ;$$

trigonal syngony, classes 3, $\bar{3}$:

$$(1 - l_3^2)^2 s_{11} + l_3^4 s_{33} + l_3^2 (1 - l_3^2)(2s_{13} + s_{44}) + 2l_2 l_3 (3l_1^2 - l_2^2) s_{14}$$
$$+ 2l_1 l_2 (3l_2^2 - l_1^2) s_{25} \ ;$$

trigonal syngony, classes $3m$, 32, $\bar{3}/m$:

$$(1 - l_3^2)^2 s_{11} + l_3^4 s_{33} + l_3^2 (1 - l_3^2)(2s_{13} + s_{44}) + 2l_2 l_3 (3l_1^2 - l_2^2) s_{44} \ ;$$

hexagonal syngony:

$$(1 - l_3^2)^2 s_{11} + l_3^4 s_{33} + l_3^2 (1 - l_3^2)(2s_{13} + s_{44}) \ ;$$

cubic syngony:

$$s_{11} + 2(s_{11} - s_{12} - \tfrac{1}{2}s_{44})(l_1^2 l_2^2 + l_2^2 l_3^2 + l_3^2 l_1^2) \ .$$

It can be seen that Young's modulus is not isotropic even for the cubic system. Its variation with the direction depends on $(l_1^2 l_2^2 + l_2^2 l_3^2 + l_3^2 l_1^2)$. This quantity is equal to zero in the direction of the $\langle 100 \rangle$ axes of the cube and has a maximum value of 1/3 in the $\langle 111 \rangle$ directions.

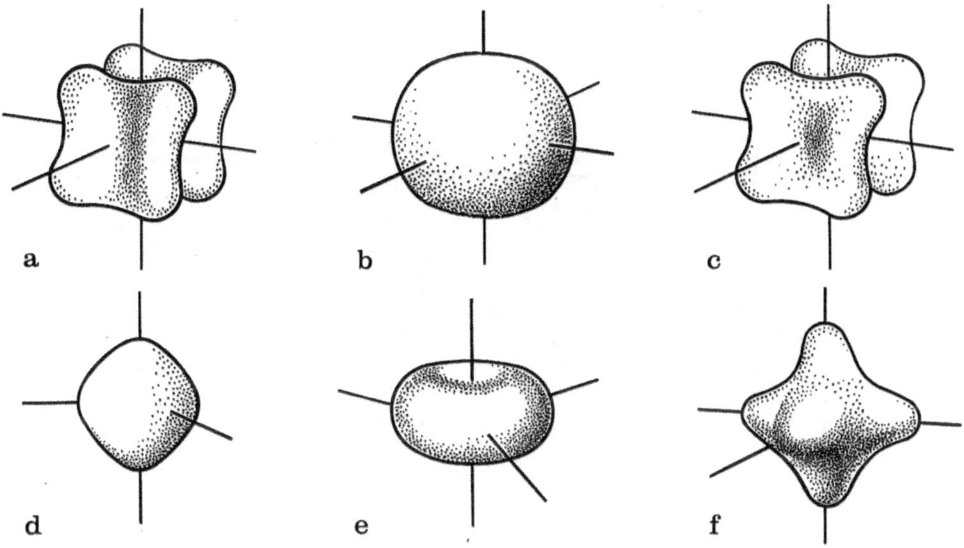

Fig. 2.11. Characteristic surfaces of Young's modulus for crystals of Au (**a**), Al (**b**), Fe (**c**), Mg (**d**), and Zn (**e**) and the torsion modulus for a crystal of Fe (**f**)

If quantity $(s_{11}-s_{12}-1/2s_{44})$ is positive, as in most of the cubic crystals, the maximum value of Young's modulus corresponds to the $\langle 111 \rangle$ directions, and the minimum value, to the $\langle 100 \rangle$ directions. Equality $(s_{11}-s_{12}-1/2s_{44}) = 0$ is the condition for the isotropy of the elastic properties of the medium.

The characteristic surfaces of s'_{33} or of Young's modulus $E = s'^{-1}_{33}$ are generally used for describing the anisotropy of the elastic properties of crystals. Sometimes the same purpose is achieved by using the characteristic surfaces of the shear modulus

$$c'_{44} = l_i l_j l_m l_n c_{44}$$

and the torsional modulus $2(s'_{44}+s'_{55})$. Figures 2.11, 12 demonstrate examples of characteristic surfaces of s'_{33}, s'^{-1}_{33}, $2(s'_{44}+s'_{55})$ and their sections for a number of crystals.

2.1.10 Adiabatic and Isothermal Values of Elasticity Coefficients

The coefficients of elasticity (elastic constants) can be determined either by static or by dynamic methods. In static measurements the temperature of the test crystal is maintained constant, and therefore the *isothermal* values of the elasticity coefficients are determined, which are sometimes denoted by superscript "T": s^T_{ij} or c^T_{ij}. In dynamic measurements of elasticity coefficients, the

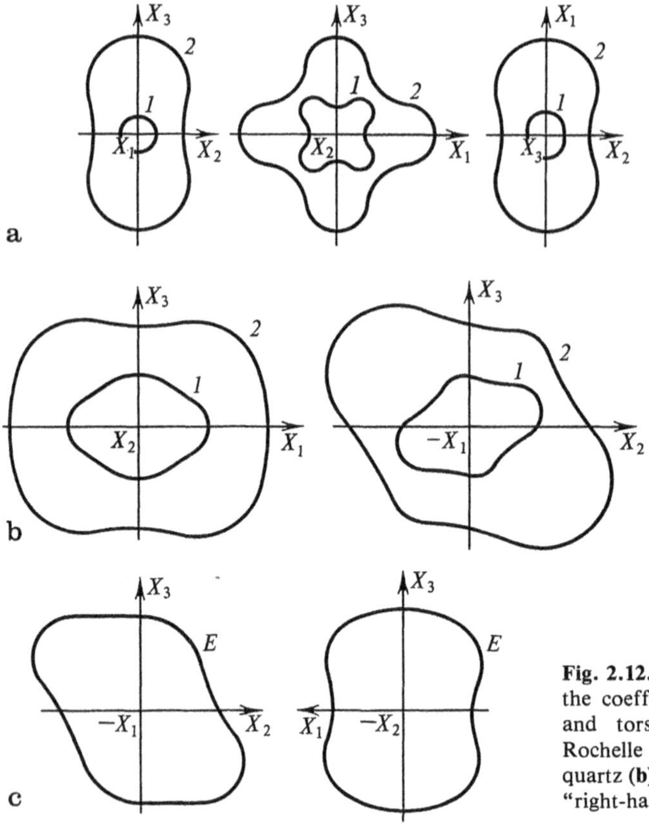

Fig. 2.12. Sections of the surfaces of the coefficients of extension s_{33} (*1*) and torsion 2 $(s'_{44}+s'_{55})$ (*2*) for Rochelle salt (**a**), "right-hand" quartz (**b**), and Young's modulus for "right-hand" quartz (**c**)

temperature of the test specimen changes, increasing in the contraction region and decreasing in the extension region. At a high vibration frequency, the temperature in the specimen bulk does not have the time to equalize, and an additional strain appears in the specimen. As a result, *adiabatic* values of the elasticity coefficients are determined in dynamic measurements; they are denoted by superscript "*S*" (*S* is the crystal entropy): s^S_{ij} or c^S_{ij}.

The relationship between the adiabatic and isothermal values of the elastic constants is found from consideration of the thermodynamics of the thermoelastic properties [2.3]. This relationship is given by

$$s^S_{ij}-s^T_{ij}=\alpha_i\alpha_k T/c^\sigma \ , \tag{2.35}$$

where α_i and α_k are the coefficients of thermal expansion, and c^σ is the heat capacity at a constant stress. Adiabatic compliances are usually less than isothermal, because the coefficients of thermal expansion are usually positive, and the heat capacity is predominantly positive.

The elasticity coefficients are also affected by electron effects [2.4].

2.1.11 Volume Compressibility of Crystals

The volume compressibility is the relative decrease in volume dV/V of a crystal under the effect of hydrostatic pressure dP, i.e.,

$$s_{iijj} = -\frac{1}{V}\frac{dV}{dP} . \tag{2.36}$$

In matrix notation, for the most general case (triclinic system) the volume compressibility has the form

$$s_{iijj} = s_{11}+s_{22}+s_{33}+2(s_{12}+s_{23}+s_{31}) . \tag{2.37}$$

For isotropic materials and cubic crystals, the volume compressibility is equal to $3(s_{11}+2s_{12})$. Its reciprocal value – the modulus of volume compressibility (triaxial compression) – is also often used:

$$k = \frac{1}{3(s_{11}+2s_{12})} = \frac{E}{3(1-2\nu)} . \tag{2.38}$$

2.1.12 Born Theory

Since the elastic strain depends on changes in lattice spacing and elastic stresses depend on particle interaction forces, the macroscopic elastic properties of crystals are defined by the rigidity of the interparticle bonds.

The relationship between the macroscopic elastic constants and the interatomic forces was first established by Born (1915), who neglected the thermal vibrations of the atoms and lattice defects and took into account only the central forces (the interaction forces directed along the lines joining the atomic or ionic centers).

For a NaCl-type crystal the interaction energy of univalent ions is calculated in [Ref. 2.5, Chap. 1]:

$$-U(r) = A\left(\frac{Me^2}{r} - M'\frac{b}{r^n}\right) . \tag{2.39}$$

The meaning of all the values appearing in this expression is also explained there. Considering that at equilibrium $r = r_0$ and $(\partial U/\partial r)_{r=r_0} = 0$, M' can be expressed through M:

$$M' = Me^2 r_0^{n-1}/nb . \tag{2.40}$$

Then

$$U(r) = -\frac{Me^2}{r}\left[1 - \frac{1}{n}\left(\frac{r_0}{r}\right)^{n-1}\right] . \tag{2.41}$$

Each of the six bonds with the nearest neighbors is associated with energy $U/6$. The value of n can be determined via the triaxial compression modulus (2.38). Indeed, since $P = -dU/dV$, then

$$k = V\frac{dP}{dV} = V\frac{d^2U}{dV^2} . \tag{2.42}$$

The volume of a gram-mole of a crystal is defined by

$$V = N\gamma r^3 ,$$

where N is the Avogadro number, and γ is a coefficient reflecting the nature of stacking of the atoms (in close-packed structures $\gamma \approx 1$). Hence,

$$\frac{dU}{dV} = \frac{1}{3N\gamma r^2}\frac{dU}{dr} , \quad \frac{d^2U}{dV^2} = \frac{1}{9N^2\gamma^2 r^2}\frac{d}{dr}\left(\frac{1}{r^2}\frac{dU}{dr}\right) . \tag{2.43}$$

Taking into account (2.41, 43), we obtain

$$n = 1 + \frac{9Vr}{e^2A}k . \tag{2.44}$$

For alkali-halide crystals, n varies from 5 for lithium salts to 11 for cesium salts. For NaCl crystals $n \simeq 9$.

2.1.13 Cauchy Ratio

If we assume that the atoms forming the crystal lattice are bound by central forces, we can additionally obtain six relations between the elastic coefficients. These so-called *Cauchy relations* are as follows:

$$c_{23} = c_{44} , \quad c_{56} = c_{14} , \quad c_{64} = c_{25} , \quad c_{31} = c_{55} , \quad c_{12} = c_{66} , \quad c_{45} = c_{36} .$$

Allowing for these relations, the number of elastic coefficients in triclinic crystals decreases from 21 to 15. From the Cauchy relations it also follows that for an isotropic material $c_{11} = 3c_{12}$ (i.e., one coefficient of elasticity remains), for a cubic crystal $c_{12} = c_{44}$, and for a hexagonal and a trigonal crystal $c_{13} = c_{44}$ and $c_{11} = 3c_{12}$, respectively.

Experience shows, however (Tables 2.3, 4), that the Cauchy relations are fulfilled more or less satisfactorily only for NaCl-type alkali-halide crystals.

According to Born's theory, the Cauchy relations are valid only when a crystal is built so that each particle in it is a center of symmetry. The nonfulfillment of the Cauchy conditions points to the necessity of taking into account the bonds in a specific crystal and, in particular, the valency electrons as independent constituent elements of the lattice.

2.1.14 Elastic Waves in Crystals

The adiabatic coefficients of elasticity find the widest application in describing the propagation of elastic waves in a crystal. In considering such waves one should use the general equations of motion of an elastic medium, equating the force of internal stresses $d\sigma_{ij}/dx_i$ to the product of acceleration $\ddot{u}$ by the mass of a unit volume of the body, i.e., by density ϱ:

$$\varrho \ddot{u}_i = \frac{\partial \sigma_{ij}}{\partial x_j} , \tag{2.45}$$

where $\sigma_{ij} = c^S_{ijkl}\varepsilon_{kl}$. Substituting (2.25) for σ_{ij} into the equation of motion (2.45) and following for the relation

$$\varepsilon_{kl} = \frac{1}{2}\left(\frac{\partial u_k}{\partial x_l} + \frac{\partial u_l}{\partial x_k}\right) ,$$

we obtain

$$\varrho \ddot{u}_i = c_{ijkl}\frac{\partial \varepsilon_{kl}}{\partial x_j} = \frac{c^S_{ijkl}}{2}\,\frac{\partial}{\partial x_j}\left(\frac{\partial u_k}{\partial x_l} + \frac{\partial u_l}{\partial x_k}\right)$$

$$= \frac{1}{2}c^S_{ijkl}\frac{\partial^2 u_k}{\partial x_j \partial x_l} + \frac{1}{2}c^S_{ijkl}\frac{\partial^2 u_l}{\partial x_j \partial x_k} .$$

Since tensor c^S_{ijkl} is symmetric with respect to subscripts k and l, by replacing summation subscript l with k (and k with l) in the second term we find that the first and second terms are identical; then

$$\varrho \ddot{u}_i = c^S_{ijkl}\frac{\partial^2 u_l}{\partial x_j \partial x_k} . \tag{2.46}$$

The equation obtained describes the law of propagation of elastic waves in a crystal. We restrict ourselves to plane monochromatic elastic waves. The expression for the displacement vector of such a wave can be written as

$$u_i = u_i^0 \exp[\mathrm{i}(\boldsymbol{kr} - \omega t)] . \tag{2.47}$$

Here $u_i^0 = \text{const}$ is the wave amplitude. Substituting (2.47) into (2.46), we get

$$\varrho \omega^2 u_i = c^S_{ijkl} k_j k_k u_l \tag{2.48}$$

(k_j and k_k are the components of the wave vector along the x_j and x_k axis, respectively). If we write $u_i = \delta_{il} u_l$ (δ_{il} is the Kronecker symbol equal to 1 at $i = l$, and 0 at $i \neq l$), then (2.48) transforms to

$$(\varrho \omega^2 \delta_{il} - c_{ijkl}^S k_j k_k) u_l = 0 \ , \tag{2.49}$$

which is a set of three homogeneous equations of the first order with respect to u_1, u_2, and u_3. Such a set, as is well known, has a nonzero solution only if a determinant composed of factors at u_l is equal to zero, i.e.,

$$|c_{ijkl}^S k_j k_k - \varrho \omega^2 \delta_{il}| = 0 \ . \tag{2.50}$$

This expression, which is a third-order equation with respect to ω^2, has three roots, different in the general case, which are a function of wave vector $\boldsymbol{k}$. Substituting each of them into (2.49), we obtain the corresponding displacement components u_i with an accuracy to a constant common multiplier.

The velocity of propagation of the wave (group velocity) is determined by the derivate of the frequency with respect to the wave vector. In an isotropic body the frequency is proportional to $|k|$, and therefore the direction of velocity $U = d\omega/dk$ coincides with the direction of $\boldsymbol{k}$. In crystals, this dependence does not hold, and the direction of propagation of the wave does not coincide with that of its wave vector. On the other hand, the wave velocity, being a function of its direction, is frequency independent while there is no dispersion of the elastic constants. Indeed, according to (2.50), ω is a homogeneous first-order function of components k_i, and hence propagation velocity $d\omega/d\boldsymbol{k}$ is a homogeneous zero-order function of k_i.

Since there are three different (for each k_i) relationships between ω and $\boldsymbol{k}$, three different elastic waves can, generally speaking, propagate along each direction in the crystal. Only in some singular directions, depending on the crystal symmetry, can the velocity of these waves coincide. For instance, for the [0001] directions of hexagonal crystals, ⟨100⟩ and ⟨111⟩ of cubic crystals, and [001] of tetragonal crystals, the velocities of the two transverse waves are identical.

In an isotropic medium, the strain in an elastic plane wave is a function of only one coordinate (for instance, x_1) and the time. For the components of displacement vector $\boldsymbol{u}$ we obtain the following expressions:

$$\frac{\partial^2 u_1}{\partial x_1^2} - \frac{1}{c_l^2} \frac{\partial^2 u_1}{\partial t^2} = 0 \ , \quad \frac{\partial^2 u_2}{\partial x_1^2} - \frac{1}{c_t^2} \frac{\partial^2 u_2}{\partial t^2} = 0 \ , \tag{2.51 a}$$

where

$$c_l = \sqrt{\frac{E(1-\nu)}{\varrho(1+\nu)(1-2\nu)}} \ , \quad c_t = \sqrt{\frac{E}{2\varrho(1+\nu)}} \tag{2.51 b}$$

(the equation for u_3 – the component of displacement along x_3 – is the same as for u_2). These are the conventional wave equations in one dimension, and c_l and c_t are the wave propagation velocities. We can see that the latter are different for component u_1, on the one hand, and components u_2 and u_3, on the other. Thus, an elastic wave in an isotropic medium actually comprises three waves propagating independently in the same direction, in one of which (u_1) the displacement is directed along the wave propagation, while in the other two (u_2, u_3) the displacement occurs in a plane perpendicular to the direction of propagation. The first wave is called longitudinal and propagates at a velocity c_l, and the second and the third are called transverse and propagate at equal velocity c_t, c_l always exceeding c_t. Velocities c_l and c_t are called the longitudinal and transverse velocities of sound. If, in an isotropic medium, to two different propagation velocities correspond pure longitudinal and pure transverse waves, then, in the crystal, to each propagation velocity corresponds a wave in which the displacement vector generally possesses components both parallel and perpendicular to the direction of propagation, i.e., along each direction three waves usually propagate, each of which is neither pure longitudinal nor pure transverse.

The propagation of elastic waves can be used for studying the anisotropy of the elastic properties of crystals. We shall now consider the *Schaefer-Bergmann* method [2.6], which is suitable for transparent materials. The method is based on investigation of the interaction of a monochromatic light beam with a system of elastic standing waves set up in a vibrating crystal. Vibrations in a crystal can be excited either directly by an electric field (for piezoelectric crystals), or, for nonpiezoelectric crystals, in some other way, for instance by means of a piezoelectric converter. As the light passes through the vibrating crystal, it is diffracted from elastic waves, and peculiar optical patterns are revealed which consist of closed curved lines (Fig. 2.13). Each of the curves is generally a section of the *surface of the inverted sound velocities* by the observation plane perpendicular to the direction of light propagation.

It is worth noting that the surfaces of the inverted sound velocities, as well as the characteristic strain surfaces, reflect the anisotropy of the elastic properties of the crystal, but there is no direct similarity between these two types of surface. The description of the elastic wave surfaces in different crystals is given by *Musgrave* [2.8].

The angle of deviation θ of the diffracted light beams from the direction of the incident beam is defined as the ratio of the wavelength of light (λ) and sound (Λ). Angle θ is often found from the formula

$$\sin\theta = \lambda/2\Lambda \ . \tag{2.52}$$

Diffraction is usually observed at frequencies of the order of tens of MHz, when angle θ reaches fractions of a degree, and $\lambda \sim 0.5$ µm.

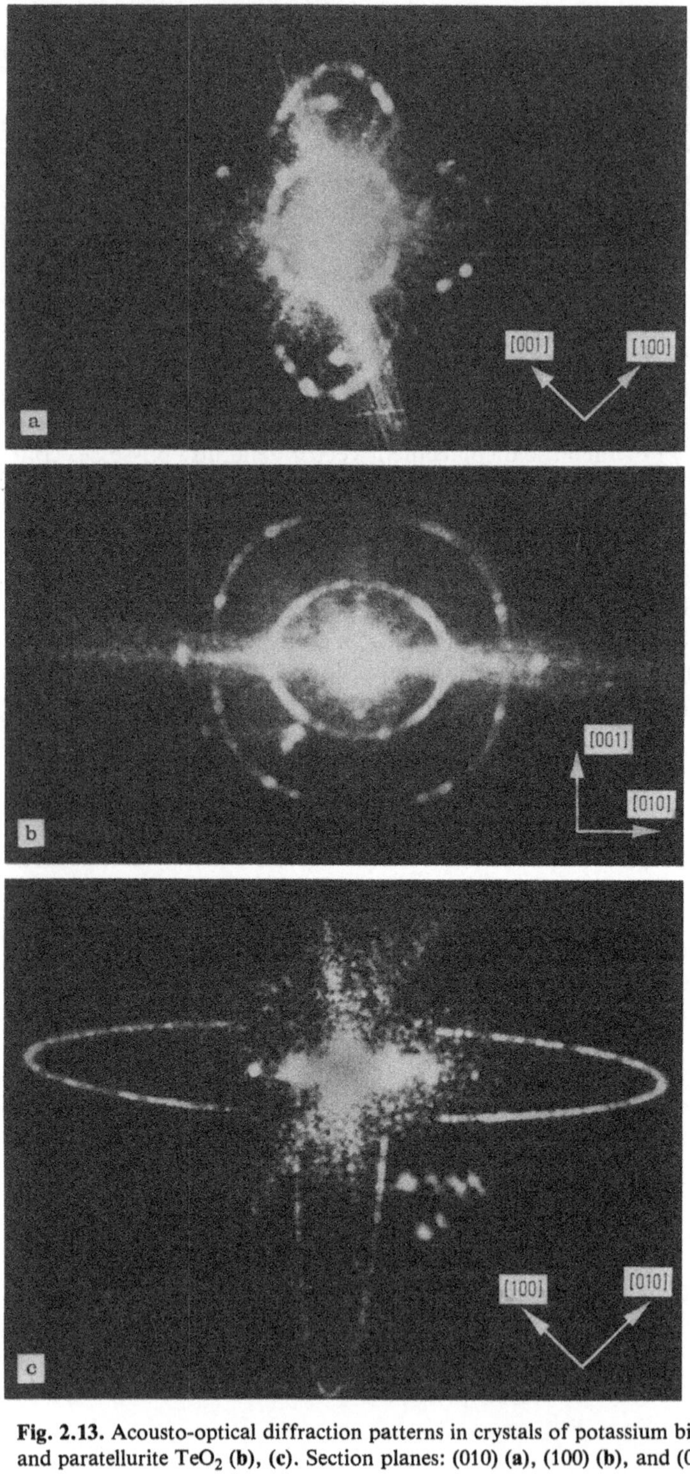

Fig. 2.13. Acousto-optical diffraction patterns in crystals of potassium biphthalate $C_6H_8O_5K$ (**a**) and paratellurite TeO_2 (**b**), (**c**). Section planes: (010) (**a**), (100) (**b**), and (001) (**c**) [2.7]

2.2 Plastic Glide Deformation

2.2.1 Elastic Deformation in Crystals

An elastic deformation in crystals, i.e., a deformation which completely vanishes after the load is removed, cannot, as a rule, exceed tenths of 1%; only in dislocation-free whiskers can it reach 3−4% [2.9]. At higher deformations (and thus higher stresses acting for sufficiently long periods) the crystal begins to "flow," acquiring a residual deformation which is preserved after unloading. The deformation remaining after the load is removed is called the *plastic deformation.* The ability to undergo plastic deformation differs from one crystal to another. In some crystals plastic deformation begins at low applied stresses [g/mm^2], and in others, at high ones (kg/mm^2). Plastic deformation may range from several percent to hundreds percent. Crystals in which only small strains are observed before failure are called *brittle.* The residual deformation, i.e., the plasticity of brittle materials, can be increased under a hydrostatic pressure. For instance, while at normal pressure corundum shows an ability to undergo macroscopically visible plastic deformation only at a temperature of about 1000 °C, at a pressure of 25,000 kgf/cm^2 it "flows" even at room temperature.

2.2.2 Translational Glide

The main method of plastic deformation of crystals at normal and low temperatures is *glide*. It is the displacement of one part of the crystal relative

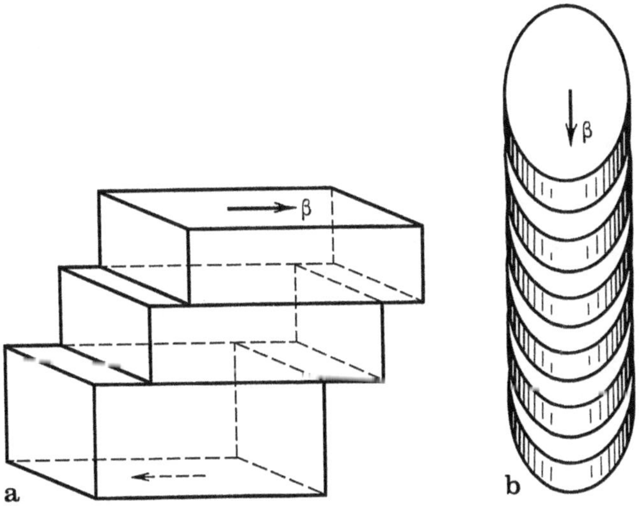

Fig. 2.14. Model of glide under the effect of shear stresses in a specimen of cubic (**a**) and cylindrical (**b**) shape

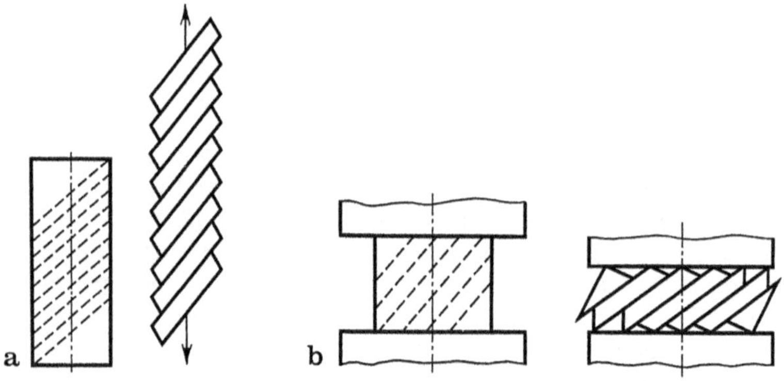

Fig. 2.15. Change in the position of glide planes in extension (**a**) and in compression (**b**). Figures on the left show the pre-deformation arrangement of the glide planes

to another without a change in crystal volume (the difference between elastic and plastic shear is explained below). Glide usually occurs over definite crystallographic planes along definite directions.

Figure 2.14 depicts a model of glide under the effect of shear stresses in specimens of various shape and also indicates the plane and direction of shear (β) in them. By gliding, parts of the crystal are displaced by a value multiple of the translation vector of the crystal lattice in the direction of glide. Therefore, glide is often termed translational. Under favorable conditions, glide may propagate throughout the cross section of the specimen.

In the cases demonstrated in Fig. 2.14, the strain under the effect of shear stresses changed only the shape of the crystal, leaving its volume and orientation unaltered. But with other methods of loading, for instance, under the effect of compressive or tensile stresses, the gliding layers rotate in the course of strain, approaching the direction of the action of the stresses (this direction is sometimes called the strain axis) in the case of extension (Fig. 2.15a) or tending to occupy a position perpendicular to the strain axis in the case of compression (Fig. 2.15b).

2.2.3 Plastic Strain Tensor

Glide results in shear strain. If a point P situated at a distance $\boldsymbol{r}$ from the origin O moves in a glide plane perpendicular to unit vector $\boldsymbol{n}$ along the direction of slipping characterized by unit vector $\boldsymbol{\beta}$, the new position of point P, i.e., point P', will be at a distance of $\boldsymbol{r}'$ from point O. Then

$$PP' = \boldsymbol{r}' - \boldsymbol{r} = \alpha(\boldsymbol{r}\cdot\boldsymbol{n})\boldsymbol{\beta} \ , \quad \text{or} \quad \boldsymbol{r}' = \boldsymbol{r} + \alpha(\boldsymbol{r}\cdot\boldsymbol{n})\boldsymbol{\beta} \ , \tag{2.53}$$

where α is the shear (displacement of Point P). If α is small, then, by analogy with (2.12, 13), it is possible to find the components of the plastic distortion tensor u_{ij}^{0} (the superscript 0 was introduced to distinguish the plastic from the elastic distortion) and the *plastic strain tensor* ε_{ij} referred to an arbitrary orthogonal system of coordinates (X_1, X_2, and X_3):

$$u_{11}^{0}=\frac{\partial u_1}{\partial x_1}=\frac{\partial}{\partial x_1}(\boldsymbol{r}'-\boldsymbol{r})=\frac{\partial}{\partial x_1}\alpha(\boldsymbol{r}\cdot\boldsymbol{n})\beta_1 \; . \tag{2.54}$$

If we write

$$\boldsymbol{r}=x_1\boldsymbol{i}+x_2\boldsymbol{j}+x_3\boldsymbol{k} \; ,$$
$$\boldsymbol{n}=n_1\boldsymbol{i}+n_2\boldsymbol{j}+n_3\boldsymbol{k} \; ,$$
$$\boldsymbol{\beta}=\beta_1\boldsymbol{i}+\beta_2\boldsymbol{j}+\beta_3\boldsymbol{k} \; ,$$

where ***i***, ***j***, and ***k*** are unit vectors, then

$$u_{11}^{0}=\frac{\partial}{\partial x_1}\alpha(\boldsymbol{r}\cdot\boldsymbol{n})\beta_1=\alpha n_1\beta_1 \; .$$

We similarly find

$$u_{23}^{0}=\frac{\partial u_2}{\partial x_3}=\frac{\partial}{\partial x_3}\alpha(\boldsymbol{r}\cdot\boldsymbol{n})\beta_2=\alpha n_3\beta_2$$

and the other components of the polar second-rank tensor u_{ij}^{0}, i.e., the plastic distortion tensor. To summarize, we can write

$$u_{ij}^{0}=\alpha\left\|\begin{matrix} n_1\beta_1 & n_2\beta_1 & n_3\beta_1 \\ n_1\beta_2 & n_2\beta_2 & n_3\beta_2 \\ n_1\beta_3 & n_2\beta_3 & n_3\beta_3 \end{matrix}\right\| \; . \tag{2.55}$$

From the orthogonality of vectors ***n*** and ***β***, it follows that

$$\alpha(n_1\beta_1+n_2\beta_2+n_3\beta_3)=u_{11}^{0}+u_{22}^{0}+u_{33}^{0}=0 \; .$$

This means the conservation of the volume by shear deformation.

A plastic distortion differs essentially from an elastic one (Fig. 2.16). In an elastic distortion the interatomic distances change, with resulting elastic deformations and lattice rotations. In the latter case the atoms move to new equilibrium positions, but the interatomic distances are preserved; owing to the translational nature of the displacements, no lattice rotations occur in plastic distortion, and only a change in the shape of the body is observed.

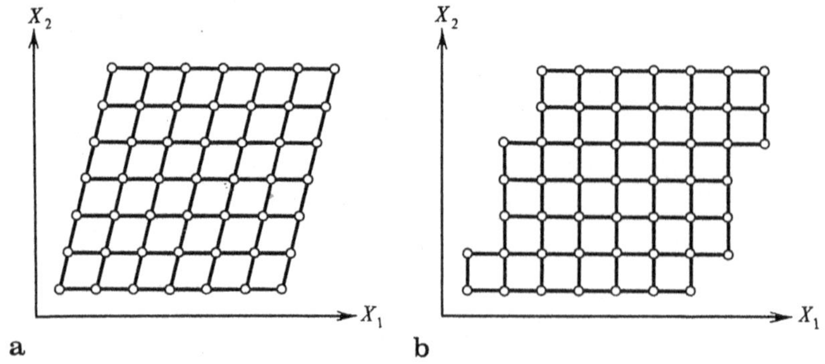

Fig. 2.16. Elastic **(a)** and plastic **(b)** distortion of the atomic net

Tensor u_{ij}^0 (2.55) can be represented as a sum of symmetric tensor ε_{ij}, describing a pure plastic deformation, and antisymmetric tensor ω_{ij}, characterizing a rotation; here

$$\varepsilon_{ij} = \left\| \begin{matrix} \alpha n_1\beta_1 & \frac{\alpha}{2}(n_1\beta_2+n_2\beta_1) & \frac{\alpha}{2}(n_1\beta_3+n_3\beta_1) \\ \frac{\alpha}{2}(n_1\beta_2+n_2\beta_1) & \alpha n_2\beta_2 & \frac{\alpha}{2}(n_3\beta_2+n_2\beta_3) \\ \frac{\alpha}{2}(n_1\beta_3+n_3\beta_1) & \frac{\alpha}{2}(n_3\beta_2+n_2\beta_3) & \alpha n_3\beta_3 \end{matrix} \right\| , \tag{2.56}$$

$$\omega_{ij} = \left\| \begin{matrix} 0 & \frac{\alpha}{2}(n_2\beta_1-n_1\beta_2) & \frac{\alpha}{2}(n_3\beta_1-n_1\beta_3) \\ -\frac{\alpha}{2}(n_2\beta_1-n_1\beta_2) & 0 & \frac{\alpha}{2}(n_3\beta_2-n_2\beta_3) \\ -\frac{\alpha}{2}(n_3\beta_1-n_1\beta_3) & -\frac{\alpha}{2}(n_3\beta_2-n_2\beta_3) & 0 \end{matrix} \right\| . \tag{2.57}$$

The significance of these tensors will be easy to understand if we perform the following operation. We choose X_1 and X_2 axes parallel to $\boldsymbol{n}$ and $\boldsymbol{\beta}$, respectively. Then expressions (2.55 – 57) transform to

$$u_{ij}^0 = \left\| \begin{matrix} 0 & 0 & 0 \\ \alpha & 0 & 0 \\ 0 & 0 & 0 \end{matrix} \right\| , \quad \varepsilon_{ij} = \left\| \begin{matrix} 0 & \alpha/2 & 0 \\ \alpha/2 & 0 & 0 \\ 0 & 0 & 0 \end{matrix} \right\| , \quad \omega_{ij} = \left\| \begin{matrix} 0 & -\alpha/2 & 0 \\ \alpha/2 & 0 & 0 \\ 0 & 0 & 0 \end{matrix} \right\| .$$

The geometric interpretation of these tensors is given in Fig. 2.7.

2.2.4 Glide Elements

The planes along which the crystal layers move are called the *glide planes,* and the directions of the displacement of the layers are called the *glide directions.* A glide plane and the glide direction lying in it are called glide elements and form a *glide system.* A set of all the crystallographically equivalent glide planes and directions constitutes a *family of glide systems.* For instance, in NaCl-type crystals belonging to class $m3m$, glide occurs over $\{110\}$ planes along $\langle 1\bar{1}0\rangle$ axes. In all, there are six planes of this type; in each of them glide can occur in the forward and backward direction along $\langle 1\bar{1}0\rangle$ axes. Thus, 12 glide systems should have been considered, but slipping off in the forward and backward direction gives one and the same result because of the centrosymmetricity of glide in these crystals. It can therefore be assumed that the family of $\{110\}$ planes and $\langle 1\bar{1}0\rangle$ directions consists of six glide systems. Table 2.5 lists the glide elements for a number of crystals.

Glide elements can be predicted from the structure of the crystal. The physical basis for this is the minimum work spent in displacement by one translation vector: the less the work, the weaker the bonds between the planes sliding relative to one another and the smaller the mutual displacement of the planes. The minimum work is required for displacing close-packed planes along close-packed directions. The reason is that close-packed planes have larger interplanar distances; therefore, the adhesion forces acting between

Table 2.5. Glide elements in some crystals[a]

Crystal	Class	Lattice type	Glide systems
fcc metals (Al, Cu) and solid solution of α-CuZn	$m3m$	F	$\langle 1\bar{1}0\rangle$, $\{111\}$
Diamond lattice: C, Si, Ge	$m3m$	F	$\langle 1\bar{1}0\rangle$, $\{111\}$ (at $T>T_m/2$)
bcc metals (Fe, Nb, Ta, W, Na, K	$m3m$	I	$\langle 1\bar{1}1\rangle$, $\{110\}$ (principal system) $\langle 1\bar{1}1\rangle$, $\{211\}$[b] $\langle 1\bar{1}1\rangle$, $\{123\}$[b] (at $T>T_m/2$ in alkali metals)
Sodium chloride type: NaCl, LiF, MgO, NaF, AgCl, NH_4I, KI, UN, LiCl, LiBr, KBr, NaBr, RbCl, KCl, NaI, AgBr	$m3m$	F	$\langle 1\bar{1}0\rangle$, $\{110\}$
TiC, UC	$m3m$	F	$\langle 1\bar{1}0\rangle$, $\{111\}$ (at $T>T_m/2$)
PbS, PbTe	$m3m$	F	$\langle 1\bar{1}0\rangle$, $\{001\}$ $\langle 001\rangle$, $\{110\}$

(Continued)

[a] The data refer to room temperature and atmospheric pressure, except when stated otherwise
[b] Systems where glide inversion causes a shear, which is crystallographically inequivalent to the forward shear

Table 2.5 (continued)

Crystal	Class	Lattice type	Glide systems
Cesium chloride type: CsBr, NH_4Cl, NH_4Br, TlBr(I), LiTl, MgTl, AuZn, CuCd, CsI	$m3m$	P	$\langle 001\rangle, \{110\}$ (principal system) $\langle 110\rangle, \{110\}$
β-CuZn	$m3m$	P	$\langle 1\bar{1}1\rangle, \{110\}$
Fluorite type: CaF_2, UO_2	$m3m$	F	$\langle 110\rangle, \{001\}$ (principal system) $\langle 110\rangle, \{110\}$ $\langle 110\rangle, \{111\}$ (at $T > T_m/2$)
Corundum α-Al_2O_3	$\bar{3}m$	R	$\langle 11\bar{2}0\rangle, \{0001\}$ (above $\langle 11\bar{2}0\rangle, \{10\bar{1}0\}$ 1000 °C) $\langle 10\bar{1}2\rangle, \{10\bar{1}1\}$ (above 1200 °C)
Graphite	$6/mmm$	P	$\langle 11\bar{2}0\rangle, \{0001\}$
Hexagonal close-packed crystals: Zn, Cd, Mg	$6/mmm$	P	$\langle 11\bar{2}0\rangle, \{0001\}$ (principal system) $\langle 11\bar{2}0\rangle, \{10\bar{1}1\}$ $\langle 11\bar{2}0\rangle, \{10\bar{1}0\}$ $\langle 11\bar{2}3\rangle, \{11\bar{2}2\}$[b]
Ti, Zr	$6/mmm$	P	$\langle 11\bar{2}0\rangle, \{10\bar{1}0\}$ (principal system) $\langle 11\bar{2}0\rangle, \{10\bar{1}1\}$ (none in Zr) $\langle 11\bar{2}0\rangle, \{0001\}$
Te	32	P	$\langle 11\bar{2}0\rangle, \{10\bar{1}0\}$ (principal system)
Be, AgMg	$6/mmm$	P	$\langle 11\bar{2}0\rangle, \{0001\}$ $\langle 010\rangle, \{001\}$
Ga	mmm	A	$\langle 010\rangle, \{102\}$ $\langle 011\rangle, \{011\}$[b]
Sphalerite type: α-ZnS, InSb	$\bar{4}3m$	F	$\langle 110\rangle, \{111\}$ (in α-ZnS, glide direction is tentative)
β-Zn	$4/mmm$	I	$\langle 001\rangle, \{110\}$ $\langle 001\rangle, \{100\}$ $\langle 101\rangle, \{101\}$[b] (systems are rare) $\langle 101\rangle, \{121\}$
Rutile TiO_2	$4/mmm$	P	$\langle 101\rangle, \{101\}$[b] $\langle 001\rangle, \{110\}$
Bi	$\bar{3}m$	R	$\langle 10\bar{1}\rangle, \{111\}$
Hg	$\bar{3}m$	R	$\langle 01\bar{1}\rangle, \{100\}$ $\langle 100\rangle, \{100\}$[b]
α-U	mmm	C	$\langle 100\rangle, \{010\}$ $\langle 100\rangle, \{001\}$ $\langle 110\rangle, \{110\}$ (systems are rare)
Spinel Al_2O_3-MgO	$m3m$	F	$\langle 1\bar{1}0\rangle, \{111\}$ $\langle 1\bar{1}0\rangle, \{110\}$ (predominates at excess of Al_2O_3)

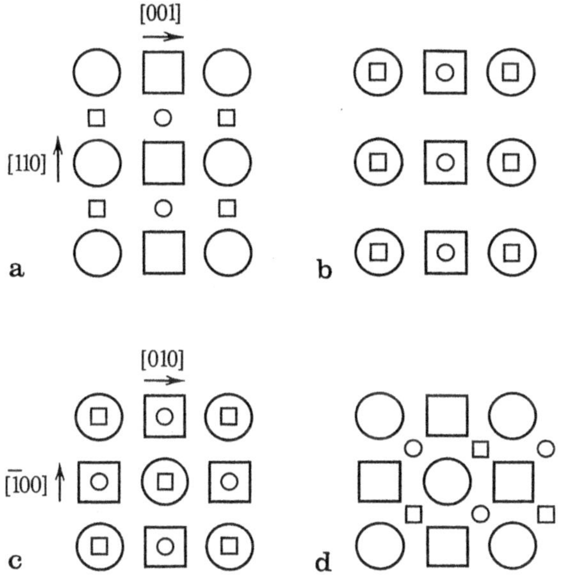

Fig. 2.17. Glide in NaCl-type crystals over systems {110} ⟨110⟩ (**a, b**) and {100} ⟨110⟩ (**c, d**): (**a, c**) before glide, (**b, d**) in the intermediate position. Large circles and squares denote cations and anions in the plane of the drawing, small circles and squares the same in the neighboring plane [2.10]

them are lower, and the vector of translation along the close-packed direction is minimal. The systems of planes corresponding to the minimum of work are called principal or primary (they are followed by secondary systems, etc.).

There are, however, many exceptions to this rule. For instance, NaCl, PbS, and TiC have identical structures, but different glide elements. Some hexagonal metals (Ti, Zr and Hf) show glide over a prism plane, and not over a basal one, as in most of the hexagonal metals. In analyzing possible glide elements, one should take into account the specifics of the interatomic bonds in the crystal, together with the structural data.

We shall use NaCl-type crystals as an illustration. According to the structure of these crystals, the most probable glide plane must be {100} as the closest-packed one, and not {110}, which is observed experimentally. But if we consider the arrangement of ions on the two sides of the glide plane after its displacement by half of the shortest translation vector ⟨110⟩ in the {110} and {100} plane, we shall see that in the former case the electrostatic attraction between the layers is conserved, while in the latter it disappears (Fig. 2.17). To overcome completely the electrostatic interaction forces, one must unquestionably spend more energy (external stresses) than for their partial overcoming. This explains the preference of glide over {110} planes in NaCl-type crystals.

If compression or extension involves glide according to one system, the plastic strain can be estimated not only by the change in specimen length, but also by the plastic shear or *specific crystallographic displacement a*.

Imagine that glide occurs in a crystal over several parallel glide planes. Suppose the shear in each glide plane is indicated by a value *da*. The dis-

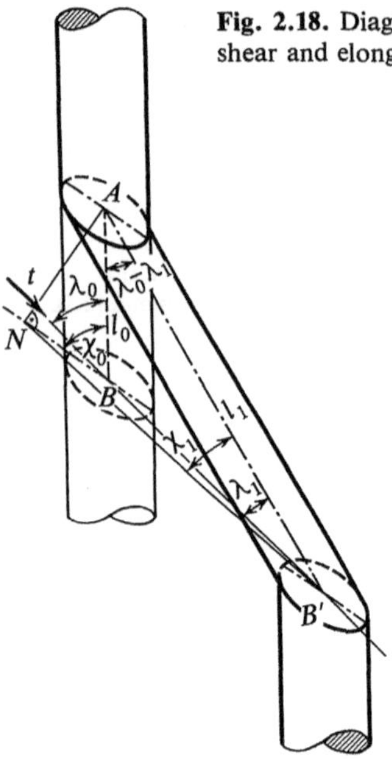

Fig. 2.18. Diagram for finding the relationship between the crystallographic shear and elongation (t: direction of glide)

placement of the crystal parts respective to the first glide plane will be, correspondingly, $2da$, $3da$, etc. The displacement along the glide direction of the crystal part located at a unit distance from the first glide plane is called the specific crystallographic displacement, or the plastic shear.

To find the relationship between the specific crystallographic displacement and the elongation (contraction), we turn to Fig. 2.18. According to this figure,

$$l_1/l_0 = 1 + e = \sin\lambda_0/\sin\lambda_1 \ ,$$

and, from triangles NAB and NAB', follows

$$l_1/l_0 = \cos\chi_0/\cos\chi_1$$

(λ_1 and χ_1, respectively, are the angles between the specimen axis and the glide direction and the normal to the plane after the deformation).
By definition, $a = BB'/l_0\cos\chi_0$, where $BB' = l_1\cos\lambda_1 - l_0\cos\lambda_0$. Hence,

$$a = \frac{(1+e)\cos\lambda_1 - \cos\lambda_0}{\cos\chi_0} = \frac{\cos\lambda_1}{\cos\chi_1} - \frac{\cos\lambda_0}{\cos\chi_0} \ . \tag{2.58}$$

Since $l_1/l_0 = \sin\lambda_0/\sin\lambda_1$, a can be expressed through the initial position of the glide elements:

$$a = \frac{1}{\sin\chi_0}\left[\sqrt{\left(\frac{l_1}{l_0}\right)^2 - \sin^2\lambda_0} - \cos\lambda_0\right] . \tag{2.59}$$

Equations (2.58, 59) were derived for glide over a single system of glide planes. Under actual conditions, an additional system of glide planes and directions may come into play owing to the change in the orientation of a crystal in the course of glide. We shall now give formulas for determining the crystallographic shear a and shear stress τ for double slip. For derivation of the formulas see *Taylor* [2.11] as well as *Bowen* and *Christian* [2.12]. In tension,

$$a = \frac{2}{h_i v_i}\ln\left[1 + \frac{h_i v_i}{|\boldsymbol{w}|}\,\frac{\sin\beta_0}{\cos\chi_0}(\cot\beta - \cot\beta_0)\right] , \tag{2.60}$$

$$\tau = \sigma\frac{|\boldsymbol{w}|}{2}\cos\beta\left[\cos\chi_0 + \frac{h_i v_i}{|\boldsymbol{w}|}\sin\beta_0(\cot\beta - \cot\beta_0)\right] , \tag{2.61}$$

where h_i are the projections of the normal to one of the glide planes onto the loading axis; v_i is the projection of the vector along the glide direction in the second glide plane onto the loading axis; $\boldsymbol{w} = \boldsymbol{u} + \boldsymbol{v}$ is the sum of unit vectors parallel to the glide directions of the first and second glide plane; χ_0 is the angle between the normal to the first glide plane and the loading axis; and β is the angle between the stress axis and $\boldsymbol{w}$, $\sin\beta/\sin\beta_0 = l_0/l$.

In compression,

$$a = \frac{2}{u_i k_i}\ln\left[1 + \frac{u_i k_i}{|\boldsymbol{p}|}\,\frac{\sin\beta_0}{\cos\lambda_0}(\cot\beta - \cot\beta_0)\right] , \tag{2.62}$$

$$\tau = \sigma\left(\frac{l}{l_0}\right)^2\frac{|\boldsymbol{p}|}{2}\cos\beta\left[\cos\lambda_0 + \frac{u_i k_i}{|\boldsymbol{p}|}\sin\beta_0(\cot\beta - \cot\beta_0)\right] , \tag{2.63}$$

where $\boldsymbol{p} = \boldsymbol{h} + \boldsymbol{k}$; $\boldsymbol{h}$ and $\boldsymbol{k}$ being the unit normals to the first and second glide plane.

2.2.5 Slip Lines

An indication that a glide has occurred in a crystal is the appearance of slip steps (striations) on the surface lateral with respect to the loading axis. The striations along traces of the slip that has taken place are called *slip lines* (single traces) or *slip bands* (groups of lines). Figure 2.19 shows photos of the slip bands on the surface of copper. Slip lines which looked like single lines on optical magnification turn out to be a group of fine slip traces when viewed with an electron microscope. By correlating the height of the slip step with the

Fig. 2.19. Slip lines (**a**) and bands (**b**) on the surface of a copper crystal [2.13]

number of fine slip traces one can estimate the shear along one plane. This shear is usually equal to 50 – 1000 lattice spacings in the direction of glide.

Slip lines are strictly straight if one glide plane is active. If several glide planes with the same shear direction can act under identical conditions (for instance, at an identical temperature or with the same specimen orientation

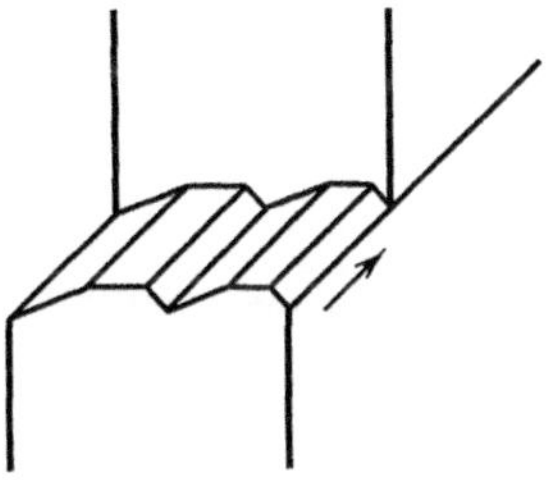

Fig. 2.20. Diagram of "pencil" glide

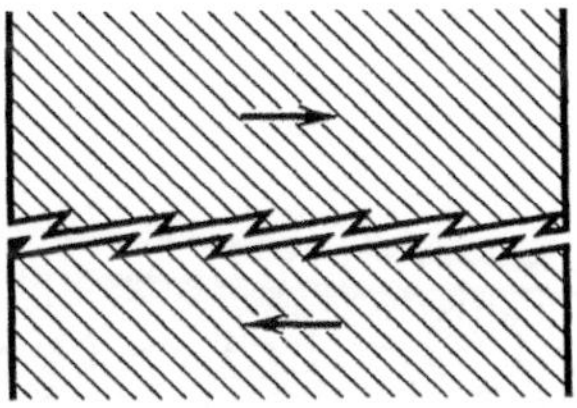

Fig. 2.21. Diagram demonstrating the cause of glide polarity in some crystals

relative to the straining force), the slip steps are not straight: a single step will be replaced by an alternation of step segments along different shear planes. This kind of strain is sometimes called "pencil" glide. Figure 2.20 illustrates such a glide.

2.2.6 Polar Glide

As a rule, glide is nonpolar, i.e., it may equally occur in the forward and backward direction. But sometimes glides in the forward and backward direction are crystallographically nonequivalent. In Table 2.5 such cases are marked with two asterisk. The *glide polarity* is then due to the asymmetric nature of the potential lattice relief in the glide plane, as is shown conventionally in Fig. 2.21. The criterion for polar glide can be formulated thus: the polar glide plane does not coincide with the symmetry plane of the crystal, while the glide direction is a polar vector in the glide plane.

A typical example of polar glide is glide over the $\{112\}\langle 1\bar{1}1\rangle$ system in bcc metals, when the above-mentioned conditions are fulfilled (Fig. 2.22).

In addition to the examples indicated in Table 2.5, polar glide is observed in a number of other crystals, for instance, $KClO_3$, $BaBr_2 \cdot 2H_2O$, and gypsum. During displacement in one direction these crystals experience glide, while during displacement in the opposite direction they undergo twinning or fracture: owing to their specific structure and bonds, the stresses for twinning or fracture are lower than those for glide in this direction.

2.2.7 Critical Shear Stress Law

Glide in crystals takes place when a sufficiently high shear stress is applied. In practice, crystals are subjected to strain predominantly under the effect of compression or extension rather than shear, and to estimate the shear stress one has to determine the component of applied pressure σ along the glide plane and direction.

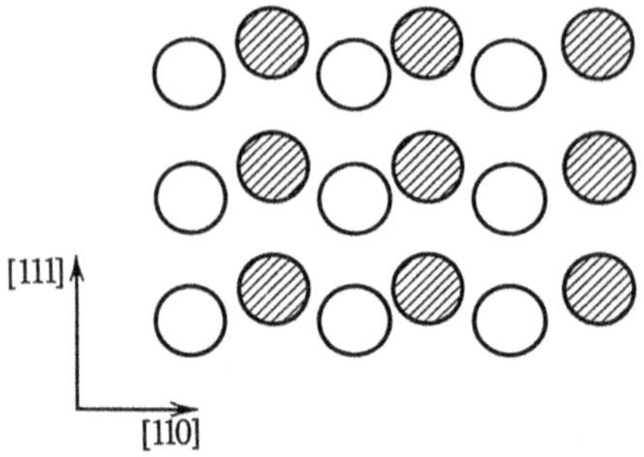

Fig. 2.22. Arrangement of atoms in the (112) plane of *bbc* metals. The shaded circles correspond to the atoms residing above the first layer

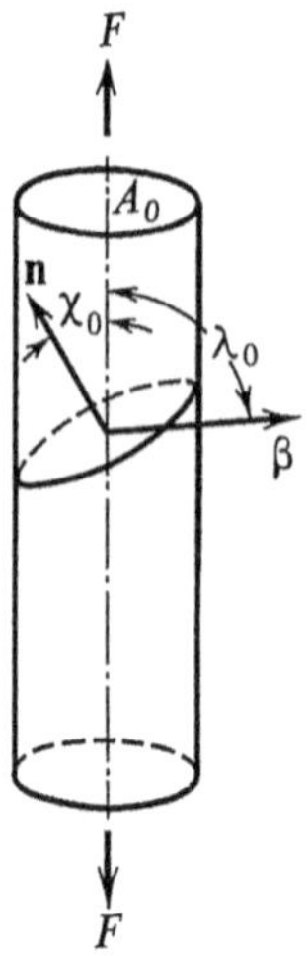

Fig. 2.23. The determination of the tangential component of the stress (shear stress)

Considering that a stress is a second-rank tensor, we use the rule for finding the tensor components (Chap. 1) to find tangential stress component τ along direction β in a plane with normal $\boldsymbol{n}$ (Fig. 2.23):

$$\tau = \sigma_{n\beta} = \sum_{ij} \alpha_{ni}\alpha_{\beta j}\sigma_{ij} = \sigma\cos\chi_0\cos\lambda_0 \ . \tag{2.64}$$

Here χ_0 and λ_0 are the angles between the specimen axis and the normal to the glide plane and the glide direction, respectively. The tangential component of stress τ is called the *shear stress,* and the factor $\cos\chi_0\cos\lambda_0$, the *Schmid factor*. In tension or compression, glide occurs at first in that system of the glide plane family in which the highest shear stress operates. In each crystal, at a definite temperature, glide begins after the shear stress reaches a definite critical value. This is the essence of the *critical shear stress law*. For crystallographically different families of glide systems, the critical shear stresses are different.

Proceeding from (2.64) (the maximum Schmid factor), it is possible to predict the active glide elements corresponding to the same family or to different families of the glide systems or specimens differently oriented with respect to the uniaxial stress. The possible glide elements can be conveniently analyzed by means of stereographic projections.

Figure 2.24 shows a standard stereographic projection of a cubic crystal onto a cube face. Letters *A, B, C,* and *D*, denote the poles of the {111} faces, and figures I–VI the ⟨110⟩ directions. The whole projection is divided into stereographic triangles. If we consider the family of {111} ⟨1$\bar{1}$0⟩ glide systems

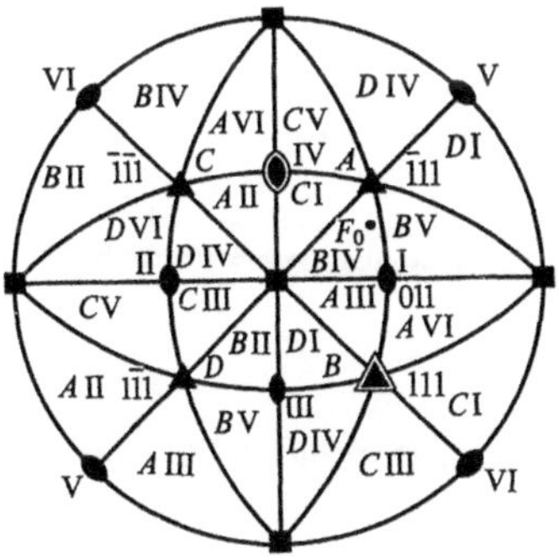

Fig. 2.24. Stereographic projection of a cubic crystal onto a cube face. The active glide elements for specimens strained along the crystallographic directions corresponding to different stereographic triangles are indicated

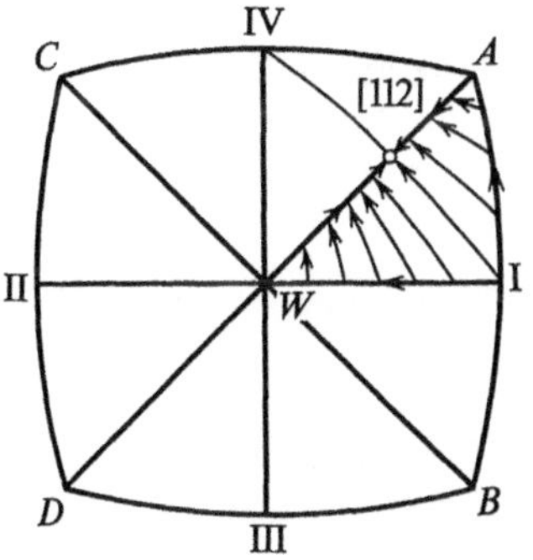

Fig. 2.25. Change in crystal orientation as a result of glide

typical of Si, Ge and fcc metals, then we can determine the most probable glide system for the direction of the specimen axis projected into each of the stereographic triangles. Consider, for instance, the case where the direction of the loading axis is projected into point F_0. Determining from the stereographic projection the angles made by F_0 with A, B, C, and $D(\chi_0)$ and with I – VI (λ_0), we see that the highest Schmid factor refers to glide systems BIV, i.e., to system (111) [101], which will be the active one.

When the direction of F_0 is inside the elementary triangle, we can use a very simple rule for finding the glide system for cubic crystals with a glide system $\{111\}$ $\langle 1\bar{1}0\rangle$ (Diehl's rule): the mirror reflection of the $\langle 110\rangle$ pole of a given triangle by the opposite side of the triangle gives the glide direction, and the mirror reflection of the $\{111\}$ pole of the same triangle by the opposite side gives the normal to the glide plane. If the loading axis is projected onto the boundary of the stereographic triangle, two glide systems will find themselves under identical conditions, and glide can occur with equal probability in either one of them or in both (double slip is discussed further in Sect. 2.2.9).

The rotation in the course of extension or contraction of the crystal regions in which the shear strain occurred (Fig. 2.14) changes angles χ_0 and λ_0, and this may result in the addition of a new glide system or replacement of the active glide system. Considering the possibility of activation of new glide systems, it is possible to predict the ultimate crystallographic orientation of the specimen after a considerable plastic deformation. In Fig. 2.25, which is a part of Fig. 2.24, the arrows indicate the paths which would be followed by the projection of the loading axis (in tension) if only one glide system BIV operated: in the long run the crystal would become oriented along the active glide direction IV. But as soon as the projection of the loading axis finds itself in the plane of rhombododecahedron WA, glide over system CI will also begin. As a result of double slip, the deformation axis is approached by the crystallographic direction, which is the bisector of the angle formed by the acting glide directions IV and I, i.e., toward the [112] direction.

A similar analysis can be carried out for deformation under the effect of the compressive stress, when the active glide plane tends to become perpendicular to the compression axis.

2.2.8 Independent Glide Systems

Tensors similar to (2.56, 57) can be composed for different glide systems. Then certain systems lead to identical strain, but to opposite rotations. The glide systems of the same family leading to different deformation are called independent. Table 2.6 lists the numbers of independent glide systems (N) for some glide elements in crystals of different symmetry.

To ensure any arbitrary plastic deformation of a crystal, at least five independent glide systems must be involved. Indeed, owing to the constancy of the crystal volume in shear strain, we have

$$\varepsilon_{11}+\varepsilon_{22}+\varepsilon_{33}=0 \ . \tag{2.65}$$

If we now take into account the symmetricity of tensor ε_{ij}, only five independent components of this tensor remain. Glide over one system may change only one component ε_{ij} irrespective of the effect on the other components, so that arbitrary assigned deformation due to glide can be ensured by at least five independent glide systems.

This condition is of great practical importance, particularly in the manufacture of polycrystalline (including ceramic) materials. Crystals having five independent glide systems each and polycrystalline materials prepared from such crystals permit any machining (rolling, stamping). Such crystals include most of the metals, and among nonmetallic substances, for instance, crystals of silver chloride and cesium iodide, which are called transparent

Table 2.6. Number of independent glide systems (N) in crystals of different classes of symmetry

Glide system	Class	N
$\langle 1\bar{1}0\rangle$, $\{111\}$	$m3m$	5
$\langle 1\bar{1}0\rangle$, $\{001\}$	$m3m$	3
$\langle 1\bar{1}0\rangle$, $\{110\}$	$m3m$	2
$\langle 1\bar{1}1\rangle$, $\{110\}$	$m3m$	5
$\langle 001\rangle$, $\{110\}$	$m3m$	3
$\langle 10\bar{1}\rangle$, $\{101\}$	$4/mmm$	4
$\langle 11\bar{2}0\rangle$, (0001)	$6/mmm$	2
$\langle 11\bar{2}0\rangle$, $\{10\bar{1}0\}$	$6/mmm$	2
$\langle 11\bar{2}0\rangle$, $\{10\bar{1}1\}$	$6/mmm$	4
$\langle 11\bar{2}3\rangle$, $\{11\bar{2}2\}$	$6/mmm$	5

metals because of their high plasticity. At the same time, NaCl and LiF crystals usually have only two independent glide systems, and ceramics prepared from them are rather brittle.

2.2.9 Cases of Complex Slipping

If, in a crystal, glide occurs over a single plane or in a single direction, it is called simple slipping to distinguish it from complex slipping, which occurs when several glide planes and/or directions are involved simultaneously. If, in the course of compression or extension, glide over different systems takes place in neighboring parts of the crystal, these parts rotate relative to each other because of the lattice rotation accompanying any glide. The greater the glide strain, the larger the misorientation angle of the lattice in the regions concerned. Reorientation of the lattice in individual regions may occur also when the glide is localized within a limited area or is nonexistent in some part, whereas it is taking place in the surrounding material.

The indicated three ways of lattice reorientation as a result of nonuniform glide are schematized in Fig. 2.26. The deformation associated with lattice rotation due to the action of different glide systems in neighboring regions is called *irrational twinning,* and the misoriented regions themselves are called irrational twins, or Brilliantov-Obreimov bands (Fig. 2.26a), because the interface between the developed regions is not a crystallographic plane and, as a rule, is only vaguely defined. As distinct from classical twinning, the misorientation angle is not constant and is a function of the degree of deformation.

Bands with localized glide (when there is no glide in the surrounding parts) are called kink bands (Fig. 2.26b), and interlayers – in which the lattice is turned about relative to the surrounding materials because they show no glide – are called deformation bands (Fig. 2.26c).

A highly specific phenomenon is glide in dynamic fatigue testing, i.e., when the crystal is subjected to sign-variable stresses for a prolonged period. During the alternating half cycles, two interchanging glide systems are active. The position of the successive slip steps on the crystal surface from one glide

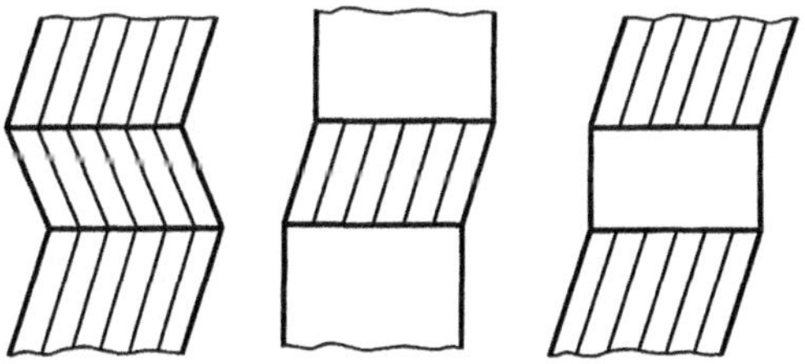

a b c

Fig. 2.26. Formation of irrational twins (**a**), deformation bands (**b**), and kink bands (**c**)

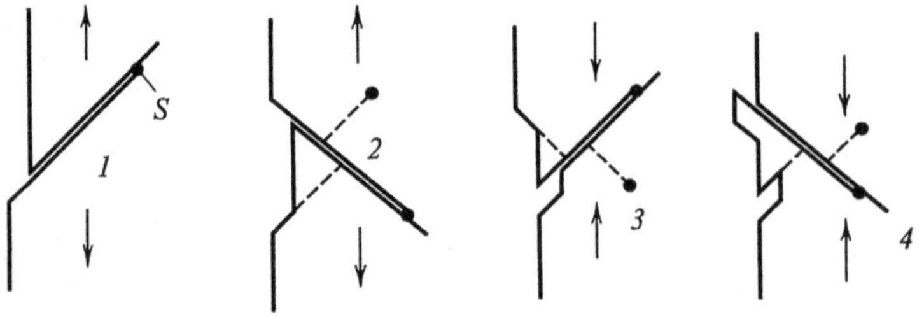

Fig. 2.27. Processes of intrusion (*1, 2*) and extrusion (*3, 4*)

system changes continuously because of interaction with another system. This process may result in tongue-shaped valleys and ridges on the surface. The appearance of ridges is called extrusion, and valleys intrusion.

Figure 2.27 illustrates extrusion and intrusion as a result of alternating action of glide systems with intersecting glide planes in cyclical longitudinal deformation (with replacement of compression by extension).

2.2.10 Stages of Glide Deformation

While an increase in strain ε in the stage of elastic deformation requires an increase in stress τ proportional to the strain, after the onset of plastic deformation the linear relationship between ε and τ is not valid any longer. If, in the course of deformation, we record the stress-strain relationship, dependence $\tau(\varepsilon)$ will give us an idea of the development of glide in the specimen. In real crystals, besides the reversible elastic strain, at comparatively low stresses still another stage of reversible strain is observed, which does not obey Hooke's law and which is called the *imperfect-elasticity stage.* This inelastic reversible strain results in the delay of strain on change in stresses, i.e., in hysterisis losses.

Figure 2.28 presents a schematic strain curve. At very low stresses, dependence $\tau(\varepsilon)$ is linear and reversible. This is the region of elastic stress, where Hooke's law is applicable (region E in Fig. 2.28). The slope of this section to axis ε is equal to Young's modulus. The stress corresponding to the termination of the elastic-deformation stage is called the *limit of elasticity* τ_e.

The stress at which the slope of the stress-strain curve changes sharply is the yield point τ_y. The yield point corresponds to the beginning of plastic (residual) deformation. If the stress-strain curve is parallel to the strain axis beyond the yield point, plastic deformation occurs without work hardening: to continue the deformation one does not have to increase the external stresses. In practice, increase in strain usually requires a higher external stress. Here we have *work hardening,* whose value depends on the slope of the stress-strain curve to the ε axis. This slope is called the *hardening coefficient,* or the plasticity modulus.

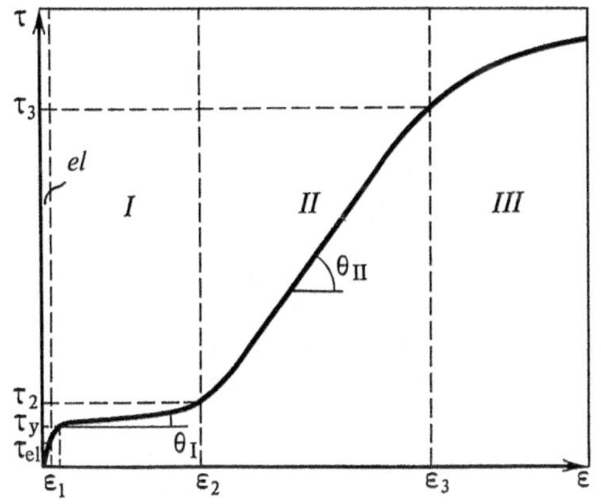

Fig. 2.28. Schematic stress-strain curve $\tau(\varepsilon)$

For most of the crystals investigated, the plastic deformation proceeds in three stages. To the initial stage (Stage I in Fig. 2.28) corresponds the near-horizontal linear section of the deformation curve. This stage is called the *easy-glide* stage, since it is characterized by very small work hardening (small angle θ_I). The next stage of deformation (Stage II in Fig. 2.28) is also a stage of linear hardening, but the corresponding section of the stress-strain curve has a much steeper slope (angle θ_{II}). The second stage is called the *fast linear hardening stage*. The last stages of plastic deformation (Stage III in Fig. 2.28 and unshown Stage IV), which are called *strain softening* (or *strain relaxation*), are described by a slightly curved, and near-horizontal section of the strain curve respectively.

Apart from hardening coefficients θ, the plastic deformation stages are characterized by other hardening parameters: shear stresses τ_y, τ_2, τ_3, and τ_4, and strains ε_1, ε_2, ε_3, and ε_4 corresponding to the beginning of the relevant stage (Fig. 2.28) (τ_4 and ε_4 are not indicated).

The hardening parameters depend on the crystal under investigation, its pre-history (the initial dislocation structure and impurity content), and also on the temperature and strain rate. Only dimensionless parameter θ_{II}/G is temperature independent and almost identical for all the fcc crystals studied ($\sim 3\times10^{-3}$).

The various stages of deformation are associated with a different character of glide. A comparison of the stages of deformation with the distribution of the slip lines showed the following: in Stage I all the crystals exhibit glide over one glide system; activization of the second glide system leads to a transition to Stage II; and the softening in Stage III and IV is the result of cross slip and the redistribution of the defects produced during deformation (dislocations, point defects) into more equilibrium configurations (annihilation of disloca-tions, polygonization, see [Ref. 2.5, Chap. 5], due to which the strain resis-tance of the crystal decreases.

Sometimes plastic deformation is preceded by a sharp drop in yield point (Fig. 2.29), after which the strain increases monotonically with the stress. This

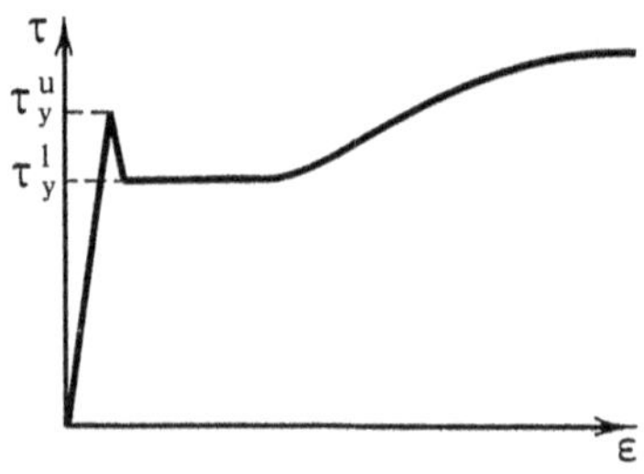

Fig. 2.29. Stress-strain curve with a yield point phenomenon

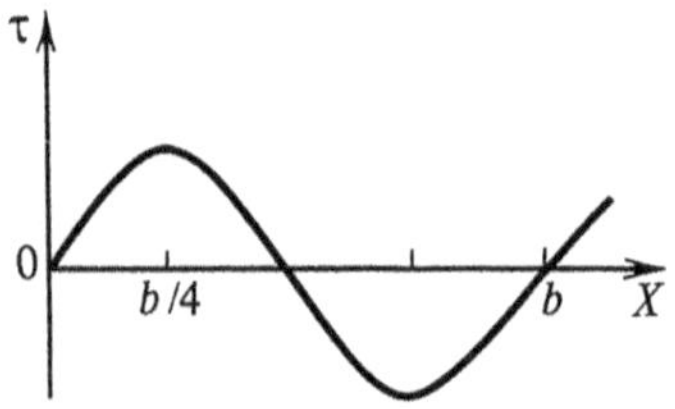

Fig. 2.30. Dependence of the shear resistance of the lattice on the displacement of atoms

phenomenon is called the *yield point phenomenon.* In the presence of a yield drop, the upper and the lower yield point are distinguished (τ_y^u and τ_y^l). The appearance of a yield point phenomenon means a sudden acceleration of plastic deformation (the reason will be considered in Sect. 2.3).

2.2.11 Theoretical Shear Strength[1]

As noted above, plastic deformation sets in after the applied mechanical stress achieves the yield point. Theoretically, this stress must correspond to the resistance of the binding forces to the shear strain.

If one part of a crystal is displaced as a whole relative to another part simultaneously over the entire glide plane, the arising shear strain resistance from the lattice will change periodically (in accordance with the lattice spacing) as the displacement of the shifting part of the crystal increases (Fig. 2.30). Assuming the sinusoidal law of variation in interatomic forces and the corresponding change in tangential stress, one can write (Frenkel's estimate)

$$\tau(x) = A \sin(2\pi x/b) \ , \tag{2.66}$$

where A is a constant, x is the displacement of atoms from the equilibrium position, and b is the period of function $\tau(x)$. For small displacements, when $\sin x \approx x$,

$$\tau(x) = A 2\pi x/b \ . \tag{2.67}$$

According to Hooke's law

$$\tau = Gx/d$$

[1] The term "strength" generally means the resistance of bodies to failure. But in theoretical consideration of glide in crystals this term is traditionally used to characterize the resistance of the lattice to shear strain.

(G is the shear modulus, and d is the distance between neighboring planes of the lattice).

Comparing the last two equations, we find

$$A = \frac{G}{2\pi} \frac{b}{d} ,$$

or

$$\tau(x) = \frac{b}{d} \frac{G}{2\pi} \sin \frac{2\pi x}{b} . \tag{2.68}$$

At $x = b/4$ the resistance is maximal, and $\tau_{\text{theor}} = b/d \cdot G/2\pi$. Assuming $b = d$, we get

$$\tau_{\text{theor}} = G/2\pi . \tag{2.69}$$

Actually, the law of variation in interatomic forces differs from the sinusoidal. For real force laws, calculation yields τ_{theor} values of the order of $G/10$ and even $G/30$ (for fcc crystals).

A near-theoretical shear strength must be exhibited either by crystals containing practically no defects (including dislocation-free crystals) [2.9] or very defective crystals, in which the plasticity is suppressed because the numerous defects hinder the development of plastic deformation. Record strengths are shown by very thin crystals, so-called whiskers [2.15] in which the high strength is due to the perfect structure, not only of the crystal itself, but also of its surface. The presence of defects at the surface (cracks, growth steps, etc.) and inside (pores, various inclusions, etc.) abruptly reduces the strength of whiskers to a much greater extent than for bulk crystals.

The shear strength of ordinary crystals is one to three orders below theoretical ($10^{-4} - 10^{-2}$ G). The discrepancy between theory and experiment in estimating the strength is partly due to the inhomogeneity of the crystal structure and the nonuniformity of the development of plastic deformation throughout the crystal and in time. The main cause for the low stress inducing plastic deformation in crystals, however, is the presence of dislocations, i.e., highly mobile line defects. Glide is actually caused by the movement of these defects in the glide plane.

2.3 Dislocation Description of Plastic Deformation by Glide

2.3.1 Dislocations as the Result of Shear Strain

In macroscopic description of glide (Sect. 2.2) we used a model of mutual displacement of rigid parts of a specimen along the entire glide plane simultaneously. But, in reality, the shear gradually "grows through" the crystal. At each instant some atoms have already displaced along the glide plane to a new position, while others still remain unmoved. Along the line lying in the glide plane and separating the displaced part from the undisplaced, the regular structure of the crystal has been disturbed: either a broken-off plane is present or the parallelism of the atomic planes is locally disturbed. The defect arising along this line is called a dislocation. A detailed description of various types of dislocation and dislocation configurations, and estimates of the dislocation-induced stresses are given in [Ref. 2.5, Chap. 5]. The present section will discuss those properties of dislocations which are directly related to plastic deformation by glide.

2.3.2 Dislocation Structure of Slip Lines and the Movement of Dislocations

If a step with a displacement by one translation vector ***b*** has appeared at the lateral surface of a crystal and the displacement has not occurred throughout the entire crystal, one dislocation appears in it. If one part of the crystal has shifted relative to another by two translation vectors and the displacement got stuck inside the crystal, the glide plane will have two dislocations. A displacement by three, four, etc. translations will cause three, four, etc. dislocations, respectively, in the crystal. A row of dislocations lying in a common glide plane form a slip line at the site of emergence of the dislocations at the surface. Figure 2.31 a gives a sketch of a slip line (it is horizontal here) made up of edge dislocations. A row of dislocations lying in the same plane is denoted by $\perp \ \perp \ \perp \ \perp \ \perp$ and called a horizontal row as distinct from a vertical row, a case where the dislocations are positioned one under another and form the boundary of misoriented blocks [Ref. 2.5, Chap. 5].

For a displacement throughout the crystal cross section, the dislocations generated in a limited area must pass through the whole glide plane. Dislocations cannot move in any arbitrary plane. They move most readily

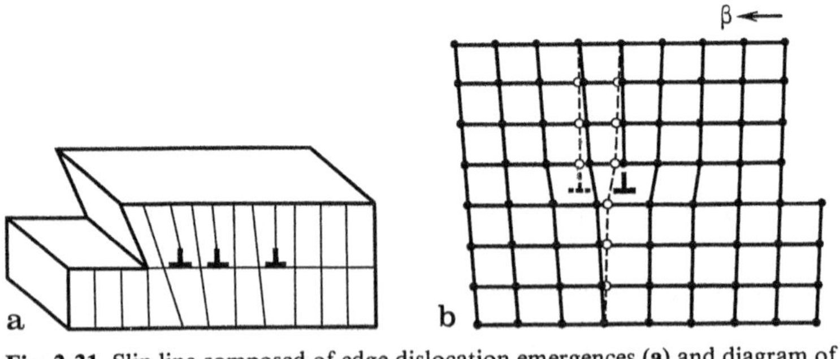

Fig. 2.31. Slip line composed of edge dislocation emergences (**a**) and diagram of movement of an edge dislocation (**b**)

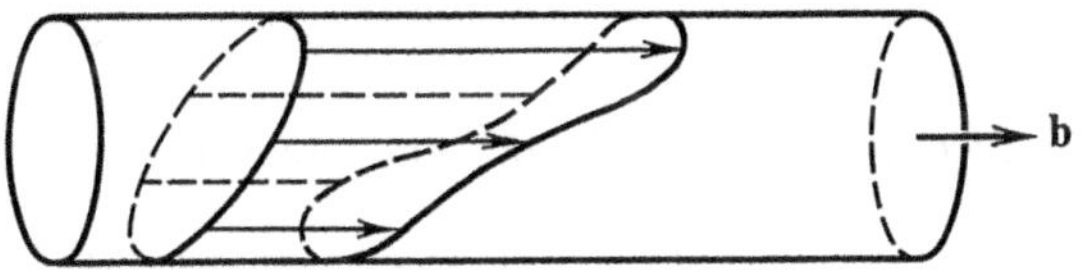

Fig. 2.32. Prismatic dislocation and its glide plane

along glide planes of the crystal (Sect. 2.2.4). Figure 2.31 b outlines the movement of an edge dislocation in a glide plane. A dislocation moves owing to successive switching of bonds between atoms, which is, of course, much easier than the displacement of a whole part of a crystal along the entire glide plane, which is considered in classical theory.

Two types of dislocation movement are distinguished: glide and climb. Dislocation *glide* is a movement involving no addition or removal of material (atoms or ions), i.e., it is a conservative movement. The dislocation glide surface is a surface containing the dislocation line, and the normal to the glide surface $\boldsymbol{n}$ is everywhere perpendicular to the translation vector *(Burgers vector)* $\boldsymbol{b}$, so that $\boldsymbol{b} \cdot \boldsymbol{n} = 0$. For a straight edge dislocation this is a glide plane, which is assigned by the dislocation line and its Burgers vector. For a screw dislocation, condition $\boldsymbol{b} \cdot \boldsymbol{n} = 0$ is satisfied by any plane through the dislocation line; therefore, screw dislocations can slide first in one and then in another plane. Transition from one glide plane to another is called *cross glide*. For a dislocation in the shape of a closed loop lying in the glide plane, the glide surface is a cylinder (or a prism, if the loop consists of straight segments) formed by lines parallel to the Burgers vector and passing through all the points of the loop (Fig. 2.32). Such loops are called *prismatic*, and the movement of prismatic loops is called *prismatic glide*. Of special interest are edge prismatic loops, which can arise, for instance, as a result of the collapse of flat pileups of vacancies or interstitials [Ref. 2.5, Chap. 5].

Another type of dislocation movement is the *climb*. Then a dislocation moves perpendicular to the glide plane. Obviously, the climb can be performed by edge and mixed dislocations and must be accompanied by extension or contraction of the extra half plane. Since the extension or contraction of the extra plane requires an influx or efflux of interstitials, the dislocation climb is a nonconservative movement. The climb is discussed in more detail in Sect. 2.4.

2.3.3 Force Acting on a Dislocation

The magnitude of the force responsible for dislocation glide depends on the work required for displacing a unit length of dislocation by a unit distance. Suppose shear stress τ acting along the glide direction makes a dislocation of unit length traverse an area of $(1 \times dx)$. Then force τdx will displace the upper part of the crystal relative to the lower by distance b and perform the work

$$dw = \tau b\,dx \ . \tag{2.70}$$

This work is equal to a change in crystal energy $-dE$, and since the force acting on the dislocation

$$F = -dE/dx \ , \tag{2.71}$$

it follows from a comparison of (2.70, 71) that

$$F = \tau b \ . \tag{2.72}$$

Let us consider a more general case, where a dislocation has the shape of an arbitrary loop lying in the glide plane, while the applied stress is randomly oriented. In accordance with (2.2) we can write down the force acting on a unit area perpendicular to $\boldsymbol{n}$ as

$$\boldsymbol{F} = \sigma \boldsymbol{n} \ . \tag{2.73}$$

Direction $\boldsymbol{n}$ is determined by the vector product of a unit vector along the dislocation line $\boldsymbol{l}$ and unit vector $\boldsymbol{d}$ in the direction of movement:

$$\boldsymbol{n} = \boldsymbol{l} \times \boldsymbol{d} \ .$$

As a result of the dislocation movement its part with negative $\boldsymbol{n}$ will have moved by $-\boldsymbol{b}$, and the work spent will be

$$w = -(\sigma \boldsymbol{n}) \cdot \boldsymbol{b} \ . \tag{2.74}$$

Owing to the symmetry of tensor σ we can write

$$w = -(\sigma \boldsymbol{b}) \cdot \boldsymbol{n} \ . \tag{2.75}$$

The dislocation energy changes by

$$-w = (\sigma \boldsymbol{b}) \cdot (\boldsymbol{l} \times \boldsymbol{d}) = (\sigma \boldsymbol{b}) \times \boldsymbol{l} \cdot \boldsymbol{d} \ . \tag{2.76}$$

On the other hand, the work of force $\boldsymbol{F}$ expended on the movement of the dislocation over distance $\boldsymbol{d}$ is expressed thus:

$$-w = \boldsymbol{F} \cdot \boldsymbol{d} \ . \tag{2.77}$$

From (2.76, 77) follows

$$\boldsymbol{F} = (\sigma \boldsymbol{b}) \times \boldsymbol{l} \ . \tag{2.78}$$

Since $\boldsymbol{l}$ is a vector parallel to the dislocation line, force $\boldsymbol{F}$ is always normal to the dislocation line.

2.3.4 Dislocation Velocity

The most important characteristic of a dislocation is the dependence of its velocity on the stress applied.

The dislocation velocity is measured experimentally when observing the movement of dislocations directly in a transmission electron microscope or by the changes in the etch pattern [Ref. 2.5, Chap. 5]. With semiconducting crystals showing a comparatively low dislocation density, x-ray topography is also used for observing the dislocation movement. Figure 2.33 demonstrates how the dislocation movement in a CsI crystal is traced by continuous etching in the course of loading. Figure 2.33 a corresponds to the case where the dislocation moved approximately uniformly during loading, and the velocity was comparable with that of the expansion of the etch patterns. The appearance of a row of flat-bottomed pits along the path of the dislocations in Fig. 2.33 b indicates that the dislocation moved jumpwise, with temporary stops. The

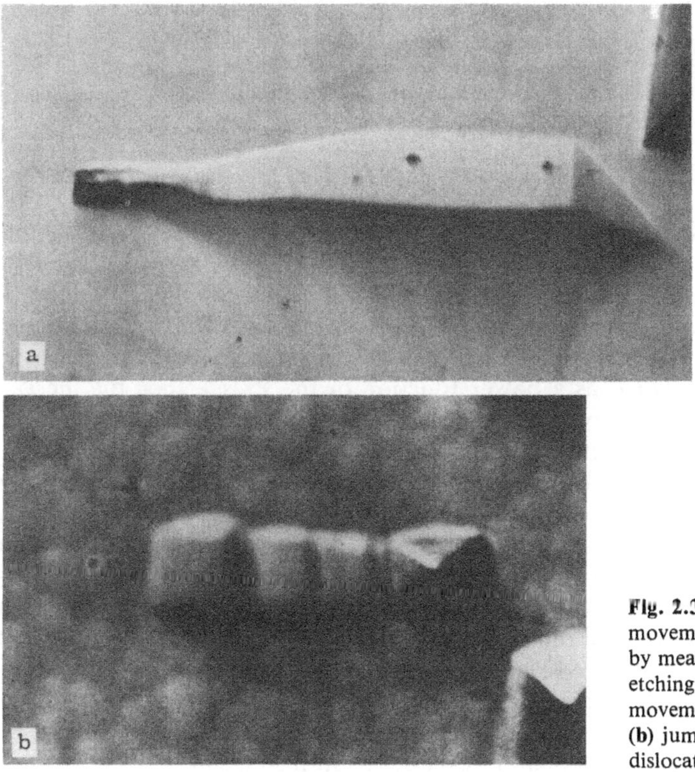

Fig. 2.33. Dislocation movement in cesium iodide by means of selective etching: (**a**) uniform movement of a dislocation, (**b**) jumpwise movement of a dislocation [2.16]

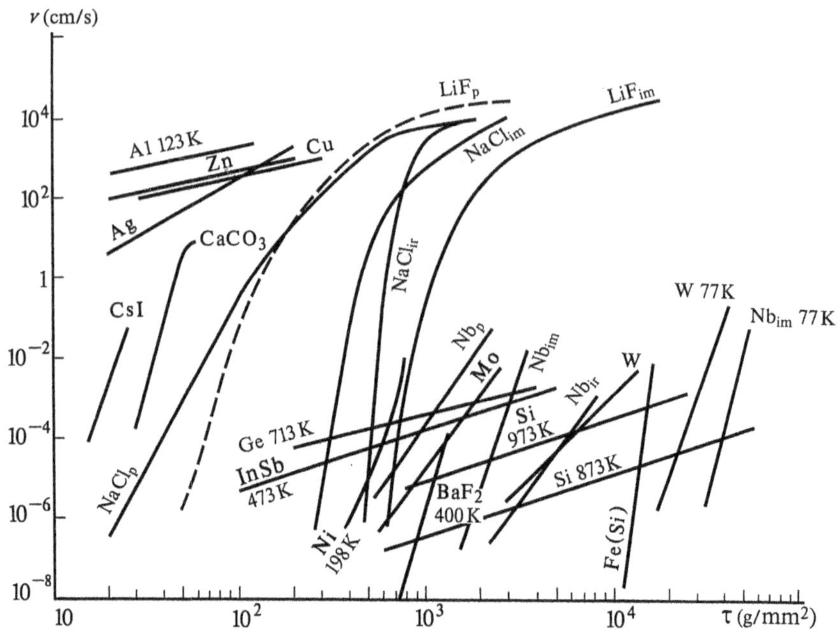

Fig. 2.34. Stress dependences of dislocation velocity in different crystals: subscript "p" denotes pure crystals, subscript "im" crystals with impurities, and subscript "ir" irradiated; where temperature is not indicated the measurements were made at room temperature (compiled by Nadgorny, see *Roitburd* [2.19])

jumpwise character of glide was first observed by *Klassen-Nekhliudova* [2.17, 18].

The dislocation velocity increases with the stress applied. Figure 2.34 depicts curves for the stress dependences of the logarithm of dislocation velocity for various crystals. At low stresses the velocity increases with the stress exponentially (or as a power function of the stress), then the increase slows down (the velocity grows linearly with the stress), and it gradually approaches the velocity of sound at very high stresses.

The deceleration of the increase in dislocation velocity at high stresses is due to the radical change in the mechanism of dislocation slowing down. The movement of dislocations in crystals requires the overcoming of various barriers associated both with the periodic structure of the crystal (Pierls-Nabarro barriers) and with the lattice defects. Slowly moving dislocations stop at these barriers and then overcome them as a result of thermal fluctuations. At high velocities of dislocations, when their kinetic energy reaches the height of the energy barriers, the slowing down of dislocations acquires a dynamic character and is limited by the pumping of the energy from the dislocations to various elementary excitations in the crystal.

Dynamic drag also plays a part during a slow movement of dislocations, affecting the kinetics of thermofluctuation overcoming of potential barriers by dislocations. In most cases, dislocation-phonon interaction plays the predominant role in the dynamic drag of dislocations. The contribution from other types of elementary excitation (electrons, excitons, etc.) manifests itself only under special conditions. For instance, in metals at low temperatures, when the phonon gas is "frozen out," the interaction of moving dislocations with conduction electrons becomes substantial.

The viscosity of the elementary excitation "gas" is defined by *drag coefficient* B, and the force acting on a unit length of a dislocation can be written as

$$F = \tau b = vB \ . \tag{2.79}$$

At room temperature, B is usually equal to $10^{-3}-10^{-4}$ Ps and falls off with decreasing temperature. In dielectrics, B drops practically to zero when 0 K is approached. In metals the viscosity coefficient is nonzero near 0 K because of scattering from electrons.

More detailed information on the theory of dynamic drag of dislocations can be obtained from reviews by *Alshitz* and *Indenbom* [2.20] and also by *Kaganov* et al. [2.21].

2.3.5 Lattice Resistance to the Dislocation Movement

The crystal lattice has a definite potential relief, which dictates the most favorable position for the dislocation lines. Ideally, dislocations must have the shape of straight lines located in the grooves of the potential relief. The dislocation energy is then minimal. Consider a dislocation line parallel to a close-packed row in a simple cubic lattice. For a dislocation to execute a translatory movement perpendicular to its line, it must shift from one groove of relief into another, overcoming a barrier of height E (Fig. 2.35). But traversing the barrier by the whole dislocation simultaneously as a rigid straight line is unlikely, because in the intermediate position a dislocation

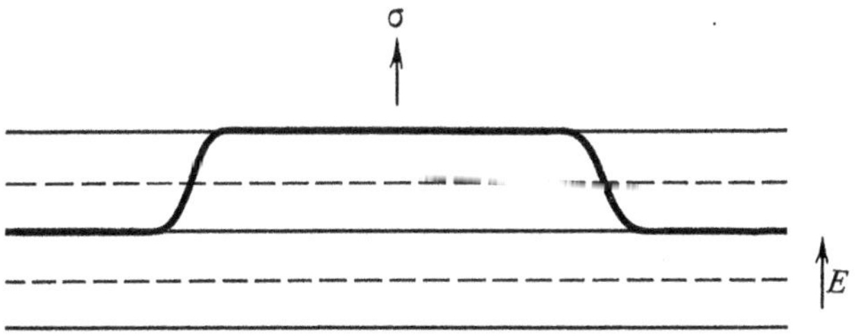

Fig. 2.35. Formation of a double kink on the dislocation line: solid lines are potential relief grooves; dashed lines are barriers between grooves

segment of macroscopic length would possess an excess energy El (l is the dislocation length). Transition of individual small sections of the dislocation line to a neighboring groove is much more probable. The segments of the dislocation line which perform transition from one groove to a neighboring one are called *kinks.* When kinks are directed to opposite sides, the length of the segment in the neighboring groove increases, and thus the entire dislocation gradually passes from one potential barrier groove to another. The movement of a kink of a dislocation line by one interatomic distance can be regarded as an elementary act of plastic glide deformation.

The lattice resistance to the dislocation movement can be estimated by using the atomic dislocation model proposed by *Frenkel* and *Kontorova* [2.22] (Fig. 2.36). The model considers the movement of a chain of atoms linked by quasi-elastic forces on a substrate of similar chains forming a periodic force field. The number of chain atoms differs by unity from that of the minima of the potential relief. According to the model, the sliding resistance, which can be interpreted as critical shear stress τ_c, is related as follows to the width of dislocation core λ [Ref. 2.5, Chap. 5], interplanar distance d, shear modulus G, and Burgers vector b:

$$\tau_c = G\frac{\pi\lambda}{3db}\exp\left(-\frac{\pi\lambda}{b}\right). \tag{2.80}$$

Quantity λ is proportional to interplanar distance d.

Frenkel and Kontorova's model takes into account only short-range interactions with the atomic chain. In actuality, each chain is under the effect of

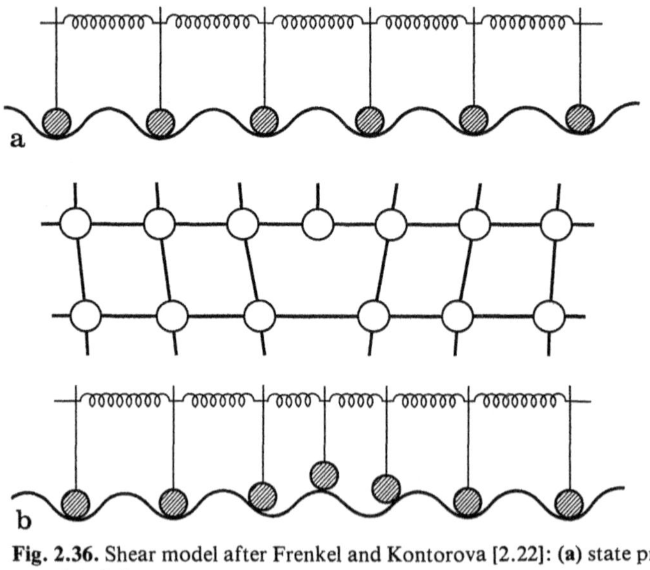

Fig. 2.36. Shear model after Frenkel and Kontorova [2.22]: (**a**) state prior to deformation, (**b**) uncompleted shear

long-range forces because of the interaction with neighboring chains. Pierls and *Nabarro* [2.23] took this into consideration and showed that the lattice resistance is associated with the critical shear stress

$$\tau_c = G \exp\left(-\frac{d}{2\pi(1-\nu)b}\right) \tag{2.81}$$

(here ν is the Poisson coefficient). This stress is called the Pierls-Nabarro stress.

The lattice resistance is not the only, nor the main cause for the slowing down of dislocations. Pierls-Nabarro stresses (forces) are significant only for a comparatively small range of crystals, for instance, semiconductors. More often, the principal resistance to the movement of dislocations is offered by other dislocations, particularly those lying in the plane intersecting the one in which the dislocation concerned is moving (fcc metals) or clusters of point defects (bbc metals, ionic crystals). Before we proceed to interactions of dislocations with each other and with point defects, it is necessary to find out in what way the number of dislocations in crystals increases.

2.3.6 Multiplication of Dislocations

In the course of deformation, dislocations move and many eventually leave the crystal altogether. In actuality, however, crystals subjected to deformation are not, as a rule, free from dislocations; on the contrary, they acquire an increasing number of dislocations. New dislocations appear because the moving dislocations produce new dislocations. At present, two cases are known in which multiplication of dislocations in the course of their movement occurs. The first was suggested by *Frank* and *Read* [2.24]. According to this mechanism, dislocations are generated as a result of the movement of the edge segment of the dislocation fixed at one or two ends. Such a fixed dislocation is called a *Frank-Read source.*

Multiplication of dislocations by means of a Frank-Read source is demonstrated in Fig. 2.37. The source consists of segment *AB* of the dislocation line (Fig. 2.37a), which is here fixed at both ends by nodes or in some other way. This segment lies in the glide plane and has a Burgers vector necessary for slipping, while segments *BC, BD, AE,* and *AF* cannot have the Burgers vectors necessary for slipping. The plane of the drawing corresponds to the glide plane. When a shear stress is applied, segment *AB* buckles (Fig. 2.37b). Since the energy of the dislocation line increases as a result of its buckling, the line acts as an elastic string and offers a resistance to this force. The tautness of the dislocation line offsets the force applied until the stress reaches the critical value

$$\tau_{\text{F-R}} \sim Gb/L \ , \tag{2.82}$$

where L is the length of segment *AB*. The further increase of the loop occurs

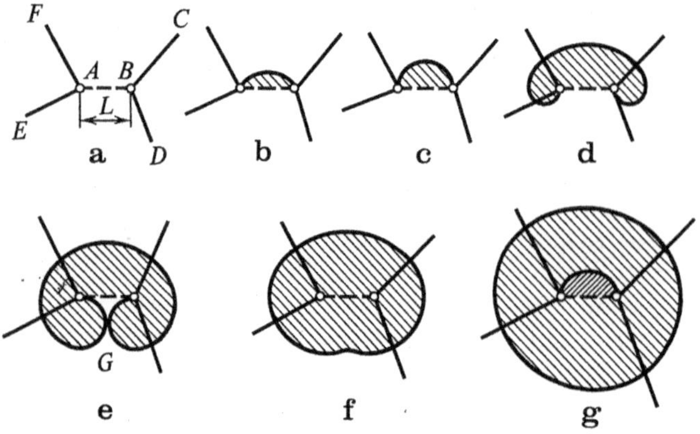

Fig. 2.37a–g. Action of a Frank-Read source

freely, because the force applied exceeds the tension of the dislocation line. In the critical state, the radius of the dislocation half loop is equal to $L/2$ (Fig. 2.37c).

With a further broadening of the dislocation loop, the configuration shown in Fig. 2.37e is achieved. Then, at point G the dislocations annihilate, leaving a closed loop, as shown in Fig. 2.37f. Finally, the process is resumed, as indicated in Fig. 2.37g.

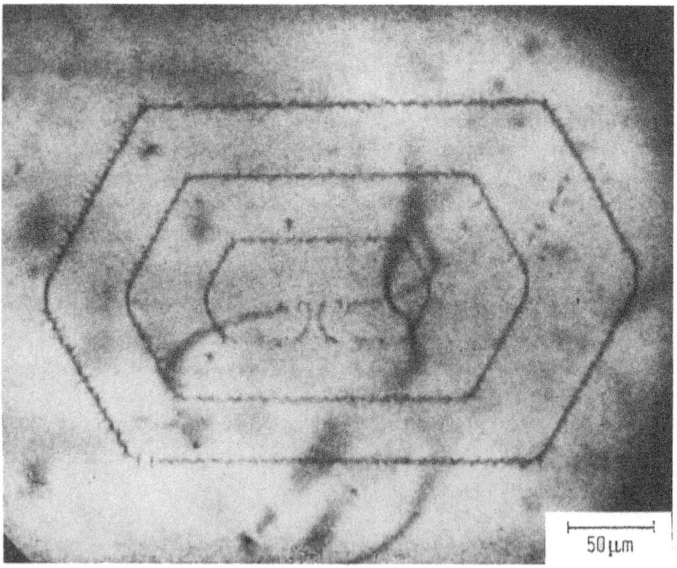

Fig. 2.38. Frank-Read source in a silicon crystal detected by x-ray topography [2.25]

Since segment *AB* of the Frank-Read source is restored each time a new loop is formed, the source can generate an unlimited number of dislocation loops, provided the nodes remain fixed in space, and the local stress acting on the source exceeds critical value $\tau_{\text{F-R}}$.

Frank-Read sources were found experimentally in silicon, potassium chloride, stainless steel, cadmium, and other materials. Figure 2.38 shows a Frank-Read source in silicon.

While Frank-Read multiplication sources ensure plastic deformation by the glide of edge dislocations in the same glide plane, the participation of cross slip of screw dislocations (double and multiple) promotes the appearance and multiplication of dislocations in neighboring planes parallel to the initial glide planes, i.e., it explains the formation of bundles of slipping lines or bands [2.26]. This second kind of multiplication is sketched in Fig. 2.39.

At first, screw dislocation *OP* moves to the left in the horizontal plane (Fig. 2.39a). Then one portion of the screw dislocation passes into the transverse glide plane, and the screw dislocation acquires segment *ABCD* lying in the transverse glide plane (Fig. 2.39b). As dislocation *OP* moves further to the left, it does not move along the whole length any more, but in portions, which results in the configuration shown in Fig. 2.39c. Portions *AB* and *CD* cannot move in the direction of movement of the initial dislocation, because, being edge dislocations (the Burgers vector is identical everywhere, regardless of the shape of the dislocation [Ref. 2.5, Chap. 5]), they must move in a plane which is not a glide plane. They can glide parallel to vector *OP*, but not perpendicular to it. Portion *BC*, together with *AB* and *CD*, which pin

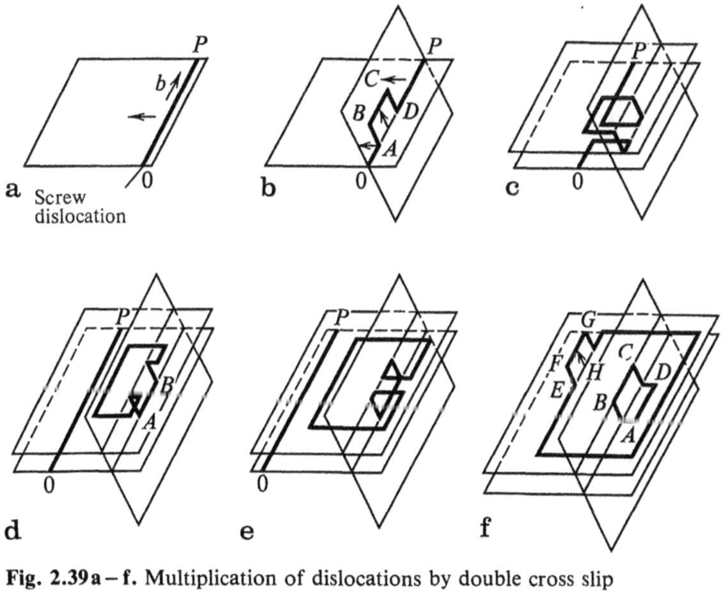

Fig. 2.39a–f. Multiplication of dislocations by double cross slip

down its ends, can act like a Frank-Read source (Fig. 2.39c – f). Similarly, *OP* can become a separate dislocation again (Fig. 2.39d, e). Finally, the loop generated by portion *BC* may experience cross slip, as *EFGH* does in Fig. 2.39f, and thus the process is resumed. At the same time portion *AD*, remaining behind *OP*, can act as a Frank-Read source in the opposite direction, because this portion has a sweeping direction opposite to that of portion *BC* (opposite mechanical sign) [Ref. 2.5, Chap. 5]. Portion *BC* returns to its initial position and can repeat the process with the formation of another loop.

By now, abundant evidence of the operation of this multiplication mechanism in real crystals (for instance, in LiF) has been obtained. Moreover, it has been shown that multiplication by *repeated cross slip* determines the development of deformation in secondary glide planes, in which the dislocations have a limited mobility.

2.3.7 Interaction Between Dislocations

As a result of the multiplication and movement of dislocations, they draw closer together. The stress fields induced by dislocations begin to affect each other. Equation (2.78), as well as [Ref. 2.5, Eqs. (5.11, 12)], make it possible to estimate the force of interaction between dislocations if we replace σ with the tensor of the stresses created by one dislocation near another.

Let us consider a few particular examples for an infinite crystal, when the edge effects can be neglected. We imply everywhere interactions of dislocations on a section of unit length.

A. Interaction of Two Parallel Screw Dislocations

Let us estimate the force with which a screw dislocation parallel to the X_3 axis (Fig. 2.40) and possessing Burgers vector $\boldsymbol{b}_1$ acts on another screw dislocation which is parallel to it, passes through point $M(r, \theta)$, and has Burgers vector $\boldsymbol{b}_2$. Since screw dislocations induce only tangential stresses, in accordance with [Ref. 2.5, Eq. (5.11)] the tangential stress in the (X_1, X_2) plane from the dislocations passing through the origin is defined by

$$\sigma_{3\theta} = Gb_1/2\pi r \ .$$

According to (2.78), the force acting on a unit length of a parallel dislocation passing through point M is perpendicular to $\sigma_{3\theta}$ and X_3 and hence is directed along radius OM, and its magnitude is defined by the formula

$$F = F_r = \frac{G\boldsymbol{b}_1 \cdot \boldsymbol{b}_2}{2\pi r} \ . \tag{2.83}$$

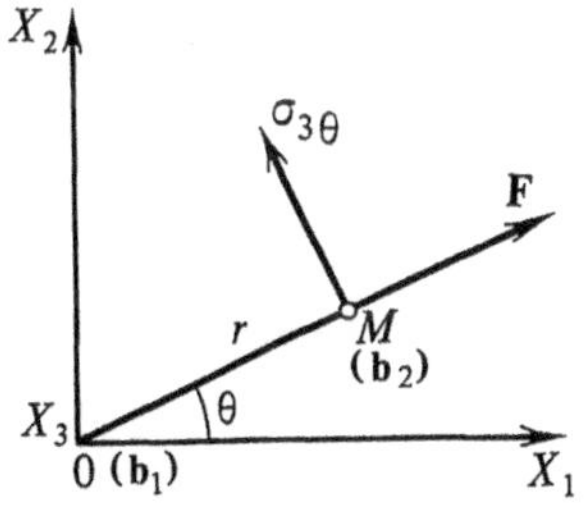

Fig. 2.40. Interaction of parallel screw dislocations

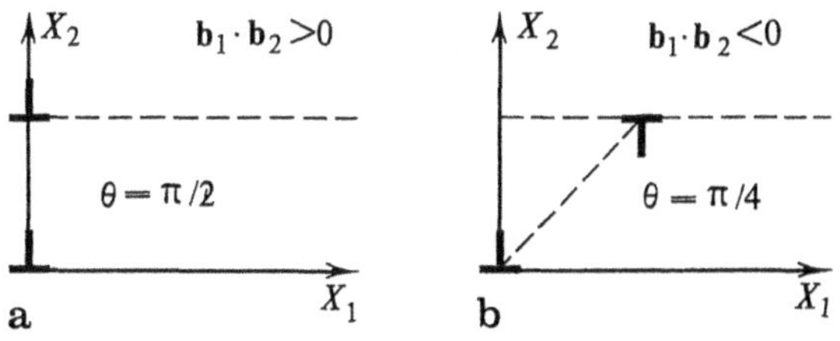

Fig. 2.41 a, b. Interaction of edge dislocations moving in parallel slip planes

This is a central force. If $\boldsymbol{b}_1 \cdot \boldsymbol{b}_2 > 0$, it draws the dislocations closer together, and if $\boldsymbol{b}_1 \cdot \boldsymbol{b}_2 < 0$, it draws them apart. There is no equilibrium position.

B. Interaction of Two Edge Dislocations with Parallel Burgers Vectors

If vectors $\boldsymbol{b}_1$ and $\boldsymbol{b}_2$ are parallel to the X_1 axis and the dislocations themselves are parallel to the X_3 axis, the first dislocation acts on the second with a force which has a radial component, F_r, and a tangential component, F_θ:

$$F_r = \frac{G\boldsymbol{b}_1 \cdot \boldsymbol{b}_2}{2\pi(1-\nu)r}\,, \tag{2.84}$$

$$F_\theta = \frac{G\boldsymbol{b}_1 \cdot \boldsymbol{b}_2 \sin 2\theta}{2\pi(1-\nu)r}\,. \tag{2.85}$$

The two above-discussed edge dislocations will be mutually attracted or repelled if they can move freely in the direction of the force. But if they can move only in their glide plane (X_1, X_3), then from consideration of the projection of the force onto the glide plane, which is equal to

$$F_r \cos\theta - F_\theta \sin\theta = \frac{G\boldsymbol{b}_1 \cdot \boldsymbol{b}_2}{2\pi(1-\nu)r} \cos\theta \cos 2\theta\,, \tag{2.86}$$

it follows that at $\boldsymbol{b}_1 \cdot \boldsymbol{b}_2 > 0$ angles $\theta = 0$, $\pi/2$ and π correspond to the stable-equilibrium positions (Fig. 2.41 a) and angles $\theta = \pi/4$ and $3\pi/4$ to unstable-equilibrium positions (at $\boldsymbol{b}_1 \cdot \boldsymbol{b}_2 < 0$ the result is opposite, Fig. 2.41 b). The pair of dislocations in Fig. 2.41 b is called a *dislocation dipole*. In deformed crystals such configurations are found rather often.

C. Interaction of Dislocations Lying in Intersecting Glide Planes

This case is of the greatest practical interest, because precisely such an interaction is the main cause of strain hardening.

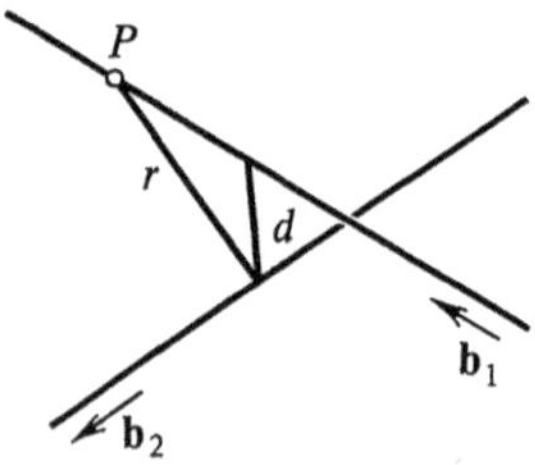

Fig. 2.42. Interaction of mutually perpendicular noncoplanar screw dislocations

Let us consider the interaction of two mutually perpendicular nonintersecting screw dislocations. Reasoning as above, we find the force with which one dislocation acts on the other at point P (Fig. 2.42):

$$F = \frac{G\boldsymbol{b}_1 \cdot \boldsymbol{b}_2}{2\pi} \frac{d}{r^2}, \tag{2.87}$$

where d is the distance between the dislocations along a normal to them. This force is directed along a normal to both dislocations. Its sign depends on the result of summation of the shear stresses induced by both dislocations at the midpoint of the line segment joining them. The dislocations will repel one another if their stress fields add together, and will attract one another if the stresses are mutually canceled.

If the moving orthogonal screw dislocations have intersected one another, then, owing to the displacement of the part of the crystal above the glide plane relative to the lower part of the crystal, both dislocations increase in length by a segment coinciding in direction and magnitude with the Burgers vector of the intersecting dislocation (a knee along $\boldsymbol{b}_2$ will appear on the dislocation with the Burgers vector $\boldsymbol{b}_1$, and vice versa). Additional sections on the dislocations are edge sections, since they are perpendicular to their Burgers vector. If the additional segment and its Burgers vector lie in the plane – the glide plane of the dislocation on which it was formed – such a knee is a *kink,* and with a further movement of the dislocation, which tends to reduce in length, it will rapidly straighten out and disappear. If the plane in which the new segment and its Burgers vector lie is not the glide plane of the dislocation containing it, the movement of such a segment, called a *jog,* is possible only by climbing, and a chain of vacancies or interstitials will trail after the moving segment (nonconservative movement).

2.3.8 Dislocation Reactions

In one type of interaction of intersecting dislocations, because of the trend toward a decrease in energy, the interacting dislocations find it profitable to join together and form a new dislocation. Such interactions of dislocations are called *dislocation reactions.*

The conditions for the joining and splitting of dislocations have been discussed and the possible dislocation reactions in an fcc lattice, a diamond lattice, and a CsCl-type lattice analyzed [Ref. 2.5, Chap. 5]. Here, we shall give an example of dislocation reactions in hexagonal close-packed crystals. These crystals have the following glide systems:

$$(0001),\ [2\bar{1}\bar{1}0]\ ,\quad \boldsymbol{b}=\tfrac{1}{3}[2\bar{1}\bar{1}0]\ ;\quad (\text{above } 200\,^\circ\mathrm{C})$$

$$(01\bar{1}0),\ [2\bar{1}\bar{1}0[\ ,\quad \boldsymbol{b}=\tfrac{1}{3}[2\bar{1}\bar{1}0]$$

$$(01\bar{1}1),\ [2\bar{1}\bar{1}0]\ ,\quad \boldsymbol{b}=\tfrac{1}{3}[2\bar{1}0]\ ;$$

$$(\bar{2}112),\ [2\bar{1}\bar{1}3]\ ,\quad \boldsymbol{b}=\tfrac{1}{3}[2\bar{1}\bar{1}3]\ .$$

The last three systems are shown in Fig. 2.43.

Thus, six Burgers vectors are possible in the basal plane (along three a axes in the forward and reverse directions: $\pm\boldsymbol{a}_i$, $i=1,2,3$), two in the plane of the prism ($\pm c$), and 12 in the plane of second-kind pyramid ($\bar{2}112$) ($\pm c$, $\pm\boldsymbol{a}_i$).

In the interaction of dislocations of all these glide systems, 21 non-equivalent reactions are possible, but, from consideration of energy gain (the dislocation energy is proportional to $\boldsymbol{b}^2$),

$$\boldsymbol{b}_3^2<\boldsymbol{b}_1^2+\boldsymbol{b}_2^2\ , \tag{2.88}$$

one can actually expect only five reactions:

$$\tfrac{1}{3}[2\bar{1}\bar{1}0]+\tfrac{1}{3}[\bar{1}2\bar{1}0]\rightarrow\tfrac{1}{3}[11\bar{2}0]\ ,$$

$$\tfrac{1}{3}[2\bar{1}\bar{1}3]+\tfrac{1}{3}[\bar{1}2\bar{1}0]\rightarrow\tfrac{1}{3}[11\bar{2}3]\ ,$$

$$\tfrac{1}{3}[2\bar{1}\bar{1}3]+\tfrac{1}{3}[2\bar{1}\bar{1}\bar{3}]\rightarrow\tfrac{2}{3}[2\bar{1}\bar{1}0]\ ,$$

$$\tfrac{1}{3}[2\bar{1}\bar{1}3]+\tfrac{1}{3}[\bar{1}2\bar{1}\bar{3}]\rightarrow\tfrac{1}{3}[11\bar{2}0]\ ,$$

$$\tfrac{1}{3}[2\bar{1}\bar{1}3]+\tfrac{1}{3}[1\bar{2}1\bar{3}]\rightarrow\quad[1\bar{1}00]\ .$$

Intersections of basal dislocations always result in glide jogs, and hardening from basal glide is small. In pyramidal glide, sessile dislocations may form, which can effectively decelerate the moving dislocations.

Condition (2.88) is called the Frank rule. It characterizes the stability of dislocations. From this condition it follows, in particular, that most of the dislocations in which the Burgers vectors are not equal to the shortest lattice vectors are unstable and decompose into two dislocations with a smaller Burgers vector. For instance, a dislocation with a Burgers vector 1/2 ⟨211⟩ in an fcc crystal decomposes into two dislocations of the type 1/2 ⟨011⟩:

$$\tfrac{1}{2}[211]\rightarrow\tfrac{1}{2}[110]+\tfrac{1}{2}[101]\ . \tag{2.89}$$

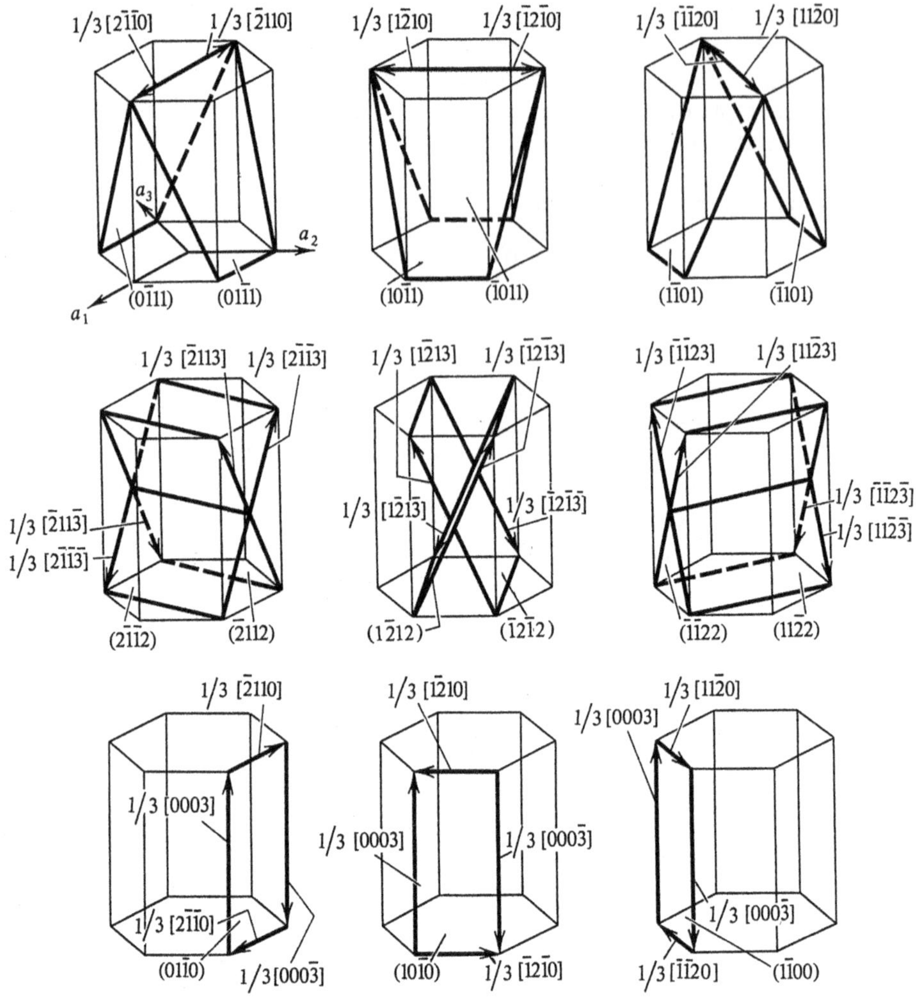

Fig. 2.43. Glide systems in close-packed hexagonal crystals [2.27]

Table 2.7 indicates the Burgers vectors of the most stable dislocations in certain Bravais lattices of high and medium symmetry.

2.3.9 Interaction of Dislocations with Point Defects

Point defects are distortions in a crystal from which elastic stresses extend by not more than several lattice spacings in any direction. A classification of point defects has been given their equilibrium concentration estimated [Ref. 2.5, Chap. 5].

Table 2.7. Stable dislocations in some Bravais lattices

Lattice	Burgers vector	Number of equivalent vectors	Square of Burgers vector
Simple cubic	$\langle 100 \rangle$	6	a^2
Bcc	$\frac{1}{2}\langle 111 \rangle$	8	$3a^2/4$
	$\langle 100 \rangle$	6	a^2
Fcc	$\frac{1}{2}\langle 110 \rangle$	12	$a^2/2$
Hexagonal	$\frac{1}{3}\langle 11\bar{2}0 \rangle$	6	a^2
	$\langle 0001 \rangle$	2	c^2
Rhombohedral $\alpha < 90°$	$\langle 100 \rangle$	6	a^2
	$\langle 110 \rangle$	6	$4a^2 \sin^2 \alpha/2$
	$\langle 111 \rangle$	6	$a^2 (1+4\sin^2 \alpha/2)$
Rhombohedral $\alpha > 90°$	$\langle 100 \rangle$	6	a^2
	$\langle 110 \rangle$	6	$4a^2 \cos^2 \alpha/2$
	$\langle 111 \rangle$	2	$9a^2 (1-4/3 \sin^2 \alpha/2)$
Simple tetragonal	$\langle 100 \rangle$	4	a^2
	$\langle 001 \rangle$	2	c^2
Bc tetragonal $c/a < \sqrt{2}$	$\frac{1}{2}\langle 111 \rangle$	8	$a^2/2+c^2/4$
	$\langle 100 \rangle$	4	a^2
Bc tetragonal $c/a > \sqrt{2}$	$\langle 001 \rangle$	2	c^2
	$\frac{1}{2}\langle 111 \rangle$	8	$a^2/2+c^2/4$
	$\langle 100 \rangle$	4	a^2

Point defects (vacancies, interstitials and impurity atoms) form upon removal, addition, or replacement of one or several atoms.

An effective way to introduce point defects into crystals is irradiation with protons, neutrons, electrons, γ-rays, x-rays, etc. If the radiation used reacts with the nuclei of crystal ions (atoms), such a reaction may generate new atoms and decay products in the crystal. Among the point defects, impurities exert the greatest influence on the movement of dislocations [2.28 – 33].

It should be noted that dislocations can themselves affect the impurity center barriers that they overcome. For instance, dislocations may be sites of nucleation of impurity precipitates (second-phase particles). The stresses arising around the precipitates may relax by emitting or absorbing vacancies, provided they are present in a sufficient number. The emitted vacancies, in turn, may help a dislocation to detach itself from the impurity center, thus promoting local (near the center) climb of the dislocation (Sect. 2.4). Dislocations increase the mass transport along their axis (tubular diffusion), causing dissipation of the impurity clusters and of other point defects [2.34].

2.3.10 Types of Interaction of Dislocations with Impurities

Single dissolved atoms (ions) of an impurity may affect the lattice spacings, the magnitude of the elasticity coefficients, the energy of the stacking faults, etc. Changes in these parameters may cause dislocation deceleration. The effect of dissolved atoms on the lattice spacings and the elasticity coefficients is estimated from the atomic-scale defect δ and that of the shear modulus η. This effect is due to interaction of the long-range field of dislocation stresses with the field of atomic displacements around the impurity center.

An atomic-size defect δ is equal to

$$\delta = \Delta r / r_0 , \tag{2.90}$$

where Δr is the difference between the radii of the matrix atom and the impurity atom, and r_0 is the radius of the intrinsic atom. It is determined experimentally from the dependence of bulk density ϱ_{bulk} or the lattice spacing a on impurity concentration c that [2.35]

$$\delta = \frac{d\varrho_{\text{bulk}}}{3\,\varrho_{\text{bulk}}\,dc} = \frac{da}{a\,dc} . \tag{2.91}$$

Knowing δ, one can find the volume change corresponding to one impurity atom

$$\Delta v_a = 3\,\Omega\delta$$

(here Ω is the volume per atom) and the dislocation-impurity binding energy

$$W_\delta = -\frac{1}{2\pi}\,\frac{1+\nu}{1-\nu}\,G|\Delta v_a| \sim G b r_0^2 \delta . \tag{2.92}$$

Here, as before, ν and G are the Poisson coefficient and the shear modulus, respectively.

The defect of the shear modulus is equal to

$$\eta = \frac{\Delta G}{G} = \frac{d\ln G}{dc} , \tag{2.93}$$

where ΔG is the difference between the shear moduli of impurity and matrix. The corresponding interaction energy

$$W_\eta = G b^2 r_0 \eta / 20 . \tag{2.94}$$

Equations (2.91 – 94) are explained in *Friedel*'s book [2.35].

From (2.93, 94) it follows that the resultant interaction energy of the dislocation and the impurity atom is proportional to the difference of the elastic coefficients and the square of dislocation-induced stresses. Depending on the sign of ΔG, the impurity is attracted to the dislocation (if G of the impurity is less than that of the matrix) or is repelled from it (if G of the impurity is larger).

Expression (2.94) is approximate. The exact expression for the dependence of the binding energy on the modulus defect has not been established yet. According to approximate estimates, the binding energy is very low, and W_η reaches more or less substantial values only for very soft impurities (with a low G).

A much more effective influence on the mobility of dislocations (particularly on their initial shift) is exerted by the appearance, around dislocations, of impurity atmosphere atoms, which fix the dislocation in place. The atmosphere may be of different origin depending on the type of interaction of the dislocations with the impurity atoms.

Four causes for the attraction of impurities to dislocations are distinguished in crystals:

1. *Elastic interaction of impurity atoms with dislocations.* In the case of edge dislocations, owing to the interaction of the elastic-stress fields of the dislocations and the impurities, the impurity atoms tend to reside in extended regions if they are larger than the matrix atoms, or in contracted regions if they are smaller than the atoms of the basic material. Screw dislocations attract only those foreign atoms which cause nonspherically symmetric distortions around them. Impurity atoms reduce the free energy of the system, but block the dislocations. Clusters of impurities around dislocations as a result of their elastic interaction are called *Cottrell atmospheres.*
2. *Chemical interaction of extended dislocations with impurity atoms.* This interaction is due to the difference in the solubility of the impurity atoms in the matrix and in the stacking fault due to extended dislocations [Ref. 2.5, Chap. 5], which has a different structure (for instance, in an fcc lattice the stacking faults have a hexagonal structure). The stacking faults attract the impurity atoms, thereby impeding the movement of the dislocations. Pileups of impurities in the stacking faults of extended dislocations are called *Suzuki atmospheres.*
3. *Impurity ordering in the dislocation field.* In metals, impurity atoms usually have several crystallographically equivalent positions in the interstices. For instance, carbon atoms in the lattice of α iron can be located on any one of the three edges of the cubic unit cell. Near dislocations, however, the impurity atoms cannot occupy all the positions, but only those where the distortion field of the impurity atom corresponds to the lattice distortion at the dislocations. In the presence of a dislocation, the carbon resides predominantly on one of the edges. Thus, structure ordering begins near the dislocation. To displace the dislocation, one has to destroy the order, which

requires additional energy expenditure and increases the yield point. The region with an ordered arrangement of the atoms at dislocations is called the *Snoek atmosphere.*

4. *Electrical interaction of impurities with dislocations.* In metals, electroneutrality must be destroyed at edge dislocations because of the difference in bulk density above and below the glide plane, since otherwise the bulk concentration of the conduction electrons does not agree with the density distribution. Owing to the constancy of the Fermi level energy, the compressed areas have a deficiency, and extended regions, an excess of electrons. Near the dislocation, an electrical linear dipole arises, which interacts with the impurity atoms. But because of the screening effect, electrical interaction in metals is negligibly small compared to the dimensional effect (see page 112). Electrical interaction of dislocations with impurities produces a stronger effect on the movement of dislocations in ionic and covalent crystals [2.29, 36]. In ionic crystals it is due to the existence, on edge and screw dislocations, of jogs with an uncompensated electric charge, which attract oppositely charged impurities.

In NaCl-type crystals, the charges have edge dislocations in the $\{111\}$ plane, which have a Burgers vector $a/2\ \langle 110\rangle$. These dislocations carry a charge $\pm e/2$ per each atomic distance [2.36]. Such dislocations possess an enormous energy, and therefore the glide over $\{111\}$ planes in NaCl-type crystals is evidently impeded. *Hirth* and *Lothe* [2.10] calculated that the charged sites are at the emergence points of edge dislocations of the ordinary glide system $\{110\}\ \langle 110\rangle$ at the $\{100\}$ plane, and the charge is equal to $\pm e/4$, whereas the emergence point of an edge dislocation at the $\{110\}$ plane is not charged.

The charging of the dislocations involved in plastic deformation is due to the appearance of charged jogs on them in the course of movement. Such jogs may form on a dislocation or at the intersection with another dislocation if the Burgers vectors of the dislocations form an angle of 60° or 120° (in NaCl-type crystals) or as a result of the departure of vacancy ions from the edge of the half plane under the effect of thermal fluctuations, or for some other reasons. Figure 2.44 shows a neutral, *A*, and a charged, *B*, jog on an edge dislocation in a NaCl-type crystal. The charged jog is half as long as the neutral. If an ion or a vacancy is precipitated on a charged jog, the jog will reverse its sign. It follows that the charge of the jog is equal to $\pm e/2$, where e is the charge of the electron.

A neutral jog may become charged if an ion or a vacancy settles on it (the dislocation may trap them in the course of movement). Then the jog splits up into two charged jogs with a total charge of $\pm e$ (jog *C* in Fig. 2.44). This origin of charged jogs is more probable than the formation of a jog of type *B*, because in that case an extra half plane of the group of ions or of vacancies of both signs must arrive at the edge. Since the mobility of ions and vacancies is limited under ordinary conditions, a mass influx of cations and anions (or their vacancies) can hardly occur.

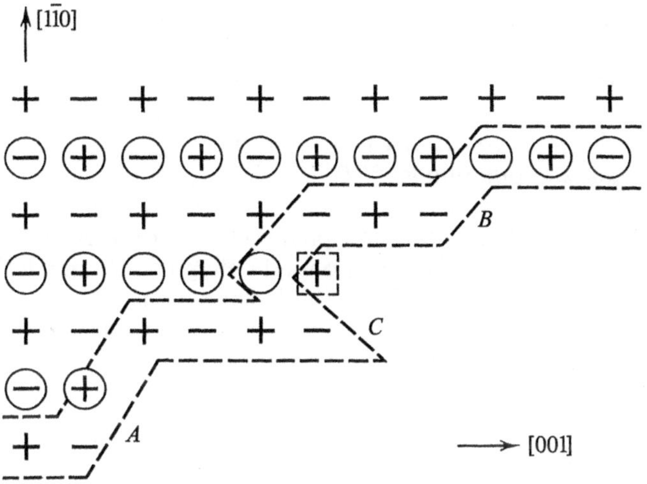

Fig. 2.44. Neutral (*A*) and charged (*B* and *C*) jogs in a NaCl-type crystal

The movement of charged dislocations through the crystal induces a potential difference on its surface. Polarization of crystals on deformation is called the *Stepanov* effect (1933) [2.37, 29]. As a rule, the charge is carried only by edge and mixed dislocations, because in the movement of edge dislocations (and the edge components in mixed dislocations) the ion signs do not alternate, and the charged jog moves together with the dislocation. In the movement of screw dislocations, ions of different signs alternate, and the sign of the charge on the jog changes all the time. A charged jog on a screw dislocation can usually move only along the dislocation. Charge transport by screw dislocations in their movement perpendicular to the Burgers vector is possible only when these dislocations can drag behind them vacancies which do not decompose into pairs of kinks.

Charged jogs attract point defects, which compensate for the jog charge. The electrostatic interaction of the jogs with these defects results in the pinning of the dislocation in the jog area. Clusters of defects around charged jogs are called *Debeye-Hückel clouds*. According to estimates [2.35] these clouds retard the movement of the dislocations much less than Cottrell atmospheres and do not affect the yield point macroscopically.

The appearance of a charge on dislocations in semiconductor crystals is due to the existence of a chain of unpaired valency electrons along the dislocation line. In n-type silicon and germanium they form a negative charge, which is the cause of attraction of point defects or impurity atoms to the dislocation. Finding themselves in the immediate vicinity of the dislocation lines, a large part of impurity atoms in a semiconductor can evidently enter readily into a peculiar chemical exchange bond with unpaired electrons. Then the binding energy of the impurity atoms with the dislocations is much higher than for a pure electrostatic interaction.

Polarization on deformation of elementary semiconducting crystals is not revealed experimentally because the charge of the dislocation line is screened owing to a rather high electrical conductivity of these crystals.

Semiconductor crystals of compounds $A^{II}B^{VI}$, which possess a considerable share of ionic bonds (ZnO, ZnS, CdS, and others), show charged dislocations. It is edge dislocations with an extra half plane $\{111\}$ that are charged, because they are composed, alternately, of oppositely charged A^{II} and B^{VI} atoms [2.36]. Dislocations at the edge of the $\{111\}$ plane made up of B^{VI} anions (α dislocations) are charged negatively, and those at the edge of the plane consisting of A^{II} cations (β dislocations), positively.

2.3.11 Precipitation Particles

If impurity atoms form large clusters (which are called precipitation particles, segregations, or precipitates), additional kinds of interactions of dislocations with impurities are observed. With precipitates and a matrix that are coherent (i.e., when the matrix and the precipitates have similar structures and a similar size of the unit cell) the dislocation can pass through a particle, displacing one of its parts relative to the other by a value of b. Then the energy of the precipitate-matrix interface increased by

$$W_\Gamma \sim \Gamma r_0 b \tag{2.95}$$

for particles of radius r_0 with specific surface energy Γ [2.38]. Here W_Γ is the energy of dislocation-particle interaction. When encountering a large incoherent particle the dislocation moves around it, leaving behind dislocation rings around the particles *(Orowan mechanism).*

Figure 2.45 shows the circumvention diagrammatically. The circumvention stress [2.39, 40] is expressed by the formula

$$\tau \sim Gb/4\pi l \ , \tag{2.96}$$

where l is the distance between the obstacles.

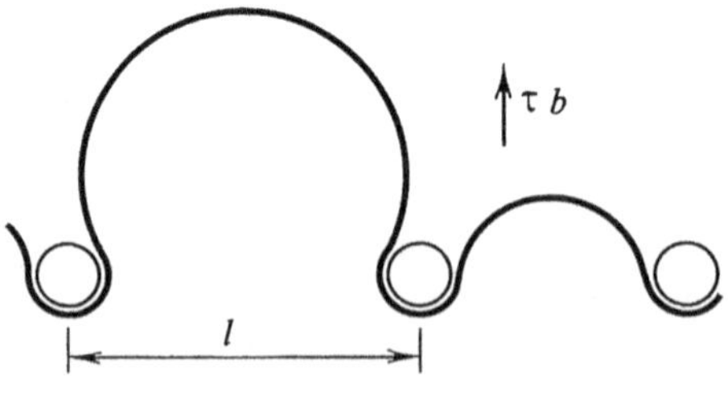

Fig. 2.45. Dislocation passing around large precipitate particles [2.38]

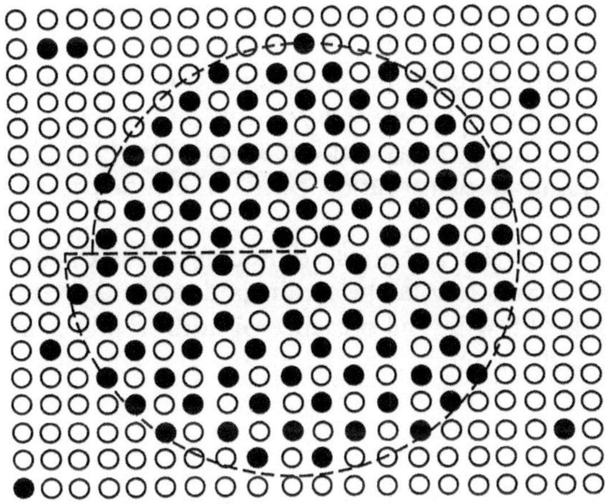

Fig. 2.46. Structure disordering in a precipitate particle along the dislocation path [2.38]

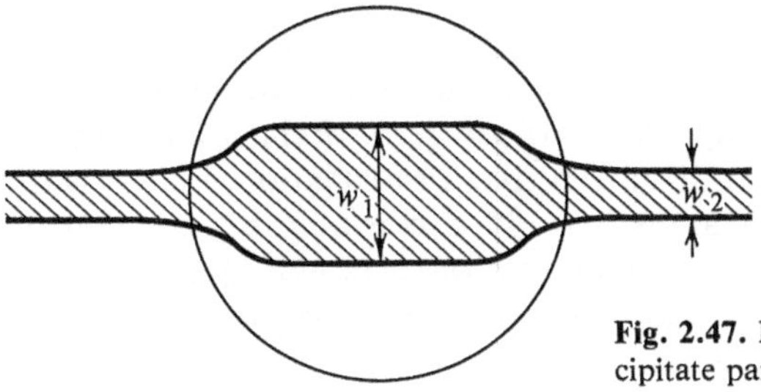

Fig. 2.47. Increase in dislocation extension in a precipitate particle [2.38]

If a precipitate particle with a radius r_0, which is in the metal with impurities, consists of an ordered alloy (Fig. 2.46) with a specific surface energy Γ', then the energy of interaction between it and the dislocation [2.38] is equal to

$$W_{\Gamma'} \sim \Gamma' r_0^2 \ . \tag{2.97}$$

This energy is spent on disturbing the order in the particle along the path of the dislocation.

The precipitate particles, in turn, can change the structure of the dislocation if it is inside the particle. For instance, if a moving dislocation splits up, the energy of the stacking fault changes from γ_1 to γ_2 and the splitting changes from w_1 to w_2 (Fig. 2.47). The corresponding interaction energy [2.38] is expressed by the formula

$$W_k \sim 2 r_0 w_1 (\gamma_1 - \gamma_2) \quad \text{for} \quad w_2 < r_0 < w_1 \ . \tag{2.98}$$

2.3.12 Internal Friction

One of the methods for analyzing the interaction of dislocations with point defects is measuring the internal friction. The *internal friction* is the ability of a solid to convert to heat the mechanical energy imparted to the body in the course of deformation and causing a disturbance of its thermodynamic equilibrium.

While in absolutely elastic bodies the relations connecting stresses and strains are time independent, in viscous media the strain is described by an equation which differs from Hooke's law by the presence of additional terms – time derivatives of the stress and strain:

$$\sigma+\tau_1\dot{\sigma}=M(\varepsilon+\tau_2\dot{\varepsilon}) , \tag{2.99}$$

where τ_1 and τ_2 are the relaxation times. The significance of coefficients τ_1 and τ_2 is as follows: if a constant load σ_0 is applied to a body instantaneously, the strain increases jumpwise to ε_0, and then grows gradually to an ultimate value of $\varepsilon(\infty)=\sigma_0/M$. As soon as load σ_0 is removed, the strain reduces sharply by a value of ε_0 and then gradually falls off to zero. Relaxation time τ_2 determines the time dependence of this aftereffect or relaxation, i.e., τ_2 is the strain relaxation time at a constant stress. If a constant strain ε_0 arises instantaneously in the body, the stress increases instantaneously to σ_0, after which it gradually diminishes to a final value of $\sigma(\infty)=M(\varepsilon_0)$. The time to the establishment of stress $\sigma(\infty)$ is stress relaxation time τ_1. Thus, τ_1 is the stress relaxation time at a constant strain. Coefficient M appearing in (2.99) characterizes the proportionality between stress and strain after the termination of the relaxation process. This coefficient is called the *relaxed elasticity modulus*. Expression (2.99) is called the *equation of a standard linear body*.

If the load varies periodically, then deformation ε is delayed in phase from the stress as a result of relaxation. Assuming $\sigma=\sigma_0\exp(\mathrm{i}\omega t)$, $\varepsilon=\varepsilon_0 \exp[\mathrm{i}(\omega t-\delta)]$ and solving (2.99), we obtain

$$(1+i\omega\tau_1)\sigma=\mathcal{M}(1+i\omega\tau_2)\varepsilon . \tag{2.100}$$

The stress and strain amplitudes are related by complex modulus $\mathcal{M}$:

$$\mathcal{M}=\frac{1+\mathrm{i}\omega\tau_2}{1-\mathrm{i}\omega\tau_1}M . \tag{2.101}$$

Relaxation results in energy dissipation, and the mechanical vibrations induced in the body gradually damp out. The tangent of the phase shift angle between stress and strain characterizes the reciprocal figure of merit of the system in which the mechanical vibrations are induced:

$$\tan\delta=Q^{-1} . \tag{2.102}$$

One of the measures of damping out of the internal friction may also be the $\Delta \bar{W}/\bar{W}$ ratio, where $\Delta \bar{W}$ is the average decrease in total mechanical energy $\bar{W}$ of the unit volume of the body within one vibration period. This ratio is called the logarithmic damping decrement Δ, which is related to Q^{-1} by the equality

$$\Delta \bar{W}/\bar{W} = \Delta = \pi Q^{-1} \ . \tag{2.103}$$

The damping out of vibrations is also observed at stresses much less than the yield point.

There are three principal sources of energy dissipation in crystals subjected to internal friction, in accordance with the three ways of internal rearrangement of a material: (1) thermal rearrangement (heat redistribution because of the nonuniform stressed state), (2) domain rearrangement (rearrangement of the domain structure of ferromagnetics and ferroelectrics, etc.), (3) atomic rearrangement (diffusion effects, relaxation along the grain boundaries in polycrystals, dislocation movement, etc.).

The internal friction depends on many factors, whose effect depends on their influence on some energy dissipation mechanisms. Each dissipation mechanism manifests itself, at constant ambient conditions (temperature, pressure, etc.), in a definite range of frequencies ω, which is reflected by the maximum on the curve $Q^{-1}(\omega)$. By varying the vibration frequency from infrasonic ($10^{-4}-10^{2}$ Hz) to hypersonic ($10^{9}-10^{11}$ Hz), one obtains a number of maxima on the curve $Q^{-1}(\omega)$ of the crystal (Fig. 2.48). These curves are called the "mechanical absorption spectrum." Changes of the specimen (its size, chemical composition, grain size) or of the environment (temperature and surrounding atmosphere) cause a characteristic change in mechanical spectrum (the number, position, and height of the maxima).

A temperature rise monotonically increases the internal friction background, as a rule. If the crystal experiences phase transformations, maxima are superimposed on background $Q^{-1}(T)$ at the appropriate temperatures. Impurities also manifest themselves by specific temperature maxima.

The internal friction is sometimes affected by the deformation amplitude. The dependence of the internal friction on deformation amplitude ε is

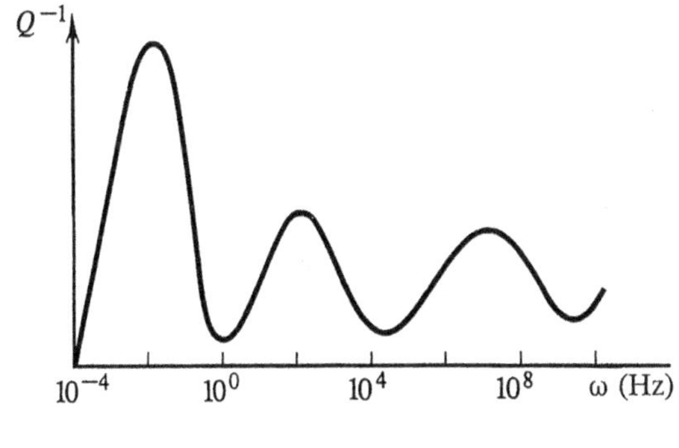

Fig. 2.48. Frequency dependence of the internal friction ("mechanical spectrum")

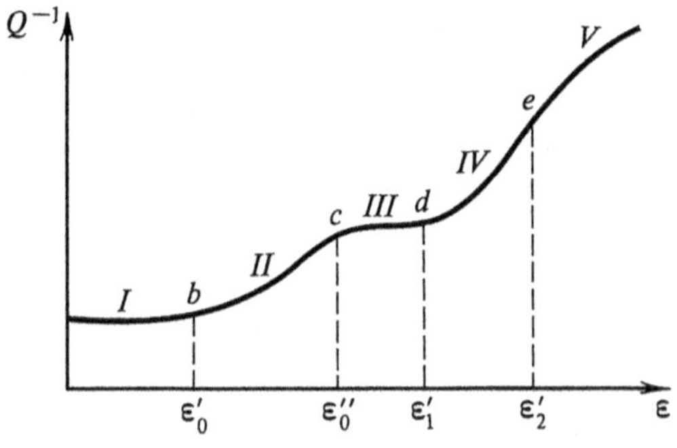

Fig. 2.49. Dependence of the internal friction on the strain vibration amplitude

schematically represented in Fig. 2.49. Regions I and III, where Q^{-1} is independent of ε, are said to be amplitude independent. The amplitudes ε_0', ε_0'', ε_1', and ε_2' correspond to the beginnings of the respective stages. Usually only the first two stages Q_{I}^{-1} and Q_{II}^{-1} are considered, which correspond to amplitude-independent and amplitude-dependent frictions.

Dislocation movement under sign-variable strains may contribute to both Q_{I}^{-1} and Q_{II}^{-1}. Amplitude-independent internal friction occurs either at very small strains (amplitude $\ll 10^{-5}$) or at very high vibration frequencies (in the ultrasonic range). In this case dislocations cannot detach themselves from the pinning centers (the forces acting on the dislocations are small, and the time of detachment of the dislocation from the pinning point exceeds the vibration period of the specimen) and vibrate about them. If, however, the deformation amplitude is high enough ($>10^{-5}$) or the vibration frequency is not too high (~ 1 Hz), so that the dislocation has time to detach itself from the pinning points, the paths of dislocation movement under the effect of opposite-sign stresses are different (dislocation hysteresis), and the internal friction becomes dependent on the deformation amplitude.

2.3.13 Anisotropy of the Internal Friction

The internal friction is among the anisotropic properties of a crystal. It is known (Sect. 2.1) that the specimen orientation may affect the elastic moduli. However, if the anisotropy of elastic moduli were the only cause for the internal friction anisotropy, the ratio of Q^{-1} to the defect of Young's modulus $\Delta E/E$ would be independent of specimen orientation, since the influence of the change in Young's modulus is determined by the defect of this modulus. Since this is not the case in experiment, one should assume the influence of the dislocation movement on the anisotropy of vibration damping. The existence of this influence is confirmed by the fact that the degree of anisotropy of internal friction is higher than that of the elastic properties, and also that the internal friction strongly depends on the previous deformation.

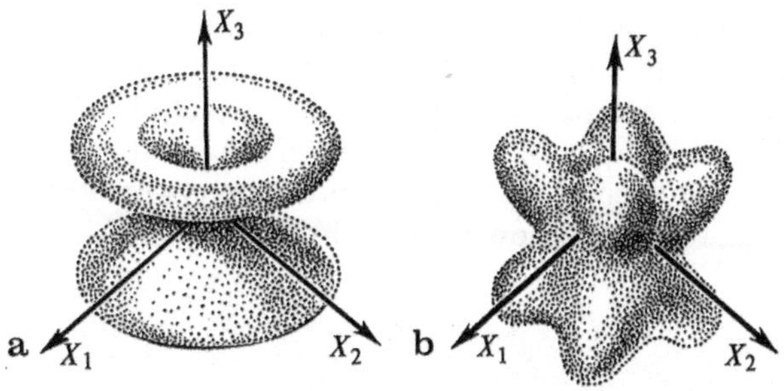

Fig. 2.50. Characteristic surface of the background of the amplitude-independent internal friction in a crystal of (**a**) zinc and (**b**) NaCl [2.41]

Figure 2.50 shows the characteristic surfaces of the background of the amplitude-independent internal friction for Zn and NaCl crystals. The length of the radius vector of the surface is proportional to the logarithmic decrement of the damping out of longitudinal vibrations in the appropriate direction. The shapes of these surfaces are in complete agreement with the law of the critical shear stress for slipping over the (0001) [11$\bar{2}$0] systems in zinc and (110) [1$\bar{1}$0] in NaCl. According to this law, the shear stress component for a given direction of load application is larger, the higher the internal friction. This proves once more that the internal friction anisotropy is related to the plasticity, rather than the elasticity, of crystals.

2.3.14 Theories of Dislocation Internal Friction

The theories of dislocation internal friction were developed by *Granato* and *Lücke* [2.42] and then by *Indenbom* and *Chernov* [2.43]. The first theory considers the contribution to the internal friction from the movement of dislocations detaching from the fixation points without the participation of thermal activation, which is true exclusively for 0 K. Idenbom and Chernov took into account the thermoactive detachment of dislocations, and this considerably increased the possibilities for obtaining information from internal-friction experiments. Their theory of dislocation hysteresis makes it possible to determine the nature of the obstacles and their distribution in the crystal on the basis of dependences $Q^{-1}(T, \tau_0)$, where τ_0 is the deformation amplitude. The theory is based on the calculation of energy losses on the internal-friction hysteresis during the vibration of the dislocation segment in an elastic field of fixed point defects, under the influence of external stresses and thermal vibrations. Indenbom's and Chernov's calculation showed that

$$Q^{-1}=\frac{Gb^2}{2T_0}L_{\min}^2\int_{L\geqslant L_{\min}} LN(L)\,dL\ , \tag{2.104}$$

where T_0 is the linear tension of the dislocation, L is the half-length of the vibrating dislocation segment, and $2L_{min}$ is its minimum length for given T and τ_0. As the temperature rises, ever-decreasing segments detach themselves from the pinning points, and segments of a given length detach themselves with a temperature increase at a lower stress. The activation energy of thermo-active detachment of dislocations with dislocation hysteresis satisfies the relation

$$H(F) \leqslant kT \ln(\nu_{eff}/\nu) \; , \tag{2.105}$$

where F is the dislocation-obstacle interaction force equal to $L_{min} b \tau_0^2$, ν_{eff} is the frequency of effective attacks of the potential barriers by a dislocation segment of length $2L$, and ν is the oscillation frequency of the external stress. From dependences $Q^{-1}(T)$, one obtains $T(\tau_0)$ for different τ_0 on sections $Q^{-1} = \text{const}$, and from $T(\tau_0)$ one finds $H(F)$. From $H(F)$ one gets the dependence of $-\partial H/\partial F$ on F by which it is possible to judge the nature of the obstacles for dislocations and of the length distribution of dislocation segments. *Indenbom* and *Chernov* [2.43], and *Savin* and *Chernov* [2.44, 45] summarize force laws for various types of obstacles.

Slightly (by a few percent) deformed fcc metals typically show an internal-friction peak in the low-temperature range (10 – 100 K) at medium and high frequencies (kHz-MHz). The position (frequency) of a peak is independent of the strain amplitude and of the introduction of impurities, and varies exponentially with the temperature. The activation energy of this effect (~0.1 eV) coincides with the energy required for displacement of dislocation segments of definite length from one potential groove to another, i.e., the internal-friction peaks reflect the vibration of dislocations around the Pierls barriers [2.46]. Therefore, they help in judging the magnitude of Pierls barriers in crystals.

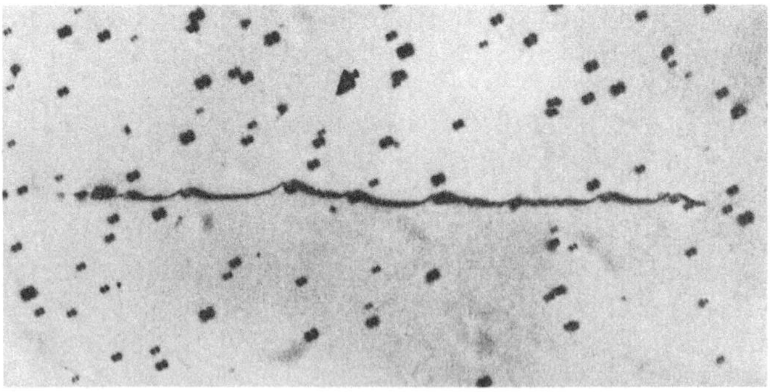

Fig. 2.51. Bending of a dislocation between precipitate particles in a crystal of germanium with a phosphorus impurity [2.47]

2.3.15 Mechanisms of Overcoming Impurity Obstacles

Because of the retardation of dislocations by impurity obstacles, impurity-free segments of dislocations buckle under the effect of the stress applied. Figure 2.51 demonstrates the buckling of a dislocation line in Ge between precipitate particles. The mechanism of detachment of dislocations from impurity centers depends on the stress level and the form of incorporation of the impurities (individual substitutional of interstitial atoms, impurity-vacancy dipoles and their complexes, segregations).

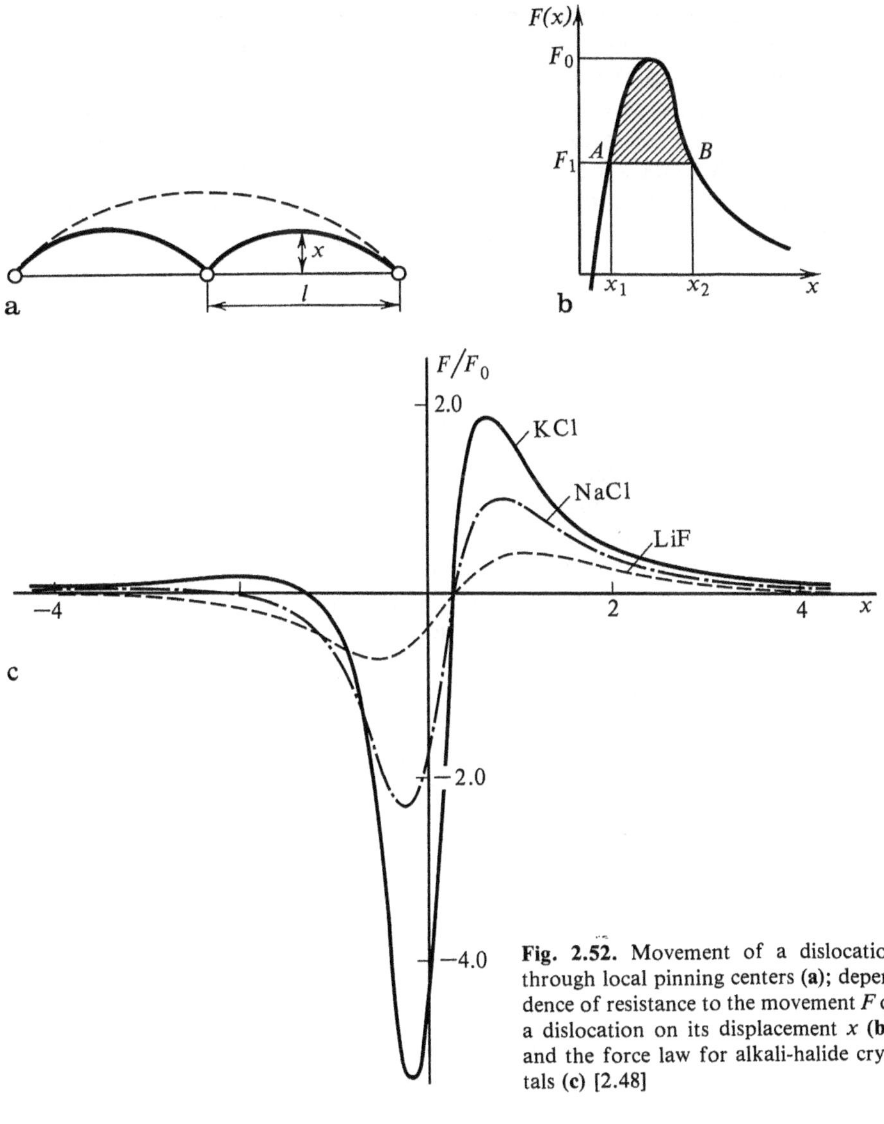

Fig. 2.52. Movement of a dislocation through local pinning centers (a); dependence of resistance to the movement F of a dislocation on its displacement x (b); and the force law for alkali-halide crystals (c) [2.48]

It can be assumed approximately that a dislocation segment is under the effect of the total force F applied to the midpoint of the segment from the side of the impurity center. The force acting on the dislocation depends on its displacement x (Fig. 2.52a) according to the law depicted in Fig. 2.52b. The plot indicates the positions of stable (x_1) and unstable (x_2) equilibrium corresponding to the stress applied τ_1 ($F_1 = \tau_1 b$). To overcome the resistance of the pinning centers, the dislocation must shift from point A to point B, which requires an energy equal in magnitude to the hatched area. This energy, which is called the activation energy of overcoming obstacle H, can be obtained from thermal fluctuations. The force law $F(x)$ can be calculated theoretically for each crystal by finding the concentration dependence of the average length l of the free segments between point centers and the corresponding change in the force of action of the dislocations on the center $F = \tau b l$ with due regard for the elastic anisotropy of the crystal, the symmetry of the center, and the orientation of the dislocation interacting with the center. According to *Barnett* and *Nix* [2.48], the force law for NaCl-type crystals in interaction of screw dislocations of glide system $\{110\}$ $\langle 110 \rangle$ with an isotropic center elongated along the $\langle 110 \rangle$ direction is calculated by the formula

$$F(x) = F_0 \beta^{3/2} \left[\frac{\beta x + 2\sqrt{2}x - 1}{(\beta x^2 + 1)^2} \right] , \tag{2.106}$$

where $F_0 = \dfrac{3\Omega \varepsilon c_{44} b}{4\pi\lambda^2}$, $\beta = \dfrac{c_{11} - c_{12}}{2c_{44}}$, $x = \dfrac{x_1}{\lambda}$, Ω is the volume of the defect, ε is the lattice strain induced by the defect, and λ is a parameter characterizing the orientation of the defect relative to the second (x_2) and the third (x_3) coordinate. Figure 2.52c shows the force laws for three alkali-halide crystals.

From Fig. 2.52b it is seen that the higher the level of external stresses, the smaller is H. If τ corresponds to $F \leqslant F_0$, where F_0 is the force corresponding to activationless detachment, the dislocation velocity v is described by

$$v = A\, \nu_{\mathrm{eff}} \exp[-H(\tau)/kT] , \tag{2.107}$$

where A is a constant, $H(\tau)$ is the activation energy dependent on stress τ, ν_{eff} is the frequency factor depending on the vibrational degrees of freedom of the dislocation, k is the Boltzmann constant, and T is the temperature. The role of these degrees of freedom amounts to the appearance in the exponent index of entropy terms:

$$\nu_{\mathrm{eff}} \sim \frac{\varphi(\tau)}{B(T)} \exp \frac{\Delta S}{k} , \tag{2.108}$$

where ΔS is the activation entropy due to the entropy difference of the dislocation in the well in front of the barrier and at the saddle point of the potential relief. Factor $1/B(T)$ (B is the coefficient of viscous deceleration of the dislocation) owes its origin to the peculiar "Brownian" motion of the dislocation under the effect of thermal fluctuations. This nature of the movement also explains the stress dependence of ν_{eff} (factor $\varphi(\tau)$).

Thus, the exponent index in the expression for the dislocation velocity contains the difference

$$\Delta G = H - T\Delta S \ , \tag{2.109}$$

which represents the free activation energy. If, within the temperature range under investigation, the impurity centers and the character of their interaction with dislocations remain unaltered ($\partial H/\partial T = 0$), then, by virtue of the thermodynamic relation,

$$T\frac{\partial \Delta S}{\partial T} = \frac{\partial H}{\partial T} = 0 \ .$$

Activation entropy ΔS and $\exp(\Delta S/k)$ are also temperature independent. Therefore, at a fixed stress τ, activation energy H can be determined from the slope of dependence $\ln(vB)$ on the reciprocal temperature. The form of dependence $H(\tau)$ permits judging the nature of the obstacles, because the force law can be reconstructed from $H(\tau)$.

Dependence $H(\tau)$ can also be obtained from the macroscopic characteristics of the deformation if the defect structure of the specimens (for instance, the concentration and distribution of the impurity centers) is preserved in the course of testing and if it is possible to pick out from the external stress τ its part τ_{eff} which directly affects the dislocation overcoming the local barrier. The remaining part of the stress τ_i applied is offset by long-range fields of stresses from various internal inhomogeneities and dislocations lying in parallel glide planes.

Internal stresses are usually estimated by the stress relaxation method (Sect. 2.4) and also by mechanical testing with a changeable strain rate. The former method is described by *Estrin* et al. [2.49] and the latter, by *Michalak* [2.50] (see also *Evans* and *Rawlings* [2.51]).

The rate of plastic strain $\dot{\varepsilon}$ is related to dislocation velocity v by the expression

$$\dot{\varepsilon} = b\varrho v \ , \tag{2.110}$$

where ϱ is the density of mobile dislocations [2.52]. Allowing for (2.107),

$$\dot{\varepsilon} = A\varrho b\nu_{\text{eff}}\exp[-H(\tau)/kT] \ . \tag{2.111}$$

Relation (2.111) allows one to determine the activation characteristics of the dislocation movement from the dependence of the yield point on the tem-

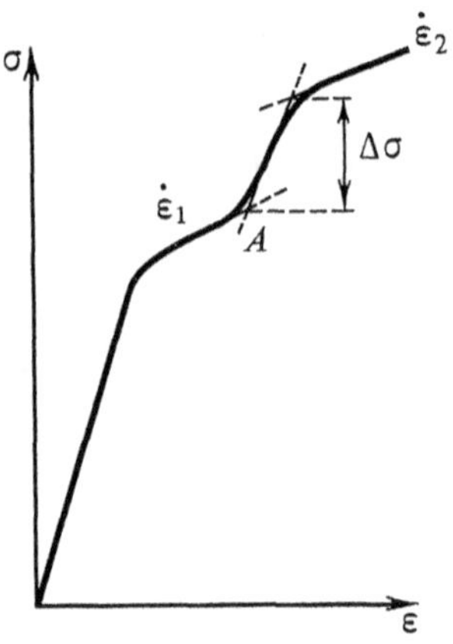

Fig. 2.53. Stress-strain curve with a change in strain rate. On change of rate $\dot{\varepsilon}_1$ to $\dot{\varepsilon}_2$ at point A, the flow stress jumped by $\Delta\sigma$

perature and deformation rate, and also from the temperature dependence of the variation in flow stress with the strain rate (step loading); the results of the corresponding experiment are schematized in Fig. 2.53 (on change of velocity $\dot{\varepsilon}_1$ to $\dot{\varepsilon}_2$ at point A, the flow stress jumped up by $\Delta\sigma$). The same problem can be solved by using the dependence of the stress relaxation rate on the stress level and the temperature. Here the starting expression is the equation

$$-\dot{\tau} = G\dot{\varepsilon} = Gb\varrho v = AGb\varrho\nu_{\text{eff}}\exp[-H(\tau)/kT] \ . \qquad (2.112)$$

According to *Frank*'s theory of impurity hardening, in alkali-halide crystals with divalent cation impurities [2.53] the stress at the yield point depends not only on the necessity of overcoming the elastic interaction of the moving dislocations with dipoles Me^{2+}-cation vacancy, but also on the dipole reorientation in the dislocation stress field so that the energy of the entire system should decrease (induced Snoek effect). If the impurity concentration is moderate and if all the divalent impurities form single dipoles with cation vacancies, then at temperatures above room temperature the yield point must be linearly proportional to the concentration of dipoles c_{d} ($\tau \sim c_{\text{d}}$). At a higher impurity concentration the dipoles join together into groups consisting of z dipoles. Usually z is not less than 3, but in NaCl it may equal 2.

Calculations show that the difference between the yield points of an impurity and an impurity-free crystal $\tau - \tau_0 = \tau_A$ is expressed by a relation of the type

$$\tau_A = p\sqrt{\alpha c_z} \ , \qquad (2.113)$$

where c_z is the concentration of complexes, and p and α are the constants characterizing this complex. Frank assumed that the yield point of an impurity-free crystal is equal to the internal stresses, which is wrong, generally speaking, because the internal stresses depend on many factors, including the presence of impurities. To take advantage of Frank's theory, τ_0 should be replaced by internal stresses determined experimentally for particular strain conditions, and, still better, for each specimen separately. According to the mass action law, the concentration of complexes c_z may be related to the concentration of individual dipoles c_d by

$$c_z = (c_d)^2 M_z \ ,$$

where M_z is a constant at a given temperature.

Using the obvious relation $c = c_d + z c_z$, we obtain

$$c = (c_z/M_z)^{1/z} + z c_z \ . \tag{2.114}$$

Equations (2.113, 114) relate τ_A to the concentration of divalent cation impurities c. In particular, at $z = 2$ the dependence of τ_A on $\sqrt{c}$ is hyperbolic with an asymptote $\tau_A = p\sqrt{\frac{\alpha}{2}}\sqrt{c} - \frac{p}{4}\sqrt{\frac{\alpha}{M_z}}$, which does not pass through the origin. At $z \geqslant 3$ (2.113, 114) are difficult to solve. However, the asymptotic relationship between τ_A and c at not-too-low impurity concentrations can easily be found in this case, too: $\tau_A = p\sqrt{\alpha c/z}$. It follows that if τ_A is determined by trimers (groups of three impurity-vacancy dipoles) and by more intricate complexes, asymptote τ_A also depends linearly on $\sqrt{c}$, but it passes through the origin in coordinates $\tau_A - \sqrt{c}$.

The influence of an impurity atmosphere depends on the level of the stress applied. If the stress exceeds that of interaction of the dislocation with the atmosphere, the dislocation will shift with respect to the cloud. At the moment of detachment of dislocations from the impurity cloud, the deformation curve shows a yield point phenomenon (Fig. 2.29). The stress required to free the dislocation from the cloud is determined by the formula [2.35]

$$\tau = c_1 |W| / b^3 \ , \tag{2.115}$$

where W is the binding energy of the cloud atoms with the dislocation, and c_1 is the limiting concentration of the impurity in the cloud. If the interaction is due to dimension factor (Sect. 2.3.10), then, according to *Friedel* [2.35],

$$\tau = G \delta c_1 / 2 \ . \tag{2.116}$$

Thermal fluctuations and stress concentration may significantly reduce the level of the stresses required for detaching the dislocation from the cloud. They help the dislocation to get rid of one or several fixation centers, after which a whole dislocation segment detaches itself from the cloud and traverses some area without stopping.

2.4 Diffusion Mechanisms of Plastic Deformation

2.4.1 Gorsky Effect

It is known that crystals may deform under the effect of mechanical stresses due to the movement of point defects (vacancies, interstitial or impurity atoms). Then one speaks of the *diffusion mechanism of deformation.*

One of the manifestations of the diffusion plasticity is the deformation of solid solutions associated with the movement of substitutional or interstitial impurities in them.

In solid substitutional solutions the components have different atomic (ionic) radii, as a rule. If such crystals are subjected to bending strain, part of the atoms (ions) of smaller radius will shift to the contraction regions, and the latter ones, to the extended parts. On bending of crystals of solid interstitial solutions, the interstitial atoms diffuse to the extended parts, as a result of which a difference in component concentration arises in the contraction and extension regions. If the crystals are kept in the bent state for a sufficiently long period, a concentration gradient, depending on changes in specimen thickness, will be established in them after equilibrium is achieved. On unloading the concentration is equalized. The shifting of atoms (ions) because of the nonuniform stressed state is called *ascending diffusion,* or the *Gorsky effect* [2.54].

At nonuniform loading, ascending diffusion may involve not only substitutional and interstitial impurities, but also intrinsic point defects of the crystal, i.e., vacancies and interstitials. The force compelling the vacancies and interstitials to move in the crystal is proportional to the pressure gradient:

$$f \sim \nabla P \ . \tag{2.117}$$

The force acting on the vacancy is directed towards the more contracted part of the crystal ($f_v \sim \nabla P$) and the force acting on the interstitial makes it move in the opposite direction ($f_i \sim -\nabla P$). Figure 2.54 presents a diagram of the movement of the interstitials and vacancies in a nonuniform stressed state. The interstitials and vacancies transport $+\Omega_0$ and $-\Omega_0$ of material, respectively. Therefore, we can write:

$$f_v = \Omega_0 \nabla P \ , \quad f_i = -\Omega_0 \nabla P \ .$$

The diffusion mobility of point defects is proportional to diffusion coefficient D, and the average drift velocity of point defects can be written

$$u = \frac{D}{kT} f \ . \tag{2.118}$$

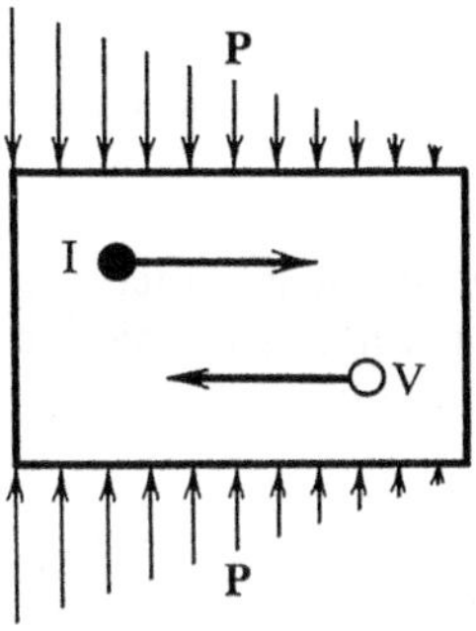

Fig. 2.54. Movement of interstitials (*I*) and vacancies (*V*) under conditions of nonuniform stress distribution [2.55]

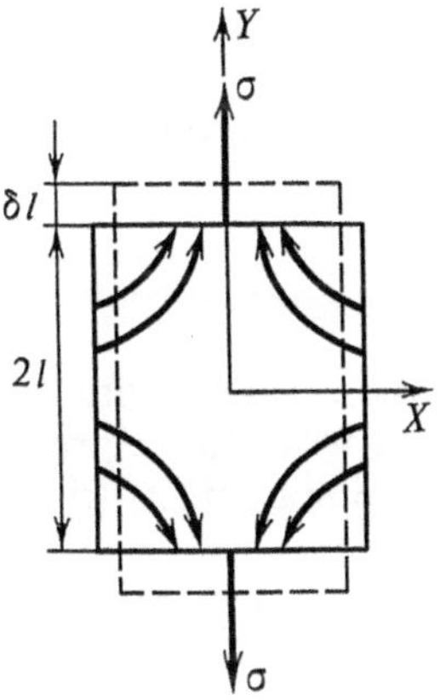

Fig. 2.55. Movement of interstitial atoms in diffusion creep due uniaxial tension [2.55]

2.4.2 Diffusion Creep

Directed diffusion of point defects may also occur in a uniform stressed state, for instance, in uniaxial tension under a constant load. Deformation under a constant load (stress) is called the creep. It can be effected by different mechanisms. If the deformation under creep is due to the movement of point defects (diffusion) alone, such a creep is called *diffusional.* Let us consider the movement of point defects under the diffusional creep.

Under a tensile stress (Fig. 2.55), the material expands in the direction of action of the stress in the region of load application. As a result, an increased concentration of vacancies and a decreased concentration of interstitial atoms arise near the loaded surface. At the lateral load-free surfaces of the specimen the concentration of vacancies (interstitials) is lower (higher) than at the loaded parts. Owing to the resulting gradient of point defect concentration, a diffusion flux of these defects is formed, which causes mass transport. The direction of the movement of interstitials under uniaxial tension is shown by the arrows in Fig. 2.55.

The mass flow density is expressed as

$$j = c_v \boldsymbol{u}_v + c_i \boldsymbol{u}_i \sim \frac{\Omega_0}{kT} (c_v D_v + c_i D_i) \nabla \sigma \ ,$$

where c_v and c_i are the equilibrium concentrations of vacancies and interstitials in the crystal; D_v and D_i are their diffusion coefficients; and $\boldsymbol{u}_v$ and $\boldsymbol{u}_i$ are their drift velocities, respectively. Since the mass flows caused by the movement of vacancies and interstitials add together, the diffusion of point defects of one type with $c = c_v + c_i$ can be considered:

$$j = cD\nabla\,(\Omega_0\sigma/kT)\ .$$

This flow density determines the velocity of the loaded surface of the specimen:

$$j = \delta l/\delta t$$

(here t is the time). Taking into account the corresponding strain increase $\delta\varepsilon = \delta l/l$, we can write the expression for the rate of strain caused by the diffusion flow of point defects (the velocity of diffusion creep):

$$\dot{\varepsilon} = \frac{\delta\varepsilon}{\delta t} = \frac{1}{l}\,\frac{\delta l}{\delta t} = \frac{j}{l} = \frac{cD}{l^2}\,\frac{\Omega_0\sigma}{kT}\ . \tag{2.119}$$

It should be noted that point defects do not necessarily begin and end their movement at the external surface. The role of sources and sinks can be played by internal surfaces, such as grain boundaries in polycrystals, the boundaries of misoriented blocks in single crystals, and also dislocations. Then, l in (2.119) corresponds to the grain or block size (or to the average distance between dislocations).

The diffusion creep is of great importance in various structures operating at high temperatures.

2.4.3 Climb of the Edge Dislocations

Diffusional plastic deformation of crystals may occur not only by diffusion of individual point defects. In their displacement, point defects encounter other disturbances – for instance, dislocations – and, interacting with them, affect their movement. If a point defect encounters an edge dislocation, it settles (condenses) at the edge of the extra half plane and produces a bump (if an interstitial was added) or a recess (if a vacancy settled on it). If a sufficient number of interstitials or vacancies have settled, the dislocation line may shift perpendicular to the glide plane because of the change in the position of the edge of the extra half plane in the area where the point defects have settled. Addition of a monatomic chain of point defects displaces the dislocation by a lattice spacing in the direction perpendicular to the glide plane.

Since the supply of material (point defects) to the dislocation line in the crystal can be effected only by diffusion, the above-described motion of dislocations is rather slow, and therefore it is called the *climb*.

The climb of dislocations plays an essential part in overcoming local obstacles (for instance, impurity centers). Dislocations which slow down at these obstacles form pileups. Stresses from a pileup of dislocations add to the external stresses and enhance their effect on the head dislocation held up by the stopper, thereby making it leave its glide plane. As a result, the head

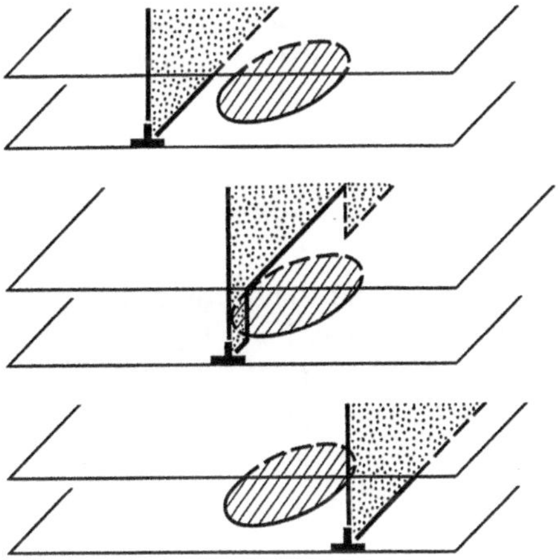

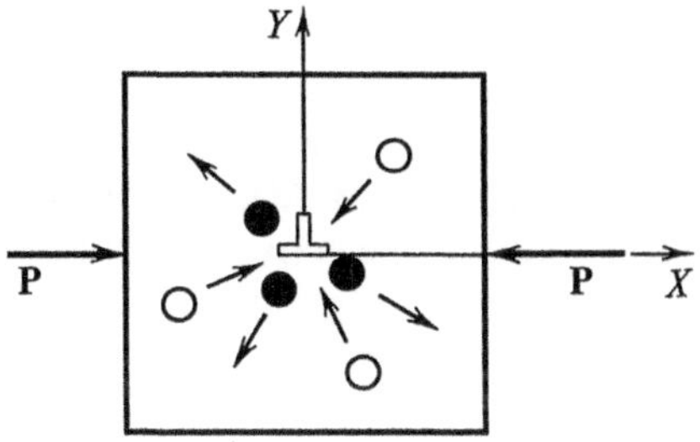

Fig. 2.56 **Fig. 2.57**

Fig. 2.56. Climbing of a dislocation segment around an obstacle (hatched)

Fig. 2.57. Movement of point defects (dark circles: interstitials, light circles: vacancies) during upward climb of a dislocation under the effect of compressive force P [2.55]

dislocation climbs to another glide plane above or below the obstacle, where its movement is not stopped and where it can pass around the obstacle. Figure 2.56 exhibits successive stages of overcoming an obstacle by the climb of the dislocation to another glide plane.

A climb may be caused by an external effect. Suppose a crystal is subjected to compressive stress P along the X axis (Fig. 2.57), which produces a force acting on the edge dislocation in the direction of the Y axis. Allowing for (2.72), the force acting on a unit length of the dislocation is defined by the expression

$$f_y \sim bP \ . \tag{2.120}$$

Here b is the modulus of the Burgers vector of the dislocation.

Owing to this force, directed fluxes of point defects appear near the dislocation: the vacancies diffuse towards the dislocation line, while the interstitials move away from the dislocation.

If an equilibrium concentration of point defects is maintained far from the dislocation [Ref. 2.5, Chap. 5], it can be assumed that the crystal has an inexhaustible source for vacancies and an infinite sink for interstitials. As a result, the process becomes stationary, with a constant climb rate of dislocations:

$$v_y \sim \frac{cD}{b}\left(\frac{\Omega_0 P}{kT}\right) \ . \tag{2.121}$$

In real cases, steady-state climb is limited not only by the mobility of point defects, but also by the scarcity of sinks and sources of point defects.

The rate of creep due to the climb of dislocations is much higher than that of the pure diffusion creep, because during the climb it is sufficient for the point defects to move over short distances.

This rate is described by an expression of the type

$$\dot{\varepsilon} \sim \sigma^n \exp(-U/kT) \ , \tag{2.122}$$

where, because of the difference between the local and the average stress and the stress dependence of the climb path, the index n may be as high as several units (usually $n = 3-5$), and the activation energy is stress independent and corresponds to the activation energy of the climb. The latter is usually equal to the activation energy of self-diffusion U_s or, at a high vacancy supersaturation, to the activation energy of vacancy migration U_m.

2.4.4 Diffusion-Dislocation Creep

Deformation under a constant load, when dislocation movement in glide planes occurs along with that of point defects, is called the *diffusion-dislocation creep.* Here the creep velocity also depends on the dislocation mobility, i.e., on the resistance of the lattice and other dislocations and point defects. This velocity is limited by the formation and further movement of jogs arising at the intersection of dislocations (Sect. 2.3.7). The jogs can migrate along dislocation lines and climb, leaving behind a trail in the form of chains of vacancies or interstitials.

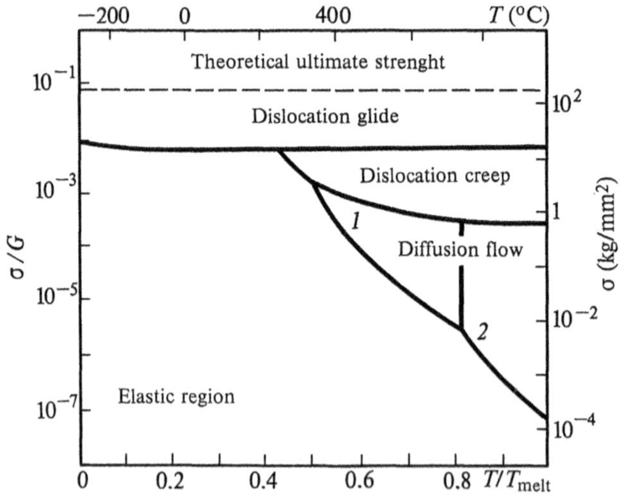

Fig. 2.58. Chart of creep mechanisms vs temperature (or vs ratio of the testing temperature to the melting point T/T_m) and stress σ (or the σ/G ratio) [2.56]: (*1*) diffusion along grain boundaries; (*2*) diffusion in the bulk

The velocity of creep due to the glide of dislocations which have climbing jogs is defined by the expression

$$\dot{\varepsilon} \sim \exp\left(-\frac{U-\gamma\sigma}{kT}\right), \tag{2.123}$$

where the activation energy is equal to U_s or U_m, reduced by the chemical potential of the vacancy at the jog, which is equal to the work of stress σ required to move a dislocation segment of length l situated between the jogs, by an interatomic distance. The quantity γ is the activation volume, equal to lb^2.

Figure 2.58 presents a chart of the creep mechanisms versus the deformation conditions.

2.4.5 Creep Stages

Creep has three stages, which are shown in Fig. 2.59 and labeled a, b, and c, respectively. The first stage is transient creep, the second steady-state creep, and the third accelerated creep, which ends in failure.

The transient creep stage includes the glide of dislocations, the formation of dislocation pileups (strain hardening), and the climb of dislocations, which permits the dislocations to leave the pileup and reduce the work hardening. In this stage, characteristic polygonization blocks are formed. The process of reduction in work hardening is called relaxation.

The stage of steady-state creep is characterized by a balance of work hardening and relaxation. In this stage the creep is limited by successive dissipation of block boundaries, beginning with the less dense ones, and by the movement of the freed dislocations to neighboring boundaries. As a result, the size of the blocks and their misorientation increase in the course of steady-state creep. As the slightly misoriented boundaries are exhausted, boundaries with a higher misorientation and dislocation density disintegrate. At sites of detachment of block boundaries and also at places of convergence

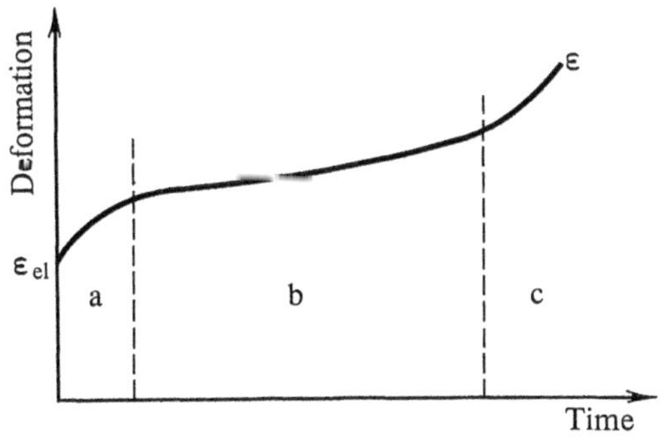

Fig. 2.59. Three-stage curve of diffusion creep

of three boundaries and at boundaries deviating from equilibrium orientation, high stresses arise which may cause cracking. In the same stage pores are formed as a result of the joining together of vacancies formed by plastic deformation.

Transition to the third, final stage of creep is associated with the increase in the opposite effect of the pores on the strain rate. Pores at the block and grain boundaries stimulate the boundary glide, and this accelerates the creep and also the failure, due to the coalescence of pores and microcracks and the formation of large cracks dangerous to the specimen.

If the dislocation glide is suppressed for some reason, the main contribution to the creep is made by the dislocations' climb. Point defects may arrive both from the crystal boundaries and from other dislocations.

A special variety of diffusion-dislocation creep is deformation due to the movement of dislocations together with the surrounding impurity atmosphere (Sect. 2.3.10). This kind of creep is called the *microcreep,* because dislocations surrounded with atmospheres move very slowly, and thus the deformation proceeds at a very low rate. The rate of microcreep is approximately described by

$$\dot{\varepsilon} \sim 2\varrho D\tau b^3/kT \ . \tag{2.124}$$

Here D is the diffusion coefficient of the impurity atoms.

We note in passing that surface defects can also take part in the creep. The involvement of surface defects (block boundaries) may be either active or passive. It is active when the creep is due to the movement of the boundaries themselves because of the climb of their individual elements. But more often the block and grain boundaries play a passive role, which amounts to the absorption and emission of vacancies in diffusion creep and to emission or retardation of dislocations during the diffusion-dislocation creep.

In the course of creep, several of the above-discussed mechanisms operate simultaneously; when the deformation conditions (temperature, stress) change, the relative participation of the particular mechanisms alters.

2.4.6 Radiation-Induced Diffusion Creep

In recent years, in connection with the development of nuclear engineering, one more variety of creep – *radiation-induced diffusion creep* – has been discovered. As noted above, irradiation of materials gives rise to additional point defects. Higher-than-equilibrium concentration of point defects, and also some heating accompanying the irradiation, promote the displacement of point defects to higher-equilibrium configurations. For instance, vacancies move from extended regions to contracted ones, and interstitials diffuse in the opposite direction.

Thus, radiation-induced diffusion creep takes place. In fissile materials or under strong irradiation of ordinary materials, so many point defects arise that they collect into flat pileups along definite crystallographic planes. Then local stresses appear near the pileups; these stresses may exceed the yield points. Then the crystal flows spontaneously even by gravity. Radiation flow is particularly pronounced and dangerous in structures where the materials are under load (for instance, in reactors).

2.4.7 Stress Relaxation

The above-considered microprocesses may occur not only in creep (under the effect of a constant stress), but also under other deformation conditions. A practically important case is *stress relaxation,* when the total deformation of the specimen is maintained constant. Then the crystals undergo plastic deformation, which leads to a monotonic decrease (relaxation) of the stresses. Stress relaxation, as well as creep, is time dependent.

Relaxation takes place not only in experiments with mechanical loading, when a constant strain is maintained, but also in heating strained (mechanically or by thermal impact) specimens. Figure 2.60 exhibits curves of the time dependence of the stresses in a crystal during relaxation after loading (***a***) and in annealing (***b***). In both cases the relaxation processes develop under the effect of internal stresses stored during preliminary deformation. The simplest example of stress relaxation is given in Fig. 2.61, which presents a photo of a germanium film with a phosphorus impurity taken in a transmission electron microscope. Label 1 denotes phosphorus inclusions, which cause elastic distortions around them; these appear on the photo as dark petals surrounding the inclusion. Label 2 denotes an inclusion, the stresses from which are relieved by the dislocation loop formed around the inclusion. Experiment shows that the relaxation rate is greater, the higher the stress (external, corresponding to the beginning of relaxation, or internal in annealing) and the temperature. Irradiation, for instance with neutrons, also accelerates the stress relaxations.

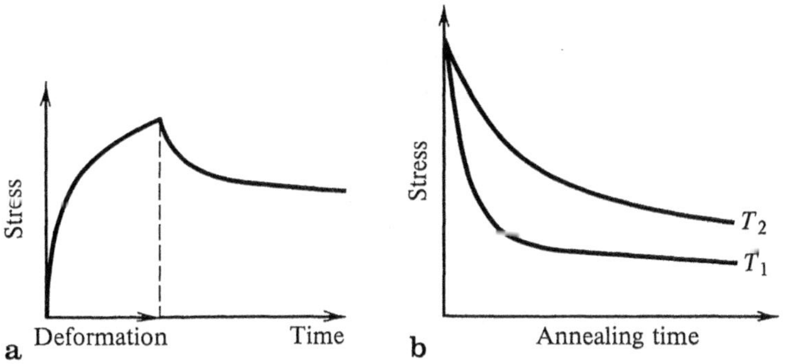

Fig. 2.60. Curves of stress relaxation after active loading (**a**) and as a result of annealing at temperatures T_1 and T_2 ($T_1 > T_2$) (**b**)

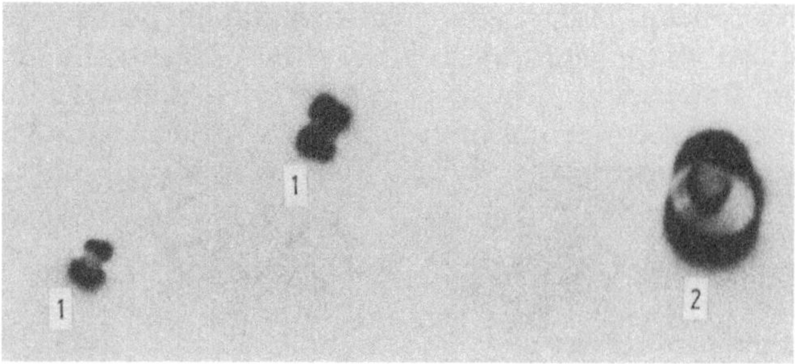

Fig. 2.61. Electron micrograph of germanium film with precipitates of phosphorus impurity (*1* and *2*) [2.47]

The stress relaxation phenomenon is widely used in practice. As noted in Sect. 2.2, the yield point usually decreases with increasing temperature. If the temperature dependence of the yield point is known, then, by selecting the annealing temperature, it is possible to achieve stress relaxation to a level close to the yield point at the annealing temperature. The internal stresses can thus be reduced: the higher the temperature and the longer the annealing, the lower the internal stresses.

2.5 Mechanical Twinning of Crystals

2.5.1 Twinning of Crystals

Twinning plays a comparatively insignificant role in strain and usually manifests itself only when the glide strain is impeded. Quartz and calcite are crystals in which twinning occurs more readily than glide. Twinning is promoted by a low temperature and impact loading, since on decrease in temperature and increase in strain rate the critical stress for glide increases more rapidly than for twinning. Twinning is inherent in all crystals, but in some it occurs more readily than in others. Among the metallic crystals, twinning proceeds most readily in hexagonal metals, slightly less readily in bcc metals, and most reluctantly in fcc metals. In some cases twinning changes the shape of the crystal (for instance, reflection twins in calcite, Fig. 2.62), and in others it does not (say, in axial twins according to the Dauphine law in quartz).

Twinning of a crystal in the process of loading is accompanied by a sharp decrease of stress in the specimen, which is reflected in the stress-strain curves as individual slumps (Fig. 2.63). Twinning may precede glide (Fig. 2.63 a) or follow it (Fig. 2.63 b).

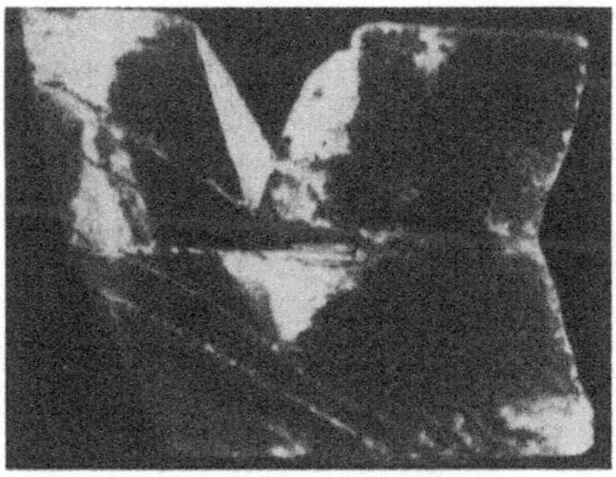

Fig. 2.62. Twinned crystal of calcite [2.57]

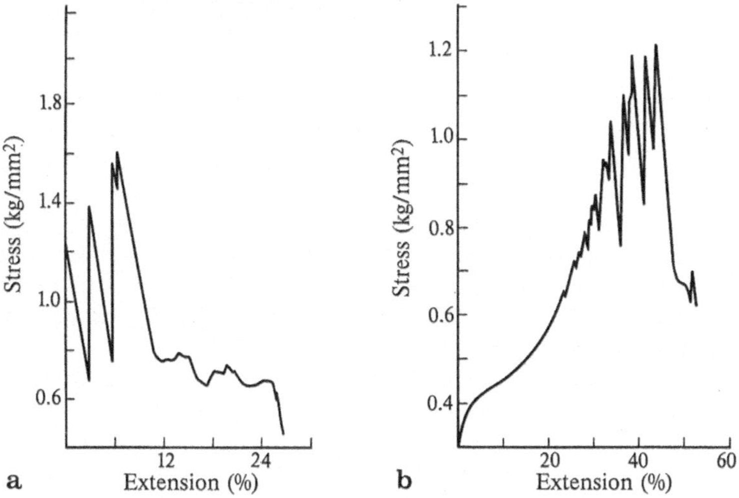

Fig. 2.63. Stress-strain curves for cadmium: **(a)** twinning precedes glide; **(b)** twinning sets in after glide strain [2.58]

2.5.2 Representation of Twinning with a Change in Shape as a Uniform Shear

If we abstract ourselves from the real movement of atoms in twinning, a comparison of the initial and final position of the crystal regions will help us to understand the rearrangement of the lattice in twinning with a change in shape. Such a rearrangement is outlined in Fig. 2.64. According to this scheme, twinning can be regarded as a successive displacement of the crystal layers from right to left parallel to the *a-a* plane, which is the mirror plane and the plane of contact between the two components of the twin. It is assumed that the atoms shift parallel to the mirror plane. In distinction to the translation glide, when the displacement is a multiple of the lattice spacing, in twinning the atoms shift by a fraction of the lattice translation vector. The shear strain in twinning is a simple shear. In a simple shear, all the atoms of a

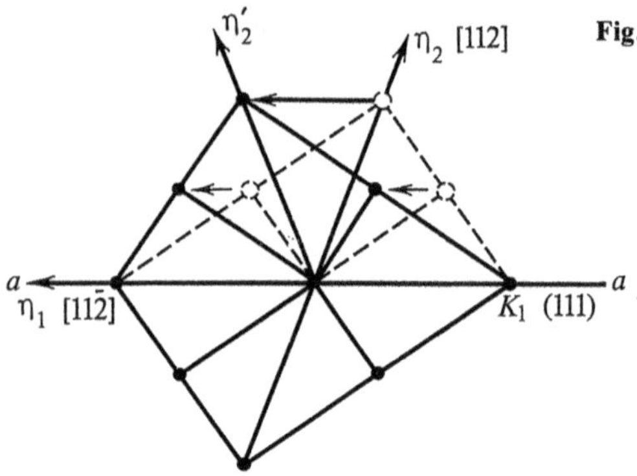

Fig. 2.64. Twinning as uniform shift of atoms

part of the crystal which have shifted to the twinning position must be displaced proportionally to their distance from the mirror plane. This plane is called the *twinning plane,* and the direction of displacement, the *twinning direction.*

The scheme of Fig. 2.64 describes the twinning of crystals with a macroscopically correct change in shape, but this scheme does not reflect quite correctly the atomic rearrangement taking place. For metals, it is true on the atomic scale as well, but for calcite $CaCO_3$, a rotation of CO_3 groups must occur in addition to the displacement of Ca atoms.

Figure 2.65 presents the projection of the calcite structure onto shear plane S. In the top right-hand corner you see part of the lattice in the twinning position. The initial position of the structural units is denoted by the dashed line. In calcite twinning all the atoms move parallel to the twinning plane, and the path lengths of the atoms are proportional to their distance from the

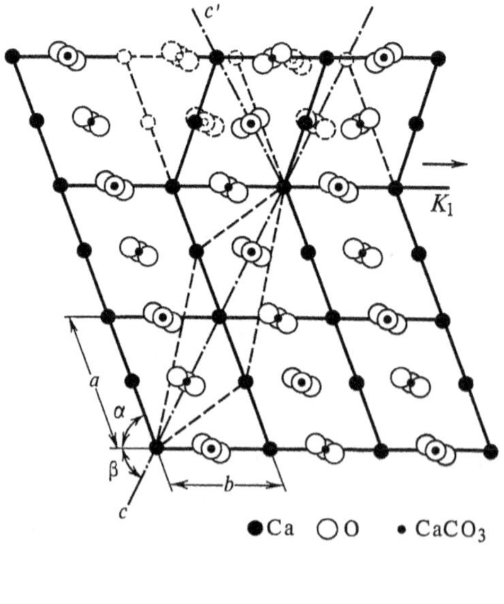

Fig. 2.65. Projection of the calcite structure onto shear plane $S(011)$. Twinning plane K_1 (101), η[010] [2.59]

twinning plane. Besides, in twinning, groups of three oxygen atoms rotate through 52° 30′ about the axis perpendicular to shear plane S passing through the center of the carbon atom.

2.5.3 Polarity of Twinning

While a glide can, as a rule, be induced by deformation in the forward and reverse direction, twinning can be achieved only if the deformation proceeds in one direction. Under deformation in the opposite direction, the twinned crystal will resume its original shape. In twinning, the displacement of the crystal layers is always polar. In some low-symmetry crystals, for instance, in monoclinic $BaBr_2 \cdot 2\,H_2O$, a strain in one direction leads to twinning, and in the opposite, to a glide.

2.5.4 Twinning Elements

As noted above, a shear in twinning with a change of shape is assigned by the shear plane and direction. A twinning shear, as well as a shear in elastic deformation (Sect. 2.1), is a uniform deformation. A *homogeneous shear deformation* is one in which the assigned macroscopic change of shape in twinning can be described by an ellipsoid to which a sphere mentally picked out in a crystal transforms. Let such a sphere (Fig. 2.66) of unit radius undergo a homogeneous shear strain along plane K_1 passing horizontally through enter O and normal to the plane of the drawing. We take as the origin the center common to the sphere and the ellipsoid formed from it due to strain. We arrange the Y axis along shear direction η_1, and choose the normal to plane K_1 as the Z axis. Since the points lying in plane K_1 are not displaced, the section of the sphere by this plane does not alter after the deformation and therefore must be one of the circular cross sections of the ellipsoid. Another circular section of the

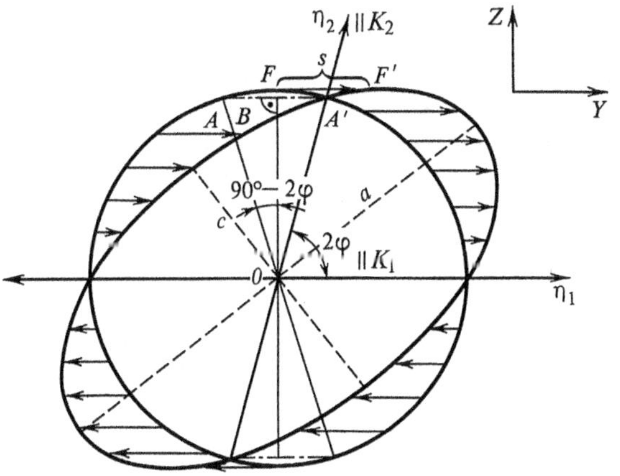

Fig. 2.66. Section of a sphere and the corresponding strain ellipsoid by shear plane S [2.58]

ellipsoid by plane K_2 is disposed symmetrically to K_1 about the major axes of the ellipsoid a and c, and passes through point A' of intersection of the circle and the ellipse, which are disposed in the plane of the drawing. The third ellipsoid axis b is perpendicular to the drawing, is equal to the sphere radius, and coincides with the line of intersection of the two circular cross sections. The plane coinciding with the drawing and defined by the ellipsoid axes (the longest a and the shortest c) is usually called the *shear plane* S. The trace of the intersection of the second circular section K_2 by shear plane S is usually designated η_2 and termed the *axis* of the main zone. The adopted measure of shear is segment s, which indicates the displacement of a point removed from plane K_1 by a unit distance. The quantity s is usually called the *crystallographic shift.* From the ellipse equation and the condition of proportionality between the displacement and the distance from the twinning plane $s:1 = AA':OB$ follows

$$s = 2/\tan 2\varphi = a - c \ , \tag{2.125}$$

where φ is the angle of inclination of the longest axis a of the ellipsoid to plane K_1.

Planes K_1 and K_2, and also directions η_1 and η_2, are called *twinning elements.* A more detailed description of twinning usually also contains shear plane S and crystallographic shear s.

Table 2.8 lists twinning elements for different crystals, in metals S always coinciding with one of the symmetry planes of the lattice. If different types of twinning are possible in a crystal, twins with a minimum shear s predominantly appear.

2.5.5 Types of Twinning

Twinning elements K_1, K_2, η_1, and η_2 do not necessarily have rational indices; they do not always pass through nodes of the Bravais lattice. Two basic cases are distinguished:

1. Here K_1 and η_2 are rational, and K_2 and η_1 are irrational. We choose, in plane K_1 (Fig. 2.67 a), two vectors $\boldsymbol{I}_1$ and $\boldsymbol{I}_2$, which remain unaltered under shear. Since η_2 is rational, a vector $\boldsymbol{I}_3$ can be chosen along it, which does not change its length after twinning ($\boldsymbol{I}_3 = \boldsymbol{I}_3'$). Vector $-\boldsymbol{I}_3$ is also a lattice vector (the lattice is centrosymmetric). It is seen from the figure that a unit cell constructed on vectors $\boldsymbol{I}_1$, $\boldsymbol{I}_2$, and $\boldsymbol{I}_3'$ is a mirror reflection of unit cell $\boldsymbol{I}_1$, $\boldsymbol{I}_2$, and $-\boldsymbol{I}_3$. Twins with rational K_1 and η_2 and irrational K_2 and η_1 (these two are not shown) are called *first-kind* or *reflection twins.*

2. Here η_1 and K_2 are rational, and K_1 and η_2 are irrational. We choose, in plane K_2 (Fig. 2.67 b), two vectors $\boldsymbol{I}_1$ and $\boldsymbol{I}_2$, and arrange vector $\boldsymbol{I}_3$ along the rational η_1 (K_1 and η_2 are not shown). Vectors $\boldsymbol{I}_1$, $\boldsymbol{I}_2$, and $\boldsymbol{I}_3$ form a unit cell which transforms to cell $-\boldsymbol{I}_1$, $-\boldsymbol{I}_2$, and $\boldsymbol{I}_3$ as a result of rotation through 180° about η_1. Such twins are called *second-type* or *rotation twins* (axial twins).

Table 2.8. Twinning elements and values of shear in different crystals

Crystal	Lattice	K_1	η_1	K_2	η_2	s
Fe, W, V, Nb, Mo, Cr	*Bcc*	(112)	$[\bar{1}\bar{1}1]$	$(11\bar{2})$	[111]	0.707
Cu, Ag, Au	*Fcc*	(111)	$[11\bar{2}]$	$(11\bar{1})$	[112]	0.707
	Hex					
Cd	$c/a = 1.886$	$(10\bar{1}2)$	$[\bar{1}011]$	$(10\bar{1}\bar{2})$	$[10\bar{1}1]$	0.171
Zn	$c/a = 1.856$	$(10\bar{1}2)$	$[\bar{1}011]$	$(10\bar{1}\bar{2})$	$[10\bar{1}1]$	1.140
Co	$c/a = 1.623$	$(10\bar{1}2)$	$[\bar{1}011]$	$(10\bar{1}\bar{2})$	$[10\bar{1}1]$	−0.130
Mg	$c/a = 1.623$	$(11\bar{2}1)$	$[11\bar{2}\bar{6}]$	(0001)	$[11\bar{2}0]$	0.614
		$(10\bar{1}2)$	$[\bar{1}011]$	$(10\bar{1}\bar{2})$	$[10\bar{1}1]$	−0.130
		$(10\bar{1}1)$	$[\bar{1}012]$	$(\bar{1}013)$	$[30\bar{3}2]$	0.137
Re	$c/a = 1.615$	$(11\bar{2}1)$	$[11\bar{2}\bar{6}]$	(0001)	$[11\bar{2}0]$	0.621
Zr	$c/a = 1.592$	$(10\bar{1}2)$	$[\bar{1}011]$	$(10\bar{1}\bar{2})$	$[10\bar{1}1]$	−0.169
		$(11\bar{2}1)$	$[\bar{1}\bar{1}26]$	(0001)	$[11\bar{2}0]$	0.63
		$(11\bar{2}2)$	$[\bar{1}\bar{1}23]$	$(11\bar{2}4)$	$[\bar{2}\bar{2}4\bar{3}]$	0.225
Ti	$c/a = 1.587$	$(10\bar{1}2)$	$[\bar{1}011]$	$(10\bar{1}\bar{2})$	$[10\bar{1}1]$	−0.175
		$(11\bar{2}2]$	$[\bar{1}\bar{1}23]$	$(11\bar{2}\bar{4})$	$[22\bar{4}\bar{3}]$	0.218
Be	$c/a = 1.568$	$(10\bar{1}2)$	$[\bar{1}011]$	$(10\bar{1}\bar{2})$	$[10\bar{1}0]$	−0.199
Graphite	$c/a = 2.722$	$(11\bar{2}1)$	$[\bar{1}\bar{1}26]$	(0001)	$[11\bar{2}0]$	0.367
Calcite $CaCO_3$	trig. *R*	(101)	[010]	(010)	[101]	0.694
Leucosapphire, Al_2O_3	ditto	$(10\bar{1}1)$	$[\bar{1}012]$	$(\bar{1}012)$	$[10\bar{1}1]$	0.202
Bi	ditto	(101)	[010]	(010)	[101]	0.118
β-Sn	tetrag., *I*	(301)	$[\bar{1}03]$	$(\bar{1}01)$	[101]	0.119
In	tetrag.	(101)	[101]	(101)	[101]	0.150
U	orthorhombic, *C*	(130)	$[3\bar{1}0]$	$(1\bar{1}0)$	[110]	0.299
		irrat.	[312]	(112)	irrat.	0.228
		(112)	irrat.	irrat.	[312]	0.228
		(121)	ditto	ditto	[311]	0.329
Plagioclase feldspar	tricl., *P*	(010)	ditto	ditto	[010]	0.151
		irrat.	[010]	(010)	irrat.	0.151

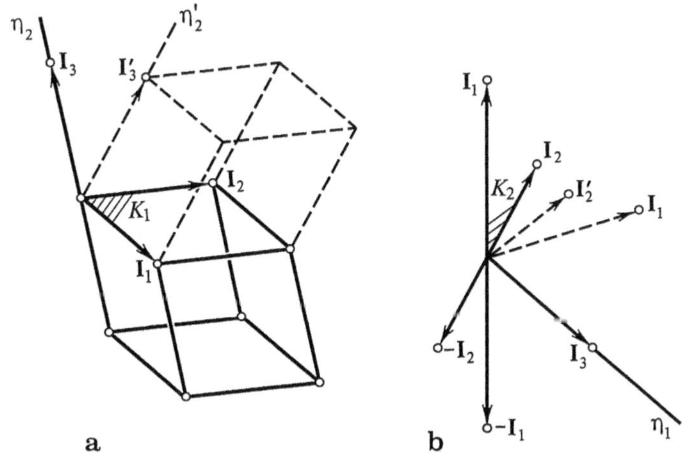

Fig. 2.67. Vector diagram of reflection (**a**) and rotation (**b**) twins [2.60]

In high-symmetry crystals all the elements K_1, K_2, η_1, and η_2 are rational, and the twins are simultaneously axial and reflection twins. These are called compound twins.

Among the compound twins one may encounter two twins with twinning elements K_1, K_2, η_1, η_2 and K_1', K_2', η_1', and η_2'; here $K_1' = K_2$, $K_2' = K_1$, $\eta_1' = \eta_2$, and $\eta_2' = \eta_1$. These are called conjugate twins.

2.5.6 Transformation of Indices of Planes and Directions in Twinning

If a number of lattice nodes formed a plane before twinning, after twinning they also lie in a plane, but the indices of this plane with respect to the axes of the twinned lattice become different. The same is true of directions. If (h, k, l) and (h', k', l') are the indices of a plane before and after twinning, we can write the following relationship between them:

$$
\begin{aligned}
h' &= 2H(Uh+Vk+Wl)-h(UH+VK+WL) \ , \\
k' &= 2K(Uh+Vk+Wl)-k(UH+VK+WL) \ , \\
l' &= 2L(Uh+Vk+Wl)-l(UH+VK+WL) \ .
\end{aligned}
\tag{2.126}
$$

Here $[UVW]$ and (HKL) are the indices of η_2 and K_1 for first-type twins, and of η_1 and K_2 for second-type. The indices of $[u\,v\,w]$ directions are transformed as follows:

$$
\begin{aligned}
u' &= 2U(Hu+Kv+Lw)-u(HU+KV+LW) \ , \\
v' &= 2V(Hu+Kv+Lw)-v(HU+KV+LW) \ , \\
w' &= 2W(Hu+Kv+Lw)-w(HU+KV+LW) \ .
\end{aligned}
\tag{2.127}
$$

2.5.7 Energy Conditions for the Formation of Twins Causing a Change in the Shape of Crystals

Let us compare the potential energy of a unit volume of a crystal in the initial and twinning orientation (we restrict ourselves to a homogeneous stress field, neglecting the appearance and displacement of twin boundaries). Suppose the crystal is under the effect of external stresses σ_{ik}. The crystal undergoes a strain ε_{ik}, which includes elastic strain $\varepsilon_{ik}^{\mathrm{el}}$. The elastic energy of the crystal is equal to the work spent on the elastic strain, which is equal to $1/2\,\sigma_{ik}\varepsilon_{ik}^{\mathrm{el}}$ (this is the area of the region bounded by the strain curve in coordinates $\sigma-\varepsilon$). The potential energy of the crystal in the field of external stresses σ_{ik} is equal to $-\sigma_{ik}\varepsilon_{ik}$, where ε_{ik} is the total strain of the crystal. The minus sign is due to the fact that the change in the potential energy of the system crystal minus the source of the external force, is a reduction in system energy because of the work of the external forces, while the elastic energy increases the crystal energy. The total potential energy of a unit volume of the crystal is equal to

$$W = \sigma_{ik}\left(\tfrac{1}{2}\varepsilon_{ik}^{\mathrm{el}} - \varepsilon_{ik}\right) , \tag{2.128}$$

which assumes summation over the repeating indices.

In the initial crystal orientation the total strain coincides with the elastic one, and therefore the total energy in this case is

$$W' = -\tfrac{1}{2}\sigma_{ik}\varepsilon_{ik}^{\mathrm{el}} = -\tfrac{1}{2}s'_{iklm}\sigma_{ik}\sigma_{lm} , \tag{2.129}$$

where s'_{iklm} are elastic compliances referred to the crystallographic axes of the initial crystal. For a crystal transformed to the twinning orientation the total strain is composed of permanent strain ε_{ik} corresponding to the change in specimen shape on twinning and elastic strain $\varepsilon_{ik}^{\mathrm{el}}$ described by elastic compliances s''_{iklm} of the twin lamellae. The total bulk density of the energy in the twin lamellae is equal to

$$W'' = -\sigma_{ik}\varepsilon_{ik}^{0} - \tfrac{1}{2}\sigma_{ik}\varepsilon_{ik}^{\mathrm{el}} = -\sigma_{ik}\varepsilon_{ik}^{0} - \tfrac{1}{2}s''_{iklm}\sigma_{ik}\sigma_{lm} . \tag{2.130}$$

By comparing the expressions for W' and W'', one can establish which of the orientations (initial or twinning) is more favorable in the field of stresses σ_{ik}. We set up the difference

$$\Delta W = W' - W'' = \sigma_{ik}\varepsilon_{ik}^{0} + \tfrac{1}{2}(s''_{iklm} - s'_{iklm})\sigma_{ik}\sigma_{lm} . \tag{2.131}$$

If $\Delta W > 0$, i.e., if $s'' > s'$, the potential energy minimum and hence the position of stable equilibrium is achieved in the twinning orientation, and the twinning component will increase in the field of stresses σ_{ik}. At $\Delta W < 0$, the twinning component will decrease. Since at low stresses σ_{ik} the linear dependence is stronger than the quadratic, for twinning with a change in shape the sign of ΔW at low stresses depends on the sign of the first term in (2.131), which depends linearly on the stress. Therefore, a change in stress sign reverses the effect, i.e., the favorable orientation becomes unfavorable (and vice versa), and the direction of the movement of the twin boundary also reverses. The twinning ellipsoid corresponding to tensor ε_{ik}^{0} determines the twinning elements.

The second term in (2.131), which is associated with the change in compliance, can be essential in twinning with a change in shape only at high stresses, when the elastic strains exceed twinning shear ε_{ik}^{0}. Here the direction of movement of the twin boundary (into the region with initial orientation or into the twinning interlayer) depends on the compliance of the crystal in two orientations.

2.5.8 Twin Nucleation

Mechanical twinning begins with the nucleation of a narrow twin interlayer in the crystal. The nucleation and growth of twin interlayers, as well as crystal growth, may proceed by different mechanisms. One of the possible ways to form twin interlayers is *layerwise (two-dimensional)* nucleation, when the nucleus of the twin interlayer is a single-layer stacking fault bounded by partial dislocations [Ref. 2.5, Chap. 5]. A further growth of the nucleus interlayer requires nucleation of new layers of stacking faults and their tangential growth along the interlayer due to the movement of partial dislocations. The nucleation of such interlayers is possible in regions where stresses sufficiently high for the appearance of stacking faults are concentrated. Most frequently it is damaged surface sites that serve as sources of nucleation of twin interlayers. If a twin interlayer has propagated throughout the crystal in a direction strictly parallel to the twinning plane, the boundary of the twin components is a geometric plane. If, however, some stacking fault layers were slowed down inside the crystal, the twin interlayer has a jogged boundary, and each jog ends in a *partial (twinning)* dislocation [Ref. 2.5, Chap. 5]. Figure 2.68 illustrates the structure of such a twin.

The formation of twins by two-dimensional nucleation requires high stresses (of the order of 0.1 G). At lower stresses, the source of twinning may include a partial dislocation in which one end is pinned in a triple node, and one or two other dislocations of the node must have a component normal to the glide plane of the moving dislocation (Fig. 2.69 a). Under the effect of the stress applied, this component will wind up into a conical spiral. The pitch of

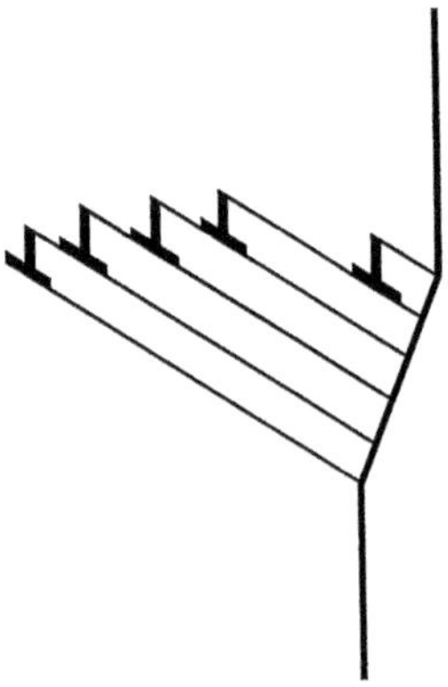

Fig. 2.68

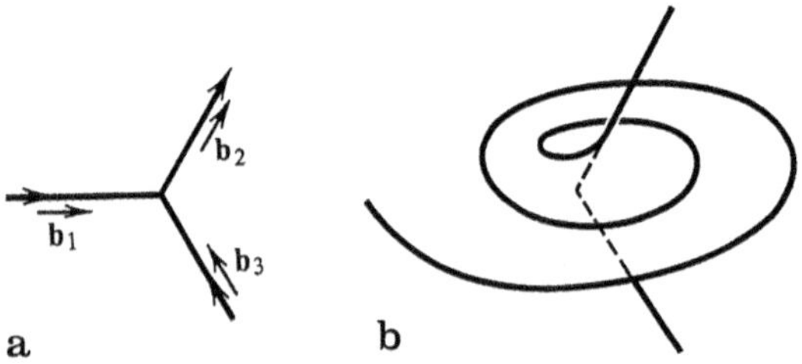

Fig. 2.69

Fig. 2.68. Twin interlayer formed as a result of two-dimensional nucleation

Fig. 2.69. Pole mechanism of nucleation of twin interlayers: **(a)** dislocation node, $\boldsymbol{b}_2$ and $\boldsymbol{b}_3$ have components perpendicular to the glide plane of the dislocation with Burgers vector $\boldsymbol{b}_1$; **(b)** action of dislocation with Burgers vector $\boldsymbol{b}_1$ as a helicoidal source

the spiral depends on the component of the Burgers vector of the dislocations forming the triple node, perpendicular to the glide plane of the moving dislocation, and each turn will produce a layer of the stacking fault (Fig. 2.69b). This mechanism of nucleation and growth of twin interlayers is called the *pole mechanism of twinning*. The minimum stress required for twin nucleation by the pole mechanism must offset stacking fault energy γ for the first period of operation of the source:

$$\sigma = \gamma/b \ .$$

There exist *elastic twin interlayers*, which disappear after the load is removed. But, in distinction to the ordinary elastic strain, elastic twinning does not obey Hooke's law and is a reversible plastic strain; the reversibility is due to the displacement of the twin interlayer by the elastic stresses stored in the crystal under deformation. As a rule, elastic twin interlayers are rather thin, no more than one micrometer in thickness, and they can be detected by means of transmission electron microscopy, while in transparent crystals they can be seen in a transmitted light owing to the appearance of interference bands of equal width (Fig. 2.70).

Twin boundaries can act as barriers for glide dislocations. Besides, intersection of glide dislocations with twin boundaries may be attended by the formation of twinning dislocations, which then penetrate into the twin interlayer.

Interaction of perfect and twinning dislocations may result in the formation of sessile (nonslipping) dislocations [Ref. 2.5, Chap. 5] which present a strong barrier for the movement of twinning dislocations and hence

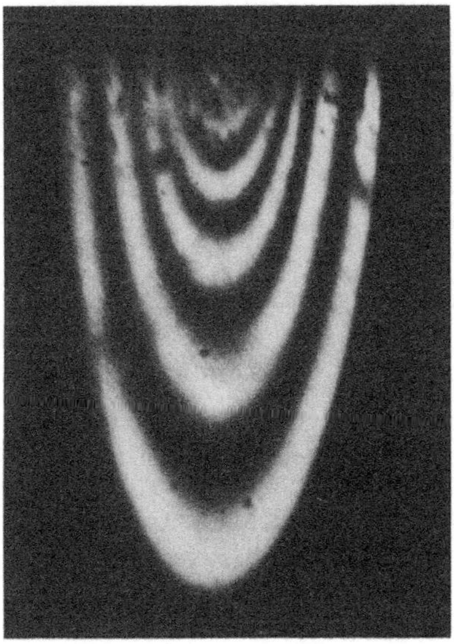

Fig. 2.70. Interference pattern on a twin interlayer of calcite [2.61]

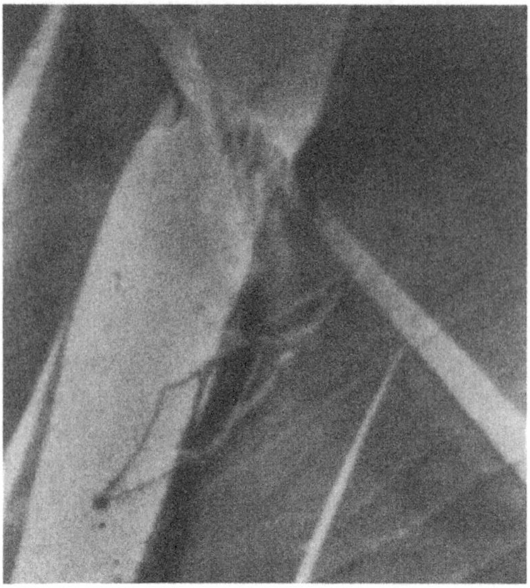

Fig. 2.71. Intersection of twin interlayers in a zinc crystal [2.62]

a source of failure, ordinary glide, or the appearance of a conjugate twin (a twin in the conjugate twinning system). An encounter of a perfect dislocation with a twin interlayer may cause the splitting of the dislocation into two twinning ones associated with a stacking fault.

If twins in two or more planes arise in a crystal, then, as the twin interlayers grow, they come in contact, and then penetrate each other. The appearance of an interlayer on the other side of the intersecting interlayer (intersection of twin interlayers) is the more difficult, the wider the intersecting interlayer. Sometimes, in the region of intersection of two interlayers, one of them absorbs the other (Fig. 2.71). Also, voids (channels), which are a source of failure, may appear along the line of intersection of the twin interlayers.

2.5.9 Twinning Without a Change in Crystal Shape

The foregoing referred to twinning which causes changes in crystal shape. In twinning without a change in shape, the very concept of residual deformation and hence the notion of the strain ellipsoid and of shear strain are inapplicable.

Let us consider the energy conditions for the formation of twins without a change in shape. Here tensor ε_{ik}^0 in (2.131) is equal to zero, and a stable orientation is shown by the twin component in which the compliance in a given stress field is the higher and, consequently, the stresses perform more work. If, however, the twin components have identical elastic constants, (2.131) is identically zero, and external mechanical forces cannot cause twin nucleation or a displacement of the twin boundary. Such twins arise only on phase transition. An example is inversion twins [Ref. 2.5, Chap. 5], characteristic, for instance, of triglycine sulphate, which thus cannot be mechanical twins.

The most typical example of twinning without a change in form is that of quartz (trigonal symmetry) according to the Dauphine law (which is named after the district of Dauphiné, France, where quartz crystals with such twins were first found). Here the components differ by a lattice rotation through 180° about the trigonal axis, which is equivalent to a change of sign of the other two principal crystallographic axes perpendicular to the optic axis. Such a transformation changes the sign of the components of the elastic compliance tensor among the indices of which those of the above two axes occur an odd number of times. We write out the matrix of the elastic compliances of quartz and mark these components:

$$\left\| \begin{matrix} s_{11} & s_{12} & s_{13} & s_{14} & 0 & 0 \\ s_{12} & s_{22} & s_{13} & -s_{14} & 0 & 0 \\ s_{13} & s_{13} & s_{33} & 0 & 0 & 0 \\ s_{14} & -s_{14} & 0 & s_{44} & 0 & 0 \\ 0 & 0 & 0 & 0 & s_{44} & 2s_{14} \\ 0 & 0 & 0 & 0 & 2s_{14} & 2(s_{11}-s_{12}) \end{matrix} \right\| .$$

Using this matrix, we write (2.131) for ΔW in the Dauphine quartz twins:

$$\Delta W = -2s_{14}[\sigma_{23}(\sigma_{11}-\sigma_{22})+2\sigma_{12}\sigma_{31}] \; . \tag{2.132}$$

Thus, the appearance of twins in quartz depends only on the tangential stresses acting on the basal plane and on the planes parallel to the optic axis. In spherical coordinates (Fig. 2.72), under uniaxial loading, (2.132) will take the form

$$\Delta W = -2s_{14}\sigma^2 \sin^3\theta \cos\theta \cos 3\varphi \; . \tag{2.133}$$

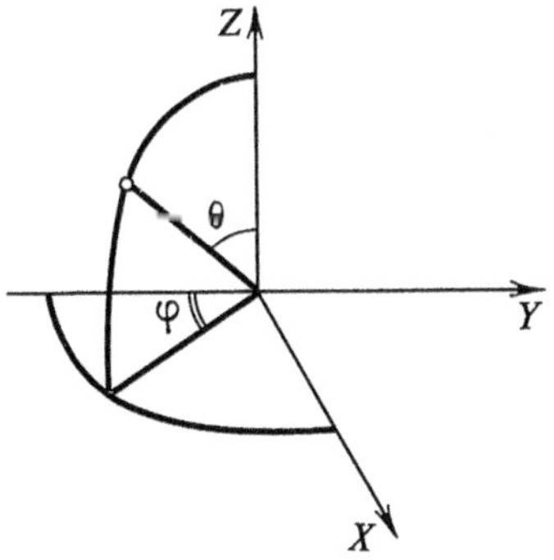

Fig. 2.72. Coordinate system adopted in calculations of the stereograms presented in Fig. 2.73

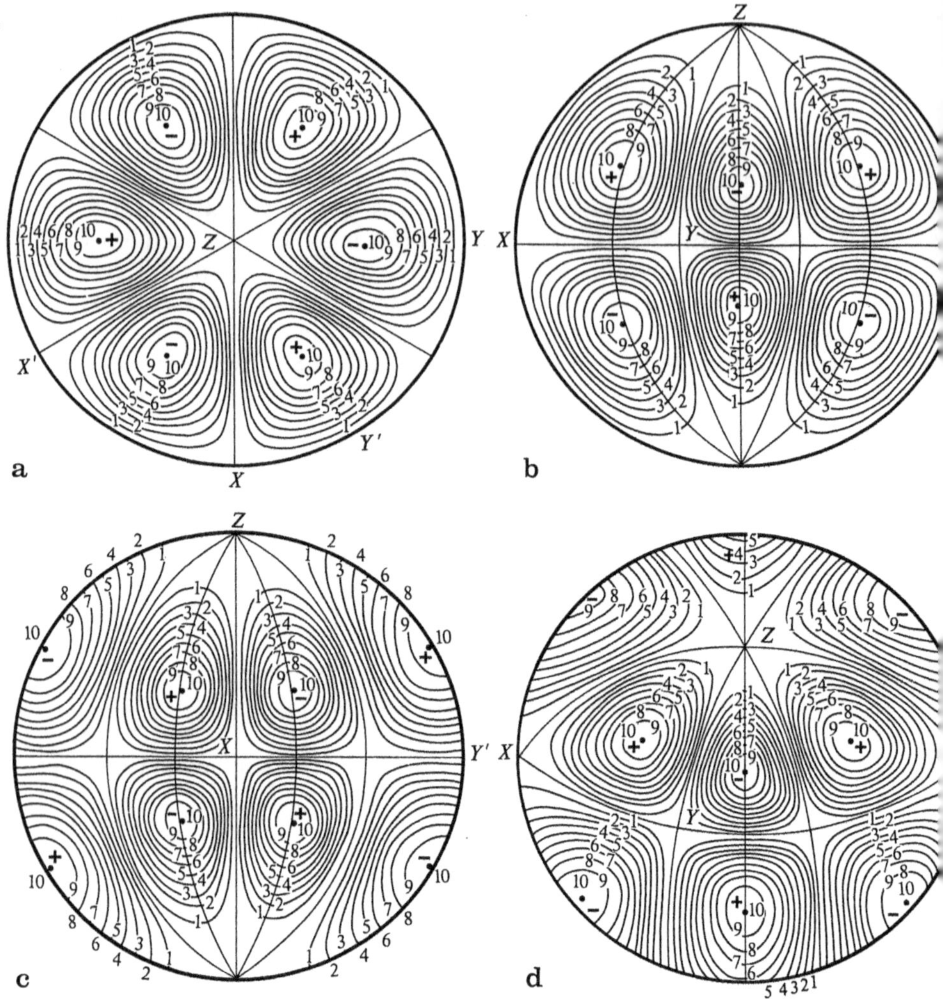

Fig. 2.73. Projections of stereograms of magnitudes $10\,\Omega$ (from 1 to 10) in quartz, onto the (0001), $(0\bar{1}10)$, $(\bar{2}110)$, and $(10\bar{1}1)$ planes (**a**, **b**, **c**, and **d**, respectively) [2.63]

For quartz, $s_{14}<0$; consequently, mechanical twinning is energetically favorable under uniaxial loading in directions for which

$$\Omega(\theta,\varphi)=\sin^3\theta\cos\theta\cos 3\varphi>0 \ .$$

Figure 2.73a–d gives the projections of stereograms of values $10\,\Omega$ onto different faces of the crystal. If the stress state differs just a little from uniaxial, Fig. 2.73 will give an approximate idea of the shape of the twin, because it fills regions with an equal degree of energy gain, i.e., regions with

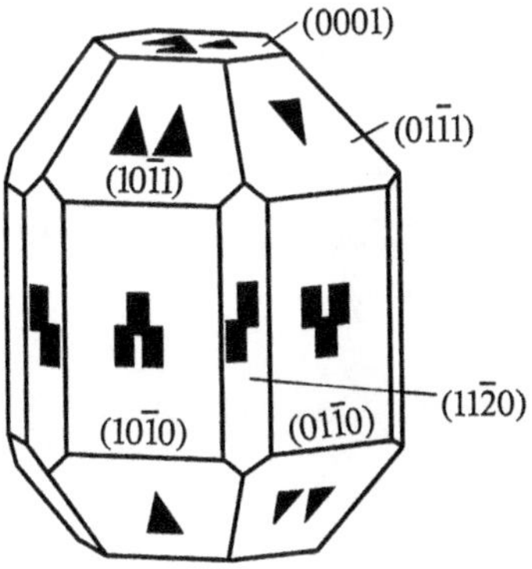

Fig. 2.74. Traces of mechanical twins on the principal faces of quartz (ball pressure) [2.64]

the same value of Ω. Figure 2.74 demonstrates twinning figures on different faces of quartz obtained under the pressure of a steel ball. Although for mechanical twinning by a concentrated load the stressed state can be considered uniaxial only very approximately, still there is a definite qualitative agreement between the stereograms and twinning figures.

Twinning in quartz can also be caused by torsion about the optic axis, when tangential stresses are induced in the basal plane. Torsion can also be used to remove twins from quartz, since they greatly deteriorate quartz products, in particular piezoelements. Experience shows that Dauphine twins are best removed by torsion about the (11$\bar{2}$0) axis [2.64].

If quartz crystals contain hydroxile groups, then during deformation these groups serve as dislocation sources. Besides twinning, usual gliding takes place in such crystals [2.65].

2.5.10 Behavior of Crystals with Polysynthetic Twins

Sometimes a deformation or, more often, a phase transition, results in the formation, not of one twin interlayer, but of a set of interlayers with alternating twinning and initial orientations. Such twins are called *polysynthetic*. Typical representatives of crystals tending to separate into polysynthetic twins are those of Rochelle salt, barium titanate, and other ferroelectrics [2.66]. In these, the components of polysynthetic twins are simultaneously ferroelectric domains. Mechanical effects, heating, or application of an electric field lead to the growth of domains of one sign at the expense of other domains [2.67, 68]. Polysynthetic twins in ferromagnetics are ferromagnetic domains; the vectors of spontaneous magnetization in neighboring twin components are oriented differently.

The trend towards polysynthetic twinning is also exhibited by some alloys (In-Cd, Cu-Mn, Cr-Mn, Au-Cd, In-Tl, etc.). Figure 2.75 shows polysynthetic twins in an In-Tl alloy. These appear under polymorphous transformation, as

Fig. 2.75. Polysynthetic twins in an In-Tl alloy (reflected polarized light) [2.69]

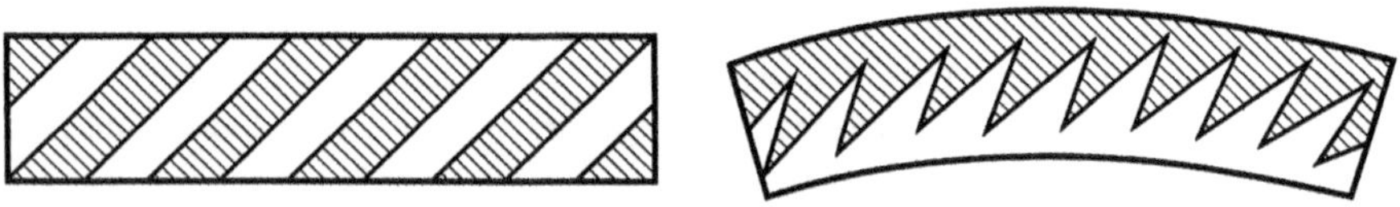

a b

Fig. 2.76. Elastic bending of an In-Tl crystal due to a shift of twin boundaries: (**a**) unstrained specimen, the interlayers have a platelet shape; (**b**) bent crystal, the twin components have acquired a wedgelike shape

in Rochelle salt. By straining twinned specimens it is possible to move twin boundaries and even make one component absorb the other.

The In-Tl alloy exhibits an interesting feature at low (below room) temperatures. If an In-Tl specimen is bent (stretched) and then the bending (or stretching) load is lifted without withdrawing the specimen from the cooling mixture, the specimen straightens out (contracts) by itself like rubber. This ability of restoring the initial shape is called *rubber elasticity*. The effect is stronger, the lower the temperature and the shorter the time of load action. Figure 2.76 depicts the shape of twin interlayers in specimen bending. The arrangement of the twin components here is in accordance with the distribution of elastic stresses: twin components of opposite sign are found on the extended and the contracted sides of the specimen. Rubber elasticity is due to the return of the twin boundaries to the initial position on unloading because of the insufficient stabilization of the new position of the twin boundaries after the deformation. If the In-Tl alloy with polysynthetic twins is strained at a higher temperature (about 20 °C), the boundaries of the twin components corresponding to the strain state manage to stabilize (evidently owing to the deposition of impurities or other defects on the boundaries), and the new distribution of the twin components is preserved after the load is removed (permanent deformation).

2.5.11 Martensitic Transformations

Twinning may also occur under the effect of stresses arising on phase transitions of a definite type. If steel is quenched from the temperature at which it is austenite (i.e., has an fcc lattice), it suddenly undergoes transformation into a very hard body-centred tetragonal phase, called *martensite.* Figure 2.77a presents martensite crystals in a Fe-Ni-Mn alloy, and Fig. 2.77b shows the scaled-up internal structure of one of the crystals, which is a polysynthetic twin.

The phase transition described is not associated with diffusion. Since a number of features of this transition (high rate, change in the shape of the crystal or of the region subjected to transition) are also inherent in other phase transitions which do not involve diffusion, diffusionless phase transitions are called *martensitic.* Martensitic transformations are characterized by a definite orientation relationship between the lattices of the initial and final phase. The most typical form of the new phase is a platelet with a low ratio of the thickness to the other linear dimensions. The habit plane of such a platelet (its flat faces or the middle plane, if the faces are strongly grooved) has a definite orientation with respect to the crystallographic axes of the maternal and the final phase. The transition changes the shape of the transformed region,

Fig. 2.77. Martensitic crystals in a Fe-33% Ni alloy: (**a**) shape and mutual arrangement of the crystals (× 500), (**b**) polysynthetic twin in one of the crystals (× 10000) [2.19]

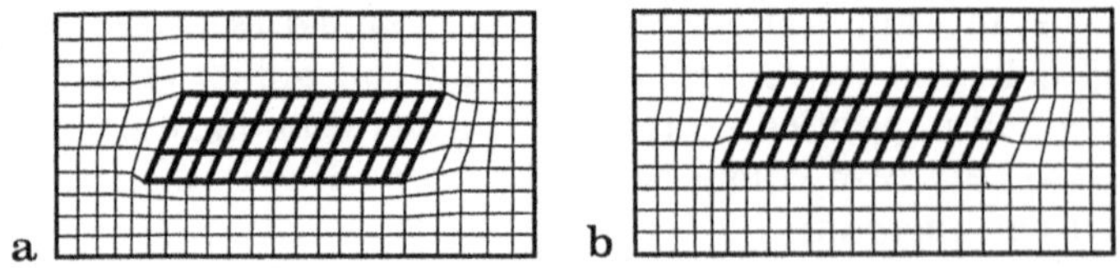

Fig. 2.78. Coherent (a) and incoherent (b) martensite crystals [2.19]

which leads to the appearance of a characteristic relief on the specimen surface.

Martensite crystals are formed not only in the course of heat treatment but also by plastic deformation at a definite (for each material) temperature. They nucleate at stress concentration sites (near dislocation pileups, interlayers, faults, and crack tops). Martensite crystals grow by consistent shifting of atoms to new positions, so that atoms which were neighbors in the initial lattice remain so in the new one. To preserve the continuity of the crystal lattice, as the nucleus grows, the stresses arising in the matrix as a result of a change in the size and shape of the nucleus must undergo continuous redistribution. At the interface of two phases in an ideal nonplastic material, one lattice continuously transforms into the other, i.e., there is a coherent relationship between the lattices of the initial and new phase (Fig. 2.78a). In actuality, however, the formation of martensitic phases is, as a rule, accompanied by the generation and movement of dislocations, and at the interface, dislocations (interphase dislocations) disturbing the coherence (Fig. 2.78b) appear.

Plastic deformation (by glide or twinning) in the parent crystal or in that of the new phase dissipates part of the elastic energy, so that the transformation proceeds irreversibly. The transformation is accompanied by an increase of the stored energy (elastic, surface energy of crystals and defect energy) and of the "friction" associated with the irreversible energy dissipation in the movement of interphase boundaries and dislocations. This phenomenon is similar to strain hardening (Sect. 2.2).

2.6 Fracture

2.6.1 Theoretical and Actual Strength

Plastic deformation in crystals develops up to a certain limit. After reaching the critical value for a given material and given strain conditions the specimen begins to disintegrate. The ability of retaining their integrity under the effect of external mechanical loads characterizes the *strength* of solids.

According to the classical theory of strength, fracture must take place after the external stresses become sufficient for breaking the bonds between

Table 2.9. Tensile strength of some crystals

Material	σ/E (%)	Ref.
Si crystals at room temperature	2	[2.70]
NaCl crystals in water (Joffe effect)	3.6	[2.71]
Si whiskers	3.6	[2.72]
Fe whiskers	4.9	–
Cu whiskers	2.8	[2.73]
Ag whiskers	4.0	–
NaCl whiskers	2.6	[2.74]
Zn crystals elongated perpendicular to the basis	0.1 – 5	[2.75]
Quartz whiskers	10 – 25	[2.76, 77]
Molten quartz	19	[2.78]

the atoms ($\sim 0.1\,E$). This value of the stress is called the theoretical strength. After it is achieved, the body must break up explosively into separate atoms or atomic layers perpendicular to the deformation axis. In reality mechanical fracture occurs much more quietly: the body disintegrates into a small number of parts at a stress which is usually much lower than the theoretical strength. The experimental material accumulated so far has made it possible to explain the discrepancy between the theoretical and real strength and to develop concepts of the mechanism of fracture of crystalline materials.

It should, however, be noted that a near-theoretical strength is still achievable in practice. As with the theoretical shear strength (Sect. 2.2), the theoretical tensile strength is achieved either in defectless crystals (for instance, in whiskers [2.15]) or in such imperfect crystals where the dislocation mobility is suppressed. While the theoretical tensile strength is close to 10 – 20% of Young's modulus, in practice a strength of the order of several percent was obtained, and 20 – 25% for quartz. Table 2.9 lists examples of high-strength crystals.

2.6.2 Cleavage Plane

Experiment shows that crystals break, as a rule, along definite crystallographic planes – *cleavage planes*. These planes are characterized by the minimum cohesion in the direction perpendicular to them.

Belov and *Klassen-Nekhlyudova* [2.79] derived some rules for determining the cleavage plane by analyzing the relationship between the structure of crystals and the cleavage plane. For homeopolar crystals in whose lattice each atom is in a symmetric environment and all the surrounding atoms are equidistant from it (diamond, sphalerite, würtzite, etc.), (1) the cleavage planes never pass between neighboring lattice planes, one of which consists of atoms of metals and the other of nonmetals; (2) cleaving propagates between the farthest-spaced planes, because the binding forces rapidly decrease with the distance.

For homeopolar crystals in which each atom is characterized by several coordination numbers (short- and long-range, graphite, molybdenite), the rules are formulated thus: (1) There is an equal number of positive and negative components between crystallographic planes along which the cleaving develops; in other words, cleavage planes bound whole molecules (for hetero- and homeopolar crystals the molecule concept is purely conventional). (2) The cleavage plane never disturbs the "short" bonds due to short-range atomic interaction.

In ionic crystals containing no radicals (rock salt, fluorite, cuprite, rutile, etc.), (1) a cleavage arises when the lattice contains neighboring pairs of parallel nets, each of which is formed by large anions and there are no cations between the two nets of the pair; (2) the cleavage is more pronounced (perfect), the larger the distance between the nets of such a pair; and (3) the most perfect cleavage is characteristic of the closest-packed planes, i.e., of planes farthest removed from one another.

In radical-containing ionic crystals (calcite, aragonite, barite, topaz, olivine, beryl, etc.), (1) the cleavage plane never separates radicals; (2) splitting along the cleavage plane occurs between pairs of planes consisting of anions, and only in the absence of such planes is the cleavage determined by the cation planes; (3) if two probable systems of cleavage planes follow from a structure, two alternatives exist: (a) if the distances between the relevant pairs of anionic planes are unequal, only a cleavage between the farthest anions arises; (b) if the distances between pairs of anionic planes are approximately equal for several families of planes, then a cleavage arises along all these planes.

Experience shows, however, that fracture does not always occur along cleavage planes. Breaks along glide and twinning planes, block boundaries, and faults are also observed.

The property of some crystals to cleavage along a system of planes whose position is controlled not by zones of weakened bonds of the ideal crystal lattice, but by the peculiarities of its distortions, is called *parting*. Examples are smooth breaking along $(10\bar{1}1)$ and $(11\bar{2}0)$ planes in crystals of synthetic corundum. Investigations show that parting along the indicated planes is associated with boundaries of twin interlayers. Smooth breaking along the (0001) plane in corundum crystals occurs along growth zones weakened by deposition of submicroscopic impurity particles.

2.6.3 Fracture Surface

In principle, during the splitting of a perfect crystal strictly along the cleavage plane the cleavage surface must be ideally smooth. In actuality this surface practically always shows irregularities. They mostly consist of cleavage steps due to the fact that a crack propagates not along a single plane, but along planes situated at different levels because of the presence of structure defects

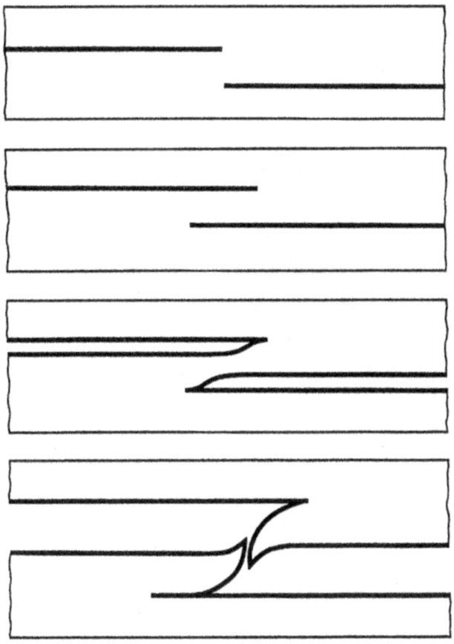

Fig. 2.79

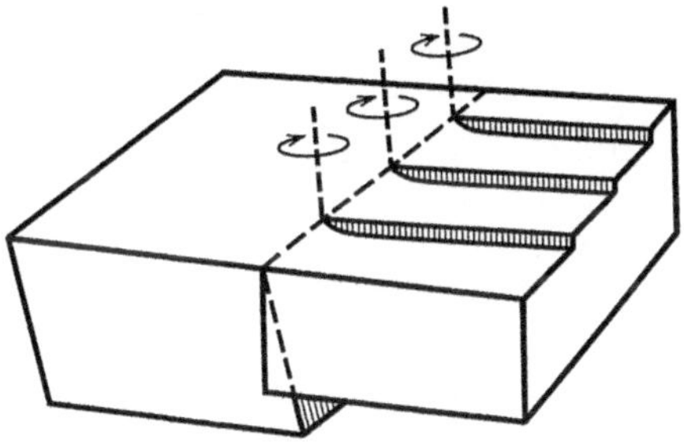

Fig. 2.80

Fig. 2.79. Formation of a cleavage step between two parallel cracks propagating at different levels

Fig. 2.80. Formation of cleavage steps on screw dislocations. The series of screw dislocations corresponds to the boundary of blocks turned relative to one another about an axis perpendicular to the boundary (torsion boundary) [2.80]

or under the effect of a cleaving force acting not strictly along the cleavage plane. A cleavage step appears at the intersection of cracks developing at different levels (Fig. 2.79). The formation of cleavage steps requires an additional energy, and therefore the cracking is impeded.

Cleavage steps are mostly caused by screw dislocations intersecting the cleavage plane. After the crack front passes through a screw dislocation the crack propagates at two different levels connected by a cleavage step with a height of the order of one Burgers vector.

If the crack intersects a block boundary consisting of a screw dislocation net, a series of cleavage jogs arises (Fig. 2.80). As the cracks continue to develop, groups of closely spaced jogs detach themselves, forming macroscopic steps. As a result, characteristic patterns, which resemble creeks joining into a major stream, are formed. They are called *creek patterns*. Figure 2.81 pictures a typical example of creek patterns on a cleavage plane of a crystal of Rochelle salt, where the "creeks" originated at the boundary of misoriented blocks.

Edge dislocations at the tilt boundaries [Ref. 2.5, Chap. 5] do not generate cleavage steps; they only change the direction of propagation of the crack.

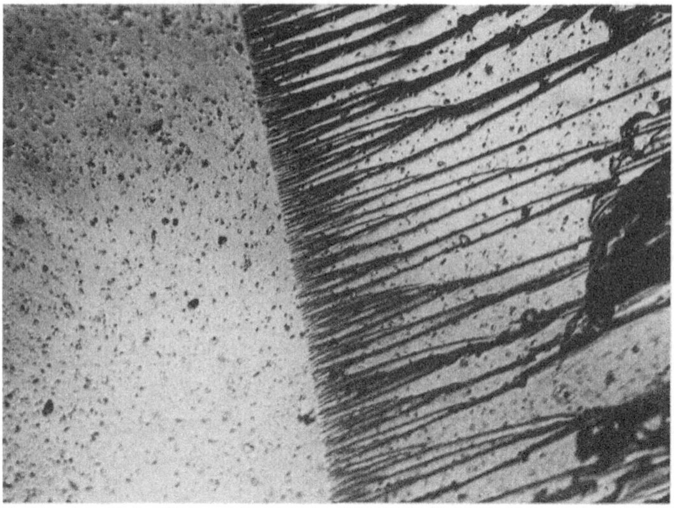

Fig. 2.81. "Creek" patterns of cleavage steps formed at the boundary of misoriented blocks in a Rochelle salt crystal; the beginning of each step corresponds to the emergence of a screw dislocation at the crystal surface [2.80]

2.6.4 Law of Critical Normal Stress

Since a fracture occurs predominantly along definite crystallographic planes (of cleavage or parting), the rupture of differently oriented specimens requires different efforts. Similar to the law of critical shear stress, which determines the stress at the beginning of the slide, there is the *law of critical normal stress.* This law states that a crystal breaks up when the critical normal stress is achieved in the direction perpendicular to some cleavage or parting plane. If N is the critical normal stress and χ is the angle between the rupture plane and the direction of tension, then stress σ which is required for fracture is defined by the formula

$$\sigma = N/\sin^2\chi \ . \tag{2.134}$$

The critical normal stress is called the ultimate strength, which, like the yield point, is not a constant of a material. Both these quantities depend on the experimental conditions: the temperature, the strain rate, the type of loading, the environment, etc. One of the causes for strength instability is also the thermal motion of atoms in solids, which changes the local interatomic bonds. At places with weakened bonds, cracks arise, whose development leads to fracture. Plastic deformation plays a still more important role in fracture.

2.6.5 Role of Plastic Deformation in the Preparation of Failure Nuclei

Some crystals do not change their pre-fracture dimensions perceptibly (quartz, chromite, etc.). Such crystals are said to be brittle. However, these crystals, too, show microscopic traces of plastic strain when subjected to thorough inspection. It turns out that absolutely brittle materials are almost nonexistent, and plastic strain may not only accompany the fracture (i.e., develop simultaneously with it in the regions surrounding the crack), but also, as first indicated by Stepanov, cause crack nucleation, if the deformation is nonuniform [2.81, 82].

The nucleus cracks are divided into force and geometric ones, according to the mechanism of their formation (Fig. 2.82) [2.56, 83]. In force cracks, the discontinuity is purely brittle; the atomic planes have detached themselves from one another without any traces of plastic strain (Fig. 2.82a). The energy expenditure for the propagation of such a crack amounts to the overcoming the adhesion forces. The formation of geometric cracks, as distinct from force cracks, involves plastic strain, which leads to sinking of lattice defects into the crack or to their ejection from the crack. In the simplest case, a geometric crack (Fig. 2.82b) corresponds to the trapping of one edge dislocation by a force crack, which requires additional energy expenditure. Owing to the trapping of the dislocation, the crack tip rounds off, and the crack becomes more stable.

The rounding of the crack tip due to the plastic strain may also result from the ejection of rows of dislocations along the glide plane, when a dislocation cross forms at the crack tip (Fig. 2.82c). Selective etching has shown that such crosses form precisely when the crack ceases to propagate.

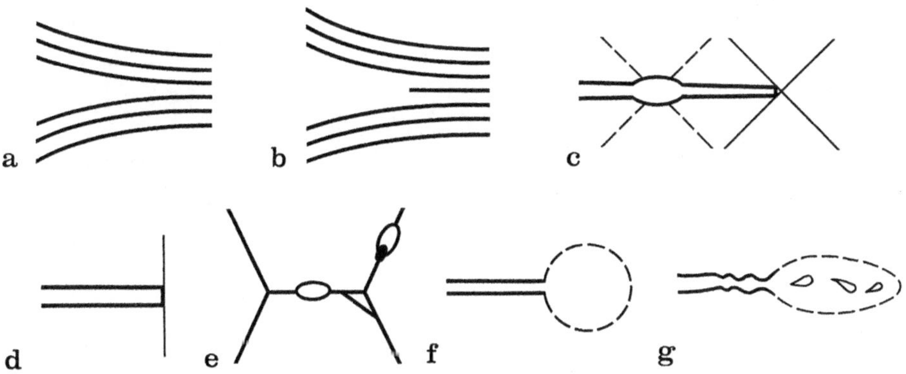

Fig. 2.82. Schematic structure of cracks: **(a)** pure force crack (the atomic planes framing the crack are shown); **(b)** geometric crack corresponding to the trapping of the dislocation by the crack **(a)**; **(c)** dislocation cross at the place of stoppage of the tip of the geometric crack; **(d)** geometric crack with a constant opening angle ensured by the plastic deformation at the tip; **(e)** pores at the grain boundary; **(f)** crack with a plastic zone; **(g)** countercracks in the plastic zone of the main crack [2.56]

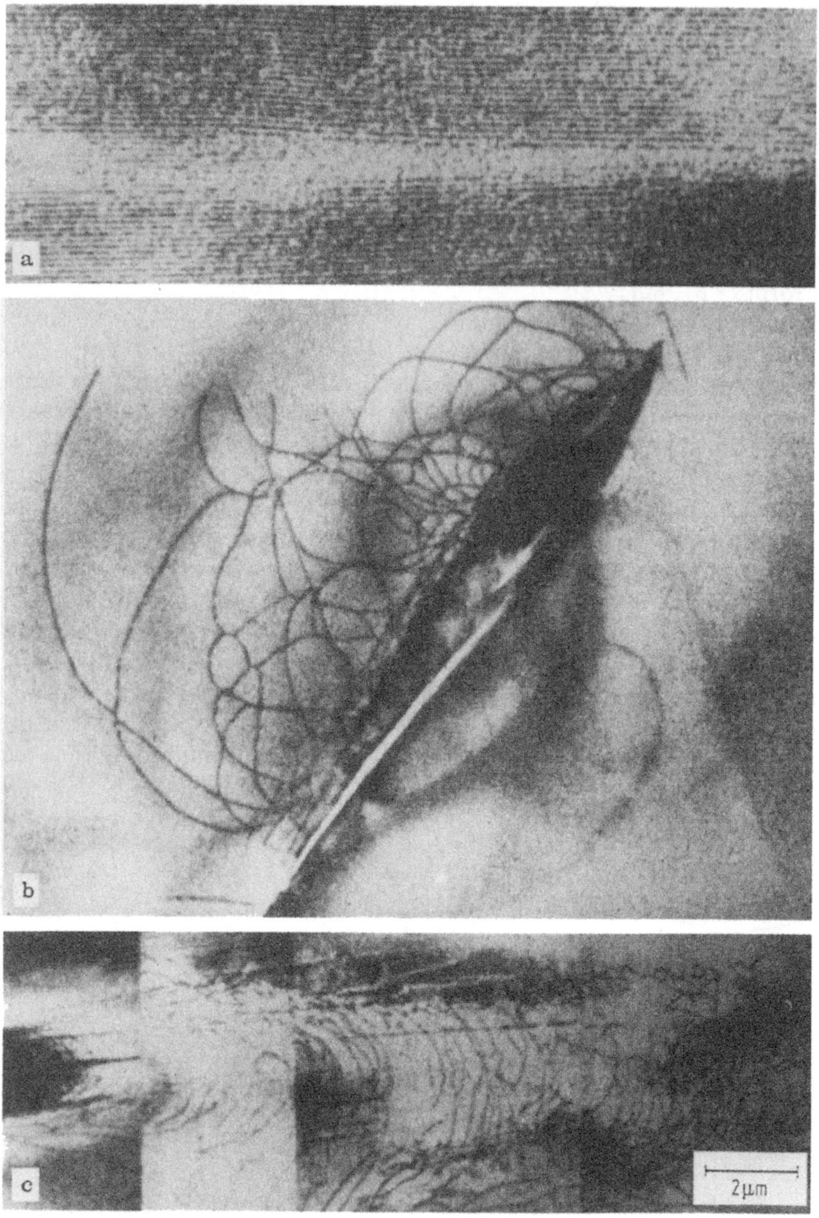

Fig. 2.83. Electron micrographs of cracks in **(a)** copper phthalocyanide, **(b)** silicon, and **(c)** iron [2.85]

When the material plasticity increases, the plastic strain at the crack tip finally turns it into a purely geometric one, which is characterized not by detachment, but by plastic separation of the crack edges (Fig. 2.82d).

A phenomenon similar to the formation of fracture nuclei is the appearance of pores at grain boundaries (Fig. 2.82e). The widening of such pores may be due either to the diffusion of vacancies from the bulk of the grains or along their boundaries, or to the near boundary glide of dislocations on slipping of the grain boundaries. If the material is highly plastic, the plastic zones at the crack tip reach macroscopic dimensions (Fig. 2.82f). Energy expenditures for the propagation of such a crack depend on the work required for displacement of the plastic zone consisting of a large number of dislocations. This work greatly exceeds the surface energy of the material. Plastic zones of macroscopic size develop independently to a large extent. New cracks often nucleate there (Fig. 2.82g) and then join together and with the master crack [2.84]. Figure 2.83 gives photos of several types of crack. In the case of copper phthalocyanine there are no dislocations at all; in low-plasticity silicon one can see a limited number of them; and in plastic crystals of iron a plastic zone consisting of a large number of dislocations is visible in front of the crack.

Crack nucleation in crystals always involves the participation of other lattice defects. Many mechanisms of crack nucleation on dislocation pileups, dislocation walls, and other dislocation aggregates are known. The corresponding diagrams are given in Fig. 2.84.

Figure 2.84a outlines the shape of a crack resulting from dislocation pileup at an obstacle. Obstacles capable of slowing down whole groups of dislocations include sessile dislocations, dislocation dipoles (pairs of parallel dislocations with opposite Burgers vectors), low-misorientation-angle block boundaries, grain boundaries, and other defects. In pileups, dislocations are so arranged under the effect of external and interaction forces that the

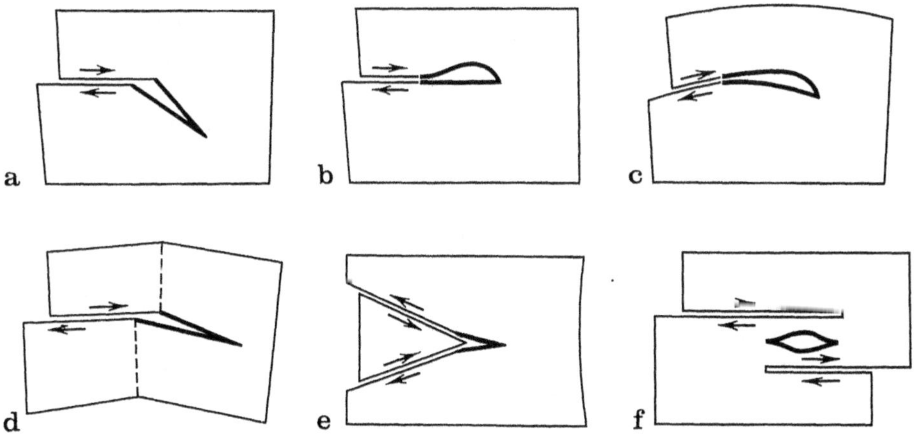

Fig. 2.84a–f. Various versions of crack nucleation [2.86]

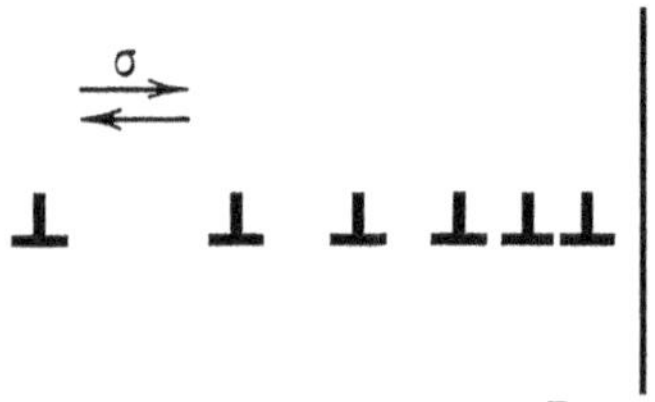

Fig. 2.85. Flat pileup of dislocations

distance between neighboring dislocations increases as they recede from the pileup (obstacle) head. Figure 2.85 shows schematically the distribution of dislocations in a pileup. In the head of a dislocation pileup, a stress concentration arises as a result of addition of stresses from all the dislocations of the pileup. Under the influence of an external stress, the number of dislocations in the pileup (and hence the stress) increases, and when the local stresses exceed the ultimate strength, a crack appears. Such a scheme of crack nucleation was proposed by *Zener* [2.87] and *Mott* [2.88]. According to this scheme a crack nucleates in front of a dislocation pileup and develops in the plane with the maximum normal stress.

Experience shows, however, that in crystals with one active glide system a crack nucleates not in front of a pileup, but in its very head, and propagates along the plane containing the dislocation pileup (Fig. 2.84b). The appearance of such cracks is attributed to the shear along the curved glide planes (Fig. 2.84c). The curving of the glide planes is due to other dislocations in the crystal, and the curvature radius of the glide planes is proportional to the dislocation density. A shear along a curved glide plane inevitably results in the appearance of normal stresses acting on the glide plane and, as a consequence, in the detachment of sliding surfaces.

The limiting case of the last scheme is a rupture of the subgrain boundary (Fig. 2.84d) either because the boundary is intersected by the active glide plane, or due to the stoppage of part of the moving boundary, or, finally, because the boundary is divided into parts when its orientation changes. Broken-off vertical rows of dislocations, as well as horizontal pileups, have increased stresses at their ends which are sufficient for crack nucleation. Fracture according to this scheme is observed in crystals in which the glide and cleavage planes coincide.

Dislocation pileups at the place of intersection of glide planes may cause a crack according to the scheme of Fig. 2.84e (examples are MgO crystals). Failure may also occur according to the scheme of Fig. 2.84f, when a nucleus crack arises in the region where tensile stresses from two dislocation pileups of opposite sign are superimposed.

Other causes for crack nucleation are also possible, of course. Special attention should be given to the fracture of high-plasticity materials in fatigue testing, i.e., when the crystal is subjected for a long period to sign-variable stresses with a strain amplitude less than the yield point in static testing.

Fatigue cracks nucleate on the surface at the very beginning of plastic deformation and gradually grow into the crystal. The nuclei of fatigue cracks are disposed in glide bands, namely, in surface depressions resulting from successive sliding in two intersecting glide bands.

2.6.6 Griffith's and Orowan's Criteria of Growth of Nucleus Cracks

Fracture (cleaving) is accompanied by the formation of two free surfaces, which requires considerable energy expenditure. Therefore cleaving, as well as gliding or twinning, cannot embrace a large surface area instantaneously. A crack develops gradually from a localized nucleus formed in one of the indicated ways. Cleaving can be regarded as a process as a result of which crack edges move apart to an approximately interatomic distance.

The specific energy of the surface formed on cleaving is equal to

$$2\gamma \sim Gb/5 \ . \tag{2.135}$$

Moving apart of the edges of a nucleus crack requires a tensile stress

$$\sigma \sim 2\gamma/b = G/5 \ . \tag{2.136}$$

This is a very high stress, which exceeds even the theoretical ultimate shear strength (0.1 G). Therefore, cleaving is possible only at high stress concentrations. As soon as a crack appears, it becomes a stress concentrator. But only large cracks can create a sufficiently high stress concentration.

The substantial role of the crack size was first pointed out by *Griffith* [2.89]. He considered stress σ_m at the ends of a flat ellipsoidal crack of diameter L in the field of tensile stress σ perpendicular to the crack. This stress is equal, in the order of magnitude, to

$$\sigma_m = \sigma\sqrt{L/r} \ , \tag{2.137}$$

where r is the distance to the end of the crack beyond its limits. If L is not too small, then, at a distance of the order of several interatomic distances from the end of the crack, σ_m may be about 100 times higher than the applied stress and thus reach the theoretical value of the strength. Then the crack will grow.

Having calculated the work spent on the elastic strain of the specimen and the strain associated with the formation of the crack, Griffith obtained the expression for the fracture stress:

$$\sigma_m = (4E\gamma/\pi L)^{1/2} \ , \tag{2.138}$$

where γ is the energy of formation of a unit surface, and E is Young's modulus.

From (2.138) follows the critical size of a crack capable of widening at a given stress:

$$L \gtrsim 4E\gamma/\pi\sigma^2 \ , \tag{2.139}$$

or (if we express E in terms of G and ν in accordance with (2.30))

$$L \gtrsim \frac{2\gamma G}{\alpha(1-\nu)\sigma^2} \ . \tag{2.140}$$

The indicated estimate of the critical size of a crack is called Griffith's criterion. Here we have considered the macroscopic theory of cracks. Griffith's criterion in the microscopic theory of cracks is discussed by *Blekherman* and *Indenbom* [2.90].

If we insert the experimental values of fracture stresses into (2.139, 140), the length of a nucleus crack will be found to exceed the specimen size severalfold. It follows that cracking involves some other process. In *Orowan*'s opinion this process may be the plastic deformation [2.91], and Griffith's criterion holds true only for ideally elastic bodies.

Suppose there is a nucleus crack on the surface of the specimen, the size of the crack being less than the critical one for the given stress. But this stress is sufficient for shear strain. Let the shear develop from point P (Fig. 2.86). If the dislocation approaches the crack end B, high stresses will arise, and the crack will grow up to point A. In the presence of a glide band this process is repeated, and the crack may gradually reach the critical size. Thus, a crack develops owing to the plastic strain, and γ in (2.140) should be replaced by the work p spent on the formation of a crack under a plastic strain. The criterion for estimating the size of a nucleus crack proposed by Orowan has the form:

$$L \geqslant \frac{2pG}{\alpha(1-\nu)\sigma^2} \ . \tag{2.141}$$

Quantity p is called the effective surface energy. It includes the true surface energy of the crack and the energy of the plastic strain near it.

It should be mentioned that the role of the plastic strain in cracking may be dual. The basic mechanism of growth of a crack is often the input of dislocations to it. Then the plastic strain determines the kinetics of the crack growth. On the other hand, dislocations near the tips of a crack often retard its propagation, rounding them out and thereby reducing the stress concentration. Similarly, the interfaces (subgrain and grain boundaries) may occasionally be channels of easy growth of cracks, and in other cases an obstacle for their propagation.

2.6.7 "Viscous" and "Brittle" Cracks

Stresses concentrated near a crack tip can be reduced by a plastic deformation: the movement of neighboring dislocations or the nucleation of new dis-

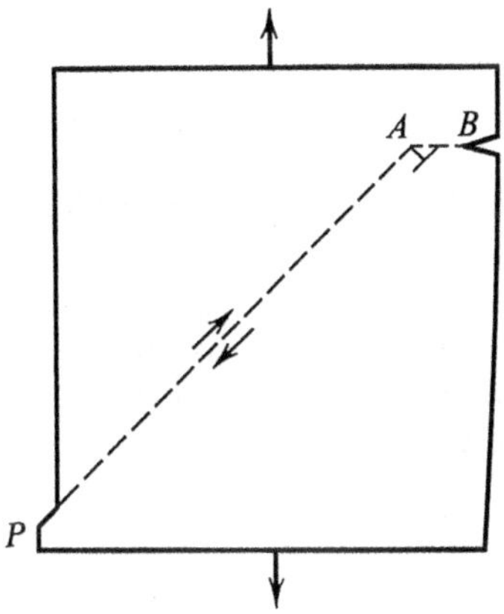

Fig. 2.86

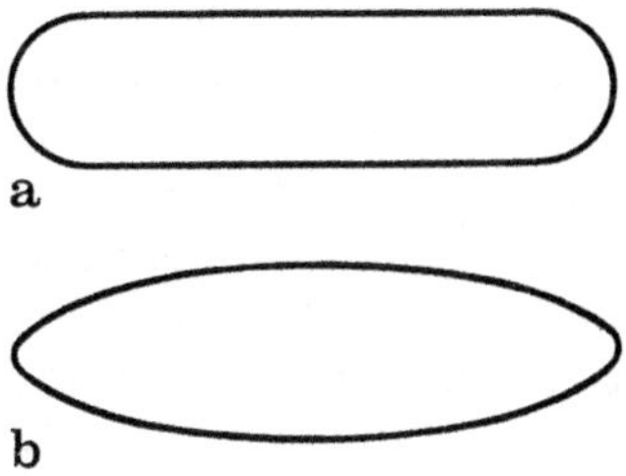

Fig. 2.87

Fig. 2.86. Development of a crack due to participation of glide [2.91]

Fig. 2.87. Shape of (**a**) a "viscous" crack and (**b**) a "brittle" crack

locations. In highly plastic materials the stress concentration at the crack tip is usually sufficiently high for activation of neighboring Frank-Read sources. If sliding on several glide systems is possible, one can pick out, near the crack tip, a cylindrical region with a sufficiently large radius, where the elastic stresses created by the crack are nearly completely relaxed. Then the dislocation sources in the above region emit loops so that some part of each loop is in the immediate vicinity of the crack tip. The resultant Burgers vector of these dislocations fully offsets the Burgers vector of the crack. A stress relaxation facilitates the accumulation of cleavage dislocations, as a result of which the crack tip rounds off, and the crack stabilizes and ceases to develop. The curvature radius at the crack tip becomes equal in the order of magnitude to the halfwidth of the crack. The edges of such cracks are parallel.

In more brittle materials the stress relaxation due to the plastic deformation is only possible by squeezing new dislocations out of the crack tip and only if the crack tip moves slowly. The rounding of the crack tips in such cases is much less pronounced, and such cracks are less stable than in high-plasticity materials; the width of the crack and its shape in the middle portion remain practically unaltered. Figure 2.87 is a schematic representation of "viscous" and "brittle" cracks.

2.6.8 Joffe Effect

If we set up obstacles to the development of cracks or at least slow down their growth, we may hope to obtain a near-theoretical strength. For instance, the theoretical strength of NaCl crystals is 200 kgf/mm^2, while in practice it usually does not exceed 0.5 kgf/mm^2. *Joffe* et al. [2.71] approached the theoretical strength of NaCl by deforming crystals and simultaneously dissolving their surface with water. Thus they succeeded in increasing the strength up to 160 kgf/mm^2.

An increase in strength due to dissolution of the surface in the course of deformation is called the Joffe effect. This effect has also been observed for metals [2.75]. The increase in tensile strength is attended by an increase in total plastic deformation. The plasticity does not increase owing to the reduction of the yield point, which remains unaltered, but does owing to the prevention of brittle fracture. According to the experimental data presently available, the Joffe effect is due to the washing off of various "harmful factors," such as surface microcracks and dislocation sources blocking each other.

2.6.9 Rebinder Effect

Discovered in 1928, Rebinder effect, as well as the Joffe effect, has to do with control of the strength of materials. The effect consists in acceleration of fine division of various solids in certain media. In distinction to corrosion, in the Rebinder effect the initial mechanical properties of solids are restored immediately after the removal of the active medium, i.e., physicochemical reversibility of the process takes place. According to the concepts of Rebinder's school, an active medium, being adsorbed on the surface of a given solid, penetrates into the surface microdefects and reduces the effort necessary for overcoming the atomic forces of adhesion. Thus, a reduction in strength is due to the decrease in surface energy (surface tension) at the interface between the failing solid and the environment. The effect of the medium is always manifested simultaneously with the mechanical load. The essence of adsorption reduction in strength is that the adsorption must occur simultaneously with bond rupture.

Surface-active media facilitate not only the failure but also the decrease in strain resistance (adsorption plasticizing) because of the decrease in the work of emergence of the dislocations at the surface. With high-plasticity materials, the adsorption increase in plasticity may even lead to increased strength [2.93].

2.7 Methods for Studying the Mechanical Properties of Crystals

The mechanical properties of crystals are evaluated by mechanical testing, which reveals certain mechanical characteristics. Although in almost any method of testing, crystals manifest elastic plastic, and strength properties, one can select deformation conditions under which some of them become most prominent. This section describes the mechanical testing methods which are usually employed in studying the mechanical properties of crystals.

2.7.1 Hardness Measurement

The fastest and simplest type of mechanical testing is hardness measurement. The hardness is not a physical constant: its value depends not only on the material studied, but also on the measuring conditions. The hardness is determined by a number of mechanical properties (elastic limit, elasticity modulus, plasticity, strength, etc.), but the role of the various mechanical characteristics is different under different conditions of hardness measurement. Therefore, the correct selection of measuring conditions plays an important part in hardness measurement. If we measure the hardness in crystals of the same class under comparable conditions, a distinct correlation between the hardness and the elastic properties will be revealed (Fig. 2.88).

There are many different methods for measuring the hardness of materials: static and dynamic indentation, scratching, abrasion, drilling, measuring the height of rebound of a penetrator dropped on the test piece, etc. The effect of each particular mechanical characteristic on the result of hardness measurement depends on the method used. Thus, in static indentation the hardness of the material mainly depends on the resistance to plastic strain. The application of the elastic rebound method permits better evaluation of the

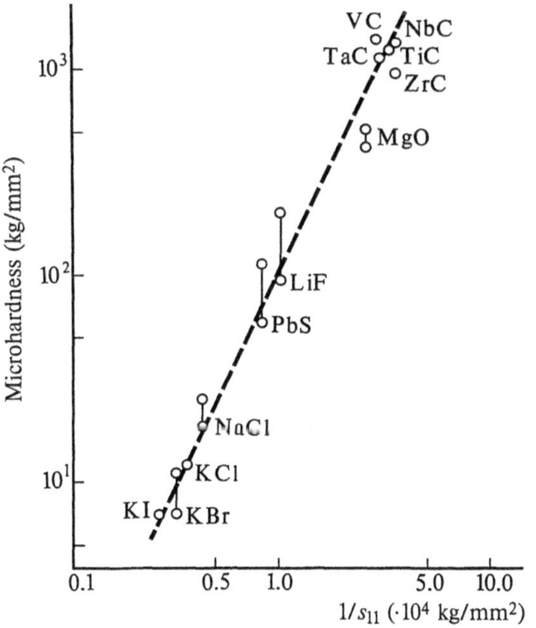

Fig. 2.88. Correlation between the microhardness and the elastic properties for crystals with a NaCl lattice [2.93]

elastic properties of the material. The results of hardness measurement by the scratching method occasionally depend on the tendency of the material toward fracture (on its brittleness).

For crystals, two groups of methods are used to estimate their plasticity and brittleness: methods of indentation (with a ball, a cone, or a pyramid) and methods of scratching with a standard point under definite conditions.

The classes of hardness are named after the inventors of different types of indenter:

1. *Brinell hardness* is determined by the impression left by a steel ball. Here, the microhardness is equal to the load applied divided by the surface area of the impression obtained, in which the diameter is measured after the load is removed.
2. *Ludwik hardness* is determined by pressing a steel cone with a vertex angle of 90° into the material. It is calculated as the load divided by the lateral surface area of the impression obtained, in which the base diameter is measured after the load is removed, while the vertex angle is conventionally assumed to be equal to the angle of the penetrator point.
3. *Rockwell hardness* is determined on a special device by pressing down a steel ball or a diamond cone of standard size. Rockwell hardness is measured in conventional units by means of different scales by the increment of the depth of penetration when passing from a small to a large standard load.
4. *Vickers hardness* is determined by pressing down a standard-sized pyramid and calculated as the standard load divided by the lateral surface of the pyramidal impression obtained. Here the base diagonal is determined after the load is removed, and the angles at the impression vertex are conventionally assumed to be equal to those at the apex of the pyramidal point. In addition to a square pyramid, trihedral Berkovich pyramids and the rhombic Knoop pyramid are used.

The hardness measured by the indentation method is usually called the *microhardness,* because the hardness in this method is evaluated within a limited region.

The results of a microhardness measurement by indentation essentially depend on the face under test (microhardness anisotropy), the choice of the load applied to the indenter, the choice of the loading time and exposure to the load, the correct pattern of impressions on the test specimen, the method for preparing the specimens for investigation, etc.

The main device used for microhardness testing in the USSR is the ПМТ-3 machine (designed by Khrushchev and Berkovich). The indenter here is a diamond pyramid with a square base and an apex angle of 136° between the opposite faces. When using such a pyramid the microhardness is calculated by the formula

$$H = 1.854\, P/d^2 \, , \qquad (2.142)$$

Table 2.10. The Mohs hardness scale

Hardness	Crystal and its formula	Test plane
1	Talc $Mg_3(OH)_2[Si_4O_{10}]$	(001)
2	Gypsum $CaSO_4 \cdot 2\,H_2O$	(010)
3	Calcite $CaCO_3$	$(10\bar{1}1)$
4	Fluorite CaF_2	(111)
5	Apatite $Ca_5(PO_4)_3$ (F, Cl, OH)	(0001)
6	Orthoclase $KAlSi_3O_8$	(001)
7	Quartz SiO_2	$(10\bar{1}1)$
8	Topaz $Al_2(F, OH)_2(SiO_4)$	(001)
9	Corundum Al_2O_3	$(11\bar{2}0)$
10	Diamond *C*	–

where P is the load applied to the indenter, and d is the impression diagonal. If P is expressed in grams and d in micrometers, then in (2.142) the microhardness will be expressed in kgf/mm^2.

The scratching hardness is measured by two methods:

1. *The Mohs hardness* is determined by the ability of a body to be scratched by a harder body or to scratch a softer one. Mohs proposed a ten-point hardness scale, which represents a succession of minerals of different hardness arranged in the ascending order. The hardnesses of these substances have been adopted as standards. It should be borne in mind that the difference between the properties of the first standard substances on the Mohs scale is much less than for the last substances, i.e., the scale is nonlinear. If the Mohs hardness is compared with the hardness of the same minerals measured by other methods, the variation in the property will be approximately parabolic. The parabola slightly smooths out if we take into account the hardness anisotropy in Mohs measurements, because the scratching hardness also depends on the face tested. Therefore, modern Mohs scale tables indicate the optimum test face in addition to the reference mineral (Table 2.10).
2. The measure of the *Martens hardness* is the width of the scratch made by a standard diamond cutter (pyramid, cone). To determine the scratching hardness (it is called the sclerometric hardness), the Martens sclerometer is used. In this device, a scratch is made by a diamond cone with an angle of 90°. The adopted measure of hardness is the load necessary for making a scratch 10 μm wide. The disadvantage of the Martens method is its comparative complexity: several scratches with different loads have to be made before plotting the graph of the scratch width versus the load in order to find the load corresponding to the scratch width of 10 μm.

It is more convenient to calculate the hardness by measuring the scratch width at a definite load. An example is *Bierbaum's method.* In his device the indenter is a trihedral diamond pyramid (a vertex of a cube), to which a

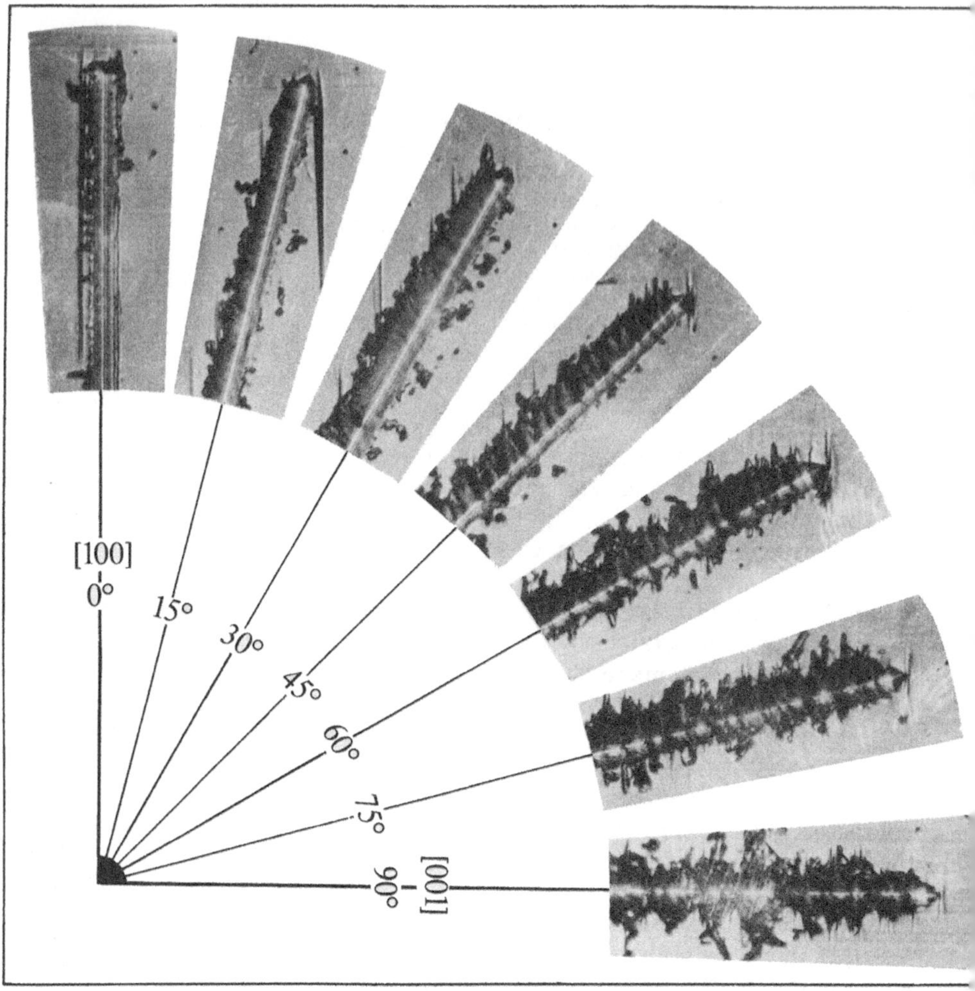

Fig. 2.89. Shape of scratches made by a diamond indenter on the (010) plane of an antimonite crystal along different directions (× 740) [2.94]

standard load (3 g) is applied. The Bierbaum hardness (K) is expressed by the formula

$$K = 10^4/d^2 \ , \tag{2.143}$$

where d is the scratch width.

For soft materials, the ПМТ-3 machine can be used for determining the sclerometric hardness by moving the specimen under a fixed indenter – a standard diamond pyramid. If the pyramid is so positioned that the scratching is done edge forward, the scratching microhardness H_{scr} can be estimated as

$$H_{scr} = 3.708\,P/d^2 \ . \tag{2.144}$$

If a trihedral pyramid is used, then in scratching edge forward

$$H_{scr} = 3.138\,P/d^2 \ . \tag{2.145}$$

The width and character of the scratch depend on the face tested and the direction of the scratch.

Figure 2.89 illustrates the anisotropy of hardness determined by scratching a face of an antimonite crystal. With this method, the closer the direction of the scratch to [001], the more cracks appear and the less plastic the crystal is.

2.7.2 Investigating the Elastic Properties of Crystals

The size and shape of the impression are different before and after the load is lifted (Fig. 2.90). This is revealed as follows. The indenter or the specimen surface is covered with a thin film of some substance. After indenting and the subsequent raising of the indenter, an impression is left on the surface, which indicates the size of the impression under load, i.e., the size of the unrestored impression. By measuring than the diagonal of the restored impression and subtracting this value from the diagonal of the unrestored impression, one obtains the elastic restoration, which gives a qualitative picture of the elastic properties of crystals [2.95, 96]. However, a quantitative estimation of the elastic properties requires the knowledge of the elastic constants of the crystal.

The methods for measuring the elastic constants are classified as static and dynamic. The static methods are based on recording the strain in compression, tension, bending, and torsion. The strain is measured by various mechanical, optical, or electrical devices (pickups). Specimens are set up in a

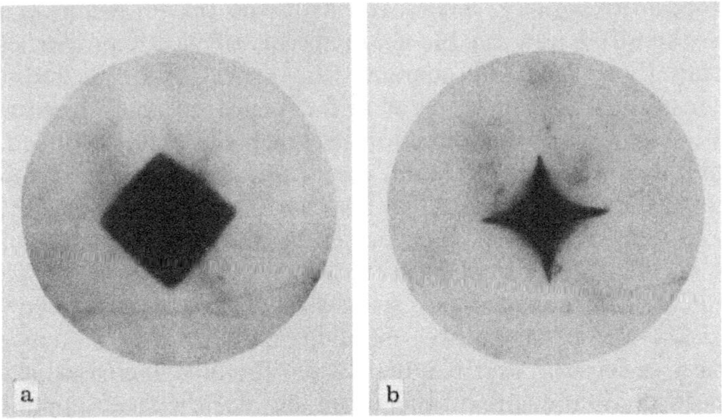

Fig. 2.90. Impression of a diamond indenter on the surface of an organic glass at the moment of loading (**a**) and after the removal of the load (**b**), $P = 20$ g ($\times$ 560) [2.94]

testing machine, the load is increased stepwise, and the strain is calculated for each loading step from the pickup readings. The elasticity moduli and the Poisson coefficient are calculated by formulas specific to each type of pickup. Static methods are mainly used in testing polycrystalline materials.

Single crystals are predominantly tested by dynamic methods, which are based on exciting wave oscillations of some type in the specimen and determining the propagation velocity of the elastic waves. The following dynamic methods are utilized:

1. *resonance methods* involving the determination of the natural oscillation frequencies of definitely oriented specimens;
2. *methods based on determining the sound velocity* in selected crystallographic directions (Schaefer-Bergmann's optical method, the wedge method, pulse methods);
3. *methods based on interaction of the crystal lattice with radiations* (the x-ray scattering method). All these methods are described in detail by *Beranek* [2.96] and *Belikov* et al. [2.97].

The region between the plastic and elastic strain, which is called the imperfect-elasticity region, is studied by measuring the internal friction. The essence of internal friction and the methods of its measurement are explained in Sect. 2.3. A comprehensive description of the techniques for measuring the internal friction can be found in *Postnikov* [2.99].

2.7.3 Investigating the Plastic Properties of Crystals

As noted above, rapid evaluation of the ability of materials to undergo plastic strain can be made by measuring the microhardness. This is mainly achieved by indentation. A disadvantage of this method is the fact that because of the extremely complicated stressed state due to the action of a concentrated load all kinds of deformation take place, including elastic and plastic strain, and also failure. A more sensitive and reliable characteristic of plastic properties alone is the yield point (Sect. 2.2). This characteristic, as well as the hardness, is not a physical constant of a given material and depends on many factors: the crystal orientation relative to the direction in which the load is applied, active glide elements, the testing temperature, the strain rate, the number and nature of defects in the specimen, etc. As in the case of microhardness, the yield point for crystals of the same class deformed in comparable conditions correlates with the coefficients of elasticity (Fig. 2.91).

The yield point is determined from stress-strain curves of the type presented in Fig. 2.28. These curves are obtained in the course of active loading (compression, extension, and bending) when the load increases at a constant rate. In deformation machines, the dependence of the stress applied on the deformation time is recorded at a constant rate. The stress is usually recorded automatically by means of strain gauges or an optical dynamometer.

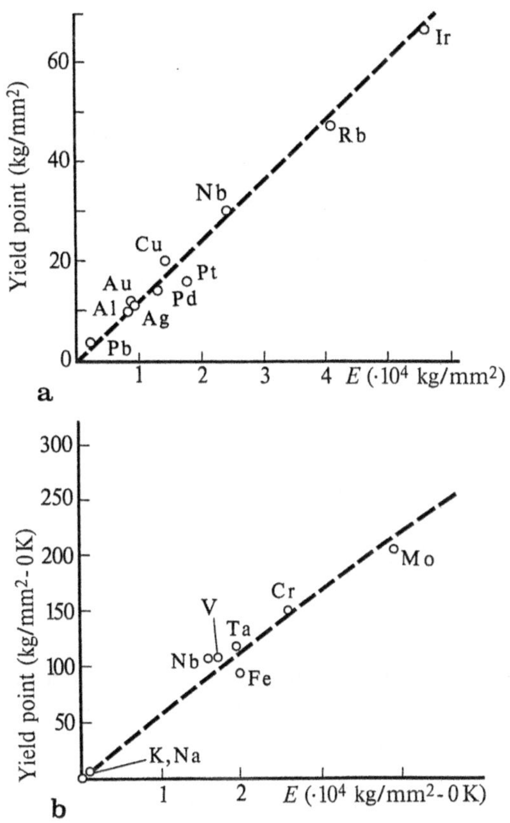

Fig. 2.91. Correlation between the yield point and the elastic properties for fcc (**a**) and bcc (**b**) metals [2.94] [2.93]

The optical dynamometer is based on the photoelasticity and the measurement of birefringence by the photoelectric method [2.100]. If an optically isotropic body (for instance, a glass prism) is subjected to mechanical effects, internal stresses arise in the body. When such a body is observed in a transmitted polarized light, birefringence arises in the stressed area. The resulting optical path difference Γ is proportional to the applied force P: $\Gamma = cP$. Proportionality factor c is called the coefficient of photoelasticity. The change in the intensity of the transmitted light caused by the birefringence is detected by a photocell and a photocurrent meter.

The loads measured by a strain gauge or an optical dynamometer (which vary at a preassigned rate in the course of loading) are recorded automatically on the recorder chart. The plotted curves (dependences of applied loads P on deformation time) are converted to the dependence of stresses $\sigma = P/S$ (S is the specimen cross section) on the degree of deformation $\varepsilon = (\Delta l/l) \times 100\%$. Dependence $\sigma(\varepsilon)$ can be converted to the dependence of shear stresses τ on shear strain a using the formulas given in Sect. 2.2. It should be taken into account that testing machines have a limited rigidity, and in straining

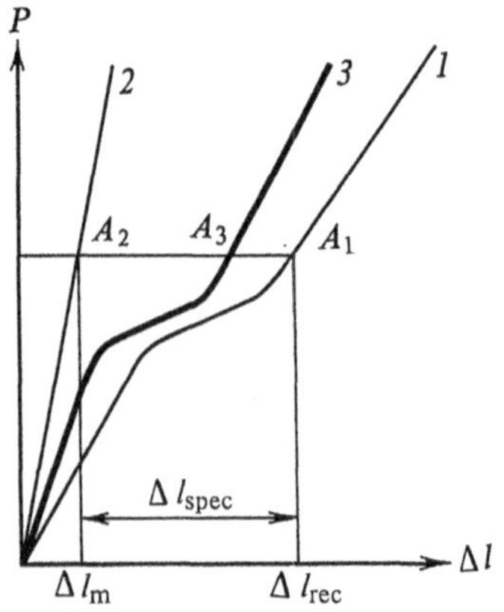

Fig. 2.92

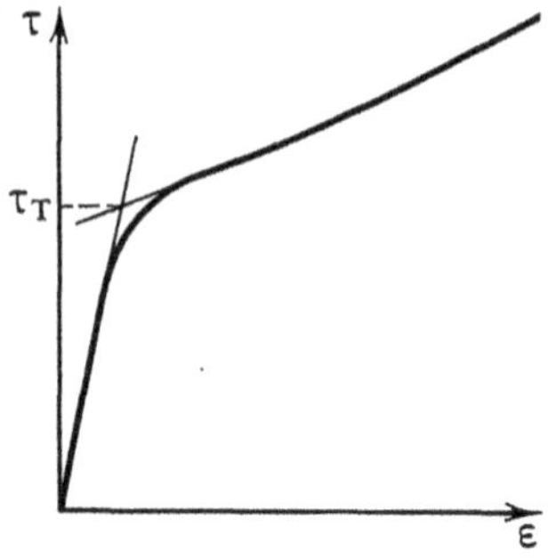

Fig. 2.93

Fig. 2.92. Construction of true stress-strain curves of a specimen: (*1*) curve taken during specimen deformation; (*2*) curve of deformation of a machine; (*3*) true specimen stress-strain curve

Fig. 2.93. Determing the yield point on the curve of the dependence of stress σ on strain ε of a LiF crystal

comparatively rigid crystals, when the ratio of the specimen and machine rigidities K_{spec}/K_{mach} ($K_{spec} = P/\Delta l_{spec}$, $K_{mach} = P/\Delta l_{mach}$) is close to unity, the true specimen strain does not agree with that recorded on the diagram. To obtain a true specimen stress-strain curve, one takes the stress-strain curve of the machine and the curve plotted during the specimen deformation (Fig. 2.92) and, making successive sections at different P, finds the true strain in the form of

$$\Delta l_{spec} = \Delta l_{rec} - \Delta l_{mach}$$

As a rule, the transition from elastic to plastic deformation occurs smoothly, and the yield point is only vaguely defined. For yield point τ_y on the stress-strain curve one takes the stress corresponding to the point of intersection of tangents to the stress-strain curve drawn on the initial steep and the following flatter section, as shown in Fig. 2.93.

The yield point in transparent crystals can be measured (with a lower accuracy, however, than from the stress-strain curves) optically by the appearance of the first birefringence fringes along the glide lines when observing the deformed specimen in a transmitted polarized light. Figure 2.94 shows a photo of birefringence fringes along the glide lines of a LiF crystal.

In the last 15 to 20 years, to achieve rapid control of the ability of a crystal to undergo plastic deformation, researchers have been studying the shape and size of dislocation rosettes formed around the indenter impression and revealed by selective etching. Figure 2.95 presents a photo of dislocation

Fig. 2.94. Birefringence fringes along the slip lines in a LiF crystal (photo taken in a transmitted polarized light, ×200)

rosettes on a cube plane of LiF, PbS, CsI, and CaF_2 crystals and on an octahedron plane of a CaF_2 crystal. The etch pits appearing at the emergence point of dislocation lines at the crystal surface are arranged in rows parallel to the lines of intersection of the test surface with the glide planes. The length of the rosette arms depends on the dislocation mobility: the more plastic the crystal, the more readily the dislocations move in it and the longer the rosette arms are; the direction of the rosette arms helps us to judge the active glide bands.

More accurate information on the mobility of dislocations in crystals can be obtained from the stress dependence of the dislocation path length or velocity. The dislocation paths are detected by selective etching (Sect. 2.3), while their average velocities are found as the path length divided by the load action time. With alkali-halide crystals, at stresses in the yield point range dislocations move at very high velocities ($\sim 10^{-3}$ cm/s), and very short loading pulses (not exceeding 10^{-6} s) are used to establish the dislocation path length with a sufficient accuracy.

2.7.4 Studying the Ability of Crystals to Undergo Failure

The brittleness can be estimated by indentation from the load causing cracks near the impression. At *Ikornikova*'s suggestion [2.101], a scale for determining the brittleness by the number of cracks has been established (Table 2.11).

The ultimate strength in compression and tension is calculated from the ratio of the specimen-breaking load to its cross-sectional area. If the strength

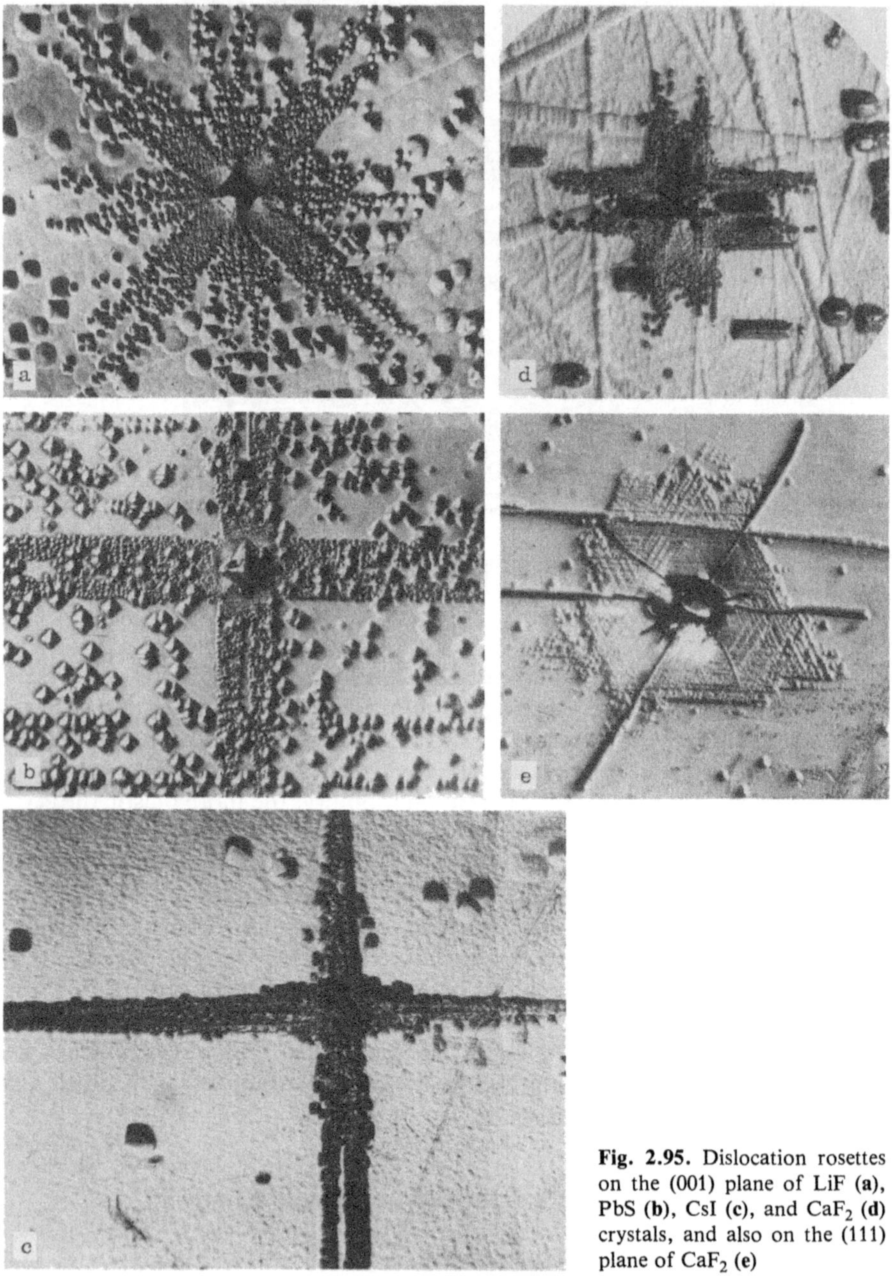

Fig. 2.95. Dislocation rosettes on the (001) plane of LiF (**a**), PbS (**b**), CsI (**c**), and CaF_2 (**d**) crystals, and also on the (111) plane of CaF_2 (**e**)

Table 2.11. Brittleness scale [2.101]

Impression character	Brittleness scale value
Impression without visible cracks and cleavages	0
A single small crack in the corner of the impression	1
A single crack not coinciding with the extension of the impression diagonal, or two cracks in adjacent corners	2
Two cracks in opposite impression corners or three cracks in different impression corners	3
More than three cracks, one or two cleavages at the sides of the impression	4
Distortion of the impression shape	5

is determined by deformation bending, the ultimate tensile stress is calculated by the formula

$$\sigma_{\text{ult}} = 3PL/2bh^2 \qquad (2.146)$$

for three-point bending and by

$$\sigma_{\text{ult}} = 3PC/bh^2 \qquad (2.147)$$

for four-point bending.

Here P is the load; b and h are the width and height of the specimen cross section, respectively; L is the distance between the cylindrical or prismatic supports carrying the specimen subjected to three-point bending (the bed with the supports on it is called a base); and C is the half-difference of the distances between the supports of the upper and lower base.

The ultimate strength in shear deformation characterizes the strength anisotropy of a crystal better than any other property. The shear is produced by two parallel cutters, the lower being fixed rigidly on a massive steel backing, and the upper in a steel prism swiveling on the upper slab of the press. As in longitudinal deformation (contraction and extension), the ultimate shear strength is calculated by the formula

$$\sigma_{\text{ult}} = P/S \ ,$$

where S is the cross-sectional area of the specimen.

The ultimate strength is occasionally determined by an impact of a pendulum or a falling load, respectively, on a Charpis or a Page impact machine, and also by torsional strain during the rotation of a shaft attached to one end of the test plate.

2.7.5 Long-Term Strength

A high ultimate strength by no means always defines the necessary strength characteristics of a material. In the present stage of technological progress, the time characteristics of failure determining the life of a machine part or a mechanism subjected to a constant load come to the foreground more and more often. These characteristics relate the time to failure (t) to the loading conditions (stress σ, temperature T, and creep rate $\dot{\varepsilon}$). The specimen lifetime under load is called longevity. Experimental data on longevity $t(\sigma, T)$ are usually well described by the following empirical formula [2.102–104]:

$$t = t_0 \exp\left(\frac{U_0 - \gamma\sigma}{kT}\right) , \tag{2.148}$$

where t_0 is independent of σ and T. Using refractory alloys as an example, the following relationship between the stationary creep rate and longevity was established:

$$t\dot{\varepsilon} = \text{const} , \tag{2.149}$$

which helps to predict the longevity of a specimen under load.

The failure process in longevity testing develops gradually as the damage accumulates. Investigations of various materials showed that damage developing in time depends on the test conditions. At high temperatures and low stresses it is determined by the nucleation and growth of cracks and pores by the vacancy mechanism. The elementary act of damage accumulation is the "swallowing" of a new vacancy by a crack or a pore. No reliable experimental data have been obtained so far as regards the damage accumulation mechanism in the range of moderate temperatures and high stresses. The experimentally revealed closeness of the activation energy of the low-temperature creep to the energy of formation of interstitials gave grounds to assume [2.104] that the determining role under these conditions is played by the movement of interstitials. The criterion of the predominance of the interstitial mechanism over the vacancy mechanism consists in the fulfillment of the condition

$$\gamma\sigma > U_s' - U_s - AkT , \tag{2.150}$$

where γ is the activation volume corresponding to the increment of plastic deformation per each interstitial produced per unit volume, U_s' is the energy of self-diffusion by the interstitial mechanism (equal to the energy of formation of an interstitial U_i), U_s is the vacancy self-diffusion energy, and A is a practically constant value of the order of several units.

In accordance with (2.150), a change in stress and temperature causes a qualitative change in the shape of curves $\log\dot{\varepsilon} - \sigma$ in connection with the

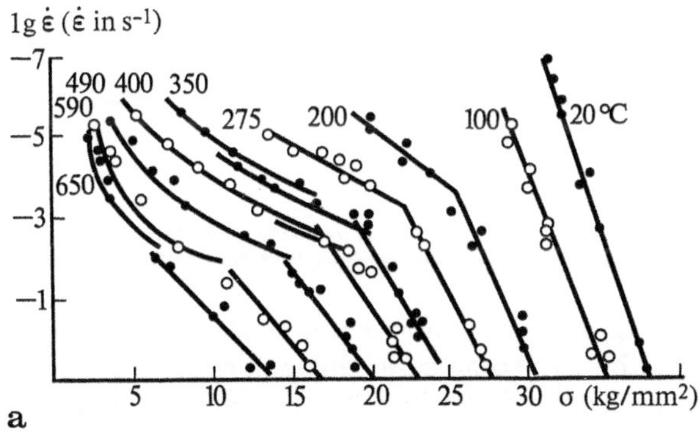

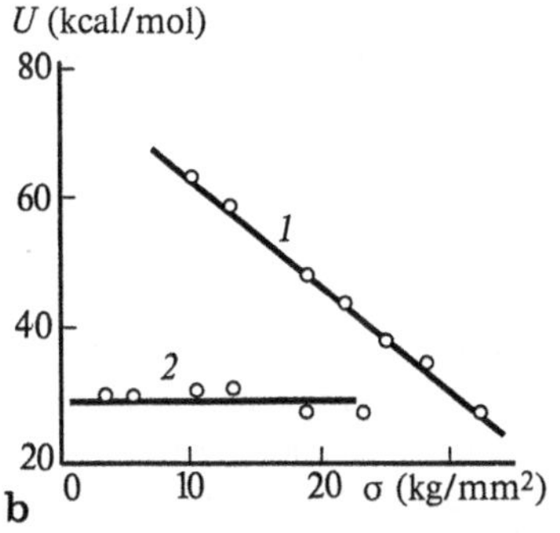

Fig. 2.96. Stress dependence of the creep rate (**a**), and stress dependcence of the activation energy (**b**) [2.56]

transition from the interstitial to the vacancy mechanism of mass transport (Fig. 2.96 a) and a sudden drop in activation energy (Fig. 2.96 b). The linear dependence of the activation energy on the stress (Curve 1) corresponds to the generation of interstitials, and the stress invariance (Curve 2) to vacancy migration. Besides, according to electron-microscopic observations of dislocation loops resulting from creep, vacancy loops prevail in the high-temperature and low-stress range, and interstitial loops at low temperatures and high stresses.

3. Electrical Properties of Crystals

The electrical properties of crystals are the phenomena associated in some way or other with electrical polarization. In some cases this polarization is spontaneous, i.e., not induced by external effects; in others it may be due to the heating of a crystal, the application of an electric field, a mechanical load, etc.

In physical crystallography the problem of the electrical properties of crystals is one of the best studied. Primary consideration here is given to phenomena closely connected with the anisotropy of dielectric crystals. For instance, dielectric constant ε is a tensor value for crystals. This is also true for the dielectric susceptibility, specific electrical conductivity, and other characteristics.

The existence of spontaneous polarization in some crystals leads to a new phenomenon unknown for the isotropic media, namely, the pyroelectric effect, which is due to a change in spontaneous polarization on heating (cooling) of a crystal. A comparatively new class of crystals – ferroelectrics – is a subclass of the pyroelectrics. The ferroelectrics are characterized by the division of a crystal into spontaneously polarized regions, domains. This explains the specifies of the electrical and other properties of the ferroelectrics (particularly in the phase transition region).

The anisotropy of crystals lies at the basis of another interesting phenomenon, the piezoelectric effect: electrical polarization under the effect of mechanical stresses (strains) as well as deformation caused by an electric field.

The electrical properties of crystals must be taken into account when studying and using them in science and technology. Pyroelectric and piezoelectric phenomena are widely used in optoelectronics, radioelectronics, hydro- and electroacoustics, energy conversion techniques, etc. This particularly refers to ferroelectrics.

3.1 Polarization, Electrical Conductivity, and Dielectric Losses

3.1.1 General

When dielectric crystals are placed in an external electric field they become polarized. The tension of the electric field $\boldsymbol{E}$ inside a dielectric can only be determined by taking into account its polarization $\boldsymbol{P}$. In addition to vectors $\boldsymbol{E}$

and $\boldsymbol{P}$ the state of a dielectric is characterized by electrical induction vector $\boldsymbol{D}$. The relationship among vectors $\boldsymbol{D}$, $\boldsymbol{P}$, and $\boldsymbol{E}$ is defined by the equalities

$$\boldsymbol{P} = \alpha \boldsymbol{E}, \quad \boldsymbol{D} = \boldsymbol{E} + 4\pi \boldsymbol{P}, \quad \boldsymbol{D} = \varepsilon \boldsymbol{E} \, (\varepsilon = 1 + 4\pi\alpha) \, , \tag{3.1}$$

where α is the polarization of the dielectric, and ε is its dielectric constant. The complete set of equations of the electrostatic field for an arbitrary set of solids containing dielectrics has the form

$$\boldsymbol{E} = -\operatorname{grad}\varphi, \quad \boldsymbol{D} = \varepsilon \boldsymbol{E}, \quad \operatorname{div}\boldsymbol{D} = 4\pi\varrho, \quad \boldsymbol{D}_{2n} - \boldsymbol{D}_{1n} = 4\pi\sigma, \tag{3.2}$$

where φ is the eletric field potential, ϱ is the volume density of free charges, $\boldsymbol{D}_{2n}$ and $\boldsymbol{D}_{1n}$ are the normal components of the induction vector at the interface of two dielectrics, and σ is the density of free charges at this interface.

Quantities ε and α for an isotropic medium are scalars. In crystals (as in all anisotropic media) the quantities α and ε, which relate polar vectors with one another, are second-rank polar tensors: α_{ij} and ε_{ij}. The relationship among vectors $\boldsymbol{D}$, $\boldsymbol{E}$, and $\boldsymbol{P}$ is evident from Fig. 3.1. We note that as a result of polarization the dielectric surface acquires charges with a sign opposite to that of the free charges. Taking into account this circumstance and the need to satisfy relations (3.1), the direction of vector $\boldsymbol{P}$ inside a dielectric must be chosen from negative to positive charges.

Tensors ε_{ij} and α_{ij} are symmetric in accordance with Onsager's principle of symmetry in static electric fields in the absence of magnetic fields.

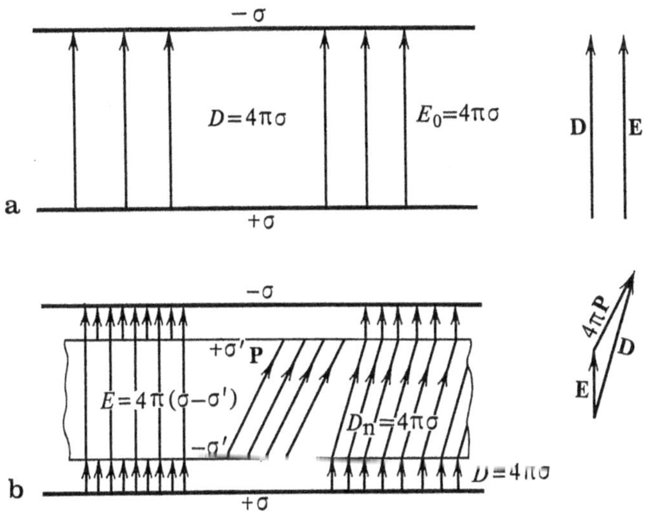

Fig. 3.1. Vectors $\boldsymbol{D}$, $\boldsymbol{E}$, and $4\pi\boldsymbol{P}$ in a parallel-side capacitor under a vacuum (**a**) and with an anisotropic dielectric plate inserted in the capacitor (**b**): σ is the density of the inducing (free) charges; σ' is the density of the polarization (bound) charges

Since the directions of vectors $\boldsymbol{E}$ and $\boldsymbol{D}$ in a crystal generally do not coincide, one can introduce, by analogy with optics, the concept of dielectric constant ε_E in the direction of vector $\boldsymbol{E}$, and of a dielectric constant ε_D in the direction of vector $\boldsymbol{D}$. Quantity ε_E indicates how many times shorter vector $\boldsymbol{E}$ is than the projection of vector $\boldsymbol{D}$ on its direction. In turn, ε_D shows how many times longer vector $\boldsymbol{D}$ is than the projection of vector $\boldsymbol{E}$ on its direction.

Experiments determine quantity ε_E, which is related by the following expression to main dielectric constants ε_1, ε_2, and ε_3 via guiding cosines c_i of vector $\boldsymbol{E}$:

$$\varepsilon_E = c_1^2\varepsilon_1 + c_2^2\varepsilon_2 + c_3^2\varepsilon_3 \,. \tag{3.3}$$

In consideration of polarization of atoms and molecules (microprocesses) the concept of *internal,* or *acting, electric field* is important. The reason is that in macroscopic consideration the field tension in a dielectric implies mean field $\boldsymbol{E}$, which neglects its atomic structure. As for the processes of polarization of molecules and atoms, they are determined not by this field, but by internal (acting) field $\boldsymbol{F}$. It is easy to see that fields $\boldsymbol{E}$ and $\boldsymbol{F}$ differ from one another at least in that calculation of tension of mean field $\boldsymbol{E}$ takes into account the field for all the charges (including those at the point concerned), whereas acting field $\boldsymbol{F}$ is calculated at a definite point, and only the field of the charges external to those at the given point are taken into account.

For weakly polarized media, in the case of isotropic dielectrics, the Lorentz approximation holds true; according to it mean field $\boldsymbol{E}$ and acting field $\boldsymbol{F}$ are related by the expression

$$\boldsymbol{F} = \boldsymbol{E} + \frac{4\pi}{3}\boldsymbol{P} \,. \tag{3.4}$$

In this approximation, the Clasius-Mossoty equation is valid, relating the dielectric constant of a dielectric to the polarization of a separate microparticle η:

$$\frac{M}{\varrho}\,\frac{\varepsilon - 1}{\varepsilon + 2} = \frac{4\pi}{3} N_0 \eta \,, \tag{3.5}$$

where M is the molecular weight, ϱ is the dielectric density, and N_0 is the Avogadro number. In the last expression the right-hand side is called the *molecular polarization.*

3.1.2 Main Types of Polarization

The various types of polarization for dielectrics/nonferroelectrics are usually reduced to four:

1) Polarization due to displacement of electrons relative to the atomic nuclei (electron-displacement polarization);
2) Polarization due to the displacement of ions of the crystal lattice relative to each other (ion-displacement polarization);
3) Polarization due to the orientation of the constant dipole moments present in the structure (thermal orientational polarization);
4) Polarization due to the motion of weakly bound ions (thermal ionic polarization).

The last two types of polarization are usually called relaxational.

Electron-displacement polarization is common to all dielectrics. It is induced by elastic displacement of the weakest-bound, mainly valency, electrons in an atom or an ion. The time to electron-displacement polarization is comparable with the light-oscillation period and equals $10^{-14}-10^{-15}$ s.

The dielectric constant ε of adielectric may, generally speaking, be due to different kinds of polarization. In the optical-frequency range, however, ε is almost entirely due to electron polarization. Here relation $n^2=\varepsilon$ (where n is the refractive index) is valid, and when considering a unit volume (3.5) takes the form

$$\frac{n^2-1}{n^2+2}=\frac{4\pi}{3}N_i\eta_i \tag{3.6}$$

(here N_i is the number of atoms of the ith sort per unit volume, and η_i is the electron polarization of the ith atom).

Using the refractive indices determined experimentally it is possible, with the aid of (3.6), to find with sufficient accuracy the values of electron polarization for many crystals, which are equal in the order of magnitude to the cube of the atomic radius, i.e., $10^{-24}\,\mathrm{cm}^3$.

The electron-displacement polarization in its pure form (in the absence of other types of polarization) occurs in crystals whose lattice is built up of nonpolar molecules (diamond, naphthalene, paraffin) and also in a number of solid polarized organic dielectrics (polystyrene, polyethylene, and others). The polarization of such dielectrics in static and low-frequency fields and also at optical frequencies is properly described by (3.5, 6). For the indicated dielectrics the relation $n^2=\varepsilon$ is fulfilled well at all frequencies.

The *ion-displacement polarization* is due to the displacement of ions relative to each other and is most pronounced in crystals with a purely or predominantly ionic bond. It should, however, be remarked that such crystals have polarization of electron displacement of ions along with ion-displacement polarization. But the former type of polarization alone cannot explain the experimentally observed value of the dielectric constants of some ionic crystals. Thus, in sodium chloride the refrative index $n=1.5$ ($n^2=\varepsilon_\infty=2.25$), while the static dielectric constant is equal to $\varepsilon_s=5.62$. This difference in the values of static and optical dielectric constants may be attributed to the ion-displacement polarization, which can be characterized by a definite value of ion polarization of an ion pair η_{ion}. The static dielectric constant ε_s, with due regard for the ion-displacement polarization in an approximation neglecting the difference between the acting and the mean field, is calculated by Born's formula

$$\varepsilon_s = n^2 + \frac{2\pi e^2}{\omega_0^2 a^3}\left(\frac{1}{m_1} + \frac{1}{m_2}\right), \tag{3.7}$$

where e is the ion charge, ω_0 is the frequency of the natural vibrations of the lattice, a is the interionic distance, and m_1 and m_2 are the ion masses. The second term in the right-hand side of the formula precisely accounts for the contribution to the dielectric constant from the ion-displacement polarization. for NaCl, for instance, this term is 2.7, which allows us to correlate properly the above values of ε_s and n for this crystal.

The term *relaxational polarization* usually implies polarization due to a change in the orientation of free or weakly bound dipoles in polar dielectrics and polarization induced by the motion of weakly bound ions. These two types of polarization are closely associated with the thermal motion of structural units; they depend on this motion and therefore are regarded as types of thermal polarization.

Some dielectric contain polar molecules or other structural units with a constant dipole moment μ_0. Dielectrics with polar molecules are represented by water, nitrobenzene, HCl, HBr, etc. If the polar molecules are free, they become, under the effect of an electric field, oriented along the field to the extent allowed by the disturbing thermal motion. Such a polarization is called the *thermal orientational polarization.*

In considering electron and ion polarization η_0 and also polarization due to dipole orientation η_{or} (neglecting the dipole interaction) the dielectric constant can be described by the Langevin-Debye formula:

$$\varepsilon - 1 = 4\pi N(\eta_0 + \mu^2/3kT), \tag{3.8}$$

where N is the number of particles per unit volume, μ is the dipole moment of the dipole ($\mu \approx 10^{-18}$ CGSE units), k is the Boltzmann constant, and T is the absolute temperature. The Langevin-Debye equation was obtained for the case where the dipole energy in an electric field is much less than the thermal motion energy ($\mu E \ll kT$). This case is encountered most often. In very strong fields, however, it is sometimes possible to orient all the dipoles along the field, and then the polarization η_{or} will be equal to μ rather than to $\mu^2/3kT$.

If in a dielectric containing polar molecules the thermal motion energy is sufficient for overcoming the potential barrier between the equilibrium positions, the molecules will be tossed from one equilibrium position to another. The external electric field creates some preferential direction in the random thermal tossing of molecules (along fixed directions), orients them, and thus polarizes the dielectric.

It has already been noted that the polarization of polar dielectrics is of a relaxational nature. The onset (or destruction) of polarization in such dielectrics is associated with the thermal motion of dipoles, is of a probabilistic character, and requires a definite time, called relaxation time τ. It is not difficult to understand that the value of τ depends on the ratio

$$\tau = \mathrm{e}^{u/kT}/2\nu\,, \tag{3.9}$$

where ν is the frequency of the natural oscillations of the elementary dipole, u is the height of the potentital barrier separating its two equilibrium positions, k is the Boltzmann constant, and T is the absolute temperature. Indeed, at a frequency ν the dipole will "attempt" twice per period to overcome the potential barrier ($\tau' = 1/2\nu$, where τ' is the time between "attempts"), and multiplication of τ' by exp (u/kT), which is the reciprocal probability of transition (according to Boltzmann's statistics), leads precisely to (3.9).

A solid dielectric often contains weakly bound ions, which may, in their thermal motion, break away from their fixed positions and move in the dielectric. The external field creates some preferential direction in the random thermal tossings and thus causes excessive tossing of ions in this direction. In a solid dielectric, the motion of weakly bound ions is strongly limited in space. Therefore, excessive tossing of ions in the direction of the field leads to asymmetry in the distribution of charges throughout the dielectric and is, after a lapse of time, offset by the back diffusion of ions. Thus, in the stationary state some electrical moment (polarization) is set up in each unit volume of the dielectric. This type of polarization, which depends on the thermal motion of the structural units of the dielectric, is called the *thermal ion polarization.*

In calculating the polarization value in thermal ion polarization, one usually proceeds from the conception that in the absence of a field a weakly bound ion can occupy two equiprobable positions 1 and 2 with identical potential energy. The probability of an ion being tossed from position 1 to 2 (or vice versa) depends on potential barrier height u and temperature and is equal to exp $(-u/kT)$. If an ion is fixed in position 1 or 2, it will vibrate with a frequency ν, determined by the elasticity of the binding forces.

A calculation of polarization due to the thermal ion polarization, based on the indicated conception, will lead to the expression

$$\eta_{\text{therm. ion}} = e^2\delta^2/12kT \tag{3.10}$$

($\delta/2$ is the free path of an ion), which is close to that for the polarization of a dipole-containing dielectric.

Thermal ion polarization usually takes place in amorphous (inorganic glasses) and polycrystalline inorganic dielectrics (ceramics). This process usually plays only an insignificant part in the polarization of pure dielectric crystals, but it may be substantial for crystals containing various impurities.

3.1.3 Electrical Conductivity

Crystals of linear dielectrics have preferential ionic conductivity (intrinsic and impurity type).

Investigations into the nature of conductivity carried out on a number of crystals show that the principal current carriers in them are those ions which have the smallest size with an equal charge, or those which have a smaller charge with similar sizes. For instance, in NaCl the principal current carriers are Na^+ and in PbCl, Cl^-. In some crystals the electrical conductivity is due to ions of both signs (for instance, in PbI_2). Besides, with a change in temperature some crystals acquire carriers of both signs in place of carrier ions of the same sign. Thus, above 600 °C the Cl^- and F^- ions, in addition to Na^+, begin to take part, though to a small extent, in charge transfer in NaCl and NaF crystals.

In strong fields, electronic conductivity is superimposed on the ionic conductivity in many crystal dielectrics. This effect has been discovered in crystals of quartz, rock salt, etc. For instance, in mica the conductivity is predominantly electronic in strong fields at low temperatures, while ionic conductivity is predominant in weak fields at sufficiently high temperatures.

If the number of particles per cm^3 capable of moving under the effect of a field is equal to n, the charge of each particle is e, and the particle mobility is $\varkappa$ then conductivity σ is defined by the relation

$$\sigma = ne\varkappa . \tag{3.11}$$

In ionic crystals, electrical conductivity may be due to the motion of the crystal lattice ions. Such conductivity is called intrinsic and is manifested most prominently at high temperatures. On the other hand, the electrical conductivity of ionic crystals may be related to the motion of relatively weakly fixed impurity ions. Such ions ensure the impurity electrical conductivity of the crystal. The electrical conductivity associated with the motion of weakly bound ions is perceptible already at rather low temperatures. More often than not, in one and the same crystal electrical conductivity is effected both by lattice ions and by weakly fixed impurity ions. Nonionic (for instance, molecular) crystals usually have impurity electrical conductivity.

The ion motion in a crystal may occur in one or two ways: (a) ions may move between lattice points, forming Frenkel defects (the interstices are free in the perfect crystal and may be occupied in crystals containing intestitial atoms); (b) ions may move by jumping over unoccupied points (Schottky defects). Such ion motion is sometimes regarded as hole conductivity.

Investigation shows that in both cases the electrical conductivity (excess and hole) is described by the expression

$$\sigma = A\,e^{-B/T} , \tag{3.12}$$

in which constant B (activation energy) is temperature independent, and A is only slightly temperature dependent. The value of B is controlled by the energy of the ion in the crystal and the energy necessary for its displacement from one stable position to another.

It has already been mentioned that in a real ionic crystal at high temperatures the conduction is usually intrinsic, and at low temperatures, impurity type (or associated with defects). Therefore, the temperature dependence of the electrical conductivity has to be described by the relation

$$\sigma = A_1 e^{-B_1/T} + A_2 e^{-B_2/T}, \tag{3.13}$$

where subscripts 1 at A and B refer to the lattice ions, and subscripts 2 to the impurity ions. It is thus clear that activation energy B_1 greatly exceeds B_2. In particular, in alkali-halide crystals the electrical conductivity is described by (3.13), and the slope of the lines $\ln \sigma = f(T)$ in the low-temperature range is equal to about one-third of that in the high-temperature range.

The electrical conductivity of nonionic crystals is usually described well by the formula

$$\ln \sigma = A - B/T, \tag{3.14}$$

which coincides with (3.12), and it is usually due to impurity ions. For quartz, the electrical conductivity along the c axis (optic axis) is higher than in the perpendicular direction (the activation energy is equal to 0.88 and 1.32 eV, respectively). The volume resistivity of quartz in the direction of the optic axis at room temperature is about 10^{14} Ωcm, and across this axis, about 10^{16} Ωcm. At 500 °C the quartz resistance decreases by about five orders of magnitude and is approximately 10^9 Ωcm in the direction of the c axis and 10^{11} Ωcm across it. The main current carriers in quartz are evidently single-charged impurity ions of Na, K, and Li.

In strong fields, at which the field work on the path of a weakly bound ion is comparable with the thermal motion energy and at which the electronic conductivity is superimposed on the ionic, the crystal may suffer a breakdown.

The breakdown of solid dielectrics has two basic forms: thermal and electrical. The *thermal breakdown* occurs because at a definite tension the dielectric leaves the state of thermal equilibrium. Its temperature begins to grow rapidly, and the dielectric breaks down. The thermal equilibrium of a dielectric may be disturbed when its electrical conductivity rises with the temperature, which is especially typical of ionic electrical conductivity.

The *electrical breakdown* is characterized by a threshold field tension, at which the electron current in a solid dielectric rises sharply.

3.1.4 Dielectric Losses

In an alternating electric field a dielectric heats up, generally speaking. The part of the energy of an alternating electric field which is converted to heat is called *dielectric losses*. Total dielectric losses (or simply losses) consist of conductivity losses corresponding to a constant applied tension, and losses due to the active component of the displacement current in the dielectric.

Dielectric losses are thus associated with onset of polarization. But rapid onset of the polariztion of electron and ion displacement does not result in any substantial dissipation of the electric field energy. The dielectric losses in crystals possessing only this kind of polarization are very small.

In dielectrics which show polarization due to the thermal motion (thermal orientational and thermal ionic orientation), polarization is associated with absorption currents. At an alternating voltage the absorption current consists of two parts. One of them (j_a) is in phase with the tension applied and is an active component of the current. The other (j_r) describes the current which leads in phasey b $\frac{\pi}{2}$ and is the reactive (capacitance) component of the current. Thus, if a dielectric shows a slow onset of polarization, then in an alternating field it suffers energy losses even in the absence of through conductivity.

Figure 3.2 exhibits a vector diagram of current densities $\boldsymbol{j}_a$ and $\boldsymbol{j}_r$ (corresponding to a slowly establishing polarization) and field tension $\boldsymbol{E}$ in a dielectric. Angle φ in the figure expresses the phase shift between the total current and voltage. Angle δ, which complements φ to $\frac{\pi}{2}$, is called the *dielectric loss angle*. The value of this angle is estimated by its tangent $\tan\delta = j_a/j_r$.

Let us first calculate the capacitance (reactive) part of the dielectric constant. Taking into account the known relationship between the reactive

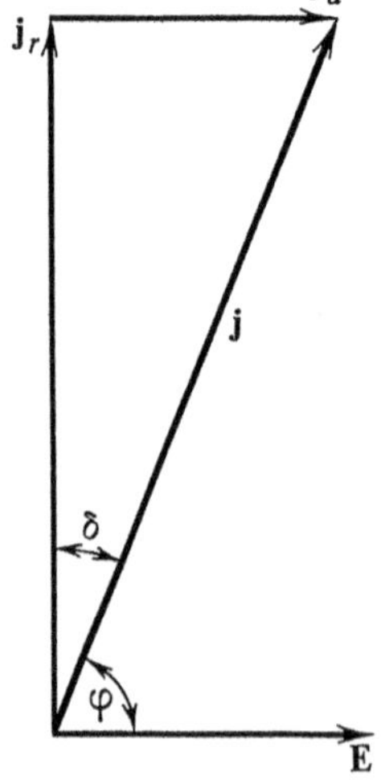

Fig. 3.2. Vector diagram of current density and field tension in a dielectric

capacitance of the capacitor C_r, current amplitude I_{rm}, and the amplitude of the voltage U_m of an alternating current of frequency ω,

$$I_{rm} = \omega C_r U_m , \tag{3.15}$$

we have the following expression for the reactive current density j_r:

$$j_r = \omega \varepsilon_r E_m / 4\pi . \tag{3.16}$$

Thus, allowing only for j_r due to the slowly establishing polarization, the reactive component of the dielectric constant is defined by the expression

$$\varepsilon_r = 4\pi j_r / \omega E_m . \tag{3.17}$$

Assuming the active current density to be due to the "active capacitance" of a dielectric possessing a dielectric constant ε_a,

$$j_a = \omega \varepsilon_a E_m / 4\pi , \tag{3.18}$$

we obtain the expression for ε_a:

$$\varepsilon_a = 4\pi j_a / \omega E_m . \tag{3.19}$$

Taking into account the phase shift of $\frac{\pi}{2}$ between ε_r and ε_a (similar to the shift between j_r and j_a) and denoting $\varepsilon_r = \varepsilon'$, $\varepsilon_a = \varepsilon''$, we can write down the dielectric constant due to the slowly establishing polarization as a complex value:

$$\varepsilon^* = \varepsilon' - \mathrm{i}\varepsilon'' . \tag{3.20}$$

The real part of (3.20), ε', characterizes the reactive (capacitance) part of the dielectric constant. On polarization of a dielectric, the displacement current associated with the charging of the capacitance characterized by ε' leads the voltage by $\frac{\pi}{2}$. The imaginary part of (3.20), ε'', characterizes the dielectric constant due to the active part of the displacement current; this current is in phase with the voltage.

The total dielectric losses are caused by all kinds of polarization (in particular, by the displacement polarization, which has been neglected so far), and also by electrical conductivity. Thus, in a real dielectric the reactive current consists of a capacitance current due to such processes as the charging of the geometric capacitance and displacement polarization, and the current induced by the relaxational polarization. Similarly, the active current is composed of the current associated with the relaxational polarization, and of the residual conductance current.

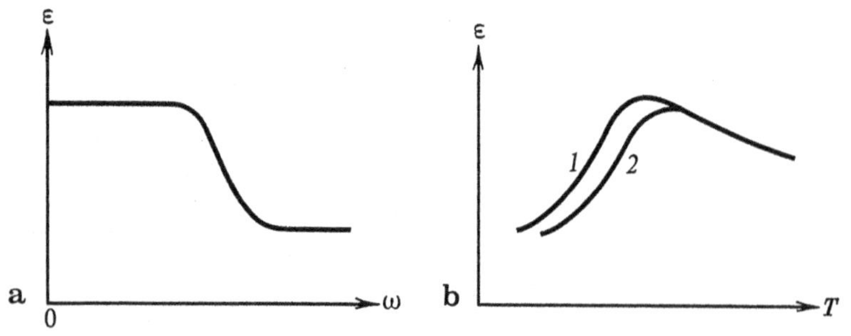

Fig. 3.3. Dependence of the total dielectric constant on the frequency (**a**) and temperature (**b**)

The capacitance part of the dielectric constant, ε, due to all the reactive currents, is frequency dependent (Fig. 3.3a). At frequencies higher than the inverted time to polarization (which determines the contribution to ε additional to that of the electron and ion polarization – the contribution from the last two polarizations is characterized by quantity ε_∞) the total dielectric polarization decreases.

The frequency dependence of $\tan\delta$ is generally rather complicated. If the dielectric conductivity is low, a clearly defined maximum is seen in the frequency variation curve of $\tan\delta$. If the conductivity is high, the maximum of $\tan\delta$ is smoothed out and may vanish altogether. With a low conductivity, i.e., with dielectric losses which are of a pure relaxational, nature, $\tan\delta$ has a maximum in the temperature variation curve. The temperature at which $\tan\delta$ is at a maximum is frequency dependent. As the frequency increases, the temperature maximum of $\tan\delta$ shifts to the right along the x axis.

At high frequencies, for dielectrics which possess relaxational polarization as well as displacement polarization, the dielectric constant will depend on the temperature as shown in Fig. 3.3b (Curve 1 corresponds to a lower frequency than Curve 2). With increasing frequency the maximum of ε shifts towards a higher temperature.

In many crystal dielectrics the dielectric losses are mainly due to conductivity. Only some dielectrics show dielectric losses of a relaxational nature (for instance, due to reorientation of polar groups in ice). In alkali-halide crystals (NaCl, LiF, KBr, etc.) the dielectric losses in a wide range of temperatures and frequencies are mainly caused by unassociated vacancies.

3.2 Pyroelectric Phenomena

3.2.1 General

It was established comparatively early that, on heating, some crystals are capable of acquiring an electric charge: positive at one end and negative at the opposite. This phenomenon was named *pyroelectricity*. For a long time attention was focused on tourmaline crystals. Later, quantitative investigations of the pyroelectric effect were conducted, its general regularities were found, and it was established that crystals of dielectrics of all the 10 polar classes must possess pyroelectric properties, namely, those of classes

1, 2, 3, 4, 6, m, $mm2$, $3m$, $4mm$, and $6mm$.

Pyroelectric properties are inherent in all the crystals showing spontaneous polarization, and the *pyroelectric effect* is a change in spontaneous polarization with the temperature.

An analysis of the thermodynamic relations describing the pyroelectric effect shows that an opposite effect must exist: a change in the temperature of the crystal on application of an electric field, which leads to a change in its spontaneous polarization. This effect was named *electrocaloric*.

The pyroelectric and electrocaloric phenomena were long regarded just as interesting physical phenomena with no practical application. The main reason was that they had formerly been studied on linear pyroelectrics.

The interest in the pyroelectric and electrocaloric effects increased after they had been studied on ferroelectrics in which both are rather pronounced, especially in the phase transition region, because of the sharp temperature dependence of the spontaneous polarization. Pyroelectric crystals are now used as extremely sensitive receivers of infrared radiation and pickups of temperature variation, as well as in other devices converting thermal to electrical energy.

3.2.2 Pyroelectric Effect

The pyroelectric effect equation describes the increment of spontaneous polarization ΔP_s of a crystal with a change in its temperature, ΔT. To a first approximation quantities ΔP_s and ΔT are related lineraly:

$$\Delta P_s = p \Delta T , \tag{3.21}$$

where p is the pyroelectric coefficient. Considering infinitely small increments of T and P_s, we obtain the differential ratio

$$\partial P_s / \partial T = p , \tag{3.22}$$

which treats pyrocoefficent p as a partial dervivative of the spontaneous polarization with respect to the temperature. Thus, on the $P_s - T$ curve the value of p for a given temperature is defined as the slope of curve $P_s = f(T)$ at a given point.

Variation in P_s with the temperature may be due to one of two factors. First of all, as the temperature changes, a free crystal changes in size, contracting or expanding because of the thermal effect. It is obvious, therefore, that even if no rearrangement in the structure of the crystal has occurred owing to temperature changes, the spontaneous polarization of the crystal alters, since the number of the charges and dipoles per unit volume (which determines the value of spontaneous polarization) also changes. Thus it is quite clear that part of the pyroelectric effect must be due to the crystal deformation. This part is usually treated as piezoelectric (Sect. 3.3). The deformational, piezoelectric part of the pyroelectric effect is called a *secondary* or *spurious* pyroelectric effect, and the pyrocoefficent characterizing it is often denoted by p''.

In the early stages of pyroelectric effect studies it was assumed that the total pyroelectric effect (characterized by coefficient p) entirely reduces to the secondary effect. It was shown, however, that a change in temperature also changes P_s, even if the crystal is clamped and does not deform. This part of the effect, which is not associated with the crystal deformation, is called the *primary* or *true* pyroelectric effect and is described by coefficient p'. The nature of the true pyroeffect is not so clear, and it is usually attributed to the changes in crystal structure with changes in its temperature. In linear pyroelectrics the true pyroeffect is small, as a rule, about 2–5% of the whole effect. (Such a relationship at room temperature exists, for instance, for tourmaline, lithium sulphate, and other pyroelectric crystals.)

The equation of the pyroeffect, which takes into account its division into primary and secondary, has the form

$$\Delta P_s = (p' + p'')\Delta T = p\Delta T\,. \tag{3.23}$$

Note that since ΔP_s is a vector, pyroelectric coefficients p, p', and p'' are also vectors. Here, however, we only give the quantitative treatment of the effect.

As a rule, at room temperature, the pyroelectric coefficient p in linear dielectrics is only slightly temperature dependent, which points to a low temperature dependence of P_s in these crystals. The absolute value of p is close to one electrostatic unit. For tourmaline, for instance, the coefficient of the total pyroeffect is -1.3 CGSE units. A vivid idea of the pyroelectric polarization of tourmaline can be obtained from the following example. A tourmaline plate cut out perpendicularly to the pyroelectric axis and having a thickness of 0.1 cm, when heated uniformly by 10 °C, acquires an electric charge of about 5×10^{-9} C/cm^2, and the potential difference between its surfaces is then about 1200 V.

Below the room temperature many crystals of linear pyroelectrics show a reduced pyrocoefficient. Near absolute zero, p is very small. In some crystals (lithium sulphate, barium nitrite, etc.) at certain temperatures the pyrocoefficient passes zero and reverses its sign, which means that P_s passes a maximum (minimum) (Fig. 3.4). An idea of the relationship between the true (p'), spurious (p''), and total (p) pyrocoefficients can be obtained from Fig. 3.4b, from

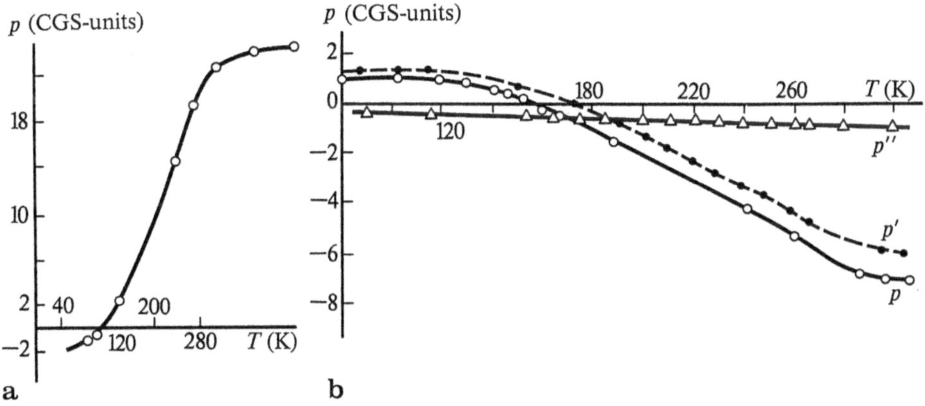

Fig. 3.4. Temperature dependence of the pyrocoefficients in lithium sulphate (**a**) and barium nitrite (**b**) crystals [3.1, 2]

which it follows that for barium nitrite the pyroeffect is mainly a true, primary one. It has been noted above that, as a rule, the secondary effect prevails in linear pyroelectrics. Barium nitrite (as well as some other crystals) is an exception to this "rule."

3.2.3 Electrocaloric Effect

This phenomenon consists in the change of the temperature (ΔT) of a pyroelectric crystal when an electric field ΔE is applied to it. Accordingly, the electrocaloric effect equation is

$$\Delta T = q\Delta E \,, \tag{3.24}$$

or, in the differential form,

$$q = \partial T/\partial E \,, \tag{3.25}$$

where q is the electrocaloric effect coefficient. Electrocaloric (as well as pyroelectric) properties are inherent only in crystals of polar classes.

It is easy to find the relationship between electrocaloric coefficient q and pyroelectric coefficient p. In thermodynamic consideration of a pyroelectric crystal with spontaneous polarization P_s, we assume that a change in this polarization will only affect the heat content of the crystal. The internal energy of the crystal, in this case too, remains unaltered, enabling us to write down several relations:

$$dU = 0 = EdP + TdS, \quad T = -EdP/dS \,, \tag{3.26}$$

$$\frac{\partial T}{\partial E} = -\frac{dP}{dS} = -\frac{\partial P}{\partial T}\,\frac{\partial T}{\partial S}\,, \tag{3.27}$$

where S is the entropy. But $\partial T/\partial E = q$, $\partial P/\partial T = p$, and it is known from thermodynamics that $dS = dQ/T$, where dQ is the heat increment; $dQ = dT\varrho cJ$ (where ϱ is the density, c is the heat capacity of the crystal, and J is the mechanical heat equivalent). Therefore, from (3.27) we finally obtain

$$q = -pT/\varrho cJ\,. \tag{3.28}$$

As would be expected, the electrocaloric and pyroelectric coefficients are porportional to one another and have opposite signs. This means that crystals with a stronger pyroelectric effect also have a stronger electrocaloric effect. Thus, if the crystal is heated (cooled), the accompanying change in P_s will lead to heat absorption (release) owing to the electrocaloric effect. As a result, the crystal will tend to retain its temperature. At $q>0$ and at a field coinciding with P_s in direction, the crystal temperature will rise (3.24), and, at $q>0$ at a field opposite to P_s in direction, it will fall off.

In accordance with the smallness of the pyroelectric effect in linear pyroelectrics, the electrocaloric effect is also small. Calculation shows, for instance, that a tourmaline plate, 1 mm thick, changes its temperature by only 5×10^{-5} deg on application of a voltage of 300 V.

3.3 Piezoelectric Effect and Electrostriction

3.3.1 Piezoelectric Effect

The *piezoelectric effect* is a set of phenomena which relate lineraly (proportionally) the mechanical stresses (strains) with the electric field (induction, polarization). Consideration of the relationship between the tensors of mechanical stresses t_{ik} and strains r_{ik} and the vectors of electric field tension $\mathfrak{U}$ and polarization $\boldsymbol{P}(\boldsymbol{P} = \boldsymbol{D}/4\pi)$ leads to the following piezoelectric-effect equations:[1]

$$\begin{aligned} P_n &= d_{nji}t_{ji}\,, & r_{ij} &= d_{mij}E_m\,,\\ P_n &= e_{nij}r_{ij}\,, & t_{ji} &= -e_{mji}E_m\,,\\ E_m &= -h_{mij}r_{ij}\,, & t_{ij} &= -h_{nji}P_n\,,\\ E_m &= -g_{mji}t_{ji}\,, & r_{ij} &= g_{nij}P_n\,. \end{aligned} \tag{3.29}$$

[1] In this chapter, in distinction to Chap. 2, mechanical stresses and strains are denoted by t and r, since σ and ε here mean the electric charge density and the dielectric constant, respectively.

Quantities d, e, g, and h are third-rank polar tensors called *piezoelectric coefficients*. The equations given in the left and right column of (3.29) describe the forward and reverse piezoeffects, respectively.

In the CGSE system, coefficients e and h have the dimensionality of electrical polarization ($\mathrm{cm}^{-1/2}\mathrm{g}^{-1/2}\mathrm{s}^{-1}$), and coefficients g and d, the reciprocal dimensionality ($\mathrm{cm}^{1/2}\mathrm{g}^{-1/2}\mathrm{s}$).

Under experimental conditions a crystal may be "short-circuited" or "open." A crystal (platelet) is "short-circuited," "closed," if its surfaces at which piezopolarization charges σ' emerge are connected by a conductor and if the crystal is simply situated in a conducting medium. The adsorbed free charge σ_0 and the bound charge compensated by it σ' are equal in magnitude and opposite in sign: $-\sigma_0=\sigma'$. A crystal is "open" if under piezoelectric polarization $P=\sigma'$ it is in an insulating medium and its surfaces with opposite charge signs are not connected electrically. Then the electrical induction $D=0$ (there are no free charges on the crystal). The field of bound charges σ' sets up a tension $E=-4\pi\sigma'/\varepsilon$ (where ε is the dielectric constant of the crystal inside the platelet).

In the reverse piezoeffect, by P is meant the surface density fo free charge σ_0. Under these conditions the crystal is open. When the external electric field E is impressed, the crystal is closed (through a battery).

Third-rank piezoelectric tensors d, e, g and h generally have 18 rather than 27 independent components owing to the symmetry with respect to two (second and third) indices. Nonzero piezocoefficients refer exclusively to crystals of noncentrosymmetric classes (with the exception of class 432, where all the tensor components of piezoelectric coefficients are zero by virtue of symmetry). There are 20 such classes: 1, 2, m, 222, $mm2$, 4, $\bar{4}$, 422, $4mm$, $\bar{4}2m$, 3, 32, $3m$, 6, $\bar{6}$, 622, $6mm$, $\bar{6}m2$, 23, $\bar{4}3m$. Strictly speaking, the crystals belonging to these classes may not be piezoelectrics as well. In practically all cases, however, the geometrical (symmetrical) and electrical polarities in crystals are equivalent, and no distinction is made between them here.

In practical calculations, the matrix form of recording (3.29) is employed, more often than not, using one subscript each at the strain and stress components and two subscripts each at the tensor components of piezocoefficients:

$$\begin{aligned}
&P_n=d_{nj}t_j\,, && r_i=d_{mi}E_m\,,\\
&P_n=e_{ni}r_i\,, && t_j=-e_{mj}E_m\,,\\
&E_m=-h_{mi}r_i\,, && t_j=-h_{nj}P_n\,,\\
&E_m=-g_{mj}t_j\,, && r_i=g_{ni}P_n
\end{aligned}\qquad (n,m=1,2,3;\quad i,j=1,2,\dots 6). \tag{3.30}$$

A comparison of coefficients d_{nji} from (3.29) and d_{ij} from (3.30) leads to the following relations:

$$
\begin{aligned}
&d_{111}=d_{11}\,, \quad d_{122}=d_{22}\,, \quad d_{133}=d_{13}\,,\\
&d_{132}=d_{123}=\tfrac{1}{2}d_{14}\,, \quad d_{131}=d_{113}=\tfrac{1}{2}d_{13}\,, \quad d_{112}=d_{121}=\tfrac{1}{2}d_{16}\,,\\
&d_{211}=d_{21}\,, \quad d_{222}=d_{22}\,, \quad d_{233}=d_{23}\,, \quad d_{232}=d_{123}=\tfrac{1}{2}d_{24}\,,\\
&d_{231}=d_{213}=\tfrac{1}{2}d_{25}\,, \quad d_{212}=d_{221}=\tfrac{1}{2}d_{26}\,, \quad d_{311}=d_{31}\,,\\
&d_{322}=d_{32}\,, \quad d_{333}=d_{33}\,, \quad d_{323}=d_{332}=\tfrac{1}{2}d_{34}\,,\\
&d_{313}=d_{331}=\tfrac{1}{2}d_{35}\,, \quad d_{312}=d_{321}=\tfrac{1}{2}d_{36}\,.
\end{aligned} \tag{3.31}
$$

Equations (3.29, 30) are generally very cumbersome. But in most of the concrete cases some of the coefficients are equal to zero or to each other because of the crystal symmetry. Indeed, as has already been noted, no piezo-effect is possible in centrosymmetric crystals. This follows from the fact that the "summing" of the symmetry elements of a crystal which has a center of symmetry and the uniform mechanical stress which also has a center of symmetry leads to a centrosymmetric group according to the Curie principle. In other words, a centrosymmetric crystal remains centrosymmetric after deformation. The presence of a center of symmetry in a deformed crystal indicates unambiguously that such a crystal has no polar directions and hence no electrical polarization.

Analytically, the effect of the symmetry on the type of piezoelectric coefficient tensors is taken into account in the same way as for tensors of any other ranks (Chap. 1). The results of such consideration for all the piezoelectric classes are given in Tables 3.1, 2. Table 3.1 shows that for 20 crystallographic classes the matrix of coefficient d_{nj} has at least one nonzero coefficient (but not more than 18, of course).

The matrices of coefficients g_{mj} coincide with those of coefficients d_{nj} listed in Table 3.1. At the same time, in accordance with the notation adopted here for the shear components of deformation tensor r_{ij} there is, for some classes, a difference between the matrices of coefficients e_{mi} and h_{mi} and the above matrices d_{nj}. The matrices of the classes for which this difference takes place are indicated inn Table 3.2. Note that in passing over from two-digit to three-digit indices for coefficients e and h in (3.31) no factors 1/2 are present.

3.3.2 Electromechanical Transformation

The energy of a crystal is generally a function of its thermal, mechanical, and electrical states. Temperature T or entropy S is used as an independent variable for describing the thermal state; the mechanical state is characterized by tension t or deformation r, and the electrical state, by electric field E or electric induction D. The elastic and dielectric properties of a crystal can be measured under quite definite conditions characterizing its thermal, mechanical, and electrical state in each particular case.

Table 3.1. Shape of the matrices of piezoelectric coefficients for all crystallographic classes. The number of nonzero coefficients is indicated in parentheses after each matrix

Triclinic syngony

Class 1

$$\begin{Vmatrix} d_{11} & d_{12} & d_{13} & d_{14} & d_{15} & d_{16} \\ d_{21} & d_{22} & d_{23} & d_{24} & d_{25} & d_{26} \\ d_{31} & d_{32} & d_{33} & d_{34} & d_{35} & d_{36} \end{Vmatrix}$$

Monoclinic syngony

Class 2, $2 \parallel Y$

$$\begin{Vmatrix} 0 & 0 & 0 & d_{14} & 0 & d_{16} \\ d_{21} & d_{22} & d_{23} & 0 & d_{25} & 0 \\ 0 & 0 & 0 & d_{34} & 0 & d_{36} \end{Vmatrix}$$

Class 2, $2 \parallel Z$

$$\begin{Vmatrix} 0 & 0 & 0 & d_{14} & d_{15} & 0 \\ 0 & 0 & 0 & d_{24} & d_{25} & 0 \\ d_{31} & d_{32} & d_{33} & 0 & 0 & d_{36} \end{Vmatrix}$$

Class m, $m \perp Y$

$$\begin{Vmatrix} d_{11} & d_{12} & d_{13} & 0 & d_{15} & 0 \\ 0 & 0 & 0 & d_{24} & 0 & d_{26} \\ d_{31} & d_{33} & 0 & d_{35} & 0 & \end{Vmatrix}$$

Class m, $m \perp Z$

$$\begin{Vmatrix} d_{11} & d_{12} & d_{13} & 0 & 0 & d_{16} \\ d_{21} & d_{22} & d_{23} & 0 & 0 & d_{26} \\ 0 & 0 & 0 & d_{24} & d_{25} & 0 \end{Vmatrix}$$

Rhombohedral syngony

Class 3

$$\begin{Vmatrix} d_{11} & -d_{11} & 0 & d_{14} & d_{15} & 2d_{22} \\ d_{22} & -d_{22} & 0 & d_{15} & -d_{14} & 2d_{11} \\ d_{31} & d_{31} & d_{33} & 0 & 0 & 0 \end{Vmatrix}$$

Class 32

$$\begin{Vmatrix} d_{11} & -d_{11} & 0 & d_{14} & 0 & 0 \\ 0 & 0 & 0 & 0 & -d_{14} & 2d_{11} \\ 0 & 0 & 0 & 0 & 0 & 0 \end{Vmatrix}$$

Class $3m$, $m \perp X$

$$\begin{Vmatrix} 0 & 0 & 0 & 0 & d_{15} & 2d_{21} \\ d_{21} & -d_{21} & 0 & d_{15} & 0 & 0 \\ d_{31} & d_{31} & d_{33} & 0 & 0 & 0 \end{Vmatrix}$$

Class $3m$, $m \perp Y$

$$\begin{Vmatrix} d_{11} & -d_{11} & 0 & 0 & d_{15} & 0 \\ 0 & 0 & 0 & d_{15} & 0 & 2d_{11} \\ d_{31} & d_{31} & d_{33} & 0 & 0 & 0 \end{Vmatrix}$$

Hexagonal syngony

Class 6

$$\begin{Vmatrix} 0 & 0 & 0 & d_{14} & d_{15} & 0 \\ 0 & 0 & 0 & d_{15} & -d_{14} & 0 \\ d_{31} & d_{31} & d_{33} & 0 & 0 & 0 \end{Vmatrix}$$

Class $6mm$

$$\begin{Vmatrix} 0 & 0 & 0 & 0 & d_{15} & 0 \\ 0 & 0 & 0 & d_{15} & 0 & 0 \\ d_{31} & d_{31} & d_{33} & 0 & 0 & 0 \end{Vmatrix}$$

Class 622

$$\begin{Vmatrix} 0 & 0 & 0 & d_{14} & 0 & 0 \\ 0 & 0 & 0 & 0 & -d_{14} & 0 \\ 0 & 0 & 0 & 0 & 0 & 0 \end{Vmatrix}$$

Class $\bar{6}$

$$\begin{Vmatrix} d_{11} & -d_{11} & 0 & 0 & 0 & 2d_{21} \\ d_{21} & -d_{21} & 0 & 0 & 0 & 2d_{11} \\ 0 & 0 & 0 & 0 & 0 & 0 \end{Vmatrix}$$

Class $\bar{6}m2$, $m \perp X$

$$\begin{Vmatrix} 0 & 0 & 0 & 0 & 0 & 2d_{22} \\ d_{22} & -d_{22} & 0 & 0 & 0 & 0 \\ 0 & 0 & 0 & 0 & 0 & 0 \end{Vmatrix}$$

Class $\bar{6}m2$, $m \perp Y$

$$\begin{Vmatrix} d_{11} & -d_{11} & 0 & 0 & 0 & 0 \\ 0 & 0 & 0 & 0 & 0 & 2d_{11} \\ 0 & 0 & 0 & 0 & 0 & 0 \end{Vmatrix}$$

(Continued)

Table 3.1 (continued)

Cubic syngony

Classes $\bar{4}3m$ and 23

$$\left\| \begin{matrix} 0 & 0 & 0 & d_{14} & 0 & 0 \\ 0 & 0 & 0 & 0 & d_{14} & 0 \\ 0 & 0 & 0 & 0 & 0 & d_{14} \end{matrix} \right\|$$

Orthorhombic syngony

Class 222

$$\left\| \begin{matrix} 0 & 0 & 0 & d_{14} & 0 & 0 \\ 0 & 0 & 0 & 0 & d_{25} & 0 \\ 0 & 0 & 0 & 0 & 0 & d_{36} \end{matrix} \right\|$$

Class $mm2$

$$\left\| \begin{matrix} 0 & 0 & 0 & 0 & d_{15} & 0 \\ 0 & 0 & 0 & d_{24} & 0 & 0 \\ d_{31} & d_{32} & d_{33} & 0 & 0 & 0 \end{matrix} \right\|$$

Tetragonal syngony

Class 4

$$\left\| \begin{matrix} 0 & 0 & 0 & d_{14} & d_{15} & 0 \\ 0 & 0 & 0 & d_{15} & -d_{14} & 0 \\ d_{31} & d_{31} & d_{33} & 0 & 0 & 0 \end{matrix} \right\|$$

Class $\bar{4}$

$$\left\| \begin{matrix} 0 & 0 & 0 & d_{14} & d_{15} & 0 \\ 0 & 0 & 0 & -d_{15} & d_{14} & 0 \\ d_{31} & -d_{31} & 0 & 0 & 0 & d_{36} \end{matrix} \right\|$$

Class 422

$$\left\| \begin{matrix} 0 & 0 & 0 & d_{14} & 0 & 0 \\ 0 & 0 & 0 & 0 & -d_{14} & 0 \\ 0 & 0 & 0 & 0 & 0 & 0 \end{matrix} \right\|$$

Class $4mm$

$$\left\| \begin{matrix} 0 & 0 & 0 & 0 & d_{15} & 0 \\ 0 & 0 & 0 & d_{15} & 0 & 0 \\ d_{31} & d_{31} & d_{33} & 0 & 0 & 0 \end{matrix} \right\|$$

Class $\bar{4}2m$, $2 \| x, y$

$$\left\| \begin{matrix} 0 & 0 & 0 & d_{14} & 0 & 0 \\ 0 & 0 & 0 & 0 & d_{14} & 0 \\ 0 & 0 & 0 & 0 & 0 & d_{36} \end{matrix} \right\|$$

Table 3.2. Shape of matrices of piezoelectric coefficients e_{ij} and h_{ij} for classes where this shape is different from that of matrices d_{ij} and g_{ij}. The number of nonzero coefficients is indicated in parentheses after each matrix

Class 3

$$\left\| \begin{matrix} e_{11} & -e_{11} & 0 & e_{14} & e_{15} & e_{21} \\ e_{21} & -e_{21} & 0 & e_{15} & -e_{14} & -e_{11} \\ e_{31} & e_{31} & e_{33} & 0 & 0 & 0 \end{matrix} \right\|$$

Class 32

$$\left\| \begin{matrix} e_{11} & -e_{11} & 0 & e_{14} & 0 & 0 \\ 0 & 0 & 0 & 0 & -e_{14} & -e_{11} \\ 0 & 0 & 0 & 0 & 0 & 0 \end{matrix} \right\|$$

Class $\bar{6}$

$$\left\| \begin{matrix} e_{11} & -e_{11} & 0 & 0 & 0 & e_{21} \\ e_{21} & -e_{21} & 0 & 0 & 0 & -e_{11} \\ 0 & 0 & 0 & 0 & 0 & 0 \end{matrix} \right\|$$

Class $3m$, $m \perp X$

$$\left\| \begin{matrix} 0 & 0 & 0 & 0 & e_{15} & e_{21} \\ e_{21} & -e_{21} & 0 & e_{15} & 0 & 0 \\ e_{31} & e_{31} & e_{33} & 0 & 0 & 0 \end{matrix} \right\|$$

Class $\bar{6}m2$, $m \perp Y$

$$\left\| \begin{matrix} e_{11} & -e_{11} & 0 & 0 & 0 & 0 \\ 0 & 0 & 0 & 0 & 0 & -e_{11} \\ 0 & 0 & 0 & 0 & 0 & 0 \end{matrix} \right\|$$

The thermal state of a crystal may either be with a constant temperature (isothermal conditions, $T = \text{const}$) or with a constant entropy (adiabatic conditions, $S = \text{const}$). Mechanically, a crystal may be free (no mechanical stresses, $t = 0$) or clamed (deformation r is constant or zero). Electrically, as indicated above, a crystal may be open (induction D is constant or zero) or closed (electric field E is constant or zero). With reference to coefficients of elastic rigidity, for instance, we can speak of quantities c^{DS}, c^{DT}, c^{ES}, and c^{ET}. The first one characterizes the rigidity measured under adiabatic conditions in the open state (superscripts D and S), the second, the rigidity measured under isothermal conditions on an open crystal, etc. Generally speaking, in measuring the dielectric characteristics it is necessary, in turn, to distinguish ε^{rS}, ε^{tS}, ε^{rT}, and ε^{tT}; ε^{rS} is measured under adiabatic conditions on a clamped crystal, and ε^{tT}, under isothermal conditions on a free crystal, etc.

In practice, adiabatic conditions are observed in measurements at high frequencies. The individual parts of the crystal have not time to exchange heat in such measurements. In static measurements, on the contrary, isothermal measuring conditions are realized.

The coefficients of elasticity (elastic rigidities c and compliances s) measured under different thermal conditions, but under equal mechanical and electrical conditions, turn out to differ only moderately, by about 1%. As a rule, adiabatic complicanes are less than isothermal (under adiabatic conditions the crystal is more rigid). Dielectric constants ε measured under different thermal conditions differ only when the crystal is a pyroelectric. The difference between ε^S and ε^T is porportional to product $p_i p_j$ (where p_i and p_j are pyrocoefficients), and the sign of this difference depends on that of p_i and p_j. Taking into consideration the relationship between coefficients c and s and also between D and E, it is easy to obtain the relationships between different piezoelectric coefficients. These have the form

$$
\begin{aligned}
d_{nj} &= \frac{\varepsilon^t_{mn}}{4\pi} g_{nj} = e_{ni} s^E_{ij}\,, \qquad g_{nj} = 4\pi \beta^t_{mn} d_{mj} = h_{ni} s^D_{ij}\,,\\
&\qquad (m, n = 1, 2, 3; \quad i, j = 1, 2, \ldots, 6), \qquad (3.32)\\
e_{nj} &= \frac{\varepsilon^r_{mn}}{4\pi} h_{mj} = d_{ni} c^E_{ij}\,, \qquad h_{nj} = 4\pi \beta^r_{mn} e_{mj} = g_{ni} c^D_{ij}
\end{aligned}
$$

β_{mn} being the tensor of the reciprocal dielectric constant.

For piezoelectric crystals the values of c^D, c^E, and (respectively) s^D, s^E are different, all other things being equal. Qualitatively, this difference can be understood by taking into account that measurement of elastic coefficients is accompanied by the piezoelectric polarization of the crystal. For an electrically open crystal ($D = \text{const}$) this means that in finding elastic coefficients one measures both the rigidity of the crystal itself and (if we can phrase it so) the "rigidity" of the electric field induced by the piezoelectric polarization in the external space. In other words, in this case part of the energy is spent on setting up an electric field. This means that $c^D > c^E$ (or $s^D < s^E$), which is actually the case. Obviously, the difference in rigidities (or compliances) is the measure of conversion from mechanical to electrical energy: the more energy spent for setting up an electric field, the higher the *electromechanical coupling coefficient k*, which is defined by the expressions

$$k^2 = \frac{s^E - s^D}{s^E} = \frac{c^D - c^E}{c^D} . \tag{3.33}$$

Finally, quantities ε^t and ε^r are distinguished for piezoelectric crystals, all other things being equal. If a crystal is free, the application of an electric field leads to mechanical deformation of the crystal in addition to polarization due to the piezoeffect. In a campled crystal, no such conversion occurs. This means that there must exist a relationship $\varepsilon^t > \varepsilon^r$. The difference between these values is also the measure of electromechanical conversion:

$$k^2 = (\varepsilon^t - \varepsilon^r)/\varepsilon^t . \tag{3.34}$$

The electromechanical coupling coefficient is an anisotropic quantity; it may have different values for different specimens performing different oscillations. The electromechanical coupling is the higher, the larger the piezocoefficients. For the ith component of field E (or of induction D) and for the jth component of tensors t and r, the following relations hold true:

$$k^2 = \frac{d_{ij}^2}{\varepsilon_{ii}^t s_{jj}^E} = \frac{h_{ij}^2}{\beta_{ii}^s c_{jj}^D}, \qquad k^2 = \frac{e_{ij}^2}{\varepsilon_{ii}^r c_{jj}^E} = \frac{g_{ij}^2}{\beta_{ii}^t s_{jj}^D} . \tag{3.35}$$

Piezoelectric coefficients d, e, g, and *h,* measured at a constant entropy S (under adiabatic conditions) and at a constant temperature T (under isothermal conditions), differ only for pyroelectric crystals. For linear pyroelectrics this difference is small and equals about 0.1%. The details of this difference for ferroelectrics in the phase transition region will be considered below.

The ability of piezocrystals to acquire polarization due to mechanical effects and to deform under the effect of electric fields makes it possible to regard them as electromechanical converters. It is clear from the foregoing that the efficiency of such a converter depends on the difference between the dielectric constants of a mechanically clamped and a free crystal and the difference between the elastic coefficients of an electrically short-circuited and an open crystal.

By the very definition, i.e., by the way piezoelectric effect equations (3.29) are written, coefficients d, g, e, and h characterize definite regimes of electromechanical conversion. Thus, from the equations for coefficient d (omitting the tensor subscripts)

$$P = dt, \qquad r = dE \tag{3.36}$$

if follows that, with a forward piezoeffect, coefficient d defines the electrical polarization (charge density or induction, since $D = 4\pi P = 4\pi\sigma$) arising under the effect of mechanical stresses. With a reverse piezoeffect, coefficient d indicates the value of the deformation due to the applied electric field. Coefficient g in a forward piezoeffect points to the potential difference due to mechanical stresses, and in a reverse piezoeffect, the value fo the deformation due to the electric charges deposited on the crystals, etc. This, in turn, means that in trying, for instance, to obtain high electric tensions under the effect of mechanical stresses one must use crystals with maximum coefficients g. Such requirements are imposed, for example, in piezoelectric receivers of sonic and ultrasonic waves. In an attempt to ensure effective deformations of a crystal under the effect of an electric field, one must select crystals with maximum coefficients d. Such requirements are envisioned in designing effective sonic and ultrasonic radiators. To ensure effective measurement of deformation by electrical methods with the aid of piezocrystals one must select crystals with maximum coefficients h and e (piezoelectric seismograph, piezoelectric adapter), since in a forward piezoeffect these coefficients characterize proportional deformations of the values of field E and polarization P.

3.3.3 Piezoelectric Properties of Linear Dielectrics

Quartz (SiO_2) is the best known and most widely used linear piezoelectric crystal. The low-temperature modification of quartz (below 573 °C), α quartz, belongs to the rhombohedral syngony (class 32, space group $D_3^4 = C3_121$). For each unit cell, which, at room temperature, has parameters $a = 4.90$ Å and $c = 5.39$ Å, there are three SiO_2 "molecules". The structure of α quartz is shown in Fig. 3.5. The motif of the structure is formed by tetrahedra of [SiO_4] enmeshed at the vertices. The tetrahedra are slightly distorted: two distances Si–O are equal to 1.61 Å, and the other two, to 1.62 Å. The enmeshed tetra-

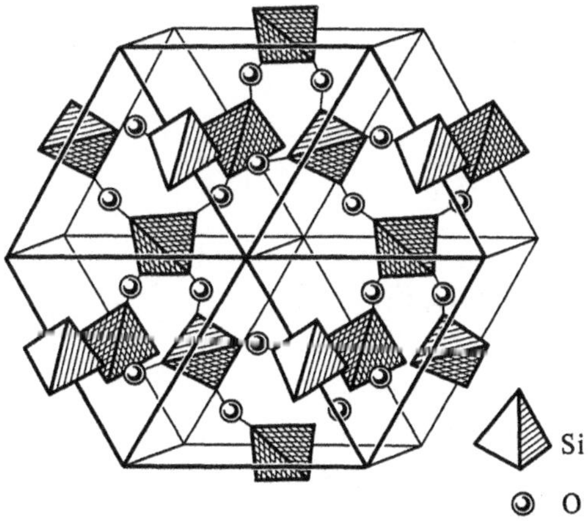

Fig. 3.5. Crystal structure of α quartz

hedra of $[SiO_4]$ form a screw motif along the *c* axis with voids at the center of a tripled hexagonal unit cell.

In the vicinity of 573 °C quartz undergoes a phase transformation and becomes hexagonal above this temperature (class 622, space group $D_6^5 = P6_122$). This modification is called β quartz; it is stable within the temperature range of 573 to 870 °C. At higher temperatures, two more modifications of SiO_2 – tridymite and christobalite – are known.

Both α quartz and β quartz possess piezoelectric properties. The equations with the use of piezomodulus d for class 32 (to which α quartz belongs) have the following form in accordance with Table 3.1:

	t_1	t_2	t_3	t_4	t_5	t_6
P_1	d_{11}	$-d_{11}$	0	d_{14}	0	0
P_2	0	0	0	0	$-d_{14}$	$-2d_{11}$
P_3	0	0	0	0	0	0

(3.37)

for the forward piezoelectric effect and

	E_1	E_2	E_3
r_1	d_{11}	0	0
r_2	$-d_{11}$	0	0
r_3	0	0	0
r_4	d_{14}	0	0
r_5	0	$-d_{14}$	0
r_6	0	$-2d_{11}$	0

(3.38)

for the reverse piezoelectric effect.

In the conventional form, the foward piezoelectric effect equations are written as follows:

$$P_1 = d_{11}t_1 - d_{11}t_2 + d_{14}t_4, \quad P_2 = -d_{14}t_5 - 2d_{11}t_6, \tag{3.39}$$

and those of the reverse:

$$\begin{aligned} &r_1 = d_{11}E_1, \quad r_2 = -d_{11}E_1, \quad r_4 = d_{14}E_1, \\ &r_5 = -d_{14}E_2, \quad r_6 = -2d_{11}E_2. \end{aligned} \tag{3.40}$$

The main peculiarity of the piezoelectric effect in quartz is that in accordance with its symmetry it does not posses piezoelectric properties in the direction of the *Z* axis (*c* axis). The simplest piezoelectric cuts of quartz are *X*

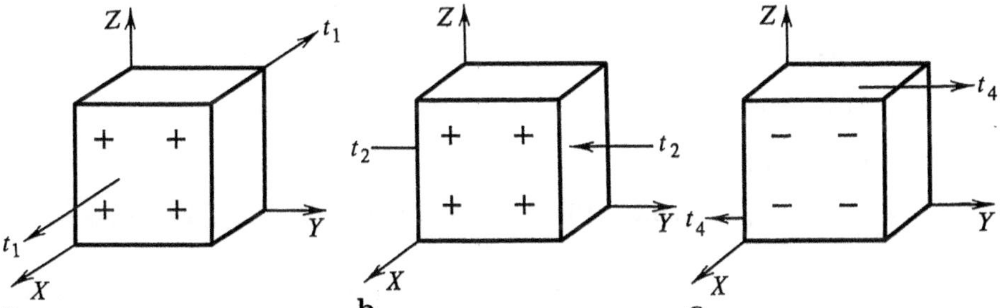

Fig. 3.6. Direct piezoelectric effect in quartz: **(a)** longitudinal, **(b)** transverse, **(c)** shear-strain-induced piezoeffects

and Y cuts perpendicular to the crystallophysical X and Y axes. Platelets that are X cut are generally used for inducing the longitudinal piezoeffect (Fig. 3.6a), and Y cut platelets, for inducing the transverse piezoeffect (Fig. 3.6b). The equation of the longitudinal piezoelectric effect in an X cut platelet has the form

$$P_1 = d_{11}\, t_1 \,, \tag{3.41}$$

and the equation of the transverse piezoelectric effect in a Y cut platelet

$$P_2 = -d_{11}\, t_2 \,. \tag{3.42}$$

The piezoelectric polarization due to shear stresses is defined by piezomodulus d_{14} (Fig. 3.6c).

The values of the quartz piezomoduli in the CGSE system are equal to

$$d_{11} = -6.76 \times 10^{-8}, \qquad d_{14} = 2.56 \times 10^{-8} \,.$$

To convey an idea of the magnitude of the piezoelectric effect in quartz, we indicate that a field of 1000 V applied to an X cut platelet of quartz with a thickness of 1 cm (in the direction of the X axis) increases the platelet thickness by 21 Å. On the other hand, a stress of 1 kgf · cm^{-2} acting on an identical platelet along the X axis sets up a potential difference of about 60 V in the direction of this axis.

The coefficients of rigidity c and compliance s of quartz, measured under adiabatic and isothermal conditions, are practically the same. The values of these coefficients at room temperature, as measured at a constant field tension, are as follows (in CGSE units):

$$s_{11}^E = 127 \times 10^{-14} \,, \qquad c_{11}^E = 86.05 \times 10^{10} \,,$$

$$s_{12}^E = -15.35 \times 10^{-14} \,, \qquad c_{12}^E = 4.85 \times 10^{10} \,,$$

$$s_{13}^E = -11.10^{-14}, \qquad c_{13}^E = 10.45 \times 10^{10},$$
$$s_{14}^E = -44.6 \times 10^{-14}, \qquad c_{14}^E = 18.25 \times 10^{10},$$
$$s_{33}^E = 95.6 \times 10^{-14}, \qquad c_{33}^E = 107.1 \times 10^{10},$$
$$s_{44}^E = 197.8 \times 10^{-14}, \qquad c_{44}^E = 58.65 \times 10^{10},$$
$$s_{66}^E = 2(s_{11}^E - s_{12}^E) = 286.6 \times 10^{-14}, \qquad c_{66}^E = (c_{11} - c_{16})/2 = 40.5 \times 10^{10}.$$

Quartz, as a rhombohedral crystal, has two independent dielectric constants $\varepsilon_1 = \varepsilon_2$, and ε_3. Their values at $t = 0$, i.e., for a mechanically free crystal, are

$$\varepsilon_1^t = 4.58, \qquad \varepsilon_3^t = 4.70.$$

Being a linear piezoelectric, quartz has a small electromechanical coupling coefficient $k = 0.1$ when X cut platelets are used and longitudinal vibrations are excited. At some shear vibrations with the use of Y cut platelets, k is slightly larger and equals 0.14.

The wide use of quartz in various devices (including piezoelectric devices) is due to its valuable properties. Thus, quartz has a Mohs hardness of 7, is water insoluble, resistant to many acids, melts at 1700 °C, and shows low thermal expansion ($\alpha_1 = 8 \times 10^{-6}$, $\alpha_3 = 13.4 \times 10^{-6}\,\mathrm{deg}^{-1}$). Quartz is a good insulator; its resistivity at room temperature is $10^{14} - 10^{15}\,\Omega$cm.

One of the most valuable properties of quartz is its ability to change its characteristics only slightly and reproduce them well over a wide range of temperatures (in the low-temperature range, down to super-low, and in the high-temperature range, almost up to the point of conversion to β quartz, 573 °C). The stability of quartz characteristics ensures its wide application, especially where a high measuring accuracy and stability of parameters are required.

As a piezoelectric, quartz is mainly used for stabilizing the generator frequency. It is also often used in high-selectivity filters, in high frequency electromechanical converters for exciting and receiving mechanical vibrations, and in various types of transmitters of mechanical quantities.

Until relatively recently, only natural quartz was used in technology. In the last few years, synthetic quartz has been grown in industrial conditions; it favorably compares with the natural mineral.

Among the linear piezoelectrics, crystals of potassium tartrate ($K_2C_4H_4O_6 \cdot 1/2\,H_2O$), ethylenediaminetartrate ($C_2H_{14}N_2O_6$), lithium sulphate ($Li_2SO_4 \cdot H_2O$), tourmaline, and ammonium dihydrophosphate ($NH_4H_2PO_4$) have been used in various devices, besides quartz. All these crystals (except the last one) possess pyroelectric properties as well.

The peculiarity of the piezoproperties of the pyroelectrics is that an electrical polarization sets up in them, in particular, under triaxial (hydrostatic) compression (extention). Therefore, pyroelectrics are used in making elements for devices recording hydrostatic compression. For instance, tourmaline (class

$3m$) platelets cut out perpendicularly to axis 3 (Z axis) are used for this purpose.

The piezoelectric effect in hydrostatic compression can be calculated by using the conventional piezoeffect equations. To do this one assumes, in the mechanical stress tensor, $t_1 = t_2 = t_3 = p$ (where p is the hydrostatic compression) and $t_4 = t_5 = t_6 = 0$. This leads (for instance, in tourmaline) to the expression for polarization

$$P_3 = (d_{33} + 2d_{31})p = d_{\text{hydro}}\, p \,, \tag{3.43}$$

from which the piezomodulus in hydrostatic compression is written as

$$d_{\text{hydro}} = d_{33} + 2d_{31} \,. \tag{3.44}$$

For crystals belonging to class 2 (lithium sulphate, etc.), d_{hydro} is determined by the following relation in accordance with Table 3.2:

$$d_{\text{hydro}} = d_{22} + d_{21} + d_{23} \tag{3.45}$$

(for the case where axis 2 is taken to be the Y axis) and by the expression

$$d_{\text{hydro}} = d_{33} + d_{31} + d_{32} \tag{3.46}$$

(for the case where axis 2 is taken to be the Z axis).

The absolute value of d_{hydro} is (in CGSE units) 7.56×10^{-8} for tourmaline, 4.1×10^{-8} for potassium tartrate, and 38.9×10^{-8} for lithium sulphate.

3.3.4 Electrostriction

The deformation of a dielectric proportional to the square of the applied field E is called *electrostriction.*

One can write the following four electrostriction equations, taking into account that stresses t and deformation r can be sued for describing the "mechanical" aspects, and field intensity E and polarization P for describing the "electrical" aspect:

$$\begin{aligned} r_{ij} &= Q_{ijmn} P_m P_n \,, & r_{ij} &= R_{ijmn} E_m E_n \,, \\ t_{ij} &= G_{ijmn} P_m P_n \,, & t_{ij} &= H_{ijmn} E_m E_n \,. \end{aligned} \tag{3.47}$$

Taking advantage of the relationship between r and t, and also between P and E, we can obtain the relationship between coefficients R, Q, G, and H. Here are some of them:

$$Q_{ijmn} = s^P_{ijkl} G_{klmn}\,, \qquad G_{ijmn} = c^P_{ijkl} Q_{klmn}\,,$$

$$R_{ijmn} = \left(\frac{1}{4\pi}\right)^2 \varepsilon_{km}\,\varepsilon_{ln}\,Q_{ijkl}\,, \qquad H_{ijmn} = \left(\frac{1}{4\pi}\right)^2 \varepsilon_{km}\,\varepsilon_{ln}\,G_{ijki}\,. \tag{3.48}$$

The electrostriction should not be confused with the reverse piezoelectric effect. The latter, as noted above, is also the result of application of an external field, but is a linear effect (the deformation is directly proportional to the field), while the electrostriction is a quadratic effect. This means, among other things, that the sign of electrostriction (which indicates whether the crystal expands or contracts under the effect of the field) is independent of the field direction, whereas in the case of the piezoelectric effect a reversal of the field changes the sign of deformation. In an alternating field, as a result of the electrostriction, the crystal vibrates with a double frequency as compared with that of the field, while in the case of the piezoelectric effect the field and deformation frequencies coincide. The electrostriction takes place in all dielectrics irrespective of their symmetry. In contrast, the piezoelectric effect occurs only in noncentrosymmetric crystals, as shown above. Quantitatively, the electrostrictive deformation due to the quadratic effect is less than the piezoelectric. Finally, in some piezoelectric crystals a field applied in definite directions does not lead to a piezoelectric deformation (Table 3.1), and in this case only an electrostrictive deformation will take place in the piezoelectric.

Each centrosymmetric dielectric crystal an, generally speaking, be transformed to a piezoelectric by applying and external field. In such an artificial piezoelectric, the value of the coefficient characterizing the piezoelectric effect (morphic effect, Chap. 1) will be proportional to the field and can be found by using one of the equations (3.47), for instance, the first:[2]

$$\frac{\partial r_{ij}}{\partial P_l} = 2\,Q_{ijll} P_l\,. \tag{3.49}$$

Comparing (3.49) with the last equation from (3.29), we have

$$g_{lij} = 2\,Q_{ijll} P_l\,, \tag{3.50}$$

which enables us to treat, for this case, the electrostriction equation (3.47) as the reverse-piezoeffect equation:

$$r_{ij} = g_{lij} P_l\,. \tag{3.51}$$

[2] It is assumed here that a centrosymmetric crystal is assigned a large free charge $\sigma_0 = P_l = D/4\pi$, after which its value varies within a narrow range, and the deformation (proportional to these small changes, σ_0) is measured.

A polarized centrosymmetric piezoelectric crystal also possesses a foward piezoeffect. In field E_l this effect is described by the equation

$$E_l = -g_{lij} t_{ij}, \tag{3.52}$$

in which g is the piezoelectric coefficient determined by (3.50). The piezoelectric effect in such a piezoelectric can also be described by other equations of the forward and reverse piezoeffect. It is clear that this morphic piezoelectric effect is very small.

A fourth-rank tensor describing the electrostriction in the most general case, contains 36, and not 81, independent coefficients because of the symmetry of the tensor with respect to the first and second pair. Allowance for the symmetry of the medium reduces the number of nonzero and independent coefficients. More often than not, crystal physics uses not the general tensor form of electrostriction equations, but the simplified, matrix form, in which one index is employed for denoting the components of the stress or strain tensor (from 1 to 6), one index for the components of the electrical vector (from 1 to 3), and two indices for the electrostriction coefficients. The matrices of thees coefficients are rather bulky, and we give only two of them, one of which corresponds to crystals of cubic classes $\bar{4}3m$, 432, $m3m$ (for instance, $BaTiO_3$ in the paraelectric modification), and the other, to all the monoclinic classes (for instance, triglycinesulphate and Rochelle salt in the ferroelectric modification):

Classes $\bar{4}3m$, 432, $m3m$ Class 2 ($2 \parallel Z$)

$$\left\| \begin{matrix} R_{11} & R_{12} & R_{12} & 0 & 0 & 0 \\ R_{12} & R_{11} & R_{12} & 0 & 0 & 0 \\ R_{12} & R_{12} & R_{11} & 0 & 0 & 0 \\ 0 & 0 & 0 & R_{44} & 0 & 0 \\ 0 & 0 & 0 & 0 & R_{44} & 0 \\ 0 & 0 & 0 & 0 & 0 & R_{44} \end{matrix} \right\| \qquad \left\| \begin{matrix} R_{11} & R_{12} & R_{13} & 0 & 0 & R_{16} \\ R_{21} & R_{22} & R_{23} & 0 & 0 & R_{26} \\ R_{31} & R_{32} & R_{33} & 0 & 0 & R_{36} \\ 0 & 0 & 0 & R_{44} & R_{45} & 0 \\ 0 & 0 & 0 & R_{54} & R_{55} & 0 \\ R_{61} & R_{62} & R_{63} & 0 & 0 & R_{66} \end{matrix} \right\| \tag{3.53}$$

For some classes of crystals the matrices of coefficents R and Q slightly differ from those of G and H in accordance with the difference in transition to one index for r and t.

The physical meaning of coefficients Q, R, G, and H is clear from (3.47). Thus, coefficient R characterizes the deformation of a free crystal on application of an electric field. The indicated coefficients characterize the operation of electrostriction converters in different regimes.

Coefficient R is determined by measuring the deformation under the effect of an inducing field, coefficient H is calculated by comparing mechanical stresses and the square of the electric field tension, etc. From a comparison of the electrostriction and the reverse piezoelectric effect it follows that the conditions for determining coefficient Q are equivalent to those for determining coefficient g. Further, to the conditions for determining coefficient R correspond the conditions for determining piezomodulus d, to G corresponds h, and to H corresponds e.

In addition to the direct method for determining electrostriction coefficients, there are also methods for indirect determination. Thus, coefficients Q and R can be determined by measuring the dependence of piezocoefficient g on the applied electric field, while coefficient R can be derived from the dependence of piezocoefficient d on the electric field, etc. (see below).

Electrostrictive deformations of crystals of linear dielectrics are extremely small. This explains the fact that until recently the scientific literature lacked the values of electrostriction coefficients of crystals.

As indicated above, a quartz crystal does not possess the piezoelectric effect if a field is applied to it along the Z axis (Table 3.1). Thus, an electric field applied in the Z direction can cause only electrostrictive deformation. Electrostriction R_{33} in this crystal was determined by establishing the dependence of the morphic piezoelectric effect, induced by a constant field applied along the Z axis, on the value of this field. In these experiments the piezomodulus at different constant fields was measured by applying an additional weak alternating field to the crystal. Using second equation of (3.47) for this case, we get

$$d_{33} = 2R_{33}E_3 . \tag{3.54}$$

The value of R_{33} found in experiments from the slope of the straight line in dependence $d_{33} = f(E)$ was equal to

$$R_{33} = 0.3 \times 10^{-12} \text{ CGSE units} .$$

It is worth noting that at a field of 100 kV/cm the value of d_{33} is 2×10^{-10} CGSE units, which is over two orders of magnitude less than d_{11} of quartz (6.8×10^{-8} CGSE units, see above). It follows that in order to obtain on a Z cut of quartz a piezomodulus equal to piezomodulus d_{11}, one would have to apply a field of the order of 10^7 V/cm.

The electrostriction coefficiencts of a NaCl crystal were measured by a similar method. The electrostriction coefficients found from dependences $d = f(E)$ are equal to (in CGSE units)

$$R_{11} = R_{22} = R_{33} = 2.7 \times 10^{-12},$$

$$R_{12} = R_{13} = R_{23} = 1.35 \times 10^{-12}, \qquad R_{44} = 0.9 \times 10^{-12} .$$

A comparison of the data for quartz and rock salt shows that the electrostrictive phenomena are much more pronounced in rock salt than in quartz. A NaCl crystal in a field of 100 kV/cm has a piezomodulus of 2×10^{-9} CGSE units.

3.3.5 Piezoelectric Textures

A *texture* is a homogeneous medium made up of particles consistently oriented in space. Examples of textures are: crystalline textures consisting of oriented crystals, fibrous materials (for instance, wood), electrets, piezoceramics, permanent polycrystalline magnets, etc.

Generally speaking, textures may be isotropic or anisotropic. Some of the anisotropic textures may possess piezoelectric properties. The possibility of existence of the piezoelectric effect in textures was predicted and proved experimentally by *Shubnikov* et al., who also conducted the first investigations of such piezoelectric textures [3.3].

Of greatest interest are textures with the symmetry of limiting groups (Chap. 1). Among the seven different types of such textures only three are piezoelectric. These are the textures described by groups ∞, ∞mm and $\infty 2$.

The piezoelectric textures of symmetry ∞ and ∞mm are polar. The texture of symmetry $\infty 2$ has no polar directions at all, but it is noncentrosymmetric, and a piezoelectric polarization can be set up in it with the aid of shear strain. The matrices of the piezoelectric coefficients of textures are given in Table 3.3. A comparison of Tables 3.1, 3, shows that the matrix of the piezocoefficients of texture ∞ coincides with that of the piezocoefficients of the crystals of classes 4 and 6; the matrix of textures ∞mm, with that of the crystals of classes $4mm$ and $6mm$; and the matrix of the piezocoefficients of texture $\infty 2$, with those of the crystals of classes 422 and 622.

The piezoelectric textures of Rochelle salt and barium titanate have been thoroughly studied. The latter textures have an extensive practical application (piezoceramics).

Textures of Rochelle salt are obtained from its melt by oriented application to some surface with a brush. After the melt solidifies (crystallizes) the texture has a symmetry $\infty 2$ and possesses the piezoelectric effect. Piezomoduli of such a texture are of the order of $(1-5)\times 10^{-7}$ CGSE units.

Unpolarized ceramics from barium titanate (or other perovskite ferroelectrics) consist of a multitude of crystallites oriented arbitrarily in space. The separate crystallites of the ceramic are divided into domains (see below)

Table 3.3. Matrices of piezoelectric moduli of textures

Symmetry ∞

$$\begin{Vmatrix} 0 & 0 & 0 & d_{14} & d_{15} & 0 \\ 0 & 0 & 0 & d_{15} & -d_{14} & 0 \\ d_{31} & d_{31} & d_{33} & 0 & 0 & 0 \end{Vmatrix}$$

Symmetry ∞mm

$$\begin{Vmatrix} 0 & 0 & 0 & 0 & d_{15} & 0 \\ 0 & 0 & 0 & d_{15} & 0 & 0 \\ d_{31} & d_{31} & d_{33} & 0 & 0 & 0 \end{Vmatrix}$$

Symmetry $\infty 2$

$$\begin{Vmatrix} 0 & 0 & 0 & d_{14} & 0 & 0 \\ 0 & 0 & 0 & 0 & -d_{14} & 0 \\ 0 & 0 & 0 & 0 & 0 & 0 \end{Vmatrix}$$

belonging to one of the symmetry groups $mm2$, $3m$, or $4mm$, depending on the temperature and composition. A set of such domains, also oriented arbitrarily in space, forms a texture belonging to symmetry group $\infty/\infty mm$. This isotropic centrosymmetric texture does not possess any piezoelectric properties.

For a ceramic to acquire piezoelectric properties it is polarized, i.e., placed in a strong electric field and held there for some time. Vectors $\boldsymbol{P}_s$ of the crystallite domains are then oriented at the minimum possible angle to the field. Owing to the complex processes of domain interaction, impurity ordering, etc., the ceramic mainly conserves such a preferred direction of $\boldsymbol{P}_s$ in the crystallites after the field is removed. In accordance with Curie's symmetry principle a polarized ceramic belongs to symmetry group ∞mm and is a piezoelectric texture.

The mean values of the piezomoduli of the $BaTiO_3$ ceramic are (in CGSE units) $d_{33} \cong 5 \times 10^{-6}$, $d_{31} \cong 2 \times 10^{-6}$, $d_{15} \cong 7 \times 10^{-6}$.

Piezoelectric properties are inherent in some rocks, in particular, those containing quartz. They are piezotextures; the quartz textures evidently have symmetry $\bar{6}m2$. The piezoelectric effect in such textures is characterized by piezomoduli equaling 0.1 to 5% of quartz piezomodulus d_{11}.

Not only textures consisting of a set of single crystals can be piezoelectric. Thus, the piezoelectric effect has been discovered and studied in wood, where piezoelectric properties are due to the oriented disposition of the fibers. The symmetry of the wood texture is $\infty 2$. The values of piezomoduli d_{14} and d_{25} are equal to about 0.5×10^{-8} CGSE units.

Comparatively recently the piezoelectric effect was discovered in the bones and sinews of man and some animals. Here textures with symmetry ∞ are found. The piezomoduli have values of the order of $10^{-8}-10^{-9}$ CGSE units.

3.4 Domain Structure and Peculiarities of the Electrical Properties of Ferroelectrics and Antiferroelectrics

3.4.1 General

In the last few decades much attention has been given to the study of ferroelectric and antiferroelectric crystals. The *ferroelectrics* are characterized by a certain temperature (Curie point) at which a spontaneous electrical polarization arises, and the crystal is thus divided into domains. The *antiferroelectrics* do not show a macroscopic spontaneous polarization, but they possess spontaneously polarized unit cells, in which the directions of spontaneous polarization are antiparallel. Two unit cells form a superstructural electrical neutral cell of the antiferroelectric, which is also divided into domains. The phase transitions and the domain structure in the ferroelectrics and antiferroelectrics exert a substantial effect on their properties (polarization, pyro- and piezoelectrical, optical, elastic, and other properties).

Investigations show that ferro- and antiferroelectrics may differ in the mechanism of spontaneous polarization. There are two such mechanisms. The first is typical of ferro- and antiferroelectrics which have the oxygen-octahedral type of structure and consists in the fact that as a result of the rearrangement of the structure the spontaneous polarization is directly due to the displacement of certain ions, its direction coinciding with that of the displacement (*displacement-type* ferroelectrics). Polarization in these is, as a rule, associated with the displacement of a cation (Ti, Nb, Ta etc.) from the center of the surrounding oxygen octahedron. The resulting dipole moments are either parallel or antiparallel, depending on the geometric characteristics of the structure (size of ions, size of voids, mutal arrangement of ions, etc.), the nature of the bond, and the electron configuration of the atoms. The most important role in the orientation processes is played by the oxygen ions. The displacement-type ferroelectrics include compounds with the structure of perovskite ($BaTiO_3$, $PbTiO_3$, and $KNbO_3$), pseudoilmenite ($LiNbO_3$, $LiTaO_3$), pyrochlor ($Cd_2Nb_2O_7$, $Pb_2Nb_2O_7$), etc.

For other ferroelectrics and antiferroelectrics the transition to the polarized state is due to the ordering of certain structure elements, which were not ordered before the transition (*ordering* ferroelectrics). In such crystals the transition is often (but not always and not necessarily) associated with the ordering of the hydrogen bond protons (for instance, in sulphates, fluoroberyllates, cyanides). In other cases, radicals become ordered as a result of their retarded rotation (sodium nitrite, dicalciumstrontiumpropionate, etc.).

3.4.2 Domains

The change in the symmetry of a crystal on appearance of spontaneous polarization can be found from P. Curie's principle of symmetry (Chap. 1). To do this, one has to consider the set of symmetry elements of the crystal in the initial (paraelectric) phase and in its spontaneous polarization. The latter is a polar vector and has a symmetry ∞mm. It is easy to see, for instance, that for class $m3m$ (the case of $BaTiO_3$) the polar vector directed along axis 4 of the initial cubic cell leads to class $4mm$; that directed along axis 2 to class $mm2$; and that directed along axis 3 to class $3m$. To these changes in symmetry corresponds the actually realized tetragonal, orthorhombic, and rhombohedral distortion of cells. The results of the change in point symmetry on ferroelectric phase transitions for crystals of the initial cubic classes are summarized in Table 3.4.

From Table 3.4 and similar tables one can thus find the point symmetry of a ferroelectric crystal in the single domain state. In such a state the crystal must necessarily belong to one of the pyroelectric classes of symmetry and be polar. If, however, the crystal is not a single-domain one, then individual domains possess such a symmetry.

If the crystal experiences several successive ferroelectric phase transitions (changing the magnitude and direction of the spontaneous polarization), one

Table 3.4. Change in the point symmetry of cubic-class crystals on appearance of P_s along different crystallographic directions. The figures in parentheses indicate the number of equivalent directions

Direction P_s	Initial phase				
	$m3m$	432	$\bar{4}3m$	$m3$	23
[100]	$4mm$ (6)	4 (6)	$mm2$ (6)	$mm2$ (6)	2 (6)
[110]	$3m$ (8)	3 (8)	$3m$ (4)	3 (8)	3 (4)
[110]	$mm2$ (12)	2 (12)	m (12)		
[hko]			1 (24)	m (12)	
[hkk]	m (24)	1 (24)	m (12)		1 (12)
[hhk]				1 (24)	
[hkl]	1 (48)		1 (24)		

can get an idea of the change in symmetry on each transition by assuming that it is realized directly from the paraelectric (initial) phase. Thus it can be said that each new ferroelectric phase transition occurs "with a call at the paraelectric phase."

In Table 3.4 the symbols of groups are followed by parenthetical figures indicating the number of directions crystallographically (and physically) equivalent in the initial phase to the direction along which the spontaneous polarization is assumed to be oriented. It is easy to understand that if the set of all domains is oriented uniformly along all the crystallographically equivalent directions, then macroscopically, on the whole, a ferroelectric crystal divided into domains has the same group of symmetry which it had in the paraelectric phase. Thus, the crystal, after being divided into domains macroscopically, "returns" to the initial symmetry, and its macroscopic properties (elastic, piezoelectric, etc.) must also possess this symmetry.

If ferroelectric domains are regarded as twins, it can be ascertained that the elements of point symmetry lost by the crystal on transition from the initial to the given ferrophase are the twinning elements of the domains.

An analysis of each particular set of crystallographically equal directions indicates which angles between domains (between the directions of their $\boldsymbol{P}_s$) can be realized and how the domain boundaries will be oriented. Considering that in a real crystal the domain-twin components are paired off, we can also find all the possible particular combinations of them, etc. In other words, this enables us to solve, for each of the crystals, all the problems referring to the geometry of the domain structure.

Thus, in crystals of class $m3m$ (the case of $BaTiO_3$) on transition $m3m \rightarrow 4mm$, domain combinations are possible which lead to 180° boundaries (antiparallel orientation of neighboring domains) and 90° boundaries (mutually perpendicular orientation with boundaries coinciding with planes of the type $\{110\}$). On transition $m3m \rightarrow 3m$, the directions of $\boldsymbol{P}_s$ in neighboring domains may equal ~71°, ~109°, and 180°, and on transition $m3m \rightarrow mm2$, 60°, 90°, 120° and 180°.

The division of a spontaneously polarized crystal into domains follows from simple energy considerations: in this way the crystal reduces its energy because of the closing of the electric field. There are other considerations, too. Domains may also arise because during a phase transition a spontaneous polarization may appear independently in different parts of the crystal and have a different (but a crystallographically equivalent) direction. These considerations are particularly important for antiferroelectrics, which have no macroscopic spontaneous polarization.

When considering possible domain boundaries in ferroelectrics, one must take into account not only geometric considerations, but also the condition of electrical neutrality of the boundary corresponding to the energy minimum of the crystal. This calls for an orientation of $\boldsymbol{P}_s$ in neighboring domains such that the projection of vector $\boldsymbol{P}_s$ onto the boundary from the side of one domain is equal in magnitude and opposite in sign to the projection of vector $\boldsymbol{P}_s$ of the neighboring domain ("head-to-tail" domain orientation).

In the presence of electron and ion polarization the main role in ferroelectrics is evidently played by electrostatic interaction between polarized ions. The interaction energies of parallel and antiparallel dipole rows in dielectrics are known to differ but slightly, and this allows one to have a narrow domain wall. Thus, the crystal anisotropy energy is the determining factor here. Owing to the high anisotropy, vector $\boldsymbol{P}_s$ in a ferroelectric does not rotate inside the wall like the spontaneous magnetization vector $\boldsymbol{M}_s$ in the case of a ferromagnetic, but simply decreases in value without a change of direction, passes the zero, and increases again, but with an opposite orientation (Fig. 3.7).

A substantial part of the wall energy in ferroelectrics may consist of elastic energy. This energy is related to the mechanical stresses due to the change in polarization inside the wall via the piezoelectric effect. The elastic energy, as a rule, favors the narrowing of the wall. It is comparable with the dipole energy in the order of magnitude.

The domain structure of ferroelectrics is studied by different methods: charged powders, etching, electroluminescence, various x-ray topography techniques, electron microscopy, and optical methods. The most important

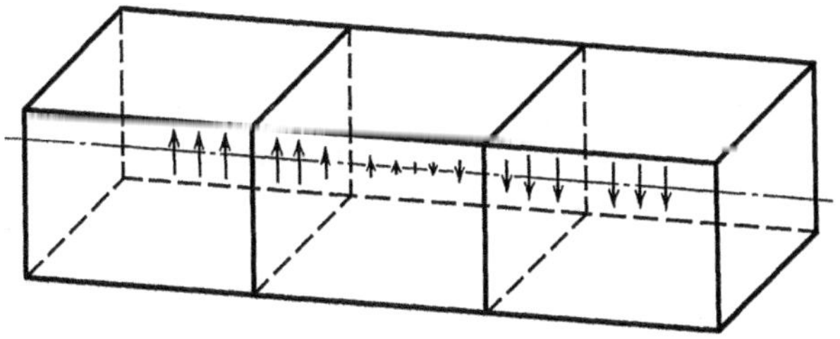

Fig. 3.7. Variation of spontaneous polarization vector at the domain boundary

information on the structure of ferroelectric domains is obtained by optical methods. Then the domains are studied under a conventional polarizing microscope. Unfortunately, this method is not always applicable: if neighboring components of twin domains have identically oriented optic indicatrices (for instance, in TGS), such domains cannot be differentiated by the optical method.

3.4.3 Polarization, Electrical Conductivity, and Dielectric Losses

Apart from the above-indicated (Sect. 3.1) types of polarization, ferroelectrics have one more (domain) type, i.e., a polarization associated with domain reorientation. As a rule, the contribution from the domain polarization greatly exceeds that from the other mechanisms, and therefore we shall henceforth consider only the features of ferroelectric polarization associated with domains.

In the absence of external fields a ferroelectric divided into domains does not show macroscopic polarization: on any crystal surface, both positive and negative charges are offset by free charges. Application of a field increases the fraction of a domain of one orientation and reduces that of the other. A change in the ratio between domains of opposite orientation is a complicated process depending mainly on the magnitude of the field and the time of its application.

Observations of the behavior of domains under a field show that repolarization of a ferroelectric is realized by two mechanisms: generation and growth of oppositely oriented nuclei. In weak fields repolarization is mainly due to the formation of nuclei of new domains not only on the surface, but throughout the crystal, and in strong fields, due to the growth of a comparatively small number of nuclei. Therefore, in weak fields the relationship between repolarization time t_n and field tension E must be of an exponential nature (statistical character of nucleation), while in high-tension fields the relation $t_n \sim 1/E$ (movement in a viscous medium) must hold.

In visual observation of a domain structure during repolarization one can see both the appearance of new domains and their lateral growth. Detailed investigations, however, show that in ferroelectrics (as distinct from ferromagnetics) nuclei grow only forward, and that the lateral movement is only apparent: a nucleus arising at the boundary, when growing forward, creates the impression of lateral movement of the wall.

Important information on the polarization of ferroelectrics can be obtained by studying them in low-tension alternating fields. In fields of tension not higher than several V · cm, one studies initial dielectric constant ε, which is only slightly associated with domain reorientation. Of greatest interest in such measurements are the temperature dependences of ε, which show its sharp increase in the phase transition region (near the Curie point). This increase in ε is due to the increase in the lability of the crystal structure as the phase transition temperature is approached. The anomalies in ε under phase transition are manifested most prominently when it is measured in the direction of the appearance of vector $\boldsymbol{P}_s$.

Experiment shows that in ferroelectrics, when the Curie point is approached from the side of the paraelectric region, the reciprocal susceptibility on the crystal varies linearly with the temperature:

$$1/\alpha = C^*(T - T_{CW}) , \tag{3.55}$$

where T_{CW} is the Curie-Weiss temperature. Taking into account relation $\varepsilon \cong 4\pi\alpha$ (for $\varepsilon \gg 1$), we have from (3.55)

$$\varepsilon = C/(T - T_{CW}) , \tag{3.56}$$

where C is the Curie constant. The last expression is called the *Curie-Weis law*.

The Curie-Weiss law holds good above transition temperature T_C (the Curie point) for many ferroelectrics. The Curie-Weiss temperature T_{CW} coincides with true transition temperature T_C (the spontaneous polarization temperature) for ferroelectrics experiencing a second-order phase transition. In a first-order transition, Curie-Weiss temperature T_{CW} is lower transition temperature T_C. Difference $T_C - T_{CW}$ is usually several degrees.

The ferroelectrics can be divided into two groups according to the absolute value of the Curie constant. In the first group, C is of the order of 10^3 deg. It includes ferroelectrics with ordering structural elements (Rochelle salt, KH_2PO_4, $NaNO_2$ etc.). The ferroelectrics of the second group are of the oxygen-octahedral type: $BaTiO_3$, $PbNb_2O_6$, etc. The absolute value of C for this group is of the order of 10^5 deg. The sharp difference in C values is due to the different types of changes in the structure of ferroelectrics in phase transitions.

The dielectric constant of ferroelectrics, which is measured in weak fields, is frequency dependent. As the frequency increases, ε reduces in the region where the field frequency is close to that of the resonance vibrations of the ferroelectric under investigation. The second region of reduction (relaxation) of ε in ferroelectrics is associated with the low response of domains, the "switching off" of the domain polarization mechanism.

The nonlinear dependence of polarization on the field, inherent in ferroelectrics, leads to dielectric hysteresis in alternating electric fields (Fig. 3.8), i.e., to a phase shift between electrical polarization P and electric field E. The dielectric hysteresis loop is one of the important characteristics of ferroelectrics and gives an idea of their dynamic susceptibility. That is why the dielectric hysteresis method is widely used for investigating the polarization of ferroelectrics in strong fields.

Consideration of hysteresis loops provides a vivid picture of the role of domain processes in polarization of ferroelectrics. This role is manifested

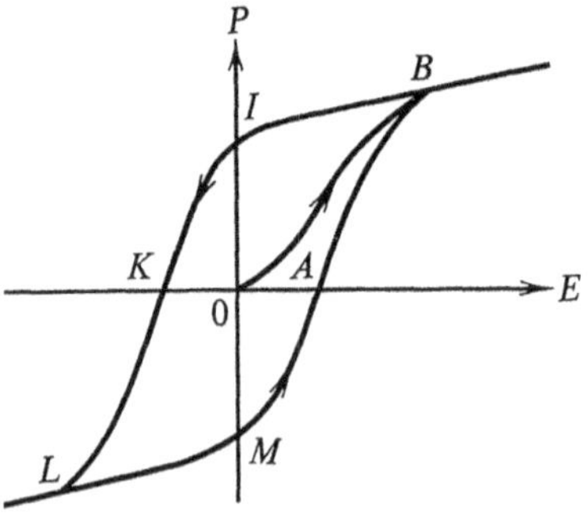

Fig. 3.8. Dielectric hysteresis loop

most prominently on intervals *MB* and *IL* of the loop (Fig. 3.8). Since repolarization depends on the field tension and the time of its application, coercive field *OK* and the loop shape will also depend both on the tension amplitude and the field frequency.

As a rule, ferroelectrics have an ionic conductivity, which is both intrinsic and impurity type (or associated with defects). This is particularly typical of ferroelectrics with ordering structural elements.

At the same time, the electrical conductivity of ferroelectrics has a number of specific features. In the first place, these specifics are due to the domain structure and phase transitions. Besides, the nonstoichiometry of the composition (mostly oxygen deficiency), which is almost always observed in oxygen-octahedral ferroelectrics, leads to electron (or hole) conductivity. This character of conductivity plays a particularly important role in the surface layers of oxygen-octahedral ferroelectrics. The nonstoichiometry of such crystals, however, is usually considerable in the bulk as well; the electron (or hole) conductivity in them exceeds the ionic, which enables us to speak in general terms of the semiconducting properties of some oxygen-octahedral pyroelectrics.

The temperature dependence of the electrical conductivity of ferroelectrics can in most cases be described by two-term equation (3.13). The most interesting and important fact is not so much that curve $\ln\sigma = f(1/T)$ has a knee as that this knee corresponds to the ferroelectric phase transition temperature. This means that the rearrangement of the structure in the phase transition region and the appearance (disappearance) of spontaneous polarization is accompanied by a change in the activation energy of the carriers. In some ferroelectrics not only a change in activation energy, but also a conductivity jump takes place in the phase transition region.

The well-known phenomenon of dielectric absorption, which consists in a slow time variation of the current flowing through the dielectric, may take place in ferroelectric crystals as well. At the same time, in ferroelectrics ordinary polarization and conductivity may be superimposed by slow processes of polarization (orientation) of domains, which will also lead to a current drop with time. Moreover, when the formation of a space charge, movement of impurities, molding, etc. are inessential, the current falloff with time will be entirely due to domain reorientation.

In weak fields ferroelectrics are polarized (to a first approximation) like linear dielectrics. Although domain processes do occur then to some extent, they do not result in a hysteresis-type dependence of polarization on the field, and the displacement of the domain boundaries can be regarded as elastic. It follows that the principal regularities of dielectric losses in ferroelectrics and weak fields are the same as in linear dielectrics with relaxational polarization.

The dielectric losses of ferroelectrics in strong fields may be of different origin, but those on dielectric hysteresis are the basic losses here. They are proportional to the hysteresis area. Note that as the temperature of the ferroelectric phase transition is approached, the dielectric losses of all types increase and pass a maximum.

The ferroelectrics possess pyroelectric properties only in the single-domain state, i.e., when the spontaneous polarization of the entire crystal is oriented in the same direction. As noted above, a ferroelectric crystal divided into domains has a symmetry of a paraelectric nonpolar phase and cannot possess pyroelectric properties. Pyroelectric properties, however, can be exhibited by a so-called unipolar crystal, i.e., one with a preferred domain orientation along one of the directions. The pyroelectric effect of unipolar crystals, however, does not present much interest, since it characterizes a particular state of the ferroelectric, while as single-domain crystal shows properties typical of the given substance.

3.4.4 Piezoelectric Properties

All ferroelectrics, at least in polar ferroelectric phases, possess piezoelectric properties. In macroscopic consideration of a crystal divided into domains as an entity, two cases may be encountered: the presence or the absence of piezoelectric properties. This depends on whether or not the given crystal possessed piezoelectric properties in the paraelectric phase before the phase transition (i.e., whether or not it had a center of symmetry in this modification). If the crystal did not have a center of symmetry, i.e., if it was a piezoelectric before the phase transition, then, in accordance with the principle of return of the crystal to the former symmetry group after being divided into domains, it will posses piezoelectric properties in the ferrophase after division into domains as well. Typical examples of such phase transitions are those with the appearance of spontaneous polarization in potassium dihydrophosphate (KDP) and Rochelle salt. Crystals which were centrosymmetric in the paraelectric phase are characterized by a centrosymmetric configuration of the domains and by the absence of the piezoelectric effect in the polydomain state. Such phase transitions occur, for instance, in triglycinesulphate (TGS).

In general, the domains in ferroelectrics are responsible, for a number of their specific piezoelectric properties. Thus, the reverse piezoeffect in ferroelectrics is associated not only with the ordinary deformation (or the appearance of stresses), but also with domain reorientation. In some crystals, in turn, domain reorientation accompanies the forward piezoeffect.

As a rule, piezoelectric properties are more pronounced in ferroelectrics than in linear piezoelectrics. This is especially true for the phase transition region. Anomalies of the piezoelectrics properties of ferroelectrics in the phase transition region constitute the second peculiarity of their piezoelectric properties as compared with those of the linear dielectrics.

As noted above, spontaneous polarization in ferroelectrics is caused by anomalies in a number of their properties (elastic, dielectric, optical, etc.) in the phase transition region. When considering the behavior of the piezoelectric properties in the phase transition region everything depends on the regime in which the piezoelectric properties of a given specimen are investigated and, accordingly, on the coefficients they are characterized by: d and e, or g and h. Piezocoefficients d and e in the transition region change very sharply, whereas coefficients g and h remain relatively constant. In ferroelectrics possessing piezoelectric properties in the paraelectric phase, g and h are approximately equal in the two regions. In ferroelectrics with a centrosymmetric (paraelectric) phase these coefficients acquire a definite value jumpwise and change only slightly even near the transformation temperature. The cause for this difference in the behavior of these pairs of coefficients is easy to understand from the conditions of their determination.

The electrostriction coefficients of single-domain ferroelectrics are close in value to the coefficients of linear dielectrics. The presence of domains, when they reorient under the effect of alternating fields, leads to a sharp increase in electrostriction coefficients. In this case the domain contribution to the electrostriction essentially depends on the reverse piezoeffect due to the double (during the field period) reorientation of the domains. Note that in practice electrocstriction oscillations are excited by application of alternating field E to the crystal. Such a regime is characterized by electrostriction coefficient R and is accompanied by domain reorientation. The application of a constant (polarizing) field to a crystal simultaneously with the alternating field leads to the switching-off of the domain mechanism and a reduction in electrostriction.

3.4.5 Antiferroelectrics

As indicated above, on phase transitions in antiferroelectrics the initial (paraelectric) cells become polarized, but the appearance of "twinning" at the microlevel (the appearance of new superstructural unit cells) makes the crystal even microscopically unpolarized. A typical representative of displacement-type antiferroelectrics is lead zirconate $PbZrO_3$, and of antiferroelectrics with ordering structure elements, ammonium dihydrophosphate $NH_4H_2PO_4$.

In weak fields, antiferroelectrics are polarized as linear dielectrics. In strong fields it is often possible to "detwin" antiparallel polarized cells. The domains then become polar: under the effect of the field the antiferroelectric experiences transition to the ferroelectric state. As a result, in high-tension alternating fields such antiferroelectrics repolarize along the double loops of dielectric hysteresis.

Antiferroelectrics evidently do not exhibit pyroelectricity in the absence of external effects. After polarization in a strong field they resume their nonpolar structure.

Antiferroelectrics may possess piezoelectric properties if they had no center of symmetry before transition to the antiferroelectric phase. An example is ammonium dihydrophosphate $NH_4H_2PO_4$. This crystal shows piezoelectric properties at room temperature (in the paraelectric phase) in accordance with its symmetry ($D_{2d} = \bar{4}2m$) and has the same properties, being an antiferroelectric (then the crystal resumes the paraelectric phase symmetry macroscopically). Centrosymmetric perovskite antiferroelectrics are not supposed to possess the piezoeffect.

3.5 Structure and Properties of Some Ferroelectrics and Antiferroelectrics

3.5.1 Barium Titanate

Barium titanate, $BaTiO_3$, has a perovskite structure (Fig. 3.9) and an ideal cubic structure above 120 °C. This paraelectric modification belongs to space group $Pm3m$ ($a \simeq 4$ Å, $Z = 1$). Each Ti ion is at the center of a regular octahedron formed by six oxygen atoms. The octahedra join at the vertices and form a skeleton containing large voids, which are occupied by Ba atoms.

At a temperature of about 120 °C, $BaTiO_3$ experiences a phase transition, and below this temperature (up to about 5 °C) it is tetragonal. The temperature of 120 °C is the Curie point for $BaTiO_3$, because below this temperature it becomes a ferroelectric. The space group of symmetry of tetragonal $BaTiO_3$ is $P4mm$; $Z = 1$, $c/a \approx 1.01$.

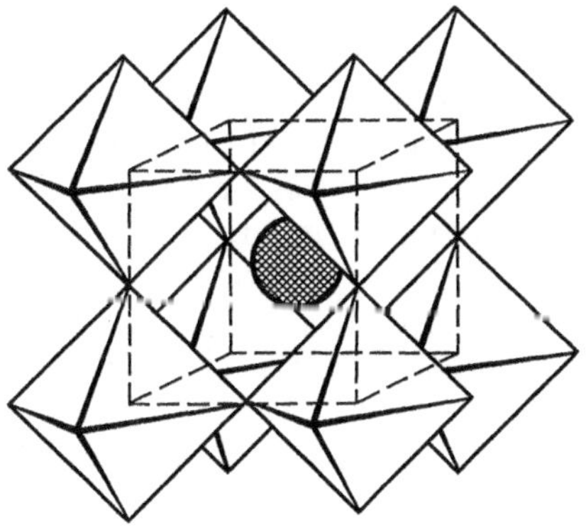

Fig. 3.9. Ideal structure of the type of perovskite ABO_3. The B atoms are at the centers of the octahedra, and the O atoms at the octahedra vertices; the A atom is hatched

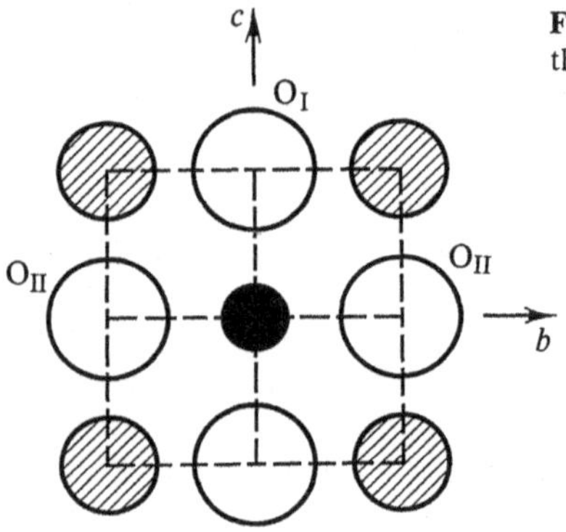

Fig. 3.10. Projection of the cubic cell of $BaTiO_3$ onto the (*bc*) plane

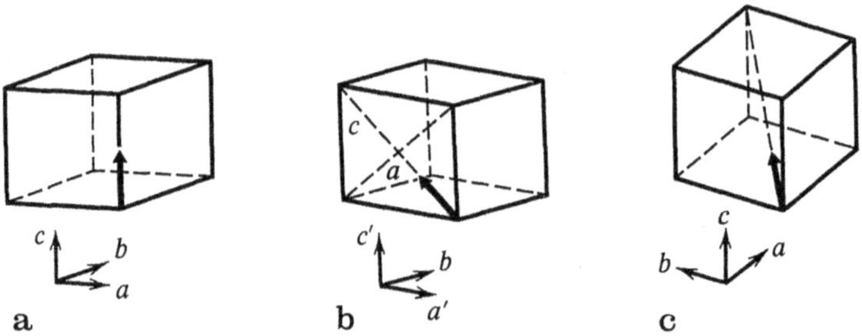

Fig. 3.11. Unit cells of three ferroelectric phases of $BaTiO_3$: (**a**) tetragonal, (**b**) orthorhombic, and (**c**) rhombohedral. The arrows indicate the direction of P_s

The oxygen octahedra forming the skeleton are distorted insignificantly. The displacement of O_I with respect to O_{II} (Fig. 3.10) is small. The titanium atoms are displaced by 0.13 Å relative to the centers of the corresponding octahedra; as a result the two bonds from O_{II} to Ti make an angle different from 180° (171°28′). The direction of spontaneous polarization in the tetragonal phase coincides with one of the fourfold symmetry axes of the initial cubic phase of the crystal (Fig. 3.11a).

With further cooling of $BaTiO_3$, which has a tetragonal symmetry, it experiences a second phase transition near the temperature of 5 °C and becomes rhombic. This structure can again be interpreted as a slightly distorted cubic one obtained by stretching in the direction of one diagonal of the face and contraction in the direction of the other. These diagonals become rhombic axes (Fig. 3.11b). The third axis retains its initial direction, but the lattice spacing along this direction changes somewhat. The space group of symmetry of rhombic $BaTiO_3$ is *Bmm*2. The unit cell referred to the new axes is face centered; axis 2 coincides with rhombic axis *c*. All the atoms are displaced in the direction of axis *c* parallel to each other.

Near −70 to −90 °C barium titanate undergoes a third phase transition, becoming rhombohedral below this temperature. A rhombohedral cell is

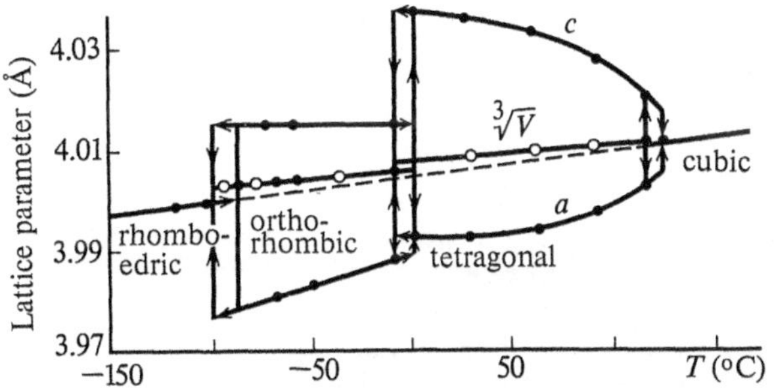

Fig. 3.12. Temperature dependences of the lattice parameters of the different phases of $BaTiO_3$ [3.4]

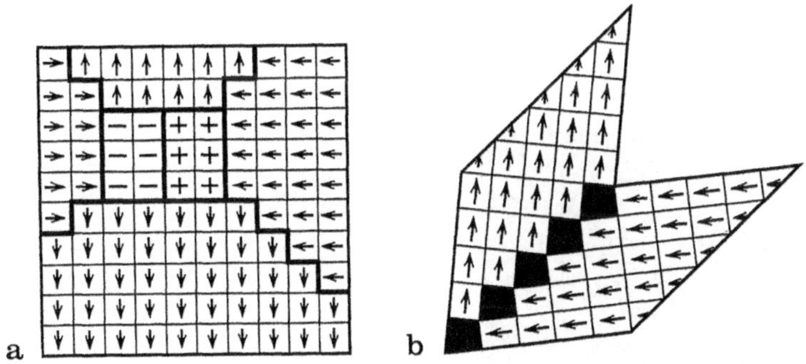

Fig. 3.13. Dipole motif of the domains (**a**) and the 90° boundary (**b**) in the tetragonal modification of $BaTiO_3$ [3.5]

formed by the stretching of the initial cube in the direction of the body diagonal (Fig. 3.11c). Its space group is $R3m$, $Z = 1$. The angle between the rhombohedral axes differs by 8′ from 90°.

The temperature variation of the parameters (a, c) and volume (V) of the unit cell for all the phases of $BaTiO_3$ is shown in Fig. 3.12. It can be seen that at the phase transition points the cell parameters and volume change jumpwise. All transitions are also accompanied by temperature hysteresis, i.e., a change in transition temperatures depending on whether the crystal is cooled or heated.

The (001) face of barium titanate in the tetragonal modification generally looks schematically as shown in Fig. 3.13. It depicts the so-called dipole motif indicating the orientation of vector $\boldsymbol{P}_s$ in the unit cells of all the domains. All the types of boundaries with angles of 180° and 90° between the directions of $\boldsymbol{P}_s$ are also visible. In the former case the direction of spontaneous polarization is parallel to the domain boundary (interface), and in the latter the boundary makes and angle of 45° with the direction of the spontaneous polarization in the separate domain.

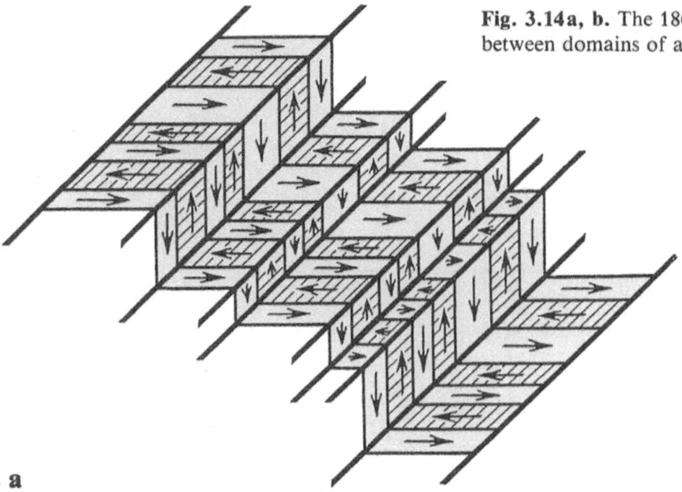

a

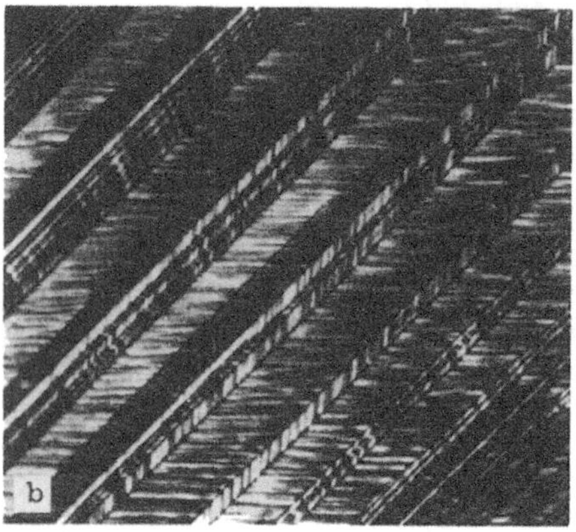

Fig. 3.14a, b. The 180° and 90° boundaries between domains of a $BaTiO_3$ crystal

Nevertheless, a real $BaTiO_3$ platelet mostly shows 90° domain boundaries, as a rule with the predominance of one component over the other. Such, for instance, is the real domain structure presented in Fig. 3.14, which was obtained during optical observations in a polarized light. Platelets where the direction of $\boldsymbol{P}_s$ lies in their plane are called *a domain* (as in Fig. 3.14). Platelets where the direction of $\boldsymbol{P}_s$ is perpendicular to their plane are called *c domain.* In neighboring *c* domains, vectors $\boldsymbol{P}_s$ are antiparallel, and in *a* domains they are at 90° to each other. An *a* domain, in turn, can be divided into antiparallel domains.

An approximate estimate taking into account the anisotropy energy and the energy of the dipole interaction in $BaTiO_3$ crystals gives, for antiparallel domains, a wall thickness of the order of one unit cell and a boundary energy of about 10 erg/cm^2. Experiment shows that the 180° boundary in $BaTiO_3$ is indeed extremely narrow, and its energy is close to the calculated one.

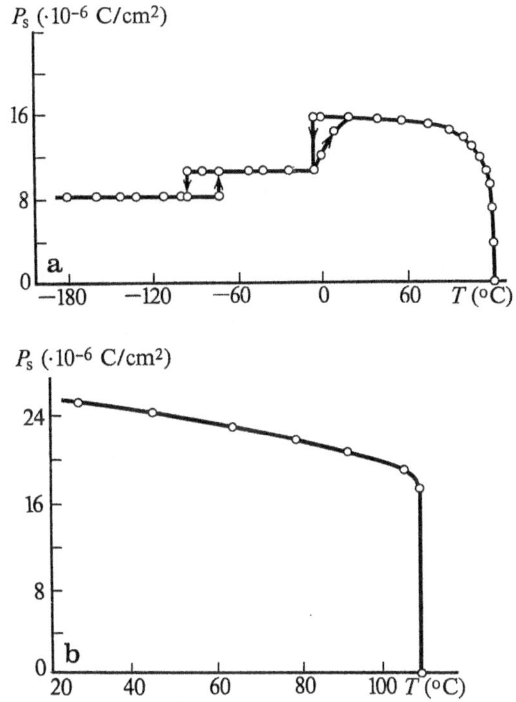

Fig. 3.15. Temperature dependence of spontaneous polarization P_s os imperfect (**a**) and perfect (**b**) crystals of barium titanate [3.6, 7]

The 90° boundary in $BaTiO_3$ has a more complicated structure and consists of slightly deformed cells. The boundary is found to be wide (up to several hundred Å), and its energy equals tens of erg/cm^2. The width of the domains in $BaTiO_3$ varies rather widely: from a micrometer or more to hundredths of a micrometer.

The domain structure of $BaTiO_3$ in the rhombic and rhombohedral modifications is more complicated: in the former case $\boldsymbol{P}_s$ may have twelve directions (along six twofold axes of the paraelectric phase) and in the latter, eight directions (along four threefold axes of the paraelectric phase).

Barium titanate, which has several equivalent polarization directions, is an example of a *multiaxial* ferroelectric.

Spontaneous polarization arises in barium titanate jumpwise at ~120 °C (Fig. 3.15). Its value at the Curie point may reach 18 μCi/cm^2. Moreover, P_s increases smoothly from the Curie point up to room temperature, reaching 26 μCi/cm^2 in good crystals, according to the most reliable measuring results. The average value of pyrocoefficient p in this temperature range is estimated at 240 CGSE units. Near T_C, p slightly exceeds this value, and at 20 °C it is slightly lower. A comparison of the deformation of the crystal lattice with the increment of P_s in the temperature range of 120 – 20 °C shows that the pyroelectric effect in this range is practically entirely secondary, spurious, i.e., is due to the change in crystal size.

The relationship between the pyroelectric and electrocaloric phenomena [Sect. 3.2, (3.28)] in ferroelectrics makes it possible, in turn, to relate the electrocaloric phenomena to the latent heat of transition (for crystals experiencing a first-order phase transition, see Sect. 3.6). The latent heat of transition is the heat required for offsetting the electrocaloric effect. The transition picture then looks approximately as follows: when the crystal being heated loses the spontaneous polarization, it tends to cool down by using the electrocaloric effect with the result that the heat supplied is spent without heating the crystal.

Let us now consider the spontaneous electrocaloric effect in barium titanate. At the Curie point the spontaneous polarization of $BaTiO_3$ arises jumpwise, thus preventing the determination of pyrocoefficient p (and hence also of electrocaloric coefficient q) at this point. However, the value of q was measured via the displacement of T_C in this crystal in the phase transition range and was found to be $q = -0.45$ CGSE units. This enables one, first, to determine the spontaneous electrocaloric effect at the Curie point and, second, to estimate by (3.28) the value of p in the phase transition region. The spontaneous polarization field in $BaTiO_3$ equals 2 CGSE units (600 V/cm), and this means that on the basis of (3.23) the spontaneous electrocaloric effect may lead to a ~1 °C change in crystal temperature. When heated from the ferroelectric region, a crystal will cool down due to the indicated effect, and when cooled from the paraelectric to the ferroelectric region it will heat up. The indicated change in transition temperature (by 1 °C) makes it possible to estimate the latent heat of transition directly from the knowledge of the crystal heat capacity. The heat capacity of $BaTiO_3$ at 120 °C is about 25 cal/(mol · deg), which yields a value of $\Delta Q = 25$ cal/mol. Estimation of p from $q = -0.45$ for $BaTiO_3$ yields $p \cong 46000$ CGSE units in the vicinity of the Curie point.

The peculiarity of repolarization of $BaTiO_3$ crystals in pulsed fields (Fig. 3.16) is that the time of repolarization on weak fields is large. This is mainly because in such fields repolarization is due to domain nucleation (this process depends on the field exponentially, Sect. 3.4). In strong fields domain germination predominates ($t_n \sim 1/E$).

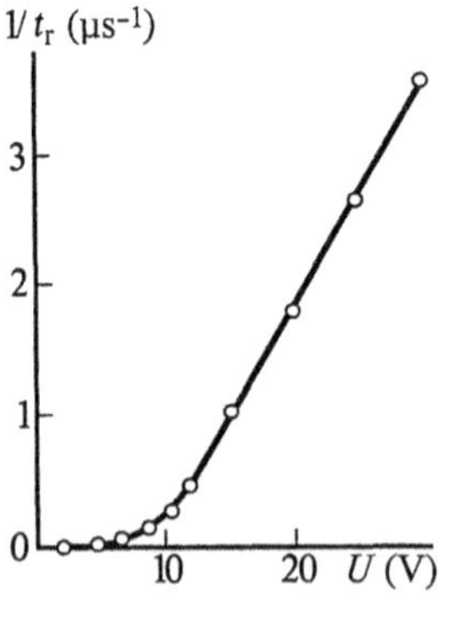

Fig. 3.16. Dependence of the inverse repolarization time $1/t_r$ of a barium titanate crystal on the field tension [3.8]

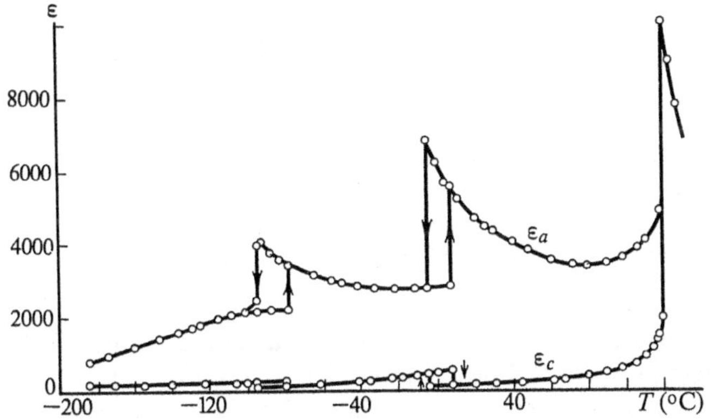

Fig. 3.17. Dependence of the dielectric constant of a single-domain $BaTiO_3$ crystal, as measured along the a and c axes (the c axis coincides with the direction of spontaneous polarization) [3.6]

The temperature dependenc of the dielectric constant in weak fields for $BaTiO_3$ is shown in Fig. 3.17. Attention is drawn to the fact that in the direction of vector $\boldsymbol{P}_s$ the dielectric constant is lower than in the perpendicular direction ($\varepsilon < \varepsilon_a$). At phase transition temperatures (particularly near T_C) dielectric constant ε_a may reach very high values. Above T_C the temperature dependence of ε_c obeys the Curie-Weiss law ($C \simeq 10^5$ deg).

In strong fields crystals of BaTiO clearly show dielectric hysteresis loops. At low frequencies coercive field E_c is equal to 500 – 2000 V/cm. In crystals consisting of c domains it is lower than in those divided into a domains.

$BaTiO_3$ crystals grown by the conventional technique possess semiconducting properties; they have an energy gap of about 3 eV and n-type conductivity. The value of the activation energy calculated from the slope of dependences $\ln \sigma = f(1/T)$ lies between 1 and 2 eV. That the activation energy of electrical conductivity is less than the width of the energy is usually attributed to the presence of impurities and (what is more important) to the oxygen deficiency of the structure. In barium titanate the donor levels due to the departure of the oxygen from the lattice lie at a depth of about 2.5 eV below the conduction band, i.e., at a distance of ~0.5 eV over the edge of the valency zone. The activation energy of the carriers in the oxygen-deficient specimens is 1.5 – 1.6 eV.

The dependence of the electrical conductivity of $BaTiO_3$ on the inverted temperature has a clearly visible knee at the phase transition point; the activation energy of the electrical conductivity in the ferroelectric phase is less than that in the paraelectric one. The knee in the curve of the dependence $\ln \sigma = f(1/T)$ for $BaTiO_3$ has also been observed in the range of transition from the tetragonal to the rhombic modifications.

The appearance of $\boldsymbol{P}_s$ at ~120 °C in a cubic centrosymmetric crystal of $BaTiO_3$ is accompanied by a spontaneous electrostrictive deformation proportional to the square of P_s. The spontaneous polarization, once arisen

Table 3.5. Dielectric and piezoelectric coefficients of a single crystal of $BaTiO_3$ in the tetragonal phase (25 °C)

ε		$d \cdot 10^6$, CGSE units	$g \cdot 10^8$, CGSE units
$\varepsilon_{11}^t = 2920$	$\varepsilon_{33}^t = 168$	$d_{15} = 11.76$	$g_{15} = 5.07$
$\varepsilon_{11}^r = 1970$	$\varepsilon_{33}^r = 110$	$d_{31} = -1.04$	$g_{31} = -7.67$
		$d_{33} = 2.57$	$g_{33} = 19.17$

jumpwise, makes the crystal a piezoelectric. With further cooling of the crystal, additional increment in P_s and additional lattice deformation take place already in the ferroelectric phase. The relationship between these two quantities is linear, and therefore the process taking place can be called a spontaneous piezoelectric effect.

In accordance with the principle explained in Sect. 3.4, a $BaTiO_3$ crystal divided into domains does not possess piezoelectric properties in any ferroelectric phase (provided, of course, that it is not unipolar) because of the presence of the center of symmetry in the paraelectric phase. Thus, for investigating (and utilizing) the piezoelectric effect the crystal must be monodomainized, which is achieved by applying a constant electric field to it.

Table 3.5 lists the absolute values of the piezoelectric coefficients and dielectric constants for the tetragonal phase of $BaTiO_3$. It should be noted that these quantities slightly differ from one crystal to another. The piezomodulus d of $BaTiO_3$ exceeds, by not less than one order of magnitude, the value of d for the best linear piezoelectrics. In accordance with Sect. 3.4, coefficients g and h are practically temperature independent, while piezocoefficients d and e greatly increase towards T_C, but all piezocoefficients reduce to zero above the Curie point.

The great difference between ε^t and ε^r points to a high electromechanical coupling coefficient k of the crystal, which reaches 67% for some modes.

Barium titanate and some other oxygen-octahedral ferroelectrics are widely used in industry as piezoelectrics, mostly in the form of piezoceramics (Sect. 3.3).

The electrostriction coefficients for $BaTiO_3$ crystals can be estimated from the jumps in unit cell parameters and the spontaneous polarization on phase transition at the Curie point. As mentioned above, the jump of P_s in this case equals about 18 $\mu C i cm^{-2}$, and x-ray measurements yield the following (also jumplike) values of spontaneous strain:

$$r_3 = 3.42 \times 10^{-3}, \quad r_1 = 1.55 \times 10^{-3}.$$

Comparing the jump of $P_s (= P_3)$ and the indicated values of r_{ij} one can write, with the aid of (3.47),

$$r_3 = Q_{33} P_3^2, \quad r_1 = Q_{13} P_3^2,$$

and then determine coefficients Q_{33} and Q_{13} (in CGSE units):

$$Q_{33} = 1.7 \times 10^{-12}, \qquad Q_{12} = 0.53 \times 10^{-12}.$$

Coefficients Q_{33} and Q_{12} describe the spontaneous electrostrictive effect. The values of these coefficients are close to those obtained by other methods for the paraelectric phase. The third electrostriction coefficient (3.53) has been found from direct measurements:

$$Q_{44} = 0.70 \times 10^{-12} \text{ CGSE units.}$$

3.5.2 Potassium Dihydrophosphate

Dihydrophosphates and dihydroarsenates of alkali metals (KH_2PO_4, RbH_2PO_4, KH_2AsO_4, RbH_2AsO_4, CsH_2AsO_4, and the corresponding deuterated compounds) are typical ferroelectrics with ordering structural elements – hydrogen bounds. Owing to the numerous x-ray and neutron diffraction investigations the detailed structure and phase transition mechanism for crystals of the type HK_2PO_4 are established clearly enough.

At room temperature KH_2PO_4 (KDP) crystallizes in tetragonal class $\bar{4}2m$ (space group $I\bar{4}2d$, $a = 7.434$ Å, $c = 6.945$ Å, $Z = 4$). The structure consists of a three-dimensional skeleton of $[PO_4]^{-4}$ groups linked by hydrogen bonds O–H...O: the two upper oxygen atoms of one $[PO_4]^{-4}$ tetrahedron are joined to the two lower oxygen atoms of the two adjacent tetrahedra, and the two lower oxygen atoms are joined to the upper oxygen atoms of the other two tetrahedra (Fig. 3.18).

At room temperature the length of the hydrogen bond is equal to 2.48 Å. This indicates a very strong bond; the hydrogen atoms are not ordered, but jump about between the two equilibrium positions spaced at ~0.5 Å along the bond line. Below the Curie point (–151 °C) the protons are ordered, i.e., they all reside either near the upper or near the lower oxygen atoms of the $[PO_4]^{-4}$ tetrahedra at a distance of 0.21 Å from the hydrogen bond center. Proton

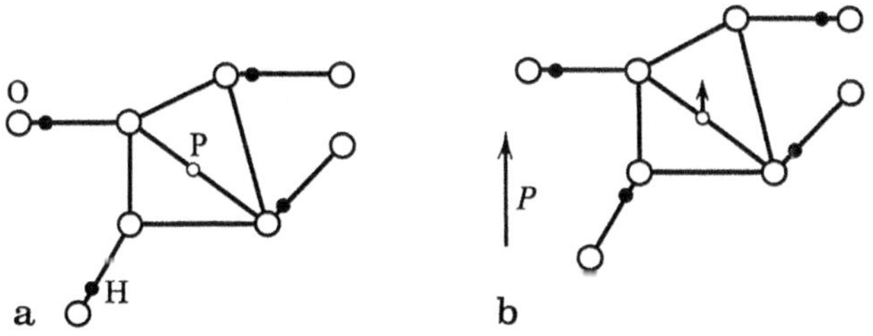

Fig. 3.18. Configurations of the H_2PO_4 group in a crystal of potassium dihydrophosphate (according to J. C. Slater) **(a)** there is no dipole moment in the direction of the c axis; **(b)** a displacement of the phosphorus atom (denoted by an arrow) sets up a dipole moment in the c direction

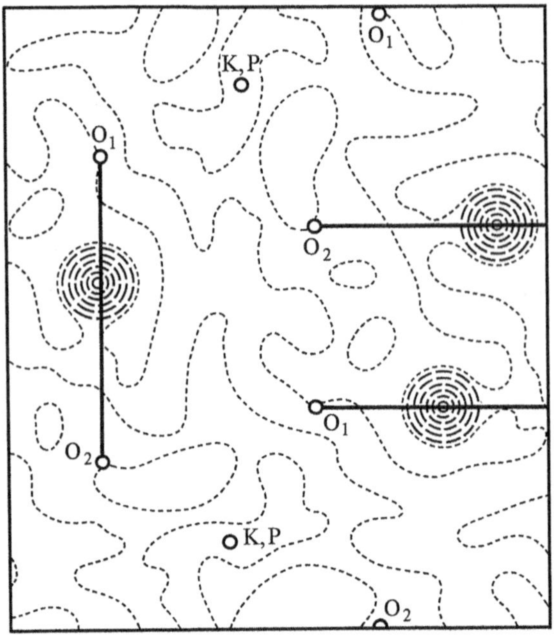

Fig. 3.19. Fourier projection of potassium dihydrophosphate at 77 K onto the (001) plane [3.9]: the dashed lines indicate the negative contours, and the dot-and-dash lines, the zero ones

ordering is shown in Fig. 3.19, which presents the projection of the neutron scattering density on the (001) plane. Such ordering disturbs the distribution of internal fields in a crystal and results in the appearance of dipole moments in the $[PO_4]^{-4}$ tetrahedra. Proton ordering is accompanied by an increase in the length of the hydrogen bond to 2.51 Å and a considerable displacement of the phosphorus and potassium atoms, which make a substantial contribution to the spontaneous polarization. The direction of the spontaneous polarization coincides with the displacement of the phosphorus in the tetrahedron and is opposite to the displacement of the potassium. The proton itself makes no contribution to the spontaneous polarization, because it moves along the hydrogen bond, which is practically perpendicular to the ferroelectric *c* axis.

Below the Curie point, $-151\,°C$, KDP belongs to rhombic class *mm*2 (space grop *Fdd*) with crystallographic axes *a* and *b* turned by 45° relative to the axes of the parelectric phase. Domain twinning brings the crystal (macroscopically) back to group $\bar{4}2m$. Figure 3.20 shows the domain structure of a cut perpendicular to the polar *c* axis revealed by the "dew" (more precisely, "hoarfrost") method. The domain boundaries are parallel to the tetragonal (100) and (010) planes; the polarization direction coincides with [001]. The domain width is $2-3\times10^{-4}$ cm, the boundary energy being about 40 erg/cm^2.

Formerly the phase transition in KDP was identified as a second-order transition, but recently it has been shown by different methods that KDP (as also practically all the ferroelectrics of this group, especially deuterated ones)

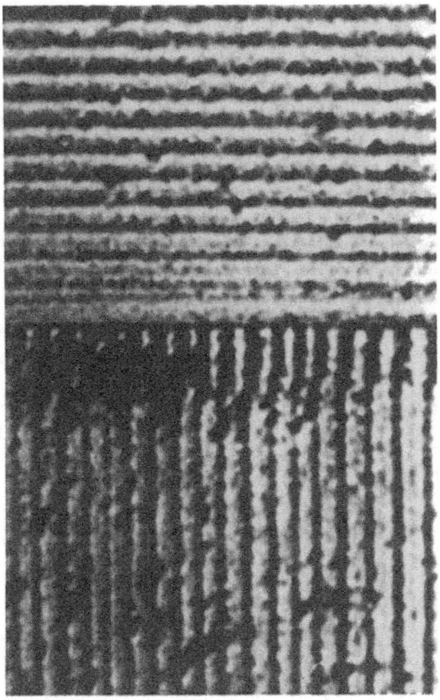

Fig. 3.20. Domain structure of a KDP crystal revealed by the "dew" method on a thin section perpendicular to the c axis [3.10]

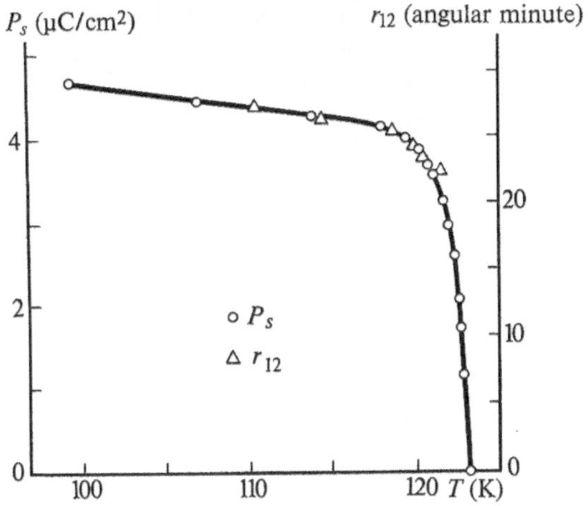

Fig. 3.21. Temperature dependence of spontaneous polarization P_s and spontaneous shear r_{12} in a KDP crystal [3.11]

experience a first-order phase transition close to the critical point. The rapid change in P_s (which is equal to about 10^4 CGSE units) in this crystal occurs within a narrow temperature range (2 °C) (Fig. 3.21). This leads to a very high value of the pyrocoefficient, about 5000 CGSE units.

Section 3.2 contains an example of the pyroelectric effect in tourmaline, which is characterized by a high potential difference at a comparatively low charge. The ferroelectrics show high values of pyroelectric coefficient p

(hundreds and thousands of times higher than for the linear pyroelectrics), which ensure relatively high charge densities (microcoulombs and even tens of microcoulombs per 1 cm^2 on a temperature change of 10 °C). At the same time, because of the high dielectric constant the potential differences are not always high here. For instance, under conditions similar to the ones mentioned for the tourmaline platelet, a KDP platelet of the same size would have a potential difference of about 300 V in the phase transition region. Electrocaloric effect coefficient q in the phase transition region is estimated at -0.06 CGSE units. Considering that in this region the tension of the spontaneous polarization field is about 1.7 CGSE units (500 V/cm), we find that on cooling from the paraelectric phase the crystal heats by about $0.06 \times 1.7 = 0.1$ °C due to the electrocaloric effect. (The crystal cools by the same value on transition from the ferro- to the paraelectric phase.) This effect is of great importance. Indeed, by using quite accessible fields of ~ 10000 V/cm it is possible to change the crystal temperature by 2 °C in the phase transition region with the aid of the electrocaloric effect. Such a change is usually called the Curie point shift under the influence of the field. The shift of the Curie point in the field towards the higher temperature is observed experimentally. The temperature dependence and anisotropy of the dielectric properties of KDP crystals are presented in Fig. 3.22.

KDP shows piezoelectric properties in the paraelectric phase in accordance with its symmetry. The crystal setting is shown in Fig. 3.23. From the matrix of class $\bar{4}2m$ piezomoduli (Table 3.1) it is seen that KDP possesses the piezoelectric effect only in shear strain. To study the effect in longitudinal strain, specimens of so-called 45° cuts are prepared. The experimental value of piezocoefficient g_{36} for KDP is 5.0×10^{-7} CGSE units.

The appearance of $\boldsymbol{P}_s$ at a temperature -150 °C in KDP is attended by a spontaneous shift $\boldsymbol{r}_{6s}$;

$$\boldsymbol{r}_{6s} = g_{36} \boldsymbol{P}_s .$$

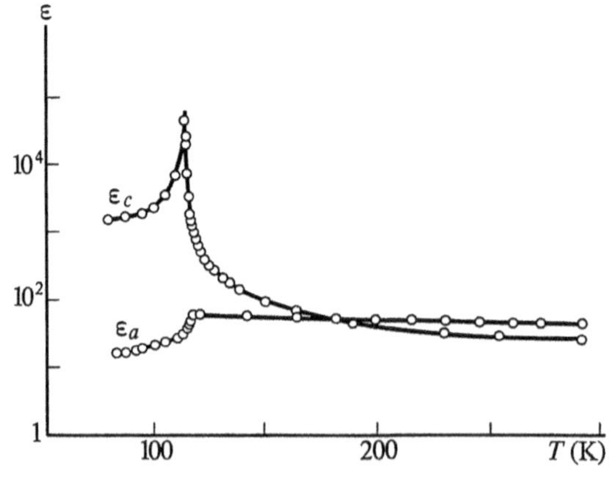

Fig. 3.22. Temperature dependence of dielectric constants ε_a and ε_c in a KDP crystal [3.12]

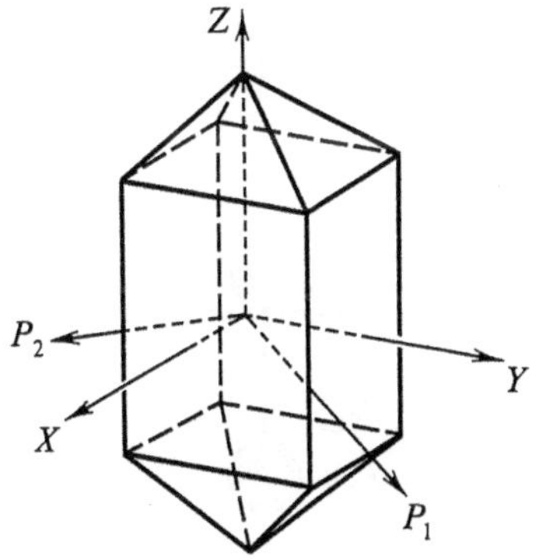

Fig. 3.23. Direction of crystallophysical axes in KH_2PO_4 and $NH_4H_2PO_4$ crystals; P_1 and P_2 denote axes parallel to the symmetry planes

Thus, the appearance of a spontaneous strain can be interpreted as a spontaneous pizeoelectric effect. A comparison of the values of r_{6s} and $\boldsymbol{P}_s$ will produce the value of coefficient g_{36} equal to 5.5×10^{-7} CGSE units, which is in good agreement with the experimental values of g_{36} obtained in the paraelectric region. It should be emphasized that the proportionally between P_s and r_6, i.e., a practical invariance of coefficient g_{36}, is retained over the entire ferroelectric temperature range. Moreover, g_{36} does not show an anomaly in passing the Curie point, which indicates that this coefficient is a true characteristic of the crystal, not associated with the anomalies of the dielectric properties. This cannot, however, be said about piezomodulus d_{36}, which increases according to the Curie-Weiss law as the Curie point is approached.

In the direction of the Z axis, the appearance of $\boldsymbol{P}_s$ is accompanied by spontaneous electrostriction. This is easy to understand, considering that in Table 3.1 modulus $d_{33} = 0$ for class $\bar{4}2m$. From the strain of the crystal along the Z axis because of the arising $\boldsymbol{P}_s$ one can directly determine coefficient Q_{33}, which was found to equal 4.3×10^{-12} CGSE units.

The piezoelectric properties of a crystal in the rhombic phase (class $mm2$) has already been described by another piezomodulus matrix (Table 3.1). The value of piezocoefficient g_{33}, estimated from the electrostrictive strain, is 8×10^{-8} CGSE units. Naturally, this coefficient, too, must not behave anomalously as the Curie point is approached.

A KDP crystal and the other ferroelectrics of this group possess high electro-optical properties and are widely used in nonlinear-optics devices (Chap. 15).

3.5.3 Rochelle Salt

Ferroelectric properties are inherent in double tartrates: $KNaC_4H_4O_6 \cdot 4\,H_2O$ (Rochelle salt), $NH_4NaC_4H_4O_6 \cdot 4\,H_2O$, $LiNH_4C_4H_4O_6 \cdot H_2O$, $LiTeC_4H_4O_6 \cdot H_2O$.Rochelle salt was the first crystal in which ferroelectric properties were discovered and studied. The space group of Rochelle salt in the high-temperature paraelectric phase is $P2_12_12_1$. The polar ferroelectric phase, bounded by

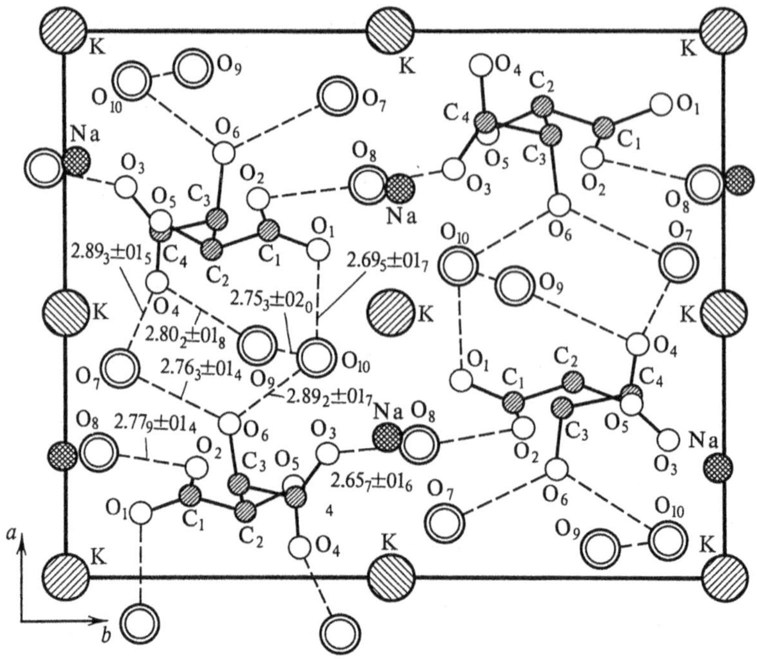

Fig. 3.24. Projection onto the (001) plane of the Rochelle salt structure in the ferroelectric phase. The dashed lines denote hydrogen bonds

two Curie points (upper $+24.5\,^\circ\mathrm{C}$ and lower $-18\,^\circ\mathrm{C}$), is monoclinic and belongs to space group $P2_1$. Monoclinic angle $\beta = 90^\circ 3'$ at $11\,^\circ\mathrm{C}$.

The Rochelle salt structure is rather complicated, as can be seen from Fig. 3.24. It is characterized by endless helical chains of hydrogen bonds O–H...O between the molecules of crystallization water and the oxygen atoms of the anions.

The numerous structural investigations of Rochelle salt have not yet fully elucidated the mechanism of spontaneous polarization in this crystal. It is, however, definitely established that the phase transition in Rochelle salt is due to the ordering of the hydrogen bonds. There are grounds to believe that the ferroelectric properties are associated with the dipoles of hydroxyl group $(OH)_5$ and, to some extent, of water molecules $(H_2O)_8$ (O_5 and O_8, respectively, in Fig. 3.24).

The domain structure of Rochelle salt is outlined in Fig. 3.25. After the phase transition the initial rhombic cell becomes monoclinic as a result of spontaneous polarization along the a axis. Domains with opposite orientations of $\boldsymbol{P}_s$ have the same sign of enantiomorphism. The starting point of the crystal symmetry (class 222) can be obtained by combining monoclinic domains (class 2). Their pairwise combinations may lead to two orientations

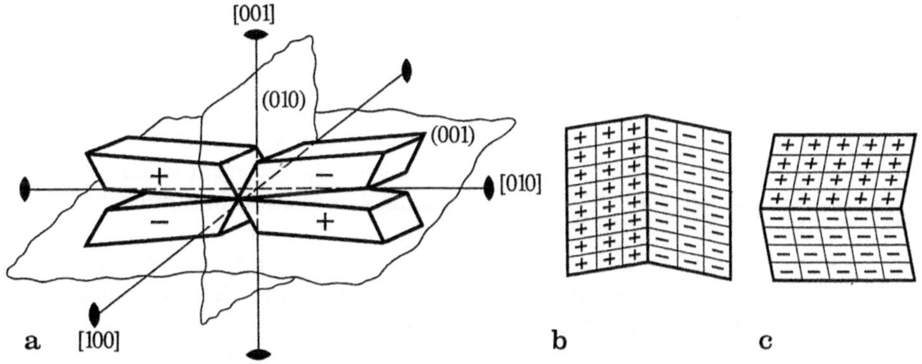

Fig. 3.25. Domains in Rochelle salt: (**a**) domain twin of symmetry 222, symmetry of one domain is 2; (**b**, **c**) dipole patterns on the (100) plane with domain boundaries along the (010) and (001) planes, respectively

Fig. 3.26. Domain structure of a plate of an *a* section of a Rochelle salt crystal

of the domain boundaries: along the (001) or (010) planes. Figures 3.25 b, c present two types of combination out of the four possible.

Domains in a real Rochelle salt crystal (Fig. 3.26) are platelets parallel to rhombic (010) and (001) planes, i.e., to the (*ac*) and (*ab*) planes. Domains with boundaries parallel to the (*ab*) plane are usually called *b* domains, and those parallel to the (*ac*) plane, *c* domains. Because of the different orientations of the unit cells in neighboring domains, during observation in a polarized light these cells become "extinguished" at different orientations of the crystal. This allows one to differentiate domains with opposite orientation of $\boldsymbol{P}_s$. Owing to the anisotropy of the crystal the energy of formation of the boundaries of *b* and *c* domains is different: *b* domains are more stable, pointing to a lower energy of the boundaries of such domains as compared with that of the boundaries of *c* domains. In Rochelle salt the width of the domain boundary is only about two spacings of the unit cell (24 Å), while its energy is several hundred erg/cm^2.

The temperature dependence of the spontaneous polarization for Rochelle salt is given in Fig. 3.27. The temperature dependence of the dielectric constant also correlates with the variation in spontaneous polarization (Fig. 3.28).

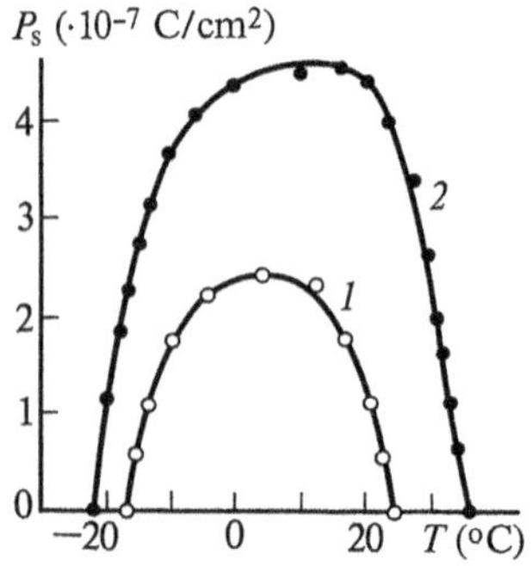

Fig. 3.27. Temperature dependence of spontaneous polarization for Rochelle salt (*1*) and deuterated Rochelle salt (*2*) [3.13]

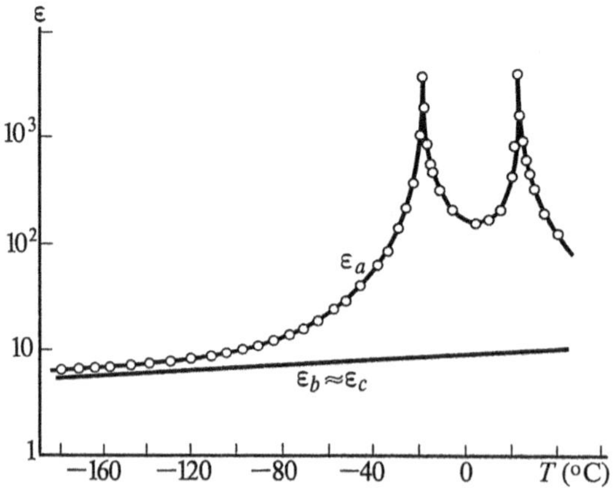

Fig. 3.28. Temperature dependence of the dielectric constant of Rochelle salt measured in weak fields

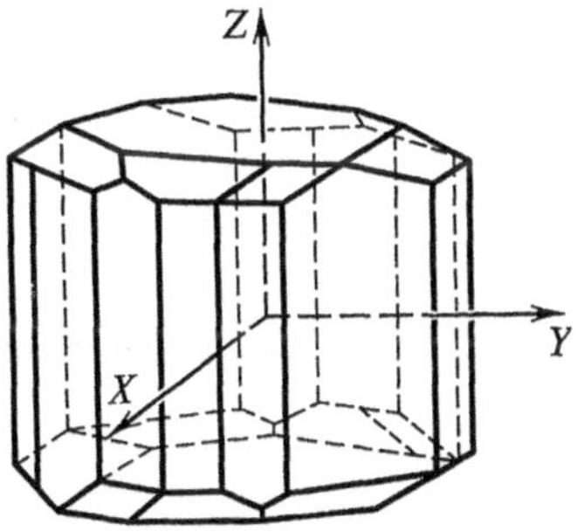

Fig. 3.29. Setting of a Rochelle salt crystal in the crystallophysical coordinate system

Outside the ferroelectric temperature range, Rochelle salt, being noncentrosymmetric, posssesses piezoelectric properties. The piezomodulus matrix has three nonzero and independent coefficients (see Table 3.1 and the crystal setting in Fig. 3.29). The spontaneous polarization along the X axis (a axis) in this crystal is thus accompanied by a spontaneous piezoelectric shear strain:

$$r_{4s} = g_{14} \boldsymbol{P}_{1s} .$$

The value of coefficient g_{14}, estimated from experimental data for $\boldsymbol{r}_s$ and $\boldsymbol{P}_s$, is about $(3-4)\times 10^{-9}$ CGSE units over a wide temperature range, including both Curie points.

It is the spontaneous deformation that makes the crystal (more precisely, the domain) in the ferrophase monoclinic, with symmetry 2. In the coordinate system of the rhombic (paraelectric) phase the piezomodulus matrix of a single-domain crystal will now have a different form (Table 3.1). The matrices of the other coefficients change similarly. Coefficient d_{11}, which is equal to zero in the rhombic phase, is $(1-1.5)\times 10^{-6}$ CGSE units in a polarized specimen in the middle of the ferroelectric temperature range. Similar measurements yield values $d_{12} = 3.8\times 10^{-7}$ and $d_{13} = -3.78\times 10^{-7}$ CGSE units.

It has been shown experimentally that piezocoefficients g_{14} and h_{14} remain practically unaltered on changes in temperature (including both Curie points) and in field tension, whereas coefficients d_{14} and e_{14} strongly depend on the temperature and the field. This points to the correctness of coefficients g_{14} and h_{14}. Summarized data on the piezo- and dielectric properties of Rochelle salt are given in Table 3.6. It can be seen that Rochelle salt possesses very high piezoelectric properties; for some vibration modes the electromechanical-coupling coefficient exceeds 50%.

Since a Rochelle salt crystal shows piezoelectric properties in the polydomain state as well, it can be repolarized under the effect of mechanical stresses (or strains). A rearrangement of the domain structure under the effect of mechanical stresses with cyclical changes of their magnitude and sign leads to the hysteresis dependence of the polarization (expressed in the ratio of the areas of oppositely oriented domains) on the mechanical stresses.

Table 3.6. Piezoelectric coefficients and dielectric constants of a Rochelle salt crystal at 20 °C (CGSE units)

Piezocoefficients				Dielectric constants	
$d \cdot 10^8$	$g \cdot 10^8$	$e \cdot 10^4$	$h \cdot 10^4$	ε^l	ε^r
$d_{14} = 1150$	$g_{14} = 31$	$e_{14} = 140$	$h_{14} = 7.58$	$\varepsilon^l_{11} = 480$	$\varepsilon^r_{11} = 220$
$d_{25} = -160$	$g_{25} = -170$	$e_{25} = -5$	$h_{25} = -5.8$	$\varepsilon^l_{22} = 12$	$\varepsilon^r_{33} = 11$
$d_{36} = 35$	$g_{36} = 44$	$e_{36} = 3.5$	$h_{36} = 4.8$	$\varepsilon^l_{33} = 10$	$\varepsilon^r_{33} = 9.8$

3.5.4 Triglycinesulphate

Triglycinesulphate $(NH_2CH_2COOH)_3 \cdot H_2SO_4$(TGS) is one of the three ferroelectric isomorphous molecular compounds; the other two are triglycineselenate and triglycinefluoroberyllate. The ferroelectric phases of these compounds belong to space group $P2_1$, and the paraelectric phases, to $P2_1/m$; in both phases $Z = 2$.

The determined x-ray and neutron-diffraction structure of triglycinesulphate is a complicated packing of glycine molecules and sulphate ions bound by hydrogen bounds O – H...O and N – H...O. Of the three glycine molecules one has the ordinary amphoteric configuration with an NH_3^+ group lying outside the plane of the other atoms, and the other two are laying in plane within the experimental error. The configurations of the glycinces are shown in Fig. 3.30.

The ferroelectric properties of triglycinesulphate are due to the ordering of the protons of the hydrogen bonds, this leading to the appearance of dipoles on the glycine molecules and sulphate ions. The main contribution to the spontaneous polarization of TGS is made by the dipole moments of the NH_3^+ groups belonging to the ions of glycine I. The ions of glycine I may occupy

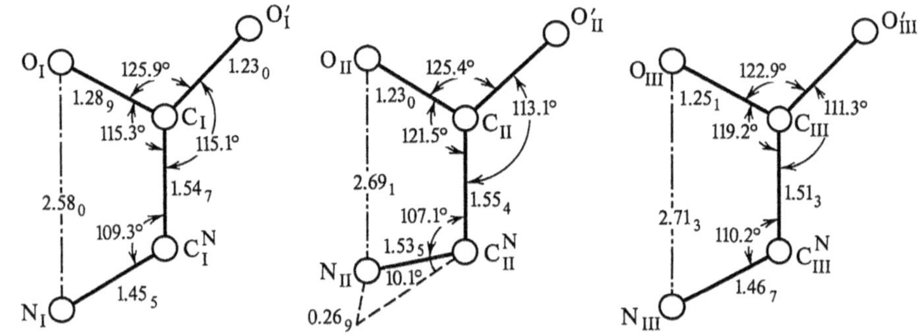

a b c

Fig. 3.30. Configurations of ions of glycine and glycinium in triglycinesulphate: (**a**, **c**) plane monoprotonated glycine ions, (**b**) amphoteric glycine ion

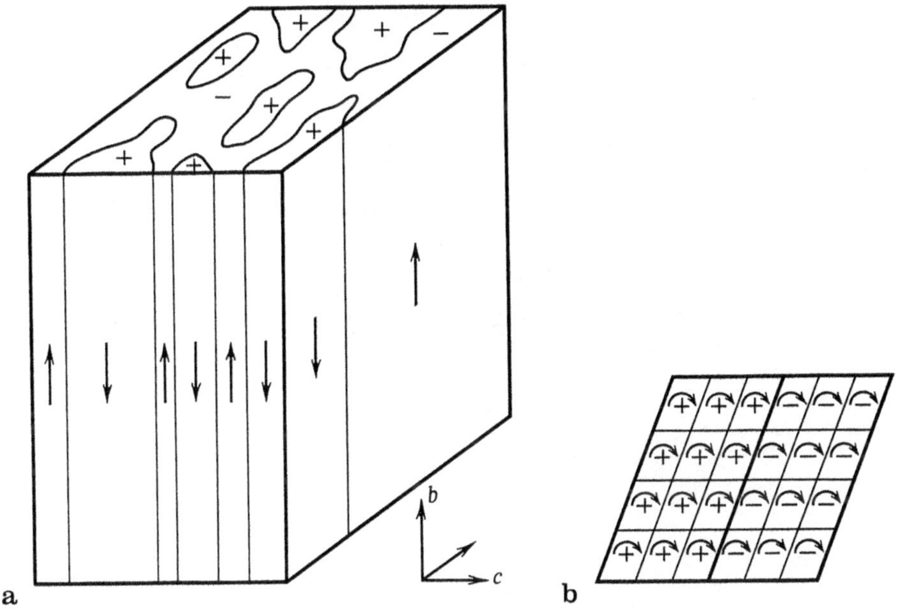

Fig. 3.31. Model of the domain structure in triglycinesulphate **(a)** and the dipole pattern in a thin section of this crystal perpendicular to the *b* axis (axis 2), along which the spontaneous polarization is oriented **(b)** [3.5]

two equivalent symmetric positions at an angle of $+12.5°$ to plane $y = 1/4$, depending on the sign of the dipole moment. The proton ordering was established by studying the diffuse scattering of x-rays near the Curie point. Interestingly, in the structure of the paraelectric phase the planes of symmetry arising perpendicular to the twofold axis are present only statistically because of the disordered orientation of the ions of glycine I.

In TGS crystals, as in KDP, the ferroelectric phase transition is of the order-disorder type. In both cases the protons of the short hydrogen bonds initiate dipole moments arising on definite groups. As for the hydrogen bond dipole moments themselves, they are almost perpendicular to the ferroelectric axis and make no contribution to the spontaneous polarization.

The domain structure of TGS is outlined in Fig. 3.31a. Domains with some one direction of $\boldsymbol{P}_s$ can conventionally be called right handed, while the other domains will be left handed. A reversal of vector $\boldsymbol{P}_s$ in each of the domains is accompanied by a change of sign of its enantiomorphism. Figure 3.31b depicts the dipole motif in the structure of two neighboring domains. The enantiomorphism of the domains is denoted by a semicircular arrow, whose direction indicates its sign.

It is seen from Fig. 3.31b that the crystal lattices of neighboring TGS domains are parallel, precluding their optical observation. The domain struc-

Fig. 3.32. Electron micrograph of the domain structure of triglycinesulphate revealed by decorating the surface of a (010) cleavage with silver chloride ($\times 4800$). The *a* axis is arranged horizontally in the plane of the drawing

ture of TGS, however, has been thoroughly investigated by the etching, electroluminescence, and charged-powder methods.

Although the domain structure of TGS greatly depends on the crystal growing conditions (particularly on the growth temperature), for most of the crystals that have not been subjected to the effect of electric and mechanical fields, like domains (of the same orientation of $\boldsymbol{P}_s$) are a kind of a matrix which contains domains of the other sign (with the opposite orientation of $\boldsymbol{P}_s$) in the form of lentil-section rods oriented along the *b* axis. The major axis of the lentil cross section is oriented along the *a* axis of the crystal (Fig. 3.31). In some cases the domain boundaries are oriented along the *a* axis of the crystal (Fig. 3.32).

The temperature dependence of the spontaneous polarization for TGS-type ferroelectrics is given in Fig. 3.33. According to this dependence, for TGS near T_C ($+49\,°C$) in the temperature range of $37-49\,°C$ the change in spontaneous polarization is about 6×10^3 CGSE units (2 μC), which enables one to estimate pyrocoefficient p in this range at 500 CGSE units. Hence, in a

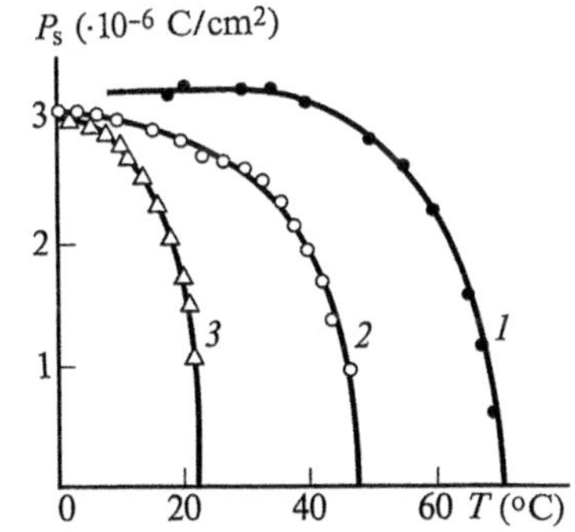

Fig. 3.33. Temperature dependence of spontaneous polarization P_s in crystals of triglycinefluoroberyllate (*1*), triglycinesulphate (*2*), and triglycineselenate (*3*) [3.14]

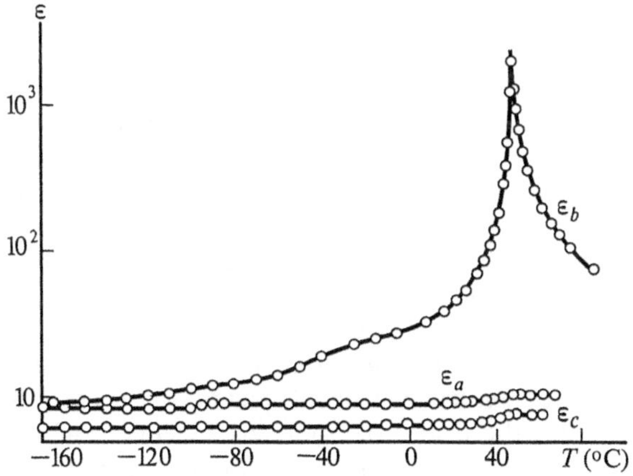

Fig. 3.34. Temperature dependence of the dielectric constant of triglycinesulphate [3.15]

TGS crystal the potential difference due to the pyroelectric effect may run into the hundreds of kilovolts because of the comparatively low ε in the transition range. Therefore, TGS plates are often broken down by the intrinsic field of spontaneous polarization due to the pyroelectric effect on rapid cooling from the paraelectric phase. Above T_C a TGS crystal possesses no piezoelectric properties, and thus the pyroelectric effect in the phase transition region is predominantly true, primary. In the temperature range of 37 – 20 °C the pyrocoefficient is about 150 CGSE units. Here TGS has a low electrocaloric coefficient $q = -0.012$ CGSE units; this means that its spontaneous electrocaloric effect is weak, and the external field only slightly affects the phase transition temperature (second-order phase transition).

The temperature dependence of the dielectric constant of TGS is presented in Fig. 3.34. The anomaly of ε_b in the range of $T_C - 49$ °C (vector P_0 is parallel to the b axis) is seen from the picture. Clearly defined loops of dielectric hysteresis (coercive field ~400 V/cm) are easily observable on TGS crystals.

Crystals of the TGS type are typical representatives of ferroelectrics with impurity ionic conductivity. In particular, the electrical conductivity of trigly-

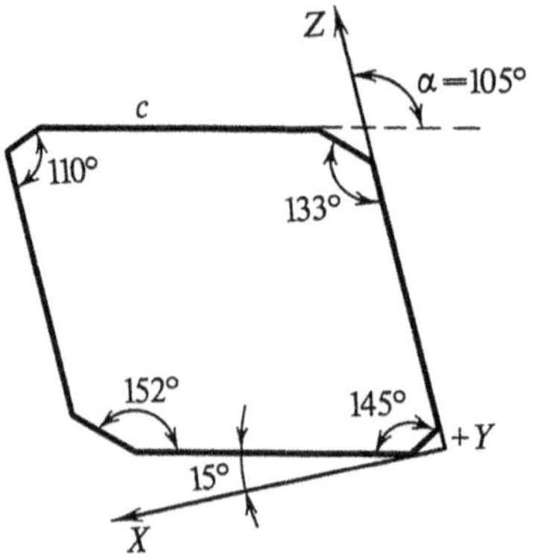

Fig. 3.35. Crystallophysical setting of a triglycinesulphate crystal

cinesulphate crystal is due to the impurity sulphate ions; TGS crystals clearly exhibit the electrical conductivity jump at the Curie point in measuring along all the three crystallophysical axes. Assuming the ionic character of the electrical conductivity, the activation energy of the carriers is 0.6 eV in the ferroelectric range and 0.4 eV in the paraelectric range. Thus, with increasing temperature in transition to the paraelectric range the activation energy may not only increase (as for $BaTiO_3$ and Rochelle salt), but also decrease.

In a TGS crystal (as well as in $BaTiO_3$), the piezoelectric effect occurs only in the ferroelectric phase. Below T_C, TGS belongs to class 2 and has (in accordance with Table 3.1) eight independent piezocoefficients. In the crystallophysical setup shown in Fig. 3.35, the values of some coefficients g, which are practically temperature independent, as noted previously, are equal to (in CGSE units)

$$g_{22} = -25.2\times10^{-8}, \quad g_{21} = -59.2\times10^{-8}, \quad g_{23} = 15.2\times10^{-8}.$$

At room temperature coefficients d are equal to

$$d_{22} = -83.8\times10^{-8}, \quad d_{21} = -44.2\times10^{-8}, \quad d_{23} = 136.7\times10^{-8}.$$

The dielectric constants measured in weak fields have the following values at room temperature:

$$\varepsilon_{11} = 8.7, \quad \varepsilon_{22} = 25, \quad \varepsilon_{33} = 4.7, \quad \varepsilon_{13} = 0.5.$$

The above data on the piezoelectric properties of TGS refer to a single-domainized specimen. It should, however, be indicated that, as a rule, natural (not field-processed) TGS crystals possess unipolarity, i.e., they have an uncompensated domain structure and hence also show piezoelectric and pyroelectric effects, although insignificant ones.

3.5.5 Antiferroelectrics

Lead zirconate $PbZrO_3$ is a typical displacement-type antiferroelectric. At temperatures above 230°C it is cubic. Below this temperature $PbZrO_3$ becomes orthorhombic with unit cell sizes equal approximately to $\sqrt{2}a_0$, $2\sqrt{2}a_0$, $2a_0$ (where a_0 is the edge of the initial cubic cell) and belongs to space group *Pbam*. Such a superstructural cell has $Z = 8$. Below the phase transition point the structure shows antiparallel displacements of atoms by about 0.2 Å along rhombic axis a (which is taken to be parallel to the subcell diagonal).

The domain structure of $BbZrO_3$ in the orthorhombic phase is similar to that of orthorhombic $BaTiO_3$. This is understandable, because in both these crystals the rhombic modification is due to the polarization in cubic paraelectric cells in the ⟨110⟩ directions in the cubic-phase axes. The difference between them is that the unit cell of orthorhombic $PbZrO_3$ is superstructural, while the unit cell of the cubic phase in $BaTiO_3$ is not multiplied.

The dielectric properties of $PbZrO_3$ have been studied thoroughly enough. In the pyroelectric phase the dielectric constant obeys the Curie-Weiss law. The Curie-Weiss temperature $T_{CW} = 190\,°C$, which is much lower than the true transition temperature (230 °C). In distinction to the antiferroelectrics, which are characterized by the ordering of the structure elements, in $PbZrO_3$ the dielectric constant reaches high values (Fig. 3.36a). the Curie constant is of the same order and magnitude as for $BaTiO_3$ ($C = 1.59 \times 10^5$ deg). The transition is of the first order (the temperature hysteresis of the transition is 4 °C).

In line with the thermodynamic theory (see below) the application of a constant field reduces the transition temperature; at 20 kV/cm, $\Delta T_C = -1.5\,°C$. In strong alternating electric fields the polarization of $PbZrO_3$ is described by double hysteresis loops (Fig. 3.36c). The critical field of a double loop F_{crit} characterizes a field at which an antiferroelectric changes to the ferroelectric state. It is temperature dependent and is the less, the closer the crystal is to the transition temperature (Fig. 3.36b).

Ammonium dihydrophosphate $NH_4H_2PO_4$ (ADP) is isomorphous to KH_2PO_4 at room temperature in the paraelectric phase and belongs to the same space group with unit cell parameters $a = 7.50$ Å and $b = 7.58$ Å. The structure is also characterized by a three-dimensional lattice of PO_4 tetrahedra bound by hydrogen bonds O–H...O of length 2.51 Å and $[NH_4]^+$ tetrahedra bound by hydrogen bonds N–H...O of length 2.88 Å with oxygen atoms of $[PO_4]^{-4}$ groups. As well as in KDP, in ADP the photons of the hydrogen bonds are ordered below the phase transition temperature ($T_C = -125\,°C$), but this ordering is much more complicated. The space group of an ADP crystal in the antiferroelectric phase is $P2_12_12_1$. The superstructure arising below the transition manifests itself as follows: the cell centers equivalent to nodes above the transition become nonequivalent to them below it. Axes 2 become the antipolarization axes along which the polarization in neighboring subcells is counteroriented.

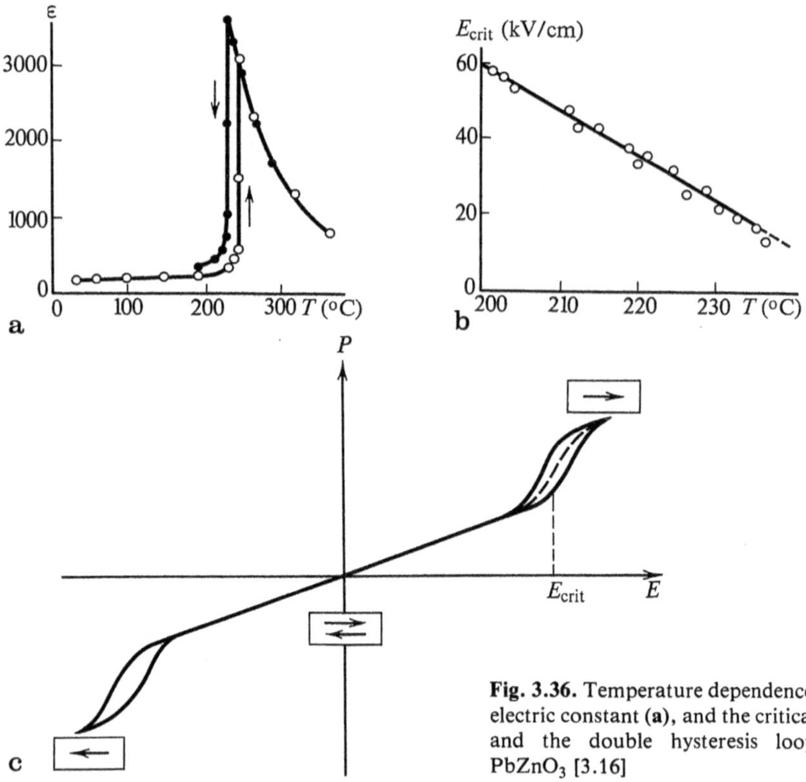

Fig. 3.36. Temperature dependence of the dielectric constant (**a**), and the critical field (**b**), and the double hysteresis loops (**c**) in $PbZnO_3$ [3.16]

The ideal superstructural cell of the antiferroelectric phase of the crystal consists of four cells of the initial phase (four subcells) (Fig. 3.37a). The domains, which make an angle of 90° (Fig. 3.37b), have opposite enantiomorphism signs: one of them is right handed, and the other left handed. That these domains must have opposite enantiomorphism signs follows from the fact that the [100] and [010] directions in the initial crystal are equal only specularly, but not compatibly. Macroscopically, the ideal set of domains of both orientations has the point symmetry of the initial (paraelectric) phase $\bar{4}2m$.

Weak-field investigations show that the dielectric constants along the antipolarization axes (a and b axes) and in the perpendicular direction (c axis) above the transition temperature obey the Curie-Weiss law with negative values of the Curie-Weiss temperature. Mechanical clamping of a crystal reduces the transition temperature.

A good piezoelectric, ADP has been used widely enough in practice. At room temperature its piezocoefficients are equal to (in CGSE units):

$$d_{36} = 148 \times 10^{-8}, \quad g_{36} = 118 \times 10^{-8}, \quad \varepsilon_{33} = 14.0 .$$

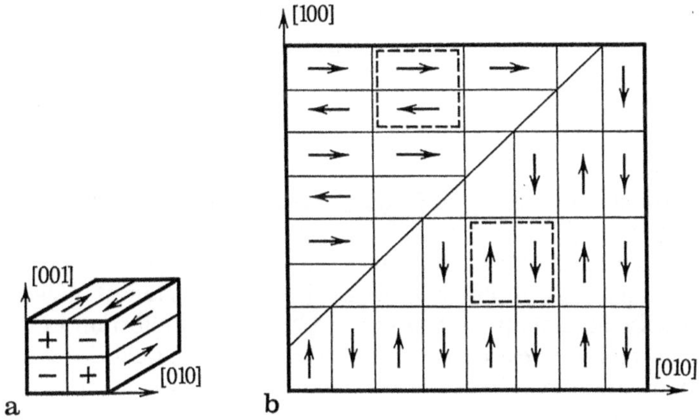

Fig. 3.37. Domains in a crystal of ammoniumdihydrophosphate [3.5]: (**a**) superstructural cell of one domain, symmetry 222; (**b**) dipole patterns in two domains and their boundary

3.6 Phase Transition in Ferroelectrics. Fundamentals of Spontaneous Polarization Theory

3.6.1 General

Spontaneously polarized states are inherent in linear pyroelectrics (tourmaline, sugar, etc.) and ferroelectrics (Rochelle salt, KDP, $BaTiO_3$, etc.). Formally, the ferroelectrics are a subclass of the pyroelectrics: the ferroelectrics are pyroelectrics which are divided into domains. Although the $\boldsymbol{P}_s$ for both crystals are of the same order of magnitude in absolute value, the origin of their spontaneous polarization is evidently different. This is indicated, in particular, by the fact that the linear pyroelectrics do not experience phase transitions. On the other hand, it is unreasonable to assume that in linear pyroelectrics the Curie temperature is very high and cannot be achieved before the melting (decomposition) of the crystal, because some ferroelectrics are known to exist only in the ferroelectric modification, and to have a domain structure (for instance, guanidealuminiumsulphate), as would be expected of a ferroelectric.

One can show qualitatively why states with spontaneous polarization exist in some crystals, i.e., are thermodynamically stable. This can be done most easily by proceeding from model considerations. For this purpose we assume that the crystal lattice is formed by point electric dipoles with a moment $\boldsymbol{p}$ (the formation and stability of a given lattice may be ensured not only by dipole forces, but also by any other forces, which are of no interest to us at the moment). Within the framework of such a model the notion of an effective, or acting, electric field $\boldsymbol{F}$, i.e., the average field acting on a separate dipole, has an exact meaning. If average macroscopic field $\boldsymbol{E}$ and polarization $\boldsymbol{P}$ are parallel (for instance, if they are directed along the symmetry axis of the lattice), then

$$\boldsymbol{F} = \boldsymbol{E} + f\boldsymbol{P}, \tag{3.57}$$

where f is the Lorentz factor (for a cubic lattice $f = 4\pi/3$). The polarization of the medium under discussion in an external electric field can be considered in the same way as the polarization of a gas consisting of dipole molecules or as magnetization of the spin gas in ferromagnetics. Direct extension of Weiss's ferromagnetism theory suggests that, at a thermodynamic equilibrium in the medium under discussion, in the absence of an external electric field $\boldsymbol{E}$ there is a spontaneous polarization $\boldsymbol{P}$ different from zero in the temperature range between $T = 0$ and Curie point T_C, and

$$T_C = fpP_\infty 3kT, \tag{3.58}$$

where P_∞ is the polarization at saturation (i.e., at $T = 0$) and k is the Boltzmann constant. At $T > T_C$, $P = 0$. Assuming $f \approx 1$, $p = 10^{-18}$, and $P_\infty \approx 10^6 - 10^5$ (CGSE units), i.e., near-real values, in the above equation, we have $T_C \approx 10^2 - 10^3$ deg.

Thus, the foregoing shows that the difference between acting field $\boldsymbol{F}$ and average macroscopic field $\boldsymbol{E}$ leads at once to the possibility (and within the framework of the selected model even to the necessity) of the existence of spontaneous polarization. It is significant that in this case (in distinction to ferromagnetism), we directly obtain the value $T \approx 10^2 - 10^3$ corresponding to the real Curie temperatures in ferroelectrics. In other words, the ordinary classical electrostatic interaction leading to values $f \approx 1$ can ensure the existence of spontaneous polarization up to $T_C \approx 10^3$. As is well known, however, in the case of ferromagnetism we must, within the framework of the Weiss theory, adopt for f the value of $10^3 - 10^4$, which cannot be due to magnetic interactions.

3.6.2 Thermodynamic Theory of Phase Transitions in Ferroelectrics

In describing processes involved in spontaneous polarization it is best to consider the free energy G of a crystal as a function of temperature T, the stresses, and polarization $\boldsymbol{P}$. The differential expression of this function has the form

$$dG = -SdT + r_i dt_i + E_m dP_m, \tag{3.59}$$

where S is the entropy, r_i are the components of the strain tensor, E_m and P_m are the components of the vector of the electric field tension and spontaneous polarization, and t is the mechanical load. Considering a free crystal ($t = 0$),

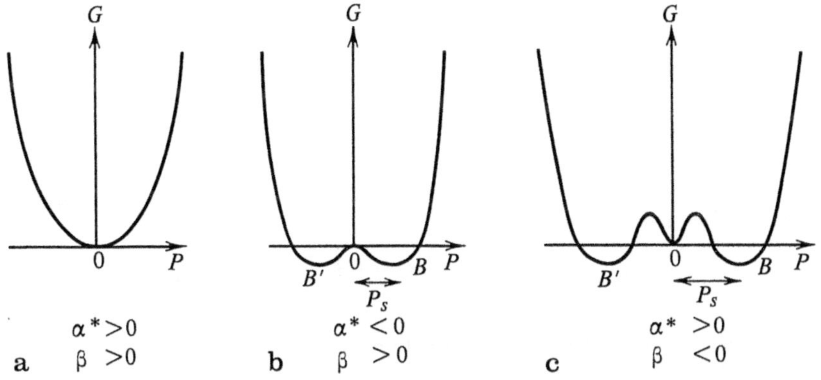

Fig. 3.38. Free energy curves for the paraelectric (**a**) and ferroelectric (**b, c**) phases: (**b, c**) second- and first-order phase transitions, respectively

function G can be expanded in degrees of polarization with temperature-dependent coefficients.

For crystals of uniaxial ferroelectrics with a single spontaneous polarization axis (Rochelle salt, TGS, KH_2PO_4, etc.) only two orientations of $\boldsymbol{P}_s$ are possible: $+P_s$ and $-P_s$. Since these two directions are equivalent, the expansion of free energy $G = G(P, t)$ at tensions t equal to zero will contain only terms with even degrees of polarization:

$$G = G_0 + \tfrac{1}{2}\,\alpha^* P^2 + \tfrac{1}{4}\beta P^4 + \tfrac{1}{6}\gamma P^6 + \dots\,. \tag{3.60}$$

The electric field tension is then defined by

$$E = \partial G/\partial P = \alpha^* P + \beta P^3 + \gamma P^5 + \dots\,. \tag{3.61}$$

The shape of the free energy curve changes substantially depending on the number, signs, and magnitude of the coefficients of expansion (3.60). If the signs of all the expansion terms are positive ($\alpha^*, \beta, \gamma \dots > 0$), the curve has a single minimum at $P_s = 0$ (Fig. 3.38a). Then the paraelectric state is stable over the entire range of P.

If one of the expansion coefficients α^* or β is negative ($\gamma > 0$ always), then, in the absence of an external field ($E = 0$), (3.61), which takes the form $\partial G/\partial P = 0$, may have, apart from solution $P = 0$, other solutions, for which $P \neq 0$, corresponding to the spontaneously polarized state. Having analyzed the free energy of such a state one can determine whether or not it will be stable.

The free energy extrema defined by the equation $\partial G/\partial P = 0$, give the first solution $P_s = 0$, which corresponds to the extremum of the function at the origin. Other extrema can be found from the equation

$$\alpha^* + \beta P^2 + \gamma P^4 + \dots = 0\,. \tag{3.62}$$

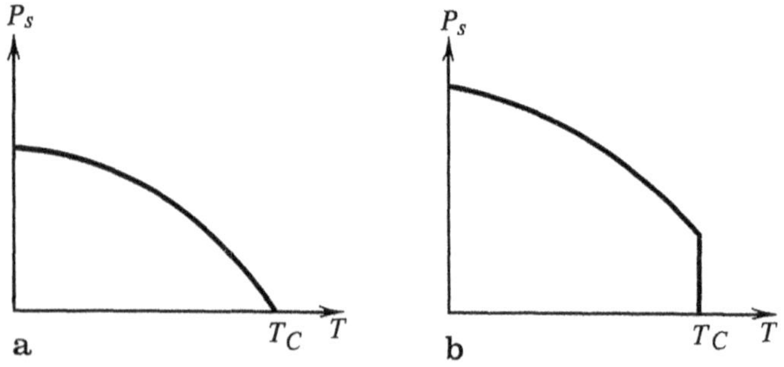

Fig. 3.39. Schematic curves illustrating the temperature dependence of spontaneous polarization at second- (**a**) and first- (**b**) order phase transitions

Let us analyze this equation in more detail. Putting at first $\gamma = 0$, $\alpha^* + \beta P^2 = 0$, we obtain the expression for the value of spontaneous polarization P_s:

$$P_s^2 = -\alpha^*/\beta . \tag{3.63}$$

The solution for P_s possesses a physical meaning if α^* and β have opposite signs. The graphic representation of function G corresponding to $\alpha^* < 0$, $\beta > 0$ is presented in Fig. 3.38b. The free energy then has a maximum at the origin and two symmetric minima B and B'. The state with spontaneous polarization is stable. A comparison of Fig. 3.38a, b shows that a continuous change of α^* from the positive to the negative value leads to a change of the stable paraelectric state ($\alpha^* > 0$) to the stable ferroelectric state ($\alpha^* < 0$, at transition temperature T_C $\alpha(T_C) = 0$). In the above-discussed case spontaneous polarization P_s is also a continuous function of the temperature (Fig. 3.39a) and reduces to zero at phase transition temperature T_C. If we assume that coefficient α^* depends linearly on the temperature [Curie-Weiss law (3.56)]

$$\alpha^* = (T - T_C)/C^* \tag{3.64}$$

(where C^* is a constant), then for spontaneous polarization P_s (at $\beta = \text{const}$) we obtain from (3.63) the parabolic temperature dependence

$$P_s^2 = k(T_C - T) , \tag{3.65}$$

where k is a constant. The crystal energy changes continuously during the transition; the transformation proceeds without any latent heat, but is accompanied by a jump in heat capacity, which is typical of second-order transitions.

Now we revert to (3.62). If coefficient β, and not α, is negative in this equation, the curve of free energy G is more complicated. The presence of a minimum at the origin ($\alpha^* > 0$) admits of the existence of $P_s \neq 0$ only if other minima satisfying condition $G_{P_s} = 0 \geqslant G_{P_s} \neq 0$. The last condition means that (3.60)

$$\tfrac{1}{2}\alpha^* P_s^2 + \tfrac{1}{4}\beta P_s^4 + \tfrac{1}{6}\gamma P_s^6 + \ldots = 0\,. \tag{3.66}$$

Equations (3.62, 65) must be simultaneous, which leads to the relations

$$P_s^2 = -\tfrac{3}{4}\beta_0/\gamma_0\,, \tag{3.67}$$

$$\alpha_0^*\,\gamma_0 = \tfrac{3}{16}\beta_0^2\,, \tag{3.68}$$

characterizing the state of the crystal at the transition temperature (this is emphasized by the zero subscripts at coefficients α^*, β, and γ). As seen from these relations, condition $\alpha^* > 0$ indicates that β and γ must have opposite signs.

The graphic representation of the free energy at a temperature close to T_C (but below it) is given in Fig. 3.38c ($\alpha^* > 0$, $\beta < 0$, $\gamma > 0$). It is seen that curve G in this case has three minima: one at the origin at $P = 0$ and two others corresponding to values $P \neq 0$ (B and B'). If the ratio of the depths of the minima is temperature dependent, at equal potential transition from the paraelectric to the ferroelectric state will occur. In distinction to the preceding case, the spontaneous polarization and energy of the crystal change jumpwise at the transition point. The latent heat of transition has a finite value, which is typical of first-order phase transitions.

In accordance with (3.59) the system entropy is defined by the expression

$$S = -\left(\frac{\partial G}{\partial T}\right)_{P,t}\,;$$

after using it we have from (3.60)

$$S = S_0 - \frac{1}{2}P^2\frac{\partial\alpha^*}{\partial T} - \frac{1}{4}P^4\frac{\partial\beta}{\partial T} - \ldots\,, \tag{3.69}$$

where S_0 is the entropy at zero polarization. Thus, if only the first term is significant in (3.69), the change in entropy on transition will be written

$$\Delta S = S - S_0 \cong -\tfrac{1}{2}P^2\,\partial\alpha^*/\partial T\,. \tag{3.70}$$

Taking into account (3.64), we can transform (3.70) to

$$\Delta S = -\tfrac{1}{2}C^* P_s^2\,. \tag{3.71}$$

From the last expression it is clear that to a jumpwise change in P_s corresponds a jumpwise change in S, which points to the existence of a latent heat of transition ΔQ (first-order transition) determined by the relation

$$\Delta Q = \frac{T_C}{2C^*} P_s^2 . \tag{3.72}$$

Similarly, in the case of second-order transitions, on which the slope of the curve of P^2 changes jumpwise, there is a jump in heat capacity. The temperature dependence of spontaneous polarization for both transition orders is outlined in Fig. 3.39.

So far we have been considering the free energy of the system neglecting the mechanical stresses and electric fields ($t = 0$, $E = 0$). Let us now take into account the effect of these quantities on the system temperature (and hence on the position of the transition temperature). For first-order transition the energies of the polarized ferroelectric ("f") and paraelectric ("p") phase in transition are equal ($G_f = G_p$); therefore, from (3.59) follows

$$0 = dG_f - dG_p = -(S_f - S_p)\,dT + (r_{if} - r_{ip})\,dt_i + (P_{mf} - P_{mp})\,dE_m , \tag{3.73}$$

where T is the transition temperature corresponding to tension T_i and field E_m. Assuming field E to be constant and pressure p to be the hydrostatic pressure, we obtain

$$\frac{\partial T}{\partial P} = \frac{\Delta V}{S_f - S_p} \tag{3.74}$$

(where ΔV is the change in volume on transition), which is the well-known Clausius-Clapeyron equation.

Taking the tensions in (3.73) to be constant, we get

$$\frac{\partial T}{\partial E_m} = \frac{P_{mf} - P_{mp}}{S_f - S_p} ; \tag{3.75}$$

this expression makes it possible to calculate the change in transition temperature under the effect of the field. In particular, by using the entropy change on transition we can, by (3.71), write down (3.75) in the form

$$q = \frac{\partial T_C}{\partial E} = \frac{\Delta P}{\Delta S} = -\frac{2C^*}{P_f} . \tag{3.76}$$

In second-order transitions the entropy in the two phases is equal. Equating S_f to S_p yields, for the case of hydrostatic pressure, Ehrenfest's equation:

$$\frac{\partial T}{\partial P} = \frac{T(\Delta_{if} - \Delta_{ip})}{C_f - C_p}, \qquad (3.77)$$

where c_f and c_p are the heat capacities at a constant pressure, and Δ_{if} and Δ_{ip} are coefficients of expansion.

Some of the thermodynamic relations given here can easily be verified from the experimental data. We note in the first place that for ferroelectrics experiencing a first-order phase transition, owing to the temperature hysteresis of the transition, T_C in a relation of the type (3.64) must be replaced by some temperature T'_C corresponding to the true transition temperature, which differs by a few degrees from T_C (for $BaTiO_3$, for instance, $T'_C - T_C = 11\,°C$). The value of coefficient α_0^* corresponding to the transformation temperature for such transitions is found from the relation

$$\alpha_0^* = (T'_C - T)/C^* . \qquad (3.78)$$

Relation (3.72) for Rochelle salt and KDP-type ferroelectrics is fulfilled fairly well in experiment. Relation (3.76) for $BaTiO_3$ ($P_s = 54 \times 10^3$ CGSE units, $\Delta S = -0.12$ cal/mol · deg) produces a value $q = \partial T_C/\partial E$ equal to -0.53 CGSE unit, which is close to the experimental value $q = -0.45$ CGSE unit. Expression (3.74) faithfully reflects the experimentally established shift of the Curie point of $BaTiO_3$ toward low temperatures in hydrostatic compression.

The general premises of thermodynamic theory given here are well developed with respect to Rochelle salt, barium titanate, and some other ferroelectrics, for which theory agrees satisfactorily with experiment.

There is no ground to believe that in all phase transitions accompanied by spontaneous polarization precisely this polarization can be taken as the parameter in which the free energy (3.60) must be expanded. Indeed, spontaneous polarization is not such a parameter when the transition is not due to the spontaneous polarization, but is only attended by it. From the symmetry standpoint, in such a transition the symmetry group of the crystal in the ferroelectric phase is a general, but no longer the highest, subgroup of the symmetry group of the crystal and of the symmetry group of the spontaneous polarization (polar) vector.

Ferroelectrics in which phase transitions are not due to spontaneous polarization are called *improper*. Examples of improper ferroelectrics are crystals in which transition to the ferroelectric phase occurs with multiplication of the cell (with the formation of a superstructure). A typical representative is gadolinium molybdate $Gd_2(MoO_4)_3$. At $T \sim 159\,°C$ this crystal experiences a ferroelectric phase transition with a change of space group $P\bar{4}2_1m$ to $Pba2$. The rearrangement of the structure is such that the volume of the unit cell is doubled, i.e., the translation in the plane perpendicular to the polar axis is multiplied.

Consideration of the ratio between the volume and correlation energies and of their role in phase transition shows that expansion (3.60) describes well the phase transition in ferroelectrics only if $(\Delta P)^2 \ll P^2$, indicating a relatively small fluctuation of polarization. A detailed consideration of the correlation effects in phase transitions in displacement-type ferroelectrics shows that these effects, which consist in interaction of the phonons of the critical low-lying branches with each other, are evidently not small and can substantially affect the transition thermodynamics. In particular, oxygen-octahedral ferroelectrics have, as a rule, first-order transitions, although they are close to second-order ones in the sense that $(T_C' - T_C) \ll T_C$. In these ferroelectrics, coefficient β is strongly temperature dependent: $\beta(T) \approx \beta_0 [1 - 9(T_C - 1)]$. A similar dependence also exists for coefficient γ. These dependences are due to anharmonic corrections, which have a relative order $T\mathscr{E}_{aT}^{-1}$ (where $\mathscr{E}_{aT} \approx 1$ eV).

The conventional phenomenological theory does not describe the temperature dependence of the dielectric constant exactly enough. Taking into account the correlation effects in the interaction of critical fluctuations at transition in displacement-type ferroelectrics makes it possible to correlate theory quite satisfactorily with the principal features of these crystals observed in phase transitions. At the same time, in a number of ferroelectrics the polarization fluctuations are small, and the behavior of such crystals in the vicinity of the Curie point is properly described by an expansion of the type (3.60), as has already been noted.

3.6.3 Macroscopic Models

To establish whether the electron polarization is high enough to transfer the crystal from the nonpolarized to the polarized state, one must bear in mind that spontaneous polarization can arise when the interaction energy of dipoles (induced by acting field F) exceeds the energy required to cause distortion of electron shells, i.e., to set up the necessary dipole moment. Consideration shows that the polarized state becomes stable when electron polarization η exceeds the critical value

$$\eta \geqslant \tfrac{3}{4}\pi N, \tag{3.79}$$

which corresponds to the so-called "$\frac{4\pi}{3}$-catastrophe".

As a rule, condition (3.79) does not hold for crystals, and there is no spontaneous polarization. But for some cubic crystals the Lorentz field has coefficients greatly exceeding $\frac{4\pi}{3}$ along some directions. If relation (3.79) is fulfilled under these conditions, spontaneous polarization may arise in the crystal at lower η as well.

If we assume that in a crystal the conditions of the "$\frac{4\pi}{3}$-catastrophe" are fulfilled, then in order to obtain the phase transition temperature we must introduce the temperature dependence of the polarization of at least one of the ions. To be able to explain the dielectric properties of displacement-type ferroelectrics (for instance, $BaTiO_3$) it is sufficient for one of the polarization η to show a weak linear temperature dependence.

To explain the properties of ferroelectrics with ordering structure elements, such as Rochelle salt, KDP, and TGS, the temperature dependence of polarization η must be of a different nature. The macroscopic susceptibility α of such ferroelectrics above the Curie point (above the upper Curie point for Rochelle salt) is known to be inversely proportional to the temperature. This suggests that in describing the spontaneous polarization in such ferroelectrics one can use a theory based on the assumed existence of constant dipoles, which can be oriented by the internal field. In polar liquids such dipoles are polar molecules. In the crystal lattice one cannot single out separate polar molecules capable of reorientation. An ion (with a charge q) capable of occupying two (or more) equilibrium positions, however, may be the equivalent of polar molecules. (An example of such an ion is the proton of the bond O–H...O in triglycinsulphate, which can occupy two stable equilibrium positions at a distance l from the bond center). This ion can be regarded as a pseudodipole with a moment $p=(ql)$, which can have two (or more) orientations. For transition from one position to another some activation energy is required. The general orientation of the pseudodipoles below the Curie point corresponds to ion ordering. With this approach an increase in spontaneous polarization of as crystal is described by a function depending on the number and distribution of the equilibrium positions of the ion to which the dipole relates.

The Lorentz field calculation for the tetragonal modification of $BaTiO_3$ shows that interaction between Ti and O_I ions is the strongest (Fig. 3.10). The coefficients describing this interaction are about 8.2 times the usual value $\frac{4\pi}{3}$, which is the cause for the considerable field increase in the region of the central ion in the perovskite structure. The shift of the titanium ion from the center of the oxygen octahedron (which is present in the tetragonal modification) greatly increases the fraction of electron polarization, mainly because of the increased polarization due to the O_I ion. The fields of the O_I ions, in turn, affect the Ti ions, with the resulting formation of chains of dipoles oriented in one and the same direction, which set up spontaneous polarization in the crystal. Calculation shows that the contribution of the titanium ions to the complete polarization is 37% (31% due to the ion shift and 6%, to the electron polarization), while the contribution of oxygen ions O_I may reach 59%. These estimates must be regarded as approximate, of course, since the calculation result greatly depends on the values of ion polarizations used and on other factors.

Detailed calculations show that the electron polarization alone does not ensure spontaneous polarization. In other words, electron polarizations of individual ions do not provide full explanation for the ferroelectric phase transition in $BaTiO_3$.

There are also other approaches to the problem of the role of electron polarization in the appearance of spontaneous polarization. In particular, the electron state of the TiO_6 octahedron is considered as a whole. In the absence of an electric field, the ground state of this electron configuration is cubically symmetric and does not possess a dipole moment. The initially degenerate energy level of the ground state splits up under the effect of interaction between the octahedra in such a way that the perturbed-state energy becomes less than that of the ground state. This explains the stability of the state with the nonzero dipole moment. Despite some drawbacks, the approach under review is of considerable interest, because it shows the fundamental possibility of existence of spontaneous polarization as a result of pure electron effects.

Spontaneous polarization in a number of crystals can be interpreted as the result of the Jan-Teller pseudoeffect. In this pseudoeffect the symmetric arrangement of the ions is unstable in the presence of two sufficiently close energy zones (valency and free) in a high-symmetry crystal. Therefore, the ions shift so that the symmetry of their arrangement is disturbed, and the edge of the valency zone lowers. Calculation shows that for the perovskites in the model taking into consideration only the interaction of the outer π states the criterion for the appearance of spontaneous polarization is written

$$\Delta < 4a^2/\varrho\omega^2 , \tag{3.80}$$

where Δ is the width of the forbidden zone, ϱ is the unit cell mass, ω is the frequency of the limiting optical vibration, and a is the electron-phonon interaction constant. To the three phase transitions in a $BaTiO_3$ crystal correspond the natural frequencies of limiting optical vibrations (degenerate in the symmetric state).

The appearance of spontaneous polarization in perovskites (as well as in some other crystal) when condition (3.80) is fulfilled is due to the interaction (mixing) of the filled and vacant zones of the crystal (in the maximally symmetric configuration). From (3.80) it follows that this interaction is the most efficient when the zones are close enough, so that the interaction between the valency and the conduction zone is the strongest.

The estimate for $BaTiO_3$ yields $a \approx 10^{-5}$ CGSE units, $\Delta \approx 1\,\text{eV}$, and $\omega \approx 10^{12} - 10^{13}\,\text{s}^{-1}$, which points to the possibility of fulfillment, for this crystal, of the criterion for the appearance of ferroelectricity (3.80). Theory also indicates that in the series of perovskites with Ti, Zr, and Hf ions, T_C must all off from Ti to Hf, and in the series of perovskites with Ca, Sr, and Ba ions, from Ba to Ca, which is actually the case.

In the model of anharmonic oscillators, a system of anharmonically vibrating ions is considered. (A system of harmonically vibrating ions cannot lead to spontaneous polarization). An analysis of the potential energy of such a system leads to an analytical expression similar to (3.79), which requires the appearance of a "$\frac{4\pi}{3}$-catastrophe" when considering internal fields. The right-hand side of this expression contains a quantity characterizing the elastic force arising on ion displacement, and the left-hand side, a quantity characterizing the force due to the appearance of an internal field or, to put it differently, the force due to interaction of the dipoles resulting from the ion shift. Thus, spontaneous polarization arises in crystals where the dipole-dipole interaction force exceeds the elastic force.

A number of authors use the Ising model for the description of phase transitions in ordering ferroelectrics. Its use is particularly effective in calculating microscopic models describing transitions with substantial fluctuations.

The Ising model for ferroelectrics is a lattice of dipoles, each of which can occupy only two positions and interact only with the nearest neighbors. In a two-dimensional lattice a dipole interacts with four neighbors, and the energy of the parallel and antiparallel dipoles is different. If we denote this difference by I, then the free energy of the system as a function of the temperature has a singularity at temperature T_C, which is determined from the ratio

$$\mathrm{sh}^2(I/kT_C) = 1 . \tag{3.81}$$

At $T < T_C$ the lattice is in the ordered state, and at $T > T_C$ in the disordered state; at $T = T_C$ a second-order phase transition occurs. The heat capacity at the transition point goes to infinity according to the law

$$c \sim -b \ln(T - T_C) \tag{3.82}$$

(where b is a constant). From the standpoint of Landau's phase transition theory, the role of the ordering parameter in the Ising model is played by middle dipole moment p_c in the lattice node. As the transition point is approached, p_c tends to zero by the law

$$p_c \sim (T - T_C)^{1/8} . \tag{3.83}$$

3.6.4 Sublattice Polarization and Phase Transitions in Antiferroelectrics

On transition to the antiferroelectric phase each paraelectric phase (subcell) in an antiferroelectric crystal becomes polarized, although the crystal does not acquire a microscopic spontaneous polarization because the dipole moments of the subcells of the superstructural unit cell cancel each other. That is why it is customary to speak of the appearance of antipolarization or polarization of antiferroelectric sublattices during phase transition.

From the phenomenological point of view the antiferroelectric phase in a dielectric appears when the state with antiparallel polarization of neighboring unit cells has a lower energy than the state with parallel polarization. Therefore, antiferroelectrics can be considered within the framework of ferroelectricity theory.

A thermodynamic description of phase transitions in antiferroelectrics can be carried out in a manner similar to the description of phase transitions in ferroelectrics. In the vicinity of the phase transition point the free energy of a crystal consisting of two sublattices with polarizations P_a and P_b, respectively, is written

$$G = \alpha_1^* (P_a^2 + P_b^2) + \alpha_2^* P_a P_b + \beta(P_a^4 + P_b^4) + \gamma(P_a^6 + P_b^6) + \dots , \tag{3.84}$$

where α_1^*, α_2^*, β, and γ are functions of the temperature (the tension of the external electric field E is taken to be zero, the anisotropy and strain being neglected). Expansion (3.84) does not take into account some of the terms of the fourth and sixth order, but the results obtained from it reflect qualitatively the basic features of transitions.

In the absence of an external electric field, relation $P_a = -P_b$ holds for an antiferroelectric, and therefore (3.84) can be rewritten

$$G - G_0 = (2\alpha_1^* - \alpha_2^*)P^2 + 2\beta P_a^4 + 2\gamma P_a^6 + \dots . \tag{3.85}$$

An analysis of the last expression suggests some conclusions as to the nature of the transition in relation to the signs of the coefficients. The results of this analysis show that if $2\alpha_1^* - \alpha_2^* > 0$, then G has a minimum at point $P_a = 0$; at $2\alpha_1^* - 2\alpha_2^* < 0$, G has a maximum at the same point, but attains a minimum at some $P_a \neq 0$ (Fig. 3.38b). Thus, if $2\alpha_1^* - \alpha_2^*$ changes from a negative to a positive value with a change in temperature, an antiferroelectric phase transition will take place.

The type of such a transition will be determined by the signs of the other coefficients. Consideration shows that at $\beta > 0$ and $\gamma > 0$ a second-order phase transition will occur. If $\beta < 0$, then at sufficiently small positive $2\alpha_1^* - \alpha_2^*$ the change in free energy with increasing P_a will be as follows: at first an increase from a minimum at zero, and then a decrease to a minimum at some $P_a \neq 0$, followed by an increase (Fig. 3.38c). At some temperature (corresponding to a small, but positive value $2\alpha_1^* - \alpha_2^*$) the minimum at $P_a \neq 0$ becomes less than the minimum at $P_a = 0$, and the system executes a jump from the state with $P_a = 0$ to the one with $P_a \neq 0$. A first-order phase transition takes place.

A further analysis shows that in a second-order antiferroelectric phase transition the dielectric constant is continuous and finite. Recall that in the case of ferroelectric transitions the dielectric constant does not experience a jump under a second-order phase transition either, but goes to infinity at the transition point, which practically means a substantial increase in dielectric constant at the transition point. At the antiferroelectric Curie point, under a second-order phase transition the dielectric constant, on the contrary, need not take large values. Under a first-order transition the dielectric susceptibility experiences a jump, but does not take large values either.

In crystals, a phase transition from the antiferroelectric state will not necessarily occur to the paraelectric phase. Transitions between the antiferroelectric and ferroelectric modifications are possible, provided there is a definite relationship between the absolute value and the signs of the expansion coefficients (3.84). First of all, it follows from the analysis that such a transition can take place only if $\alpha_1^* < 0$. Further, it turns out that with the following relationship between the coefficients,

$$0 \geqslant \alpha_2^* > \alpha_1^* , \tag{3.86}$$

the ferroelectric phase is stable, while the antiferroelectric one is metastable. The situation is reversed when

$$0 < \alpha_2^* < -\alpha_1^* . \tag{3.87}$$

A change in α_1^* and α_2^* with the temperature and pressure may result in replacement of one of the last two relations by the other, which will ensure a phase transition between the antiferroelectric and ferroelectric modification.

It has already been noted that an applied external electric field increases the temperature of the ferroelectric-paraelectric transformation (it is assumed that the paraelectric modification exists at a higher temperature). Such a behavior of a ferroelectric is easy to understand, if only from the fact that the application of a field increases the crystal polarization, which can be destroyed by the thermal motion only at a higher temperature. The application of an electric field to an antiferroelectric below the Curie point promotes transition to the ferroelectric structure (for which P_a has the same direction as P_b). This means that the application of a field reduces the temperature of the antiferroelectric-ferroelectric phase transition when the ferroelectric phase is at a higher temperature. If, however, the high-temperature modification for an antiferroelectric is paraelectric, the arguments about the nature of the Curie point shift are not so obvious. But regarding the antiferroelectric phase as more ordered than the ferroelectric one and proceeding from the above statement that the application of a field brings the antiferroelectric modification closer to the ferroelectric one, it can be inferred that the temperature of the antiferroelectric-paraelectric transition falls off under the effect of the field. This is actually observed in experiment.

By applying an external field it is possible, generally speaking, to transfer an antiferroelectric to the ferroelectric state, i.e., to orient the dipole moments of both sublattices in the same direction (Sect. 3.4). This can be done even when the high-temperature modification is paraelectric instead of ferroelectric. The field at which this transition occurs is called the critical field. (If the high-temperature phase is ferroelectric, the transition from the antiferroelectric to the ferroelectric state can be interpreted as the result of a shift of the Curie point to a lower-temperature range under the effect of the field). Thermodynamic consideration shows that the value of critical field E_{crit} is expressed via the coefficients in the expansion of the function of free energy by

$$E_{\mathrm{crit}} = \frac{4}{3\sqrt{6\beta}} (\alpha_2^* - \alpha_1^*)^{3/2} . \tag{3.88}$$

As indicated in Sect. 3.4, the presence of a critical field in antiferroelectrics leads, in strong alternating fields, to their polarization, which has the form of double hysteresis loops in dependence $P = P(E)$ (Fig. 3.36c).

The value of the critical field for a given antiferroelectric depends on the crystal temperature, more precisely, on how much below the transition point the crystal temperature is. The critical field determines the field-induced transition of the antiferroelectric to the ferroelectric modification. This is a first-order phase transition accompanied by the release of the latent heat of transition.

The effect of the hydrostatic pressure on the phase transition temperature in ferroelectrics can be perceived by considering pyroelectric and electrocaloric phenomena (Sects. 3.2, 5). The relevant qualitative considerations for the antiferroelectrics amount to the following. Suppose hydrostatic compression increases the dipole moment of the antiferroelectric subcells. This increase also means an increase in the stability of the antiferroelectric state. Here the antiferroelectric-ferroelectric and antiferroelectric-paraelectric transition temperature will increase. If the hydrostatic compression reduces the subcell polarization, both temperatures of the respective transitions will decrease.

The above conceptions of spontaneous polarization in various ferroelectrics, based on microscopic models, can, in principle, be extended to the antiferroelectrics. The achievement of the antiferroelectric state requires that a dipole moment appear in the subcells, and the latter must be so arranged as to cancel each other. The first of these conditions is equivalent to a "polarization catastrophe" (in its interpretation via electron or ion polarization) within a single subcell. The second condition holds when the dipole interaction energy of the chains of antiparallel dipoles is lower than the interaction energy of the chains of parallel dipoles (the energy of the charge-dipole electrostatic interaction makes a negative contribution to the crystal energy).

3.6.5 Crystal Lattice Dynamics and Phase Transitions in Ferroelectrics

The termal vibrations of atoms in a crystal can be regarded as a combination of normal vibrations of the crystal lattice. The vibrations representing a displacement of the sublattices of the crystal as a whole are called *limiting*. If the limiting vibrations are associated with a change in macroscopic dipole moment, they are called *dipole* vibrations.

The equation of motion for the normal coordinate in electric field $E\exp(\mathrm{i}\omega t)$ (neglecting the damping) has the form

$$\ddot{u} + \omega_0^2 u = \frac{e^*}{m} E \mathrm{e}^{\mathrm{i}\omega t} \ . \tag{3.89}$$

This is the equation of motion of a harmonic oscillator. Here ω_0 is the frequency of normal vibrations, e^* is the effective dynamic charge of the given mode (the ion charge, in the simplest case), and m is the reduced mass for the given normal vibration.

The polarization associated with normal coordinate u is equal to

$$P = e^* N u \ . \tag{3.90}$$

Considering that the polarization will also be a periodic function of the frequency ($P = P_0 \mathrm{e}^{\mathrm{i}\omega t}$), (3.89) can – with due regard for (3.90) – be reduced to

$$\omega^2 P_0 + \omega_0^2 P_0 = e^* NE/m \ . \tag{3.91}$$

Contribution ε_u to dielectric constant ε due to the normal vibrations, as calculated from the last expression, is

$$\varepsilon_u = 4\pi \frac{\partial P_0}{\partial E} = \frac{f}{\omega_0^2 - \omega^2} \ , \quad \text{where} \tag{3.92}$$

$$f = e^{*2} N/m \ .$$

For several limiting vibrations the expression for ε has the form

$$\varepsilon = \varepsilon_\infty + \sum \frac{f_k}{\omega_k^2 - \omega^2} \ .$$

Here ε_∞ is the high-frequency dielectric constant (for $\omega \gg \omega_k$), ω_k are the limiting vibration frequencies, and f_k is the oscillator strength. If damping constant γ_k is assigned to each mode, the last expression takes the form

$$\varepsilon = \varepsilon_\infty + \sum \frac{f_k}{\omega_k^2 - \omega^2 + 2\mathrm{i}\omega\gamma_k} \ . \tag{3.93}$$

It is seen that a sharp increase in ε at low frequencies in some temperature range (which is typical of the ferroelectric phase transition) may be attributed to an abrupt change in the frequency ω_k of one of the limiting vibrations ($\gamma_k \gg \omega_k$). The vibration whose frequency sharply falls off as $T \to T_\mathrm{C}$ is usually called the *ferroelectric vibration*, or the *soft mode*. The frequency of this vibration has – with due regard for the Curie-Weiss law (3.56) – a temperature dependence of the type

$$\omega_k \sim (T - T_\mathrm{C})^{1/2} \ . \tag{3.94}$$

The approach to phase transitions in ferroelectrics from the standpoint of crystal lattice dynamics was first developed by Ginzburg, Anderson, Cochran, and others. In the past decade many investigations have been carried out which demonstrate the fruitfulness of such an approach to phase transitions in many ferroelectrics.

As is well known, the dispersion ratio (the dependence of frequency ω on wave number k) for a vibrating chain of two atoms with masses m_1 and m_2, spaced at $a/2$, has the form

$$\omega^2 = \frac{M(m_1+m_2)}{m_1 m_2} \pm \frac{M}{m_1 m_2}\,[(m_1+m_2)^2 - 4m_1 m_2 \sin^2(\pi q a)]^{1/2}\,, \qquad (3.95)$$

where M is the quasielastic coupling coefficient. Two solutions obtained from (3.95) correspond to two frequency branches: acoustic $\omega_-(k)$ and optical $\omega_+(k)$. As wave number k decreases (an increase in wavelength), the frequency of vibrations in the acoustic branch reduces to zero and in the optical, increases and reaches a maximum. This is associated with the difference in the nature of vibrations of neighboring particles (atoms). In the acoustic branch neighboring particles vibrate in phase, and to long waves fully correspond the displacements of the unit cells. In the optical branch neighboring particles vibrate in counterphase. To long waves then correspond the vibrations of the particles inside the unit cells. These can be regarded as mutual vibrations of two translation lattices consisting of particles of different sorts. In the acoustic branch the particles vibrate in phase and with equal amplitudes. In the optical branch the particles vibrate in counterphase, and the vibration amplitude is inversely proportional to the particle masses. For the maximum value of wave number k, the heavier particles in the acoustic branch are at rest.

The theory of vibrations of a one-dimensional chain can be generalized for the three-dimensional case. For a diatomic crystal (containing two sorts of atoms) there are six branches of function $\omega(k)$. Three of them correspond to acoustic vibrations, and the other three to optical.

Investigations into the possibility that some crystals can, under certain conditions, be unstable to at least one mode of normal vibrations were carried out at first for diatomic cubic crystals (NaCl, CsCl). Such instability may lead to a rearrangement of the crystal with the resulting spontaneous polarization, after which the crystal regains its stability. Then (3.95) takes the form

$$\mu\omega_{\mathrm{T}}^2 = R_0' - 4\pi(\varepsilon_e+2)(Z'e)^2/9V\,, \qquad (3.96)$$

where $\mu = m_1 m_2/(m_1+m_2)$, R_0' is the short-range interaction constant, ε_e is the high-frequency dielectric constant, V is the volume of the unit cell, and $Z'e$ is its effective charge.

From (3.96) it follows that the frequency ω_{T} of the transverse optical vibration mode in the crystals under consideration can actually vanish provided that

$$R_0' = 4\pi(\varepsilon_e+2)(Z'e)^2/9V\,. \qquad (3.97)$$

It can be shown that this condition is equivalent to that of "polarization catastrophe". From the Lidden-Sax-Teller relation

$$\omega_L^2/\omega_T^2 = \varepsilon_s/\varepsilon_e \ , \tag{3.98}$$

where ω_L is the longitudinal mode frequency, and ε_s is the static dielectric constant, it follows that to the vanishing of frequency ω_T corresponds an increase of static dielectric constant ($\varepsilon_s = 0$) to infinity. This stems from the fact that the crystal experiences a ferroelectric phase transition as a result of its instability to the soft vibration mode.

In real diatomic alkali-halide crystals the left- and right-hand sides of (3.96) are of the same order of magnitude, but R_0' is approximately twice as large as the right-hand side. Investigation of barium titanate shows that in this crystal transition due to lattice instability is possible. Account is taken of the anharmonicity of the lattice vibrations leading to the temperature dependence of R_0'. The transition is determined by the relationship between the value of elastic energy F_{el} (as a function of shift u)

$$F_{el} = R'u + Bu^3 + B'u^5 \tag{3.99}$$

and the Coulomb force (3.97)

$$F_{Coul} = 4\pi(\varepsilon_e + 2)Z'eu/9V \ . \tag{3.100}$$

Both these forces are represented graphically in Fig. 3.40. (The curvature the elastic force plot is exaggerated, B being assumed negative.) In accordance with (3.96) the value of ω_T^2 is proportional to the difference of the slopes of the two lines at the origin, while ε_s is inversely proportional to the slope difference. With decreasing temperature the slope of the line determining the displacing force increases until the curves intersect at a definite value of u_{int}. In the vicinity of u_{int}, stable but anharmonic vibrations of the two lattices are

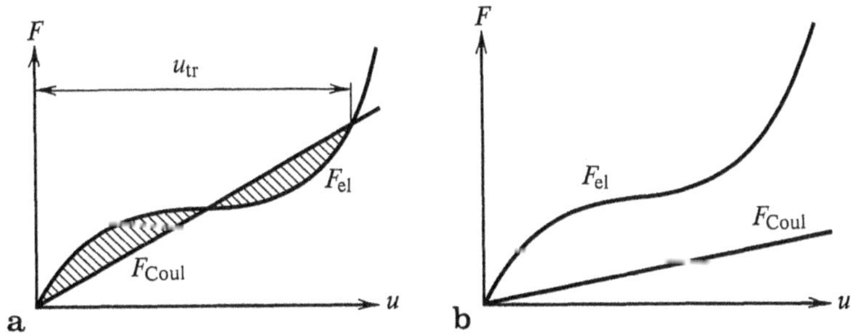

Fig. 3.40. Dependence of the restoring force F_{el} and the displacing coulomb force F_{Coul} on the value of displacement at (**a**) and above (**b**) the transition temperature [3.17]

possible. The indicated case corresponds to the phase transition of $BaTiO_3$ to the tetragonal modification with spontaneous polarization. The transition, however, will not occur until the free energies of the two phases become equal. Graphically, such an equality corresponds to that of the hatched areas in Fig. 3.40a. The crystal is stable above T_C, metastable within a certain range (between T'_C and T_{CW}), and unstable below T_{CW}. It is because of this instability that the crystal transfers to the tetragonal modification.

It seems possible to relate the macroscopic parameters α^*, β, and γ used in expansions of the type (3.60) to the microscopic parameters. By appropriate choice of the macroscopic parameters, one can obtain in $BaTiO_3$ three consecutive phase transitions with spontaneous polarization directed along [100], [110], and [111], respectively, and describe its properties in good agreement with experiment.

In the foregoing, only mode ω_T was taken into consideration (which is unjustified in the case of anharmonicity, when the vibration modes cease to be normal). Owing to the allowance for the delay effects, limiting frequency ω_T corresponds not to an infinite wavelength ($k = 0$), but to a wavelength smaller than the crystal, but larger than the unit cell.

As indicated by experiment, in barium titanate the soft (transverse optical) mode lies below the acoustic brances with the exception of very small wave vector values. Quantitatively, this means that interaction between the acoustic and the transverse optical branch can take place only in the immediate vicinity of the phase transition temperature. Such an interaction must occur with a higher confidence below the Curie point, because the ferroelectric modification is piezoelectric.

We have already given some attention to the concepts which place proton ordering at the basis of phase transitions in KDP-type ferroelectrics. Recently, however, difficulties were revealed in correlating theory with experiment without taking into account the interaction of ordering processes with lattice vibrations. It may be assumed as established by now, both theoretically and experimentally, that at least in some ferroelectrics experiencing transition with ordering of the structural elements, the lattice dynamics plays an important role: the long-range coulomb forces resulting from the dynamic motion of charges in crystals (as has been described above with reference to ionic crystals) here, too, may lead to the weakening of the short-range forces and to lattice instability. This behavior of ordering ferroelectrics is caused by the fact that, at least in some of them, ordering is accompanied by displacement of sublattices as a whole (for instance, the displacement of K and P ions along the Z axis in KDP). It is these displacements that are responsible for spontaneous polarization. As indicated above, while for the Curie point one observes a considerable isotopic effect ($T_C = 122$ K for KH_2PO_4, and $T_C = 213$ K for KD_2PO_4), for spontaneous polarization and the Curie constant the isotopic effect is almost nil. A considerable change in Curie point on deuteration indicates the primary role of the proton subsystem in the phase transition. In contrast, the absence of substantial changes in spontaneous polarization and

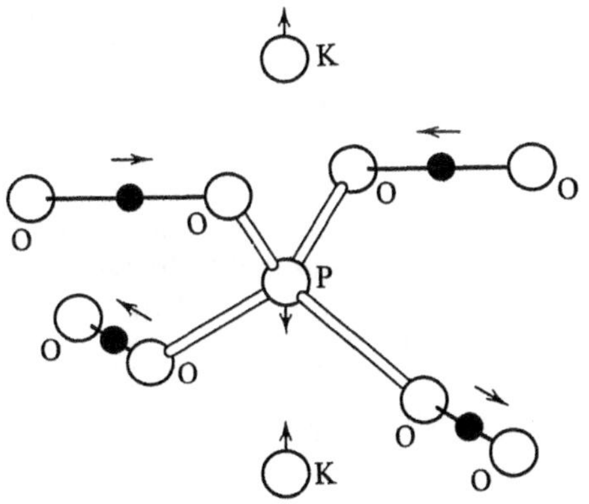

Fig. 3.41. Vibration corresponding to the ferroelectric mode in KDP [3.17]

Curie constant, as well as the fact that spontaneous polarization is almost completely due to the displacement of ions, requires taking into account, in theory, the degrees of freedom corresponding to the ion displacements.

In investigations on the lattice dynamics in such crystals as KDP, one proceeds from the premise that in these crystals the ferroelectric mode corresponds to proton ordering in which two "upper" proton move in the direction of the corners of the tetrahedron concerned, while the two "lower" ones move away from them (Fig. 3.41). It is easy to see that this model is based on the conceptions used by Slater for describing the proton ordering in KDP (Fig. 3.18). It should, however, be borne in mind that these models resemble one another only outwardly, since, according to Slater, in the paraelectric phase the hydrogen atoms are statistically disordered with respect to the bonds, and in this model they execute continuous vibrations.

In other models it is assumed, in describing the phase transition in KH_2PO_4, that the excitation associated with proton tunneling strongly interacts with the optical vibration of the $K-PO_4$ along the Z axis, and the frequency of the bound mode tends to zero at the transition point. The form of the optical vibration involved in the phase transition is similar to the one presented in Fig. 3.41. Such a model explains the isotopic effects for the Curie point and its almost complete absence for the Curie constant and spontaneous polarization.

The last model suggests that KH_2PO_4 belongs to the "mixed" type of ferroelectrics, since the phase transition in it is associated both with the ordering of the proton system and with the instability of the limiting optical vibration.

Consideration of the model of phase transition in KDP shwos that the mixed mode, corresponding to interaction between the proton ordering and the optical vibrations of the lattice, is "acoustic" rather than "optical." In KDP, the depolarizing field evidently does not effect the crystal spectrum.

Owing to the strong bond between the optical and acoustic branches, with a change in temperature the crystal may become unstable relative to the transverse acoustic mode even before it becomes unstable with respect to the transverse optical mode. In thermodynamic theory such a situation is reflected by the fact that the condition of the phase transition depends not only on the behavior of the dielectric constants, but also on the elastic and piezoelectric constants.

Thus, some ferroelectric transitions which were formerly regarded as "order-disorder" transitions must actually be considered more complicated. In these crystals the phase transitions are closely associated with the behavior of the optical and acoustic vibration modes of lattice. Examples of such ferroelectric transitions are offered by crystals of KDP, $K_4Fe(CN)_6 \cdot 3\,H_2O$, and Rochelle salt.

Phase transitions in antiferroelectrics can also be attributed to the instability of the crystal to some normal vibration modes. In particular, theoretical consideration of transitions in some perovskites ($PbZrO_3$, $NaNbO_3$ etc.) shows that the average displacement of atoms from their equilibrium positions in a cubic lattice can be expressed in terms of static displacements corresponding to vibrations with some wave number $k = k_{oi}$.

The numerous attempts, on various ferroelectrics, to discover the soft mode – the limiting dipole transverse optical vibration with a frequency decreasing anomalously as T_C is approached – include the study of infrared spectra and Raman light scattering spectra, thermal and diffusion x-ray and electron scattering, and inelastic neutron scattering. Some data on the lattice dynamics can also be obtained from investigations of the Mössbauer effect, and the NMR and EPR spectra.

The primary objects of investigation were displacement-type ferroelectrics ($BaTiO_3$, $SrTiO_3$, $KTaO_3$, $LiNbO_2$, etc.). Investigations into Raman spectra of scattering on $BaTiO_3$ (in the tetragonal phase) revealed a soft ferroelectric mode, which is damped out. The frequency of this mode in the undamped state is 36 cm^{-1}, which is also in good agreement with calculations by (3.98).

Investigations on inelastic neutron scattering on $BaTiO_3$ (in the cubic phase) revealed a strong quadrupole neutron scattering due to the polarization fluctuations of a relaxational nature. The scattering has a large anisotropy, and the strongest correlation of fluctuations takes place along the [100] axis perpendicular to the direction of $\boldsymbol{P}_s$.

Investigation into inelastic neutron scattering on a single-domain crystal of $BaTiO_3$ in the tetragonal phase led to the discovery of a very strong energy anisotropy in the damping of the soft optical mode in relation to the direction of the unit polarization vector $\boldsymbol{e}$. The lowest-frequency transverse optical branch with a wave vector $\boldsymbol{k} \parallel [110]$ are $\boldsymbol{e} \parallel [1\bar{1}0]$ damps out only slightly, with the exception of small k values (< 0.05 Å^{-1}), below which it becomes overdamped. The energy of this soft mode grows rapidly with k, and its extrapolation to zero values of k yields 4.5 MeV, which agrees well with the Raman scattering data. The other soft modes with $\boldsymbol{e} \parallel [100]$ and $k \parallel [010]$ have an unusually low energy and high damping constants for the entire zone. They lie below some acoustic branches and cause certain anomalies in their dependence on the temperature and on k when intersecting these branches.

According to the results of investigation into the diffusion x-ray scattering in $BaTiO_3$ crystals, cubic $BaTiO_3$ is disordered, and separate ordered chains of Ti ions are uniformly oriented along the fourfold axes of symmetry. It can, however, be shown that the anisotropic distribution of the optical phonon frequencies also leads to diffusion scattering layers and to atomic displacements, strongly correlated along the [100] directions. The indicated correlation will be instantaneous rather than static (as is assumed in diffusion scattering investigations). Thus, these experiments do not allow one to draw an unambiguous conclusion and regard cubic $BaTiO_3$ as a disordered phase. At the same time, ordering and displacement-type ferroelectrics appear to represent two limiting states, between which all the ferroelectric compounds are distributed.

At a very low temperature ($\sim$4 K), $KTaO_3$ experiences a ferroelectric phase transition. This crystal was one of the first on which the soft ferroelectric mode was discovered by different methods and the principal premises of the dynamic theory of ferroelectricity were verified. According to investigations into inelastic neutron scattering, in this crystal the ferroelectric transverse optical mode has a phonon energy of 10.7 ± 0.3 MeV (86.3 ± 2.5 cm^{-1}) at $k = 0$. The temperature dependence of the square of its energy (Fig. 3.42) corresponds to the temperature dependence of $1/\varepsilon$ with a transition temperature of 15 K (the extrapolated transition temperature $T_C' \approx 2-4$ K). The deviations of the temperature dependence of ε from the Curie-Weiss law (at a very low and a very high temperature) exactly correspond to the temperature dependence of the soft-mode energy. The last dependence showed anomalies at 4.3 K, whose analysis suggests that $KTaO_3$ experiences a structural phase transition at 10 K due to the interaction of the longitudinal acoustic mode with

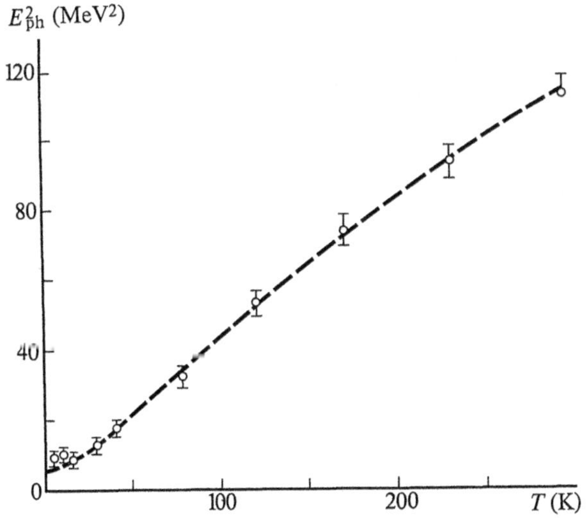

Fig. 3.42. Temperature dependence of the energy of the transverse optical mode at $k = 0$ for $KTaO_3$ [3.18]

the transverse optical (ferroelectric) one. In the temperature range of 4 – 30 K, a strong dependence of the damping of ultrasonic (longitudinal) waves has been revealed. Investigations into the damping of ultrasonic waves supply data on the soft mode in $KTaO_3$ which completely agree with those obtained by other methods.

Investigations into inelastic neutron scattering on $KTaO_3$ also revealed a strong relationship between the transverse acoustic and the soft transverse optical mode. In additional to the expected "softening," the soft transverse optical mode shows a strong temperature dependence of the energy. The cross section of neutron scattering on the phonons of the soft transverse acoustic mode also exhibits an unexpectedly strong temperature dependence. The relationship between the modes discussed here strongly depends on the direction of vector $\boldsymbol{k}$.

The anomalous behavior of the optical mode in ferroelectrics must reduce the intensity of the Mössbauer line in the paraelectric range as the Curie temperature is approached. This effect was actually observed in investigations on the temperature dependence of the Mössbauer effect in the solid solutions of $Ba(Ti_{1-x}Sn_x)O_3$ enriched in ^{119}Sn. It was shwon that the probability of the Mössbauer effect indeed passes a minimum at the Curie point.

Investigations into the Raman light spectra in KDP crystals resulted in the discovery of a broad low-frequency ferroelectric mode, which can be approximated by a harmonically damped oscillation function. The characteristic frequency of mode ω_0 tends to zero as T_C is approached from the side of the paraelectric phase. At a high temperature this frequency is $\sim 99\ cm^{-1}$, and the band is $\sim 170\ cm^{-1}$ wide. The approach of ω_0 to zero near T_C corresponds to the conceptions based on the soft mode and on cooperative tunneling of protons (associated with frequencies of 85 and 99 cm^{-1}, respectively).

Study of the Brillouin scattering spectrum of KDP, found that the frequency of the acoustic mode in KDP reduces to zero at 122 K at a finite wavelength. As $T \to T_C$, from the side of the paraelectric phase the soft mode must intersect the acoustic one, and since the crystal is a piezoelectric, the soft mode is linearly related to the transverse acoustic mode. On extrapolation, the optical mode frequency vanishes at 117 K. Thus, from a comparison of the above data, it follows that in KDP the acoustic mode becomes unstable before the optical. The difference in T_C values for the two modes corresponds to the difference in T_C values for clamped and free crystals.

3.6.6 Incommensurate Phase Transitions in Ferroelectrics

Structural phase transitions, which are vividly exemplified by the best-studied transitions in ferroelectrics, can be divided into two broad classes: commensurate and incommensurate phase transitions. For definition we will speak of displacement-type transitions. The change in symmetry on such transitions is due to atomic displacements corresponding to a normal coordinate, i.e., to a static "displacement wave" of the atoms. In most of the known cases the period of this displacement wave is 1, 2, 3, or 4 times that of the crystal lattice of the initial (more symmetric) phase. Such transitions are called commensurate. In recent years incommensurate phase transitions have been commonly observed in various compounds, including ferroelectrics; at these transitions, the wave period of the arising atomic displacements is not a multiple of, or commensurate with, the crystal lattice period of the more symmetric phase. As a result, the crystal loses its periodicity along a definite direction in the incommensurate phase, and this precisely points to a very large size of the unit cell.

The difference between commensurate and incommensurate phase transitions manifests itself in the different nature of disperison of the soft branch of the crystal phonon spectrum. At a commensurate phase transition, the soft mode is localized in the k space; that is, the soft-branch minimum is at a definite point of the Brillouin zone, which does not change its position with temperature. Such points are determined by symmetry and correspond to values $ka/2\pi = 1, 2, 3, 4$, where a is the lattice parameter [3.19].

At an incommensurate phase transition, the position of the soft-branch minimum is not determined by crystal symmetry and therefore changes with temperature. The value $ka/2\pi$, corresponding to the soft mode, is an arbitrary (generally speaking, irrational) number, i.e., the values k and $2\pi/a$ are incommensurate.

An example of the case in Fig. 3.43a is the phase transition in barium titanate (the soft mode is triple degenerate at $k = 0$). The thermodynamical theory of such transitions, which is based on Landau's general theory of phase transitions, was first developed by *Ginzburg* [3.20]. In this theory, the order parameter is polarization, which is proportional to the soft-mode coordinate. The Ginzburg theory explains the dielectric anomalies and other properties of barium titanate and many other ferroelectrics.

To the case in Fig. 3.43b corresponds the ferroelectric phase transition in gadolinium molybdate. Here polarization is not an order parameter and arises on phase transition exclusively because it is proportional to the quadratic combination of the coordinates of the doubly degenerate soft mode at $k = a/\pi$. The dielectric anomalies in such ferroelectrics, called improper, possess a number of specific features. The Curie-Weiss law is not valid here, and a second-order phase transition is not eliminated by the electric field as it is in proper ferroelectrics, which are described by the Ginzburg theory.

The case in Fig. 3.43c is associated with an incommensurate phase transition, for instance, in ferroelectric sodium nitrate. This crystal exhibits two successive transitions on temperature decrease: first, to an incommensurate phase, and then to an ordinary (commensurate) polar phase. The structure of the incommensurate phase can be characterized by a polarization wave with a period exceeding the crystal lattice spacing by one or more orders. The position of this polarization wave is not fixed in an infinite crystal, i.e., the wave can shift without any resistance in a perfect, defect-free crystal. This explains

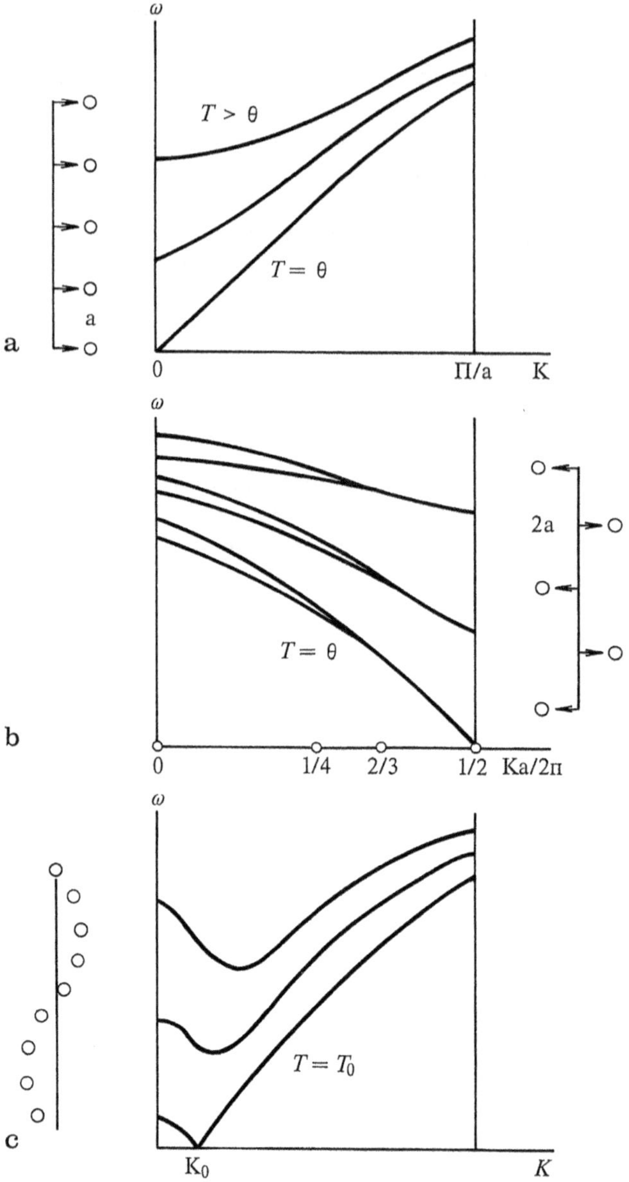

Fig. 3.43. Soft optical branches of the phonon spectrum of crystal vibrations for the following cases: commensurate phase transition **(a)** without a change in translational symmetry of the crystal, **(b)** with doubling of the unit cell period; **(c)** incommensurate phase transition

the existence of a new branch of acoustic-type vibrations, the so-called phason branch, which was predicted by theory. In distinction to acoustic phonons, long-wave phasons are overdamped and should show neither in infrared-absorption nor in light scattering spectra. Nevertheless, experiments on vibrations corresponding to the phason branch and possible at sufficiently large $k \sim k_0$ are of definite interest. Also of interest are experiments on observing, in the incommensurate phase of ferroelectrics, the macroscopic quadrupolar moment (which is anomalously large because of the polarization wave) by generation of the second light harmonic (also anomalously large); by studying NMR spectra (which have peculiarities of their own); by measuring the anomalous properties of dielectric susceptibility, heat capacity; etc. An incommensurate phase is characterized by a strong influence of defects on its properties. This influence is associated with the possibility of fixing the displacement wave on defects.

The thermodynamic theory of incommensurate superlattice for magnetics was first constructed by *Dzyaloshinskii* [3.21]. A similar approach was applied to incommensurate phase transitions in ferroelectrics [3.22]. One of the possible macroscopic (i.e., symmetry-generated) causes for incommensurate transitions was actually indicated in *Lifshitz*'s paper as early as 1941 [3.19]. The presence of a gradient invariant of the type $\eta \partial \xi / \partial \chi - \xi \partial \eta / \partial \chi$, which is called Lifshitz's gradient invariant, for the components of the order parameter η, ξ forbids the location of the minimum of the branch of normal crystal vibrations at the k-space point associated with the η, ξ mode. The branch minimum shifts to an arbitrary, not symmetry-determined, point of the k space, and a second-order phase transition can proceed only to a phase with an incommensurate structure. *Indenbom* [3.23] was the first to point out the necessity of taking into account Lifshitz's gradient invariant when considering phase transitions in ferroelectrics. Inclusion of Lifshitz's gradient invariant in the thermodynamic potential enables one to describe the sequence of transitions – initial to incommensurate to commensurate (polar) phase – and explains the nature of the anomalies of physical quantities which are observed at these transitions in ammonium fluoroberyllate, rubidium trihydroselenite, potassium selenate and related crystals, and also in a number of other ferroelectrics.

In sodium nitrate and in thiourea, the polar phase is equitranslational with the initial (high-temperature) one and hence the order parameter describing transition between these phases transforms according to the representation of the point symmetry group of the initial phase of the crystal D_{2h}, while all the representations of this group are unidimensional. Therefore, no Lifshitz's gradient invariant exists. Nevertheless, the symmetry of the initial phase admits specific interaction of two distinct modes of the crystal vibration spectrum, expressed by a gradient invariant which outwardly resembles Lifshitz's invariant. Now, however, η and ξ are not the coordinates of the double-degenerate soft mode, but those of two different modes, of which only one is soft. Inclusion of such a gradient invariant in the thermodynamic potential enables us to describe a sequence of phase transitions: initial to incommensurate to commensurate (polar) phase. Consequently, the symmetry-generated relationships of the order parameter η with the other variable ξ via the gradient invariant can also be regarded as a macroscopic cause for the appearance of the incommensurate phase.

4. Magnetic Properties of Crystals

As noted above, crystallography widely uses the ideas of symmetry as a basis for crystal investigations. In this chapter we shall consider the magnetic phenomena precisely from this standpoint. Our attention will be focused on magnetic symmetry and the application of the symmetry principles in explaining the anisotropy of the physical properties and describing the structures of magnetic crystals. In discussing the anisotropy of the physical properties of ferromagnetics, we use the tensor method, which is traditional for crystallography. The previous courses of crystallography practically ignored the anisotropy of ferromagnetic crystals.

Restricting ourselves to the specific "crystallographic" aspects and to the minimum necessary information from the theory of magnetism, we naturally cannot dwell on some other important problems relating to the magnetic properties of crystals. This chapter completely omits the resonance phenomena, problems associated with the behavior of magnetically ordered crystals in alternating fields, and optical and other "nonmagnetic" properties. Nor does it fully cover questions concerning the wide use of magnetic materials in modern technology, although they are partially discussed in passing. All this information can be found in the specialized textbooks and monographs on magnetism which are listed in the Bibliography.

4.1 Disordered Magnetics

4.1.1 Basic Relations Characterizing the Behavior of a Substance in a Magnetic Field

By the magnetic properties of a substance is meant the ability of bodies to interact with a magnetic field. It is well known that any moving electrically charged body or particle produces a magnetic field of its own and thus interacts with the external magnetic field. It follows that magnetism is a universal property of any substance, because matter is inconceivable without moving charged particles.

A substance placed in a magnetic field $\boldsymbol{H}$ is magnetized. It acquires a resultant magnetic moment consisting of the elementary magnetic moments of the separate particles. A characteristic, or a measure, of the magnetized state of a substance is *magnetization* $\boldsymbol{I}$, or the magnetic moment of a unit volume. In moderate fields, there exists the following simple relationship between $\boldsymbol{I}$ and field $\boldsymbol{H}$:

$$\boldsymbol{I} = \varkappa \boldsymbol{H} , \tag{4.1}$$

where $\varkappa$ is the *volume magnetic susceptibility* of the substance. In addition to volume susceptibility $\varkappa$, the *molar susceptibility* ($\varkappa_{\mathrm{mol}}$) and the *specific susceptibility* ($\varkappa_{\mathrm{spec}}$) are sometimes used. Their relationship is given by expressions $\varkappa_{\mathrm{spec}} = \varkappa/\varrho$, $\varkappa_{\mathrm{mol}} = \varkappa M/\varrho$, where ϱ and M are the density and the molecular weight. Accordingly, *molar* and *specific magnetization* are distinguished. The latter is usually denoted by σ.

Inside the magnetized substance an intrinsic field $4\pi \boldsymbol{I}$ is induced; therefore, another vector, the *magnetic induction*, is introduced in addition to vectors $\boldsymbol{I}$ and $\boldsymbol{H}$:

$$\boldsymbol{B} = \boldsymbol{H} + 4\pi \boldsymbol{I} . \tag{4.2}$$

Substituting (4.1) into (4.2), we obtain

$$\boldsymbol{B} = \boldsymbol{H}(1 + 4\pi\varkappa) = \mu \boldsymbol{H} . \tag{4.3}$$

The quantity $\mu = 1 + 4\pi\varkappa$ is called the magnetic permeability of the substance.

If a body of an open shape is subjected to magnetization, "magnetic charges" arise on the surface of the body in the direction of the external field and induce an additional magnetic field $\boldsymbol{H}_{\mathrm{d}}$ directed inside the body oppositely to the external field. Field $\boldsymbol{H}_{\mathrm{d}}$, which is called the *demagnetizing field*, is, to a first approximation, proportional to the magnetization:

$$\boldsymbol{H}_{\mathrm{d}} = -N\boldsymbol{I} , \tag{4.4}$$

where N is the so-called *demagnetizing factor*, which depends on the shape of the body. Taking into account the demagnetizing field, the total true field inside the body, which actually causes magnetization $\boldsymbol{I}$, will be equal to

$$\boldsymbol{H} = \boldsymbol{H}_{\mathrm{e}} + \boldsymbol{H}_{\mathrm{d}} = \boldsymbol{H}_{\mathrm{e}} - N\boldsymbol{I} , \tag{4.5}$$

where $\boldsymbol{H}_{\mathrm{e}}$ stands for the external magnetic field. Allowance for the demagnetizing field is essential only for strong magnetic substances (for instance, ferromagnetics), for which the value of $\boldsymbol{I}$ is comparable with the external field. Note that field $\boldsymbol{H}_{\mathrm{d}}$ is homogeneous only for elliptical bodies.

Substituting (4.5) into (4.1), we get

$$\boldsymbol{I} = \varkappa \boldsymbol{H} = \frac{\varkappa}{1 + \varkappa N} \boldsymbol{H}_{\mathrm{e}} = \varkappa_0 \boldsymbol{H}_{\mathrm{e}} , \tag{4.6}$$

where $\varkappa_0$ can be called the susceptibility of the body, as distinct from the susceptibility of the substance $\varkappa$, since the former depends not only on the nature of the material, but also on the shape of the body. The magnetic permeability of the body μ_0 can be defined similarly.

One feature associated with the symmetry of vectors $\boldsymbol{H}$, $\boldsymbol{B}$, and $\boldsymbol{I}$ should be emphasized. From the electromagnetic field equation curl $\boldsymbol{H} = 4\pi \boldsymbol{j}/c$, where $\boldsymbol{j}$ is the polar vector of the electric current density, and c is the light velocity, it follows that $\boldsymbol{H}$ is an axial vector with symmetry ∞/m (Chap. 1). The above equation shows that closed lines of force are induced around each line of current. Relations (4.1, 2) indicate that $\boldsymbol{B}$ and $\boldsymbol{I}$ are also axial vectors. This means that magnetic field $\boldsymbol{H}$, the magnetic moments, and vectors $\boldsymbol{B}$ and $\boldsymbol{I}$ should, strictly speaking (Chap. 1), be depicted as rotating line segments of rotating cylinders (Fig. 4.2b, d, e). But this is not always convenient, and therefore magnetic moments and fields will hereafter be denoted by traditional arrows (Figs. 4.1, 4.2c, and further), bearing in mind, however, that this does not reflect the true symmetry of the axial vector.

4.1.2 Tensors of Diamagnetic and Paramagnetic Susceptibility

If we confine ourselves to an isotropic magnetic, then, according to (4.1), magnetization vector $\boldsymbol{I}$ is collinear with vector $\boldsymbol{H}$, and magnetic susceptibility $\varkappa$ is a scalar quantity. If $\varkappa < 0$, magnetization is antiparallel to the external field, and if $\varkappa > 0$, $\boldsymbol{I}$ is parallel to $\boldsymbol{H}$. In the former case the substance is said to be *diamagnetic*, and in the latter, *paramagnetic*. The absolute values of $\varkappa$ for diamagnetic and paramagnetic substances lie between 10^{-4} and 10^{-6}. A diamagnetic substance can be distinguished from a paramagnetic one by placing it in an inhomogeneous magnetic field: a diamagnetic will be pushed out of the field, whereas a paramagnetic will be pulled into the region of highest intensity.

For crystals, i.e., for an anisotropic medium, $\varkappa$ and μ are second-rank polar tensors. Therefore, (4.1, 3) must be written in tensor form:

$$I_i = \varkappa_{ij} H_j \ , \qquad B_i = \mu_{ij} H_j \ . \tag{4.7}$$

Since $\boldsymbol{I}$, $\boldsymbol{B}$, and $\boldsymbol{H}$ in (4.7) transform as components of axial vectors, $\varkappa_{ij}$ and μ_{ij} transform as components of a second-rank polar tensor. The features and form of the second-rank tensor for crystals of different symmetry have been considered in Chap. 1.

The magnetic susceptibility of a crystal can generally be characterized by three values of the principal susceptibilities $\varkappa_1$, $\varkappa_2$, and $\varkappa_3$, when tensor $\varkappa_{ij}$ is transformed to the principal axes (Chap. 1).

4.1.3 Classification of Magnetic Substances

Because of the diversity of manifestations of magnetism in nature and, often, a fantastic relationship between various types of magnetism, it is difficult to find a unified principle for classifying the magnetics. The drawback of the traditional classification of magnetics based on the susceptibility sign ($\varkappa < 0$ for diamagnetics, $\varkappa > 0$ for all the others) and the value of susceptibility ($|\varkappa| \approx 10^{-4}$ to 10^{-6} for weakly magnetic substances: diamagnetics and paramagnetics; $\varkappa \gg 1$ for ferromagnetics) is that an ever-increasing number of substances are being discovered which can hardly be definitely assigned to any one of the above-listed types of magnetics. Suffice it to say that some crystals are known to be paramagnetic along one axis and diamagnetic along the other. Some crystals are magnetized as ferromagnetics along one axis and as weakly magnetic substances along the other. As regards their behavior in a magnetic field, the antiferromagnetic crystals could be classified as weakly magnetic substances, although antiferromagnetism is the closest to ferromagnetism in its nature. From the standpoint of crystallography, which uses symmetry as a method, it is best to apply a classification based on the absence or presence of ordering of the magnetic moments of the atoms in crystals. All the crystals can be divided into two groups according to this feature: ordered and disordered magnetics.

The following sections consider magnetics without magnetic ordering. They include diamagnetics and paramagnetics.

4.1.4 Diamagnetism

The diamagnetics are substances whose atoms and molecules have no magnetic moments. In such substances the electron shells are closed. It is known from the theory of electromagnetism that any change in an external magnetic field intersecting an electric circuit induces in it a current so directed that its magnetic field opposes the external change (Lenz's law). Therefore, when a diamagnetic is placed in a magnetic field the external magnetic field is "screened" by the arising internal field directed oppositely to the external. The opposition to the external field is manifested as some slowing down of the angular velocity of the orbital motion of the electrons. Magnetic field $\boldsymbol{H}$ makes the orbit precess around the field direction, which induces an additional moment directed opposite to the field:

$$\Delta I = -\frac{e^2 Q}{4\pi m c^2} H \; . \tag{4.8}$$

Here Q is the area of the projection of the orbit onto a plane perpendicular to the magnetic field, and m and e are the mass and charge of the electron, respectively.

For closed shells with a spherical symmetry, Q can be replaced by $\pi\overline{\varrho^2}$, where $\overline{\varrho^2}$ is the average value of the square of the projection of the distance from the atomic nucleus to the electron onto a plane perpendicular to the magnetic field. For a spherically symmetric shell (all the orbit orientations are equiprobable), $\overline{\varrho^2} = 2\overline{r^2}/3$, where $\overline{r^2}$ is the mean square radius of the orbit. Then the total diamagnetic moment of the set of N isolated atoms containing Z electrons each will be defined by the relation

$$\Delta I = -\frac{Ne^2H}{6mc^2}\sum_{i=1}^{i=Z}\overline{r_i^2}\,, \tag{4.9}$$

whence the diamagnetic susceptibility is expressed as

$$\varkappa = \frac{\Delta I}{H} = -\frac{Ne^2}{6mc^2}\sum_{i=1}^{i=Z}\overline{r_i^2}\,. \tag{4.10}$$

Equation (4.10) is called the Langevin-Pauli formula. From (4.10) it is seen that diamagnetism depends on the radius of the electron orbits and is temperature independent. The Langevin-Pauli formula holds good for the description of diamagnetism of inert gases and ions in solutions, because it was derived for a set of isolated atoms. Putting $\overline{r_i^2} \sim 10^{-16}$ cm^2, we find that $\varkappa \sim 10^{-6}$, which agrees, as regards the order of magnitude, with the experimental results.

The diamagnetic susceptibility of a crystal body consists, to a first approximation, of the atomic susceptibilities of its constituent atoms, but the nature of the chemical bond must be taken into account. In many compounds, for instance, in ionic crystals of the type KCl, NaCl, and KBr, diamagnetism coexists with so-called *polarization paramagnetism*. This paramagnetism (which is often called *Van Vleck* paramagnetism) is due to the mutual deformation of the ions, which disturbs the spherical symmetry of the electron shells. The susceptibility of polarization paramagnetism is also practically temperature independent.

The diamagnetic elements include antimony, carbon, tellurium, arsenic, bismuth, mercury, zinc, gold, silver, and copper. The strongest diamagnetic is bismuth. Plant and animal tissues, many minerals, water, and some glasses are also diamagnetics. The diamagnetic bodies include a large number of chemical compounds, in particular, almost all the organic compounds and petroleum.

Metals and semiconductors are a special case of diamagnetism; along with the diamagnetism of the atomic skeletons, they show so-called *diamagnetism of conduction electrons*. This kind of diamagnetism is associated with a change in the orbital free electron motion arising in the external magnetic field. The diamagnetism of the electron gas is exceeded severalfold by the paramagnetism caused by the conduction electrons having an intrinsic spin magnetic moment. The diamagnetism and paramagnetism of the conduction electrons are pure quantum effects. From quantum theory it follows that both diamagnetism and paramagnetism of the electron gas are temperature independent.

In conclusion it should be added that diamagnetism is inherent in all the substances without exception. The universality of magnetism is due precisely to diamagnetism. But it may be obscured by stronger magnetic effects, compared to which it can be neglected.

4.1.5 Paramagnetism

As indicated above, the specific feature of a paramagnetic is the positive value of its magnetic susceptibility. A necessary condition for the paramagnetic state of a substance, as distinct from the diamagnetic, is that the atoms have constant magnetic moments irrespective of the external magnetic field.

Let us consider the nature of the atomic magnetic moments. The magnetic moment of an atom or an ion is determined by the total angular momentum of the electrons. Each electron in an atom possesses an angular momentum related to the orbital motion, and an intrinsic angular momentum (spin). According to the Russel-Saunders model, in order to obtain the *total angular momentum* $\boldsymbol{J}$ of a multielectron atom, one should first add vectorially the orbital angular momenta of the individual electrons and obtain *total orbital angular momentum* $\boldsymbol{L}$, then add up all the spin angular momenta of the electrons and obtain the total spin angular momentum $\boldsymbol{S}$ of the atom. Total angular momentum $\boldsymbol{J}$ will be equal to the vector sum of momenta $\boldsymbol{S}$ and $\boldsymbol{L}$; $\boldsymbol{J}=\boldsymbol{L}+\boldsymbol{S}$. Momenta $\boldsymbol{L}$, $\boldsymbol{S}$, and $\boldsymbol{J}$ obey definite quantization rules and are characterized by the appropriate *orbital* L, *spin* S, and *total angular* J quantum numbers [4.1]. The following magnetic moments are related to the orbital and spin angular momenta

$$\boldsymbol{\mu}_L = -\mu_{\mathrm{B}}\boldsymbol{L}\ , \qquad \boldsymbol{\mu}_S = -2\mu_{\mathrm{B}}\boldsymbol{S}\ , \tag{4.11}$$

where value $\mu_{\mathrm{B}}=\hbar|e|/2mc=0.927\times 10^{-20}$ erg/Oe is called the Bohr magneton and is a kind of magnetic moment "quantum" of the system of electrons. It is worth noting here that the term "spin" means the intrinsic angular (mechanical) momentum, rather than the magnetic moment, although these notions are often erroneously identified which each other.

From (4.11) it is seen that ratio μ_S/S is twice as large as the corresponding magnetomechanical ratio for the orbital motion of electrons. Because of this gyromagnetic "anomaly" the resultant magnetic moment $\boldsymbol{\mu}_J$ of the electron shell will not lie on the same axis with the angular momentum of the atom $\boldsymbol{J}$: $\boldsymbol{\mu}_J = -\mu_{\mathrm{B}}(\boldsymbol{L}+2\boldsymbol{S})$. In the presence of an external magnetic field one is usually interested only in the time-average projection $\boldsymbol{\mu}_J$ onto axis $\boldsymbol{J}$. Such an effective magnetic moment of the atom is calculated by the equation

$$|\mu_{\rm eff}| = g_J \mu_B \sqrt{J(J+1)} \quad \text{or} \quad p = \mu_{\rm eff}/\mu_B = g_J \sqrt{J(J+1)} \ , \qquad (4.12)$$

where

$$g_J = 1 + \frac{J(J+1) + S(S+1) - L(L+1)}{2J(J+1)}$$

is the so-called Landé g factor of the electron shell. Quantity p, which characterizes the effective number of Bohr magnetons, can be used conveniently for experimental determination of the effective magnetic moment. The maximum positive value of the projection μ_J onto the direction of the field (the saturation moment) is defined by the relation

$$\mu_{\max} = g_J \mu_B J \ . \qquad (4.13)$$

The magnitude of the g factor depends on how the orbital and spin angular momenta of the electrons add together, forming the total angular momentum. If the total momentum of the atom is due only to the spin momentum ($S \neq 0$, $L = 0$), then $g_j = 2$. In this case the orbital momentum is said to be completely "quenched." The effective magnetic moment $\mu_{\rm eff}$ of the atom is then expressed as

$$\mu_{\rm eff} = 2\,\mu_B \sqrt{S(S+1)} \ , \quad p = 2\sqrt{S(S+1)} \ . \qquad (4.14)$$

A comparison of the experimental values of p with those calculated by (4.12, 14) shows that the magnetic moments of the ions of the iron group (Ti^{3+}, V^{3+}, Cr^{3+}, Mn^{3+}, Co^{2+}, Ni^{2+}, and Cu^{2+}) are due almost exclusively to the spin momenta, since the experimental values of p agree better with those calculated by (4.14), i.e., at $g_J = 2$. A similar comparison for ions of rare-earth metals (Ce^{3+}, ... Lu^{3+}) shows that orbital momenta ($1 < g_J < 2$) take an active part in the magnetism of these ions.

The difference in the nature of the magnetic moments for the indicated two groups of ions results because in the iron-group crystals the effect of the crystal field on the orbital momentum is so strong that the latter acquires a fixed orientation in the lattice (it is "quenched"). The coupling beween L and the lattice in the iron-group crystals is so strong because the unfilled $3d$ layer, responsible for the magnetic moment of the atom, is an outer one. In the crystals or rare-earth ions the magnetic moment is due to the unfilled $4f$ layer, which is located in the "interior" of the electron shell. Here the crystal field induces only a partial effect of "quenching" of the orbital momentum.

When considering the basic laws governing the paramagnetics, one should bear in mind that all the paramagnetic bodies possess so-called *orientational paramagnetism* in distinction to the Van Vleck polarization paramagnetic effect encountered in diamagnetics (see above).

Let us consider a medium containing N atoms per unit volume, each of the atoms possessing a magnetic moment μ_A. In the absence of an external field the directions of these moments are distributed randomly (therefore, we classify the paramagnetics as disordered magnetics). The application of an external field causes some orientation of the magnetic moments in the direction of the field ($\varkappa > 0$) and the magnetization of the unit volume $I = N\mu_A$ (4.1). The orienting effect of the field is resisted by the thermal motion, and therefore orientational paramagnetism is essentially temperature dependent. The dependence of orientational paramagnetic magnetization I on H and absolute temperature T is described by the Brillouin-Debye formula:

$$I = N g_J \mu_B J B_J(x) \ , \tag{4.15}$$

where

$$B_J(x) = \frac{2J+1}{2J} \coth\left(\frac{2J+1}{2J} x\right) - \frac{1}{2J} \coth \frac{x}{2J} \ , \tag{4.16}$$

$$x = \frac{g_J \mu_B J H}{kT} \ . \tag{4.17}$$

Equations (4.15 – 17) contain the quantities introduced previously, with the exception of Boltzmann constant k. For weak fields

$$\mu_B H \ll kT \ , \quad B_J(x) \approx (J+1)x/3J \ , \quad \text{and}$$
$$\varkappa = \frac{I}{H} = \frac{N\mu_B^2 g_J^2 J(J+1)}{3kT} = \frac{C}{T} \ . \tag{4.18}$$

Equation (4.18) is the well-known Curie law (C is the Curie constant) discovered experimentally by Pierre Curie in 1895. The Curie law holds true only in media where the interaction between the magnetic moments is low (gases, dilute salt solutions). For the vast majority of solids in which interaction of magnetic moments takes place the so-called *Curie-Weiss law* (discovered by Weiss in 1907) is valid:

$$\varkappa = C/(T - \Delta) \ . \tag{4.19}$$

Constant Δ takes into account the interaction between the magnetic moments. It will be shown in Sect. 4.2 that the Curie-Weiss law is valid for the magnetic susceptibility of most of the ferromagnetics above the Curie point. Recall (Chap. 3) that the same law is followed by the dielectric susceptibility of most of the ferroelectrics above the Curie point.

In the other limiting case $H/T \to \infty$ (superstrong fields or $T \to 0$), the value of $I \to N g_j \mu_B J = N\mu_{max}$. Then (4.1) fails, and the paramagnetic begins to show so-called saturation, i.e., a state in which all the magnetic moments are parallel to the external field (a kind of an artificial ferromagnetic).

Among the typical paramagnetic crystals are those containing ions of the iron- and rare-earth-group elements. The nature of the magnetic moment of these ions has been discussed above.

4.2 Ordered Magnetics

4.2.1 Different Types of Magnetic Structures in Crystals. Ferromagnetism, Antiferromagnetism, and Ferrimagnetism

We now proceed to crystals in which cooling below a definite temperature causes ordered orientation of the magnetic moments of the atoms.

To describe magnetic structure we have to introduce the concept of *magnetic unit cell*, by which we mean the least set of groups of atoms of a crystal lattice, with periodic repetitions (translations) of this set resulting in the entire magnetic structure of the crystal. A crystal unit cell may either coincide with the magnetic one or be its multiple. But, as will be shown below, this condition may not hold in helocoid structures.

All the magnetically ordered crystals are divided into two groups: those with a nonzero and those with a zero resultant (total) magnetic moment of the unit magnetic cell. A nonzero macroscopic resultant magnetic moment per unit volume is called *spontaneous* magnetization and denoted by I_S. Crystals with $I_S \neq 0$ are called *ferromagnetics*, and crystals with $I_S = 0$, *antiferromagnetics*. It will be shown below that in addition to the classification of the magnetically ordered crystals according to the presense or absence of I_S, a more detailed classification can be made which would reflect the space distribution patterns for the magnetic moments of the atoms.

Figure 4.1 outlines the different types of magnetic structure. Figure 4.1a corresponds to a simple, "ordinary" ferromagnetic structure: all the magnetic moments of the atoms (Fig. 4.1a depicts atoms of the same sort) are mutually parallel. It is obvious that in such a structure $I_S \neq 0$. An example of crystals with such a magnetic order are the "classical" ferromagnetics: iron, nickel, and cobalt.

Figure 4.1b shows a simple antiferromagnetic structure. The neighboring sites of the lattice are occupied by atoms which have equal but opposite magnetic moments (hence the prefix "anti"). In such a structure the magnetic moments of the atoms cancel each other, i.e., $I_S = 0$. The axis along which the antiferromagnetically ordered magnetic moments are arranged is called the *antiferromagnetic axis*.

Crystallographically, all the atoms with the same direction of the magnetic moments in the structure can be unified into so-called *magnetic sublattices*. In our case all the atoms whose magnetic moments are directed "upward" form one sublattice, and the atoms with the "downward" direction of the magnetic moments form the other. These two sublattices are formed by atoms in crys-

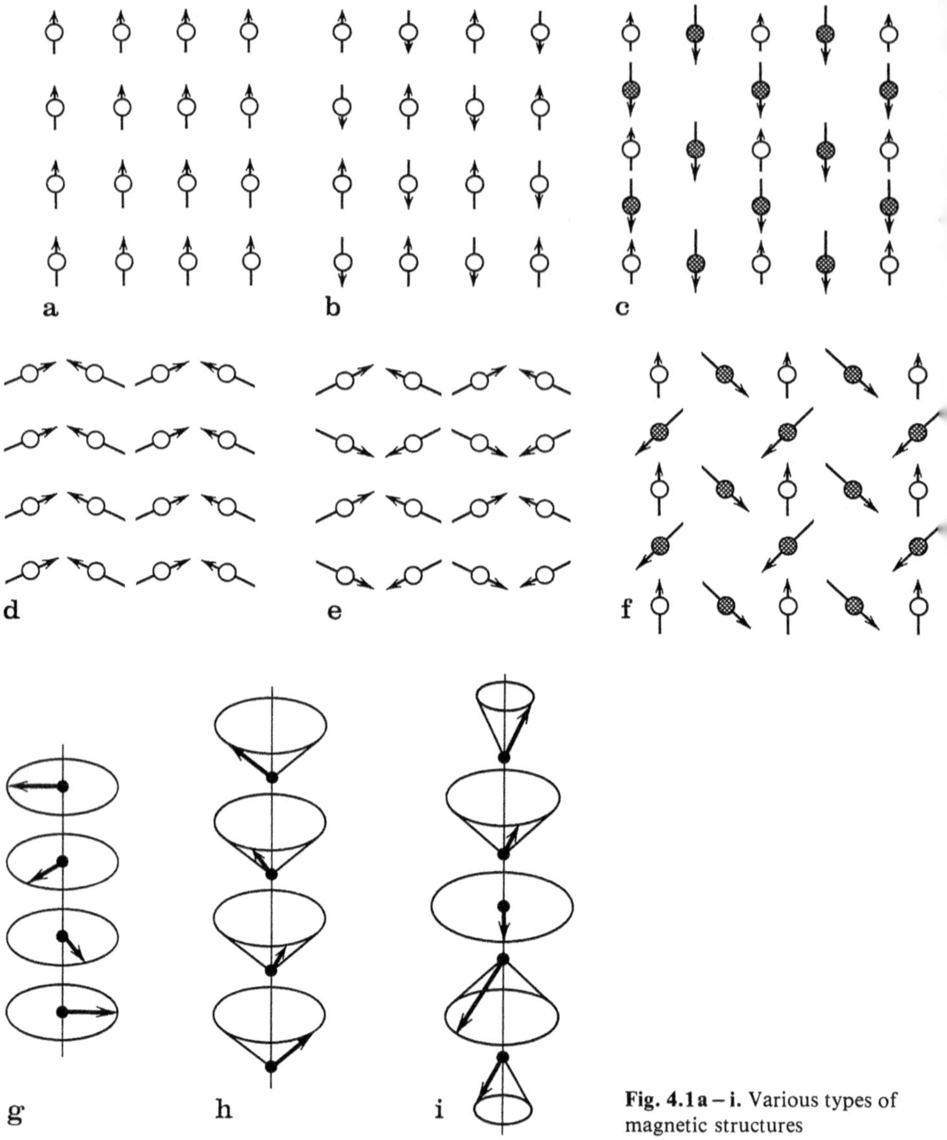

Fig. 4.1 a – i. Various types of magnetic structures

tallographically equivalent positions (two equivalent magnetic sublattices). Generally, the magnetic structure can contain several sublattices formed by atoms in crystallographically nonequivalent positions. Strictly speaking, a magnetic sublattice is the set of all the atomic magnetic moments obtainable from each other by parallel translations over distances multiple of the magnetic unit cell spacings. Typical representatives of crystals with an antiferromagnetic structure are oxides of transition metals MnO, NiO, CoO, FeO, many fluorides, chlorides, sulphides, selenides, etc.

In a *ferrimagnetic* collinear magnetic structure (Fig. 4.1 c) neighboring atoms also have an antiparallel orientation, but the total magnetic moment of the cell is different from zero. Such a structure will possess spontaneous magnetization, because the magnetic moments of the ions of different sublattices are uncompensated. The incomplete compensation may be attributed to several causes. First, a unit magnetic cell may contain a different number of ions belonging to two sublattices (the magnetic moments of the ions may be the same). Second, the magnetic moments of the ions for two different sublattices may differ in magnitude. In most cases both causes are present, as shown in Fig. 4.1 c. The ferrimagnetic substances include so-called ferrites (Sect. 4.5), hence the term "ferrimagnetism". Ferrimagnetism is sometimes called *uncompensated antiferromagnetism*, which is a little longer, but reflects the nature of the phenomenon more accurately. If we do not particularly wish to emphasize the character of magnetic ordering in a ferrimagnetic crystal, it will not be a grave terminological error if we call it ferromagnetic, because it is also among the substances which possess spontaneous magnetization (some inconsistency in the classification of magnetically ordered crystals is still unavoidable).

The indicated types of magnetic structure are *collinear* magnetic structures. There are also various types of *noncollinear* magnetic structures; the most typical are the following.

Weakly noncollinear magnetic structure (Fig. 4.1 d), inherent in so-called *weak ferromagnetics*, is characterized by a small resultant magnetic moment (it is directed upward in Fig. 4.1 d due to a slight canting of the directions of the antiferromagnetic ordering of the sublattice magnetic moments. The weak ferromagnetics (canted antiferromagnetics) include crystals of α-Fe_2O_3 (hematite), $FeBO_3$, FeF_3, carbonates $MnCO_3$, $CoCO_3$, $NiCO_3$, so-called orthoferrites $RFeO_3$, and orthochromites $RCrO_3$ (R being the ion of a rare-earth element).

At the same time, a *weakly noncollinear antiferromagnetic structure* (Fig. 4.1 e) is possible, which does not possess a resultant moment.

The *strongly noncollinear* magnetic structures include the *triangular* (angular) structure shown in Fig. 4.1 f. Here the magnetic sublattice formed by the shaded atoms is subdivided into two sublattices, whose magnetic moments, directed at an angle to each other, produce, as a result of addition, a moment antiparallel to that of the third sublattice. Such a structure occurs in a number of ferrites and can be regarded as a particular case of ferrimagnetic structure.

A special and very interesting group consists of crystals with *screw* or *helicoid* magnetic ordering. Such ordering is observed in some hexagonal rare-earth metals. The simplest screw structure is the antiferromagnetic helicoid (Fig. 4.1 g), which was found in dysprosium and holmium at certain temperatures. The atomic magnetic moments in such structures are perpendicular to the sixfold axis (i.e., they lie in the basal plane), but in each basal plane they are turned by a certain angle with respect to the direction of the moment in the neighboring basal plane. Thus, the termini of the vectors of these moments describe a helical line about the sixfold axis, and the magnetic cell as a whole (corresponding to the screw pitch) does not possess a resultant magnetic moment. A more complicated picture of ordering was found in holmium (Fig. 4.1 h). The magnetic moments of holmium atoms in the ferromagnetic state are inclined by a certain constant angle to the sixfold axis: the projections of the magnetic moment in the basal plane are ordered helicoidally, as in the preceding case. Spontaneous magnetization in such a structure is due to the magnetic moment component parallel to the hexagonal axis. The structure described is called a *ferromagnetic helicoid*[1]. A very complicated magnetic structure is observed in erbium in the antiferromagnetic state. The magnitude and direction of the magnetic moment component along the hexagonal axis vary sinusoidally from one layer to another (Fig. 4.1 i); such magnetic ordering is called cycloidal. Examples of magnetic structures are also given in [Ref. 2.5, Fig. 1.40 a, b and Fig. 1.41 a – d].

Helicoid magnetic structures have been found in some ferrites with a hexagonal structure.

The direct method for determining the atomic magnetic structure is the neutron diffraction method [Ref. 1.1, Chap. 4]. In some cases sufficiently detailed information on the magnetic structure can be obtained with the aid of the Mössbauer effect and nuclear magnetic resonance.

4.2.2 Magnetic Symmetry

The symmetry of disposition and orientation of magnetic moments in a crystal is described by *magnetic space groups*. The magnetic symmetry takes into account the space distribution of the density of magnetic moment $\mu(x, y, z) \equiv \mu(\boldsymbol{r})$ induced by the electrons in the crystal. Function $\mu(\boldsymbol{r})$ contains a so-called R transformation consisting in the reversal of the magnetic moment. Under this transformation (it is equivalent to a change of sign of time, "time inversion") the invariance of the quantum mechanics equations is preserved. The symmetry which uses operation R along with ordinary operations is called the magnetic symmetry. Because of the specifics of operation R, the magnetic symmetry is sometimes called the space-time symmetry.

Operation R is isomorphous to the recoloring operation in the antisymmetry. Therefore, the point groups of the magnetic symmetry are isomorphous to the point groups of antisymmetry, and the space groups, to the space Shubnikov's, groups of antisymmetry [Ref. 1.1, Chap. 2].

The relationship between the magnetic classes of crystals and the antisymmetry classes is shown in Table 4.1. The parentheses indicate the number of the respective space groups. Of the total number of 122 classes, 32 contain

[1] For reasons not yet quite clear, in helicoid structures the magnetic unit cell spacing (the screw pitch) is not always a multiple of the crystallographic unit cell parameters.

Table 4.1. Relationship between the classes of antisymmetry and those of magnetic symmetry. The classes in which $I_S \neq 0$ is admissible are underscored

G Polar (single-color)	G' Mixed-polarity (two-color)	$G1'$ Neutral (gray)	Total number of classes
$13 + 19 = 32(230)$	$18 + 40 = 58(1191)$	32(230)	122(1651)
Total 90(1421)			

operation R explicitly (classes $G1'$). In crystals belonging to classes $G1'$ the time-average value of the magnetic moment density is equal to zero, because transformation R taken separately means $\mu(r) = -\mu(r) = 0$. Such crystals have no magnetic structure (diamagnetics and paramagnetics).

All the ordered magnetics belong to classes $(G + G')$. From Table 4.1 it is seen that the total number of magnetic classes describing magnetically ordered crystals is 90 (they are listed in Table 4.2). Among these, 32 classes contain no operation R at all (conventional crystallographic classes G). The remaining 58 classes contain operations R in conjunction with other operations. Among the 90 magnetic classes, 31 classes $(13 + 18)$ are those for which crystals can display a nonzero spontaneous magnetization (ferromagnetics, ferrimagnetics, and weak ferromagnetics). These classes are underscored in Table 4.2. The remaining 59 classes $(19 + 40)$ of the 90 magnetic classes describe antiferromagnetic crystals. They include crystals belonging to the magnetic classes of the cubic syngony, whereas these classes cannot (according to Table 4.2) contain crystals showing spontaneous magnetization. The antiferromagnetics may, in principle, include crystals belonging to any one of the 90 magnetic classes, including 31 classes of crystals potentially capable of possessing spontaneous magnetization.

Why is it necessary to use magnetic groups of symmetry? From the simple example depicted in Fig. 4.2 one can see that the crystallographic symmetry generally does not reflect the magnetic structure to the full extent. The four (nonmagnetic) atoms depicted in Fig. 4.2a form a configuration which is characterized by a fourfold axis in the direction of the Z axis (perpendicular to the plane of the drawing). If all the atoms are assigned magnetic moments and aligned ferromagnetically along the Z axis (Fig. 4.2b), this axis, as is easy to see, will remain axis 4. If, however, with the parallelism of the moments retained, they are "flipped" in some direction in the XY plane, as in Fig. 4.2c, the Z axis will become a twofold axis of antisymmetry $2'$: in order to superimpose the moments of atoms 1, 3, and 2, 4 it is necessary, in addition to rotation through 180°, to reverse the direction of the moments simultaneously, i.e., to apply operation R in combination with a 180° rotation. (The terms "antisymmetry axis", "anticenter", etc., which will be used for describing magnetic structures, are borrowed from antisymmetry theory [Ref. 1.1, Chap. 2], where they denote operations accompanied by recoloring of

Table 4.2. Distribution of 90 magnetic classes ($G'+G$) among the crystallographic syngonies. The symmetry elements containing operation R are primed. The classes in which $I_S \neq 0$ is admissible are underscored

Syngony	G'	G
Triclinic	$\bar{1}'$	$\underline{1}$, $\underline{\bar{1}}$
Monoclinic	$\underline{2'}$, $\underline{m'}$, $\underline{2'/m'}$, $2/m'$, $2'/m$	$\underline{2}$, $\underline{m}$, $\underline{2/m}$
Orthorhombic	$\underline{22'2'}$, $\underline{m'm'2}$, $\underline{m'm2'}$ $\underline{mm'm'}$, $m'm'm'$, $m'mm$	222, $mm2$, mmm
Tetragonal	$4'$, $\bar{4}'$, $4'/m$, $4/m'$, $4'/m'$ $4'22'$, $\underline{42'2'}$, $4'mm'$, $\underline{4m'm'}$ $\bar{4}'2m'$, $\bar{4}'2'm$, $\underline{\bar{4}2'm'}$, $4'/mmm'$ $\underline{4/mm'm'}$, $4/m'm'm'$, $4/m'mm$ $4'/m'mm'$	$\underline{4}$, $\underline{\bar{4}}$ $\underline{4/m}$ 422 $4mm$, $\bar{4}2m$ $4/mmm$
Trigonal	$\bar{3}'$, $\underline{32'}$, $\underline{3m'}$, $\underline{\bar{3}m'}$ $\bar{3}'m'$, $3'm$	$\underline{3}$, $\underline{\bar{3}}$, 32 $3m$, $\bar{3}m$
Hexagonal	$6'$, $\bar{6}'$, $6'/m'$, $6/m'$, $6'/m$ $6'22'$, $\underline{62'2'}$, $6'mm'$, $\underline{6m'm'}$ $\bar{6}'m'2$, $\bar{6}'m2'$, $\underline{\bar{6}m'2'}$ $6'/m'mm'$, $\underline{6/mm'm'}$, $6/m'm'm'$ $6/m'mm$, $6'/mmm'$	$\underline{6}$, $\underline{\bar{6}}$ $\underline{6/m}$, 622 $6mm$, $\bar{6}m2$ $6/mmm$
Cubic	$m'3$, $4'32'$, $\bar{4}'3m'$ $m3m'$, $m'3m'$, $m'3m$	23, $m3$ 432, $\bar{4}3m$, $m3m$

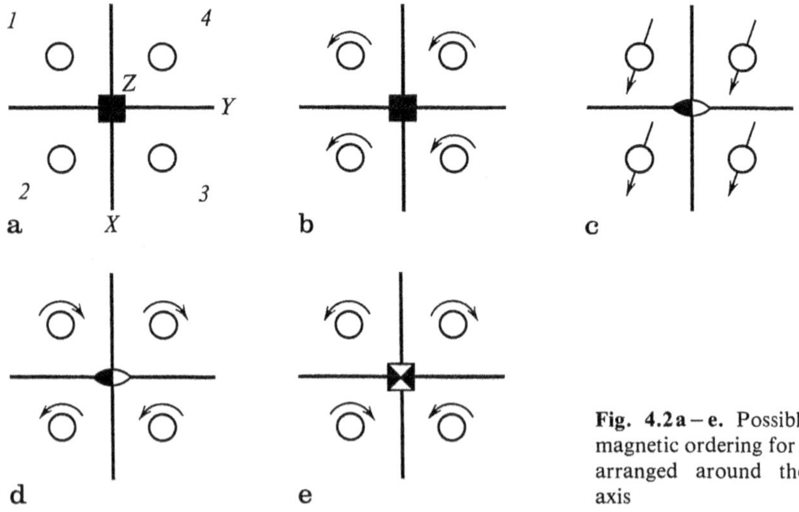

Fig. 4.2a–e. Possible types of magnetic ordering for four atoms arranged around the fourfold axis

figures. Magnetic symmetry has no special term for symmetry elements containing operation R.)

Figure 4.1 d shows that axis $2' \| Z$ may also correspond to antiferromagnetic ordering of the moments along the Z direction. The antiferromagnetic ordering shown in Fig. 4.2 e is characterized by a fourfold antisymmetry axis $4' \| Z$.

The above example illustrates the following important symmetry rules:

1) With a given crystal structure a crystal may have several types of magnetic structure.
2) The appearance of magnetic ordering may only lead to a reduction (or retention, at least) of the symmetry of the crystal as compared to its nonmagnetic state.

These rules reflect one of the fundamental principles of crystal physics, the Curie principle (Chap. 1). Examples of particular magnetic structures for whose description magnetic symmetry is used are given in Sect. 4.5.

4.2.3 Basic Types of Interaction in Ordered Magnetics

Atomic interaction in magnetically ordered crystals may be either of an electrostatic or of a magnetic nature. Estimation of the magnetic interaction of atoms shows that it is not sufficiently large to ensure a magnetic order within a wide temperature range. Moreover, an attempt to reveal experimentally a strong internal magnetic field which could cause magnetic ordering showed that no such field actually exists (Dorfman's experiments on the passage of electrons through a magnetized nickel foil, 1927). Therefore, the cause for the ordering of magnetic moments should be sought in the electrostatic interaction of atoms.

Along with the electrostatic energy of coulomb interaction, ordered magnetics also have an additional electrostatic energy, which is called the *exchange energy* (E_{exch}), of a purely quantum origin. The expression for the exchange energy of interaction of i and j atoms possessing spin momenta $\boldsymbol{s}_i$ and $\boldsymbol{s}_j$ has the form

$$E_{\text{exch}} = -2I_{ij}\boldsymbol{s}_i\boldsymbol{s}_j \ , \tag{4.20}$$

where I_{ij} is the so-called *exchange integral*, which takes into account the overlapping of the atomic wave functions p. The exchange energy promotes the establishment of an ordered distribution of the spin magnetic moments. If exchange integral I_{ij} is negative, an antiparallel spin orientation takes place (antiferromagnetism). If, however, $I_{ij} > 0$, the spins are parallel to each other (ferromagnetism). The sign of the exchange integral depends, apart from other factors, on the interatomic distance. The exchange energy is invariant to permutation of spins i and j (it is symmetric with respect to the spins, $I_{ij} = I_{ji}$).

Direct and *indirect exchange interactions* (superexchange) and distinguished. The direct exchange interaction discussed so far is associated with the overlapping of the wave functions of the electrons of neighboring atoms; it is realized mainly in ferromagnetic metals and their alloys, where all the nearest atoms possess magnetic moments. Indirect exchange interaction occurs when atoms possessing magnetic moments are separated by magnetically neutral atoms, for instance, by oxygen anions in magnetic oxides. Then the exchange interaction takes place through the electrons of nonmagnetic atoms. In some alloys containing rare-earth elements, in which the distances between magnetic atoms are too large for realization of direct exchange interaction, magnetic ordering is due to the indirect exchange interaction of electrons of neighboring atoms through the conduction electrons.

The exchange energy in the form of (4.20) is isotropic: when the whole spin system is rotated through any angle with respect to the crystal lattice, the exchange energy of the system remains unaltered (in recent years calculations of so-called quadrupole interaction have led to the concept of anisotropic exchange energy, which will not be considered here). It is known from experiment, however, that the directions of the magnetic spin moments and of the spontaneous magnetization vector in a crystal are always related to its symmetry. Spontaneous magnetization vector $\boldsymbol{I}_s$ in a ferromagnetic crystal is always bound (although not "firmly", as in the spontaneous polarization in ferroelectrics) to a definite symmetry axis, which is called the *axis of easy magnetization* (or preferred direction). The meaning of this term will become clear when we consider the magnetization processes in Sect. 4.3. To deflect vector $\boldsymbol{I}_s$ from this axis (by applying the external field), a certain amount of energy must be spent. This energy is called the *magnetocrystalline anisotropy* energy, or the *magnetic anisotropy* energy. That vector $\boldsymbol{I}_s$ "senses" the crystal lattice is due to the magnetic interaction of the magnetic moments of the atoms. In addition to the pure dipole magnetic interaction of magnetic moments, which can be identified with the interaction of a system of magnetic arrows, the relationship between magnetic moments and the lattice may be effected via the so-called spin-orbital coupling. It has been noted in Sect. 4.1 that the orbital angular momentum of the atom $\boldsymbol{L}$ acquires a fixed orientation with respect to the crystal lattice because of partial or complete "quenching". The coupling with the lattice is transferred to the spin magnetic moments of the atoms by means of spin-orbital interaction.

The part of the free energy of the crystal which is due to magnetic interaction, i.e., the magnetic anisotropy energy, can be represented as an expansion in the powers of the components of vector $\boldsymbol{I}_s$ (or in the powers of its direction cosines). The anisotropy energy must be invariant to operation R, whereas the magnetization changes sign in this operation. If follows that the anisotropy energy, being a scalar quantity, must be an even function of the magnetization components. The expression for the anisotropy energy has the form

$$V = k'_{mn} I_m I_n + k'_{mnop} I_m I_n I_o I_p + k'_{mnopqr} I_m I_n I_o I_p I_q I_r + \dots \,, \tag{4.21}$$

where $m, n, \dots r$ take the values 1, 2, and 3 in accordance with the three crystallophysical axes X_1, X_2, and X_3; $I_m \dots I_r$ stand for the components of vector

I_s along the X_1, X_2, and X_3 axes; and k'_{mn}, k'_{mnop}, and k'_{mnopqr} are second-, fourth-, and sixth-rank tensors, whose type depends on the crystal symmetry in the paramagnetic phase (Sect. 4.4).

As will be shown in Sect. 4.4, the expression of the magnetic anisotropy energy for crystals of cubic ferromagnetics is written

$$V_{\text{cube}} = K_1(\alpha_1^2\alpha_2^2 + \alpha_2^2\alpha_3^2 + \alpha_3^2\alpha_1^2) + K_2\alpha_1^2\alpha_2^2\alpha_3^2 + \dots \, . \tag{4.22}$$

The components of the magnetization vector in (4.22) are replaced by the direction cosines; K_1 and K_2 are the so-called first and second *magnetic anisotropy constants*.

The magnetic anisotropy energy for uniaxial crystals (neglecting the anisotropy in the basal plane) can be written

$$V_{\text{uni}} = K_1 \sin^2\theta + K_2 \sin^4\theta \ , \tag{4.23}$$

where θ is the angle between I_s and a highest-order axis. Anisotropy constants K_1 and K_2 in (4.23) are not equal to constants K_1 and K_2 in (4.22).

The above-listed (exchange and magnetic) interactions in crystals ensure various types of magnetic ordering. In each particular case these interactions give rise to some equilibrium-ordered structure of the magnetic moments.

4.2.4 Molecular Field Theory. The Curie and Néel Points

We now consider the effect of the temperature on the properties of ordered magnetics. At a certain temperature, which is called the *Curie temperature* (or *Curie point*) T_C for ferromagnetic crystals and the Néel temperature T_N for antiferromagnetics, the magnetic moment ordering, which was "ideal" at temperature $T=0$, completely disappears. At these points the exchange energy responsible for the ordering of the magnetic moments becomes equal to the average energy of the thermal motion of the atoms, which achieves a value sufficient to destroy the ordered structure. Above points T_N and T_C the crystal becomes paramagnetic. Cases are known where, with increasing temperature, crystals of some rare-earth metals first change from the ferromagnetic to the antiferromagnetic phase, and only then to the paramagnetic.

To describe the temperature dependence of I_s we shall use Weiss's molecular field theory. It was said in Sect. 4.1 that in strong fields with $T \to 0$ a paramagnetic approaches the state of saturation and resembles a ferromagnetic, since all the atomic magnetic moments become field oriented. Assume that there is some strong field inside the ferromagnetic at $T < T_C$ (we shall it a *molecular field*), which causes the ordering of the magnetic moments. Molecular field theory, which was proposed by Weiss in 1907, long before the introduction of the theory of exchange interaction, postulated the presence of such an internal magnetic field (molecular field H^M), leading to the ordering of the atomic magnetic moments (in fact, there is no such field, and therefore H^{M} must be regarded as some equivalent measure of exchange interaction). The magnitude of this field, which takes into account the interaction of the magnetic moments of the atoms, is assumed to be proportional to the magnetization

$$H^{\mathrm{M}} = \lambda I \ . \tag{4.24}$$

Constant λ is called the *molecular field constant*. In this case the consideration of a ferromagnetic is reduced to that of a paramagnetic in which external magnetic field H is replaced by some effective field equal to the sum $(H + H^{\mathrm{M}})$.

Spontaneous magnetization is obtained with the aid of Brillouin-Debye equation (4.15), provided we substitute effective field $(H + H^{\mathrm{M}})$ into it and take external field H to be zero. Substituting (4.24) into (4.15) and assuming $H = 0$, we get

$$I = NJg_J\mu_{\mathrm{B}}B_J(x) \ , \tag{4.25}$$

$$x = Jg_J\mu_{\mathrm{B}}\lambda I/kT \ . \tag{4.26}$$

Originally in Weiss's theory a classical Langevin function $L(x) = \coth x - 1/x$ was used instead of Brillouin function (4.16). As $J \to \infty$, B_J changes to $L(x)$. Condition $J \to \infty$ means that the magnetic moments, being in the state of thermal motion, can assume any orientation relative to the direction of the field.

To solve (4.25), in which I appears both in the argument of the Brillouin function and in the left-hand side of the equation, one uses the graphic method shown in Fig. 4.3. The most important result of Weiss's theory is that at $H = 0$; when Line 2 expressing the dependence of I on x (4.26) passes through the origin, there exists a nonzero solution for I. The magnetization corresponding to $H = 0$ has the meaning of spontaneous magnetization. The value of $I = I_{\mathrm{S}}$ is found as the intersection point of Curve 1 (4.25) with Line 2.

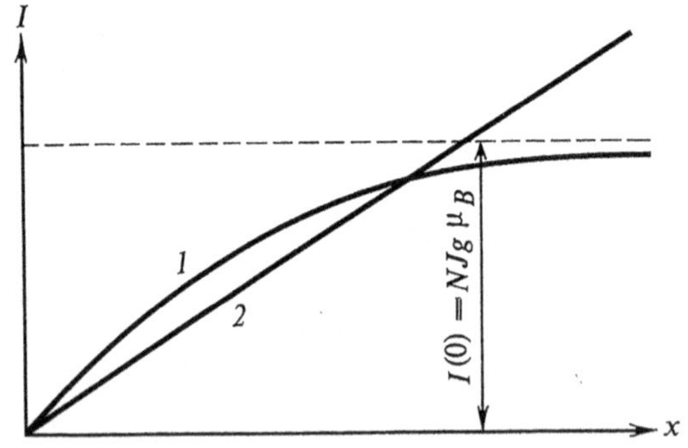

Fig. 4.3. Graphical solution of the equation of state in Weiss's molecular-field theory

As follows from (4.26), the slope of Line 2 is proportional to T. When $T \to 0$, $x \to \infty$, and $I = I_S$ tends to its maximum value

$$I(0) = NJ g_J \mu_B \ . \tag{4.27}$$

In Fig. 4.3 this limiting value is denoted by the dashed line. With increasing temperature, I_S decreases and diappears when the slope of Line 2 is equal to the derivative of function $B_J(x)$ at the origin. Thus, the initial slope of Curve 1 corresponds to Curie point T_C. At $T > T_C$, magnetization I appears only in the external field.

Introducing the external field, parameter x can be written as

$$x = \frac{J g_J \mu_B (H + \lambda I)}{kT} \ . \tag{4.28}$$

Neglecting the higher-order terms in the expansion of the Brillouin function with respect to parameter $x \ll 1$, as in deriving (4.18), we obtain

$$I = \frac{N g_J^2 \mu_B^2 J(J+1)}{3kT} (H + \lambda I) \ . \tag{4.29}$$

If we single out value $\varkappa = I/H$ from (4.29), we get an expression similar to Curie-Weiss law (4.19), where $\Delta = C\lambda$, and

$$C = \frac{N g_J^2 \mu_B^2 J(J+1)}{3k} \ . \tag{4.30}$$

Quantity $C\lambda$, which has the temperature dimensionality, corresponds to the temperature at which Line 2 in Fig. 4.3 will touch Curve 1 at the origin, i.e., $\Delta = C\lambda = T_C$. The Curie-Weiss law can now be written

$$\varkappa = \frac{I}{H} = \frac{C}{T - C\lambda} = \frac{C}{T - T_C} \ . \tag{4.31}$$

The relationship of constants λ and C with the microscopic characteristics can be readily found by comparing (4.31, 29).

Variation of spontaneous magnetization with the temperature can conveniently be considered with respect to magnetization at absolute zero $I(0)$ (4.27), and temperature variation, with respect to T_C. The expression for the temperature dependence of magnetization in such "reduced" coordinates is easily obtained by expressing λ in (4.26) via T_C/C. Then, with due regard for (4.27), we get

$$x = \frac{3J}{J+1} \frac{T_C I(T)}{T I(0)} \ . \tag{4.32}$$

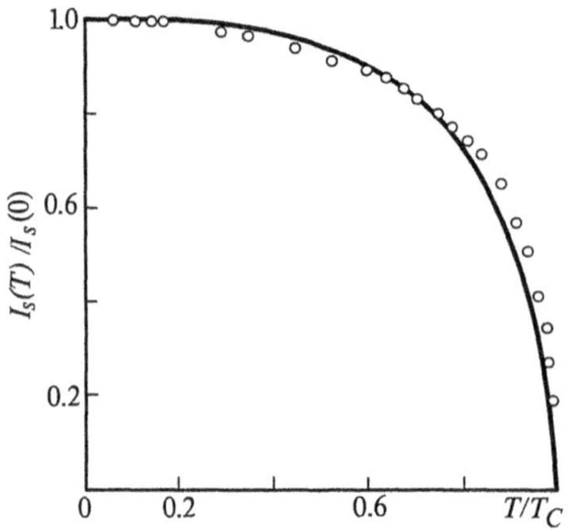

Fig. 4.4. Temperature dependence of spontaneous magnetization: solid line (4.34); points, experimental results for Ni

Here $I(T)$ and $I(0)$ at $H=0$ have the meaning of spontaneous magnetization at T and absolute zero, respectively. Then the expression for the temperature dependence of I_S in the reduced coordinates will, in accordance with (4.32, 25), have the form

$$\frac{I_S(T)}{I_S(0)} = B_J\left[\frac{3J}{J+1}\,\frac{I_S(T)/I_S(0)}{T/T_C}\right] . \tag{4.33}$$

If there is only one uncompensated spin per atom, then $J=1/2$ and $g_J=2$, and (4.33) simplifies to

$$\frac{I_S(T)}{I_S(0)} = \tanh\frac{I_S(T)/I_S(0)}{T/T_C} . \tag{4.34}$$

A comparison of (4.33) with experiment shows that the best agreement for iron and nickel is obtained at low J and $g_J=2$. This is in line with the atomic electron configurations of these metals and indicates that the principal carriers of magnetism in Ni and Fe are electron spins. In Fig. 4.4 the solid curve denotes dependence (4.34), and the dots, the experimental data for Ni.

Although the molecular field theory is simple, it gives a true qualitative picture of the fundamental properties of a ferromagnetic, predicting spontaneous magnetization and explaining its temperature dependence. But a detailed comparison of theory with experiment reveals a divergence, which is particularly pronounced in the region $T\approx 0$ and $T\approx T_C$. At T close to absolute zero the temperature dependence of I_S in most of the ferromagnetics obeys the so-called Bloch's $T\,3/2$ law,

$$I_S(T) = I_S(0)(1-\alpha T^{3/2}) , \tag{4.35}$$

following from the spin-wave theory (α is some constant depending on the lattice type and the exchange integral). Molecular field theory products a much faster drop in I_S with the temperature. Divergences with dependence $I_S(T)/I_S(0) \sim (T_C - T)^{1/2}$, which is predicted by molecular field theory, have also been found at temperatures near the Curie point.

At the Curie point, as well as at the ordinary phase transition point between the polymorphous modifications of a crystal, specific anomalies of various "nonmagnetic" physical properties (heat capacity, coefficient of thermal expansion, etc.) manifest themselves. For most of the ferromagnetics (including the ferrimagnetics) the phase transition at the Curie point is a second-order transition, although first-order transitions are known.

Molecular field theory is also fruitful when the magnetic structure of the crystal is characterized by several sublattices (ferrimagnetism). Such a theory was developed for ferrites by Néel. In the case of n sublattices one can introduce the concept of sublattice magnetization. Since the atoms of a given ferrimagnetic sublattice are surrounded by neighbors, some of which refer to the same sublattice, and the others to other sublattices, the molecular field acting on a given atom depends on the magnetizations of all the sublattices, i.e., it is equal to

$$H_i^{\mathrm{M}} = \sum_{j=1}^{n} \lambda_{ij} I_i \ . \tag{4.36}$$

Molecular field theory has been modified many times and is still used successfully for interpreting experimental data [4.2].

The effect of the temperature on the magnetization of each sublattice may prove different depending on the interaction among the spins within a single sublattice and among the spins of different sublattices. This does not, however, mean that each sublattice has a Curie point of its own, since the magnetization of each sublattice is associated with the magnetizations of the other sublattices. For the simplest system of two sublattices a and b the following particular cases may take place:

1. The magnetizations of sublattices $\boldsymbol{I}_a$ and $\boldsymbol{I}_b$ differ in magnitude (ferrimagnetism), but the effect of the temperature on either sublattice is equal and obeys the law (4.33). In this case the temperature dependence of spontaneous magnetization, which is a vector sum of $\boldsymbol{I}_a$ and $\boldsymbol{I}_b$, does not differ in the least from the temperature dependence for pure ferromagnetism and is of a "Weissian" nature, i.e., also obeys the law (4.33).
2. The crystal is ferrimagnetic, but each of the sublattices has its "own" temperature dependence of magnetization $\boldsymbol{I}_a$ and $\boldsymbol{I}_b$. Then the temperature dependence of spontaneous magnetization $\boldsymbol{I}_S = \boldsymbol{I}_a + \boldsymbol{I}_b$ is anomalous, i.e., fails to obey the law (4.33). In a particular case typical of ferrimagnetism, at a certain temperature spontaneous magnetization may pass through zero (so-called *compensation point*) or through the maximum, as shown in Fig. 4.5. Compensation points have been found in a ferrite with a composition

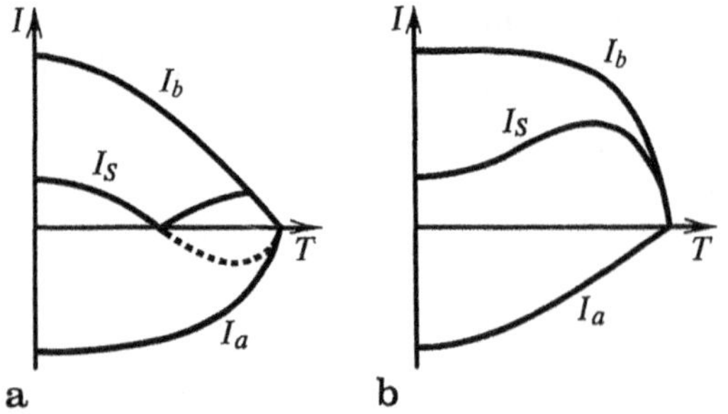

Fig. 4.5 a, b. Examples of anomalous dependence $I_S(T)$ for ferrimagnetics

$Li_{0.5}Cr_{1.25}Fe_{1.25}O_4$, in some rare-earth ferrites with a garnet structure, and in orthoferrites (Sect. 4.5). Magnetization measurements do not reveal the change of sign of I_s (dashed line in Fig. 4.5 a); experiment indicates the dependence shown in Fig. 4.5a by a solid line. Theoretically, differently shaped curves $I_S(T)$ are possible for ferrimagnetics. Here we give only two most typical cases.

3. Sublattice magnetizations are identical and cancel each other, $I_a + I_b = 0$. The sublattices are fully equivalent, and the temperature effect on them is also identical. This is a case of "pure" antiferromagnetism. At the Néel point T_N antiferromagnetics show the maximum of susceptibility $\varkappa$. At $T < T_N$ the magnitude and temperature variations of $\varkappa$ strongly depend on the direction of the external field with respect to the antiferromagnetic ordering axis.

Measurements of the temperature dependence of I_S in ferromagnetics do not yield any information on the temperature dependence of the magnetizations of individual sublattices. To study the temperature dependence of each sublattice one uses methods for measuring local magnetic fields on nuclei (nuclear magnetic resonance, the Mössbauer effect). Determination of this dependence is based on proportionality between the local field on the nucleus and the time-average magnetic moment of ions forming a given sublattice.

We should note one important feature which helps to differentiate ferro-, ferri-, and antiferromagnetics above the Curie (Néel) point. Figure 4.6 exhibits the dependence of reciprocal susceptibility $1/\varkappa$ on T for the magnetics indicated. If the Curie-Weiss law (4.19, 31) holds good, the above dependence is expressed by a straight line intersecting with the x axis at $T = T_C$, and for ferromagnetics T_C must coincide with the ferromagnetic phase transition temperature, as shown in Fig. 4.5. This requirement stems from the theory of second-order phase transitions. But experiment shows that the Curie point, as determined from the law (4.31), is always above the temperature at which I_S disappears with heating. This difference sometimes reaches several tens of degrees. Therefore, we distinguish between the *paramagnetic Curie point* Θ and the *ferromagnetic Curie point* T_C in ferromagnetics. The difference between Θ and T_C is usually attributed to the effect of fluctuations of the short-range magnetic order. The Curie-Weiss law is valid for ferromagnetics

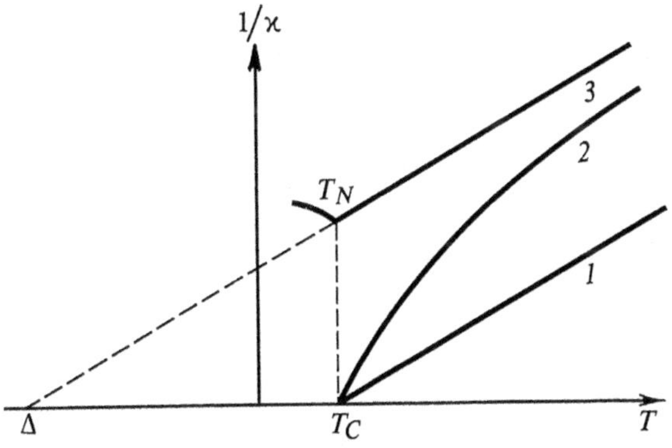

Fig. 4.6. Dependence of $1/\varkappa$ on T for ferromagnetic (*1*), ferrimagnetic (*2*), and antiferromagnetic (*3*) crystals

over a wide temperature range far removed from the Curie point, with the exception of the indicated peculiarity.

The Curie Weiss law is also fulfilled for the antiferromagnetics in the range $T > T_N$, but the value of Δ (4.19) is negative, which is directly associated with the negative sign of exchange integral (4.20).

For the ferrimagnetics, the dependence of $1/\varkappa$ on T does not obey the ordinary Curie-Weiss law. It has a characteristic convex shape approaching the hyperbola. This feature, according to the Néel theory, is due to the complicated relationship between sublattice magnetizations and molecular fields (4.36).

4.3 Domain Structure of Ferromagnetic Crystals and Magnetization Processes

4.3.1 Change of Symmetry at the Curie Point. Symmetry Aspects of Splitting of a Ferromagnetic into Domains

According to the most general considerations based on Curie's symmetry principle the appearance of spontaneous magnetization I_S in a crystal at the Curie temperature must be accompanied by a change in its symmetry, i.e., must be a phase transition. We saw this when discussing the configuration shown in Fig. 4.2. The same figure indicates that the point group of the magnetic symmetry of a crystal below T_C will be different depending on the direction in which I_S arises in the paramagnetic crystal.

Table 4.3 shows to which magnetic class the paramagnetic crystal transfers on appearance of spontaneous magnetization in it [1.14]. Holohedral classes of all systems are taken is initial classes. This is sufficient to embrace practically all the known phase transitions in ordered magnetics. It is seen, in particular, that if I_S appeared in the [100] direction of the initial cubic crystal

Table 4.3. Change in magnetic point symmetry on appearance of $\boldsymbol{I}_S$ in crystals of holohedral classes

$\boldsymbol{I}_S$ direction	Initial class $m3m$	$\boldsymbol{I}_S$ direction	Initial class $6/mmm$	$\bar{3}m$	$\boldsymbol{I}_S$ direction	Initial class $4/mmm$	$\boldsymbol{I}_S$ direction	Initial class mmm	$2/m$	$\bar{1}$
$\langle 100\rangle$	$4/mm'm'$	$\langle 0001\rangle$	$6/mm'm'$	$\bar{3}/m'$	$\langle 001\rangle$	$4/mm'm'$	$\langle 001\rangle$	$mm'm'$	$2/m$	1
$\langle 111\rangle$	$\bar{3}m'$	$\langle 11\bar{2}0\rangle$[b]	$mm'm'$	$2/m$	$\langle 100\rangle$	$mm'm'$	$\langle 010\rangle$	$mm'm'$	$2'/m'$	1
$\langle 110\rangle$	$mm'm'$	$\langle 10\bar{1}0\rangle$[a]	$mm'm'$	$2'/m'$	$\langle 110\rangle$	$mm'm'$	$\langle 100\rangle$	$mm'm'$	$2'/m'$	1
$\langle hk0\rangle$	$2'/m'$	$\langle hki0\rangle$	$2'/m'$	$\bar{1}$	$\langle hk0\rangle$	$2'/m'$	$\langle hk0\rangle$	$2'/m'$	$2'/m'$	1
$\langle hkk\rangle$	$2'/m'$	$\langle hh2\bar{h}l\rangle$	$2'/m'$	$\bar{1}$	$\langle h0l\rangle$	$2'/m'$	$\langle h0l\rangle$	$2'/m'$	$\bar{1}$	1
$\langle hhk\rangle$	$2'/m'$	$\langle h0\bar{h}l\rangle$[a]	$2'/m'$	$2'/m'$	$\langle hhl\rangle$	$2'/m'$	$\langle 0kl\rangle$	$2'/m'$	$\bar{1}$	1
$\langle hkl\rangle$	$\bar{1}$	$\langle hkil\rangle$	$\bar{1}$	$\bar{1}$	$\langle hkl\rangle$	$\bar{1}$	$\langle hkl\rangle$	$\bar{1}$	$\bar{1}$	1

[a] For initial class $\bar{3}m$ these directions are along the vertical symmetry plane.
[b] For initial class $\bar{3}m$ these directions coincide with one of the twofold axes.

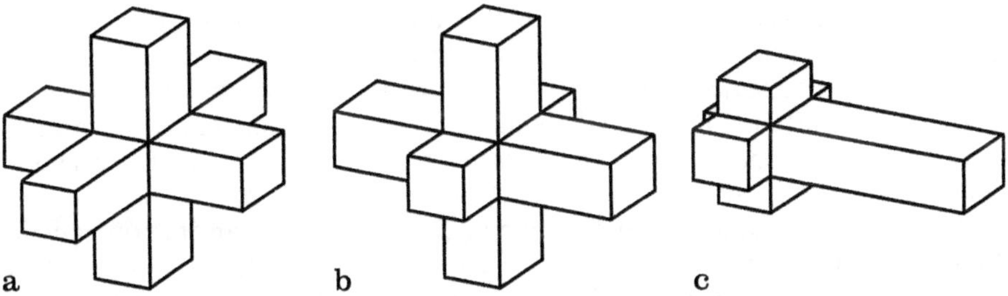

Fig. 4.7. Schematic representation of a domain structure corresponding to **(a)** the ideally demagnetized state of a crystal, **(b)** the state with a magnetic texture, and **(c)** the state with residual magnetization

belonging to class $m3m$, the symmetry of the crystal reduces to tetragonal ($4/mm'm'$). If $I_s \| [111]$, the symmetry of the crystal becomes trigonal ($\bar{3}m'$). The first example refers to iron, and the last to nickel and most of the cubic ferrites with a spinel structure. Generally, when I_S is directed arbitrarily relative to the symmetry axes of the initial crystal of any class, the resultant symmetry is the lowest, triclinic. The last version is never realized at the Curie point, because I_S always arises along definite symmetry axes of the initial phase (these are easy-magnetization axes), but it can be realized in the external field or under reorientation transitions between magnetic classes (Sect. 4.5).

If we include transitions not only from holohedral, but also from the other crystallographic classes in Table 4.3, we shall obtain the classes (31 in all) underscored in Table 4.2.

Let us consider again, as an example, a cubic crystal of the iron type. Since a cubic crystal has three fourfold axes, there are, from the crystallographic and energy standpoints, six equivalent ⟨100⟩ directions along which I_S may arise. The appearance of spontaneous magnetization in a phase transition realizes this "possibility" and, indeed, at $T < T_C$, I_S is disposed along all the indicated directions. Then the crystal volume is divided into separate regions – domains – a kind of component of a polysynthetic twin, inside which I_S is parallel to one of the six ⟨100⟩ directions of the initial cubic phase. In accordance with Table 4.3 each of the domains will be tetragonal ($4/mm'm'$). But the space distribution of the domains is such that the symmetry of a polydomain crystal (pseudosymmetry) actually remains cubic ($m3m$) This is illustrated in Fig. 4.7a where six groups of tetragonal domains (each group occupying 1/6 of the crystal volume in the ideal case) distributed ove the ⟨100⟩ axes, form a figure with the symmetry of the initial paramagnetic phase, $m3m$. This figure corresponds to the "ideally demagnetized" state, since the domains with mutually opposite directions of I_S occupy equal portions of the crystal volume. On the whole, such a crystal does not show any macroscopic magnetization (the absence of such magnetization was actually the cause for Weiss's introduction of the concept of separate regions magnetized along different directions).

If, however, one or several groups of domains occupy a larger volume compared with the others, a so-called *magnetic texture* arises (Fig. 4.7b), which may, in particular, bring about a state with residual magnetization (Fig. 4.7c). In this state the crystal will exhibit macroscopic magnetization (it may serve as a permanent magnet). In the cases shown in Fig. 4.7b, c the pseudo-symmetry of the crystal can no longer be considered cubic.

In addition to domains in ferromagnetics, there exist antiferromagnetic domains in antiferromagnetic crystals. Their existence is most obvious when antiferromagnetic ordering arises in cubic crystals. The appearance of domains in this case is due to the fact that the initial crystal has several crystallographically and physically equivalent axes, along which antiferromagnetic ordering can arise. Therefore, cubic crystals can be called *magnetically multiaxial,* and the crystals of all the other syngonies, in which the ordering of the magnetic moments can occur only along one of the crystallographic axes, *magnetically uniaxial.*

It has been shown by *Shuvalov* [1.14] that the symmetry aspects of splitting of a ferro- or antiferromagnetic into domains are similar to those considered in Chap. 3 for ferroelectrics and antiferroelectrics.

Adjacent domains in a crystal are separated by *domain walls* (bondaries). From Fig. 4.7a it follows that in a Fe-type crystal the directions of $\boldsymbol{I}_S$ in neighboring domains may differ by 90° or 180°. Thus, 90° and 180° domain walls are distinguished. It is easy to see that Ni-type crystals (with easy-magnetization axes ⟨111⟩) can have ~71°, ~109°, and 180° walls. From the foregoing it follows that the modified symmetry of a crystal on appearance of $\boldsymbol{I}_S$ implies the symmetry of a separate domain. The symmetry of the domain will coincide with that of the whole crystal magnetized to saturation along one of the directions of $\boldsymbol{I}_S$ in the domain (the easy axis). Such a magnetized crystal will be a large single domain.

In considering the properties of ferromagnetics one may wonder whether a ferromagnetic crystal must be regarded as tetragonal or cubic (for a Fe-type crystal). The solution of this problem depends on whether the physical property concerned is determined by the individual properties of the domain or by the space distribution of the domains. For instance, in "cubic" ferrite garnets (point group $m3m$ above T_C) one can observe magnetic rotation of the plane of light polarization (this phenomenon is utilized for studying the domain structure), whereas according to Neumann's principle this phenomenon is forbidden for $m3m$-class crystals. The Faraday effect is observed precisely because each domain has as lower symmetry than $m3m$. On the other hand, as regards such properties as the electrical conductivity, a pseudocubic multidomain (nontextured) ferromagnetic behaves as an isotropic medium, i.e., as would be expected of a cubic crystal.

To avoid terminological confusion when describing the symmetry of magnetically ordered crystals we shall use the term crystallographic (nonmagnetic) for those groups (classes) which describe the symmetry of a given crystal in the paramagnetic phase, i.e., without magnetic ordering. Since there is usually no need to take into account the reduction of the symmetry of a crystal under its transformation to a magnetically ordered state, it is customary to allocate a crystal, even in the magnetically ordered phase, to the same class as for the paramagnetic phase (the usual 32 crystallographic classes and the corresponding 230 space groups). Therefore, expressions of the type "cubic ferromagnetic," which appear sometimes, should not mislead the reader.

If, on the other hand, it is necessary to emphasize the symmetry due to a magnetic order, the crystal (domain) should be allocated to one of the 90 magnetic classes, and the magnetic structure will then be described by one of the magnetic space groups of symmetry (it may, of course, turn out that the magnetic class coincides with the crystallographic).

When describing the structures of particular crystals in Sect. 4.5, we shall indicate their respective crystallographic and magnetic classes whenever possible.

4.3.2 Spontaneous Magnetostriction. Magnetoelastic Energy

The change in the symmetry of a crystal due to $\boldsymbol{I}_S$ is associated with the deformation of the crystal. Spontaneous magnetization is always accompanied by so-called spontaneous deformation, i.e., a distortion of the crystal structure, which depends on the direction and magnitude of the spontaneous magnetization. This deformation is called *spontaneous magnetostriction.* It consists of the isotropic and the anisotropic part (Sect. 4.4). The anisotropic part is responsible for the linear magnetostriction of the ferromagnetics, i.e, for the relative change in linear dimensions on magnetization (Fig. 4.15b).

In most of the ferromagnetics the relative spontaneous deformation at temperatures considerably lower than the Curie point is of the order of 10^{-5}. Near the Curie point distortions will be still less distinct because of the smallness of quantity $\boldsymbol{I}_S$ itself; x-ray diffraction does not always reveal such small lattice distortions. But in some ferromagnetics the spontaneous deformation of a domain can be detected by x-ray diffraction, for instance, in cobalt ferrite, where the anisotropic spontaneous deformation may be of the order of 10^{-4}. That of the lattice is particularly large (on the order of 10^{-3}) in some rare-earth metals (Tb, Dy, Ho, and Er). A gigantic relative linear magnetostriction at room temperature is observed in alloys of rare-earth metals with iron, for instance in $TbFe_2$ ($\sim 10^{-3}$). The magnitude of such magnetostriction is vividly demonstrated by the fact that a 1m rod is elongated by 1 mm on magnetization.

The magnetostriction phenomenon is widely utilized in the production of various devices such as magnetostrictive gauges, transducers, resonators, ultrasonic instruments, etc.

The splitting of a crystal into domains causes internal mechanical stresses of magnetostrictive origin in the crystal. This is explained in Fig. 4.8. Stresses arise because neighboring domains with boundaries different from 180° (for

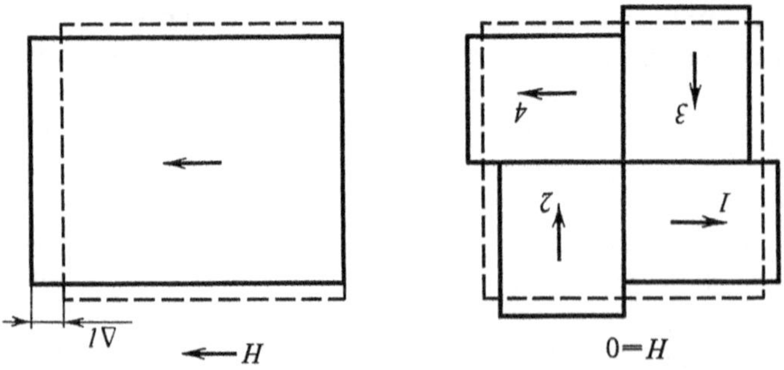

Fig. 4.8. Diagram explaining the occurrence of magnetostrictive stresses

instance, 1 and 2, 2 and 4) are deformed in different directions. These stresses play an important part in magnetization. The relationship between the domain structure and the stresses manifests itself in the fact that the domain structure can be changed by applying external stresses.

Figure 4.8 shows that if we magnetize a crystal by applying an external field directed along one of the "easy" axes, the crystal will be transferred into a single-domain state and change its length in the direction of the field by Δl.[2] If a crystal is magnetized at a certain angle $\varphi \neq \pi/2$ to the "easy" axis, the magnitude of the magnetostriction will be different.

To take into account the relationship between the deformations and the direction of $\boldsymbol{I}_S$ in ferromagnetic crystals, the so-called *magnetoelastic energy* is introduced, which can be written as the tensor series

$$E_{\mathrm{me}} = M_{ijkl}\,\varepsilon_{ij}\,I_k\,I_l + M_{ijklmn}\,\varepsilon_{ij}\,I_k\,I_l\,I_m\,I_n + \ldots = V^0_{ij}\,\varepsilon_{ij}\ , \tag{4.37}$$

where ε_{ij} are the components of the deformation tensor. The magnetoelastic energy can also be expressed in terms of stress component σ_{ij}:

$$E_{\mathrm{me}} = N_{ijkl}\,\sigma_{ij}\,I_k\,I_l + N_{ijklmn}\,\sigma_{ij}\,I_k\,I_l\,I_m\,I_n + \ldots\ . \tag{4.38}$$

Equations (4.37, 38) contain even powers of the magnetization components for the same reasons as (4.21) for the magnetic anisotropy energy. The coefficients forming tensors N and M can be called the *magnetostrictive*, or *magnetoelastic coefficients*. The difference between N and M is about the same as between stiffness c and compliance s in the theory of elasticity (Chap. 2). For the sake of the following presentation it is important to emphasize that the symmetry properties of tensors $[N_{ijkl}]$ and $[N_{ijklmn}]$ fully coincide with

[2] The spontaneous deformation in the direction of magnetization is not necessarily positive, which is demonstrated, in an exaggerated manner, in Fig. 4.8. The domain may be contracted in the direction of $\boldsymbol{I}_S$, i.e., it may have a negative spontaneous striction. The sign and magnitude of the spontaneous deformation depend on the nature of the interatomic forces.

those of tensors $[M_{ijkl}]$ and $[M_{ijklmn}]$. As in the case of magnetic anisotropy energy V, these properties depend on the crystal symmetry of the initial paramagnetic phase of the crystal.

Energy E_{me} can be regarded as a part of the free energy of a ferromagnetic crystal, represented as an expansion into a series with respect to deformation (this can be done because of the smallness of the magnetostrictive deformations):

$$V' = V^0 + V^0_{ij}\,\varepsilon_{ij} + 1/2\; V^0_{ijkl}\,\varepsilon_{ij}\,\varepsilon_{ki} + \dots \;, \tag{4.39}$$

$$V^0 = k_{mn}\,I_m I_n + k_{mnop}\,I_m I_n I_o I_p + k_{mnopqr}\,I_m I_n I_o I_p I_q I_r \;. \tag{4.40}$$

The zero superscripts indicate that the expansion coefficients are calculated for an underformed crystal. The first term in (4.39) corresponds to the anisotropy energy at zero deformations, and the meaning of V^0_{ij} is clear from (4.37). Tensors k_{mn}, k_{mnop}, and k_{mnopqr} in (4.40) and k'_{mn}, k'_{mnop}, and k'_{mnopqr} in (4.21) are identical from the standpoint of symmetry limitations, but the former reflect the anisotropy at zero stresses, and the latter, at zero deformations. The relationships between these two types of magnetic anisotropy will be considered in Sect. 4.4.

The third term in (4.39), taking into account the additional contribution to the elastic energy and associated with the change in symmetry on lattice deformation (so-called "morphic" effect), is usually neglected because of its smallness.

4.3.3 Energy Aspects of Domain Structure. Domain Walls

A rigorous theory of splitting of a finite-size ferromagnetic crystal into domains was developed by *Landau* and *Lifshitz* [4.3]. This theory includes also the symmetry aspects considered above.

Since a finite-size crystal has a demagnetizing field $\boldsymbol{H}_{\mathrm{d}}$, the crystal possesses a magnetic (magnetostatic) energy $NI^2/2$, which will reach a maximum if the crystal is a single magnetized region (a single domain), as indicated for a magnetic uniaxial crystal in Fig. 4.9a. The minimum of the magnetic energy will correspond to a state in which the crystal is divided into separate domains (Fig. 4.9b, c). Calculation taking into account the competition of the exchange forces and the demagnetizing action of the surface shows that domain size l is related to specimen size L by the approximate expression $l \sim \sqrt{L}$. To the most "favorable" domain structure of a magnetic uniaxial ferromagnetic corresponds the structure shown in Fig. 4.9c, d, which has so-called "closure domains," with magnetization perpendicular to the direction of $\boldsymbol{I}_{\mathrm{S}}$ of the principal domains. For such a structure, the magnetic flux closes inside the crystal (hence the term "closure domains"), and the magnetic energy associated with the demagnetizing field is equal to zero. The direction of

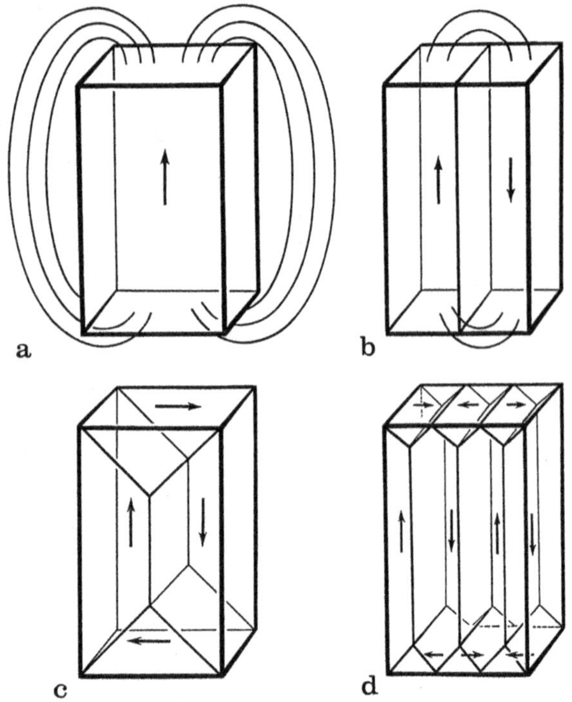

Fig. 4.9a–d. Splitting of a uniaxial crystal into domains [4.3]

I_S in such "clossure" domains does not coincide with the easy-magnetization axis.

If I_S is small or if the crystal shape does not promote the formation of closure domains, the uniaxial crystal will consist of parallel-sided domains (so-called stripe structure). When domains arise, *domain walls* form between them, which separate domains with different directions of the magnetic moments. The magnetic moments in the wall must change their direction "fanwise." The exchange energy tends to make the change in the directions of the moments take place less abruptly, because a sharp change in the direction of the moment in the wall is unfavorable as regards the exchange energy. But an excessively wide wall is unfavorable from the standpoint of the magnetic anisotropy energy. As a result, an equilibrium transition layer is established, whose width δ is determined by relation $\delta = (E_{\text{exch}}/Va)^{1/2}$, where a is of the order of the unit cell parameter, E_{exch} is the exchange energy, and V is the anisotropy energy of the unit volume.

In various ferromagnetic materials the width of the walls may fluctuate within a rather considerable range (from tens to hundreds of atomic distances). The formation of domain walls requires the spending of energy. The splitting of the crystal into domains will continue until the energy expenditure for the formation of the new boundaries exceeds the energy gain due to the reduction in magnetic energy. Thus, the equilibrium domain struc-

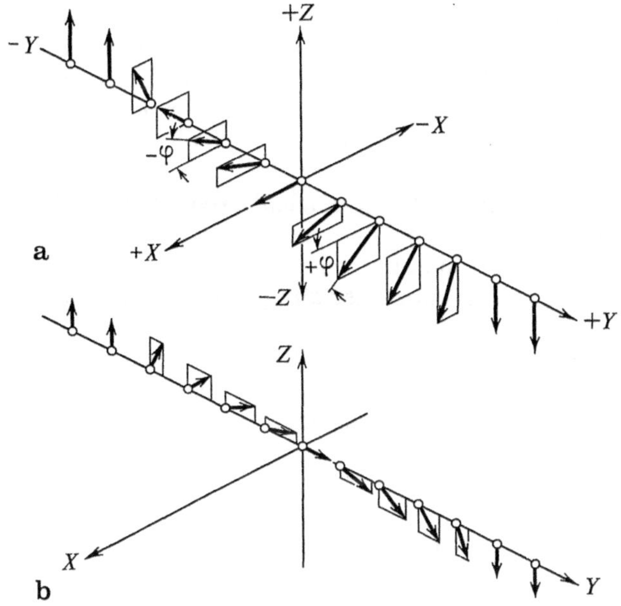

Fig. 4.10. Two types of 180° walls: Bloch wall (**a**) and Néel wall (**b**)

ture results from the competition among the various types of energy – namely, the exchange energy, the magnetostatic energy, and the magnetic anisotropy energy – and corresponds to the minimum of the free energy of the crystal, which consists of the above-listed types.

It is worth noting that the antiferromagnetics lack the energy factors favoring the division into domains associated with the magnetostatic energy ($I_S = 0$). Nevertheless, an antiferromagnetic also possesses a domain structure, whose appearance can be understood at least from symmetry considerations (Sect. 4.3.1).

A rotation of the magnetic moments in the domain wall leads to a local reduction in magnetic symmetry as compared with that of the domain.

Two extreme cases of domain boundaries – *Bloch* and *Néel* walls – are distinguished, depending on the nature of the rotation of the magnetic moment. Figure 4.10 shows models of 180° walls of these two types. In a Bloch wall (Fig. 4.10a) the magnetic moment rotates in planes parallel to the XZ plane of the boundary layer. Here the wall is not "magnetically charged" normally to its plane. In a Néel-type wall the magnetic moment rotates in the YZ plane perpendicular to that of the boundary layer; it is "charged" in the direction of the normal.

The dependence of rotation angle φ (Fig. 4.10a) on the coordinate reckoned from the center of the wall along its normal is given for a 180° Bloch wall by the expression

$$\sin\varphi = \tanh(y/\delta) \ , \tag{4.41}$$

where δ is a parameter which has the meaning of the wall width. From (4.41) it follows that the domain wall should not be represented as something having a strictly definite beginning and end; therefore, the concept of *transitional layer* suits it better. From the symmetry standpoint it is interesting to note that walls may be right and left handed, because from the energy viewpoint it makes no difference in what direction (clockwise or anticlockwise) the rotation of the magnetic moment begins. To distinguish "right-handed" from "left-handed" walls, the notion of wall polarization is sometimes introduced, which depends on the direction of the magnetic moment at the center of the domain wall. For instance, the wall shown in Fig. 4.10a is polarized along the positive direction of the X axis.

In an ideally demagnetized ferromagnetic crystal it is natural to assume the existence of an equal number of right- and left-handed domain walls.

A moving wall has inertial properties, and therefore it can be assigned some effective mass.

4.3.4 Methods for Observing the Domain Structure. Examples of Domain Structures

A domain structure can be observed by several methods. The powder pattern method is the most widespread. In this method a thin layer of suspension of a ferromagnetic powder in a liquid is applied on a thoroughly polished crystal surface. On settling, the powder collects at sites with the largest gradients of the scattering field, i.e., at the domain walls. This enables one to observe the domain walls under a microscope. The domain structure can also be observed with the aid of magneto-optic methods by using Kerr's and Faraday's effects (see Chap. 7). The former is based on the fact that a polarized light reflected from the crystal surface has different (in magnitude and direction) rotation of the plane of polarization as a result of different orientations of I_S in different domains. To observe domains in very thin films or in transparent ferromagnetics, Faraday's effect can be used. Methods for observing the domain structure with the aid of neutron and electron beams are being developed, and x-ray topography is used successfully.

Figure 4.11 demonstrates an example of a domain structure, revealed by the powder pattern method, for a hexagonal cobalt crystal in which the sixfold axis (c axis) is the easy-magnetization axis. The photo of Fig. 4.11a shows the $(10\bar{1}0)$ plane containing the c axis; the domain walls extend along the c axis with antiparallel directions of magnetization I_S (indicated by the arrows). This structure resembles the one depicted in Fig. 4.10a except that instead of the "closure domains" the photograph shows wedgelike domains with inverse (with respect to the principal domain) magnetization. The wedgelike domains are caused by the demagnetizing field. In the basal plane perpendicular to the c axis one observes a peculiar structure in the form of stars (Fig. 4.11b), formed by the extended domains "emerging" at the crystal surface. In a hexa-

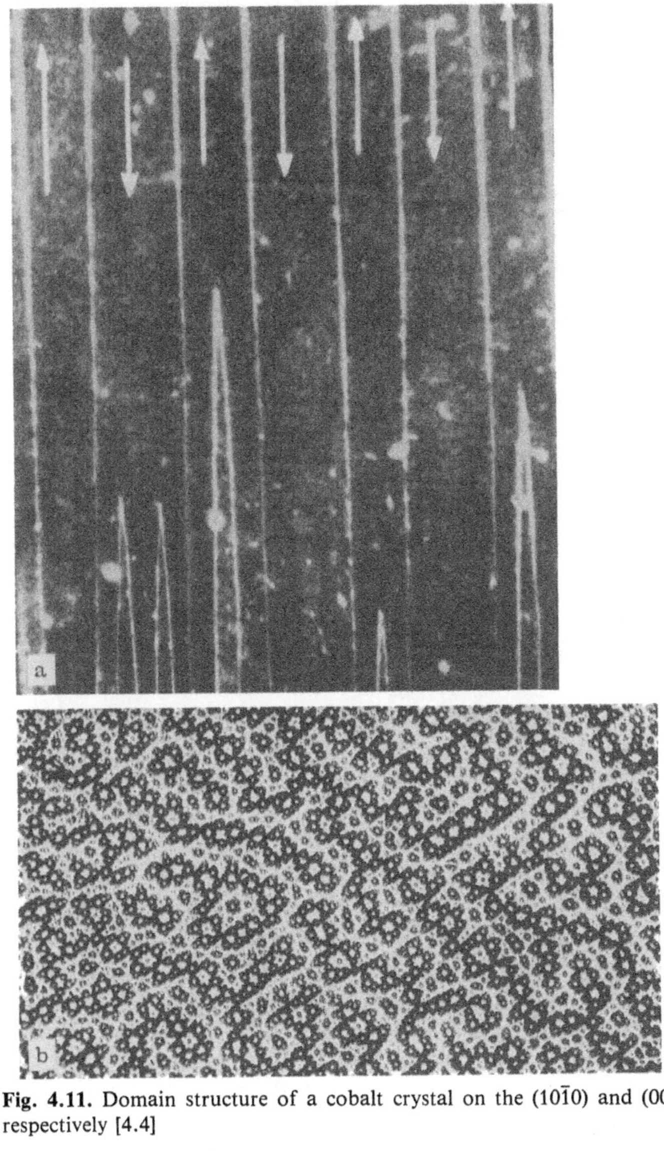

Fig. 4.11. Domain structure of a cobalt crystal on the (10$\bar{1}$0) and (0001) planes, (**a**) and (**b**), respectively [4.4]

gonal crystal, not only the c axis can serve as the easy-magnetization axis. Crystals are known in which the easy magnetization directions lie in the basal plane, forming the *easy-magnetization plane* (when all the directions for I_S in the basal plane are almost equivalent energetically), or form some angle with the c axis (*easy-magnetization cone*). A more complex domain structure will be observed in such crystals.

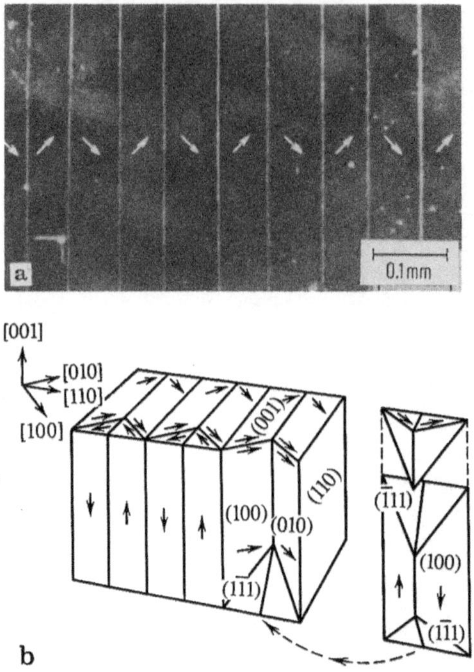

Fig. 4.12. Domain structure on the (001) surface of a silicon-iron crystal (**a**) and its interpretation (**b**) [4.4]

In cubic crystals, either tetragonal ⟨100⟩ or trigonal ⟨111⟩ axes usually serve as easy-magnetization axes. Figure 4.12a is an example of a domain structure on the (001) surface of a silicon-iron crystal, for which the ⟨100⟩ axes are easy-magnetization axes. The horizontal direction on the photo coincides with the [110] axis. This domain structure is explained in Fig. 4.12b. The directions of $\boldsymbol{I}_S$ in adjacent domains on the photo form angles of 90° and alternate from [100] to [010]. Figure 4.12b also shows the closure domains, although they are not seen on the photo. As noted above, in the general case 180°, ∼09°, ∼71°, and 90° domain walls can exist in neighboring domains of a cubic crystal, depending on the type of easy-magnetization axes (⟨100⟩ or ⟨111⟩) and hence on the angles between the vectors of $\boldsymbol{I}_S$. In Fig. 4.12b for instance, one can see 180° and 90° walls. In the above-discussed case of a uniaxial hexagonal cobalt crystal there are only 180° walls. If, however, the domains in a hexagonal crystal are arranged along twofold axes in the basal planes, domain walls separating domains with angles of 30°, 60°, 90°, 120°, and 180° between $\boldsymbol{I}_S$ can exist.

Figure 4.13 portrays the domain structure of a thin (0.05 mm) platelet of an $YFeO_3$ (yttrium orthoferrite) crystal revealed with the aid of the Faraday effect. This orthorhombic magnetouniaxial crystal exhibits a so-called strip (labyrinthine) domain structure with walls parallel to the easy-magnetization axis (it is perpendicular to the plane of the photo and coincides with one of the twofold axes). It can be seen that some domains have a closed cylindrical

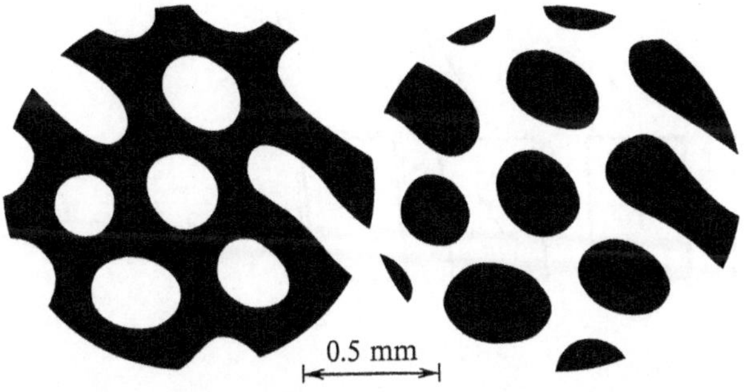

Fig. 4.13. Ferromagnetic domains in a $YFeO_3$ crystal observed with the aid of Faraday's effect. The contrast is reversed by a turn of the microscope analyzer [4.4]

shape. Such separate cylindrical or bubble domains can also be created in crystal platelets of some other ferrites (for instance, in garnets). The absence or presence of a cylindrical domain at a given place of a crystal can be used as two stable states. Magnetic materials possessing such a domain structure acquire an ever-increasing importance as elements of integral circuits in computer technology. Single-crystal platelets and films of magnetouniaxial ferrites may contain up to $10^6 - 10^7$ bits (mobile cylindrical domains) capable of moving controllably in weak gradient fields. Techniques have already been developed which make it possible to move, generate, destroy, and record cylindrical domains reliably, i.e., to perfom all the necessary operations over the information contained in a crystal.

The character of a domain structure is affected by a great number of factors associated with the structure of a real crystal. In addition to such factors as the shape and size of the crystal, the specific features of the magnetic anisotropy energy, and the crystallographic orientation of the crystal surface, the domain structure and particularly the movement of domain walls are substantially affected by the internal stresses, the crystal lattice defects, and nonuniformity (inclusions, voids, and dislocations).

4.3.5 Magnetization Processes

In the absence of an external field, the most energetically favorable state of a ferromagnetic is when domains with different directions of I_S occupy equal portions of the crystal volume, as shown in Fig. 4.7a. Such a domain distribution corresponds to the demagnetized state in which the crystal as a whole does not show magnetization.

Let us now consider the changes occurring in a crystal on application of an external field. Magnetization of a polydomain ferromagnetic consists essentially of two processes: the displacement of the domain walls and the rotation of vector I_S of the domains. We have encountered wall displacement in ferro-

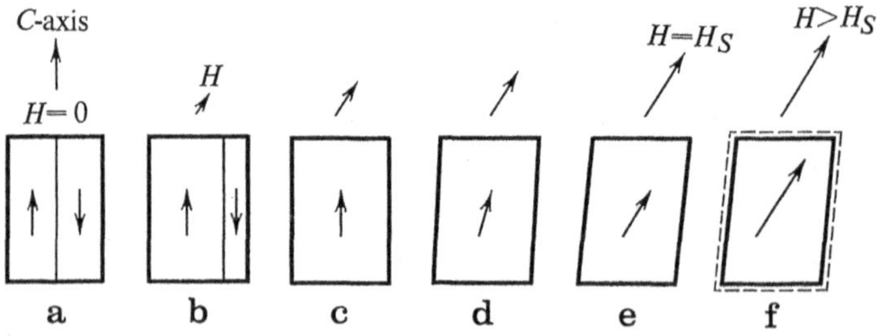

Fig. 4.14a – f. Magnetization processes

electrics, but rotation is characteristic only of ferromagnetics; in ferroelectrics the spontaneous polarization vector cannot deviate from a definite symmetry axis of the crystal. Displacement and rotation are illustrated in Fig. 4.14 which shows a magnetouniaxial (for instance, hexagonal) crystal with a c axis which is the easy-magnetization axis.

Suppose the magnetizing field is directed at some arbitrary angle to the c axis. This corresponds to the most general case, when magnetization I_S in domains and domain walls makes angles different from 0 or $\pi/2$ with the external field. In weak fields, wall displacement predominates (Fig. 4.14b): the volume of the domains, whose vectors I_S make the least angle with the direction of the magnetic field, increases at the expense of the neighboring domains, and the direction of I_S remains unaltered, only the position of the domain walls changing (the wall shifts and disappears, Fig. 4.14c). In stronger fields vector I_S (Fig. 4.14d) turns towards the direction of the field, and at a certain field H_S saturation takes place, when the crystal magnetization in the direction of the field becomes equal to the magnetization of a separate domain (i.e., I_S, Fig. 4.14e). With a further field increase, I_S does not change its direction, but increases slightly in absolute value (Fig. 4.14f), tending, as $H \to \infty$, to absolute saturation corresponding to I_S at $T = 0$, when the disordering effect of the temperature on the spin orientation is absent. Thus, this additional increase in magnetization, which is called *true magnetization*, is associated with the overcoming of the disordering action of the thermal motion. The traditional term for this process in the Russian literature is the *paraprocess*, because this phenomenon is similar to the magnetization of a paramagnetic, which has been considered in Sect. 4.1.

We have discussed magnetization processes for the case of a uniaxial crystal, when the external field is directed at an arbitrary angle to the easy-magnetization axis. In the general case we can only speak of the preponderance of some magnetization mechanism or other (displacement, rotation, or paraprocess) in definite fields, since these mechanisms are often superimposed upon each other. In cubic (magnetomultiaxial) ferromagnetics, magnetization processes are always superimposed, because they exhibit a complicated distribution of angles between the external field and domain magnetization.

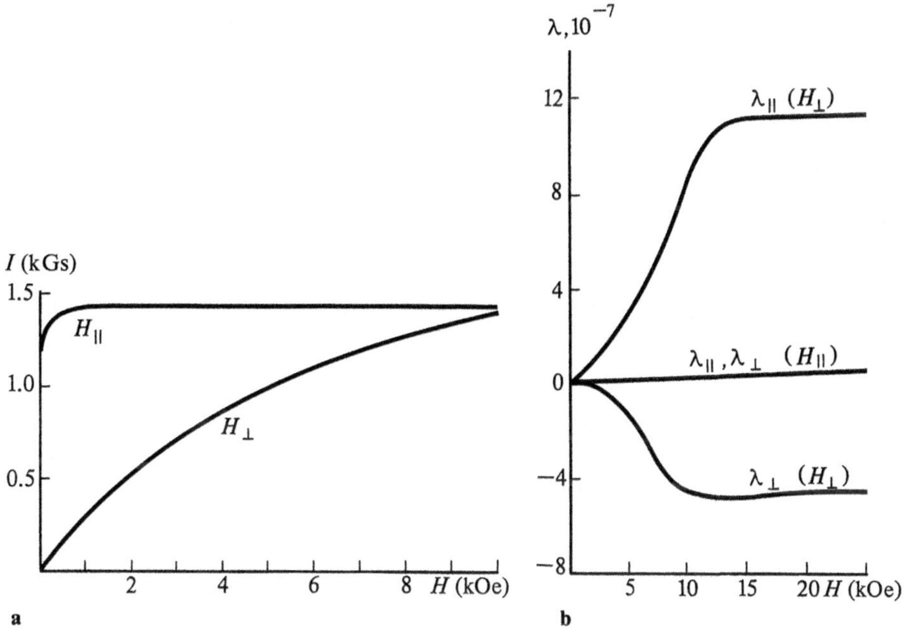

Fig. 4.15. Magnetization (**a**) [4.5] and magnetostriction (**b**) [4.6] curves for a single crystal of cobalt at room temperature: $H_{\parallel}$ external field is parallel to the c axis; $H_{\perp}$ perpendicular to the c axis; $\lambda_{\parallel}$ magnetostriction along the c axis; $\lambda_{\perp}$ magnetostriction perpendicular to the c axis

It is, however, possible to choose directions of the external field for a uniaxial crystal in such a way that magnetization will be due either to rotation alone or to wall displacement. For instance, if we magnetize a uniaxial ferromagnetic crystal, shown in Fig. 4.14a along the "easy" c axis, such a crystal will be magnetized in comparatively low fields to saturation exclusively by displacement of 180° domain walls, which will correspond to the curve $\boldsymbol{H} \| \boldsymbol{c}$ in Fig. 4.15a demonstrating magnetization curves (i.e., the I-H dependence) for a cobalt single crystal. The displacement of 180° walls is not accompanied by deformation (magnetostriction) of the crystal, because the spontaneous deformation of the domain will remain unaltered after the reversal of $\boldsymbol{I}_S$.

If a crystal is magnetized perpendicular to the easy-magnetization direction, the magnetization process (curve $\boldsymbol{H} \perp \boldsymbol{c}$ in Fig. 4.15a) will be due exclusively to the rotation of vectors $\boldsymbol{I}_S$ because the directions of $\boldsymbol{I}_S$ in the domains make equal 90° angles with the field. The magnetic anisotropy forces will tend to "keep" $\boldsymbol{I}_S$ in the direction of the c axis, and the forces of interaction with the external field, to turn $\boldsymbol{I}_S$ in the direction of the field. The total energy (per 1 cm^3) will then be equal to

$$F = V - \boldsymbol{H}\boldsymbol{I}_S \cos(\boldsymbol{H}, \boldsymbol{I}_S) = V - \boldsymbol{H}\boldsymbol{I}_S \sin\theta \ , \qquad (4.42)$$

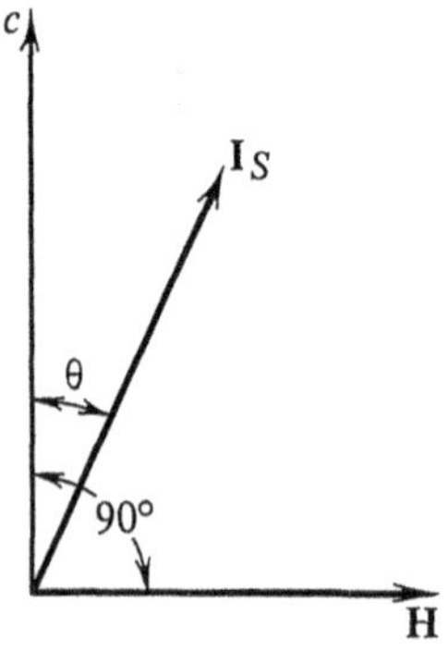

Fig. 4.16. Calculation of the magnetization curve of a uniaxial crystal perpendicular to the easy-magnetization axis (c axis)

where V is the anisotropy energy defined by (4.23). The mutual orientation of I_S and $\boldsymbol{H}$ relative to the c axis are shown in Fig. 4.16. From the equilibrium conditions $\partial F/\partial\theta = 0$ and, taking into account only the first ansiotropy constant in (4.23), we find

$$\sin\theta_0 = HI_S/2K_1 \ . \tag{4.43}$$

The susceptibility in the direction $\boldsymbol{H}\perp\boldsymbol{c}$ is equal to the component of magnetization I_S along $\boldsymbol{H}$ divided by by H:

$$\varkappa(\perp c) = I_S \sin\theta_0/H \ .$$

From (4.43) we find that the susceptibility

$$\varkappa = I_S^2/2K_1 \tag{4.44}$$

is a constant value, i.e., the magnetization curve of the crystal at $\boldsymbol{H}\perp\boldsymbol{c}$ must be a straight line up to saturation $\theta_0 = \frac{\pi}{2}$, which is achieved in the field

$$H = 2K_1/I_S \ , \tag{4.45}$$

and which is called anisotropy field H_a. The deviation of the magnetization curve for Co at $\boldsymbol{H}\perp\boldsymbol{c}$ from the straight line (Fig. 4.15a) is due to the effect of constant K_2 (4.23).

The rotation of I_S is associated with a change in crystal symmetry (Table 4.3) and the appearance of magnetostrictive deformations, which is schematized in Fig. 4.14d, e. From the magnetization curves in Fig. 4.15a it follows that the terms "easy-magnetization axis" and "hard-magnetization axis" efficiently express the essence of the matter: at $\boldsymbol{H}\|\boldsymbol{c}$ the magnetization of the crystal is "easy," and at $\boldsymbol{H}\perp\boldsymbol{c}$, "hard." From the area between the curves $\boldsymbol{H}\|\boldsymbol{c}$ and $\boldsymbol{H}\perp\boldsymbol{c}$ one can estimate the magnetic anisotropy energy.

If the field makes an angle different from zero or from $\pi/2$ with the c axis (the general case discussed by us at the beginning of this section), the magnetization curve will occupy an intermediate position between $\boldsymbol{H} \| \boldsymbol{c}$ and $\boldsymbol{H} \perp \boldsymbol{c}$ in Fig. 4.15 a.

Figure 4.15 b shows so-called magnetostriction curves $\lambda = \Delta l/l$ (the relative change in linear dimensions) for magnetization $\boldsymbol{H} \| \boldsymbol{c}$ and $\boldsymbol{H} \perp \boldsymbol{c}$ of cobalt crystals [in the experiment a strain gauge was glued to the crystal surface, and the deformation either along the c axis ($\lambda_{\|}$) or perpendicular to it (λ_l) was measured]. As indicated above, when a crystal is magnetized parallel to the c axis, the magnetostriction along any direction is practically equal to zero: only a slight magnetostriction, depending linearly on the field and due to the paraprocess, is observed. In the case of $\boldsymbol{H} \perp \boldsymbol{c}$ magnetization the crystal deforms because of the rotation of $\boldsymbol{I}_S$; the magnetostriction measured along c axis and perpendicular to it has a different sign and magnitude. In fields of the order of 1200 Oe the magnetostriction reaches saturation. A further field increase only causes magnetostriction of the paraprocess, and curves $\lambda(H)$ at $H > 12000$ Oe have the same slope as $\boldsymbol{H} \| \boldsymbol{c}$, when the magnetostriction is due exclusively to the paraprocess. This last circumstance indicates that the magnetostriction of the paraprocess is of a "volume" isotropic nature, which is shown schematically in Fig. 4.14 f (see Note on p. 294).

4.3.6 Magnetic Hysteresis

It is well known that if a ferromagnetic material is magnetized in an external field changing from some positive value $+H$ to a negative one $-H$ and back, the dependence of magnetization I on H [or on magnetic induction $B = H + 4\pi I$, see (4.2)] will follow a so-called *hysteresis loop*. As an example, Fig. 4.17 shows a hysteresis loop for molybdenum permalloy (an alloy of 79 wt.% Ni, 17% Fe, and 4% Mo with a cubic crystal structure subjected to special heat treatment).

Figure 4.17 exhibits a family of hysteresis loops corresponding to different values of the maximum external magnetic field. With an increase in external field amplitude the loop width increases and its shape changes. But after a definite value of the field is achieved the loop ceases to expand (this state is almost achieved for the loop shown in Fig. 4.17). The loops inside such a major loop are called *minor cycles*. The locus of the apices of the minor loops is called the *virgin magnetization curve*. The major hysteresis loop is used to determine the basic characteristics of the ferromagnetic material: *saturation magnetization* (or saturation induction B_S), remanent magnetization I_r (a romanent induction R_r), coercive field (force) H_c[3] (i.e., the field necessary for reducing the remanent magnetization to zero).

[3] The coercive field, determined from the induction loop, does not coincide with H_c, which is determined from the hysteresis loop for magnetization I, because $B = H + 4\pi I = 0$, $I = -H/4\pi \neq 0$. Nevertheless, this difference is substantial only for materials with a high H_c.

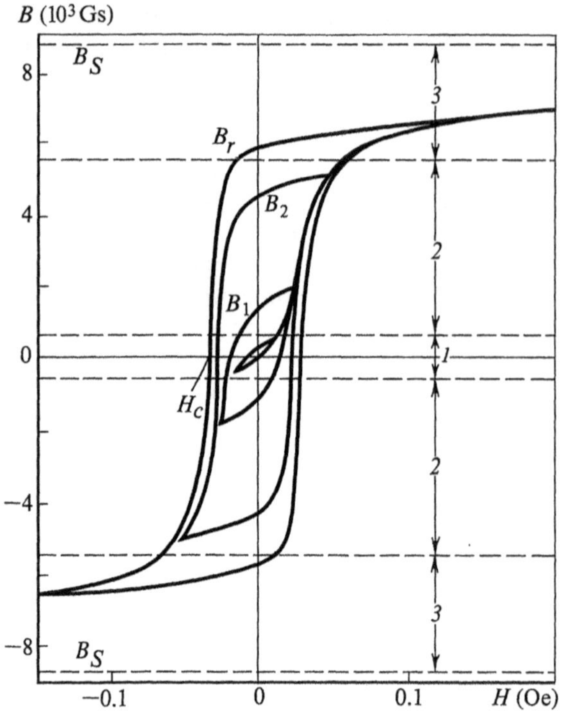

Fig. 4.17. Hysteresis loops of molybdenum permalloy. Portions *1*, *2*, and *3* correspond to the different magnetization processes [4.7]

Hysteresis is an extremely structure-sensitive property. Various types of magnetic hysteresis due to various mechanisms are observed in ferromagnetic materials. Three basic mechanisms can be singled out among them: (a) hysteresis due to the delay in the displacement of domain walls (irreversible displacement), (b) hysteresis due to the delay in the growth of remagnetization nuclei, and (c) hysteresis due to irreversible rotation.

The hysteresis due to irreversible wall shift is typical of inhomogeneous magnetic materials, where the moving wall can be delayed on various inhomogeneities (elastic stress gradients, inclusions, dislocations, etc.). The effect of irreversible displacement processes on the nature of hysteresis can be illustrated on the loop family of Fig. 4.17. In the weak field region (Region 1) one mainly observes and elastic reversible displacement of the walls, which does not result in a noticeable hysteresis during the change in the magnitude and direction of the magnetic field (when the field is switched off the wall resumes its original position). Portion 1 corresponds to the region of so-called *initial permability* $\mu_{\text{unit}} = B/H$.

In stronger fields corresponding to the steep portion of the magnetization curve (Region 2) the irreversible shift of the walls is the determining process: when the field is switched off the wall is "fixed" in the new position (from the energy standpoint it "finds" its new potential well). If the boundary was on the "enery hump" at a certain field strength, it may move spontaneously to a new potential well without any further field increase, which will lead to a so-called irreversible *Barkhausen jump*. To Portion 2, in which Barkhausen jumps occur, corresponds the *maximum permeability* μ_{max}. Because of the irreversibility of the processes, when the field in Region 2 is switched off, the domain structure will not resume its initial state, but will move to a state with some remanent induction (points B_1 and B_2 in Fig. 4.17).

In higher fields close to saturation (Region 3), magnetization processes are due to the rotation of vectors $\boldsymbol{I}_S$. If the magnetic field is reduced to zero after saturation, the magnetization vectors of the domains will rotate from the direction of the field to the closest easy-magnetization direction (in the initial state the vectors were equiprobably distributed among all the easy-magnetization directions), which will lead to a state with remanent induction B_r corresponding to the major hysteresis loop. The state with remanent magnetization is depicted schematically in Fig. 4.7c.

The next hysteresis mechanism – the delay in the growth of remagnetization nuclei – takes place to some extent in all ferromagnetics, but manifests itself in its purest form in uniaxial crystals. Let us consider the crystal whose magnetization curves are given in Fig. 4.15a. When $\boldsymbol{H} \| \boldsymbol{c}$, the main role in the remagnetization of such a crystal is played by *remagnetization nuclei* similar to wedgelike domains of inverted magnetization, which are visible on the surface of a cobalt crystal in Fig. 4.11. With a change in the sign of the field and its further increase, at a certain value of the magnetic field the nuclei begin to grow rapidly, absorbing the whole volume of the crystal by displacement of the walls. The hysteresis loop in such uniaxial crystals must have a rectangular shape at $\boldsymbol{H} \| \boldsymbol{c}$, as in Fig. 4.18. The magnitude of the coercive force, which is equal to the value of the field H_c determining the beginning of remagnetization, depends on the nucleation of "inverted" domains and on the delays in the displacement of domain boundaries due to various distortions of the crystal lattice (inclusions, internal stresses, and dislocations). The origin of the remagnetization nuclei is not sufficiently understood as yet.

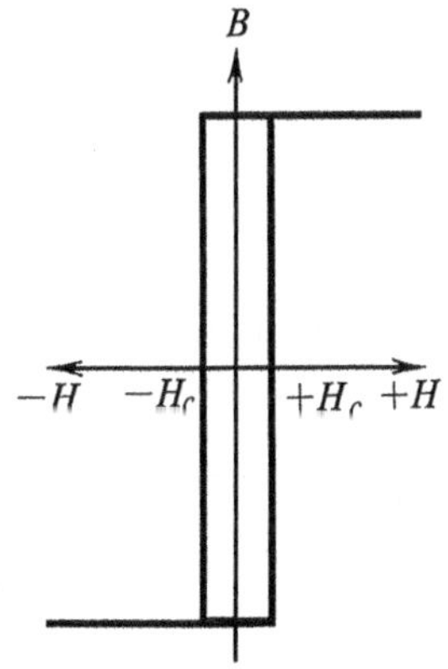

Fig. 4.18. Rectangular hysteresis loop

The third cause for hysteresis – irreversible rotation processes – is almost never encountered in large crystals. Such a hysteresis can occur in heterophase ferromagnetic materials, thin films, or fine particles of a ferromagnetic powder, where each particle is a single domain, this precluding irreversible displacement of the walls and the appearance of remagnetization nuclei. In the case of single crystals irreversible rotation can manifest itself as hysteresis during magnetization in rotating magnetic fields.

In conclusion, we note that the energy spent on remagnetization of a ferromagnetic is converted to heat. A ferromagnetic subjected to cyclical remagnetization, such as a transformer core, is heated not only due to so-called eddy currents, but also hysteresis losses: the higher the latter, the larger the area of the hysteresis loop. Therefore, materials with a high magnetic permeability (steep magnetization curve) and a narrow hysteresis loop (low coercive force) are used for work in alternating fields. Such materials are called *magnetically soft*. An example of such materials is molybdenum permalloy, whose hysteresis loop is shwon in Fig. 4.17. The coercive force of this material is of the order of 0.05 Oe, as follows from Fig. 4.17. In contrast to magnetically soft materials, *magnetically hard* materials have high values of the coercive force (several hundred oersted) and remanence. In technology, these materials are used as permanent magnets. The hysteresis loop of magnetically hard materials has a large area.

Classical magnetically hard materials include some grades of steels and alloys based on systems Ni-Al-Fe (alni), Ni-Al-Co-Fe (alnico), Co-V-Fe (vicalloy), Cu-Ni-Fe (cunife), Cu-Ni-Co (cunico), and Fe-Co-Ni-Al (ticonal). Permanent magnets are also made of some highly coercive ferrites. In recent years the optimum permanent magnet characteristics have been found in intermetallic compounds of rare-earth metals with transition metals of the iron group, mainly with cobalt (for instance, the $SmCo_5$ alloy). To convey an idea of the power of such a magnet, it will suffice to mention that an $SmCo_5$ magnet from a single-crystal sphere, 3 mm in diameter, can hold a load about 1000 times its own weight.

4.4 Anisotropy of Ferromagnetic Crystals

4.4.1 Peculiarities of the Tensor Description of Ferromagnetic Crystals

If follows from Sect. 4.3 that as vector $\boldsymbol{I}_S$ rotates in a ferromagnetic crystal (which may result, for instance, from rotation of the crystal in a sufficiently strong magnetic field), the symmetry of the crystal will vary, "following" the direction of $\boldsymbol{I}_S$. The crystal will resemble a rolling liquid drop. This peculiarity causes a number of difficulties in describing the anisotropic properties of ferromagnetic crystals. Probably because of this, the anisotropy of ferromagnetic crystals has not been considered in crystal physics courses until now.

It is shown in Chap. 1 that the relationship between two vector properties $\boldsymbol{A}$ and $\boldsymbol{B}$ in an anisotropic medium has the tensor form

$$A_i = \varepsilon_{ik} B_k \ , \tag{4.46}$$

and the type of tensor ε_{ik} depends on the class of symmetry of the crystal. From the change in the symmetry of a ferromagnetic crystal on rotation of vector $\boldsymbol{I}_S$ it follows that in such a case the type of tensor ε_{ik} must depend on the direction of vector $\boldsymbol{I}_S$. When vector $\boldsymbol{I}_S$ is directed arbitrarily with respect to the symmetry axes, one must take into account all the six components of tensor ε_{ik} (the properties for which condition $\varepsilon_{ik} = \varepsilon_{ki}$ is fulfilled are to be considered), because the symmetry of the crystal reduces to triclinic (Table 4.3). The effect of the symmetry on the type of tensor ε_{ik}, due to the change in the direction of $\boldsymbol{I}_S$, is taken into account by introducing the dependence of components of ε_{ik} on components I_1, I_2, and I_3 of vector $\boldsymbol{I}_S$. The crystallophysical axes of the initial paramagnetic crystal ($\boldsymbol{I}_S = 0$) are chosen as the coordinate axes X_1, X_2, and X_3.

Considering the smallness of the lattice distortions on change in the direction of $\boldsymbol{I}_S$, the dependence of components of ε_{ik} on components of $\boldsymbol{I}_S$ can be represented as the following tensor series:

$$\varepsilon_{ik} = N_{iklm} I_l I_m + N_{iklmno} I_l I_m I_n I_o + \dots \ , \tag{4.47}$$

where *i, k, l, m, n,* and *o* take values 1, 2, and 3. Expansion is carried out in even powers, because, as in the case of magnetic anisotropy, we consider tensor properties invariant to operation R. The properties of crystals described by tensor ε_{ik} are among the so-called *even effects* unaffected by the reversal of the direction of $\boldsymbol{I}_S$. At a certain constant temperature components I_1, I_2, and I_3 can be expressed in terms of direction cosines α_1, α_2, and α_3 with respect to axes X_1, X_2, and X_3:

$$I_1 = I_S \alpha_1 \ , \quad I_2 = I_S \alpha_2 \ , \quad I_3 = I_S \alpha_3 \ .$$

We shall now see which particular properties can be characterized by tensors ε_{ik}.

Suppose ε_{ik} in (4.46, 47) have the meaning of the deformation tensor components which characterize the change in length and position of radius vector $\boldsymbol{B}_k$ drawn to a certain point inside the crystal. Then (4.47) is nothing other than the spontaneous magnetostrictive deformation on change in the direction (rotation) of $\boldsymbol{I}_S$ in the crystal. The dependence (4.47) for magnetostrictive deformations can also be obtained by differentiating magnetoelastic energy (4.38) with respect to the stress components: $\varepsilon_{ij} = \partial E_{me}/\partial \sigma_{ij}$. If we now assume that (4.46) expresses the relationship between the electric field and current vectors, then ε_{ik} will be the components of the resistivity tensor. In this case (4.47) describes the change in electrical resistance during rotation of $\boldsymbol{I}_S$.

Similarly, if we have in mind the relationship between the temperature gradient and the arising strength of the thermal-emf field, then ε_{ik} in (4.47) will characterize the so-called even thermomagnetic effect, i.e., the change in thermal emf during rotation of $\boldsymbol{I}_S$. In the last two cases we can, by analogy with the term "spontaneous magnetostrictive deformation," speak of "spontaneous change in electrical resistance" and "spontaneous change in thermal emf." The formal description of all the effects listed is similar. The anisotropy of the above-mentioned even effects can be traced most vividly for the case of magnetostriction.

To obtain the expression for the spontaneous magnetostriction (we denote it by ε_S) measured along a certain direction in the crystal, we change the coordinate system so that the new axis X_3' coincides with the direction of interest. Then, according in Chap. 1, the deformation concerned will be equal to

$$\varepsilon_S = \varepsilon_{33}' = \sum_i \sum_j \varepsilon_{ik} \beta_l \beta_m = \beta_1^2 \varepsilon_{11} + \beta_2^2 \varepsilon_{22} + \beta_3^2 \varepsilon_{33} + 2\beta_2 \beta_3 \varepsilon_{23} + 2\beta_3 \beta_1 \varepsilon_{13} + 2\beta_1 \beta_2 \varepsilon_{12} \,, \quad (4.48)$$

where β_1, β_2, and β_3 are the cosines of the angles between new axis X_3' and old ones X_1, X_2, and X_3, i.e., the direction cosines of magnetostriction measurement direction relative to the crystallophysical axes of the crystals.

By determining the dependence of components of ε_{ik} on the magnetization direction for a crystal of a particular symmetry with the aid of (4.47) and substituting the resulting expression into (4.48), we can obtain the so-called anisotropy formula, which will characterize the spontaneous magnetostriction (or any other of the even effects listed above) in an arbitrary direction (β_1, β_2, and β_3) at an arbitrary direction of magnetization (α_1, α_2, and α_3). In the next section we shall consider the properties of tensors N_{iklm} and N_{iklmno} and obtain expressions for the magnetostriction anisotropy of ferromagnetic crystals of different symmetry.

We shall, however, emphasize one feature distinct from the other even effects, a feature which should be borne in mind when deriving the magnetostriction formulas according to the scheme under consideration.

The dependence of ε_{ik} on the direction of $\boldsymbol{I}_S$ in the form (4.47) assumes that the crystal deforms freely due to magnetostriction. Actually, free deformation is hindered by the crystal elasticity. The equilibrium deformation components ε_{ik}^* are found from the condition

$$\frac{\partial(U+V')}{\partial \varepsilon_{ij}} = 0 \,, \quad (4.49)$$

where $U = \frac{1}{2} c_{ijkl} \varepsilon_{ij} \varepsilon_{kl}$ is the elastic energy (Chap. 2), and V' is the free energy in the form of (4.39) neglecting the morphic effect (Sect. 4.3.2). Differentiating (4.39), we find

$$-c_{ijkl}\,\varepsilon^*_{kl} = V^0_{ij} \ , \tag{4.50}$$

where function V^0_{ij} is defined in (4.37).

Replacing the stiffness coefficients c_{ijkl} by compliances s_{ijkl}, we can write (4.50) as

$$\varepsilon^*_{ij} = -s_{ijkl}\,V^0_{kl} \ . \tag{4.51}$$

Comparing (4.51) with Hooke's law (Chap. 2), we can see that function V^0_{ij} can be given the meaning of magnetostrictive stresses offset by the elasticity forces, which produce some equilibrium values of deformation ε^*_{ij}. Thus, coefficients N in (4.47) should, strictly speaking, be replaced by the products of the respective components of s by M from (4.37). This, however, will not change the general form of the functional dependence of components of ε_{ij} on components of $\boldsymbol{I}_S$. Therefore, we shall not take into account the elastic energy to avoid the loss of generality in derivation of anisotropy formulas for all the even effects when considering the magnetostriction of crystals of different syngonies.

4.4.2 Magnetostriction Anisotropy in Ferromagnetics of Different Symmetry

Although expression (4.47) is written for a crystal in the ferromagnetic state, tensors N_{iklm} and N_{iklmno} must satisfy the conditions of symmetry of the "initial" (paramagnetic) phase of the crystal, in which spontaneous magnetization arises. The reason is that the spontaneous magnetostrictive deformations leading to a reduction in crystal symmetry can be considered small. Owing to the intrinsic symmetry (symmetry with respect to the suffixes) of tensors N_{iklm} and N_{iklmno}, they can be written in matrix form $N_{\alpha\beta}$, and $N_{\alpha\beta\gamma}$, where α, β, and γ take the values 1, 2, 3, 4, 5, and 6 and are related to the suffixes of the tensor components as follows: 1 = 11, 2 = 22, 3 = 33, 4 = 23, 5 = 13, and 6 = 12 (Chap. 1). Tensor $N_{iklm} = N_{\alpha\beta}$ is similar in symmetry to the tensors of the electrostriction (Chap. 3) and piezo-optical (Chap. 7) coefficients. Tensor $N_{\alpha\beta}$ differs from the elastic-constants tensor as it does not admit mutual permutation of suffixes α and β and therefore generally has 36 independent components, and not 21 components as does the elastic-constants tensor (Chap. 2).

Sixth-rank tensor $N_{\alpha\beta\gamma}$ generally has 90 independent components. Detailed derivation of the type of tensor for crystals of different symmetry would take up much space; therefore, we shall give the final matrices, whose derivation the reader can find in [4.8, 9]

a. Cubic Crystal

The matrix of tensor $N_{iklm} = N_{\alpha\beta}$ for a crystal of cubic symmetry $m3m$ (nickel, iron, and ferrite spinels) has the form [4.7]:

$$N_{\alpha\beta}=\begin{Vmatrix} N_{11} & N_{12} & N_{12} & 0 & 0 & 0 \\ N_{12} & N_{11} & N_{12} & 0 & 0 & 0 \\ N_{12} & N_{12} & N_{11} & 0 & 0 & 0 \\ 0 & 0 & 0 & N_{44} & 0 & 0 \\ 0 & 0 & 0 & 0 & N_{44} & 0 \\ 0 & 0 & 0 & 0 & 0 & N_{44} \end{Vmatrix} . \tag{4.52}$$

When calculating ε_{ik} according to (4.47) the sought-for components ε_{11}, ε_{22}, ε_{33}, ε_{23}, ε_{13}, and ε_{12} must be placed to the left of the matrix (4.52) (vertically downward) in accordance with the number of the line, and the magnetization components $I_1 I_1 = I_S^2 \alpha_1^2$, $I_2 I_2 = I_S^2 \alpha_2^2$, $I_3 I_3 = I_S^2 \alpha_3^2$, $I_2 I_3 = I_3 I_2 = I_s^2 \alpha_2 \alpha_3$, $I_1 I_3 = I_3 I_1 = I_S^2 \alpha_1 \alpha_3$ and $I_1 I_2 = I_2 I_1 = I_S^2 \alpha_1 \alpha_2$ must be placed above (horizontally from left to right).

The matrix of the sixth-rank tensor $N_{iklmno} = N_{\alpha\beta\gamma}$ for crystals of the same symmetry can conveniently be written as follows [4.8]:

$$N_{\alpha\beta\gamma}=\begin{Vmatrix}
\alpha\beta 1 = \begin{Vmatrix} N_{111} & N_{112} & N_{112} & 0 & 0 & 0 \\ N_{112} & N_{112} & N_{123} & 0 & 0 & 0 \\ N_{112} & N_{123} & N_{112} & 0 & 0 & 0 \\ 0 & 0 & 0 & N_{441} & 0 & 0 \\ 0 & 0 & 0 & 0 & N_{551} & 0 \\ 0 & 0 & 0 & 0 & 0 & N_{551} \end{Vmatrix} & \alpha\beta 2 = \begin{Vmatrix} N_{112} & N_{112} & N_{123} & 0 & 0 & 0 \\ N_{112} & N_{111} & N_{112} & 0 & 0 & 0 \\ N_{123} & N_{112} & N_{112} & 0 & 0 & 0 \\ 0 & 0 & 0 & N_{551} & 0 & 0 \\ 0 & 0 & 0 & 0 & N_{441} & 0 \\ 0 & 0 & 0 & 0 & 0 & N_{551} \end{Vmatrix} \\
\alpha\beta 3 = \begin{Vmatrix} N_{112} & N_{123} & N_{112} & 0 & 0 & 0 \\ N_{123} & N_{112} & N_{112} & 0 & 0 & 0 \\ N_{112} & N_{112} & N_{111} & 0 & 0 & 0 \\ 0 & 0 & 0 & N_{551} & 0 & 0 \\ 0 & 0 & 0 & 0 & N_{551} & 0 \\ 0 & 0 & 0 & 0 & 0 & N_{441} \end{Vmatrix} & \alpha\beta 4 = \begin{Vmatrix} 0 & 0 & 0 & N_{123} & 0 & 0 \\ 0 & 0 & 0 & N_{112} & 0 & 0 \\ 0 & 0 & 0 & N_{112} & 0 & 0 \\ N_{123} & N_{112} & N_{112} & 0 & 0 & 0 \\ 0 & 0 & 0 & 0 & 0 & N_{456} \\ 0 & 0 & 0 & 0 & N_{456} & 0 \end{Vmatrix} \\
\alpha\beta 5 = \begin{Vmatrix} 0 & 0 & 0 & 0 & N_{112} & 0 \\ 0 & 0 & 0 & 0 & N_{123} & 0 \\ 0 & 0 & 0 & 0 & N_{112} & 0 \\ 0 & 0 & 0 & 0 & 0 & N_{456} \\ N_{112} & N_{123} & N_{112} & 0 & 0 & 0 \\ 0 & 0 & 0 & N_{456} & 0 & 0 \end{Vmatrix} & \alpha\beta 6 = \begin{Vmatrix} 0 & 0 & 0 & 0 & 0 & N_{112} \\ 0 & 0 & 0 & 0 & 0 & N_{112} \\ 0 & 0 & 0 & 0 & 0 & N_{123} \\ 0 & 0 & 0 & 0 & N_{456} & 0 \\ 0 & 0 & 0 & N_{456} & 0 & 0 \\ N_{112} & N_{112} & N_{123} & 0 & 0 & 0 \end{Vmatrix}
\end{Vmatrix} . \tag{4.53}$$

Components of ε_{ik} for each of the small matrices are calculated as in the case of the fourth-rank tensor, the only difference being that each of the coefficients in the small matrices is multiplied additionally by the respective components $I_n I_o = I_\gamma$ in accordance with the third suffix γ.

As can be seen from (4.52, 53, tensors N_{iklm} and N_{iklmno} for a cubic crystal of the *m*3*m* class, which generally had 36 and 90 independent components, respectively, now retain only 3 and 6 components: N_{11}, N_{12} N_{44} and N_{111},

$N_{112}, N_{123}, N_{441}, N_{551}, N_{546}$. The components of ε_{ik} calculated with the aid of (4.52, 53) are equal to

$$\begin{aligned}
\varepsilon_{11} &= N_{12} I_S^2 + (N_{11}-N_{12})\,\alpha_1^2 I_S^2 + N_{112}(\alpha_2^4+\alpha_3^4)\, I_S^4 + 3N_{123}\,\alpha_2^2\alpha_3^2 I_S^4 \\
&\quad + 3N_{112}(\alpha_3^2\alpha_1^2+\alpha_1^2\alpha_2^2)\, I_S^4 ,\\
\varepsilon_{22} &= N_{12} I_S^2 + (N_{11}-N_{12})\,\alpha_2^2 I_S^2 + N_{112}(\alpha_3^4+\alpha_1^4)\, I_S^4 + 3N_{123}\,\alpha_3^2\alpha_1^2 I_S^4 \\
&\quad + 3N_{112}(\alpha_1^2\alpha_2^2+\alpha_2^2\alpha_3^2)\, I_S^4 ,\\
\varepsilon_{33} &= N_{12} I_S^2 + (N_{11}-N_{12})\,\alpha_3^2 I_S^2 + N_{112}(\alpha_1^4+\alpha_2^4)\, I_S^4 + 3N_{123}\,\alpha_1^2\alpha_2^2 I_S^4 \\
&\quad + 3N_{112}(\alpha_2^2\alpha_3^2+\alpha_3^2\alpha_1^2)\, I_S^4 ,\\
\varepsilon_{23} &= N_{44}\,\alpha_2\alpha_3 I_S^2 + (N_{441}+N_{123}+2N_{456})\,\alpha_1^2\alpha_2\alpha_3 I_S^4 \\
&\quad + (N_{551}+N_{112})(\alpha_2^3\alpha_3+\alpha_3^3\alpha_1)\, I_S^4 ,\\
\varepsilon_{13} &= N_{44}\,\alpha_3\alpha_1 I_S^2 + (N_{441}+N_{123}+2N_{456})\,\alpha_2^2\alpha_3\alpha_1 I_S^4 \\
&\quad + (N_{551}+N_{112})(\alpha_3^3\alpha_1+\alpha_1^3\alpha_2)\, I_S^4 ,\\
\varepsilon_{12} &= N_{44}\,\alpha_1\alpha_2 I_S^2 + (N_{441}+N_{123}+2N_{456})\,\alpha_3^2\alpha_1\alpha_2 I_S^4 \\
&\quad + (N_{551}+N_{112})(\alpha_1^3\alpha_2+\alpha_2^3\alpha_3)\, I_S^4 .
\end{aligned}$$

In calculating, it was taken into account that the direction cosines in the rectangular system of coordinates are related by the following expression:

$$\alpha_1^2+\alpha_2^2+\alpha_3^2 = 1 . \tag{4.54}$$

Using this relation and replacing:

$$\alpha_2^4+\alpha_3^4 = 1-\alpha_1^4-2S(\alpha_1^2\alpha_2^2) , \qquad \alpha_2^2\alpha_3^2 = \alpha_1^4-\alpha_1^2+S(\alpha_1^2\alpha_2^2) , \tag{4.55}$$

where $S(\alpha_1^2\alpha_2^2) = \alpha_1^2\alpha_2^2+\alpha_2^2\alpha_3^2+\alpha_3^2\alpha_1^2$ (similar rearrangements are also made for other components), the expressions for ε_{ik} can, after some rearrangements, be written as follows:

$$\begin{aligned}
\varepsilon_{11} &= h_0+h_1\,\alpha_1^2+h_4\,[\alpha_1^4+\tfrac{2}{3}\,S(\alpha_1^2\alpha_2^2)]+h_3\,S(\alpha_1^2\alpha_2^2) ,\\
\varepsilon_{22} &= h_0+h_1\,\alpha_2^2+h_4\,[\alpha_2^4+\tfrac{2}{3}\,S(\alpha_1^2\alpha_2^2)]+h_3\,S(\alpha_1^2\alpha_2^2) ,\\
\varepsilon_{33} &= h_0+h_1\,\alpha_3^2+h_4\,[\alpha_3^4+\tfrac{2}{3}\,S(\alpha_1^2\alpha_2^2)]+h_3\,S(\alpha_1^2\alpha_2^2) ,\\
\varepsilon_{23} &= h_2\,\alpha_2\alpha_3+h_5\,\alpha_1^2\alpha_2\alpha_3 ,\\
\varepsilon_{13} &= h_2\,\alpha_3\alpha_1+h_5\,\alpha_2^2\alpha_3\alpha_1 ,\\
\varepsilon_{12} &= h_2\,\alpha_1\alpha_2+h_5\,\alpha_3^2\alpha_1\alpha_2 .
\end{aligned} \tag{4.56}$$

The values h_0, h_1, h_2, h_3, h_4, and h_5 in (4.56) are called magnetoelastic coupling constants. They are related in the following way to the components of matrices (4.52, 53) and spontaneous magnetization I_S:

$$\begin{aligned}
h_0 &= N_{12} I_S^2 + N_{112} I_S^4 \ , \\
h_1 &= (N_{11} - N_{12}) I_S^2 + (3N_{112} - 3N_{123}) I_S^4 \ , \\
h_2 &= N_{44} I_S^2 + (N_{551} + N_{112}) I_S^4 \ , \\
h_3 &= [N_{123} + \tfrac{2}{3}(N_{112} - N_{111})] I_S^4 \ , \\
h_4 &= (N_{111} - 4N_{112} + 3N_{123}) I_S^4 \ , \\
h_5 &= (N_{441} + N_{123} + 2N_{456} - N_{551} - N_{112}) I_S^4 \ .
\end{aligned} \tag{4.57}$$

Substituting (4.56) into (4.48) and bearing in mind that $\beta_1^2 + \beta_2^2 + \beta_3 = 1$, we obtain

$$\begin{aligned}
\varepsilon_S = {} & h_0 + h_1 S(\alpha_1^2 \beta_1^2) + 2h_2 S(\alpha_1 \alpha_2 \beta_1 \beta_2) + h_3 S(\alpha_1^2 \alpha_2^2) \\
& + h_4 S(\alpha_1^4 \beta_1^2 + \tfrac{2}{3} \alpha_1^2 \alpha_2^2) \ .
\end{aligned} \tag{4.58}$$

Here we have introduced operators S for brevity, which denote summation of the parenthetical values by cyclic permutation of the suffixes [for instance, $S(\alpha_1^4 \beta_1^2 + \frac{2}{3} \alpha_1^2 \alpha_2^2) = \alpha_1^4 \beta_1^2 + \alpha_2^4 \beta_2^2 + \alpha_3^4 \beta_3^2 + \frac{2}{3} (\alpha_1^2 \alpha_2^2 + \alpha_2^2 \alpha_3^2 + \alpha_3^2 \alpha_1^2)$].

Equation (4.58) shows that the appearance of spontaneous magnetizatin in a cubic crystal is accompanied by spontaneous magnetostrictive deformation, which consists of the isotropic part h_0 and the anisotropic one depending on the direction of measurement of deformation and on the orientation of $\boldsymbol{I}_S$. If we take into account the elastic energy, constants $h_0, h_1, \ldots, h_5$ in (4.58) will contain compliances s_{ij}, but the general form of the equation will remain unaltered.

In deriving (4.58), for the state with zero deformation, one adopts a state in which there is no spontaneous magnetization, i.e., some state of the crystal above the Curie temperature. When measuring the magnetostriction it is often more convenient to take, as a reference state, the demagnetized state of the specimen at the temperature of measurement of magnetostriction. This enables one to use (4.58) in measuring deformation corresponding to magnetization of the crystal from $H = 0$ (demagnetized state) to saturation, in different directions in the crystal at some constant temperature.

For cubic ferromagnetic crystals it is possible to distinguish two principal cases corresponding to the demagnetized state: vectors $\boldsymbol{I}_S$ of the domains are arranged along either the $\langle 111 \rangle$ or the $\langle 100 \rangle$ easy-magnetization axis. The ideally demagnetized state of a crystal with $\langle 111 \rangle$ easy-magnetization axes (nickel) is characterized by a uniform distribution of the magnetization vectors in all the domains along four axes (eight directions): $[111]$, $[\bar{1}\bar{1}\bar{1}]$, $[1\bar{1}\bar{1}]$, $[\bar{1}11]$, $[\bar{1}\bar{1}1]$, $[11\bar{1}]$, $[\bar{1}1\bar{1}]$, $[1\bar{1}1]$ (four groups of domains with an antiparallel direction of I_S in each). Such a state is realized at the following values of the direction cosines:

$$\begin{aligned} \pm\alpha_1 &= \pm\alpha_2 = \pm\alpha_3 = 1/\sqrt{3}\ , \qquad \mp\alpha_1 = \mp\alpha_2 = \mp\alpha_3 = 1/\sqrt{3}\ , \\ \pm\alpha_1 &= \mp\alpha_2 = \mp\alpha_3 = 1/\sqrt{3}\ , \qquad \mp\alpha_1 = \pm\alpha_2 = \mp\alpha_3 = 1/\sqrt{3}\ . \end{aligned} \tag{4.59}$$

Substituting (4.59) into (4.58), multiplying each of the solutions by the "weight" of the group (1/4), and adding them together, we get the expression

$$h_0+\tfrac{1}{3}h_1+\tfrac{1}{3}h_3+\tfrac{1}{3}h_4\ , \tag{4.60}$$

which characterizes the spontaneous deformation on transition of a Ni-type crystal from the nonferromagnetic to the ideally demagnetized state. In the case of a crystal with three ⟨100⟩ easy-magnetization axes (for instance, iron) there are three groups of domains which are characterized by the direction cosines:

$$\begin{aligned} &\alpha_1 = \pm 1\ , \quad &&\alpha_2 = 0\ , \quad &&\alpha_3 = 0\ , \\ &\alpha_1 = 0\ , &&\alpha_2 = \pm 1\ , &&\alpha_3 = 0\ , \\ &\alpha_1 = 0\ , &&\alpha_2 = 0\ , &&\alpha_3 = \pm 1\ . \end{aligned}$$

A similar calculation leads to the expression

$$h_0+\tfrac{1}{3}h_1+\tfrac{1}{3}h_4\ . \tag{4.61}$$

Subtracting (4.60, 61) from (4.58), we obtain the necessary equations describing the anisotropy of magnetostriction caused by the magnetization to saturation for Ni- and Fe-type crystals (it is more convenient to consider relative deformation $\Delta l/l = \lambda$):

$$\begin{aligned} \lambda(\mathrm{Ni}) &= h_1 S(\alpha_1^2\beta_1^2-\tfrac{1}{3})+2h_2 S(\alpha_1\alpha_2\beta_1\beta_2)+h_3 S(\alpha_1^2\alpha_2^2-\tfrac{1}{3}) \\ &\quad +h_4 S(\alpha_1^4\beta_1^2+\tfrac{2}{3}\alpha_1^2\alpha_2^2-\tfrac{1}{3})+2h_5 S(\alpha_1^2\alpha_2\alpha_3\beta_2\beta_3)\ , \\ \lambda(\mathrm{Fe}) &= h_1 S(\alpha_1^2\beta_1^2-\tfrac{1}{3})+2h_2 S(\alpha_1\alpha_2\beta_1\beta_2)+h_3 S(\alpha_1^2\alpha_2^2) \\ &\quad +h_4 S(\alpha_1^4\beta_1^2+\tfrac{2}{3}\alpha_1^2\alpha_2^2-\tfrac{1}{3})+2h_5 S(\alpha_1^2\alpha_2\alpha_3\beta_2\beta_3)\ . \end{aligned} \tag{4.62}$$

Expressions for magnetostriction in the form of (4.62) were obtained by *Becker* and *Döring* in 1939 [4.10].

If we put $N_{\alpha\beta\gamma} = 0$, i.e., if we take into consideration only the quadratic terms of magnetization in (4.47), then h_3, h_4, and h_5 will also vanish, and (4.62) will change to the two-constant equation:

$$\begin{aligned} \lambda &= h_1(\alpha_1^2\beta_1^2+\alpha_2^2\beta_2^2+\alpha_3^2\beta_3^2-\tfrac{1}{3}) \\ &\quad +2h_2(\alpha_1\alpha_2\beta_1\beta_2+\alpha_2\alpha_3\beta_2\beta_3+\alpha_3\alpha_1\beta_3\beta_1)\ . \end{aligned} \tag{4.63}$$

This expression was obtained by *Akulov* in 1939 [4.11]. It is more usual to apply (4.63) in the form

$$\lambda = \tfrac{3}{2}\lambda_{100}(\alpha_1^2\beta_1^2+\alpha_2^2\beta_2^2+\alpha_3^2\beta_3^2-\tfrac{1}{3}) + 3\lambda_{111}(\alpha_1\alpha_2\beta_1\beta_2+\alpha_2\alpha_3\beta_2\beta_3+\alpha_3\alpha_1\beta_3\beta_1)\ . \qquad (4.64)$$

Equations (4.63, 64) are often sufficient for approximate description of the magnetostriction of cubic ferromagnetic crystals. Constants λ_{100} and λ_{111} correspond to the longitudinal (deformation along the direction of the external field) magnetostriction along the $\langle 100\rangle$ axes ($\alpha_1=\beta_1=1$) and along the $\langle 111\rangle$ axes ($\alpha_1=\alpha_2=\alpha_3=\beta_1=\beta_2=\beta_3=1/\sqrt{3}$) on magnetization of a crystal from $H=0$ to the saturation field at which $I_S \| H$. For instance, for crystals of magnetite (Fe_3O_4) the values of the magnetostriction constants at room temperature are $\lambda_{100} = -20\times10^{-6}$, $\lambda_{111} = +78\times10^{-6}$.

b. Hexagonal Crystal

In considering hexagonal, tetragonal, and orthorhombic crystals we shall restrict ourselves to the quadratic terms in (4.47). The matrix of tensor $N_{\alpha\beta}$ for crystals of the hexagonal class $6/mm$ (cobalt, hexagonal ferrites) has the form [4.9]

$$N_{\alpha\beta} = \left\| \begin{matrix} N_{11} & N_{12} & N_{13} & 0 & 0 & 0 \\ N_{12} & N_{11} & N_{13} & 0 & 0 & 0 \\ N_{31} & N_{31} & N_{33} & 0 & 0 & 0 \\ 0 & 0 & 0 & N_{44} & 0 & 0 \\ 0 & 0 & 0 & 0 & N_{44} & 0 \\ 0 & 0 & 0 & 0 & 0 & N_{11}-N_{12} \end{matrix} \right\| . \qquad (4.65)$$

Calculating the components of tensor ε_{ik} as for the cubic crystal and substituting them into (4.48), we get

$$\varepsilon_S = I_S^2(N_{33}-N_{31})\,\alpha_3^2\beta_3^2 + I_S^2(N_{11}-N_{13})(\alpha_1^2\beta_1^2+\alpha_2^2\beta_2^2) + I_S^2(N_{12}-N_{13})(\alpha_2^2\beta_1^2+\alpha_1^2\beta_2^2) + 2I_S^2(N_{11}-N_{12})\,\alpha_1\alpha_2\beta_1\beta_2 + 2I_S^2N_{44}\,\alpha_3\beta_3(\alpha_2\beta_2+\alpha_1\beta_1) + N_{13}I_S^2 + I_S^2(N_{31}-N_{13})\,\beta_3^2\ . \qquad (4.66)$$

The demagnetized state of a hexagonal crystal (for instance, cobalt) with a magnetization axis coinciding with the hexagonal one is characterized by a single group of antiparallel domains, i.e., is determined from the condition

$$\alpha_3 = \pm 1\ , \quad \alpha_1=\alpha_2=0\ . \qquad (4.67)$$

Substituting (4.67) into (4.66), we obtain

$$\varepsilon_S = I_S^2(N_{33}-N_{13})\,\beta_3^2 + N_{13}I_S^2\ . \qquad (4.68)$$

Subtracting (4.68) from (4.66), we get the following expression for the magnetostriction caused by magnetization to saturation:

$$\lambda = I_3^2 [(N_{11}-N_{12})(\alpha_1\beta_1+\alpha_2\beta_2)^2+(N_{12}-N_{13})(1-\beta_3^2)(1-\alpha_3^2) + (N_{31}-N_{33})(1-\alpha_3^2)\beta_3^2+2N_{44}(\alpha_1\beta_1+\alpha_2\beta_2)\alpha_3\beta_3] \ . \tag{4.69}$$

In practice, it is more convenient to use (4.69) is a somewhat rearranged form:

$$\lambda = \lambda_A [(\alpha_1\beta_1+\alpha_2\beta_2)^2-(\alpha_1\beta_1+\alpha_2\beta_2)\alpha_3\beta_3]+\lambda_B[(1-\alpha_3^2)(1-\beta_3^2) - (\alpha_1\beta_1+\alpha_2\beta_2)^2]+\lambda_C[(1-\alpha_3^2)\beta_3^2-(\alpha_1\beta_1+\alpha_2\beta_2)\alpha_3\beta_3] + 4\lambda_D(\alpha_1\beta_1+\alpha_2\beta_2)\alpha_3\beta_3 \ , \tag{4.70}$$

where constants λ_A, λ_B, λ_C, and λ_D are determined from experiment at the following orientations of the direction of measurement of magnetostriction and magnetization: $\lambda_A(\alpha_1=\beta_1=1)$, $\lambda_B(\alpha_1=1,\ \beta_2=1)$, $\lambda_C(\alpha_1=1,\ \beta_3=1)$, $\lambda_D(\alpha_1=\beta_1=1/\sqrt{2},\ \alpha_3=\beta_3=1/\sqrt{2})$. They are related to components $N_{\alpha\beta}$ and spontaneous magnetization as follows:

$$\lambda_A=(N_{11}-N_{13})I_S^2 \ , \quad \lambda_B=(N_{12}-N_{13})I_S^2 \ , \quad \lambda_C=(N_{31}-N_{33})I_S^2 \ ,$$
$$\lambda_D=[\tfrac{1}{4}(N_{11}-N_{13})+\tfrac{1}{4}(N_{31}-N_{33}+2N_{44})]I_s^2 \ .$$

For a cobalt crystal the magnetostriction constants at room temperature are equal to $\lambda_A=-45\times10^{-6}$, $\lambda_B=-95\times10^{-6}$, $\lambda_C=+110\times10^{-6}$, and $\lambda_D=-100\times10^{-6}$. Constants λ_A and λ_C are readily found from the magnetostriction curves shown in Fig. 4.15b. They are equal to the saturation magnetostriction for cases $\lambda_\perp$, $H_\perp$, and $\lambda_\|$, and $H_\perp$, respectively.

c. *Tetragonal and Orthorhombic Crystals*

The matrix of tensor $N_{\alpha\beta}$ for tetragonal class $4/mm$ has the form [4.8]

$$N_{\alpha\beta}=\begin{Vmatrix} N_{11} & N_{12} & N_{13} & 0 & 0 & 0\\ N_{12} & N_{11} & N_{13} & 0 & 0 & 0\\ N_{31} & N_{31} & N_{33} & 0 & 0 & 0\\ 0 & 0 & 0 & N_{44} & 0 & 0\\ 0 & 0 & 0 & 0 & N_{44} & 0\\ 0 & 0 & 0 & 0 & 0 & N_{66} \end{Vmatrix} \ . \tag{4.71}$$

For orthorhombic crystals of class mmm [4.9]

$$N_{\alpha\beta}=\begin{Vmatrix} N_{11} & N_{12} & N_{13} & 0 & 0 & 0\\ N_{12} & N_{22} & N_{23} & 0 & 0 & 0\\ N_{31} & N_{32} & N_{33} & 0 & 0 & 0\\ 0 & 0 & 0 & N_{44} & 0 & 0\\ 0 & 0 & 0 & 0 & N_{55} & 0\\ 0 & 0 & 0 & 0 & 0 & N_{66} \end{Vmatrix} \ . \tag{4.72}$$

As a result of calculations, which we shall not dwell upon, we get the following formulas for the magnetostriction caused by magnetization to saturation.

For a tetragonal crystal

$$\lambda = h_1(1-\alpha_3^2)\beta_3^2 + h_2(\alpha_1^2\beta_1^2+\alpha_2^2\beta_2^2) + h_3(\alpha_2^2\beta_1^2+\alpha_1^2\beta_2^2) + 2h_4\alpha_1\alpha_2\beta_1\beta_2 + 2h_5(\alpha_2\beta_2+\alpha_1\beta_1)\alpha_3\beta_3 , \quad (4.73)$$

where

$$h_1 = (N_{31}-N_{33})I_S^2 , \quad h_2 = (N_{11}-N_{13})I_S^2 , \quad h_3 = (N_{12}-N_{13})I_S^2 ,$$
$$h_4 = N_{66}I_S^2 , h_5 = N_{44}I_S^2 .$$

For an orthorhombic crystal

$$\lambda = h_1\alpha_1^2\beta_1^2 + h_2\alpha_2^2\beta_1^2 + h_3\alpha_1^2\beta_2^2 + h_4\alpha_2^2\beta_2^2 + h_5\alpha_1^2\beta_3^2 + h_6\alpha_2^2\beta_3^2 + 2h_7\alpha_2\alpha_3\beta_2\beta_3 + 2h_8\alpha_3\alpha_1\beta_3\beta_1 + 2h_9\alpha_1\alpha_2\beta_1\beta_2 , \quad (4.74)$$

where

$$h_1 = (N_{11}-N_{13})I_S^2 , \quad h_4 = (N_{22}-N_{23})I_S^2 , \quad h_7 = N_{44}I_S^2 ,$$
$$h_2 = (N_{12}-N_{13})I_S^2 , \quad h_5 = (N_{31}-N_{33})I_S^2 , \quad h_8 = N_{55}I_S^2 ,$$
$$h_3 = (N_{21}-N_{23})I_S^2 , \quad h_6 = (N_{32}-N_{33})I_S^2 , \quad h_9 = N_{66}I_S^2 .$$

For the demagnetized state, in both cases the equiprobable domain distribution along the positive and negative direction of the $X_3(Z)$ axis was chosen.

d. Polycrystal

Let us consider the ideal polycrystal with an equiprobable orientation of the domain magnetization vectors along all possible directions. There is no need to take into account the demagnetized state here, since all the magnetization directions are equiprobable.

The matrix of tensor $N_{\alpha\beta}$ for a polycrystal has the form

$$N_{\alpha\beta} = \left\| \begin{matrix} N_{11} & N_{12} & N_{12} & 0 & 0 & 0 \\ N_{12} & N_{11} & N_{12} & 0 & 0 & 0 \\ N_{12} & N_{12} & N_{11} & 0 & 0 & 0 \\ 0 & 0 & 0 & N_{11}-N_{12} & 0 & 0 \\ 0 & 0 & 0 & 0 & N_{11}-N_{12} & 0 \\ 0 & 0 & 0 & 0 & 0 & N_{11}-N_{12} \end{matrix} \right\| . \quad (4.75)$$

Components ε_{ik} are equal to

$$\varepsilon_{11} = N_{12} + (N_{11}-N_{12})\alpha_1^2 , \quad \varepsilon_{23} = (N_{11}-N_{12})\alpha_2\alpha_3 ,$$
$$\varepsilon_{22} = N_{12} + (N_{11}-N_{12})\alpha_2^2 , \quad \varepsilon_{13} = (N_{11}-N_{12})\alpha_3\alpha_1 ,$$
$$\varepsilon_{33} = N_{12} + (N_{11}-N_{12})\alpha_3^2 , \quad \varepsilon_{12} = (N_{11}-N_{12})\alpha_1\alpha_2 .$$

For the magnetostriction in the direction characterized by cosines β_1, β_2, and β_3 (the rectangular coordinate system is chosen arbitrarily) during magnetization in the direction of α_1, α_2, and α_3 we obtain

$$\lambda = N_{12}+(N_{11}-N_{12})\,S(\alpha_1^2\beta_1^2)+2(N_{11}-N_{12})\,S(\alpha_2\,\alpha_3\,\beta_2\,\beta_3)\ .$$

Denoting the angle between the direction of λ measurement and the direction of magnetization by θ, and bearing in mind that $\cos\theta = \alpha_1\beta_1+\alpha_2\beta_2+\alpha_3\beta_3$, we get

$$\lambda = M_1+M_2\cos^2\theta\ ,$$

where $M_1 = N_{12}$ and $M_2 = N_{11}-N_{12}$.

In a Ni-type cubic crystal the constants M_1 and M_2 of the polycrystal are related to the constants h_1, h_2, h_3, h_4, and h_5 of the single crystal in the following way:

$$M_1 = -\tfrac{1}{5}\left(\tfrac{2}{3}\,h_1+h_2+\tfrac{2}{3}\,h_3+\tfrac{4}{7}\,h_4+\tfrac{1}{7}\,h_5\right)\ ,$$
$$M_2 = \tfrac{1}{5}\left(2\,h_1+3\,h_2+\tfrac{12}{7}h_4+\tfrac{3}{7}\,h_5\right)\ .$$

4.4.3 Magnetic Anisotropy Energy Corresponding to Zero Strains and Zero Stresses

At zero strains (at a constant volume, $\varepsilon_{ij}=0$) the sum of the elastic and magnetic energy is equal, by (4.39), to

$$U+V' = V^0\ . \tag{4.76}$$

The expression for V^0 is given in (4.40). Energy V^0 is called the *free magnetic anisotropy energy of an underformed crystal.* At zero stresses, strain components ε_{ij} will have equilibrium values ε_{ij}^*, which are defined by (4.51), and therefore

$$U+V' = -\tfrac{1}{2}\,V_{ij}^0\,\varepsilon_{ij}^*+V^0+V_{ij}^0\,\varepsilon_{ij} = V^0+\tfrac{1}{2}\,V_{ij}^0\,\varepsilon_{ij}\ . \tag{4.77}$$

Expression (4.77) practically corresponds to the anisotropy energy measured at atmospheric pressure, i.e., to energy V (4.22), which can now be written as

$$V = V^0+\tfrac{1}{2}\,V_{ij}^0\,\varepsilon_{ij}^*\ . \tag{4.78}$$

Thus, the density of magnetic anisotropy energy consists of two parts: the energy corresponding to the underformed crystal, and the contribution from the magnetoelastic energy. Consideration of the magnetoelastic contribution ultimately amounts to the fact that anisotropy constants K_1 and K_2 in the final expression for V – for instance, (4.22, 23) – become equal: $K_1 = K_1^0 + K_1'$ and $K_2 = K_2^0 + K_2'$, where the constants with the zero superscripts refer to the energy V^0 of the undeformed crystal, and the primed constants take into account the contribution from the magnetoelastic energy. As noted in Sect. 4.3, from the symmetry standpoint, tensors k_{mn}, k_{mnop}, k_{mnopqr}, and k'_{mn}, k'_{mnop}, k_{mnopqr} in (4.21, 40) are identical, respectively, but the anisotropy energy equations for crystals of different systems are more conveniently derived under the assumption of zero strains. We abandon the zero superscripts and primes and assume that the anisotropy constants obtained in the final expression correspond to those available from experiment, i.e., we assign to them the meaning of constants K_1 and K_2 in (4.22, 23).

Let us see what V will look like for crystals of a particular symmetry.

a. Cubic Crystal

To find the magnetic anisotropy energy it is necessary, as in the case of magnetostriction, to know the specific type of tensors k_{mn}, k_{mnop}, and k_{mopqr} for a given class of symmetry. Magnetic anisotropy tensors k_{mnop} and k_{mnopqr} can be obtained with the aid of matrices (4.52, 53) subject to an additional condition that all the terms with suffixes m, n, o, p, q, and r can be interchanged in any possible way (tensor k_{mnop} can easily be obtained on the same condition from a tensor similar to that of elastic constants). Tensor k_{mn}, similar to that of the dielectric constant, has only one component for cubic crystals (it is degenerated to a scalar). The tensors under consideration are written, as in the case of tensors $N_{\alpha\beta}$ and $N_{\alpha\beta\gamma}$, more conveniently in matrix form k_α, $k_{\alpha\beta}$, and $k_{\alpha\beta\gamma}$. The relationship between suffixes m, n, o, p, q, r, and α, β, γ has been considered on p. 311. For a cubic crystal, the number of components of tensor $k_{\alpha\beta}$ (provided it is possible to permute all the subscripts) reduces from three [tensor $N_{\alpha\beta}$ (4.52)] to two (k_{11} and k_{12}), because $k_{44} = k_{12}$, and for the sixth-rank tensor, from six [tensor $N_{\alpha\beta\gamma}$ (4.53)] to three (k_{111}, k_{112}, and k_{123}), since $k_{112} = k_{551}$, $k_{123} = k_{441} = k_{456}$. Replacing the components of the spontaneous magnetization vector by the direction cosines, as before, and appropriately combining the components of tensors k_α, $k_{\alpha\beta}$, and $k_{\alpha\beta\gamma}$, one can obtain the following expression for the magnetic anisotropy energy of a cubic ferromagnetic crystal:

$$V = K_0 + K_1 (\alpha_1^2 \alpha_2^2 + \alpha_2^2 \alpha_3^2 + \alpha_3^2 \alpha_1^2) + K_2 \alpha_1^2 \alpha_2^2 \alpha_3^2 \; . \tag{4.79}$$

Magnetic anisotropy constants K_1 and K_2 in (4.79) are related to the spontaneous magnetization in the following manner:

$$K_1 = a I_S^4 + b I_S^6 \; , \quad K_2 = c I_S^6 \; ,$$

where a and b are some linear combinations of components k_{11}, k_{12}, and k_{111}, respectively, and quantity c is a combination of the components of the sixth-rank tensor k_{123}, k_{112}, and k_{111}. Term K_0 (additive term) includes combinations of the tensor components independent of the direction of I_S in the crystal, i.e., making no contribution to the anisotropy; K_0 includes, in particular, the only component k_α of k_{mn}, which takes into account the quadratic interaction of the magnetization components. Owing to (4.54), this interaction does not lead to anisotropy.

The form of representation of the magnetic anisotropy energy – for instance, as (4.79) – is not unique. As independent invariants one can choose, instead of combinations of the direction cosines standing at K_1 and K_2 in (4.79), a single independent invariant of the form $\alpha_1^{2n}+\alpha_2^{2n}+\alpha_3^{2n}$, where n may have values 2, 3, 4, or 5 with an allowance for higher magnetization degrees [for $n=1$ we obtain (4.54)].

Sometimes it is found preferable to represent the magnetic anisotropy energy not as a power series, but as an expansion in harmonic polynomials.

b. *Hexagonal, Tetragonal, and Orthorhombic Crystals*

For a hexagonal crystal of class $6/mmm$, second-rank tensor K_α has two components k_1 and k_3, and fourth-rank tensor $k_{\alpha\beta}$, four components k_{11}, k_{12}, k_{13}, and k_{33} instead of the six components for the magnetostriction tensor (4.65) of instead of five components for the elastic constants tensor [when using a tensor of the form of (4.65) it should be borne in mind that the component in the lower right-hand corner of the matrix must have the form $2(k_{11}-k_{12})$, as in the elastic constant tensor]. If we restrict ourselves to the second and fourth powers of the magnetization constants in (4.21), i.e., if we take into consideration only the second- and fourth-rank tensors, then after rather simple calculations using (4.54, 55), expression (4.21) for a hexagonal crystal transforms to

$$V=k_1 I_S^2+k_{11} I_S^4+(k_{11}+k_{33}-3\,k_{13})\,I_S^4\,\alpha_3^4 \\ +[(k_3-k_1)\,I_s^2+(3\,k_{13}-2\,k_{11})\,I_S^4]\,\alpha_3^2\ . \tag{4.80}$$

The first two terms in (4.80) are similar to additive term K_0 in (4.79), and the last two terms show that the magnetic anisotropy energy for a hexagonal crystal depends in this approximation exclusively on α_3, i.e., on the orientation of the magnetization vector with respect to the hexagonal axis. Note that for a hexagonal crystal the allowance for the squares of magnetization in (4.21) leads to anisotropy [k_1 and k_3 in (4.80)] in distinction to cubic crystals, where tensor k_α degenerated into a scalar and did not contribute to the anisotropy.

For magnetouniaxial crystals, it is more convenient to replace direction cosines α_1, α_2, and α_3 by spherical coordinates

$$\alpha_1=\sin\theta\cos\varphi\ ,\qquad \alpha_2=\sin\theta\sin\varphi\ ,\qquad \alpha_3=\cos\theta\ ,$$

where angle θ defines the orientation of the vector of spontaneous magnetization I_S relative to the hexagonal axis, and angle φ is reckoned in the basal plane from the twofold axis. Then, bearing in mind that $\alpha_3^2 = \sin^2\theta - 1$, $\alpha_3^4 = 1 - 2\sin^2\theta - \sin^4\theta$, we can reduce expression (4.80) to

$$V = K_0 + K_1 \sin^2\theta + K_2 \sin^4\theta \ , \tag{4.81}$$

where K_0, K_1, and K_2 are the magnetic anisotropy constants of a hexagonal crystal (4.24). Expressions (4.80, 81) show that taking into account the second and fourth powers of magnetization in (4.21) does not lead to anisotropy in the basal plane (α_2, α_3 or coordinate φ are nonexistent). To make anisotropy "appear" in the basal plane we must take into consideration the sixth powers of the magnetization components and use sixth-rank tensor $k_{\alpha\beta\gamma}$. In calculation, the anisotropy energy expression will include five components of tensor $k_{\alpha\beta\gamma}$: k_{333}, k_{113}, k_{133}, k_{111}, and k_{222}. Since the calculations are cumbersome, we give only the final result, which can be written as

$$V = K_0 + K_1 \sin^2\theta + K_2 \sin^4\theta + K_3' \sin^6\theta + K_3 \sin^6\theta \cos 6\varphi \ , \tag{4.82}$$

where K_1, K_2, K_3', and K_3 are the magnetic anisotropy constants of a hexagonal crystal. Here K_3' and K_3 have different physical meanings (the former, like K_1 and K_2, defines the relationship between vector I_S and the c axis, and the latter takes into account the anisotropy in the basal plane depending on angle φ), but they stand at the sixth-power terms and are therefore denoted by identical subscripts.

Similar calculations of the magnetic anisotropy energy for tetragonal and orthorhombic crystals yield the following expressions [4.9]:

$$\begin{aligned} V(\text{tetragonal}) &= K_0 + K_1 \sin^2\theta + K_2 \sin^4\theta + K_3 \sin^4\theta \sin^2\varphi \cos^2\varphi \ , \\ V(\text{orthorhombic}) &= K_0 + (K_1 \cos^2\varphi + K_2 \sin^2\varphi) \sin^2\theta \\ &\quad + (K_3 \cos^4\varphi + K_4 \sin^2\varphi \cos^2\varphi + K_5 \sin^4\varphi) \sin^4\theta \\ &\quad + (K_6 \cos^2\varphi + K_7 \sin^2\varphi) \sin^2\theta \cos^2\theta \ . \end{aligned} \tag{4.83}$$

Angle θ in both cases is taken from the $X_3-(Z)$ axis, and angle φ, from the twofold axis in the plane perpendicular to the X_3 axis.

4.4.4 Equilibrium Directions of Spontaneous Magnetization

From the equations for the magnetic anisotropy energy one can obtain conditions at which I_S in the crystal is directed along definite easy-magnetization axes. As an example, we consider a hexagonal crystal for which the anisotropy energy (neglecting the sixth powers of magnetization and the isotropic term) is written as (4.23, 81)

$$V = K_1 \sin^2\theta + K_2 \sin^4\theta \ .$$

The equilibrium directions of I_S, defined by angle θ_0, must correspond to the minimum of energy V. From the condition of the minimum

$$dV/d\theta = \cos\theta \sin\theta (2K_1 + 4K_2 \sin^2\theta) = 0 \ , \quad (4.84)$$

we find that for a hexagonal crystal there are three solutions corresponding to the equilibrium directions of I_S:

$$\sin\theta_0 = 0 \ , \quad (4.85)$$

$$\cos\theta_0 = 0 \ , \quad (4.86)$$

$$\sin^2\theta_0 = -K_1/2K_2 \ , \quad (4.87)$$

Solution (4.85) means that the equilibrium direction of I_S coincides with the sixfold axis (c axis). It is fulfilled at $\theta = 0°$ and $\theta = 180°$; this means that the crystal has one group of domains with mutually opposite directions of I_S. As can be shown by comparing the energies corresponding to the condition of the minimum $d^2V/d\theta > 0$, solution (4.85) is achieved when

$$K_1 + K_2 > 0 \qquad K_1 > 0 \ . \quad (4.88)$$

If, however,

$$0 < -K_1 < 2K_2 \ , \quad (4.89)$$

then to the equilibrium direction of I_S corresponds solution (4.87). If we neglect the anisotropy in the basal plane [the term with K_3 in (4.82)], each generatrix of cone of revolution with an axis coinciding with the c axis and with cone angle $2\theta_0$ will be the equilibrium direction of I_S.

At any other relations between K_1 and K_2 solution (4.86) will be achieved, and any direction in the basal plane (easy-magnetization plane) will be the equilibrium direction of I_S.

Strictly speaking, the anisotropy energy in the basal plane must always depend on angle φ (4.82). Therefore, in the case of (4.87) six "easiest" axes can be singled out in a cone. Then the crystal is divided into six groups of anti-parallel domains, and in each group half of the domains have an I_S directed "upward" relative to the c axis, an the other half, and I_S directed "down-ward." In the case of the "easy-magnetization plane" all the "easy" directions of I_S will be in the basal plane.

A similar consideration can be carried out for tetragonal and orthorhombic crystals.

If we assume that in orthorhombic crystals I_S is located in one of the symmetry planes (for instance, in the XZ plane), then (4.83 b) ($\varphi = 0$) reduces to the form

$$V = K_1 \sin^2\theta + K_3 \sin^4\theta \ , \tag{4.90}$$

similar to the one considered above for hexagonal crystals: the equilibrium position of I_S may either coincide with the twofold axis ($\theta_0 = 0°$, $180°$; $\theta_0 = 90°$, $270°$) or be positioned in the XZ plane at a certain angle θ_0 different from 0° or 90°. In the former case there will be one group of antiparallel domains and in the latter, two groups of domains. The symmetry of the latter (as well as of the hexagonal crystal domains corresponding to the easy-magnetization cone) will differ from that of domains with $\theta_0 = 0$ and from the symmetry of domains with $\theta_0 = 90°$ (Table 4.3).

It should be emphasized that anisotropy energy V reflects the symmetry of the original (paramagnetic) phase. The various solutions corresponding to the minimum of V show that the crystal can be divided into domains in different ways, retaining its initial symmetry as a whole. The relationship between the symmetry of a domain and that of a crystal was considered in Sect. 4.3.

In cubic crystals, there can exist ⟨100⟩, ⟨110⟩, and ⟨111⟩ easy-magnetization axes. The relations between constants K_1 and K_2 in (4.79), for which the directions of I_S coincide with the indicated axes, are represented in Fig. 4.19. When the term with K_1 in (4.79) is predominant, then at $K_1 < 0$ the ⟨111⟩ axes are the easy-magnetization directions, and at $K_1 > 0$, the ⟨100⟩ axes are.

To different temperatures may correspond different conditions of the minimum of V. Then a change in crystal temperature causes a phase transition with a change in the orientation of the easy-magnetization axis, which is accompanied by a change of symmetry and a rearrangement of the domain structure. Such phase transitions were discussed at the end of Sect. 4.5.

4.4.5 Magnetic Anisotropy Measurement

The simplest and most reliable method for investigating the magnetic anisotropy is the so-called torque method. A crystal prepared in the shape of a sphere or a disc is gradually rotated in a certain crystallographic plane in a strong homogeneous magnetic field. During the rotation the torque balancing the crystal in the magnetic field is measured. The torque is referred to a unit volume of the crystal [erg/cm^3]. The measurement of the magnetic anisotropy amounts to determining anisotropy constants K_1, K_2, etc., from the torque curves.

Suppose we have to measure constants K_1 and K_2 for a cobalt crystal. If we prepare from it a disc with a plane coinciding with $(10\bar{1}0)$ and suspend this

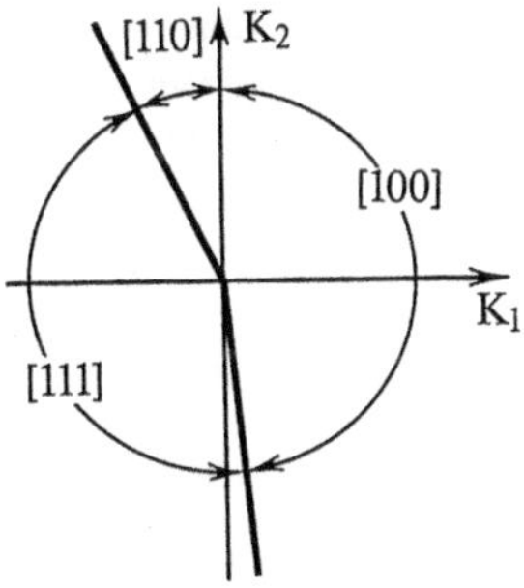

Fig. 4.19. Relationship between anisotropy constants K_1 and K_2 and the easy-magnetization direction for cubic crystals

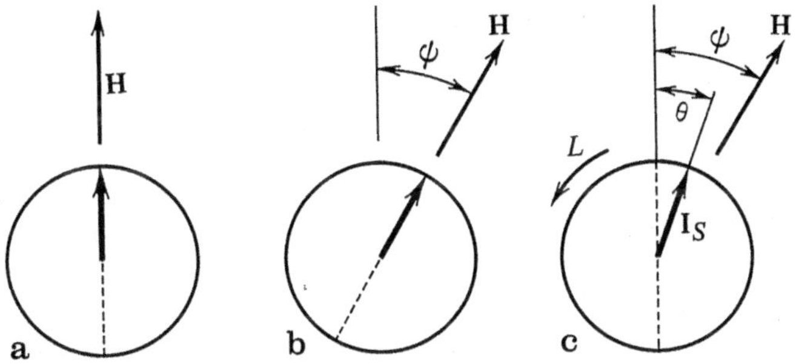

Fig. 4.20 a – c. Principle of measuring the magnetic anisotropy energy by the torque method

disc freely in a homogeneous magnetic field of sufficiently high intensity, the crystal will occupy a position in the field in which the easy-magnetization direction coincides with that of the field (this property is used in the method for orienting ferromagnetic crystals in a magnetic field). This is schematized in Fig. 4.20a (the direction of the c axis is shown by a dashed line). If we change the direction of the field by angle ψ, the unattached crystal will rotate together with the field (Fig. 4.20b). To return the crystal to position a without changing the direction and magnitude of the field, a certain torque L must be applied.

The torque applied will be offset by the oppositely directed torque defined from the equilibrium condition (4.42):

$$L = -dV/d\theta = -HI_S \sin(\psi - \theta) \ . \tag{4.91}$$

As a result, vector I_S will deviate from the c axis by a certain angle θ towards the field (Fig. 4.20c), while the energy expended on rotation of the crystal and its "retention" in the magnetic field will balance off with the anisotropy energy, which resists the deviation of vector I_S from the direction of the "easy" c axis. The expression for L will be obtained from (4.84) by passing to double angles:

$$L = -(K_1 + K_2)\sin 2\theta - \frac{K_2}{2}\sin 4\theta \ . \tag{4.92}$$

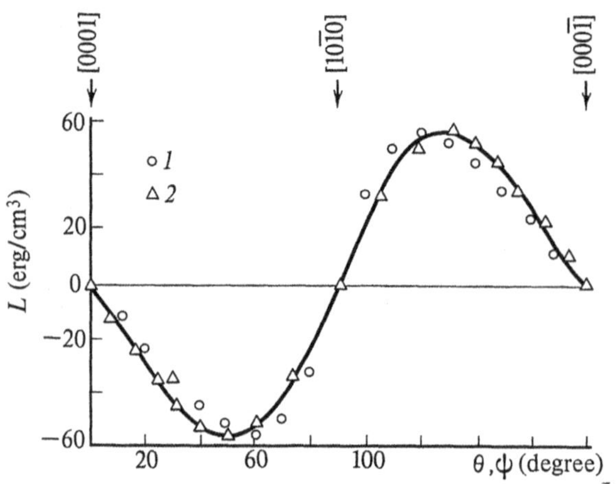

Fig. 4.21. Torque curves L for a cobalt crystal in the (10$\bar{1}$0) plane at room temperature: (*1*) $L(\psi)$, (*2*) $L(\theta)$ [4.6]

Field H is usually taken large enough for condition $\theta \approx \psi$ to be satisfied. By measuring L in relation to angle $\theta \approx \psi$, one can obtain the so-called torque curve and calculate K_1 and K_2 in (4.92) by the least squares method.

If, however, condition $\theta \approx \psi$ is not fulfilled (as shown in Fig. 4.20c), it is necessary to introduce a correction for the angle of "lagging" $\psi - \theta$ of vector I_S from the direction of field $\boldsymbol{H}$. This can be done with the aid of (4.91), provided the values of $\boldsymbol{H}$ and I_S are known. Figure 4.21 demonstrates the torque curve for a single-crystal disc of cobalt [(10$\bar{1}$0) plane] at room temperature [4.6]. The circles denote the values of L in relation to angle ψ. The triangles denote the same values of L, but are related to angle θ by means of (4.91). The solid line is drawn according to (4.92) for $K_1 = 4.3 \times 0^6$ erg/cm^3, $K_2 = 1.2 \times 10^6$ erg/cm^3. At $\psi = \theta = 0$ the crystal is in stable equilibrium and the torque is zero, because $\boldsymbol{H}$ coincides in direction with the c axis. The choice of signs of L is such that to the stable position of the crystal corresponds point $L = 0$ on the section of curve $L(\theta)$ with $dL/d\theta < 0$. At pount $L = 0$ and $\theta = 90°$, where $dL/d\theta > 0$, magnetization I_S is oriented in the basal plane, which is the hard-magnetization plane. The crystal is then in unstable equilibrium: at any slight deviation of $\boldsymbol{I}_S$ from $\theta = 90°$ it tends to occupy a position in which I_S will make the least angle with the easy-magnetization direction.

It has been indicated in Sect. 4.3.5 that the magnetic anisotropy energy can also be estimated from the area between the magnetization curves of the crystal along different axes.

4.5 Structure of Some Magnetically Ordered Crystals and Reorientation Transitions

4.5.1 Ferrites

The most interesting crystals from the standpoint of the magnetic atomic structure are those of iron and rare-earth oxides. This class of substances includes ferrites. The most typical of these are ferrites with a spinel and garnet structure and also hexagonal ferrites.

As regards their electric properties, the ferrites are semiconductors: their conductivity lies between 10^{-10} ohm^{-1} cm^{-1} and 10^2 ohm^{-1} cm^{-1}, depending on their structure and composition. Owing to their low electrical conductivity, ferrites are widely used in various high-frequency devices, mainly because eddy current losses in them are much lower than in magnetic alloys.

Considering that the ferrite structure has been described in sufficient detail in the literature [4.12, 13] we shall supply only the most essential information on it. Our attention will be focused on the structure of weak ferromagnetics as representatives of ordered magnetics, the properties of which are successfully described by the magnetic symmetry.

a. Ferrites with a Spinel Structure

A crystallographic space group of spinel above the Curie point is $O_h^7 = Fd3m$. The chemical formula of ferrites with a structure similar to that of minerals spinel ($MgAl_2O_4$) and magnetite (Fe_3O_4) is written as $MeFe_2O_4 = MeO \cdot Fe_2O_3$. Here Me stands for one of the divalent ions of transition metals Mn, Fe, Co, Ni, Cu, and Zn, and also for Mg and Cd ions. A combination of ions with average valency 2 is possible. To the spinel unit cell correspond 8 molecules of $MeFe_2O_4$. It contains 32 oxygen atoms which cubic close packing. Of the 64 tetrahedral and 32 octahedral sites available in such a cell, only 8 tetrahedral (*a* sites) and 16 octahedral (*b* sites) are occupied in the spinel structure. The structure of spinel-type oxides is shown in [Ref. 2.5, Fig. 2.17].

Two extreme cases of spinel structure are distinguished: *normal spinel* and *inverse spinel*. In normal spinel, 8 divalent metal ions Me^{2+} occupy 8 *a* sites and 16 trivalent ions Fe^{3+}, 16 *b* sites. The formula of normal spinel can be represented as

$$\{Me^2 + [Fe^{3+}\, Fe^{3+}]O_4\} \cdot 8 \ .$$

Here the ions occupying the *a* sites are written in front of the square brackets, and *b* ions, in the brackets. The formula of inverse spinel has the form

$$\{Fe^{2+}[Me^{2+}\, Fe^{3+}]O_4\} \cdot 8 \ .$$

Table 4.4. Spontaneous magnetic moment at 0 K and the Curie temperatures of some spinel ferrites

Ferrite	T_C, °C	A ion	B ions	Magnetic moment (in Bohr magnetons)		Total magnetic moment	
				A ion	B ions	Theory	Experiment
Fe_3O_4	585	Fe^{3+}	$Fe^{2+}+Fe^{3+}$	5	4+5	4	4.1
$CoFe_2O_4$	520	Fe^{3+}	$Co^{2+}+Fe^{3+}$	5	3+5	3	3.7
$NiFe_2O_4$	585	Fe^{3+}	$Ni^{2+}+Fe^{3+}$	5	2+5	2	2.3
$CuFe_2O_4$	455	Fe^{3+}	$Cu^{2+}+Fe^{3+}$	5	1+5	1	1.3
$MgFe_2O_4$	440	Fe^{3+}	$Mg^{2+}+Fe^{3+}$	5	0+5	0	1.1
$Li_{0.5}Fe_{2.5}O_4$	670	Fe^{3+}	$Li^{1+}_{0.5}+Fe^{3+}_{1.5}$	5	0+7.5	2.5	2.6

Completely normal or completely inverse structures are the extreme cases, as noted above. Generally, the cation distribution can be written as follows (for one formula unit):

$$Me^{2+}_{\delta}\,Fe^{3+}_{1-\delta}\,[Me^{2+}_{1-\delta}\,Fe^{3+}_{1+\delta}]\,O_4\ .$$

Parameter δ can be regarded as the degree of inversion of the spinel: for normal spinel $\delta=1$, and for inverse $\delta=0$.

The magnetic ions at a and b sites form two magnetic sublattices, respectively. In most of the ferrites, ions of different sorts at b sites are distributed statistically. The magnetic moments of a and b sublattices show antiparallel orientation: exchange integral I_{ab} for ab interaction is negative and considerable exceeds the positive exchange integrals I_{aa} and I_{bb}. Inside each of the a and b sublattices the magnetic moments show parallel orientation. At $T=0$ the magnetic structure will represent two saturated a and b sublattices with antiparallel magnetic moments, as shown in Fig. 4.1c. The resultant spontaneous magnetization can readily be found as the difference between the magnetizations of the a and b sublattices (the latter is usually the higher). Such a structure demonstrates the simplest case of ferrimagnetism.

As an illustration, Table 4.4 lists experimental and theoretical values of the spontaneous magnetic moment at 0 K [quantity p, (4.12)], and also of the Curie points for a number of ferrites with a spinel structure. It also indicates the ion distribution among the a and b sublattices.

For inverse spinel, the resultant magnetization is equal to the magnetic moment of an ion of a divalent metal in the b sublattice.

It is seen from Table 4.4 that the two-sublattice model correctly (in principle) reflects the features of the magnetic structure of the spinels. In some spinel ferrites, however, more complicated types of ferrimagnetic structure may arise (Fig. 4.1f).

Most of the ferrites with a spinel structure have an easy-magnetization axis of the ⟨111⟩ type (the magnetic class below T_C is $\bar{3}m'$).

b. Ferrites with a Garnet Structure

The formula for garnet ferrites can be written as $3\,R_2O_3 \cdot 5\,Fe_2O_3$ or $R_3^{3+}Fe_2^{3+}Fe_3^{3+}O_{12} = R_3Fe_{15}O_{12}$, where R^{3+} is one of the trivalent ions of the rare-earth elements. Many ways exist for replacement of R and Fe ions. The crystallographic space group of garnets above T_C is $O_h^{10} = Ia3d$. The ferrimagnetic garnets are similar in structure to the natural mineral spessartite $Mn_3Al_2Si_3O_{12}$. The crystal structure of garnet is shown in [Ref. 2.5, Fig. 2.18].

The garnet structure has three sorts of intermediate sites occupied by three sorts of cations: 24 tetrahedral sites (*d* sites, point symmetry of the environment $\bar{4}$) with fourfold coordination, 16 octahedral sites (*a* sites, environment symmetry $\bar{3}$) with sixfold coordination, and 24 sites (*c* sites, environment symmetry 222) surrounded by eight oxygen ions forming a dodecahedron – a deformed cube. It is easy to see that the unit cell contains 8 formula units of $R_3Fe_5O_{12}$ or 4 units of $3\,R_2O_3 \cdot 5\,Fe_2O_3$. To explain the magnetic properties of garnet ferrites, a model consisting of three magnetic sublattices is used (in distinction to spinels, where two sublattices were considered). Two of them (*a* and *d*) contain Fe^{3+}, the number of tetrahedral sites (24 *d*) exceeding the number of octahedral ones (16 *a*). The third sublattice (24 *c* sites) contains R^{3+}.

According to Néel's theory, there is a strong negative exchange interaction between the *a* and *d* sublattices with resulting magnetization due to the excess of "tetrahedral" (*d*) magnetic moments. The R^{3+} ions are magnetized by the weak field from the *a* and *d* sublattices. As a result, the moments of the ions at the *c* sites are directed opposite to the magnetization in the *d* sublattice.

The arrangement of the sublattice magnetizations can be outlined as follows:

$$\underset{R^{3+}}{\overset{c}{\longrightarrow}}\ \underset{Fe^{3+}}{\overset{d}{\longleftarrow}}\ \underset{Fe^{3+}}{\overset{a}{\longrightarrow}} \ .$$

This scheme makes it possible to calculate the total magnetic moment of garnet ferrites (per mole of $3\,R_2O_3 \cdot 5\,Fe_2O_3$) at 0 K. Thus, for gadolinium garnet $3\,Gd_2O_3 \cdot 5\,Fe_2O_3$ we obtain (considering that the magnetic moments of Gd^{3+} and Fe^{3+} are 7 and 5 Bohr magnetons, respectively)

$$6 \cdot 7\,\mu_B - (6 \cdot 5\,\mu_B - 4 \cdot 5\,\mu_B) = 32\,\mu_B \ .$$

Experiment yields 30 μ_B. For yttrium garnet (the Y^{3+} ion is nonmagnetic) we have $6 \times 5\,\mu_B - 4 \times 5\,\mu_B = 10\,\mu_B$, which also agrees with experiment.

The magnetic class of most of the garnet ferrites is the same as for spinel, $\bar{3}m'$.

Owing to the competition between teh exchange interactions of the ions in the three magnetic sublattices, the garnets exhibit a large variety of $I_S - T$ dependences, which are discussed in item 2 on p. 287 (for instance, the appearance of compensation points).

c. *Hexagonal Ferrites*

From the theory of closest packing [Ref. 2.5, Chap. 1] it is known that the sequence of layers of the type *ABABA*... leads to a hexagonal structure, and the sequence *ABCABCA*... to the cubic closest packing [Ref. 2.5, Fig. 1.63]. The crystal structure of the hexagonal ferrites is built according to the principle of alternating blocks of cubic and hexagonal packings of oxygen ions. In some layers part of the oxygen ions are replaced by barium ions, which have a slightly larger ionic radius than the oxygen ions.

The simplest of the hexagonal structures, which is denoted by ***M***, is similar to the structure of the mineral magnetoplumbite. The sequence of the layers in the structure has the form *AC'ACBAB'ABCAC'ACB*.... The primed layers contain barium ions. The above sequence indicates that the unit cell of a crystal with the ***M*** structure contains 10 layers of oxygen ions. Above the Curie point the crystallographic space group of a crystal with the ***M*** structure is $P6_3/mmc$. The chemical formula of ***M*** structure compounds is BaO $\cdot\, 6\,Fe_2O_3 = BaFe_{12}O_{19}$ (magnetoplumbite contains Pb atoms instead of Ba). The unit cell contains two formula units of $BaFe_{12}O_{19}$.

The cross section of a hexagonal unit cell of the ***M*** structure along a plane containing sixfold screw axes (6_3), sixfold inversion axes ($\bar{6}$), and a twofold screw axis (2_1) is shown schematically in Fig. 4.22. Such a structure is conveniently represented as an alternation of blocks lettered ***S*** and ***R*** in Fig. 4.22. The construction of block ***S*** is similar to a spinel with the [111] axis directed vertically. Block ***S*** corresponds to the formula unit of 2 ($MeO \cdot Fe_2O_3$), where in the ***M*** structure Me = Fe^{3+}. It is easy to see that the electroneutrality condition is not fulfilled for the above formula (there is an excess charge +2). But the ***M*** structure is electroneutral as a whole, because the two oxygen ions in the block ***R*** = $(BaFe_6O_{11})^{-2}$ are replaced by divalent barium cations. In Fig. 4.22 ***R**** and ***S**** correspond to blocks rotated by 180° about axis 2_1 with respect to blocks ***R*** and ***S***. The crystal structure shown in Fig. 2.22 can be denoted by ***RSR*S****.

The Fe^{3+} ions in an ***M*** structure occupy five nonequivalent positions (lettered a, b, c, d, and e in Fig. 4.22), forming five magnetic sublattices. In addition to the previously encountered octahedral and tetrahedral sites, hexagonal ferrites have a new type of environment (sublattice e) consisting of five oxygen ions, whose centers form a trigonal bipyramid. All the Fe^{3+} ions belonging to the same sublattice are joined in Fig. 4.22 by dashed lines passing through the centers of symmetry of the structure.

The distribution of ions among the five sublattices in the ***M*** structure is shown in Table 4.5, which also demonstrates that the total magnetic moment

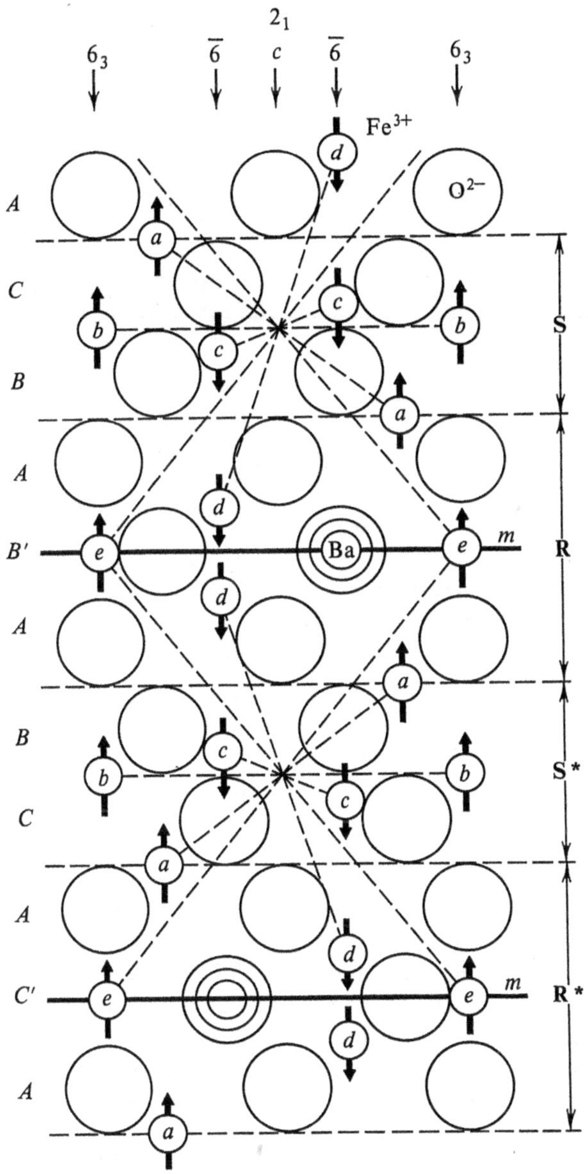

Fig. 4.22. Cross section of the *M* structure along the (11$\bar{2}$0) plane [4.12]

is easily calculated as the difference between the number of magnetic moments directed (conventionally) "upward" and the number of moments directed "downward" (taking into account that the magnetic moment of Fe^{3+} ions is equal to 5 μ_B):

$$(6+1-2-2+1)\cdot 5\,\mu_B = 20\,\mu_B \;.$$

Table 4.5. Distribution of Fe^{3+} ions among the sublattices in an ***M*** structure and total magnetization at $T=0$

Sublattice type	Environment	Number of Fe^{3+} ions	Direction of magnetic moment of sublattice
a	octahedron	6	↑
b	octahedron	1	↑
c	tetrahedron	2	↓
d	octahedron	2	↓
e	bipyramid	1	↑

Table 4.6. Main types of structures of hexagonal ferrites

Formula	Designation	Sequence of blocks	Number of layers per unit cell	Crystallographic space group
$BaO \cdot 6\,Fe_2O_3$	M	$\boldsymbol{RSR^*S^*}$	10	$P6_3/mmc$
$BaO \cdot 2\,MeO \cdot 8\,Fe_2O_3$	$W(MS)$	$\boldsymbol{RS_2R^*S_2^*}$ ($W = M+2S$)	14	$P6_3/mmc$
$2\,BaO \cdot 2\,MeO \cdot 6\,Fe_2O_3$	Y	$(\boldsymbol{TS})_3$	$3\times6=18$	$R\bar{3}m$
$3\,BaO \cdot 2\,MeO \cdot 12\,Fe_2O_3$	$Z(MY)$	$\boldsymbol{RSTSR^*S^*T^*S^*}$ ($Z = M+Y$)	22	$P6_3/mmc$
$2\,BaO \cdot 2\,MeO \cdot 14\,Fe_2O_3$	$X(M_2S)$	$(\boldsymbol{RSR^*S_2^*})_3$	$3\times12=36$	$R\bar{3}m$
$4\,BaO \cdot 2\,MeO \cdot 18\,Fe_2O_3$	M_2Y	$\boldsymbol{RSR^*S^*TS^*}$	16	$R\bar{3}m$

Besides the ***R*** and ***S*** blocks of which the structures of hexagonal ferrites are built, there is also the so-called ***T*** block consisting of four oxygen layers stacked according to the type $AB'A'B$. Two internal neighboring layers in this block contain barium ions. The ***T*** block can be regarded as a combination of the internal layers of two ***R*** blocks. The chemical composition of a ***T*** block corresponds to the hypothetical compound $2\,\{BaO \cdot 2\,Fe_2O_3\}$. Magnetically, the ***T*** block is "antiferromagnetic," because the magnetic moments of the ions cancel each other.

The ***R***, ***S***, and ***T*** blocks are those "bricks" of which the structures of hexagonal ferrites are built. Table 4.6 gives the main types of ferromagnetic oxides with a hexagonal structure, their traditional designations, and principal characteristics. In addition to crystals with the structures listed in Table 4.6, crystals with structures of the type $(\boldsymbol{TS}_n)\boldsymbol{T}$ have been synthesized, where n can, in principle, be increased to infinity. Such structures exemplify polytypy in magnetic substances. The unit cell parameters along the c axis in such structures increase to several thousand angstrom units.

4.5.2 Weakly Ferromagnetic Crystals (Canted Antiferromagnetics)

There is a group of crystals with almost antiferromagnetic ordering of the magnetic moments, which, however, show a slight spontaneous magnetization resulting from the disturbance (canting) of the strict collinearity in the arrangement of the magnetic moments. In Sect. 4.2 such crystals were called weak ferromagnetics. They are the most prominent examples of crystals whose magnetic properties were not fully understood until magnetic symmetry was used.

A necessary (but not a sufficient) condition for the appearance of weak ferromagnetism in a ferromagnetic crystal is that it should belong to one of the 31 ferromagnetic classes (Table 4.2) for which crystals can display $I_S \neq 0$.

Let us consider two isomorphous antiferromagnetics α-Fe_2O_3 (hematite) and Cr_2O_3. In the paramagnetic phase both crystals have a corundum-type structure [Ref. 2.5, Fig. 1.72] and belong to crystallographic class $\bar{3}m$. The rhombohedral unit cell contains four magnetic ions arranged on the threefold axis (Fig. 4.23). The spacing of the rhombohedral cell is denoted by c in Fig. 4.23.

For α-Fe_2O_3, at $T < 260$ K the sequence of the magnetic moments in atoms along the threefold axis is such (Fig. 4.23 a) that the structure retains the usual center of symmetry (its position is denoted by a cross in Fig. 4.23). The magnetic symmetry class coincides with crystallographic class $\bar{3}m$. In Cr_2O_3 (Fig. 4.23 c) the sequence of magnetic moments in the magnetically ordered phase is such that the center of symmetry binds together atoms with opposite directions of the magnetic moments, i.e., the inversion operation at the point is combined with operation R (center of antisymmetry). The corresponding magnetic class from among G' in Table 4.2 is class $\bar{3}'m'$; $\bar{3}m$ and $\bar{3}'m'$ are not among the classes for which a spontaneous magnetic moment is

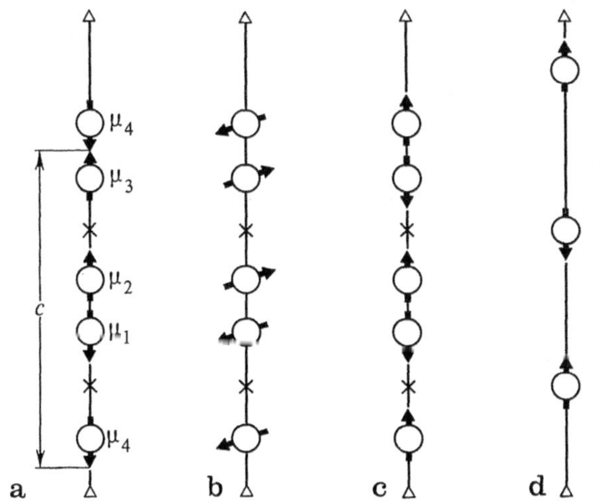

Fig. 4.23 Various types of magnetic ordering in corundum-type structures: (**a**) α-Fe_2O_3 ($T < 260$ K), (**b**) α-Fe_2O_3 ($T > 260$ K), (**c**) Cr_2O_3, (**d**) $FeCO_3$

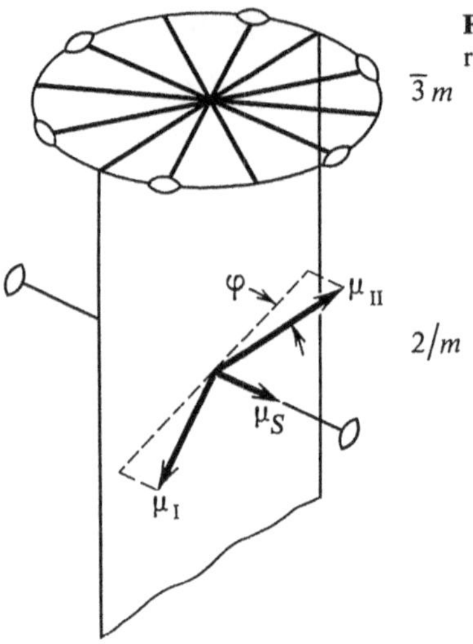

Fig. 4.24. Diagram explaining the possibility of weak ferromagnetism in hematite

permissible. Thus, the symmetry "forbids" the existence of weak ferromagnetism in them.

On heating to $T = 260$ K hematite undergoes a phase transition (so-called Morin point), in which the magnetic moments rotate (Fig. 4.23b), moving into a plane perpendicular to the threefold axis (into the basal plane). It is assumed that the rotation of the magnetic moments occurs in one of the three longitudinal planes of symmetry. A deviation of the antiferromagnetic axis from the threefold axis reduces the symmetry: it must now correspond to class $2/m$, which is among the magnetic classes for which crystals can exhibit a spontaneous moment along axis 2. It has been known from experiment that at $T > 260$ K hematite possesses a small spontaneous moment perpendicular to the threefold axis. The symmetry cause for the admissibility of weak ferromagnetism here consists in the possibility of changing the directions of the magnetic moments in such a way as to create a nonzero total magnetic moment in the unit cell while still preserving the symmetry of the crystal.

Figure 4.24 shows a spin configuration corresponding to class $2/m$. The four magnetic moments are here replaced by two issuing from the same point: $\mu_I = \mu_1 + \mu_4$ and $\mu_{II} = \mu_2 + \mu_3$ (two-sublattice model). It is seen from Fig. 4.24 that the noncollinearity characterized by angle φ does not disturb symmetry $2/m$. It is accompanied by the appearance a spontaneous moment μ_S directed along the twofold axis.

Since a crystal of α-Fe_2O_3 has three twofold axes (upper part of Fig. 4.24), it will be divided into three groups of domains in accordance with the principles described in Sect. 4.3; and 60°, 120°, and 180° walls may arise between the separate domains.

The symmetry approach to the explanation of the weak ferromagnetism of hematite was first used by *Dzyaloshinsky* [4.13]. Earlier it was supposed that the weak ferromagnetism is due to nonideality of the crystal, and therefore called "parasitic" ferromagnetism. From the energy standpoint, the noncollinearity in the arrangement of the spins of atoms 1 and 2 is taken into account by introducing an interaction of the form

$$E_d = -2D_{12}[S_1 \times S_2]$$

into the free energy.

This part of the free energy decreases when S_1 and S_2 tend to be perpendicular to one another and to some vector $\boldsymbol{D}$ (Dzyaloshinsky vector). Contrary to exchange energy E_{exch} [4.20], energy E_d is antisymmetric with respect to spins $\boldsymbol{D}_{12} = -\boldsymbol{D}_{21}$.

The value of E_d is usually one order of magnitude less than E_{exch} and is a second-order correction to it; therefore, in weak ferromagnetics angle φ is very small ($\sim 0.5°$).

The above-discussed phase transition in hematite $\bar{3}m \rightarrow 2/m$, which is accompanied by the appearance of moment μ_s parallel to axis 2, is one of the transitions listed in Table 4.3. It is seen there that, in principle, transition $\bar{3}m \rightarrow 2'/m'$ is possible, in which μ_S is disposed in the plane of symmetry, perpendicularly to the threefold axis.

Similar reasoning can be applied in describing the magnetic structures of antiferromagnetic carbonates $MnCO_3$, $CoCO_3$, $NiCO_3$, and $FeCO_3$. The symmetry of these crystals above T_N is described by the same crystallographic space group as for α-Fe_2O_3 and Cr_2O_3 ($D_{3d}^6 = R\bar{3}c$), but the rhombohedral cell has only two metal ions on the threefold axis, as shown in Fig. 4.23 d. In $FeCo_3$, the antiferromagnetic axis is arranged along the threefold [111] axis, and the magnetic symmetry class, as in α-Fe_2O_3 at $T < 260$ K, coincides with the crystallographic ($\bar{3}m$), which rules out weak ferromagnetism. In crystals of $MnCO_3$, $CoCO_3$, and $NiCO_3$ the antiferromagnetic axis does not coincide with [111], and therefore they belong to a lower-symmetry magnetic class ($2/m$ or $2'/m'$), for which noncollinearity of magnetic moments is permissible. Experiment shows that carbonates $CoCO_3$, $NiCO_3$, and $MnCO_3$ are indeed weak ferromagnetics with $\mu_S \perp$ [111]; $FeBO_3$ and FeF_3 are also trigonal weak ferromagnetics.

One more interesting group of antiferromagnetics with weak ferromagnetism is represented by crystals of fluorides NiF_2, MnF_2, FeF_2, and CoF_2, which have a rutile-type structure (symmetry above T_N $4/mmm$). The rutile structure is shown in [Ref. 2.5, Fig. 2.15]. The determination of the magnetic structure showed that below T_N NiF_2 belongs to magnetic class $mm'm'$, whereas MnF_2, FeF_2, and CoF_2 belong to $4'/mmm'$.

Figure 4.25 shows the magnetic structures corresponding to these two classes. In $4'/mmm'$ the antiferromagnetic axis is parallel to the fourfold axis. For crystals of this class the spontaneous moment μ_S is forbidden, and therefore MnF_2, FeF_2, and CoF_2 cannot be weak ferromagnetics. In a NiF_2 crystal the antiferromagnetic axis is parallel to axis $2'$. Superimposing magnetic ions I

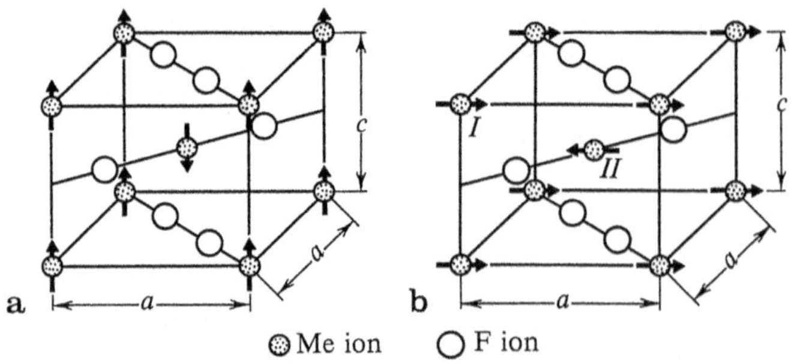

Fig. 4.25. Magnetic structure of fluorides: (**a**) MnF_2, FeF_2, CoF_2 ($4'/mmm'$); (**b**) NiF_2 ($mm'm'$)

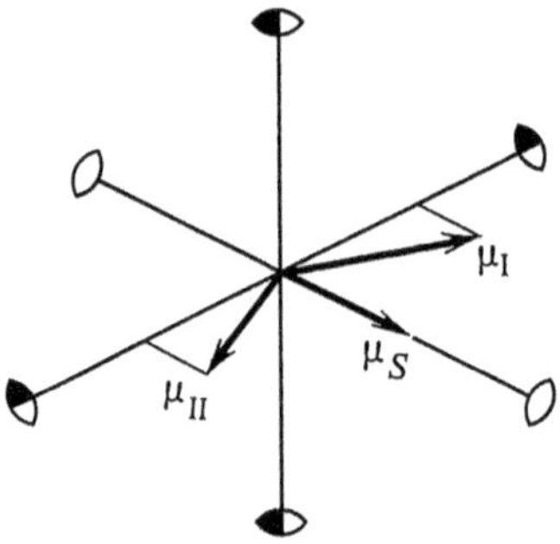

Fig. 4.26. Configuration composed of two magnetic moments corresponding to point group $mm'm'$. Only axis 2 and two axes 2′ are shown

and II, we obtain a spin configuration corresponding to point group $mm'm'$. From Fig. 4.26 it is seen that in NiF_2 noncollinearity of μ_I and μ_{II} is admissible with the appearance of μ_S along the simple twofold axis. The remaining two axes are twofold antisymmetry axes which reverse the direction of the magnetic moments on rotation through 180° and thus leave the entire configuration unaltered. One can see from Tables 4.2, 3, that $mm'm'$ is among those classes for which crystals can display ferromagnetism.

From the above-discussed symmetry considerations it becomes clear why, of all the fluorides, NiF_2 alone shows weak ferromagnetism.

We now proceed to a large group of weak ferromagnetics with formula RMO_3, where R^{3+} is a rare-earth ion. These compounds are called orthoferrites ($M = Fe^{3+}$) or orthochromites ($M = Cr^{3+}$), depending on the type of the M ion. They crystallize in the orthorhombically distorted structure of perovskite with four molecular units per unit cell. The space group of these compounds in the paramagnetic phase is $D_{2h}^{16} = Pbnm$. Figure 4.27 depicts the crystal structure of $RFeO_3$. Inside the orthorhombic unit cell (dot-and-dash lines) one can single out two monoclinic pseudocells with a monoclinicity angle $\beta \approx 92-93°$ (dashed lines). Each monoclinic pseudocell actually represents a distorted perovskite structure, which explains the allocation of the orthoferrites to the group of perovskite-like compounds. Sometimes the perovskite structure is portrayed as a cube with R^{3+} ions located at the

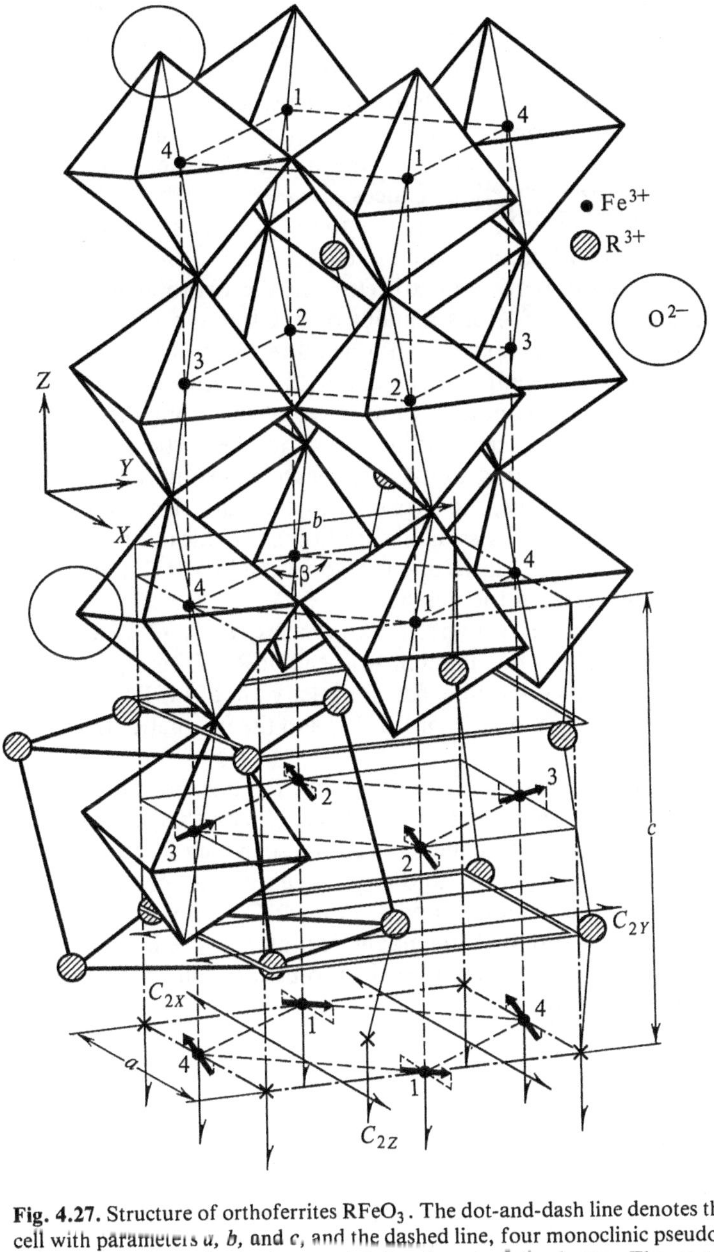

Fig. 4.27. Structure of orthoferrites $RFeO_3$. The dot-and-dash line denotes the orthorhombic unit cell with parameters a, b, and c, and the dashed line, four monoclinic pseudocells. The perovskite cube is singled out in the bottom left-hand part of the figure. The centers of symmetry are denoted by crosses, and the mirror planes of symmetry by double lines. The orientation of the magnetic moments corresponds to configuration Γ_4 shown in Fig. 4.28b

vertices and the oxygen anions O^{2-} at the face centers. The distorted cube depicted in the bottom left-hand part of Fig. 4.27 corresponds to such a cell (the perovskite structure can also be seen in [Ref. 2.5, Fig. 2.16]).

The perovskite structure in rare-earth orthoferrites is distorted because the radii of R ions are small as compared with those required for realization of the ideal perovskite structure. As the radius of the R ion increases (from Lu to La), the degree of structure distortion reduces.

In undistorted cubic perovskite structure the O^{2-} ions surrounding the Fe^{3+} ions form octahedra whose axes are directed along the edges of the unit cube. The distortion of the perovskite structure in the orthoferrites is such that the environment of the Fe^{3+} ions remains octahedral as a whole, but the octahedra axes deviate from the above-mentioned directions. The regularity in the alternation of oxygen octahedra can be understood from Fig. 4.27, where the pattern of the structure is extended by one more period along the $c(Z)$ axis for clarity. From the figure it is seen, for instance, that the vertical axes of the chain of octahedra arranged along the Z axis (passing through ions 1, 2, 1, 2... or 3, 4, 3, 4...) form zigzags, the axes of the octahedra surrounding ions 1 and 3, 2 and 4 being parallel. A similar situation exists along the X and Y directions. Along the Z axis, a zigzag line is also formed by R^{3+} ions, which are located in the voids between the octahedra.

Space group *Pbnm* has three twofold screw axes, C_{2X}, C_{2Y}, and C_{2Z}, which are shown in Fig. 4.27. The magnetic unit cell of $RFeO_3$ is equivalent to the crystallographic one. We assume for simplicity that the R^{3+} ions are diamagnetic (for instance, Y^{3+}, La^{3+}, or Lu^{3+}), and the magnetic structure is formed only by Fe^{3+} ions.

Let us consider, from the symmetry standpoint, the possible arrangement of the magnetic moments in the structure of $RFeO_3$. To do this, it will suffice to use the generating symmetry elements, for which we choose, for group *Pbnm*, twofold screw axes C_{2X}. C_{2Y}, and C_{2Z}. It can be seen from Fig. 4.27 that screw axis C_{2X} interchanges ions 2 and 3, 1 and 4; C_{2Y}, ions 1 and 3, 2 and 4; and C_{2Z}, ions 1 and 2, 3 and 4. If the magnetic moments are inclined to the X, Y, and Z axes (general positions), their projections on these axes are related by the following expressions:

$$\begin{aligned} \mu_{1X} &= -\mu_{2X} = -\mu_{3X} = \mu_{4X} , \\ \mu_{1Y} &= -\mu_{2Y} = \mu_{3Y} = -\mu_{4Y} , \\ \mu_{1Z} &= \mu_{2Z} = -\mu_{3Z} = -\mu_{4Z} . \end{aligned} \tag{4.93}$$

From (4.93) it is seen that the projections of the magnetic moments in each direction cancel one another, i.e., for the structure with simple twofold symmetry axes (point group *mmm*) only antiferromagnetism is possible. Spin configuration (4.93), which is denoted by symbol Γ_1, is shown in Fig. 4.28a.

Configuration Γ_1 does not exhaust the possible types of magnetic structures which can be realized in orthoferrites, with their allocation to crystal-

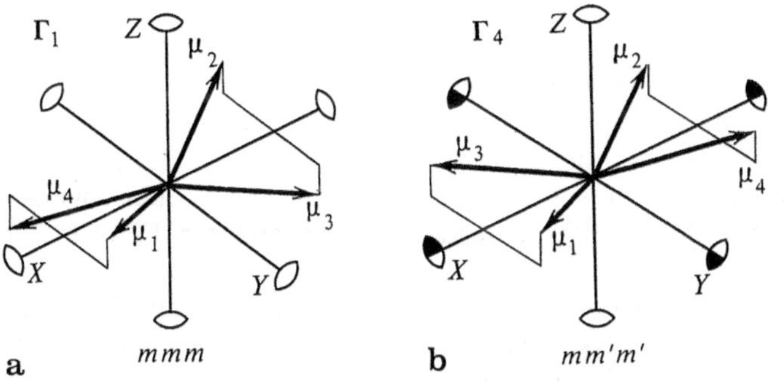

Fig. 4.28. Spin configuration of the type Γ_1 (**a**) and Γ_4 (**b**) corresponding to point groups *mmm* and *mm'm'*, respectively

lographic class *mmm* retained. To make the consideration complete, we must use the magnetic symmetry. According to Table 4.2, the following magnetic classes correspond to class *mmm: m'm'm'*, *m'mm*, and *mm'm'*. Among the symmetry elements classes *m'm'm'* and *m'mm* contain a center of antisymmetry, which contradicts the presence of a magnetic moment in Fe^{3+} ions, because they occupy a centrosymmetric position in the structure (Fig. 4.27). Hence, classes *m'm'm'* and *m'mm* must be excluded from consideration. Class *mm'm'* contains an ordinary center of symmetry. In considering the possible magnetic structures it is sufficient to use, as before, the operations corresponding to axes 2, 2′, 2′ in this case. Then there are three possible combinations of simple axes and antisymmetry axes within one and the same class *mm'm'*: $2_X, 2'_Y, 2'_Z$ (configuration Γ_2); $2'_X, 2_Y, 2'_Z$ (Γ_3); $2'_X, 2'_Y, 2_Z$ (Γ_4). The projections of the magnetic moments for configuration Γ_4 are related by the expressions

$$\begin{aligned} \mu_{1X} &= -\mu_{2X} = \mu_{3X} = -\mu_{4X} \,, \\ \mu_{1Y} &= -\mu_{2Y} = -\mu_{3Y} = \mu_{4Y} \,, \\ \mu_{1Z} &= \mu_{2Z} = \mu_{3Z} = \mu_{4Z} \,. \end{aligned} \tag{4.94}$$

Graphically, spin configuration (4.94) is depicted in Fig. 4.28 b.

From (4.94) and Fig. 4.28 b it is seen that configuration Γ_4 (as distinct from Γ_1) is appropriate for the appearance of a spontaneous moment along the Z axis, which agrees with experiment (almost all the orthoferrites have a magnetic structure with a weak ferromagnetic moment $\mu_S \| Z$ at room temperature and higher). Configurations Γ_1 ($\mu_S = 0$) and Γ_2 ($\mu_S \| X$) have also been found in some orthoferrites within definite temperature ranges.

Thus, in the case of orthoferrites we ascertain once more that in describing the magnetic properties of weak ferromagnetics we cannot restrict ourselves to the conventional symmetry.

4.5.3 Reorientation Transitions

At the end of Sect. 4.4 it was shown that the equilibrium direction $\boldsymbol{I}_S$ in a crystal (easy axis) is realized at definite relationships between magnetic anisotropy constants K_1 and K_2. If these conditions are changed, the easy magnetization axis changes its direction. We now turn again to a crystal of cobalt, whose magnetic properties have already been considered in this chapter. Figure 4.29 shows the temperature dependence of anisotropy constants K_1 and K_2 for cobalt. It is seen that near 520 K constant K_1 passes through zero and becomes negative. Below 520 K, K_1 and K_2 satisfy condition $K_1+K_2>0$ and $K_1>0$ (4.88), and in the range of 520 – 580 K, condition $0<-K_1<2K_2$ (4.89). This means that at $T<520$ K the c axis is the easy-magnetization axis, and at $T>520$ K the equilibrium direction of $\boldsymbol{I}_S$ begins to deviate from the c axis by angle θ_0, which is defined by (4.87). A polydomain crystal (see p. 324) must then retain, on the whole, the symmetry of the initial paramagnetic phase; therefore, at $520<T<580$ K the directions of $\boldsymbol{I}_S$ in the separate domains are distributed along the generatrices of a cone with a cone angle $2\theta_0$ (the cone axis coincides with the c axis). Accurate anisotropy measurements show that not all the directions coinciding with the cone generatrix are energetically equivalent: a rotation of $\boldsymbol{I}_S$ in domains occurs in one of the longitudinal planes of symmetry.

Transition in cobalt at 520 K and any other transition in a ferromagnetic accompanied by a change in the direction of the easy-magnetization axis is a phase transition from one magnetic class (from among the classes underscored in Table 4.2) to another. Knowing the direction of $\boldsymbol{I}_S$, one can establish the magnetic class with the aid of Table 4.3. It is easy to see that the above-discussed case of transition in cobalt is a transition from class $6/mm'm'$ to $2'/m'$. At $T\approx 580$ K, constant K_1 in cobalt becomes equal to $-2K_2$ (Fig. 4.29). According to (4.87) this means that θ_0 reaches 90°, and $\boldsymbol{I}_s$ is disposed in the

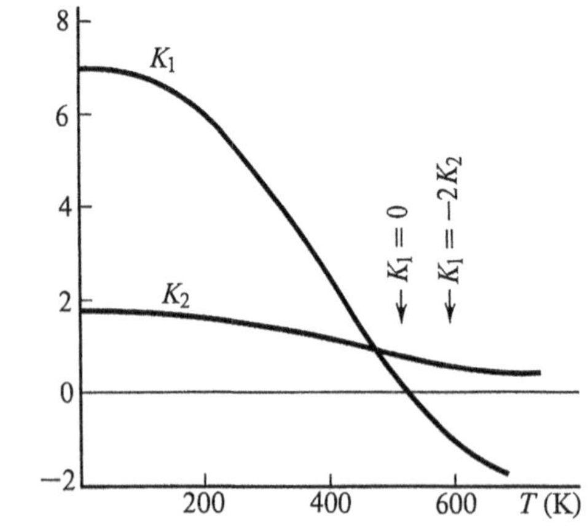

Fig. 4.29. Temperature dependence of anisotropy constants K_1 and K_2 for a cobalt crystal [4.15]

basal plane (more precisely, coincides with one of the twofold axes). A "cone" → "plane" transition is a phase transition of the type $2'/m' \rightarrow mm'm'$. Thus, when cobalt is heated, the following chain of magnetic phase transitions is observed: $6/mm'm' \rightarrow 2'/m' \rightarrow mm'm'$. We note that at $T \approx 690$ K cobalt undergoes a phase transition of a nonmagnetic nature, in which it transfers to the cubic modification.

The peculiarities of the magnetic anisotropy during the heating of cobalt were studied as far back as the early thirties. Many other cases are known where a change in the temperature or composition of a ferromagnetic crystal changes the easy-magnetization directions. It is only in recent years, in connection with the successful utilization of the principles of symmetry in describing the magnetic properties of crystals (especially orthoferrites), that attention has been given to the fact that this process involves phase transitions. Phase transitions associated with a change in magnetic symmetry and a change in the equilibrium direction of $\boldsymbol{I}_S$ in a crystal are called *reorientation transitions*. The meaning of this term is clear: a change in the direction of $\boldsymbol{I}_S$ is associated with a change in the orientation of the magnetic moments with respect to the symmetry elements of the initial paramagnetic phase. By reorientation transitions, in a broad sense, one can imply any phase transitions involving a change in the orientation of the magnetic moments, taking place at temperatures below the Curie or Néel point (such phase transitions are called "order-order" ones, as distinct from the "order-disorder" transitions at the Curie or Néel point). Then transitions from the ferromagnetic (or weakly ferromagnetic) to the antiferromagnetic state can also be classified among the reorientation transitions.

Figure 4.30 shows examples of reorientation transitions in some magnetically ordered crystals belonging to different syngonies. A crystal of hematite, whose properties were discussed in the preceding section, has been chosen as a trigonal crystal. Orthorhombic orthoferrites and orthochromites $RFeO_3$ and $RCrO_3$ hold the first place as regards the diversity of reorientation transitions. Figure 4.30 presents only transitions most typical for these compounds. It is interesting to note that the configurations Γ_2, Γ_3, and Γ_4, between which the phase transitions take place, belong to one and the same magnetic class $mm'm'$ (they naturally belong to different magnetic space groups).

These transitions are usually continuous and occur in a certain temperature range by a smooth rotation of the axis of antiferromagnetism in one of the symmetry planes; the weak ferromagnetic moment then also rotates in this plane and changes its direction from one symmetry axis to another. In the course of reorientation, the magnetic symmetry of each domain reduces to monoclinic $2'/m'$. Such transitions are observed in $SmFeO_3$ at above-room temperatures and also in $SmCrO_3$ and $GdCrO_3$ at low temperatures (Fig. 4.30). Some of the orthochromites and orthoferrites ($DyFeO_3$, $ErCrO_3$) transfer at low temperatures to the antiferromagnetic state, as in hematite at 260 K. In the range of helium temperatures, in $RFeO_3$ and $RCrO_3$, where R^{3+} has a magnetic moment, one observes additional phase transitions due to the

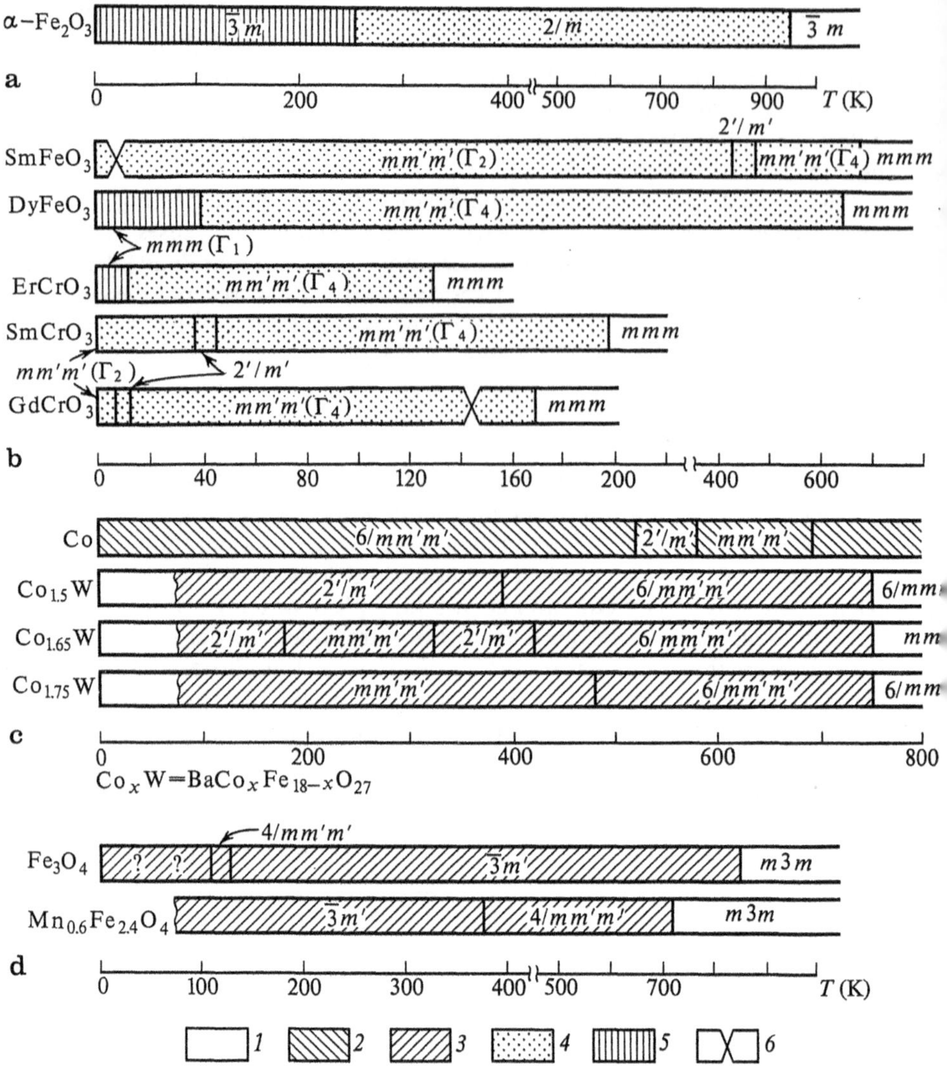

Fig. 4.30. Examples of reorientation transitions in magnetically ordered crystals of various symmetry: (**a**) trigonal, (**b**) orthorhombic, (**c**) hexagonal (at $T > 695$ K cobalt undergoes allotropic transformation and becomes cubic), (**d**) cubic crystals; (*1*) paramagnetic phase, (*2*, *3*) ferro- and ferrimagnetism, respectively, (*4*) weak ferromagnetism, (*5*) antiferromagnetism, (*6*) compensation point

appearance of a magnetic order in the rare-earth sublattice (second Néel point). These transitions are not shown in Fig. 4.30.

An interesting feature of orthoferrites and orthochromites is the presence of compensation points in some of them. This is associated with the nature of interaction between sublattices containing iron and rare-earth metal ions. But the compensation points do not correspond to the phase transitions, because the passage through zero by the total magnetic moment is not associated with a rearrangement of the magnetic structure.

The reorientation transitions in hexagonal crystals shown in Fig. 4.30 include the above-discussed case of metallic cobalt. Transitions of the type "easy axis" ($6/mm'm'$) → "cone" ($2'/m'$) → "easy plane" ($mm'm'$) are observed not only in cobalt. Figure 4.30 gives examples of such transitions observed in some hexagonal ferrites with substitution of Co^{2+} ions (data on anisotropy at $T < 77$ K are unavailable for them).

In cubic crystals one usually observes transitions between classes $\bar{3}m'$ (easy-magnetization axis of the ⟨111⟩ type) and $4/mm'm'$ (⟨100⟩ type axis). Figure 4.30 includes examples of transitions of crystals of pure magnetite Fe_3O_4 and magnetite with partial replacement of iron ions by manganese. On cooling below 119 K, magnetite crystals undergo a phase transition due to the ordering of Fe^{2+} and Fe^{3+} in the b sublattice. The magnetic class of the low-temperature modification of mangetite has not yet been precisely established.

In conclusion, we note that reorientation transitions may be either of the first or the second order. For instance, gradual reorientation in orthoferrites and orthochromites of the type $\Gamma_2 \rightleftarrows \Gamma_4$ through the intermediate phase $2'/m'$ proceeds as two successive second-order phase transitions. The same kind of transition should be expected for transitions of the type "axis" ⇄ "cone" ⇄ "plane" in hexagonal crystals, because the phenomenological description of magnetic anisotropy in such transitions is similar to that of the reorientation in orthoferrites. On the other hand, transitions from the weakly ferromagnetic phase to the paramagnetic phase occur jumpwise as a first-order phase transition (transition at 260 K in hematite or similar transitions in $DyFeO_3$ and $ErCrO_3$). Transitions between classes $\bar{3}m'$, and $4'/mm'm'$ in cubic crystals are evidently also first-order phase transitions. The nature of phase transitions may change in the presence of an external magnetic field.

4.6 Piezomagnetic and Magnetoelectric Effects

By the piezomagnetic effect is meant the appearance of a macroscopic magnetic moment on crystal deformation. The piezomagnetic effect is described by the relation

$$I_i = Q_{ijk}\,\varepsilon_{jk} \ .$$

Since tensor Q_{ijk} relates axial vector I_i to polar tensor ε_{jk}, it is a third-rank axial tensor. All the components of tensor Q_{ijk} must reverse their signs under time inversion operation R, because vector I_i possesses this property, while tensor ε_{jk} does not. From the above-enumerated properties of tensor Q_{ijk} it follows that it is identically equal to zero in crystals belonging to those magnetic classes which contain operation R in combination with a symmetry center (i.e., the antisymmetry center). These classes are as follows (there are 21 in all):

$$\bar{1}', 2/m', 2'/m, m'm'm', m'mm, 4/m', 4'/m', 4/m'm'm', 4'/m'mm', 4/m'mm, \bar{3}', \bar{3}'m', \bar{3}'m, 6'/m, 6/m', 6/m'mm, 6/m'm'm', 6'/mmm', m'3, m'3m', m'3m \ .$$

Because the deformation tensor is symmetric ($\varepsilon_{jk} = \varepsilon_{kj}$), tensor Q_{ijk} is equal to zero for three more classes of the cubic syngony: $4'3m$, 432, and $m3m$. Thus, the piezomagnetic effect can be observed in crystals belonging to the remaining ($90-21-3=66$) magnetic classes. These classes are listed in Table 4.7 which also includes the matrices of the coefficients of the piezomagnetic tensor for 66 magnetic classes in which piezomagnetism is possible. Tensors Q_{ijk} are written in matrix form (Q_{iA}), where $i = 1, 2, 3$, and $A = 1, 2, 3, 4, 5, 6$.

Despite the large number of magnetic classes admitting piezomagnetic effect, it has been observed experimentally only in three antiferromagnetics: CoF_2, MnF_2 [4.16], and $FeCO_3$. Since CoF_2 and MnF_2 belong to magnetic class $4'/mmm'$, the piezomagnetic effect (Table 4.7) must be characterized by two coefficients Q_{14} and Q_{36}. The piezomagnetic coefficients ($Gs \cdot kg^{-1}\, cm^2$) have the following values at 20.4 K: $Q_{14} = 2.1 \times 10^{-3}$, $Q_{36} = 0.8 \times 10^{-3}$, and $Q_{14} \simeq 10^{-5}$, $Q_{36} \approx 0$ for CoF_2 and MnF_2, respectively. It is seen that the piezomagnetic effect in MnF_2 is about 100 times weaker than in CoF_2.

By the magnetoelectric effect is meant the appearance of a macroscopic magnetic moment I_i on application of electric field E_j:

$$I_i = Q_{ij} E_j \ .$$

Tensor Q_{ij} is a second-rank axial tensor. Its components (as in tensor Q_{ijk}) must reverse their signs on operation R.

Tensor Q_{ij} is equal to zero for 11 centrosymmetric classes from group G (Table 4.2) –

$$\bar{1}, 2/m, mmm, 4'/m, 4/mmm, \bar{3}, \bar{3}m, 6/m, 6/mmm, m3, m3m \ –$$

and for 10 classes from group G' containing structures with an ordinary center of symmetry –

$$2'/m', mm'm', 4'/m, 4/mm'm', 4'/mmm', \bar{3}m', 6'/m', 6/mm'm', 6'/m'mm', m3m' \ .$$

Table 4.7. Matrices of coefficients of tensor Q_{ik} for 66 magnetic classes in which the piezomagnetic effect is possible

Classes 1, 1′

$$\begin{Vmatrix} Q_{11} & Q_{12} & Q_{13} & Q_{14} & Q_{15} & Q_{16} \\ Q_{21} & Q_{22} & Q_{23} & Q_{24} & Q_{25} & Q_{26} \\ Q_{31} & Q_{32} & Q_{33} & Q_{34} & Q_{35} & Q_{36} \end{Vmatrix}$$

Classes 2, *m*, 2/*m*

$$\begin{Vmatrix} 0 & 0 & 0 & Q_{14} & Q_{15} & 0 \\ 0 & 0 & 0 & Q_{24} & Q_{25} & 0 \\ Q_{31} & Q_{32} & Q_{33} & 0 & 0 & Q_{36} \end{Vmatrix}$$

Classes 2′, *m*′, 2′/*m*′

$$\begin{Vmatrix} Q_{11} & Q_{12} & Q_{13} & 0 & 0 & Q_{16} \\ Q_{21} & Q_{22} & Q_{23} & 0 & 0 & Q_{26} \\ 0 & 0 & 0 & Q_{34} & Q_{35} & 0 \end{Vmatrix}$$

Classes 222, *mm*2, *mmm*

$$\begin{Vmatrix} 0 & 0 & 0 & Q_{14} & 0 & 0 \\ 0 & 0 & 0 & 0 & Q_{25} & 0 \\ 0 & 0 & 0 & 0 & 0 & Q_{36} \end{Vmatrix}$$

Classes 22′2′, *m*′*m*′2, *m*′*m*2′, *mm*′*m*′

$$\begin{Vmatrix} 0 & 0 & 0 & 0 & Q_{15} & 0 \\ 0 & 0 & 0 & Q_{24} & 0 & 0 \\ Q_{31} & Q_{32} & Q_{33} & 0 & 0 & 0 \end{Vmatrix}$$

Classes 4, $\bar{4}$, 4/*m*, 6, $\bar{6}$, 6/*m*

$$\begin{Vmatrix} 0 & 0 & 0 & Q_{14} & Q_{15} & 0 \\ 0 & 0 & 0 & Q_{15} & -Q_{14} & 0 \\ Q_{31} & Q_{31} & Q_{33} & 0 & 0 & 0 \end{Vmatrix}$$

Classes 4′, $\bar{4}$′, 4′/*m*

$$\begin{Vmatrix} 0 & 0 & 0 & Q_{14} & Q_{15} & 0 \\ 0 & 0 & 0 & -Q_{15} & Q_{14} & 0 \\ Q_{31} & -Q_{31} & 0 & 0 & 0 & Q_{36} \end{Vmatrix}$$

Classes 422, 4*mm*, $\bar{4}2m$, 4/*mmm*, 622, 6*mm*, $\bar{6}m2$, 6/*mmm*

$$\begin{Vmatrix} 0 & 0 & 0 & Q_{14} & 0 & 0 \\ 0 & 0 & 0 & 0 & -Q_{14} & 0 \\ 0 & 0 & 0 & 0 & 0 & 0 \end{Vmatrix}$$

Classes 4′22′, 4′*mm*′, $\bar{4}'2m'$, $\bar{4}'2'm$, 4′/*mmm*′

$$\begin{Vmatrix} 0 & 0 & 0 & Q_{14} & 0 & 0 \\ 0 & 0 & 0 & 0 & Q_{14} & 0 \\ 0 & 0 & 0 & 0 & 0 & Q_{36} \end{Vmatrix}$$

Classes 42′2′, 4*m*′*m*′, $\bar{4}2'm'$, 4/*mm*′*m*′, 62′2′, 6*m*′*m*′, $\bar{6}m'2'$, 6/*mm*′*m*′

$$\begin{Vmatrix} 0 & 0 & 0 & 0 & Q_{15} & 0 \\ 0 & 0 & 0 & Q_{15} & 0 & 0 \\ Q_{31} & Q_{31} & Q_{33} & 0 & 0 & 0 \end{Vmatrix}$$

Classes 3, $\bar{3}$

$$\begin{Vmatrix} Q_{11} & -Q_{11} & 0 & Q_{14} & Q_{15} & -2Q_{22} \\ -Q_{22} & Q_{22} & 0 & Q_{15} & -Q_{14} & -2Q_{11} \\ Q_{31} & Q_{31} & Q_{33} & 0 & 0 & 0 \end{Vmatrix}$$

Classes 32, 3*m*, $\bar{3}m$

$$\begin{Vmatrix} 0 & 0 & 0 & Q_{14} & 0 & -2Q_{22} \\ -Q_{22} & Q_{22} & 0 & 0 & -Q_{14} & 0 \\ 0 & 0 & 0 & 0 & 0 & 0 \end{Vmatrix}$$ [a]

Classes 32′, 3*m*′, $\bar{3}m'$

$$\begin{Vmatrix} Q_{11} & -Q_{11} & 0 & 0 & Q_{15} & 0 \\ 0 & 0 & 0 & Q_{15} & 0 & -2Q_{11} \\ Q_{31} & Q_{31} & Q_{33} & 0 & 0 & 0 \end{Vmatrix}$$ [a]

Classes 6′, $\bar{6}$′, 6′/*m*′

$$\begin{Vmatrix} Q_{11} & -Q_{11} & 0 & 0 & 0 & -2Q_{22} \\ -Q_{22} & Q_{22} & 0 & 0 & 0 & -2Q_{11} \\ 0 & 0 & 0 & 0 & 0 & 0 \end{Vmatrix}$$

Classes 6′22′, 6′*mm*′, 6′*m*′2, 6′*m*2′, 6′/*m*′*mm*′

$$\begin{Vmatrix} 0 & 0 & 0 & 0 & 0 & -2Q_{22} \\ -Q_{22} & Q_{22} & 0 & 0 & 0 & 0 \\ 0 & 0 & 0 & 0 & 0 & 0 \end{Vmatrix}$$ [a]

Classes 23, *m*3, 4′32′, $\bar{4}'3m'$, *m*3*m*′

$$\begin{Vmatrix} 0 & 0 & 0 & Q_{14} & 0 & 0 \\ 0 & 0 & 0 & 0 & Q_{14} & 0 \\ 0 & 0 & 0 & 0 & 0 & Q_{14} \end{Vmatrix}$$

[a] The *Y* axis coincides with the 2 axis

A detailed consideration of the properties of tensor Q_{ij} shows that it is equal to zero in 11 more classes:

$$\bar{6}, 6', 6'/m, \bar{6}m2, \bar{6}m'2', 6'm'm, 6'22', 6'/mmm', \bar{4}3m, 4'32', m'3m \ .$$

Thus, the magnetoelectric effect is possible in crystals belonging to $90-11-10-11=58$ magnetic classes. Table 4.8 lists 58 magnetic classes in

Table 4.8. Matrices of the coefficients of tensor Q_{ij} for 52 magnetic classes in which the magnetoelectric effect is possible

Classes 1, 1′	Classes 2, m', $2/m'$	Classes 2′, m, $2'/m$
$\begin{Vmatrix} Q_{11} & Q_{12} & Q_{13} \\ Q_{21} & Q_{22} & Q_{23} \\ Q_{31} & Q_{32} & Q_{33} \end{Vmatrix}$	$\begin{Vmatrix} Q_{11} & Q_{12} & 0 \\ Q_{21} & Q_{22} & 0 \\ 0 & 0 & Q_{33} \end{Vmatrix}$	$\begin{Vmatrix} 0 & 0 & Q_{13} \\ 0 & 0 & Q_{23} \\ Q_{31} & Q_{32} & 0 \end{Vmatrix}$
Classes 222, $m'm'2$, $m'm'm'$	Classes $22'2'$, $2mm$, $m'm2'$, $m'mm$	Classes 4, $\bar{4}'$, $4/m'$, 3, $\bar{3}'$, 6, $\bar{6}'$, $6/m'$
$\begin{Vmatrix} Q_{11} & 0 & 0 \\ 0 & Q_{22} & 0 \\ 0 & 0 & Q_{33} \end{Vmatrix}$	$\begin{Vmatrix} 0 & Q_{12} & 0 \\ Q_{21} & 0 & 0 \\ 0 & 0 & 0 \end{Vmatrix}$	$\begin{Vmatrix} Q_{11} & Q_{12} & 0 \\ -Q_{12} & Q_{11} & 0 \\ 0 & 0 & 0 \end{Vmatrix}$
Classes $4'$, $\bar{4}$, $4'/m'$	Classes 422, $4m'm'$, $\bar{4}'2m'$, $4/m'm'm'$, 32, $3m'$, $\bar{3}'m'$, 622, $6m'm'$, $\bar{6}'m'2$, $6/m'm'm'$	Classes $4'22'$, $4'mm'$, $\bar{4}2m$, $\bar{4}2'm'$, $4'/m'mm'$
$\begin{Vmatrix} Q_{11} & Q_{12} & 0 \\ Q_{12} & -Q_{11} & 0 \\ 0 & 0 & 0 \end{Vmatrix}$	$\begin{Vmatrix} Q_{11} & 0 & 0 \\ 0 & Q_{11} & 0 \\ 0 & 0 & Q_{33} \end{Vmatrix}$	$\begin{Vmatrix} Q_{11} & 0 & 0 \\ 0 & -Q_{11} & 0 \\ 0 & 0 & 0 \end{Vmatrix}$
Classes $42'2'$, $4mm$, $\bar{4}'2'm$, $4/m'mm$, $32'$, $3m$, $\bar{3}'m$, $62'2'$, $6mm$, $\bar{6}'m2'$, $6m'mm$	Classes 23, $m'3$, 432, $\bar{4}'3m'$, $m'3m'$	
$\begin{Vmatrix} 0 & Q_{12} & 0 \\ -Q_{12} & 0 & 0 \\ 0 & 0 & 0 \end{Vmatrix}$	$\begin{Vmatrix} Q_{11} & 0 & 0 \\ 0 & Q_{11} & 0 \\ 0 & 0 & Q_{11} \end{Vmatrix}$	

which the magnetoelectric effect is possible and gives the matrices of the coefficients of tensor Q_{ij}.

The previous note concerning the insufficient experimental data on the detection of the piezomagnetic effect fully refers to the magnetoelectric effect as well. The magnetoelectric effect was first observed experimentally [4.17] in a crystal of Cr_2O_3 (magnetic class $\bar{3}'m'$). The crystal was affected by an alternating electric field with simultaneous recording of changes in magnetization. Later, also on Cr_2O_3 crystals, the existence of the reciprocal effect was proved, namely, the appearance of electrical polarization under application of a magnetic field.

A special case of magnetoelectric effect takes place where the crystal shows spontaneous magnetization and spontaneous polarization simultaneously. This rare property is inherent in a crystal of Ni-I boracite ($Ni_3B_7O_{13}I$). At $T < 64$ K this crystal belongs to magnetic class $m'm2'$ and is simultaneously a ferroelectric and a weak ferromagnetic; the spontaneous polarization and magnetization are directed along different symmetry axes. As follows from symmetry considerations ($m'm2'$ is among the classes listed in Table 4.8), Ni-I boracite possesses a linear magnetoelectric effect at $T < 64$ K.

Note that the detection of a magnetoelectric (or piezomagnetic) effect may often serve as one of the methods for diagnosing the class of magnetic symmetry of a crystal. Thus, as mentioned above, Cr_2O_3 belongs to crystallographic class $\bar{3}m$, in which the magnetoelectric effect is forbidden. But the fact that Cr_2O_3 shows the magnetoelectric effect proves that this antiferromagnetic crystal belongs to magnetic class $\bar{3}'m'$.

5. Semiconducting Crystals

5.1 Basic Properties of Semiconductors

5.1.1 Metals, Semiconductors, and Insulators

Electronic semiconductors are crystals and amorphous solids whose electrical conductivity is intermediate between metals ($\sigma = 10^4 - 10^5\ \Omega^{-1} \cdot \mathrm{cm}^{-1}$) and insulators ($\sigma = 10^{-10} - 10^{-20}\ \Omega^{-1} \cdot \mathrm{cm}^{-1}$). In distinction to the ionic conductors, the conduction in metals and semiconductors is effected by the drift of electrons and holes.

The elements of the zone theory for crystals presented in [Ref. 2.5, Chap. 3] are manifested most vividly for semiconductors whose physical properties are closely related to the structure of energy bands. In accordance with the zone theory, electrons in crystals, as well as in free atoms, can occupy only definite quantum states, i.e., can possess only a definite energy. The system of discrete energy levels of an isolated atom turns in a crystal into a system of bands of finite width, which are called *energy bands.* A cubic centimeter of a crystal contains $10^{22} - 10^{23}$ atoms. The number of energy levels in these bands is of the same order of magnitude (each atom introduces its own level into the band).

The width of energy bands is several electron-volts; therefore, the difference between the energies of neighboring levels in a band is on the order of 10^{-22} eV, i.e, the levels within the bands form a practically continuous spectrum.

The zone theory makes it possible to formulate the criterion for dividing the solids into two classes: metals and semiconductors (insulators). The metals are characterized by the presence in the valence band of free levels, which can be occupied by electrons with additional energy due to acceleration in an electric field. Such levels form when free atoms constituting the crystal have unfilled shells of valence electrons. Thus, a Na or a Li atom has one valence *s* electron, while a filled *s* shell must have two electrons. Therefore, the valence band of such a crystal is only half-filled. In alkali-earth elements (Be, Mg, Ca etc.) the valence band is filled completely, but overlaps with the conduction band. In both cases, the electrons may be accelerated and transferred to free levels, i.e., take part in the electrical conduction under the effect of an external electric field, even the weakest one. Thus, the distinguishing feature

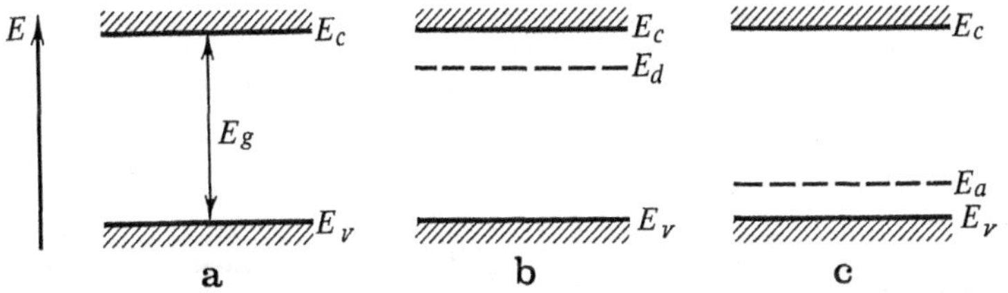

Fig. 5.1. Energy bands of intrinsic (**a**), electron-type (**b**), and hole-type (**c**) semiconductors; E_d and E_a are the donor and acceptor impurity levels, respectively

of the metals is that in the ground unexcited state (at 0 K) they have conduction electrons, i.e., those which are involved in the ordered motion under the effect of an external electric field.

In semiconductors and insulators at 0 K, the valence band is fully populated, while the conduction band is separated from it by a forbidden band (Fig. 5.1a) and contains no carriers. Therefore, a not-too-strong electric field cannot accelerate the electrons of the valence band and transfer them to the conduction band. In other words, such crystals must be ideal insulators at 0 K. When such a crystal is heated or irradiated, the electrons can absorb quanta of thermal or radiant energy sufficient for transition to the conduction band. Such electrons already can take part in electrical conduction. On this transition some holes appear in the valence band; they can also take part in the electricity transport. The probability of transition of an electron from the valence to the conduction band is proportional to $\exp(-E_g/kT)$, where E_g is the width of the forbidden band. At a large E_g ($\sim 2-3$ eV) this probability is very low.

Thus, the division of the crystals into metals and nonmetals is well grounded. In contrast, the division of the nonmetallic crystals into semiconductors and insulators is purely conventional. Formerly, it was believed that crystals with $E_g \approx 2-3$ eV could be referred to as insulators (e.g., [5.1]), but later it was established that many such crystals were typical semiconductors (Table 5.1). Moreover, it was shown that one and the same crystal can be a semiconductor or an insulator depending on the concentration of impurities or excessive (above the stoichiometric composition) atoms of one of the components. This is true, for instance, for crystals of diamond ($E_g = 5.4$ eV), zinc oxide, gallium nitride, etc. Even such typical insulators as barium and strontium titanates, as well as titanium dioxide, acquire semiconductor properties on partial reduction, which is due to the appearance of excess metal atoms playing the part of impurities. On the other hand, it is known that many insulators and semiconductors reveal metallic conduction at sufficiently high pressures [5.2]. In other words, the same crystals can have either metallic or nonmetallic properties depending on the temperature, pressure, impurity concentration, etc. Therefore, strictly speaking, one should distinguish between the metallic and the semiconducting states, rather than between the

Table 5.1. Energy gap width E_g for some semiconductors (at 300 K)

Semiconductor group	Crystal	E_g, eV	Semiconductor group	Crystal	E_g, eV
A^{IV}	C (diamond)	5.4	$A^{II}B^{VI}$	ZnO	3.20
A^{IV}	Si	1.1	$A^{II}B^{VI}$	ZnS	3.60
A^{IV}	Ge	0.67	$A^{II}B^{VI}$	ZnSe	2.80
$A^{IV}B^{IV}$	SiC(α)	3.10	$A^{II}B^{VI}$	ZnTe	2.30
$A^{III}B^{V}$	GaN	3.60	$A^{II}B^{VI}$	CdS	2.42
$A^{III}B^{V}$	GaP	2.32	$A^{II}B^{VI}$	CdSe	1.85
$A^{III}B^{V}$	GaAs	1.52	$A^{II}B^{VI}$	CdTe	1.51
$A^{III}B^{V}$	GaSb	0.80	$A^{IV}B^{VI}$	PbS	0.40
$A^{III}B^{V}$	InP	1.40	$A^{IV}B^{VI}$	PbSe	0.27
$A^{III}B^{V}$	InAs	0.43	$A^{IV}B^{VI}$	PbTe	0.30
$A^{III}B^{V}$	InSb	0.22			

metals and semiconductors. But if we take crystals at normal conditions, the division into metals and nonmetals is expedient.

The division of the nonmetallic crystals into semiconductors and insulators is also rational, since quite a number of crystals are known whose electrical conductivity cannot be notably increased either by introducing impurities or by illumination or heating. This resistance to increased conductivity is associated either with the very short lifetime of photoelectrons or with the existence of deep traps in crystals, which capture the electrons formed due to illumination and, finally, with the very low mobility of electrons, i.e., the extremely low velocity of their drift in an electric field. Such crystals include, in particular, those of alkali halides, whose chemical bonds are typically ionic. In these crystals an electron polarizes the surrounding ions by its electric field. The electric field due to this polarization acts, in turn, on the electron, reducing its potential energy. Thus, the electrons produce a polarization well, the electron escape out of this well being energetically disadvantageous. The application of an external electric field makes the electron move together with the well. The low velocity of polarization motion severely limits the mobility of the electron.

Thus, the role of charge carriers in crystals with a tendency to polarization is played not by free electrons or holes, but by more complex systems, which include electrons and polarization regions. Such systems were termed *polarons* on the suggestion of *Pekar* [5.3], who had developed the appropriate theory.

Along with the *electronic semiconductors*, in which the electricity is transported by electrons or holes, there is a group of crystals whose electrical conduction is associated with the ion drift. These are so-called *ionic-conductors* or *solid electrolytes* [5.4]. The last term reflects the fact that the charge carriers in such crystals are of the same nature as in the ordinary electrolytes. A particularly high ionic conductivity ($\sim 0.2\ \Omega^{-1}\cdot\mathrm{cm}^{-1}$) is exhibited by crystals with a statistical (disordered) distribution of the ions among the lattice points. The number of ions is less than that of the points, which leaves many points unoccupied. This greatly increases the probability of an ion jumping over from one point to another. Such crystals include AgI, Ag_4RbI_5, Ag_2S, Li_3N, etc.

Solid electrolytes have a high energy capacity (~300 W per 1 kgwt). Storage batteries made of such materials are widely used in technology. In particular, they are extremely convenient for use in the city transport. Unfortunately, crystals with a sufficiently high ionic conductivity at room temperature (so-called *superionic conductors*) have not been found yet. And it is not quite convenient to heat a battery to 300–400 °C, say, in a bus. At present, many laboratories are searching for superionic conductors.

5.1.2 Intrinsic and Extrinsic Conductivity

On thermal excitation of carriers from the valence to the conduction band (this is observed in high-purity crystals) the concentration of electrons is always equal to that of the holes. The conductivity of a crystal due to the drift of these carriers is called *intrinsic*. The concentration of intrinsic carriers is related to the energy gap by the equation

$$n_i = p_i = CT^{-3/2} \exp(-E_g/kT) . \tag{5.1}$$

Here n_i and p_i are the electron and hole concentrations, C is a constant, E_g is the energy gap, and k is the Boltzmann constant.

Intrinsic conductivity is usually observed at rather high temperatures. The reason is that the energy gap of most of the semiconducting crystals is ~0.5–2.0 eV, while the thermal excitation energy is ~3/2 kT, i.e., it is equal to 0.03 eV at 300 K. Intrinsic carrier concentrations for a number of semiconductors are listed below.

Crystal:	Si	Ge	InSb	GaAs
n_i, cm^{-3}	1.4×10^{10}	2.4×10^{13}	1.7×10^{16}	1.1×10^{7}
ϱ_i, Ωcm	2.3×10^{5}	47	4.5×10^{-3}	6.4×10^{7}

It is well known that the physical properties of semiconductors depend not only on the structure of the energy bands, but also on impurities and defects. Moreover, during the operation of most of the semiconductor devices, one deals with fluxes of carriers excited from shallow (i.e., low-excitation-energy) impurity levels arising in the forbidden band of the crystal as a result of doping (Fig. 5.1 b, c). In particular, the discovery in 1948 of the transistor effect in germanium, which brought about a scientific and technological revolution in electronics and radioengineering, became possible only after high-purity crystals had been grown and assigned amounts of required impurities introduced into them.

In the case of germanium and silicon such impurities are elements of groups V and III of the periodic table. The elements of group V have five valence electrons each, and the formation of bonds with four neighbors in the Ge or Si lattice requires only four electrons. Thus, on replacement of the main lattice atom by an impurity atom, one extra electron remains. It moves in the field of the ionic residue and the other valence electrons, i.e., in the field of effective charge $+e$, to form a hydrogenlike system. Under the effect of thermal excitation this electron can easily transfer to the conduction band, since its binding energy with the impurity center, i.e., the donor ionization energy, is ~ 0.01 eV. Such impurities are called *donors,* while the semiconductors are said to be electronic (n type).

Similarly, replacement of the main atom by an atom of a group-III element gives rise to a system of an effective charge $-e$ and an electron vacancy moving around it (a positive hole). Owing to thermal excitation a hole bound with the impurity center can also be delocalized, i.e., instead of a hole bound with the given center another hole may appear at a different place of the lattice; this hole will move through the crystal as a result of successive transition of other electrons of the bond into it. The process is similar to the detachment of a bound electron from a donor and can be called the transfer of a hole to the free state. Impurities of this kind are called *acceptors*, and the respective semiconductors, *hole* (p-type) semiconductors.

The ionization energy of the impurity atoms in a crystal can be determined, for instance, from the spectral dependence of the impurity photoconductivity or from the dependence of the Hall coefficient on $1/T$ at a low temperature [5.5]. The corresponding values for group-III and group-V elements in Ge and Si, found from Hall's measurements, are given in Table 5.2. As we can see, all the ionization energies are much less than the energy gap of germanium (~ 0.79 eV) and silicon (1.1 eV). Therefore, practically all the impurity centers are fully ionized in Ge already at a temperature between ~ 5 and 10 K.

The conductivity of a crystal due to the drift of free carriers delocalized from the impurity centers is called *impurity conductivity.* The concentration of free electrons in impurity semiconductors is related to the ionization energy of the donors by the equation

$$n_0 = 2(NN_c)^{1/2} \exp(-E_d/kT) \,, \tag{5.2}$$

where N is the total concentration of the impurity centers, $N_c = (2\pi m_e kT/h^2)^{2/3}$ is the density of states in the conduction band (the total number of allowed energy states in this band), and E_d is the donor ionization energy.

A similar equation is valid for the hole concentration:

$$p_0 = 2(NN_v)^{1/2} \exp(-E_a/kT) \,,$$

Table 5.2. Ionization energy of impurities in germanium and silicon

Impurity	Donor or acceptor	Ionization energy, eV	
		in Ge	in Si
B	A	0.0104	0.045
Al	A	0.0102	0.057
Ga	A	0.0108	0.065
In	A	0.0112	0.160
P	D	0.0120	0.044
As	D	0.0127	0.049
Sb	D	0.0096	0.039
Bi	D	–	0.069

where N_v is the density of states in the valence band, and E_a is the ionization energy of the acceptors.

It should be noted that the solubility of group-III and group-V elements in Ge and Si is rather high (in particular, that of Ga, In, and P in Ge approaches 10^{21} atoms per cm^3). This enables one to vary the electron and hole concentration in these crystals over a very wide range and thus design various devices.

One can estimate the effective size of the region occupied by an electron or a hole associated with the corresponding impurity center. This estimate yields $\sim 80\,a_0$, where a_0 is the radius of the first Bohr orbit of hydrogen ($a_0 = 53 \times 10^{-8}$ cm). In other words, the wave function of an impurity electron embraces a region containing an enormous number of main atoms of the crystal.

Thus, there are three basic conductivity mechanisms in semiconducting crystals:

1) *Intrinsic conductivity,* in which equal numbers of electrons and holes are involved;
2) *Electron conductivity,* which is effected by electrons excited from donor centers;
3) *Hole conductivity,* in which the current is transported by free holes left in the valence band after part of the electrons have transferred to acceptor centers.

5.1.3 Electrical Conductivity of Semiconductors

In the absence of an external field the electrons and holes perform a random thermal motion in the crystal, which closely resembles the motion of the molecules in gases. The electron then has a definite free path length l and relaxation time τ equal to the average time between two successive collisions.

Under the effect of an external electric field the electrons acquire some additional velocity, which is directed along the field. If τ were not limited, the electron velocity in the direction of the field would increase indefinitely, i.e., the electron would move absolutely freely in the crystal, which is impossible, except in the ideal crystal.

In real crystals there are always disturbances in perfect periodicity (impurities, defects, thermal vibrations of the atoms, etc.), on which electrons are scattered. An electron moves with an acceleration only over free path length l; then it experiences a collision, and the entire process begins all over again. If the velocity of the thermal motion of the electrons is v, then $\tau \approx l/v$. Thus, on application of an external electric field, the electrons acquire some additional velocity along the field, which results in the appearance of electric current, i.e., an ordered motion (drift) of the electrons along the field. The mean velocity of this motion is called drift velocity v_d.

The electron motion equation has the form

$$\boldsymbol{F} = m_n \dot{v} = -e\boldsymbol{E} ,$$

where m_n is the effective mass of an electron. From this equation it follows that the drift velocity of an electron is proportional to field tension $\boldsymbol{E}$:

$$v_d = \mu_n \boldsymbol{E} , \quad \text{where} \tag{5.3}$$

$$\mu_n = e\tau/m_n \tag{5.4}$$

is the drift velocity of an electron in an unit field, which is called the *electron mobility*.

If there are carriers of one sort only, electrons, for instance, the current density

$$j = env_d = en\mu_n \boldsymbol{E} , \tag{5.5}$$

where n is the electron concentration. At the same time, in accordance with Ohm's law,

$$j = \sigma \boldsymbol{E} , \tag{5.6}$$

where σ is the specific conductivity. Comparing these two expressions, we obtain the well-known equation

$$\sigma = en\mu_n . \tag{5.7}$$

If the conduction involves not only electrons, but also holes, then

$$\sigma = e(n\mu_n + p\mu_p) , \tag{5.8}$$

where p is the concentration of the holes and μ_p their mobility. Accordingly,

$$j = e(n\mu_n + p\mu_p)\boldsymbol{E} . \tag{5.9}$$

The foregoing is true only for isotropic semiconductors, which are exclusively amorphous, and polycrystalline substances and cubic crystals. Generally, however, the electrical conduction of semiconductors is anisotropic and is described (as will be shown in the next chapter) by a symmetric second-rank polar tensor σ_{ik}. An example of a semiconductor with anisotropic conduction is tellurium, whose electrical conductivity along the hexagonal axis is almost half that in the perpendicular direction:

$$\sigma_x = \sigma_y = 1.95\,\sigma_z .$$

Here the ellipsoid of effective mass is that of revolution compressed along the Z axis.

It is usually assumed that the electrical conductivity is a scalar only in semiconductors with spherical isoenergetic surfaces and a scalar effective mass of charge carriers [5.6]. It will be shown in the next chapter that cubic crystals are isotropic from the standpoint of electrical conductivity. As for isoenergetic surfaces – for instance, in germanium and silicon – they are not spherical: in Ge they are ellipsoids of revolution which are elongated along four $\langle 111 \rangle$ directions, while in Si they are ellipsoids elongated along $\langle 100 \rangle$ [5.5]. In the case of silicon, to each ellipsoid corresponds an effective mass tensor, which has two nonzero components m_L and m_T when reduced to the main axes. The values of m_L and m_T are the same for all ellipsoids, although they correspond to different directions in the crystal.

If we write the equation for the current density $j = \sigma E$, where $\sigma = en\mu_c$, then μ_c, which can be called the "ohmic" mobility, is related to the components of the effective mass tensor by the equation

$$\mu_c = (2\mu_T + \mu_L)/3 ,$$
where
$$\mu_T = e\tau/m_T \quad \text{and} \quad \mu_L = e\tau/m_L . \tag{5.10}$$

Accordingly, the "ohmic" effective mass

$$\frac{1}{m_c} = \frac{1}{3}\left(\frac{2}{m_T} + \frac{1}{m_L}\right).$$

To put it differently, we have here the averaging of the mobility of electrons residing at different minima of the band, as a result of which the "ohmic" mobility is equal for all the directions.

If, however, ellipsoids of equal energy do not coincide with the crystallographic axes (as in Ge, for instance), then for each ellipsoid there will be a conductivity tensor of the general type, i.e., with nonzero nondiagonal components. But on averaging over all the ellipsoids, the nondiagonal components of the conductivity tensor disappear, and all the diagonal ones are equal, i.e., the electrical conductivity tensor degenerates into a scalar. Thus, the nonsphericity of isoenergetic surfaces does not mean that the electrical conductivity of the crystal is necessarily anisotropic.

Let us estimate some quantities characterizing the properties of semiconductors. In pure Ge at 300 K, $\mu_n = 4000\ \mathrm{cm^2/V\,s}$, $m \approx 0.3$ m, i.e., $e/m_n = 1.8 \times 10^{18}$ CGSE, and $\tau \approx 6 \times 10^{-13}$ s. The thermal (mean-square) velocity of an electron can be found by the equation $mv^2/2 = 3kT/2$ $\bar{v} = 2.5 \times 10^7$ cm/s) or $l = \bar{v}\tau = 1.5 \times 10^{-5}$ cm; that is, the free path length contains hundreds of interatomic distances. The drift velocity of an electron is $\sim 4 \times 10^3$ cm/s, which is four orders of magnitude less than the thermal velocity (at $E \approx 1$ V/cm).

One of the features distinguishing the semiconductors from the metals is that their electrical conductivity increases with the temperature. From the foregoing (5.1, 2, 9) it follows that this is due to the increased concentration of the carriers. As for the carrier mobility temperature dependence, it is not so unambiguous. The principal factors limiting the carrier mobility are charged impurity centers, thermal vibrations of the atoms, and structural defects.

Theory shows that for carrier scattering on charged centers in covalent crystals $\mu \sim T^{3/2}$. Here we observe a rare case where the mobility of electrons of holes increases with the temperature. This dependence can be explained as follows: the higher the temperature, the higher the kinetic energy of the electron and its velocity, i.e., the shorter the time of its flight near the charged center and the lower the scattering probability. The indicated dependence, however, is observed only in the low-temperature range. At a high temperature the thermal vibration amplitude of the atoms increases abruptly, and carrier scattering on thermal vibrations becomes predominant. In covalent crystals, then, $\mu \sim T^{-3/2}$, and in ionic ones $\mu \sim T^{-1/2}$ at $T > \Theta$ and $\mu \sim T^{\Theta/T}$ at $T < \Theta$, where Θ is the Debye temperature of the crystal.

5.1.4 Hall Effect

Discovered in 1879, Hall effect makes it possible to investigate the properties of charge carriers in crystals in the simplest and most effective way. Measurement of the Hall emf enables one to determine their sign, concentration, and mobility.

We shall show in the next chapter that the Hall effect in crystals is described by a third-rank axial tensor and shall consider the dependence of the number of tensor nonzero components on the symmetry of the crystal. In particular, we will see that in cubic crystals the Hall coefficient tensor contains only one nonzero component and degenerates into a scalar. To find the relationship between the Hall coefficient and the charge carrier concentration we shall consider this simplest case.

It is well known that an electron moving in a magnetic field is acted upon by the Lorentz force perpendicular to electron velocity and to the magnetic

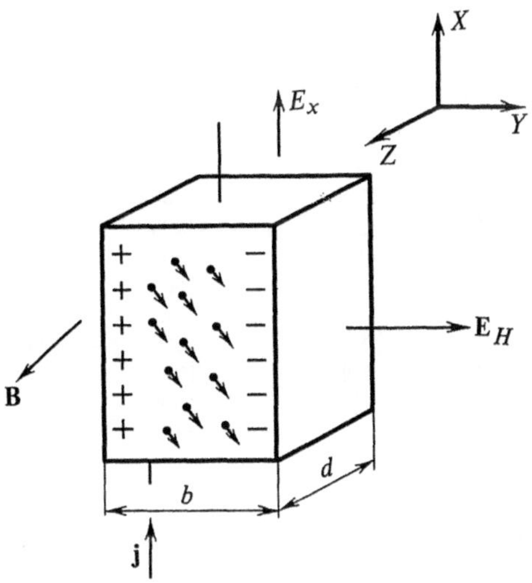

Fig. 5.2. Origin of the Hall emf

field vector $\boldsymbol{B}$. Therefore, if a conductor through which current $\boldsymbol{j}$ is flowing is placed in magnetic field $\boldsymbol{B}$ normal to the electric current, an emf (Hall emf) arises in the direction perpendicular to the plane containing vectors $\boldsymbol{B}$ and $\boldsymbol{j}$.

Obviously, in the stationary state this emf must balance off the Lorentz force. If the current flows along the X axis and the magnetic field is applied along the Z axis, the Lorentz force acts on the electron along the Y axis (Fig. 5.2) and is equal to $-eBv_x$.

The electric field which must offset this force is defined from the condition

$$eE_y = -eBv_x = -Bj_x/n \quad (\text{since } j_x = env_x) \, .$$

Hence,

$$E_y = -Bj_x/en = RBj_x, \quad \text{where} \tag{5.11}$$

$$R = -1/ne \tag{5.12}$$

is the Hall coefficient.

If the thickness of the specimen is d and its width b, then $j_x = I_x/bd$, and therefore from (5.11)

$$E_y = RBI_x/bd \, . \tag{5.13}$$

Thus, the Hall emf $E_H = E_y b$ will be defined by

$$E_H = RBI/d \, . \tag{5.14}$$

Table 5.3. Mobility of electrons and holes ($cm^2/V \cdot s$) at 300 K for some semiconducting crystals

Crystal	μ_n	μ_p	Crystal	μ_n	μ_p
C (diamond)	1800	1400	ZnO	180	–
Si	1300	500	ZnSe	260	15
Ge	3800	1800	ZnTe	1000	100
SiC	220	50	CdS	300	–
GaN	100	–	CdSe	500	–
GaP	300	100	CdTe	1500	600
GaAs	8800	400			
InP	4600	150	PbS	600	600
InAs	33000	460	PbSe	1000	900
InSb	78000	730	PbTe	1600	600

By measuring the values of E_H, B, I, and d, we can determine the value of R, which is related to the carrier concentration by (5.12). If we simultaneously find $\sigma = en\mu_n$, we can obtain the carrier mobility

$$\mu = R\sigma . \tag{5.15}$$

Table 5.3 gives the carrier mobilities of a number of semiconductors. The sign of the carriers depends on that of the Hall coefficient. Thus, the measurements of the Hall effect and of the conductivity of crystals yield the basic parameters of the charge carriers: their concentration, sign, and mobility.

The Hall effect lies at the basis of a number of devices widely used for measurement of constant and variable magnetic fields and investigation of their distributions, for addition and multiplication performed by computers, etc. [5.7]. The operation of these devices is based on the fact that the Hall emf is proportional to the current flowing through the Hall generator and to the magnetic field tension (5.14). Consequently, by measuring E_H and I, we can find B. The generator can be made very small; for instance, the linear dimensions of thin-film generators obtained by evaporation of thin films through a mask may be as small as $\sim 10-20$ μm. Using such a generator, we can measure a magnetic field practically at one point and thus obtain the field distribution in space, determine the presence of field gradients, etc.

If we place a generator in a strongly inhomogeneous field, a change in its position will change the Hall emf. Therefore, a generator can be used to detect the displacements or vibrations of objects placed into this field [5.7]. A Hall generator with a system inducing a magnetic field (such as an induction coil) is the simplest multiplying element. Indeed, if one of the signals $U = U_0 \sin \omega_1 t$ is fed to the current electrodes of the generator and another signal $I = I_0 \sin(\omega_2 t + \varphi)$, to the electrodes of the field-inducing coil, we obtain at the Hall contacts an emf proportional to the amplitudes of the input signals $E_H \approx U_0 I_0$. Consequently, a signal will be generated at the output with an amplitude proportional to the product of the amplitudes of the incoming signals.

5.1.5 Photoconductivity

In distinction to equilibrium carriers, which are formed in a conductor as a result of thermal excitations, carriers resulting from other excitations are called nonequilibrium carriers. The simplest method for producing nonequilibrium carriers consists in illuminating the semiconductor with a light of appropriate wavelength. On appearance of nonequilibrium carriers the total electron concentration will naturally increase, thereby increasing the conductivity of the semiconductor. This phenomenon is called *photoconductivity*. If the energy of light quanta $\hbar\omega \geqslant E_g$, illumination makes an electron rush from the valence to the conduction band. Here we deal with *intrinsic photoconductivity*. If, however, $E_d \leqslant \hbar\omega < E_g$, where E_d is the ionization energy of the donor, the absorption of a quantum excites an electron (or a hole) from the impurity centers (*extrinsic photoconductivity*).

Optical generation of electrons and holes is accompanied by strong light absorption. Here *intrinsic absorption* (at $\hbar\omega \geqslant E_g$) and *extrinsic absorption* (at $\hbar\omega < E_g$) are distinguished. Absorption in the intrinsic-band range is usually much higher than extrinsic absorption. Thus, by taking the absorption spectrum of some semiconductor, one can determine the value of E_g from the edge of the intrinsic band.

The additional conductivity associated with the optical generation of electron-hole pairs is written

$$\Delta\sigma = e(\Delta n \mu_n + \Delta p \mu_p)\,. \tag{5.16}$$

At a relatively low concentration of equilibrium carriers, illumination may change the conductivity of a crystal by hundreds or thousands of times. This lies at the basis of photoresistors used for detection and measurement of the light signal intensity.

It is clear that under illumination the electron and hole concentration cannot increase without bounds. Having found itself near a hole in the valence band or near a vacant impurity center, an electron may fill these vacancies, giving up the excess energy to the lattice or emitting it in the form of a quantum of the corresponding frequency. This process is called carrier *recombination*, and if the excess energy is given up to the lattice, the recombination is called *nonradiative*, while if it is emitted in the form of a light quantum, *radiative*. Radiative transitions form the basis of the operation of *light diodes* and also of *semiconductor lasers*.

The photoconductivity differs from the ordinary or dark conductivity only by the method of charge carrier generation. As for the mechanism of current passage through the crystal and, in particular, the relationship between the vectors of current density $\boldsymbol{j}$ and field tension $\boldsymbol{E}$, they remain the same as for ordinary conductivity. This means that from the standpoint of photoconductivity cubic crystals are isotropic, whereas uniaxial crystals have two and triaxial crystals three main coefficients of photoconductivity. The anisotropy of photoconductivity is observed, in particular, in CdS and CdSe crystals.

In making photoresistors it is customary to use crystals of compounds A^{II} B^{VI} (ZnO, ZnS, ZnSe, ZnTe, CdS, CdSe, CdTe) and A^{IV} B^{VI} (PbS, PbSe, PbTe, and $Pb_xSn_{1-x}Te$), which possess a high photosensitivity in the visible (group one) and the ir (group two) range of the spectrum.

5.2 Electron-Hole Junctions

5.2.1 Electron Energy Distribution

In analyzing many phenomena in semiconductors and describing the operating principle of semiconductor devices, it is important to know the law of distribution of electrons and holes among the energy levels. To describe this law one introduces the so-called distribution function $f(\varepsilon)$ characterizing the probability of the occupancy of a level with a given energy ε.

According to quantum statistics, distribution function $f(\varepsilon)$ depends only on the energy and is equal to

$$f(\varepsilon)=\left[\exp\left(\frac{\varepsilon-\varepsilon_F}{kT}\right)+1\right]^{-1}, \tag{5.17}$$

where ε_F is the mean thermodynamic potential per electron (this quantity is more often called the level of chemical potential, or the *Fermi level*).

Let us consider this function in more detail. If $\varepsilon>\varepsilon_F$, then, as $T\to 0$, $f(\varepsilon)$ also tends to zero; if $\varepsilon<\varepsilon_F$, then, as $T\to 0$, $(\varepsilon-\varepsilon_F)/kT\to -\infty$ and $f(\varepsilon)\to 1$; at $\varepsilon=\varepsilon_F$, $f(\varepsilon)$ jumps from 1 to 0 and hence can be represented as a polygonal (broken) line (Fig. 5.3). This means that at $T=0$ K all the states below the Fermi level are filled, and those above it are vacant. Thus, in metals at 0 K the Fermi level passes across the middle of the valence band. If $T\neq 0$ K, then $f(\varepsilon)=0$ at $(\varepsilon-\varepsilon_F)/kT\gg 0$ and $f(\varepsilon)=1$ at $(\varepsilon-\varepsilon_F)/kT\ll 0$, but for ε values close to ε_F the situation is different. Now at $\varepsilon=\varepsilon_F$, $f(\varepsilon)=1/2$; at $\varepsilon<\varepsilon_F$, $1/2<f(\varepsilon)<1$; and at $\varepsilon>\varepsilon_F$, $0<f(\varepsilon)<1/2$. Thus, function $f(\varepsilon)$ now varies smoothly from 1 to 0 in the range $\varepsilon\sim 4\,kT$ (Fig. 5.3). This means that with an increase in temperature some of the electrons pass from the lower half of the band to the upper. Transitions occur in the energy range of $\sim 4\,kT$, which, at room temperature, equals only 0.1 eV. Thus, on the average, the energy of all the free electrons in the metal remains practically unaltered on changes in T. The state of the electrons when their energy is temperature independent is called *degenerate*.

In the case of semiconductors the situation is essentially different. The number of free electrons in semiconductors fluctuates between 10^{13} and $10^{18}\,\mathrm{cm}^{-3}$, while the total number of electron levels, or states, in the conduc-

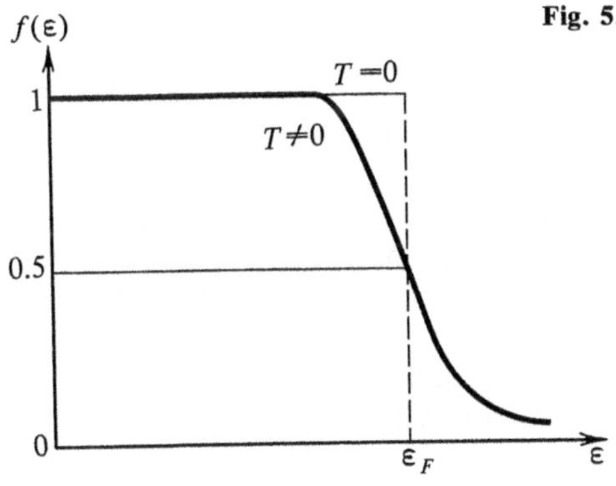

Fig. 5.3. Distribution functions $f(\varepsilon)$

tion band is equal to $2N\cdot g$, where N is the number of atoms, and g is the number of electrons at the atomic level of which the band was formed. One cubic centimeter of the crystal contain $\sim 10^{22}$ atoms, and hence the number of states will be of the same order. Thus, the portion of electron levels occupied by electrons is negligibly small here, i.e., $f(\varepsilon) \ll 1$ or, what is the same, $(\varepsilon-\varepsilon_F)/kT \gg 1$, i.e., the chemical potential level must pass below the bottom of the conduction band. Then we can neglect the unity in (5.17); hence,

$$f(\varepsilon) = \exp\left(\frac{\varepsilon_F-\varepsilon}{kT}\right). \tag{5.18}$$

As is well known, this expression is called the Maxwell-Boltzmann distribution, and the state of the electron gas which is described by (5.18) is called *nondegenerate*.

From the foregoing it follows that the degree of degeneracy of electrons is higher, the more electrons reside in the conduction band (i.e., the higher the impurity concentration and the temperature). In an intrinsic semiconductor the carrier concentration depends on E_g. Accordingly, in semiconductors with small E_g, degeneracy sets in at a relatively low temperature (5.1). All this is also true for hole semiconductors.

In nondegenerate semiconductors the kinetic energy of the electrons is equal to $3/2\,kT$ on the average (~ 0.03 eV at 300 K). In other words, electrons moving among the crystal atoms exchange energy with them as a result of collisions, i.e., are in thermal equilibrium with the lattice. The velocity of the thermal random motion of an electron is $\sim 10^7$ cm/s.

Quite a different picture is observed in the metals. The width of the valence band in them is several eV. If in such a band all the states are filled to a level of at least 1 eV (due to the high concentration of electrons), the thermal vibrations of the atoms cannot effect the energy distribution of the electrons (the energy of the thermal vibrations of atoms is ~0.1 eV even at 1000 K).

The Fermi level is of great importance for the description of many phenomena in semiconductors. If two or several semiconductors are brought in contact, i.e., have a chance to exchange free carriers, they reach, sooner or later, a state of thermodynamic equilibrium. Then their chemical potential, i.e., the Fermi level, reckoned from one and the same energy level, is equal. This fact is widely used in analyzing contact phenomena and the operation of semiconductor devices with p–n junctions.

The concentrations of electrons in the conduction band and of holes in the valence band of an intrinsic semiconductor are related to the Fermi level by equations

$$n_i = 2\left(\frac{2\pi m_n kT}{h^2}\right)^{3/2} \exp(\varepsilon_F/kT) = N_c \exp(\varepsilon_F/kT)\,, \tag{5.19}$$

$$p_i = 2\left(\frac{2\pi m_p kT}{h^2}\right)^{3/2} \exp\left(-\frac{\varepsilon_F+E_g}{kT}\right) = N_v \exp\left(-\frac{\varepsilon_F+E_g}{kT}\right), \tag{5.20}$$

where N_c and N_v are the densities of states in the conduction and valence bands (5.2) and, m_n and m_p are the effective masses of the electrons and holes. Equating $n_i = p_i$, we find ε_F:

$$\varepsilon_F = -\frac{E_g}{2} + \frac{3kT}{4}\ln\frac{m_p}{m_n}\,. \tag{5.21}$$

At $m_p = m_n$ the Fermi level passes exactly across the middle of the forbidden band. In semiconductors such as InSb, where $m_p/m_n \approx 20$, $E_g = 0.2$ eV, it is shifted appreciably toward the conduction band. Substituting the value of ε_F into (5.19, 20), we get

$$n_i = p_i = 2(2\pi kT/h^2)^{3/2}(m_n m_p)^{3/4} \exp(-E_g/2kT)\,. \tag{5.22}$$

Since in an intrinsic semiconductor $n_i = p_i$, by multiplying (5.19) and (5.20) we obtain the well-knwon expression for the carrier concentration:

$$n_i = (N_c N_v)^{1/2} \exp(-E_g/2kT)\,. \tag{5.23}$$

Equation (5.21) can also be expressed in terms of N_c and N_v:

$$\varepsilon_F = -\frac{E_g}{2} - \frac{1}{2}kT\ln\left(\frac{N_v}{N_c}\right). \tag{5.24}$$

In the case of impurity semiconductors, similar equations hold true, which contain, in place of the energy gap, the ionization energies of the donors E_d and acceptors E_a, and also the concentrations of donor and acceptor centers N_d and N_a:

$$n = (N_d N_c)^{1/2} \exp(-E_d/2kT), \qquad \varepsilon_F = -E_d/2 - (kT/2) \ln\left(\frac{N_v}{N_d}\right).$$

At $T = 0$ the chemical potential level in an impurity electron semiconductor is located between the bottom of the conduction band and the donor levels. An increase in temperature causes transition of electrons to the conduction band and their scattering ("evaporation" from the band bottom). When the first process predominates, the electron gas density in the conduction band increases, and the Fermi level rises. If the second process predominates, the Fermi level goes down, and after half of the impurity electrons have transferred to the conduction band, it coincides with the impurity levels; then it goes further down, approaching the middle of the energy gap, and the zone of intrinsic conduction (Fig. 5.4). A similar process takes place in a hole semiconductor.

5.2.2 Semiconductor-Metal Contact

When two different semiconductors or a semiconductor and a metal are in contact, potential barriers often arise, which play an important part in the operation of semiconductor devices. Consider, for example, the contact of a metal with an n-type semiconductor. Figure 5.5a shows their energy diagram before contact and Fig. 5.5b, c their diagrams after contact is formed. Here E_0 is the energy level of an electron in a vacuum, E_c and E_v are the band boundaries in the semiconductor, and ε_m and ε_s are the positions of the Fermi levels in the metal and the semiconductor.

If the work function in the metal is lower than in the semiconductor, then after the contact is formed more electrons will transfer from metal to semiconductor than vice versa. In the opposite case the situation will be reversed. At equilibrium we get the pattern exhibited in Figs. 5.5b, c. The band edges are slightly bent, and the electron concentration in the contact layer of the semiconductor will be higher (Fig. 5.5b) or lower (Fig. 5.5c) than in the bulk. Therefore, a potential barrier eU_c arises at the interface, where U_c is the potential difference due to the different electron concentration in the semiconductor bulk and in the contact layer (depleted or enriched). From the condition for the constancy of the Fermi level at equilibrium it follows that the height of the potential barrier arising at the contact is

$$eU_c = \varepsilon_s - \varepsilon_m , \tag{5.25}$$

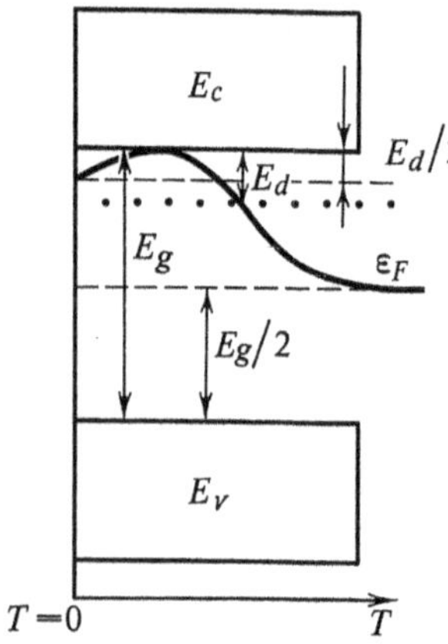

Fig. 5.4. Temperature dependence of the position of the Fermi level of an electron semiconductor

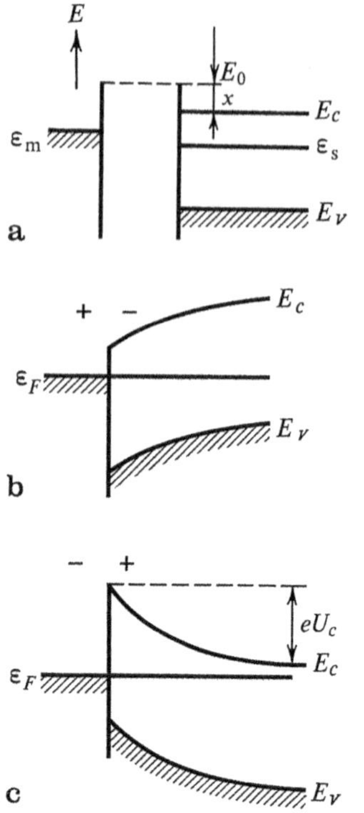

Fig. 5.5. Energy bands of a metal and a semiconductor before **(a)** and after **(b, c)** the formation of the contact

where ε_s and ε_m are the positions of the Fermi levels in the semiconductor and the metal before they came in contact. Taking advantage of the relationship between the chemical potential and thermionic work function Φ – $\Phi = E_0 - \varepsilon_F$, where E_0 is the energy of an electron in a vacuum – we can write expression (5.25) as

$$eU_c = \psi_m - \Phi_s \,. \tag{5.26}$$

If, on contact between a metal and a semiconductor, a carrier-enriched layer arises in the semiconductor, such a contact is *ohmic*.

For current rectification purposes the formation of a depletion layer is of interest. Let us consider it in more detail. We estimate the number of electrons which have to transfer from semiconductor to metal in order to set up an equilibrium potential difference U_c. Suppose the difference in work functions $eU_c = 1$ eV and metal-semiconductor gap $d = 10^{-7}$ cm. The field in the gap is then $E = 10^7$ V/cm, and the surface charge density necessary for inducing such a field, $q = E/4\pi$. Hence, $n = q/e = E/4\pi e \approx 10^{12}$ electrons will transfer from the semiconductor to the metal.

If the concentration of the donor centers in the semiconductor is $\sim 10^{16}\,\mathrm{cm}^{-3}$, then to induce such a charge one must ionize all the donors at a

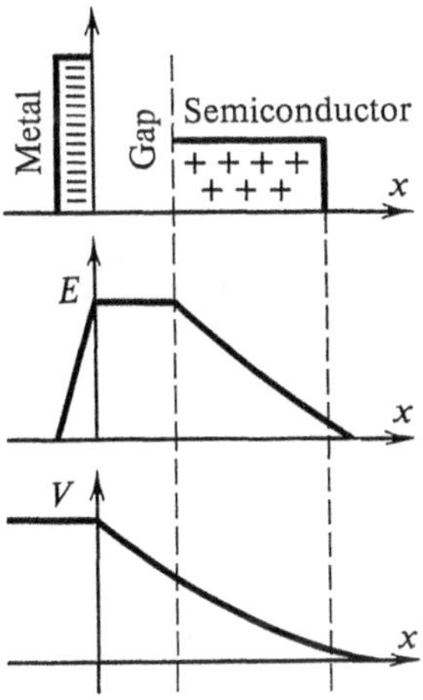

Fig. 5.6. Charge, field tension, and potential distribution in the contact layer of an electron semiconductor (contact with a metal)

distance of $\sim 10^{-4}$ cm from the surface. This is about 1000 times the gap width d (~ 10 Å). Consequently, the semiconductor acquires a layer of space charge (because of the escape of electrons) of thickness ~ 1 µm, in which the charge density is approximately constant, the field increases linearly, and the potential quadratically (Fig. 5.6). The main potential drop here occurs not in the gap, but in the space-charge layer. This layer, which offers a very high resistance, is called the *blocking layer*, and the corresponding contact, the *rectifying contact*.

The passage of the current through the rectifying contact depends on the sign of the external (applied) potential difference V. If it adds to the contact potential difference, the thickness of the blocking layer increases with the external voltage, and the current through the barrier reduces:

$$I_1 = I_s \exp(-eV/kT) \quad , \tag{5.27a}$$

where I_s is the saturation current. In the opposite case the barrier height decreases, and the current rises:

$$I_1 = I_s \exp(eV/kT) \quad . \tag{5.27b}$$

Here the following expression is obtained for the voltage-current characteristic:

$$I = I_s [\exp(eV/kT) - 1] \quad . \tag{5.28}$$

From (5.28) it follows that at a positive potential of the semiconductor with respect to the metal the current increases rapidly with the voltage. Already at eV equal to several kT the current increases exponentially (the

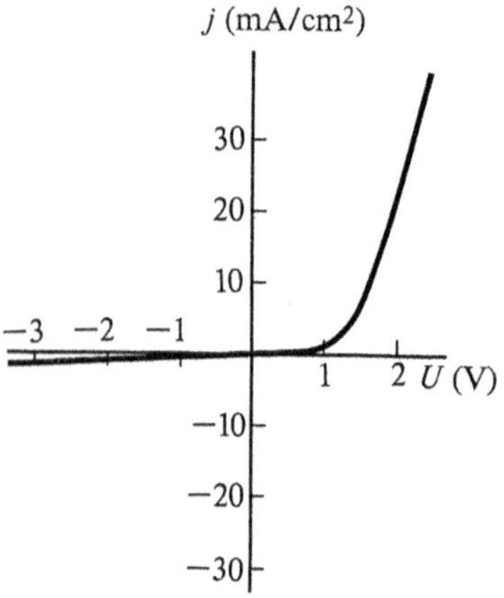

Fig. 5.7. Voltage-current characteristic of an Ag-ZnO contact

unity can be neglected). If $V<0$, the exponent decreases rapidly, and at $eV \approx kT$, $I = -I_s$ (Fig. 5.7), i.e., current reaches saturation.

The unidirectional conduction of metal-semiconductor contacts is used extensively in semiconductor rectifiers. In recent years such contacts have found wide application in making integral circuits on Si plates and also in constructing various devices on crystals of ZnO, CdS, CdSe, and other compounds in which hole conductance or p-n junctions cannot be obtained.

5.2.3 Semiconductor Diodes with p-n Junctions

If one part of a semiconducting crystal contains donor-type impurities and the other, acceptor-type, an electron-hole junction (p-n junction) arises at the interface between these parts (Fig. 5.8a).

Suppose that these parts of the crystal are brought in contact. The electrons will begin to pass from the n region, where there are many of them, to the p region, where there are few of them, and the holes will similarly move from the p region to the n region. This transition cannot last indefinitely: indeed, when the electrons escape from the n region, they leave behind uncompensated donor centers (positive ions), while the p region acquires negatively charged acceptor centers. Some of the carriers, however, will transfer to neighboring regions, so that the n-type material adjoining the interface will acquire a positive charge, while the p-type material, negative.

The transition of electrons and holes will continue until the resulting potential difference φ_0 offsets the difference in the work functions of the electron and hole semiconductor and the Fermi levels of these parts equalize (Fig. 5.8b). Then an electric field appears in the transition region, which prevents further diffusion of the main carriers across the interface. Only carriers with a sufficiently high energy can overcome the field countereffect and cross the transition region. But the same number of carriers moves in the opposite

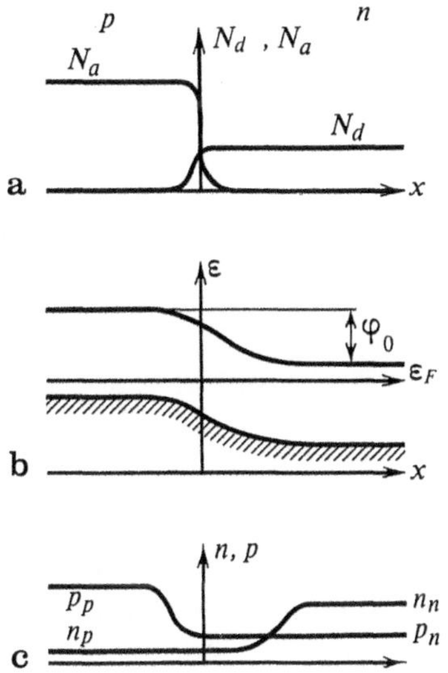

Fig. 5.8. Distribution of the donor (N_d) and acceptor (N_a) impurities (**a**), of the potential (**b**), and of the free carriers (**c**) in a crystal with a p-n junction

direction under the effect of the field. In other words, under equilibrium conditions the fluxes of carriers in both directions are the same.

The transition layer has almost no free carriers, which cannot linger in it and fly through it rapidly. This carrier-depleted layer (of thickness ~1 μm) (Fig. 5.8c) offers a very high resistance as compared with the other parts of the crystal. Therefore, when a crystal with a p-n junction is switched into the circuit, practically all the voltage is concentrated on the p-n junction.

If a carrier concentration Δn_0 excessive in comparison with the equilibrium carrier concentration arises at some point of a semiconductor (for instance, under illumination), these carriers will immediately begin to diffuse (dissipate), and their concentration will decrease exponentially, by recombination, with the distance from the indicated point

$$\Delta n(x) = \Delta n_0 \exp(-x/L) ,$$

where constant L is called the *diffusion length* (distance at which the nonequilibrium concentration decreases by a factor of e). It is related to the coefficient of diffusion and carrier lifetime τ by the Einstein equation:

$$L = \sqrt{D\tau} . \tag{5.29}$$

We now assume that the p-n junction is subjected to an external voltage in the direction of transmission (+ on the p region). Then the holes will rush into the n region, where they will become minority carriers and will recombine with the electrons. However, owing to the finite value of τ_p this recombination will occur not at once, but after passing some distance beyond the p-n junction,

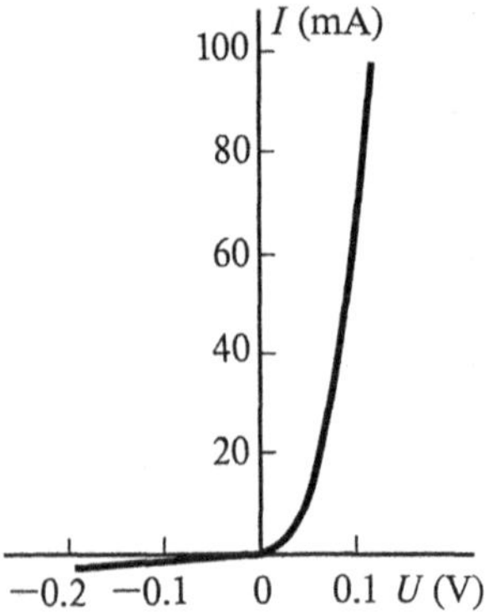

Fig. 5.9. Current-voltage characteristic of a Ge diode with a p-n junction

where the hole concentration will remain above the concentration of equilibrium holes, p_n. At the same time the electron concentration in the n region will also increase, since additional electrons will emerge from the electrode to compensate the space charge of the arriving holes. The same thing will happen to the electrons in the n region. Under the effect of the applied field part of these electrons will pass into the p region, recombining with the holes, whereas from the electrode additional holes will enter the volume of the p region to offset the electron charge. This phenomenon is called the *injection* of minority carriers.

Injection results in a sharp nonlinearity of the current-voltage characteristic of a p-n junction, which is described by the same equation as the characteristic of the rectifying metal-semiconductor contact. A positive external voltage leads to hole injection from the p to the n regions and to the holes' intensive recombination with the electrons. The same is true for electrons injected from then region to the p region. These recombination currents may be high even at low voltages, because the concentration of the main carriers is high enough, and the injection is not accompanied by the appearance of an appreciable space charge.

At a negative voltage on the p region the barrier height increases, and the junction can be crossed only by minority carriers, whose concentration is much lower than that of the main carriers. Therefore, the reverse current of p-n junctions are low (Fig. 5.9).

Owing to the nonlinearity of the current-voltage characteristic, p-n junctions are widely used for rectifying alternating currents, and detecting and transforming signals. In powerful rectifiers the area of the p-n junction is made large to increase the current, and the thickness of the transition layer is increased to raise the breakdown voltage. This is done by making three-layer structures whose intermediate layer has intrinsic conduction (p-i-n structures).

In producing semiconductor diodes extensive use is made of crystals of Ge and Si, as well as GaP, GaAs, InAs, InSb, and other compounds and solid solutions of the type $GaAs_{1-x}P_x$, $Ga_{1-x}In_xAs$, etc. Recently, p-n junctions on diamond crystals have been designed.

5.2.4 Semiconductor Triode (Transistor)

The invention of the transistor – a device designed to amplify and generate electric signals – has been one of the most important discoveries in the field of crystal physics. This invention stimulated the rapid development of semiconductor physics and microelectronics, computer technology, and cybernetics. A schematic design of the transistor is presented in Fig. 5.10. It consists of a semiconductor plate in which a narrow ($\sim 1\ \mu m$) n-type region separates two p-type regions. The left-hand part of the device is called the emitter, the middle part the base, and the right-hand part the collector (*E*, *B* and *C* in Fig. 5.10). The hole concentration in the emitter is usually made about two orders of magnitude higher than that of the electrons in the base. If a direct voltage is fed to the left-hand junction, holes will be injected through it. Because the base is thin, most of the holes will reach the right-hand junction, which is switched in reverse. Under the effect of the field of the right-hand junction, the holes will be dumped into the collector. Hence, when the left-hand junction is open, almost the same current as flows through the left-hand junction will flow through the right-hand junction instead of the weak back current, and a considerable voltage drop $V = IR_c$, determined by the emf of the storage battery, will occur on resistor R_c. Since the direct current through the p-n junction depends very strongly on the voltage, the current in the transistor changes very sharply upon small changes in the voltage across the left-hand junction. Thus, a bipolar transistor is similar to a vacuum triode, in which the emitter plays the role of the cathode, the collector that of the anode, and the base that of the grid.

In the production of diodes and transistors p-n junctions were first made either by adding the necessary dopants to the melt from which the single crystal was pulled, or by subsequent diffusion of the dopants from the gas phase (in sealed-off ampoules), or else from a small drop of a metal deposited on the semiconductor surface. Later on an epitaxial technique was widely used for this purpose, i.e., growing thin single-crystal layers of semiconductors with the required type and concentration of dopant on the substrate. This technique is extremely promising, and at present it is used extensively in the

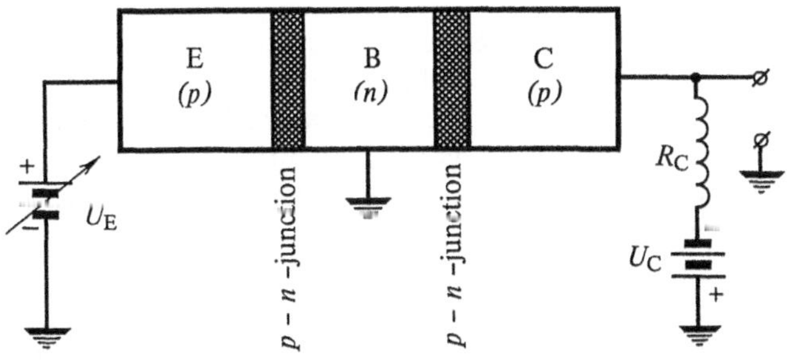

Fig. 5.10. p-n-p transistor circuit

industry, particurlarly in making integral circuits, when hundreds of thousands of micron-size devices are built on the same single plate, and also in the manufacture of multilayer structures, for instance, those used in semiconductor lasers.

In recent years, the molecular-beam epitaxy method has come into use; it is based on the condensation of molecular beams in a vacuum. By using several different evaporators it produces very thin (monomolecular, in the limit) layers of different composition and so-called superlattices, i.e., multilayer structures whose composition changes periodically along the normal to the surface (for instance, 10 – 50 Å layer of GaAs and Ge). Such artificial "superstructures" must show a number of interesting properties [5.8]. The idea of epitaxial growth of single-crystal films for producing p-n junctions and multilayer structures was first proposed at the Institute of Crystallography, USSR Academy of Science in 1954 [5.9].

Lately, semiconductor devices and integral circuits are also manufactured by the ion implantation method, i.e., by incorporating into a thin semiconductor surface layer the necessary ions with an energy from several tens of keV to several hundred keV obtained on special accelerators [5.10].

5.2.5 Semiconductor Laser

Stimulated radiation can be obtained in semiconducting and dielectric crystals only by achieving a state *with a negative temperature.* As mentioned above, under normal conditions an equilibrium electron and hole energy distribution sets up in a semiconductor, which is described by the Fermi function (5.17). Then the number of electrons (for definiteness sake, we shall operate only with electrons) at the lower levels is much larger than at the upper. These levels exchange electrons continuously; the transfer of particles from the lower to the upper levels results from absorption of energy, while the transfer from the upper to the lower level is accompanied by its emission. The probability of intensifying the radiation by such a system is very low, since the concentration of electrons at the lower levels is much higher than at the upper. In other words, a crystal in the state of thermal equilibrium cannot intensify the incident radiation.

To create conditions under which a semiconductor could intensify the radiation or become a source of stimulated radiation, one has to disturb the equilibrium distribution of the electrons, i.e., to ensure a distribution at which the number of electrons at the upper levels exceeds that at the lower. To put it differently, one has to achieve an *inverse population of the levels,* which is described by the same equation (5.17) as the equilibrium distribution, provided that T is prefixed with a minus. Therefore, the state with an inverse population of the levels is often called the state *with a negative temperature.*

In principle, two methods for obtaining the state with a negative temperature in semiconducting crystals are possible:

1. Disturbing the *intraband equilibrium*. If the electrons are transferred from the lower part of a band (for instance, the conduction band) to the upper with the aid of some power source, the restoration of the disturbed equilibrium takes very little time (10^{-13} s). Therefore, this method for achieving an *inverted* population of the levels is impractical.

2. Disturbing the *interband equilibrium*. If electrons are transferred from the valence band to the conduction band, the relaxation time of such interband transitions is much longer than in the former case; it varies between 10^{-3} and 10^{-9} s. Therefore, this method for achieving an inverted population of the levels is quite feasible.

Since the equilibrium inside the bands is restored within $\sim 10^{-13}$ s, it may be assumed that all the electrons transferred to the conduction band have time enough to reach a thermal equilibrium with the lattice, i.e., they become concentrated in a thin layer near the bottom of the conduction band; similarly, all the holes will concentrate near the ceiling of the valence band. Consequently, in semiconducting crystals it is possible to intensify or generate a radiation whose energy of quanta is approximately equal to the energy gap. The state with a negative temperature can then be achieved only by disturbing the interband equilibrium.

With dielectric lasers the principal method for achieving an inverse population of the levels is *optical pumping*. The disadvantage of this method is the low conversion factor due to the broad spectrum of the pumping sources. The shortwave radiation is absorbed in the thin surface layer of the crystal, turning to heat, while the long-wave radiation is not absorbed by the crystal at all, i.e., it takes no part in the excitation. The *electron-beam excitation* method is more promising, since the electrons are completely absorbed in solids, but great disadvantages of this method are the formation of radiation defects in the crystal irradiated and the relative bulk of the pumping devices: electron guns and their power sources. For these reasons electron-beam pumping is used only in producing lasers with such crystals in which p-n junctions cannot be made (CdS, CdSe, ZnO, GaN, etc.). The most promising method for achieving an inverted population density in semiconducting crystals is the *injection of electrons and holes* through a p-n junction.

Let us consider this method briefly. Imagine a crystal with a p-n junction whose n region is strongly doped with donors, and the p region with acceptors. The conduction band of the n region will then have a sufficient number of electrons, and the electrons of the valence band will reside at the acceptor levels, i.e., a sufficient number of vacant sites will have formed in this band to which electrons from the conduction band could move. If these two parts of the crystal were not separated by the p-n junction, the electrons of the conduction band could "drop" into the valence band and recombine with the holes, as a result of which one could obtain a bright pulse of light of recombination radiation (Fig. 5.11). Hence, if it were possible to remove the potential barrier of the p-n junction, one could obtain an inverted population density of the levels.

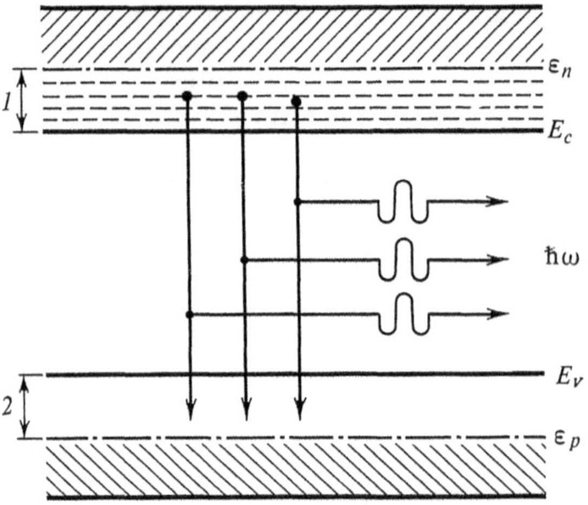

Fig. 5.11. Interband inverse level population in a semiconductor crystal (ε_n and ε_p are the Fermi quasilevels of electrons and holes): (*1*) region of filled state, (*2*) region of vacant state

As noted above, the potential barrier of a p-n junction can be reduced by applying a voltage in the forward direction. With high levels of doping, when both parts of the crystal are practically degenerate, the Fermi quasilevels[1] of the electrons and holes are disposed inside the bands. If we apply a voltage V to such a crystal in the forward direction, whose value satisfies condition $eV>E_g$, the barrier of the p-n junction will practically disappear, and a strong current will flow through the crystal. As a result of the disappearance of the potential barrier, the electrons from the conduction band will move to the valence band of the p region and actively recombine with the holes. In some semiconductors this recombination is accompanied by intensive photon radiation, and such semiconductors (GaAs, InAs, InSb, etc.) are used in laser manufacture.

Recombination of minority carriers is naturally attended by depletion of the respective bands. Therefore, in order to reproduce the inverted population density one has to offset the escape of nonequilibrium carriers by continuous injection of electrons into the n region and holes into the p region. This injection is performed by an external current source, which plays the part of the pumping source here.

It is well known that in setting up the generation mode one has to ensure a positive feedback between radiation and absorption. This feedback is achieved in ordinary lasers by means of a *Fabri-Perot resonator* with parallel-side mirrors. In semiconductor lasers the specimen itself serves as resonator,

[1] When nonequilibrium carriers appear in the crystal bands, Fermi level ε_F splits into quasilevels of electrons ε_n and holes ε_p, each of them shifting towards its own band. The higher the carrier concentration, the closer ε_n and ε_p approach the bottom of the conduction band and the valence band ceiling. In degenerate semiconductors the quasilevels are disposed inside the respective bands.

because semiconductors have a high refractive index ($n = 3-4$), and the semiconductor-air interface can reflect up to 30% of the radiation. The parallelism of such "mirrors" is ensured automatically by providing cleavage faces perpendicular to the p-n junction. The high amplification factor of semiconductor lasers ($\sim 2000\ \mathrm{cm}^{-1}$) makes it possible to use crystals of very small sizes (fractions of a millimeter) for making lasers.

The early injection lasers were constructed on GaAs crystals doped with zinc (p region) and selenium (n region). In the first devices the threshold density of the current through a p-n junction at which the laser radiation became coherent was several thousand amperes per cm^2, and therefore they could operate only at low temperature (4 – 20 K). (If the current density is below the threshold level, the radiation is incoherent, i.e., the laser operates like an ordinary light diode). At present the threshold current density has been reduced to $100\ \mathrm{A/cm}^2$, and semiconductor lasers can operate at room temperature. This improvement in the parameters is due to the perfection of the laser manufacturing techniques, primarily by using multilayer heterojunctions in crystals of solid solutions of the type $Ga_{1-x}Al_xAs$ [5.11]. The sizes of injection lasers are very small because of the high amplification factor and the high density of the threshold current. Here the radiating area is $\sim 10^{-4}\ \mathrm{cm}^2$, radiation power ~ 10 W, and efficiency 70 – 80% [5.12].

5.2.6 Photocells with a p-n Junction

Soon after the advent of copper-oxide rectifiers it was established that their illumination induces a current in the rectifier circuit without applying an external emf. This served as a basis for the manufacture of blocking layer photocells designed for the conversion of light signals to electrical ones. Such devices are used, for instance, in exposure meters. They are not employed for power generation, since their efficiency in only fractions of one percent. The situation changes appreciably after germanium and silicon photocells with p-n junctions were made.

The efficiency of such photocells is 10 – 12%, and in devices with heterojunctions it may be as high as 30% [5.11]. This opens up the way for utilizing photocells in direct conversion of solar to electrical energy.

Let us consider briefly the operating principle of a photocell with a p-n junction. Suppose a light with a quantum energy sufficient for generating electron-hole pairs falls on the surface of a crystal with a p-n junction (Fig. 5.12). For simplicity we assume that the entire light is absorbed in the n region, and that the distance of the p-n junction from the illuminated surface is less than the diffusion length of the holes. Then all the light induced holes will reach the p-n junction and transfer to the p region under the effect of the contact potential difference eU_c. The light-induced electrons are the majority carriers for the n region; they, unlike the holes, will be delayed by the contact field of the p-n junction and remain in the n region.

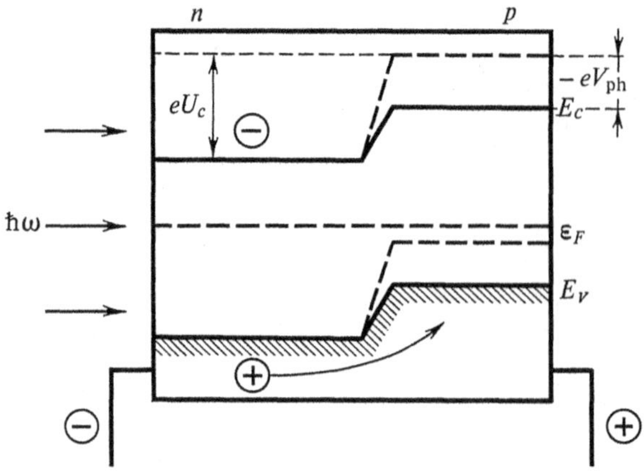

Fig. 5.12. Generation of a photo-emf in a photocell

Thus, under the effect of illumination, a photocurrent

$$I_{ph} = eRL_p \tag{5.30}$$

will flow through the p-n junction (R is the number of electrons induced by the light per 1 s, and L_p is the diffusion length of the holes). As a result of escape of the holes from the n region, this region will be charged negatively and the p region positively, and a potential difference eV_{ph} (so-called photo-emf) will arise between them (Fig. 5.12). This emf reduces the contact potentials difference eU_c, and the so-called leakage current will flow through the junction in the opposite direction:

$$I_l = I_s[\exp(eV_{ph}/kT)-1] \quad .$$

At a certain value of V_{ph} a stationary state must set in, at which the photocurrent will be equal to the opposing leakage current. The stationarity condition has the form

$$eRL_p - I_s[\exp(eV_{ph}/kT)-1] = 0 \quad ,$$

whence we obtain the following expression for the photocell voltage:

$$V_{ph} = \frac{kT}{e}\ln\left(1+\frac{eRL_p}{I_s}\right) \quad . \tag{5.31}$$

At low levels of generation of minority carriers V_{ph} is proportional to R and hence to the intensity of the incident light. At a high light intensity, V_{ph} increases until the p-n junction barrier disappears. The maximum voltage obtainable at a p-n junction is equal to the voltage resulting from the different positions of the Fermi levels in n- and p-type semiconductors. It is approximately equal to the voltage corresponding to the energy gap E_g.

If we connect several photocells in series, their emf add together. For instance, if we have a series of 100 Si photocells, we obtain a storage battery with a voltage on the order of 100 V. Such batteries are widely used in spacecraft and satellites.

When using a photocell for light signal detection it is advisable to switch it into the circuit with the emf source in the blocking direction. Then a very weak reverse current due to the drift of the minority carriers will flow through the p-n junction. Illumination of such a diode with a light of photon energy $\hbar\omega > E_g$ sharply increases the concentration of minority carriers and hence the current through the p-n junction. By detecting this increase of the current in some way one can detect or record light signals. A photocell operating in this mode is called a *photodiode*. The distinguishing feature of the photodiodes is that they show a high photosensitivity and a very high response, and are therefore extensively used for detecting short weak signals. These devices are common in optics, electronics, automation, and computer technology. Photodiodes based on semiconductors with a narrow energy gap (PbS, PbSe, PbTe, $Pb_{1-x}Sn_xTe$, $Hg_{1-x}Cd_xTe$, a.o.) are used as infrared light detectors ($\lambda = 3-10\ \mu m$).

6. Transport Phenomena in Crystals

Transport phenomena are associated with the electric charge, heat, and mass flow in electrical and thermal conductivity and diffusion [1.4]. They also include thermoelectric effects in crystals in the presence of electric potential and temperature gradients [1.5], and galvano- [6.1, 6.2] and thermomagnetic effects [1.11; 6.3; 6.4] which arise in crystals when an electric current or a heat flux passes through a crystal located in a magnetic field. This chapter briefly considers all these effects.

6.1 Electrical Conductivity of Crystals

6.1.1 Electrical Conductivity and Resistivity Tensors

An electric field induces a current in conducting crystals. At weak fields the relationship between current density j and field intensity E is mostly linear (Ohm's law). In the case of homogeneous isotropic conductors the linearity reduces to the simple proportionality

$$j = \sigma E \ . \tag{6.1}$$

Coefficient σ, which depends on the conductor type and temperature, is called *conductivity*. This coefficient is related to the current carrier concentration and mobility by the well-known equation

$$\sigma = en\mu = enl\tau/m^* \ , \tag{6.2}$$

where e is the charge of the carriers, n is their concentration, μ is the mobility, m^* is the effective mass, τ is the time between two successive collisions, and l is the mean free path.

In single crystals, the directions of vectors j and E generally do not coincide, and the linear relationship between them is expressed by generalized Ohm's law

$$j_i = \sigma_{ik} E_k \ , \tag{6.3}$$

where σ_{ik} is a symmetric second-rank polar tensor, in which $\sigma_{ik} = \sigma_{ki}$. It is called the *tensor of electrical conductivity coefficients.* The crystal conductivity anisotropy results from the anisotropy of the *reciprocal effective mass* of the electrons (holes) m^*, [6.5] which is defined by the symmetry of the isoenergetic surfaces and is a symmetric second-rank tensor [Ref. 2.5, Chap. 3]. It should be mentioned that (6.3) could also have a free term: the crystal symmetry allows for its existence [6.6]. Then the relationship between j_{ik} and E_k would have the form

$$j_i = \sigma_{ik} E_k + j_i^0 \ .$$

The presence of a constant vector j_i^0 would mean that an electric field could exist in a conducting crystal in the absence of a current ("pyroelectricity" of a conductor). We shall prove this to be impossible.

According to the Joule-Lenz law the amount of heat released in 1 cm^3 of a crystal per 1 s during the passage of a current is defined by the equation

$$jE = \sigma E^2 = j^2/\sigma \ .$$

The release of heat increases the entropy S of the system. With the release of heat $dQ = jE\,dv$, the entropy dS of volume element dv increases by dQ/T. The rate of change in the total entropy of the system is thus equal to

$$\frac{dS}{dt} = \int \frac{jE}{T} dv \ . \tag{6.4}$$

By the law of conservation of energy this derivative must be strictly positive. In the meantime, term j_i^0 in the integrand of (6.4) could have either sign, and as a result, dS/dt could become negative. Hence, $j_i^0 = 0$. Substituting $j = \sigma E$ into (6.4), we see that conductivity σ must be positive.

Each of the coefficients σ_{ik} has quite a definite physical meaning. Thus, if a unit electric field is applied along the X_2 axis, σ_{22} is equal to the current density along this axis, while σ_{12} and σ_{32} are the current densities along the X_1 and X_3 axes [6.7]. If tensor σ_{ik} is reduced to the main axes, (6.3) takes the simple form

$$j_1 = \sigma_1 E_1 \ , \quad j_2 = \sigma_2 E_2 \ , \quad j_3 = \sigma_3 E_3 \ .$$

The second-order characteristic surface for the tensor of conductivity coefficients is described by the equation

$$\sigma_{ik} x_i x_k = 1 \ , \tag{6.5}$$

which, on passing to the main axes, takes the form

$$\sigma_1 x_1^2 + \sigma_2 x_2^2 + \sigma_3 x_3^3 = 1 \ . \tag{6.6}$$

As noted above, coefficients σ_1, σ_2, and σ_3 are always positive, and therefore (6.6) is the ellipsoid equation. This is the characteristic ellipsoid of the conductivity coefficients. Its shape and orientation must correspond to the symmetry of the crystal (Chap. 1).

Tensor σ_{ik} can often be conveniently replaced by resistivity tensor ϱ_{ik}. Then (6.3) transforms to

$$E_i = \varrho_{ik} j_k \ . \tag{6.7}$$

Coefficients ϱ_{ik} also form a symmetric second-rank polar tensor, the tensor of *resistivity coefficients*. If this tensor is written in the form of a matrix, the latter will obviously be the reciprocal of that of the conductivity coefficients. This does not mean, however, that all the components ϱ_{ik} are the reciprocal values of the respective components σ_{ik}, for instance, $\varrho_{12} \neq 1/\sigma_{12}$ (Chap. 1).

Coefficients ϱ_{12} and σ_{12} are determined in various experiments. Thus, σ_{12} is found in the measurement of the component of the current along X_1, when an electric field is applied along X_2, i.e.,

$$j_1 = \sigma_{12} E_2 \ .$$

In distinction to this, coefficient ϱ_{12} is determined by measuring the field tension along the X_1 axis, when the current vector is directed along X_2, i.e.,

$$E_1 = \varrho_{12} j_2 \ .$$

Since $\sigma_{ik} = \sigma_{ki}$, then also $\varrho_{ik} = \varrho_{ki}$. If tensor ϱ_{ik} is reduced to the main axes (they coincide with the main axes of tensor σ_{ik}), then (6.7) takes the simple form

$$E_1 = \varrho_1 j_1 \ , \quad E_2 = \varrho_2 j_2 \ , \quad E_3 = \varrho_3 j_3 \ , \tag{6.8}$$

whence follow the relations

$$\varrho_1 = 1/\sigma_1 \ , \quad \varrho_2 = 1/\sigma_2 \ , \quad \varrho_3 = 1/\sigma_3 \ . \tag{6.9}$$

When reduced to the main axes, the characteristic surface (ellipsoid) of the tensor of resistivity coefficients is decribed by the equation

$$\varrho_1 x_1^2 + \varrho_2 x_2^2 + \varrho_3 x_3^2 = 1 \ . \tag{6.10}$$

As for the dependence of the number of independent components of tensors σ_{ik} and ϱ_{ik} on the symmetry elements, it is, evidently, the same as for any symmetric second-rank polar tensor. Hence, from the conductivity standpoint all the crystals can be divided into three groups:

Table 6.1. Electrical conductivities of metallic crystals (at 20 °C)

Crystal	Syngony	$\varrho_1 = \varrho_2$, 10^{-8} ohm · m	ϱ_3, 10^{-8} ohm · m
Tin	Tetragonal	9.9	14.3
Bismuth	Trigonal	109	138
Cadmium	Hexagonal	6.8	8.30
Zinc	Hexagonal	5.91	6.13
Tellurium	Trigonal	29×10^8	59×10^8
Tungsten	Cubic		5.48
Copper	Cubic		1.51

i) crystals of *cubic syngony,* which have a single main conductivity coefficient (i.e., they are isotropic);
ii) crystals of *intermediate syngonies,* which have two main conductivity coefficients each;
iii) crystals of *lower syngonies,* which have three main conductivity coefficients each.

Examples of crystals with anisotropic electrical conductivity are tin, bismuth, cadmium, and zinc (Table 6.1). At room temperature the conductivity of tellurium crystals along the hexagonal axis is almost twice as low as in the perpendicular direction:

$$\sigma_1 = \sigma_2 = 1.95\,\sigma_3 \ .$$

Crystals which are built up of weakly bound close-packed packets show a considerable conductivity anisotropy. In such crystals the conductivity along the packets ($\sigma_{\|}$) may exceed that in the perpendicular direction $\sigma_{\perp}$ by a few orders of magnitude [6.8]. The anisotropy is particularly high when layers of molecules of some other substance, say, pyridine Py, are sandwiched between the packets (for instance, in TaS_2), which results in so-called intercalated structures (STaSPySTaS...).

6.1.2 Crystal Conductivity in a Given Direction

Since vectors $\boldsymbol{j}$ and $\boldsymbol{E}$ are generally nonparallel, the concept of "crystal conductivity in a given direction" requires an exact definition. As mentioned in Chap. 1, by quantity T, characterizing the tensor property in a given direction, is meant the ratio of the projection of vector $\boldsymbol{a}$ onto this direction to vector length $\boldsymbol{b}$ laid off in the same direction ($b_i = T_i a_j$). Consequently, if a field $\boldsymbol{E}$ is applied to a crystal, then electrical conductivity σ in the direction of $\boldsymbol{E}$ is equal to the ratio to E, of the current density component $j_{\|}$ parallel to vector $\boldsymbol{E}$, i.e., $\sigma = j_{\|}/E$. Let us now consider, for instance, a thin crystal plate

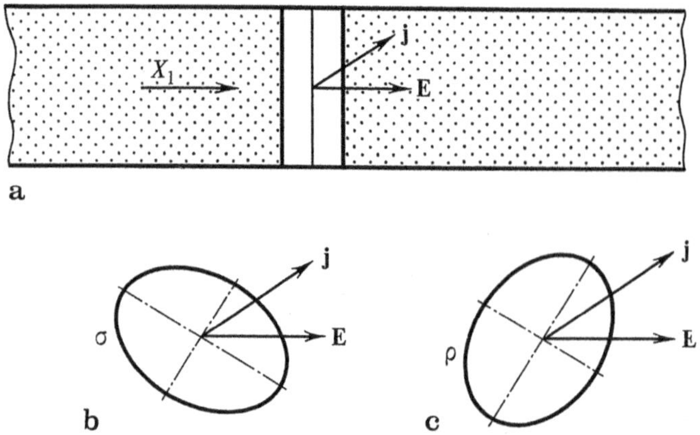

Fig. 6.1. Flow of a current through a thin crystal plate (**a**) and the corresponding ellipsoids of the electrical conductivity σ (**b**) and resistivity ϱ (**c**) coefficients

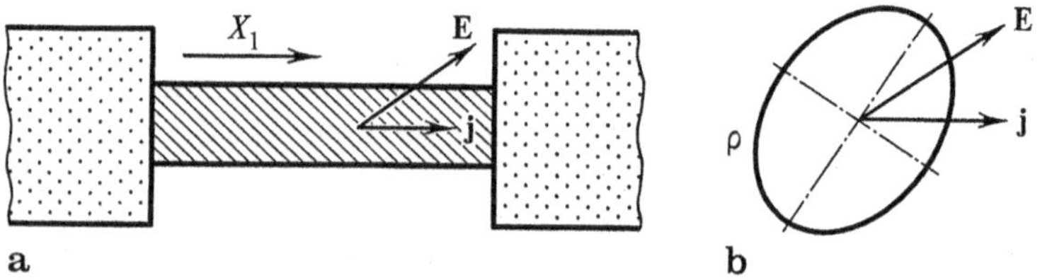

Fig. 6.2. Flow of a current through a long rod (**a**) and the ellipsoid of the resistivity coefficients ϱ (**b**)

sandwiched between two flat electrodes, to which electric field $\boldsymbol{E}$, perpendicular to the large faces of the plate, is applied. In such a plate current density vector $\boldsymbol{j}$ will generally have some other direction. The arrangement of vectors $\boldsymbol{j}$ and $\boldsymbol{E}$ and of the respective ellipsoids of the conductivity and resistivity coefficients is given in Fig. 6.1. Since here the direction of vector $\boldsymbol{E}$ is given, to find the conductivity of the plate in the direction of the normal to its surface, one must use (6.3):

$$j_1 = \sigma_{11}E \ , \quad j_2 = \sigma_{21}E \ , \quad j_3 = \sigma_{31}E \ .$$

If, instead of a thin plate, a thin crystal rod is placed between the metal electrodes, then the current density vector is parallel to the rod axis, and the electric field vector is directed at some angle to the rod axis (Fig. 6.2). Here the direction of the vector, $j = [j_1, 0, 0]$, is given, and hence it is more convenient to use (6.7):

$$E_1 = \varrho_{11}j_1 \ , \quad E_2 = \varrho_{21}j_1 \ , \quad E_3 = \varrho_{31}j_1 \ .$$

In this case the field component is best measured along the rod axis, E_1, whose ratio to current density j_1 will yield ϱ_{11}. The length of the radius vector of the resistivity ellipsoid in the direction of the X_1 axis is equal to $1/\sqrt{\varrho_{11}}$ (Fig. 6.2b). By measuring the conductivity or resistivity in different directions whose orientations relative to the crystallographic axes can be determined, for instance, by x-ray methods, one can find the main conductivity and resistivity coefficients.

6.2 Thermal Conductivity of Crystals

6.2.1 Tensors of the Thermal Conductivity and Thermal Resistivity Coefficients

The descriptions of electrical and thermal conductivities of crystals are absolutely identical. The role of the vector of electric field tension is played here by the vector of the temperature gradient, and the role of the current density vector, by the vector of the heat flux density.

If there is a temperature difference between two points or parts of a solid, one observes heat transport between them, i.e., equalization of the temperature, which is achieved by thermal conductivity. The amount of heat passing through a cross section of some rod per unit time is called the heat flux. By analogy with current density vector $\boldsymbol{j}$, we introduce heat flux density vector $\boldsymbol{q}$. This vector characterizes the amount and direction of the heat flux: it is parallel to the heat flux, and if a unit area is perpendicular to vector $\boldsymbol{q}$, then $\boldsymbol{q}$ is the density of the heat flux through this area. The components of vector $\boldsymbol{q}$ – q_1, q_2, and q_3 – are the amounts of heat passing per unit time through unit areas perpendicular to the X_1, X_2, and X_3 axes.

In isotropic substances the thermal conductivity is described by an equation similar to (6.1):

$$\boldsymbol{q} = -\chi \operatorname{grad} T , \tag{6.11a}$$

where χ is the thermal conductivity coefficient.

Thus, the heat flux is proportional to the temperature gradient. Equation (6.11a) can evidently be written

$$q_i = -\chi \frac{\partial T}{\partial x_i} . \tag{6.11b}$$

In crystals, vector $\boldsymbol{q}$ is generally nonparallel to grad T, and (6.11a) is replaced by the generalized thermal-conductivity equation similar to (6.3):

$$q_{ij} = -\chi_{ij} \frac{\partial T}{\partial x_j} . \tag{6.12}$$

Coefficients χ_{ij} relate two polar vectors, i.e., they form a second-rank polar tensor, called the *tensor of thermal conductivity coefficients*. The minus sign in (6.11 a, 12) is due to the fact that heat is transported from hotter to cooler parts. From (6.12) it follows that each component of vector $\boldsymbol{q}$ depends on all the three components of the temperature gradient, rather than on one of them, as in the case of isotropic substances.

Each of the coefficients χ_{ij} has quite a definite physical meaning. For instance, if there is a temperature gradient $\partial T/\partial x_1 = 1$ grad/cm along the X_1 axis, the heat flux along this axis is equal to χ_{11}, and along the X_2 and X_3 axes to χ_{21} and χ_{31}. In accordance with Onsager's principle (Sect. 6.2.2)

$$\chi_{ij} = \chi_{ji} , \tag{6.13}$$

i.e., tensor χ_{ij} is a symmetric second-rank polar tensor. If this tensor is reduced to the main axes, (6.12) takes the form

$$q_1 = -\chi_1 \frac{\partial T}{\partial x_1} , \quad q_2 = -\chi_2 \frac{\partial T}{\partial x_2} , \quad q_3 = -\chi_3 \frac{\partial T}{\partial x_3} ; \tag{6.14}$$

here χ_1, χ_2, and χ_3 are the main coefficients of thermal conductivity.

The characteristic surface for the thermal conductivity coefficients on passing to the main axes is described by the equation

$$\chi_1 x_1^2 + \chi_2 x_2^2 + \chi_3 x_3^2 = 1 . \tag{6.15}$$

If we solve (6.12) for $\partial T/\partial x_i$, we get

$$\frac{\partial T}{\partial x_i} = -r_{ij} q_j , \tag{6.16}$$

where r_{ij} are the thermal resistance coefficients, which form a symmetric second-rank polar tensor, the *tensor of thermal resistance coefficients*.

All that has been said above concerning coefficients σ_{ik} and ϱ_{ik} is true for coefficients χ_{ij} and r_{ij}. Tensors r_{ij} and χ_{ij} for crystals of different classes have the same form as the dielectric constant tensor. In particular, cubic crystals are isotropic from the standpoint of thermal conductivity. Table 6.2 lists the values of the thermal conductivity coefficients of a number of crystals.

6.2.2 Onsager's Principle

Considering the thermal conductivity of crystals we assumed, in accordance with Onsager's principle, that $\chi_{ij} = \chi_{ji}$. The thermodynamics of irreversible processes, from which the indicated principle stems, was developed by Onsager in 1931. The conclusion about the symmetricity of the tensor of the

Table 6.2. Thermal conductivity coefficients of some crystals (J/m · s · deg)

Crystal	Syngony	T, °C	$\chi_1 = \chi_2$	χ_3
Quartz	Trigonal	30	6.5	11.3
Calcite	Trigonal	30	4.2	5.0
Bismuth	Trigonal	14	9.2	6.6
Graphite	Hexagonal	30	355	89
Aluminium	Cubic	30	208	
Copper	Cubic	0	410	

thermal conductivity coefficients had been drawn much earlier on the basis of measurements of the thermal conductivity of crystals. To understand the meaning of these measurements, we first assume that $\chi_{ij} \neq \chi_{ji}$. We now resolve tensor χ_{ij} into the symmetric and antisymmetric parts. Then we rotate the coordinate axes until they coincide with the main axes of the symmetric tensor. The transformations of the axes leaves the antisymmetric tensor unaltered. As a result we get

$$\left\| \begin{matrix} \chi_{11} & 0 & 0 \\ 0 & \chi_{22} & 0 \\ 0 & 0 & \chi_{33} \end{matrix} \right\| + \left\| \begin{matrix} 0 & \chi_{12} & -\chi_{31} \\ -\chi_{12} & 0 & \chi_{23} \\ \chi_{31} & -\chi_{23} & 0 \end{matrix} \right\| = \left\| \begin{matrix} \chi_{11} & \chi_{12} & -\chi_{31} \\ -\chi_{12} & \chi_{22} & \chi_{23} \\ \chi_{31} & -\chi_{23} & \chi_{33} \end{matrix} \right\| .$$

Suppose the crystal has a three-, four-, or sixfold axis of symmetry. Let this axis coincide with X_3. If the temperature gradient is directed along X_3, the vector of the heat flux must also be parallel to X_3. This means that $\chi_{23} = \chi_{31} = 0$. Besides, the presence of a three-, four-, or sixfold axis of symmetry leads to $\chi_{11} = \chi_{22}$. Consequently, tensor χ_{ij} takes the form

$$\left\| \begin{matrix} \chi_{11} & \chi_{12} & 0 \\ -\chi_{12} & \chi_{11} & 0 \\ 0 & 0 & \chi_{33} \end{matrix} \right\| . \tag{6.17}$$

Let us cut from a crystal of this symmetry a round plate perpendicular to the main axis. Suppose there is a point heat source at the center of the plate O (Fig. 6.3). It can be proved that tensor (6.17) will not change on any rotation of the coordinate system about the X_3 axis. Hence, the equations describing the heat transport must have a circular symmetry relative to the X_3 axis. This means that the isotherms on the flat surface of the plate must be circular in shape.

The equations for heat fluxes at the point K situated on the X_1 axis have the form

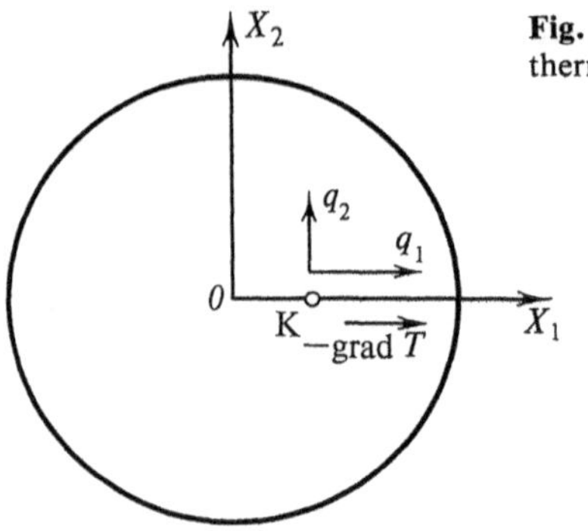

Fig. 6.3. Proof of the symmetricity of the tensor of the thermal conductivity coefficients

$$q_1 = -\chi_{11}\frac{\partial T}{\partial x_1}\ , \quad q_2 = -\chi_{21}\frac{\partial T}{\partial x_1} = \chi_{12}\frac{\partial T}{\partial x_1}\ , \quad q_3 = 0\ .$$

Thus, vector $\boldsymbol{q}$ makes with the X_1 axis an angle φ defined by the expression

$$\tan\varphi = q_2/q_1 = -\chi_{12}/\chi_{11}\ .$$

Because of the circular symmetry, vector $\boldsymbol{q}$ must form the same angle with any radius vector. In other words, the heat flux must propagate along an unwinding spiral with a constant tilt of the tangent relative to the radius vector. Such an unexpected result made many scientists carry out experiments to ascertain such "helical" flow of heat in crystals. All these experiments, however, ended in a failure. Therefore, it was concluded that, at least within the experimental error, the heat from a point source flows along straight lines, and hence (6.13) holds true.

Let us now consider Onsager's principle. Let $x_1, x_2, \ldots, x_a$ be some parameters characterizing the state of a system. We introduce the values

$$X_a = -\frac{\partial s}{\partial x_a}\ ,$$

where s is the entropy of a unit volume of the system and the derivative is taken at a constant energy of this volume. In a near-equilibrium state, x_a are close to their equilibrium values, and X_a are small. Then the system will tend to return to the equilibrium state, and the rate of change in x_a, i.e., $\partial x_a/\partial t$, at each point of the system will be a function of X_a at the same points. Expanding these rates into a series in powers of x_a and restricting ourselves to the linear terms, we obtain equations of the type

$$\frac{\partial x_a}{\partial t} = -\sum \gamma_{ab} X_b\ . \tag{6.18}$$

In thermodynamics of irreversible processes the values $\partial x_a/\partial t$ are called *fluxes,* the values X_a *forces* conjugate to the respective fluxes, and γ_{ab} kinetic coefficients. In accordance with Onsager's principle, coefficients γ are symmetric, i.e.,

$$\gamma_{ab} = \gamma_{ba} \quad . \tag{6.19}$$

When using this principle we should choose the fluxes and the respective forces correctly. This problem is usually solved with the aid of an equation determining the time variation of the total system entropy [6.6]:

$$\frac{dS}{dt} = -\int \sum_a X_a \frac{\partial x_a}{\partial t} dv \quad , \tag{6.20}$$

where integration is performed over the entire volume of the system. In particular, during the passage of a current along a conductor we have

$$\frac{dS}{dt} = \int \frac{jE}{T} dv \quad .$$

Hence, current density j must be taken as flux $\partial x_a/\partial t$, and the components of vector $-E/T$ as force X_a conjugate to it. From (6.3, 18) it follows that the role of the kinetic coefficients is played by the T-fold components of the conductivity tensor, whose symmetricity thus follows from (6.19).

6.3 Thermoelectric Effects

6.3.1 Definition of Thermal Effects

The heat and electricity transports in crystals have so far been considered independently of one another. In isotropic conductors or crystals, however, both a temperature gradient and a potential gradient are often present simultaneously. This produces a number of effects, called *thermoelectric* [6.9]. They include the Zeebeck, Peltier, and Thomson effects. The Bridgman effect is also possible in crystals.

In the Zeebeck effect, an emf (thermal emf) arises in a closed circuit consisting of two different conductors, whose junctions are maintained at a different temperature. This emf is proportional to the temperature difference:

$$d\varphi = \alpha dT \quad ; \tag{6.21}$$

here $\alpha = d\varphi/dT$ is the thermal emf referred to a unit change in temperature and called the *differential thermal* emf. Along with the direct thermoelectric effect, there exists a reciprocal, *Peltier effect:* a junction of two conductors is heated or cooled (depending on the direction of the current) on passage of an electric current through it. The rate of heat release or absorption is proportional to the current density

$$Q = \Pi_{ab} j \; ; \tag{6.22}$$

here Π_{ab} is the Peltier coefficient.

In the *Thomson effect,* heat is released or absorbed on passage of a current in a homogeneous conductor with a temperature gradient along it. The amount of this heat is proportional to the current density and the temperature gradient:

$$dQ = \tau j \, dT \; , \tag{6.23}$$

where τ is the Thomson coefficient. The relationship among the coefficients α, Π, and τ is expressed by the *Thomson relations:*

$$\Pi = \alpha T \; , \quad \tau_1 - \tau_2 = T \frac{d\alpha}{dT} \; . \tag{6.24}$$

6.3.2 Thermoelectric Effects in an Isotropic Medium

To make a correct choice of fluxes and the respective forces, we first consider thermoelectric effects in an isotropic medium. We assume that the medium is inhomogeneous, i.e., that its properties change from point to point. During the passage of a current the Peltier heat will be distributed throughout the medium (the Peltier effect at a contact is only a particular case).

When temperature and electrical potential gradients are present simultaneously, the equations for the electricity and heat fluxes have the form

$$j = -a \,\mathrm{grad}\, \varphi - b \,\mathrm{grad}\, T \; , \tag{6.25 a}$$

$$q = -c \,\mathrm{grad}\, \varphi - d \,\mathrm{grad}\, T \; . \tag{6.25 b}$$

Let us write the Gibbs equation for a small volume element of the medium

$$T ds = du - \varphi de \; ; \tag{6.26}$$

here s is the specific entropy, u is the specific energy, e is the electric charge, and φ is the potential. From (6.26) it is seen that the entropy, energy, and electric current are related by the following expression

$$s = \frac{1}{T} u - \frac{\varphi}{T} j \; ;$$

consequently,

$$\frac{ds}{dt} = u\,\mathrm{grad}\,\frac{1}{T} - j\,\mathrm{grad}\left(-\frac{\varphi}{T}\right), \tag{6.27}$$

because $\mathrm{div}\,u = \mathrm{div}\,j = 0$.

But $u = q + \varphi j$ and therefore

$$\frac{ds}{dt} = (q+\varphi j)\,\mathrm{grad}\,\frac{1}{T} + j\,\mathrm{grad}\left(-\frac{1}{T}\right) = (q+\varphi j)\,\mathrm{grad}\,\frac{1}{T}$$

$$+ j\,\frac{-\mathrm{grad}\,\varphi}{T} - \varphi j\,\mathrm{grad}\,\frac{1}{T} = q\,\mathrm{grad}\,\frac{1}{T} - j\,\frac{\mathrm{grad}\,\varphi}{T}\ . \tag{6.28}$$

Comparing (6.28, 20) we see that if q and j are chosen as fluxes, the corresponding forces will be $\mathrm{grad}\,(1/T) = T^{-2}\mathrm{grad}\,T$ and $-\mathrm{grad}\,(\varphi/T) = E/T$. Thus, (6.25 a, b) takes the form

$$j = aT\,\frac{-\mathrm{grad}\,\varphi}{T} + bT^2\,\mathrm{grad}\,\frac{1}{T}\ , \tag{6.29 a}$$

$$q = cT\,\frac{-\mathrm{grad}\,\varphi}{T} + dT^2\,\mathrm{grad}\,\frac{1}{T}\ . \tag{6.29 b}$$

According to Onsager's principle we have

$$bT^2 = cT \quad \text{or} \quad bT = c\ .$$

Expressing $-\mathrm{grad}\,\varphi$ from (6.29 a) in terms of j and $\mathrm{grad}\,T$ and substituting it into (6.29 b), we obtain the following expression for the heat flux:

$$q = \frac{b}{a}\,jT - \left(d - \frac{b^2T}{a}\right)\mathrm{grad}\,T \tag{6.30}$$

or

$$q = \frac{b}{a}\,jT - \chi\,\mathrm{grad}\,T\ , \tag{6.31}$$

where $\chi = (d - b^2T/a)$ is the thermal conductivity coefficient.

We shall now find the expression for the rate of heat release. In the presence of an electron flux the total energy flux

$$u = q + \varphi j\ .$$

The rate of heat release per unit volume of the medium in the vicinity of a given point is equal to the divergency of the energy flux with the opposite sign:

$$-\operatorname{div}(q+\varphi j) = -\operatorname{div}\left(\frac{b}{a}T+\varphi\right)j+\operatorname{div}(\chi \operatorname{grad} T)$$

$$= -j\operatorname{grad}\varphi-\frac{b}{a}j\operatorname{grad}T-Tj\operatorname{grad}\frac{b}{a}+\operatorname{div}(\chi\operatorname{grad}T) \ .$$

But as follows from (6.25),

$$-\operatorname{grad}\varphi = j/a+\frac{b}{a}\operatorname{grad}T \ ,$$

hence,

$$-\operatorname{div}(q+\varphi j) = jj/a - Tj\operatorname{grad}(b/a)+\operatorname{div}(\chi\operatorname{grad}T) \ . \tag{6.32}$$

The first term in the right-hand side of this expression represents the Joule heat, and the last one, the heat dissipated by conduction, while the second,

$$\dot{q} = -Tj\operatorname{grad}(b/a) \ , \tag{6.33}$$

is the Thomson heat released owing to thermoelectric effects. We write

$$\alpha = b/a \ . \tag{6.34}$$

In a homogeneous medium, α is only a function of the temperature; hence,

$$\operatorname{grad}\alpha = \frac{d\alpha}{dT}\operatorname{grad}T \ .$$

Then (6.33) transforms to

$$\dot{q} = -Tj\frac{d\alpha}{dT}\operatorname{grad}T \ .$$

It is seen that $-Td\alpha/dT$ is nothing else than Thomson coefficient τ (6.24), and α is the thermal emf coefficient. In the general case, (6.33) contains both the Thomson and the Peltier heat. Indeed, in an inhomogeneous medium, coefficient α is a function not only of the temperature, but also of the coordinate (it may change with changes in the chemical composition of the medium or when the intermetallic contact is crossed by the flux).

Consequently, $\operatorname{grad}\alpha$ contains two terms, and the expanded form of (6.33) will be

$$\dot{q} = -Tj\,\mathrm{grad}\,\alpha = -Tj_i\frac{\partial\alpha}{\partial x_i} = -j_i\left(\frac{\partial\alpha}{\partial x_i}\right)_T - j_iT\left(\frac{\partial\alpha}{\partial T}\right)\frac{\partial T}{\partial x_i} \ . \tag{6.35}$$

The first term in the right-hand side of this equation is the Peltier heat, proportional to the current density, and the second, the Thomson heat, proportional to the current density and the temperature gradient.

6.3.3 Thermoelectric Effects in Crystals

For crystals, equations relating the electron and heat fluxes to the respective forces have the same form as (6.29), except that the scalar coefficients *a, b, c,* and *d* are replaced by second-rank polar tensors a_{ik}, b_{ik}, c_{ik}, and d_{ik}:

$$j_i = a_{ik}T\left(-\frac{1}{T}\frac{\partial\varphi}{\partial x_k}\right) + b_{ik}\frac{\partial}{\partial x_k}\left(\frac{1}{T}\right) , \tag{6.36a}$$

$$q_i = c_{ik}T - \left(\frac{1}{T}\frac{\partial\varphi}{\partial x_k}\right) + d_{ik}T^2\frac{\partial}{\partial x_k}\left(\frac{1}{T}\right) , \tag{6.36b}$$

From the Onsager relations we have

$$c_{ik} = b_{ik}T \ . \tag{6.37}$$

Expressing $E_i = -\partial\varphi/\partial x_k$ in terms of j, $\partial/\partial x_i(1/T)$ from (6.36a) and substituting (6.37) into (6.36b), we get

$$E_i = a_{ik}^{-1}j_k + a_{ik}^{-1}b_{lk}\frac{\partial T}{\partial x_k}$$

or

$$q_i = b_{il}a_{lk}^{-1}Tj - (d_{ik} - b_{il}a_{lm}^{-1}b_{mk}T)\frac{\partial T}{\partial x} ,$$

$$E_i = \varrho_{ik}j_k - \alpha_{ik}\frac{\partial T}{\partial x_k} , \quad q_i = -\beta_{ik}j_k - \chi_{ik}\frac{\partial T}{\partial x_k} \ . \tag{6.38}$$

In these equations, ϱ_{ik} is the electrical resistance tensor and χ_{ik} the thermal conductivity tensor. Tensors α_{ik} and β_{ik} are bound by Thomson's relation $\beta_{ik} = T\alpha_{ik}$. They can be called tensors of the thermoelectrical properties of crystals. In particular, χ_{ik} is the tensor of the thermal emf coefficients. Both these tensors are second-rank polar tensors. They are generally asymmetric.

For the rate of heat release per unit volume of the crystal, we obtain an equation similar to (6.32) for an isotropic medium:

$$-\frac{\partial u}{\partial x_k} = \frac{j_i j_k}{a_{ik}} - T\frac{\partial}{\partial x_k}(j_i \alpha_{ik}) + \frac{\partial}{\partial x_k}\left(\chi_{ki}\frac{\partial T}{\partial x_i}\right) . \tag{6.39}$$

As in (6.32), here the first term in the right-hand side is the Joule heat, the second is the heat released owing to thermoelectric effects, and the third, the heat dissipated by conduction. Let us write separately the term responsible for the thermoelectric effects:

$$\dot{q} = -T\frac{\partial}{\partial x_k}(j_i \alpha_{ik}) . \tag{6.40}$$

For heat release at a contact between two crystals or between two parts of one and the same crystal, we get

$$\dot{q} = -T\Delta_n(j_i \alpha_{ik}) = -\Delta_n(j_i \alpha_{ik} T) , \tag{6.41}$$

where Δ_n is the jump of the normal component of vector $j_i \alpha_{ik}$. Obviously, $\alpha_{ik} T$ are Π_{ik}, i.e., Peltier coefficients. Thus,

$$\dot{q} = -\Delta_n(j_i \Pi_{ik}) . \tag{6.42}$$

6.3.4 Dependence of the Tensor of the Thermoelectric Coefficients on the Crystal Symmetry

In the general case (triclinic syngony), tensors α_{ik} and Π_{ik}, related by expression $\Pi_{ik} = T\alpha_{ik}$, have nine independent components each. The presence of symmetry elements in crystals reduces the number of independent components, as usual. Table 6.3 lists the matrices of tensors α_{ik} for various syngonies.

6.3.5 Peltier, Thomson, and Bridgman Effects

Generally, both longitudinal and transverse thermoelectric effects can be observed. Let us consider, for instance, a long parallelepiped carved out of a crystal with a noncubic symmetry. Suppose an electric current $\boldsymbol{j} = [j00]$ is flowing along such a crystal (X_1 axis). Vector $\boldsymbol{j}_i \cdot \alpha_{ik}$ will be nonparallel to $\boldsymbol{j}$, because there are transverse components $j \cdot \alpha_{12}$ and $j \cdot \alpha_{13}$ along with the longitudinal component $j \cdot \alpha_{11}$. On the upper face, perpendicular to X_2, there is a jump of the normal component $j_i \alpha_{ik}$ equal to $-j\alpha_{12}$. In accordance with (6.41), an amount of heat equal to $Tj\alpha_{12}$ will be released here. The same amount of heat must be absorbed on the opposite (lower) face.

To obtain more detailed information on the nature of the heat released owing to thermoelectric effects, we differentiate the right-hand side of (6.40),

Table 6.3. Matrices of thermoelectric coefficients α_{ik} for crystals of different syngonies

Syngony / classes	Matrix
Triclinic syngony	$\begin{Vmatrix} \alpha_{11} & \alpha_{12} & \alpha_{13} \\ \alpha_{21} & \alpha_{22} & \alpha_{23} \\ \alpha_{31} & \alpha_{32} & \alpha_{33} \end{Vmatrix}$
Monoclinic syngony	$\begin{Vmatrix} \alpha_{11} & \alpha_{12} & 0 \\ \alpha_{21} & \alpha_{22} & 0 \\ 0 & 0 & \alpha_{33} \end{Vmatrix}$
Orthorhombic syngony	$\begin{Vmatrix} \alpha_{11} & 0 & 0 \\ 0 & \alpha_{22} & 0 \\ 0 & 0 & \alpha_{33} \end{Vmatrix}$
Tetragonal, trigonal, and hexagonal syngonies: Classes 4, $\bar{4}$, $4/m$, 3, $\bar{3}$, 6, $\bar{6}$, $6/m$	$\begin{Vmatrix} \alpha_{11} & \alpha_{12} & 0 \\ -\alpha_{12} & \alpha_{11} & 0 \\ 0 & 0 & \alpha_{33} \end{Vmatrix}$
Classes 422, $4mm$, $\bar{4}2m$, $4/mmm$, 32, $3m$, $\bar{3}m$, 622, $6mm$, $\bar{6}m2$, $6/mmm$	$\begin{Vmatrix} \alpha_{11} & 0 & 0 \\ 0 & \alpha_{11} & 0 \\ 0 & 0 & \alpha_{33} \end{Vmatrix}$
Cubic syngony	$\begin{Vmatrix} \alpha_{11} & 0 & 0 \\ 0 & \alpha_{11} & 0 \\ 0 & 0 & \alpha_{11} \end{Vmatrix}$

bearing in mind that α_{ik} depends on x_k in a dual way. First, α_{ik} varies with the coordinate because of the change in crystal composition or orientation (it crosses the contact or the twin boundary). Second, α_{ik} may vary with the coordinate because of the nonuniform temperature distribution.

We ultimately get

$$\dot{a} = -j_i T \left(\frac{\partial \alpha_{ik}}{\partial x_k}\right)_T - j_i T \left(\frac{\partial \alpha_{ik}}{\partial T}\right) \frac{\partial T}{\partial x_k} - T \alpha_{ik} \frac{\partial j}{\partial x_k} \ . \tag{6.43}$$

The first term in the right-hand side of (6.43) is the Peltier heat due to the change in crystal properties with the coordinate (at a constant temperature). Along with the longitudinal Peltier effect, a *transverse Peltier effect* is also possible in crystals, because not only the temperature gradient in the direction of the current, but also other gradients, make a contribution here.

The second term is the Thomson heat, which is proportional to the current density and the temperature gradient. This amount of heat released per unit volume of the crystal is defined by the expression

$$-j_i T \left(\frac{\partial \alpha_{ik}}{\partial T}\right) \frac{\partial T}{\partial x_k} = \tau_{ik} j \frac{\partial T}{\partial x_k} \ , \tag{6.44}$$

where τ_{ik} is the tensor of the Thomson coefficients, which has the form

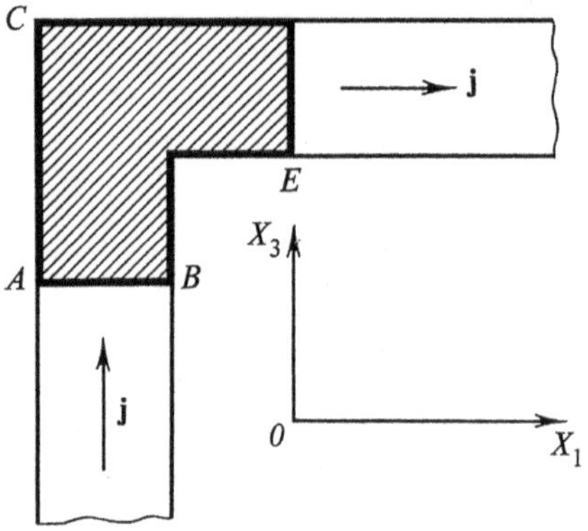

Fig. 6.4. Scheme of observation of the Bridgman effect

$$\tau_{ik} = -T\left(\frac{\partial \alpha_{ik}}{\partial T}\right)_x . \tag{6.45}$$

The last term in (6.43) describes an effect which can be observed only in crystals. In the simplest case of a cubic crystal, only the diagonal components of tensor α_{ik} are nonzero. They are equal to each other, and therefore the last term of (6.43) can be represented as

$$-T\alpha\frac{\partial j}{\partial x_i} . \tag{6.46}$$

In the case of stationary currents, $\partial j/\partial x_i = 0$. Hence, this term leads to a heat release due to the nonuniform current distribution. This is the so-called *Bridgman effect.*

To demonstrate this effect, we consider an L-shaped crystal with a unit cross section and a symmetry $4mm$ (Fig. 6.4). The crystallographic orientation and temperature of the two parts of the crystal are the same. Suppose an electron flux j flows along the X_3 axis and then takes a new course parallel to X_1 axis. Not to complicate the picture by the transverse Peltier effect, we orient the crystal main axis parallel to X_3. We apply (6.40) to the hatched part of the crystal. Components α_{ik} are constant, and the value $-q/T$ is equal to divergency $j_i\alpha_{ik}$ integrated over the singled-out volume. By the Gauss theorem such an integral is equal to the flux of normal components $j\alpha_{ik}$ directed outwards through the crystal faces bounding the volume. The components of the flux through AB and DE are equal to $-j\alpha_{33}$ and $j\alpha_{11}$, respectively (we take the components directed towards the outward normals). Consequently,

$$\dot{q} = Tj(\alpha_{33} - \alpha_{11}) . \tag{6.47}$$

This is the expression for the Bridgman heat, which follows here from the change in the direction of current j in the crystal. Generally, any nonuniformity in the distribution of current densities in an anisotropic crystal leads to the release of Bridgman heat.

6.3.6 Thermoelectric Effects in Technology

Thermoelectric effects are widely used in the manufacture of thermoelectric generators and refrigerators, i.e., appliances designed for direct conversion of heat to electricity and for refrigeration by passing an electric current across a junction of two semiconductors (Peltier effect). The idea of making such appliances was conceived by academician *Ioffe* [6.9].

A thermogenerator consists of several series-connected thermoelements (thermocouples), which are usually made of semiconductors with a high thermal emf coefficient, a sufficiently high electrical conductivity, and a low thermal conductivity of the lattice (Bi_2Te_3, Sb_2Te_3, PbTe, etc.). The efficiency of such generators may be as high as 10 – 15%. It is defined by

$$Z = (\alpha_1^2 + \alpha_2^2)/(\sqrt{\chi_1 \varrho_1} + \sqrt{\chi_2 \varrho_2})^2 ,$$

where α_1 and α_2 are the thermal emf of the branches, χ_1 and χ_2 are the thermal conductivities of the lattices, and ϱ_1 and ϱ_2 are the resistivities of the semiconductors.

6.4 Galvano- and Thermomagnetic Effects

6.4.1 Onsager's Principle in the Presence of a Magnetic Field

An external magnetic field exerts a substantial effect on the electrical, thermal, and thermoelectrical phenomena in conducting crystals. A phenomenological description of this effect is carried out with the aid of the same system of tensors as for thermoelectric effects, except that the tensor coefficients of (6.38) of the preceding section are here functions of magnetic field $\boldsymbol{H}$.

In the general form, the equations describing the propagation of an electric current and a heat flux in a crystal located in a magnetic field have the form

$$E_i = \varrho_{ik}(H) j_k + \alpha_{ik}(H) \operatorname{grad} T , \tag{6.48a}$$

$$q_i = -\beta_{ik}(H) j_k + \chi_{ik}(H) \operatorname{grad} T . \tag{6.48b}$$

In the presence of a magnetic field, Onsager's principle (6.19) takes the form

$$\gamma_{ik}(\boldsymbol{H}) = \gamma_{ki}(-\boldsymbol{H}) . \tag{6.49}$$

This follows from the time symmetry principle for mechanics equations, on which the proof of symmetry of kinetic coefficients is based. This principle states that if the time and velocities of all the particles of the system are reversed simultaneously, the particles will move along their former trajectories. In other words, the symmetry of mechanics equations for the motion of individual particles with respect to time reversal is conserved. This symmetry, however, changes somewhat for a uniformly rotating body and for bodies in an external magnetic field. Here the symmetry of the equations of motion with respect to time sign reversal takes place only if the sign of rotation velocity Ω or the direction of magnetic field $\boldsymbol{H}$ changes as well. Therefore, for the kinetic coefficients, which depend on $\boldsymbol{H}$ as on a parameter, (6.49) holds true.

Thus, we can write the following expressions for the coefficients of (6.48):

$$\varrho_{ik}(\boldsymbol{H}) = \varrho_{ki}(-\boldsymbol{H}) \ , \quad \chi_{ik}(\boldsymbol{H}) = \chi_{ki}(-\boldsymbol{H}) \ , \quad \alpha_{ik}(\boldsymbol{H}) = T\beta_{ki}(\boldsymbol{H}) \ . \tag{6.50}$$

This means that the tensors of the coefficients of electrical resistivity and thermal conductivity, which are symmetric in the absence of a magnetic field, become asymmetric.

6.4.2 Hall, Righi-Leduc, Nernst, and Ettingshausen Effects

It is well known that any second-rank tensor can be represented as a sum of the symmetric and antisymmetric parts. In particular, for tensor ϱ_{ik} we can write

$$\varrho_{ik} = s_{ik} + a_{ik} \ , \tag{6.51}$$

where s_{ik} and a_{ik} are the symmetric and antisymmetric tensors. By definition

$$s_{ik}(\boldsymbol{H}) = s_{ki}(\boldsymbol{H}) \ , \quad a_{ik}(\boldsymbol{H}) = -a_{ki}(\boldsymbol{H}) \ . \tag{6.52}$$

From (6.50, 52) it follows that

$$\begin{aligned} s_{ik}(\boldsymbol{H}) &= s_{ki}(-\boldsymbol{H}) = s_{ik}(-\boldsymbol{H}) \ , \\ a_{ik}(\boldsymbol{H}) &= a_{ki}(-\boldsymbol{H}) = -a_{ik}(-\boldsymbol{H}) \ . \end{aligned} \tag{6.53}$$

This means that the symmetric part of s_{ik} is an even function of $\boldsymbol{H}$, and the antisymmetric a_{ik}, an odd one. The antisymmetric part a_{ik} can be represented as an axial vector $\boldsymbol{a}$, whose components are defined by the relations

$$a_1 = a_{23} \ , \quad a_2 = -a_{13} \ , \quad a_3 = a_{12} \ . \tag{6.54}$$

In the absence of a temperature gradient we have from (6.48a)

$$E_i = \varrho_{ik} j_k = s_{ik} j_k + a_{ik} j_k \ ,$$

or, using axial vector $\boldsymbol{a}$,

$$E_i = s_{ik} j_k + [\boldsymbol{j} \times \boldsymbol{a}]_i \ . \tag{6.55}$$

To obtain the Joule heat we must take the scalar product $\boldsymbol{j} \cdot \boldsymbol{E}$. Since vectors $[\boldsymbol{j} \times \boldsymbol{a}]$ and $\boldsymbol{j}$ are mutually perpendicular, their scalar product is equal to zero, and we get from (6.55)

$$\boldsymbol{j} \cdot \boldsymbol{E} = s_{ik} j_i j_k \ .$$

Hence, the Joule heat is defined only by the symmetric part of the tensor of the resistivity coefficients ϱ_{ik}. The presence of the term $[\boldsymbol{j} \times \boldsymbol{a}]$ in (6.55) means that the magnetic field perpendicular to the electric current induces an electric field perpendicular to the direction of the magnetic field and current. This is called the Hall effect.

Let us represent the components of the tensor of resistivity coefficient $\varrho_{ik}(\boldsymbol{H})$ as a series in increasing powers of $\boldsymbol{H}$. Restricting ourselves to the terms of the second order with respect to $\boldsymbol{H}$, we obtain

$$\varrho_{ik}(H) = \varrho_{ik}^{(0)} + \varrho_{ikl} H_l + \varrho_{iklm} H_l H_m \ . \tag{6.56}$$

Obviously, the symmetric part of the tensor must contain terms of an even power, and the antisymmetric, of an odd power with respect to H, i.e.,

$$s_{ik}(H) = \varrho_{ik}^{(0)} + \varrho_{iklm} H_l H_m \ , \tag{6.57}$$

$$a_{ik}(H) = \varrho_{ikl} H_l \ . \tag{6.58}$$

The Hall emf is proportional to the first power of the magnetic field tension. In other words, the Hall effect must be described by the antisymmetric part of the tensor of the resistivity coefficients which is represented by (6.58). Tensor ϱ_{ikl} is called the Hall coefficients tensor. It relates antisymmetric second-rank polar tensor $a_{ik}(H)$ and axial vector $\boldsymbol{H}$, i.e., is a third-rank axial tensor.

The second term in the right-hand side of (6.57) reflects the variation of the electric resistance of the crystal in a magnetic field. In other words, it describes the magnetoresistance effect. Tensor ϱ_{iklm} is called the magnetoresistance tensor. It relates the symmetric second-rank polar tensor with the product of vectors $\boldsymbol{H}_l \cdot \boldsymbol{H}_m$. We may consequently assume that it relates two symmetric second-rank polar tensors, i.e., is a symmetric fourth-rank polar tensor.

We shall now consider the Hall effect and other similar phenomena for an isotropic medium in more detail. Later on we shall see that cubic crystals can be allocated to such a medium. In the case of an isotropic medium, the ordinary tensor of the resistivity coefficients $\varrho_{ik}^{(0)}$ and the tensor of the Hall coefficients ϱ_{ikl} are isotropic, i.e., have one nonzero component each, whereas the tensor of the magnetoresistance coefficients has two such components. The axial vector, defined by (6.54), is parallel to $\boldsymbol{H}$, and the Hall emf vector is perpendicular to the current and magnetic induction vector. The relationship between ϱ, a, and $\boldsymbol{H}$ can be written

$$\varrho(\boldsymbol{H}) = a(\boldsymbol{H}) = -R\boldsymbol{H} \ ,$$

where R is the nonzero component of the Hall coefficients tensor, which is called the Hall coefficient.

The following vectors, linear in $\boldsymbol{j}$ and quadratic in $\boldsymbol{H}$, can be formed from vectors $\boldsymbol{j}$ and $\boldsymbol{H}$: $\boldsymbol{H}^2\boldsymbol{j}$ and $(\boldsymbol{j}\cdot\boldsymbol{H})\cdot\boldsymbol{H}$. Thus, the general equation for vector $\boldsymbol{E}$ containing terms of second order with respect to $\boldsymbol{H}$ can be written

$$\boldsymbol{E} = \varrho_0\boldsymbol{j} + R[\boldsymbol{H}\times\boldsymbol{j}] + \beta_2(\boldsymbol{j}\cdot\boldsymbol{H})\cdot\boldsymbol{H} \ . \tag{6.59}$$

If the magnetic field is directed only along the X_3-axis, we obtain from this equation

$$\begin{aligned} E_1 &= (\varrho_0 + \beta_1 H_3^2)j_1 - RH_3 j_2 \ , \\ E_2 &= RH_3 j_1 + (\varrho_0 + \beta_1 H_3^2)j_2 \ , \\ E_3 &= [\varrho_0 + (\beta_1 + \beta_2)H_3^2]j_3 \ . \end{aligned} \tag{6.60}$$

Here coefficient $(\varrho_0 + \beta_1 H_3^2)$ is the resistivity measured in the direction of the current when the magnetic field is perpendicular to the current, i.e., it is the longitudinal effect in a transverse magnetic field ($\varrho_\perp$). Term RH_3 yields the transverse effect in a transverse field; this is actually the Hall effect. Finally, the term in the square brackets gives the longitudinal effect in a longitudinal field and is denoted by $\varrho_\parallel$. The quantities $(\varrho_\perp - \varrho_0)/\varrho_0 = \Delta\varrho_\perp/\varrho_0$ and $(\varrho_\parallel - \varrho_0)/\varrho_0 = \Delta\varrho_\parallel/\varrho_0$ are called the coefficients of transverse and longitudinal magnetoresistances.

We can consider the effect of the field on the tensor of thermal conductivity coefficients χ_{ij} in much the same way as we considered the electric resistivity tensor. The first-order effect, i.e., the term proportional to $\boldsymbol{H}$, is an analog of the Hall effect. It consists in the appearance of a transverse temperature gradient when a conductor with a heat flux is subjected to the action of a transverse magnetic field. This is the so-called *Righi-Leduc effect.* The terms proportional to the second power of the magnetic field correspond to the magnetic resistance and describe the *magnetothermal conductivity* (variation of thermal conductivity in a magnetic field).

Tensor α_{ik} in (6.48 a) relates electric field E_i to temperature gradient $\partial T/\partial x_k$ in the absence of an electric current ($j=0$). In the presence of a magnetic field, it is naturally a function of this field. From (6.48 a) at $j = 0$ we have

$$E_i = \alpha_{ik}(H) \frac{\partial T}{\partial x_k} . \qquad (6.61)$$

Similarly, β_{ik} relates heat flux q_i with electric current j in the absence of a temperature gradient:

$$q_i = -\beta_{ik}(H) j_k . \qquad (6.62)$$

Onsager's relation (6.49) for tensors α_{ik} and β_{ik} differs from the respective relations for ϱ_{ik} and χ_{ik}. As a result, the symmetric and antisymmetric part of tensors α_{ik} and β_{ik} are no longer an even and an odd function of $\boldsymbol{H}$. In the case of a heat flux, we can write for α_{ik} and β_{ik} expressions similar to (6.55), which means that the antisymmetric parts of these tensors produce transverse effects. But since the antisymmetric parts are not equivalent to the odd terms of the functions, we cannot assert that the first-order effects caused by these parts are pure transverse effects, as in the case of tensors ϱ_{ik} and χ_{ik}.

By analogy with (6.56) we can expand functions $\alpha_{ik}(H)$ into a series in H:

$$\alpha_{ik}(H) = \alpha_{ik}^{(0)} + \alpha_{ikl} H_l + \alpha_{iklm} H_l H_m . \qquad (6.63)$$

Substituting (6.63) into (6.61), we get

$$E_i = (\alpha_{ik}^{(0)} + \alpha_{ikl} H_l + \alpha_{iklm} H_l H_m) \frac{\partial T}{\partial x_k} . \qquad (6.64)$$

It is seen that tensor α_{ikl} is similar to the tensor of Hall coefficients, but, as noted above, it does not represent a pure transverse effect. This is also true of β_{ikl}. The respective transverse effects in an isotropic medium are called the *Nernst effect* and the *Ettingshausen effect*.

The Nernst effect consists in the appearance of a transverse potential difference in a conductor with a heat flux placed in a transverse magnetic field. It can be assumed that tensor α_{ikl} describes this effect in crystals, but this description is not accurate owing to the above-mentioned circumstance. Tensor α_{ikl} is called the tensor of Nernst coefficients. Component α_{ikl} reflects the variation in the ith component of the thermal emf due to the kth component of the temperature gradient under the effect of the lth component of the magnetic field.

The Ettingshausen effect consists in the appearance of a temperature difference in the direction normal to the longitudinal electric current and magnetic field. Tensor β_{ikl} approximately describes this effect in crystals and is called the tensor of the Ettingshausen coefficients. Tensors α_{ikl} and β_{ikl} relate the second-rank polar tensors to the axial vector, and hence are third-rank axial tensors.

The respective second-order effects are described by tensors α_{ikl} and β_{ikl}. Each of them relates a second-rank polar tensor to a vector product of the type $\boldsymbol{H}_l \cdot \boldsymbol{H}_m$ and is a fourth-rank polar tensor. Tensor α_{iklm} can be called a magnetothermoelectric tensor. It describes the variation of the thermal emf in a magnetic field.

The effects due to the action of the magnetic field on the electrical and thermoelectrical properties in the absence of temperature gradients are called galvanomagnetic. They include the Hall effect, the magnetoresistance effect, and the Ettingshausen effect. The effects due to the action of the magnetic field on the thermal and thermoelectric properties in the absence of an electric current are called thermomagnetic. They include the Nernst effect, the magnetothermal emf, the Righi-Leduc effect, and magnetothermal conductivity. There are also effects of higher orders, but they are not comsidered here.

Thus, the transport phenomena in crystals can be classified as follows:

1. *Zero-order effects,* whose coefficients are denoted by second-rank polar tensors. They include the electrical and thermal conductivity phenomena, and also thermoelectrical phenomena occurring in the absence of a magnetic field. They are described by tensors ϱ_{ik}, χ_{ik}, α_{ik}, and β_{ik}.
2. *First-order effects* proportional to the first-power magnetic field vector. They include the Hall, Righi-Leduc, Nernst, and Ettingshausen effects. The tensors of the Hall and Righi-Leduc coefficients ϱ_{ikl} and χ_{ikl} are third-rank axial tensors; because of the internal symmetry they are equivalent to second-rank polar tensors. The tensors of the Nernst and Ettingshausen coefficients α_{ikl} and β_{ikl} are third-rank axial tensors relating second-rank polar tensors to axial vectors.
3. *Second-order effects* proportional to the second power of the magnetic field tensor. They are described by fourth-rank polar tensors ϱ_{iklm}, χ_{iklm}, α_{iklm}, and β_{iklm}. These are the magnetoresistance, magnetothermal conductivity, and magnetothermal emf (the fourth effect has no name).

6.4.3 Tensors of the Hall, Righi-Leduc, Nernst, and Ettingshausen Coefficients and of Magnetoresistance for Crystals of Different Classes

As noted above, the Hall effect is described by the antisymmetric part of the tensor of the coefficients of resistivity a_{ij} (6.58). By definition, $a_{ij} = -a_{ji}$ and $a_{ii} = 0$. Hence, the components of the Hall coefficients tensor ϱ_{ikl}, in which the first two subscripts are identical, are equal to zero, and for general-type components equality $\varrho_{ikl} = -\varrho_{kil}$ holds true.

If α_{ip} are the direction cosines of the angles between the new (crystallophysical) and old (arbitrary) coordinate axes, the law of transformation of tensor ϱ_{ikl} will be written as follows in accordance with (1.17):

$$\varrho_{ikl} = \alpha_{ip}\,\alpha_{kq}\,\alpha_{lr}\,\varrho_{pqr} \;.$$

Let us consider the respective transformations for classes 2 and $\bar{3}m$.

In class 2, if the twofold axis is parallel to X_3, the axes transformation matrix will be written

	X_1	X_2	X_3
X_1'	-1	0	0
X_2'	0	-1	0
X_3'	0	0	1

As has already been mentioned before, $\varrho_{ikl} = -\varrho_{kil}$, $\varrho_{iil} = 0$. We shall now consider the possible nonzero coefficients:

$$\varrho'_{121} = \alpha_{14}\alpha_{22}\alpha_{11}\varrho_{121} = (-1)(-1)(-1) \quad = -\varrho_{121} = 0 \ ,$$

$$\varrho'_{131} = \alpha_{11}\alpha_{33}\alpha_{14}\varrho_{131} = (-1)(1)(-1) \quad = \varrho_{131} \quad \neq 0 \ ,$$

$$\varrho'_{231} = \alpha_{22}\alpha_{33}\alpha_{14}\varrho_{231} = (-1)(1)(-1)\varrho_{231} = \varrho_{231} \quad \neq 0 \ .$$

It is seen that only those coefficients whose subscripts contain a single numerical 3 are different from zero. Consequently, for the first part of tensor ϱ_{ik1} related to the component of the magnetic field vector H_1, we get the matrix

$$\begin{Vmatrix} 0 & 0 & \varrho_{131} \\ 0 & 0 & \varrho_{231} \\ -\varrho_{131} & -\varrho_{231} & 0 \end{Vmatrix} .$$

We can similarly establish that the second and third matrices associated with components H_2 and H_3 have the form

$$\begin{Vmatrix} 0 & 0 & \varrho_{132} \\ 0 & 0 & \varrho_{232} \\ -\varrho_{132} & -\varrho_{232} & 0 \end{Vmatrix} , \quad \begin{Vmatrix} 0 & \varrho_{123} & 0 \\ -\varrho_{123} & 0 & 0 \\ 0 & 0 & 0 \end{Vmatrix} .$$

In class $\bar{3}m$ the symmetry elements include axis $\bar{3}$ parallel to X_3, a twofold axis parallel to X_1, and a center of symmetry. The transformation of the coordinate axes by the twofold axis is described by the matrix

$$\begin{Vmatrix} 1 & 0 & 0 \\ 0 & -1 & 0 \\ 0 & 0 & -1 \end{Vmatrix} .$$

Applying it to the components of the Hall coefficients, we obtain

$$\varrho'_{121} = \alpha_{11}\alpha_{22}\alpha_{11}\varrho_{121} = -\varrho_{121} = 0 \ ,$$

$$\varrho'_{131} = \alpha_{11}\alpha_{33}\alpha_{11}\varrho_{131} = -\varrho_{131} = 0 \ ,$$

$$\varrho'_{231} = \alpha_{22}\alpha_{33}\alpha_{11}\varrho_{231} = \varrho_{231} \neq 0 \ .$$

It follows that only those components whose subscripts contain a single unity can be nonzero. Hence, the matrix of the Hall coefficients will be:

$$\begin{array}{ccc} H_1 & H_2 & H_3 \\ \left\Vert\begin{matrix} 0 & 0 & 0 \\ 0 & 0 & \varrho_{231} \\ 0 & -\varrho_{231} & 0 \end{matrix}\right\Vert , & \left\Vert\begin{matrix} 0 & \varrho_{122} & \varrho_{132} \\ -\varrho_{122} & 0 & 0 \\ -\varrho_{132} & 0 & 0 \end{matrix}\right\Vert , & \left\Vert\begin{matrix} 0 & \varrho_{123} & \varrho_{133} \\ -\varrho_{123} & 0 & 0 \\ -\varrho_{133} & 0 & 0 \end{matrix}\right\Vert . \end{array}$$

The transformation by axis $\bar{3}$ has a matrix corresponding to that of the transformation by axis 3 and by the center of symmetry, i.e.,

	X_1	X_2	X_3
X'_1	1/2	$-\sqrt{3}/2$	0
X'_2	$\sqrt{3}/2$	1/2	0
X'_3	0	0	-1 .

We must now consider alternately those nonzero coefficients which were found with the aid of the matrix of transformation by the threefold axis (bearing in mind that the center of symmetry requires a minus sign in the expression for ϱ'_{ikl}):

$$\begin{aligned} \varrho'_{231} &= \alpha_{21}\alpha_{33}\alpha_{11}\varrho_{131} + (-)\alpha_{22}\alpha_{33}\alpha_{11}\varrho_{231} + (-)\alpha_{21}\alpha_{33}\alpha_{12}\varrho_{132} \\ &\quad + (-)\alpha_{22}\alpha_{33}\alpha_{12}\varrho_{232} \\ &= 0 + \tfrac{1}{4}\varrho_{231} + (-\tfrac{3}{4})\varrho_{132} + 0 = \varrho_{231} \ , \end{aligned}$$

since, according to Neumann's principle,

$$\varrho_{231} = -\varrho_{132} \ ;$$

$$\begin{aligned} \varrho'_{122} &= \alpha_{11}\alpha_{22}\alpha_{22}\varrho_{122} + (-)\alpha_{12}\alpha_{21}\alpha_{22}\varrho_{212} \\ &= \tfrac{1}{8}\varrho_{122} - \tfrac{3}{8}\varrho_{122} = -\tfrac{1}{4}\varrho_{122} = \varrho_{122} = 0 \ ; \end{aligned}$$

$$\varrho'_{123} = \alpha_{11}\alpha_{22}\alpha_{33}\varrho_{123} + (-)\alpha_{12}\alpha_{21}\alpha_{33}\varrho_{213} = \tfrac{1}{4}\varrho_{123} + \tfrac{3}{4}\varrho_{123} = \varrho_{123} \neq 0 \ ;$$

$$\varrho'_{133} = \alpha_{11}\alpha_{33}\varrho_{133} = \tfrac{1}{2}\varrho_{133} = \varrho_{133} = 0 \ .$$

As a result we get the following matrices of the Hall coefficients:

$$\overset{H_1}{\left\| \begin{matrix} 0 & 0 & 0 \\ 0 & 0 & \varrho_{231} \\ 0 & -\varrho_{231} & 0 \end{matrix} \right\|}, \quad \overset{H_2}{\left\| \begin{matrix} 0 & 0 & -\varrho_{231} \\ 0 & 0 & 0 \\ \varrho_{231} & 0 & 0 \end{matrix} \right\|}, \quad \overset{H_3}{\left\| \begin{matrix} 0 & \varrho_{123} & 0 \\ -\varrho_{123} & 0 & 0 \\ 0 & 0 & 0 \end{matrix} \right\|}.$$

Suppose the direction of the current coincides with the X_1' axis, the magnetic field is parallel to X_3', and the component along X_2 is determined for the Hall emf, i.e., we are dealing with Hall coefficient ϱ_{123}'. To find its expression in terms of the respective known components of the Hall coefficients tensor we use the transformation

$$\varrho_{123}' = \alpha_{1p}\alpha_{2q}\alpha_{3r}\varrho_{pqr} ,$$

where ϱ_{pqr} stands for all the nonzero ϱ_{ikl}, and α_{ik} are the coefficients of the matrix of transformation of axes $X_1X_2X_3$.

To illustrate, we consider a thin bar of a bismuth crystal whose length is perpendicular to the threefold axis and whose normal to the main face is inclined at an angle θ to the threefold axis. We direct the X_1' axis along the length of the bar and the X_3' axis along the normal to the main face. Then the transformation matrix will have the form

	X_1	X_2	X_3
X_1'	0	1	0
X_2'	$-\cos\theta$	0	$\sin\theta$
X_3'	$\sin\theta$	0	$\cos\theta$.

Let the external magnetic field be directed along the X_3' axis. Then the respective Hall coefficient will be ϱ_{123}'. Its dependence on the angle can be obtained as follows: using the matrix ϱ_{ikl} found above for bismuth, we write

$$\varrho_{123}' = \alpha_{12}\alpha_{23}\alpha_{31}\varrho_{231} + \alpha_{12}\alpha_{21}\alpha_{33}\varrho_{213} = \sin^2\theta \cdot \varrho_{231} + (-)\cos^2\theta \cdot (-\varrho_{123})$$
$$= \sin^2\theta \cdot \varrho_{231} + \cos^2\theta \cdot \varrho_{123} .$$

Thus, if $\theta = 0°$, then $\varrho_{123}' = \varrho_{123}$, and if $\theta = 90°$, then $\varrho_{123}' = \varrho_{231}$.

Let us consider, in conclusion, a thin bar cut out of a bismuth crystal parallel to the basal plane (i.e., perpendicular to X_3). Let the long side of the bar make angle θ with the twofold axis X_1 of the crystal. We shall now find the dependence of the Hall coefficient on angle θ.

Table 6.4. Matrices of the Hall and Righi-Leduc coefficients for crystals of different syngonies

Triclinic syngony

$$\left\|\begin{matrix} 0 & \varrho_{121} & \varrho_{131} \\ \varrho_{121} & 0 & \varrho_{231} \\ -\varrho_{131} & -\varrho_{231} & 0 \end{matrix}\right\| \quad \left\|\begin{matrix} 0 & \varrho_{122} & \varrho_{132} \\ -\varrho_{122} & 0 & \varrho_{232} \\ -\varrho_{132} & -\varrho_{232} & 0 \end{matrix}\right\| \quad \left\|\begin{matrix} 0 & \varrho_{123} & \varrho_{133} \\ -\varrho_{123} & 0 & \varrho_{233} \\ -\varrho_{133} & \varrho_{233} & 0 \end{matrix}\right\|$$

Monoclinic syngony

$$\left\|\begin{matrix} 0 & 0 & \varrho_{131} \\ 0 & 0 & \varrho_{231} \\ -\varrho_{131} & -\varrho_{231} & 0 \end{matrix}\right\| \quad \left\|\begin{matrix} 0 & 0 & \varrho_{132} \\ 0 & 0 & \varrho_{232} \\ -\varrho_{132} & \varrho_{232} & 0 \end{matrix}\right\| \quad \left\|\begin{matrix} 0 & \varrho_{123} & 0 \\ -\varrho_{123} & 0 & 0 \\ 0 & 0 & 0 \end{matrix}\right\|$$

Orthorhombic syngony

$$\left\|\begin{matrix} 0 & 0 & 0 \\ 0 & 0 & \varrho_{231} \\ 0 & -\varrho_{231} & 0 \end{matrix}\right\| \quad \left\|\begin{matrix} 0 & 0 & \varrho_{132} \\ 0 & 0 & 0 \\ -\varrho_{132} & 0 & 0 \end{matrix}\right\| \quad \left\|\begin{matrix} 0 & \varrho_{123} & 0 \\ -\varrho_{123} & 0 & 0 \\ 0 & 0 & 0 \end{matrix}\right\|$$

Tetragonal, trigonal, and hexagonal syngonies:

Classes 4, $\bar{4}$, $4/m$, 3, $\bar{3}$, 6, $\bar{6}$, $6/m$

$$\left\|\begin{matrix} 0 & 0 & \varrho_{131} \\ 0 & 0 & -\varrho_{132} \\ -\varrho_{131} & \varrho_{132} & 0 \end{matrix}\right\| \quad \left\|\begin{matrix} 0 & 0 & \varrho_{132} \\ 0 & 0 & \varrho_{131} \\ -\varrho_{132} & -\varrho_{131} & 0 \end{matrix}\right\| \quad \left\|\begin{matrix} 0 & \varrho_{123} & 0 \\ -\varrho_{123} & 0 & 0 \\ 0 & 0 & 0 \end{matrix}\right\|$$

Classes 422, $4mm$, $\bar{4}2m$, $4/mmm$, 32, $\bar{3}m$, 622, $6mm$, $\bar{6}m2$, $6/mmm$

$$\left\|\begin{matrix} 0 & 0 & 0 \\ 0 & 0 & \varrho_{231} \\ 0 & -\varrho_{231} & 0 \end{matrix}\right\| \quad \left\|\begin{matrix} 0 & 0 & -\varrho_{231} \\ 0 & 0 & 0 \\ \varrho_{231} & 0 & 0 \end{matrix}\right\| \quad \left\|\begin{matrix} 0 & \varrho_{123} & 0 \\ -\varrho_{123} & 0 & 0 \\ 0 & 0 & 0 \end{matrix}\right\|$$

Cubic syngony and isotropic medium

$$\left\|\begin{matrix} 0 & 0 & 0 \\ 0 & 0 & -\varrho_{132} \\ 0 & \varrho_{132} & 0 \end{matrix}\right\| \quad \left\|\begin{matrix} 0 & 0 & \varrho_{132} \\ 0 & 0 & 0 \\ -\varrho_{132} & 0 & 0 \end{matrix}\right\| \quad \left\|\begin{matrix} 0 & -\varrho_{132} & 0 \\ \varrho_{132} & 0 & 0 \\ 0 & 0 & 0 \end{matrix}\right\|$$

It is clear that the axes transformation matrix has the form

	X_1	X_2	X_3
X_1'	$\cos\theta$	$\sin\theta$	0
X_2'	$-\sin\theta$	$\cos\theta$	0
X_3'	0	0	1

The sought-for Hall coefficient is ϱ_{123}'. Using nonzero coefficients ϱ_{ikl} for bismuth, we get

Table 6.5. Matrices of the Nernst and Ettingshausen coefficients for crystals of different syngonies

Triclinic syngony

$$\begin{Vmatrix} \alpha_{111} & \alpha_{121} & \alpha_{131} \\ \alpha_{211} & \alpha_{221} & \alpha_{231} \\ \alpha_{311} & \alpha_{321} & \alpha_{331} \end{Vmatrix} \quad \begin{Vmatrix} \alpha_{112} & \alpha_{122} & \alpha_{132} \\ \alpha_{212} & \alpha_{222} & \alpha_{232} \\ \alpha_{312} & \alpha_{322} & \alpha_{332} \end{Vmatrix} \quad \begin{Vmatrix} \alpha_{113} & \alpha_{123} & \alpha_{133} \\ \alpha_{213} & \alpha_{223} & \alpha_{233} \\ \alpha_{313} & \alpha_{323} & \alpha_{333} \end{Vmatrix}$$

Monoclinic syngony

$$\begin{Vmatrix} 0 & 0 & \alpha_{131} \\ 0 & 0 & \alpha_{231} \\ \alpha_{311} & \alpha_{321} & 0 \end{Vmatrix} \quad \begin{Vmatrix} 0 & 0 & \alpha_{132} \\ 0 & 0 & \alpha_{232} \\ \alpha_{312} & \alpha_{322} & 0 \end{Vmatrix} \quad \begin{Vmatrix} \alpha_{113} & \alpha_{123} & 0 \\ \alpha_{213} & \alpha_{223} & 0 \\ 0 & 0 & \alpha_{333} \end{Vmatrix}$$

Orthorhombic syngony

$$\begin{Vmatrix} 0 & 0 & 0 \\ 0 & 0 & \alpha_{231} \\ 0 & \alpha_{321} & 0 \end{Vmatrix} \quad \begin{Vmatrix} 0 & 0 & \alpha_{132} \\ 0 & 0 & 0 \\ \alpha_{312} & 0 & 0 \end{Vmatrix} \quad \begin{Vmatrix} 0 & \alpha_{123} & 0 \\ \alpha_{213} & 0 & 0 \\ 0 & 0 & 0 \end{Vmatrix}$$

Tetragonal and hexagonal syngonies:

Classes 4, $\bar{4}$, $4/m$, $\bar{6}$, 6, $6/m$

$$\begin{Vmatrix} 0 & 0 & \alpha_{131} \\ 0 & 0 & \alpha_{231} \\ \alpha_{311} & \alpha_{321} & 0 \end{Vmatrix} \quad \begin{Vmatrix} 0 & 0 & \alpha_{231} \\ 0 & 0 & \alpha_{131} \\ \alpha_{311} & \alpha_{321} & 0 \end{Vmatrix} \quad \begin{Vmatrix} \alpha_{113} & \alpha_{123} & 0 \\ \alpha_{123} & \alpha_{113} & 0 \\ 0 & 0 & \alpha_{333} \end{Vmatrix}$$

Classes $4mm$, $\bar{4}2m$, 422, $4/mmm$, $\bar{6}m2$, 622, $6/mmm$

$$\begin{Vmatrix} 0 & 0 & 0 \\ 0 & 0 & \alpha_{231} \\ 0 & \alpha_{321} & 0 \end{Vmatrix} \quad \begin{Vmatrix} 0 & 0 & \alpha_{231} \\ 0 & 0 & 0 \\ \alpha_{321} & 0 & 0 \end{Vmatrix} \quad \begin{Vmatrix} 0 & \alpha_{123} & 0 \\ \alpha_{123} & 0 & 0 \\ 0 & 0 & 0 \end{Vmatrix}$$

Trigonal syngony:

Classes 3, $\bar{3}$

$$\begin{Vmatrix} \alpha_{111} & \alpha_{121} & \alpha_{131} \\ \alpha_{121} & -\alpha_{111} & \alpha_{231} \\ \alpha_{311} & \alpha_{321} & 0 \end{Vmatrix} \quad \begin{Vmatrix} -\alpha_{121} & -\alpha_{111} & -\alpha_{231} \\ -\alpha_{111} & -\alpha_{121} & \alpha_{131} \\ -\alpha_{321} & \alpha_{311} & 0 \end{Vmatrix} \quad \begin{Vmatrix} \alpha_{113} & \alpha_{123} & 0 \\ -\alpha_{123} & \alpha_{113} & 0 \\ 0 & 0 & \alpha_{333} \end{Vmatrix}$$

Classes $3m$, 32, $\bar{3}m$

$$\begin{Vmatrix} \alpha_{111} & 0 & 0 \\ 0 & -\alpha_{111} & \alpha_{231} \\ 0 & \alpha_{321} & 0 \end{Vmatrix} \quad \begin{Vmatrix} 0 & -\alpha_{111} & -\alpha_{231} \\ -\alpha_{111} & 0 & 0 \\ -\alpha_{321} & 0 & 0 \end{Vmatrix} \quad \begin{Vmatrix} 0 & \alpha_{123} & 0 \\ -\alpha_{123} & 0 & 0 \\ 0 & 0 & 0 \end{Vmatrix}$$

(Continued)

Table 6.5 (continued)

Cubic syngony:

Classes 23, $m3$

$$\begin{Vmatrix} 0 & 0 & 0 \\ 0 & 0 & \alpha_{231} \\ 0 & \alpha_{321} & 0 \end{Vmatrix} \quad \begin{Vmatrix} 0 & 0 & \alpha_{321} \\ 0 & 0 & 0 \\ \alpha_{231} & 0 & 0 \end{Vmatrix} \quad \begin{Vmatrix} 0 & \alpha_{231} & 0 \\ \alpha_{321} & 0 & 0 \\ 0 & 0 & 0 \end{Vmatrix}$$

Classes $\bar{4}3m$, 432, $m3m$; isotropic medium

$$\begin{Vmatrix} 0 & 0 & 0 \\ 0 & 0 & \alpha_{231} \\ 0 & -\alpha_{231} & 0 \end{Vmatrix} \quad \begin{Vmatrix} 0 & 0 & -\alpha_{231} \\ 0 & 0 & 0 \\ \alpha_{231} & 0 & 0 \end{Vmatrix} \quad \begin{Vmatrix} 0 & \alpha_{231} & 0 \\ -\alpha_{231} & 0 & 0 \\ 0 & 0 & 0 \end{Vmatrix}$$

$$\varrho'_{123} = \alpha_{11}\alpha_{22}\alpha_{33}\varrho_{123} + \alpha_{12}\alpha_{21}\alpha_{33}\varrho_{123} = \cos^2\theta \cdot \varrho_{123} + \sin^2\theta \cdot \varrho_{123} = \varrho_{123} \ .$$

Thus, for a thin bar whose large face is perpendicular to X_3, the Hall coefficient is independent of θ.

The matrices of the Hall and Righi-Leduc coefficients for crystals of different syngonies are given in Table 6.4, and the matrices of the Nernst and Ettingshausen coefficients in Table 6.5.

The schemes of the nonzero components of magnetoresistance tensor ϱ_{iklm} and magnetothermal conductivity tensor χ_{iklm} are identical and have the same form as the photoelasticity tensor.

In conclusion, we note that the Hall and the magnetoresistance effects in semiconducting crystals and thin films are widely used in science for determining the principal parameters of charge carriers and in technology for designing a number of applicances and devices [5.7, 6.1].

7. Optical Properties of Crystals

7.1 Plane Electromagnetic Waves in an Anisotropic Medium

7.1.1 Dielectric Constants of Crystals

The properties of an anisotropic continuous medium with respect to electromagnetic waves are described by Maxwell's electrodynamic equations, and the relationship between the induction $\boldsymbol{D}$ and tension $\boldsymbol{E}$ of the electric field, on the one hand, and the induction $\boldsymbol{B}$ and tension $\boldsymbol{H}$ of the magnetic field, on the other, is defined, in distinction to an isotropic medium, by the tensors of the dielectric constant and magnetic permeability, which depend on frequency ω: $\varepsilon_{ik}(\omega)$, $\mu_{ik}(\omega)$. The coupling equations here have the form

$$D_i = \varepsilon_{ik}(\omega) E_k \ , \quad B_i = \mu_{ik}(\omega) H_k \ . \tag{7.1}$$

If a body is not in an external magnetic field, the generalized principle of the symmetry of kinetic coefficients requires, as is well known, that the tensor be symmetric: $\varepsilon_{ik} = \varepsilon_{ki}$ [7.1].

The electromagnetic energy flux density is expressed by the Umov-Poynting vector

$$\boldsymbol{S} = \frac{c}{4\pi} [\boldsymbol{E} \times \boldsymbol{H}] \ , \tag{7.2}$$

and the change in energy per unit volume in unit time is calculated as follows:

$$-\operatorname{div} \boldsymbol{S} = \frac{c}{4\pi} (\boldsymbol{E} \operatorname{curl} \boldsymbol{H} - \boldsymbol{H} \operatorname{curl} \boldsymbol{E}) = \frac{1}{4\pi} \left(\boldsymbol{E} \frac{\partial \boldsymbol{D}}{\partial t} + \boldsymbol{H} \frac{\partial \boldsymbol{B}}{\partial t} \right) .$$

For a monochromatic wave, it is convenient to represent $\boldsymbol{E}$ and $\boldsymbol{H}$ by complex values $\boldsymbol{E}_0 \exp(-\mathrm{i}\omega t)$, $\boldsymbol{H}_0 \exp(-\mathrm{i}\omega t)$, and then, after time averaging, we have for the electric losses:

$$Q = \frac{\mathrm{i}\omega}{8\pi} (\varepsilon_{ik}^* - \varepsilon_{ki}) E_i E_k^* \ . \tag{7.3}$$

In the absence of absorption $\varepsilon_{ik}^* = \varepsilon_{ki} = \varepsilon_{ik}$, i.e., the polar tensor of the dielectric constant is not only symmetric, but also real here. As shown in Chap. 1, to such a tensor one can compare the ellipsoid ($\boldsymbol{r}\varepsilon\boldsymbol{r} = 1$, $\boldsymbol{r}$ being the radius vector), which is called the *Fresnel ellipsoid* in crystal optics.

By choosing the appropriate (principal) system of coordinates, we can reduce the ellipsoid equation to its conventional form:

$$\varepsilon_{11}x^2 + \varepsilon_{22}y^2 + \varepsilon_{33}z^2 = 1 \ . \tag{7.4}$$

The directions of the coordinate axes of such a system are called the principal directions, and the values $\varepsilon_{11} = \varepsilon_x$, $\varepsilon_{22} = \varepsilon_y$, $\varepsilon_{33} = \varepsilon_z$, the principal values of tensor ε_{ik}.

Recall now the conditions imposed by the symmetry of a crystal on the type of the symmetric second-rank polar tensor (Chap. 1). First, the components of such a tensor remain unaltered on inversion transformation, and therefore, out of the 32 point symmetry classes of crystals, we have to consider only 11, which possess a center of symmetry. Further reduction in the number of classes under consideration results because the type of symmetric tensor ε_{ik} is the same for all the classes which have three-, four-, or sixfold symmetry axes. For crystals of cubic classes, tensor ε_{ik} turns into a scalar. As a result, five distinct tensors for different crystal syngonies remain:

$$\begin{Vmatrix} \varepsilon_{11} & \varepsilon_{12} & \varepsilon_{13} \\ \varepsilon_{12} & \varepsilon_{22} & \varepsilon_{23} \\ \varepsilon_{13} & \varepsilon_{23} & \varepsilon_{33} \end{Vmatrix}, \quad \begin{Vmatrix} \varepsilon_{11} & \varepsilon_{12} & 0 \\ \varepsilon_{12} & \varepsilon_{22} & 0 \\ 0 & 0 & \varepsilon_{33} \end{Vmatrix}, \quad \begin{Vmatrix} \varepsilon_{11} & 0 & 0 \\ 0 & \varepsilon_{22} & 0 \\ 0 & 0 & \varepsilon_{33} \end{Vmatrix},$$

$$\begin{Vmatrix} \varepsilon_{11} & 0 & 0 \\ 0 & \varepsilon_{11} & 0 \\ 0 & 0 & \varepsilon_{33} \end{Vmatrix}, \quad \begin{Vmatrix} \varepsilon_{11} & 0 & 0 \\ 0 & \varepsilon_{11} & 0 \\ 0 & 0 & \varepsilon_{11} \end{Vmatrix}.$$

For the first three types of tensors in the cases of triclinic, monoclinic, and orthorhombic syngonies, the characteristic surface is a triaxial ellipsoid. For the trigonal, tetragonal, and hexagonal syngonies it is an ellipsoid of revolution, and finally, for the cubic syngony the tensor ellipsoid degenerates into a sphere.

In crystals of the triclinic syngony the triad of mutually perpendicular axes of the principal directions of the dielectric constant tensor may have any orientation, depending on the values of the tensor components (axes dispersion). In monoclinic crystals, only one of the principal tensor axes is always directed along the twofold axis or is perpendicular to the plane of symmetry. In crystals of orthorhombic or higher symmetry there is no dispersion of the axes, and their direction is completely determined by the symmetry elements of the crystal.

7.1.2 Plane Waves in Transparent Crystals

We first consider an optically transparent anisotropic and nonmagnetic medium [7.2 – 5]. The relationship between the intensities and inductions of the electric and magnetic field must be written as

$$D_i = \varepsilon_{ik} E_k \ , \quad B_i = H_i \ , \tag{7.5}$$

where ε_{ik} is a real symmetric tensor with positive main values. For plane monochromatic waves with a frequency ω and a wave vector $\boldsymbol{k}$, $\boldsymbol{E} = \boldsymbol{E}_0 \exp[\mathrm{i}(\omega t - \boldsymbol{kr})]$, where $\boldsymbol{k} = (\omega/c)\boldsymbol{n}$ ($\boldsymbol{n}$ is the wave normal); from Maxwell's equations

$$\operatorname{curl} \boldsymbol{H} = \frac{1}{c}\,\frac{\partial \boldsymbol{D}}{\partial t}, \quad \operatorname{curl} \boldsymbol{E} = -\frac{1}{c}\,\frac{\partial \boldsymbol{H}}{\partial t}$$

we have

$$\boldsymbol{H} = [\boldsymbol{nE}] \ , \quad \boldsymbol{D} = -[\boldsymbol{nH}] \ .$$

Thus, vectors $\boldsymbol{n}$, $\boldsymbol{E}$, and $\boldsymbol{D}$, lie in the same plane perpendicular to $\boldsymbol{H}$ and, besides, $\boldsymbol{D} \perp \boldsymbol{n}$ (Fig. 7.1). Eliminating vector $\boldsymbol{H}$ from the last two equations, we find

$$\boldsymbol{D} = n^2 \boldsymbol{E} - \boldsymbol{n}(\boldsymbol{nE}) \ . \tag{7.6}$$

Using coupling equation (7.5), we obtain three linear homogeneous equations for the components of vector E_i:

$$(n^2 \delta_{ik} - n_i n_k - \varepsilon_{ik}) E_k = 0 \ , \quad \delta_{ik} = \begin{cases} 0 \ , & i \neq k \ , \\ 1 \ , & i = k \ . \end{cases} \tag{7.7}$$

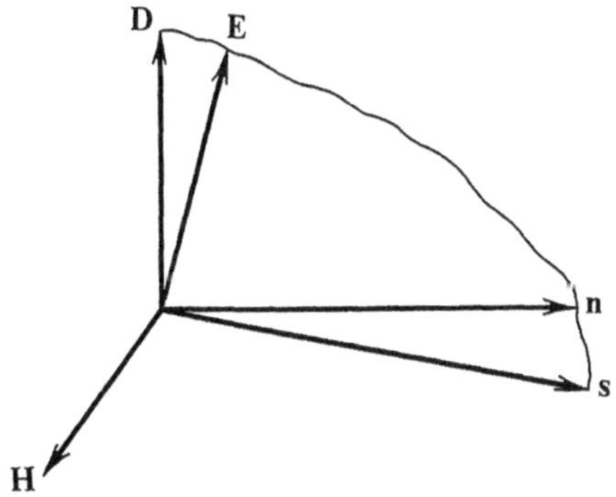

Fig. 7.1. Arrangement of vectors ***E, D, H, n,*** and ***s*** of a light wave in a crystal (***s*** is the beam vector, $\boldsymbol{ns} = 1$)

As is well known, the condition for the compatibility of a set of linear homogeneous equations is that the determinant of the system must be equal to zero. In the Cartesian system of coordinates, whose axes coincide with the principal directions of tensor ε_{ik}, this condition leads to the main equation of crystal optics *(Fresnel equation)*:

$$n^2(\varepsilon_x n_x^2+\varepsilon_y n_y^2+\varepsilon_z n_z^2)-[n_x^2\varepsilon_x(\varepsilon_y+\varepsilon_z)+n_y^2\varepsilon_y(\varepsilon_x+\varepsilon_z)+n_z^2\varepsilon_z(\varepsilon_x+\varepsilon_y)]$$
$$+\varepsilon_x\varepsilon_y\varepsilon_z=f(k_x,k_y,k_z)=0\ , \tag{7.8}$$

or in the symmetric form:

$$\frac{(n_x^0)^2}{1/n^2-1/\varepsilon_x}+\frac{(n_y^0)^2}{1/n^2-1/\varepsilon_y}+\frac{(n_z^0)^2}{1/n^2-1/\varepsilon_z}=0\ . \tag{7.9}$$

If the principal values ε_x, ε_y, ε_z of tensor ε_{ik} as a function of frequency ω are known, the Fresnel equation defines the absolute value of vector $\boldsymbol{n}$, provided its direction is given by unit vector $\boldsymbol{n}^0$. Generally, to each direction of the wave vector correspond two different values of *refractive index n* and two directions of induction vector $\boldsymbol{D}$ ($\boldsymbol{D}'$ and $\boldsymbol{D}''$ are the directions of light oscillations). Thus, in a crystal, in distinction to an isotropic medium, two plane, linearly polarized waves with two different direction-dependent phase velocities can propagate in each direction. In determining the direction of oscillations of these waves it is convenient to choose, instead of the main system of coordinates, a new one with the Z' axis directed along vector $\boldsymbol{n}^0$. From (7.6) one can obtain the following two equations for the transverse components of vector $\boldsymbol{D}$:

$$(\varepsilon_{\alpha\beta}^{-1}n^2-\delta_{\alpha\beta})D_\beta=0 \tag{7.10}$$

with a zero determinant, which actually define the direction of vector $\boldsymbol{D}$ [in system (7.10), α, $\beta=X'$, Y', and summation is carried out with respect to β].

It is also easy to see from (7.10) that the directions of vectors $\boldsymbol{D}'$ and $\boldsymbol{D}''$, corresponding to two values of n (roots of the Fresnel equation) are mutually perpendicular (Fig. 7.2).

Let us now consider the energy flux vector of Umov-Poynting

$$\boldsymbol{S}=\frac{c}{4\pi}[\boldsymbol{E}\times\boldsymbol{H}]=\frac{c}{4\pi}[\boldsymbol{n}E^2-\boldsymbol{E}(\boldsymbol{n}\boldsymbol{E})]\ . \tag{7.11}$$

Vector $\boldsymbol{S}$ lies in the plane of vectors $\boldsymbol{D}$, $\boldsymbol{E}$, and $\boldsymbol{n}$, is perpendicular to the vector $\boldsymbol{E}$ of the electric field intensity of the wave, and generally does not coincide with vector $\boldsymbol{n}$ in direction. It can be proved that vector $\boldsymbol{S}$ is directed along group velocity vector $\partial\omega/\partial\boldsymbol{k}$. We designate as *beam vector* $\boldsymbol{s}$ the one which is directed along $\boldsymbol{S}$ and whose absolute value is such that $\boldsymbol{ns}=1$. We

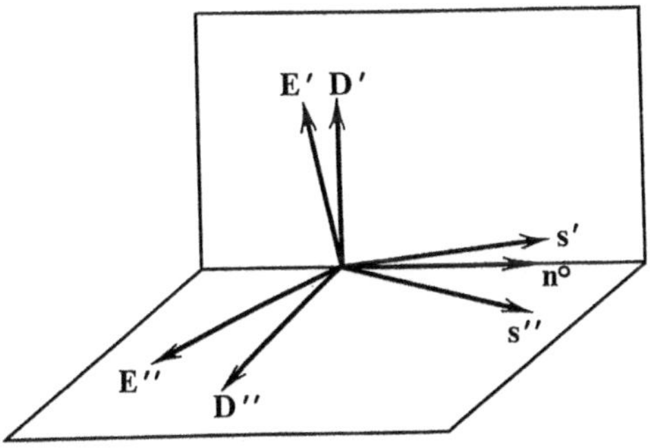

Fig. 7.2. Arrangement of vectors $\boldsymbol{E}'$, $\boldsymbol{D}'$, $\boldsymbol{D}'$, s' and $\boldsymbol{E}''$, $\boldsymbol{D}''$, s'' of two plane, linearly polarized waves in a crystal

now have $\boldsymbol{s}\boldsymbol{E}=0$; $\boldsymbol{s}\boldsymbol{H}=0$ (7.11) and further $[\boldsymbol{s}\times\boldsymbol{H}]=[\boldsymbol{s}\times[\boldsymbol{n}\times\boldsymbol{E}]]=-\boldsymbol{E}$, $[\boldsymbol{s}\times\boldsymbol{D}]=-[\boldsymbol{s}\times[\boldsymbol{n}\times\boldsymbol{H}]]=\boldsymbol{H}$.

Comparing the two series of equations obtained,

$$D_i=\varepsilon_{ik}E_k\ ,\qquad \boldsymbol{D}=-[\boldsymbol{n}\times\boldsymbol{H}]\ ,\quad \boldsymbol{H}=[\boldsymbol{n}\times\boldsymbol{E}]\ ,\quad \boldsymbol{n}\boldsymbol{s}=1\ , \tag{7.12a}$$

$$E_i=\varepsilon_{ik}^{-1}D_k\ ,\qquad \boldsymbol{E}=-[\boldsymbol{s}\times\boldsymbol{H}]\ ,\quad \boldsymbol{H}=[\boldsymbol{s}\times\boldsymbol{D}]\ ,\quad \boldsymbol{n}\boldsymbol{s}=1\ , \tag{7.12b}$$

we see that by replacements $\varepsilon_{ik}\rightarrow\varepsilon_{ik}^{-1}$, $\boldsymbol{D}\rightarrow\boldsymbol{E}$, $\boldsymbol{n}\rightarrow\boldsymbol{s}$ from the equations of the first series for $\boldsymbol{D}$, $\boldsymbol{H}$, and $\boldsymbol{n}$, we obtain equations of the second series for $\boldsymbol{E}$, $\boldsymbol{H}$, and $\boldsymbol{s}$. This is the essence of what is called the *duality principle* in crystal optics. For instance, to find the absolute value of beam vector $\boldsymbol{s}$ from its direction $\boldsymbol{s}^0$, we obtain from (7.9) the equation

$$\frac{(s_x^0)^2}{1/s^2-1/\varepsilon_x^{-1}}+\frac{(s_y^0)^2}{1/s^2-1/\varepsilon_y^{-1}}+\frac{(s_z^0)^2}{1/s^2-1/\varepsilon_z^{-1}}=0\ .$$

7.1.3 Optical Surfaces

Relations (7.10) can be assigned a simple geometric meaning by using the following constructions. Let us consider the ellipsoid of tensor ε_{ik}^{-1} with semi-axes $\sqrt{\varepsilon_x}$, $\sqrt{\varepsilon_y}$, $\sqrt{\varepsilon_z}$, which is called the *optic indicatrix.* Now we choose some direction of wave vector $\boldsymbol{k}$. In the section of this ellipsoid by a plane perpendicular to the chosen direction $\boldsymbol{n}^0$ we obtain an ellipse in the general case. Equation (7.9) shows that the sizes of the principal semiaxes of the section ellipse are equal to the values of refractive indices n, and their directions coincide with those of the induction vectors $\boldsymbol{D}'$ and $\boldsymbol{D}''$ of the two waves in a crystal with a given $\boldsymbol{n}^0$. The replacement rule (duality principle) leads to a similar construction for determining the directions of intensity vectors $\boldsymbol{E}'$ and $\boldsymbol{E}''$ of waves with a given direction of beam vector $\boldsymbol{s}$.

The dependence of the refractive indices on the direction of the light wave in a crystal is characterized most vividly by the surface of the wave vectors, the absolute values of whose radius vectors in the given direction $\boldsymbol{n}^0$ are equal to

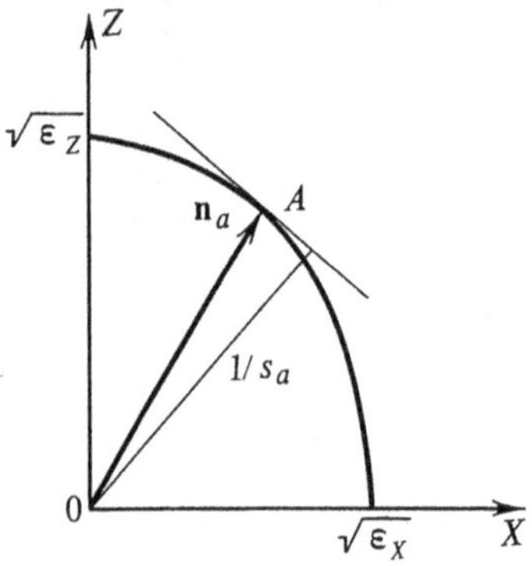

Fig. 7.3. Derivation of the geometric relationship between a beam vector and a wave vector in a crystal

the refractive indices n determined by the Fresnel equation. A similar fourth-order surface can be constructed for beam vectors $\boldsymbol{s}$. Later on we shall consider the form of such surfaces for crystals of different classes.

Let $f(k_x, k_y, k_z, \omega) = 0$ be the equation of the wave-vector surface. The components of the vector of group velocity $\partial\omega/\partial k_i = -(\partial f/\partial k_i)/(\partial f/\partial\omega)$ (7.8) are proportional to derivatives $\partial f/\partial n_i$, and therefore the beam vector is parallel to grad$f$, i.e., is directed along the normal to the wave-vector surface. Suppose now that $\boldsymbol{n}_a$ is the radius vector of some point of the wave-vector surface, and $\boldsymbol{s}_a$, the respective beam vector. The equation of the plane tangent to the wave-vector surface at point A (Fig. 7.3) will be $\boldsymbol{s}_a(\boldsymbol{n}-\boldsymbol{n}_a) = 0$, and since $\boldsymbol{n}_a\boldsymbol{s}_a = 1$, then $\boldsymbol{s}_a\boldsymbol{n} = 1$. Hence, the length of a perpendicular dropped from the origin to the plane tangent to the wave-vector surface is $1/s_a$. Similarly, the length of a perpendicular dropped from the origin to the plane tangent to the beam vector surface is $1/n_a$. Thus, the geometric relationship is established between the beam and wave vectors of the light waves in a crystal.

7.2 Uniaxial and Biaxial Crystals

7.2.1 Uniaxial Crystals

Consideration of the types of dielectric constant tensors shows that all the crystals can be divided into three groups according to the number (1, 2, 3) of distinct principal values of the dielectric constant tensor (Sect. 7.1).

For cubic crystals, tensor ε_{ik} degenerates into scalar $\varepsilon = n^2$, and hence such crystals do not differ from isotropic solids in their optical properties (Table 7.1). For crystals of the trigonal, tetragonal, and hexagonal syngonies, tensor ε_{ik} has two principal values $\varepsilon_z = \varepsilon_\parallel = n_e^2$, $\varepsilon_x = \varepsilon_y = \varepsilon_\perp = n_0^2$ (Table 7.2), and the respective characteristic surface is an ellipsoid of revolution with an axis parallel to the highest-order axis of symmetry. For uniaxial crystals, the

Table 7.1. Refractive indices n of some cubic crystals [7.6]

Crystal	n	Crystal	n
Fluorite CaF_2	1.434	Diamond C	2.419
Rock salt NaCl	1.544	Cuprite Cu_2O	2.705
Spinel $MgAl_2O_4$	1.718	Germanium Ge	~4.0
Pyrope $Mg_3Al_2Si_3O_{12}$	1.705		

Table 7.2. Refractive indices n_0 and n_e of some uniaxial crystals [7.6]

Crystal	n_0	n_e	Crystal	n_0	n_e
Ice H_2O	1.309	1.313	Corundum Al_2O_3	1.768	1.760
Quartz SiO_2	1.544	1.553	Scheelite $CaWO_4$	1.918	1.934
Beryl $Be_3Al_2\ Si_6O_{18}$	1.568	1.564	Zircon $ZrSiO_4$	1.926	1.985
Saltpeter $NaNO_3$	1.587	1.336	Rutile TiO_2	2.616	2.903
Calcite $CaCO_3$	1.658	1.486	Cinnabar HgS	2.854	3.201

Fresnel equation in the main system of coordinates decomposes into two quadratic equations:

$$n^2 - \varepsilon_\perp = 0 , \quad \frac{n_z^2}{\varepsilon_\perp} + \frac{n_x^2 + n_y^2}{\varepsilon_\parallel} = 1 . \tag{7.13}$$

Thus, in a uniaxial crystal two waves can propagate in each direction of the wave vector: an *ordinary* wave with a refractive index $n_0 = \sqrt{\varepsilon_\perp}$ independent of the direction (as its name implies) and an extraordinary one with a refractive index n dependent on the slope, θ, of the $\boldsymbol{n}$ vector to the Z axis, which is directed along the higher-symmetry axis in the crystal:

$$\frac{1}{n^2} = \frac{\sin^2\theta}{\varepsilon_\parallel} + \frac{\cos^2\theta}{\varepsilon_\perp} . \tag{7.14}$$

In one special direction, when angle $\theta = 0$, the refractive indices of the two waves coincide: $n_0 = n = \sqrt{\varepsilon_\perp}$, and in the crystal, as well as in an isotropic body, the waves propagate at the same velocity; such a direction in a crystal is called the *optic axis*. Therefore, crystals of the trigonal, tetragonal, and hexagonal syngonies are called uniaxial. To determine the difference of the refractive indices of waves with the same direction of the wave vector, we easily obtain, by assuming the principal birefringence (the difference $\sqrt{\varepsilon_\perp} - \sqrt{\varepsilon_\parallel} = n_0 - n_e$) to be small,

$$n - n_0 \approx (n_e - n_0) \sin^2\theta . \tag{7.15}$$

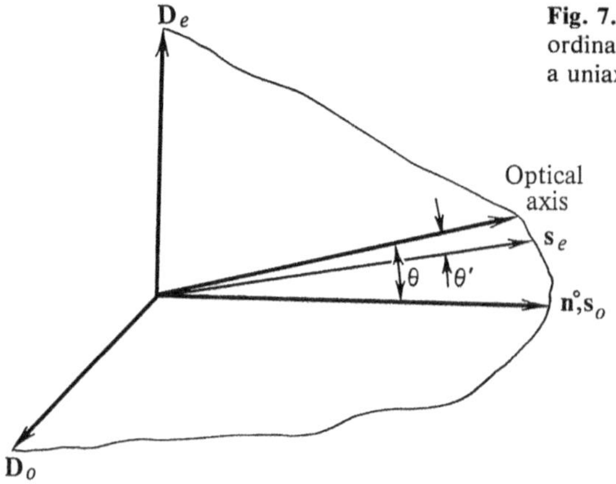

Fig. 7.4. Direction of light oscillations in an ordinary wave and an extraordinary wave in a uniaxial crystal

The solution of the set of Eqs. (7.7) defining the directions of vectors $\boldsymbol{E}$, and hence $\boldsymbol{D}$, by (7.1) shows that the direction of light oscillations in an ordinary wave is perpendicular to the plane containing the optic axis and the wave vector. Such a plane is called the *principal section* (Fig. 7.4). In an extraordinary wave, on the contrary, the direction of oscillations lies in the plane of the main section. The beam vector of an ordinary wave coincides with wave vector $\boldsymbol{n}$ in direction and forms an angle θ with the optic axis of the crystal. The beam vector of an extraordinary wave lies naturally in the plane of the main section (vectors $\boldsymbol{n}$, $\boldsymbol{D}$, $\boldsymbol{s}$ and $\boldsymbol{E}$ are always coplanar), but does not coincide with the direction of wave vector $\boldsymbol{n}$ and forms a different angle θ' with the optic axis, which can be found from the condition

$$\tan\theta' = \frac{\varepsilon_\perp}{\varepsilon_\parallel}\tan\theta \ . \qquad (7.16)$$

As can be seen from the Fresnel equation, the wave-vector surface in uniaxial crystals decomposes into two surfaces: spherical for the ordinary wave, and an ellipsoid of revolution for the extraordinary one. The two surfaces touch one another at two points lying on the optic axis. If $n_o < n_e$, uniaxial crystals are said to be positive, and if $n_o > n_e$, negative (Fig. 7.5).

7.2.2 Biaxial Crystals

When constructing the wave-vector surface for triclinic, monoclinic, and orthorhombic crystals, we proceed from the triaxial ellipsoid $\varepsilon_{ik}^{-1}x_i x_k = 1$ with three different principal values. To construct the section of the wave-vector surface perpendicular to the Y axis of the ellipsoid ($\varepsilon_x < \varepsilon_y < \varepsilon_z$), we proceed as follows. We cut the tensor ellipsoid by the YZ plane; we lay off lengths equal to $\sqrt{\varepsilon_y}$ and $\sqrt{\varepsilon_z}$ on the X axis. Rotating the section plane about the Y axis in the ellipsoid section, we obtain ellipses with one constant semiaxis $\sqrt{\varepsilon_y}$

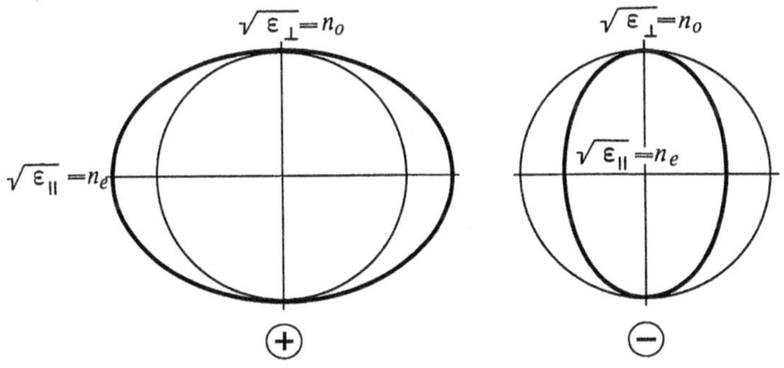

Fig. 7.5. Surfaces of wave vectors for positive and negative crystals

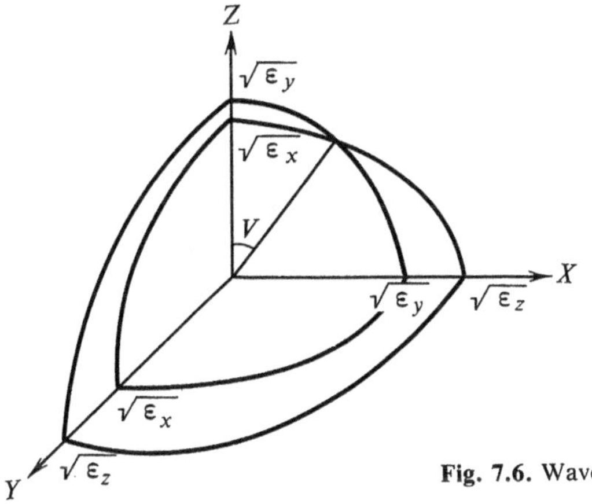

Fig. 7.6. Wave surfaces in a biaxial crystal

and another, variable, semiaxis, which takes values from $\sqrt{\varepsilon_z}$ to the minimal $\sqrt{\varepsilon_x}$. Thus, in the section of the wave-vector surface we obtain a circle of radius $\sqrt{\varepsilon_y}$ and an ellipse with semiaxis $\sqrt{\varepsilon_z}$ and $\sqrt{\varepsilon_x}$; in a similar way we construct the other two sections perpendicular to the X and Z axes of the ellipsoid of tensor ε_{ik}^{-1}. The general view of the surface is given in Fig. 7.6.

The surface is formed by two shells touching at four points and possesses a center of symmetry. In two directions, called *optic axes* or *binormals*, issuing from the origin towards these points, the refractive indices of the two waves coincide, and there is no birefringence (to these directions correspond circular cross sections of the tensor ellipsoid). Crystals of the triclinic, monoclinic, and orthorhombic syngonies are therefore biaxial. The optic axes make an angle V with the Z axis, which can be found by simultaneous solution of the equations of circles $x^2+z^2=\varepsilon_y$ and ellipse $x^2/\varepsilon_z+z^2/\varepsilon_x=1$. Hence,

$$\tan V = \left(\frac{\varepsilon_z(\varepsilon_y-\varepsilon_x)}{\varepsilon_x(\varepsilon_z-\varepsilon_y)}\right)^{1/2} . \tag{7.17}$$

Table 7.3. Refractive indices of some biaxial crystals [7.6]

Crystal	n_1	n_2	n_3	$2V$, deg
Saltpeter KNO_3	1.3328	1.4988	1.4994	6
Ammonia saltpeter NH_4NO_3	1.411	1.605	1.6296	35
Gypsum $CaSO_4 \cdot 2H_2O$	1.520	1.523	1.530	58
Aragonite $CaCO_3$	1.530	1.680	1.685	18
Muscovite $KAl_2[(OH, F)(AlSi_3O_{10})]$	1.572	1.611	1.615	30
Celestine $SrSO_4$	1.622	1.624	1.631	51
Barite $BaSO_4$	1.636	1.637	1.648	37.5
Potassium chromate $K_2Cr_2O_7$	1.715	1.762	1.892	64

Table 7.3 lists the values of n and V for some biaxial crystals.

Analytically, the values of refractive indices of the waves in relation to the direction of the wave vector are defined by Fresnel equations (7.8, 9). For simplicity, the direction of the wave vector is given not by the direction cosines n_x^0, n_y^0, and n_z^0 of the angles with the main axes of the dielectric constant tensor, as before, but by two angles φ_1 and φ_2, which the wave vector makes with the optic axes of the crystal. Using such assignment of the direction of the wave vector it is possible, for instance, to obtain a simple expression defining the difference between the refractive indices n' and n'' of the waves propagating in this direction in the crystal

$$\frac{1}{(n')^2}-\frac{1}{(n'')^2}=\left(\frac{1}{\varepsilon_x}-\frac{1}{\varepsilon_z}\right)\sin\varphi_1\sin\varphi_2 \ . \tag{7.18}$$

The problem of the direction of wave oscillations in a biaxial crystal is solved according to the *Fresnel theorem*. The directions of the oscillations of light waves, i.e., vectors $\boldsymbol{D}$ corresponding to vector $\boldsymbol{n}$, are the bisectors of the angles between the traces of two planes, each of which contains vector $\boldsymbol{n}$ and one of the optic axes, on a plane perpendicular to vector $\boldsymbol{n}$.

Indeed, suppose n' and n'' are the semiaxes of the ellipse of the section of the ellipsoid of tensor ε_{ik}^{-1} by a plane perpendicular to vector $\boldsymbol{n}$ (Fig. 7.7a), and diameter $A'A'$ is the line of intersection of this section with the circular section of the ellipsoid. We draw, in the plane of the elliptic section, a diameter $N'N'$ perpendicular to $A'A'$. It is clear that $\boldsymbol{n}$, N_0, and $N'N'$ lie in the same plane perpendicular to $A'A'$ (Fig. 7.7b). A similar construction can be made for the other circular section and the other optic axis. After such a construction we have, in the plane of the elliptic section, two equal diameters, $A'A'$ and $A''A''$ (as diameters of circular sections), lying on the different sides of ellipse semi axis n'. Hence, n' bisects the angle between them, and

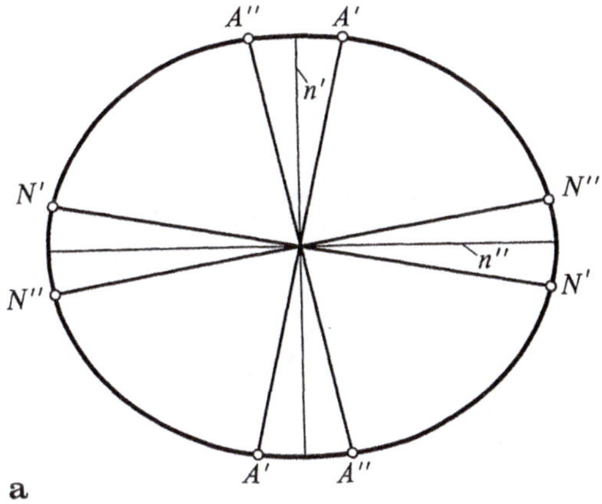

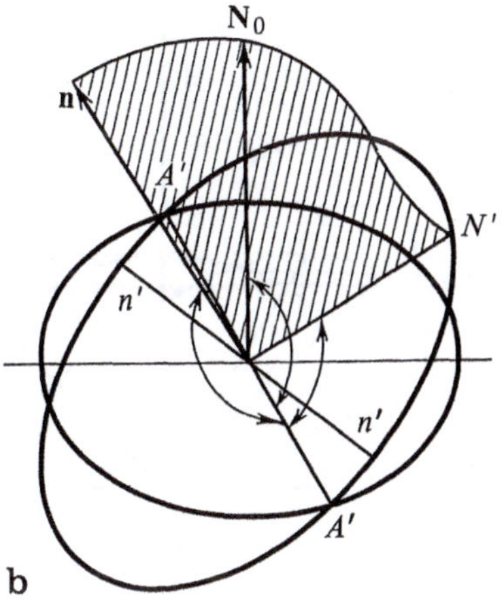

Fig. 7.7a, b. Derivation of the Fresnel theorem on the polarization of light waves in a biaxial crystal

since $A'A' \perp N'N'$ and $A''A'' \perp N''N''$, n' is also the bisector of the angle between $N'N'$ and $N''N''$.

According to the duality principle the beam-vector surface is constructed either on the basis of the ellipsoid of tensor ε_{ik} or on the basis of the wave-vector surface. The beam-vector surface also has four points of intersection, through which the optic axes of the beams (*biradials*) inclined to the Z axis at an angle

$$\tan \gamma = \tan V \sqrt{\varepsilon_x/\varepsilon_z}$$

pass.

To each such direction, i.e., the biradial (Fig. 7.8a), corresponds an infinite number of wave vectors whose directions fill a whole cone (the cone of *external conical refraction*). Triangle AOB in Fig. 7.8a is a section of this cone. Similarly, to the direction of the optic axis (binormal) responds a whole cone of beam vectors. Triangle $A'O'B'$ (Fig. 7.8b) is a section of the cone (cone of *internal conical refraction*).

The angle ψ of the cone of internal conical refraction can be found from the equation $\tan \psi = [(\varepsilon_x - \varepsilon_y)(\varepsilon_y - \varepsilon_z)/\varepsilon_x \varepsilon_z]^{1/2}$, and for the external one we obtain, according to the duality principle, $\tan \chi = (\sqrt{\varepsilon_x \varepsilon_z}/\varepsilon_y) \tan \psi$.

When observing the external conical refraction under the microscope (Fig. 7.9a), a converging light beam is directed towards a plate cut out of a biaxial crystal perpendicular to one of the biradials in such a way that the beam focus is on the surface of the plate. The other surface is covered by a diaphragm with an aperture opposite the focus of the incident beam. The converging rays of the incident beam always include one whose wave vector will move in the crystal along direction OA after refraction (Fig. 7.8a).

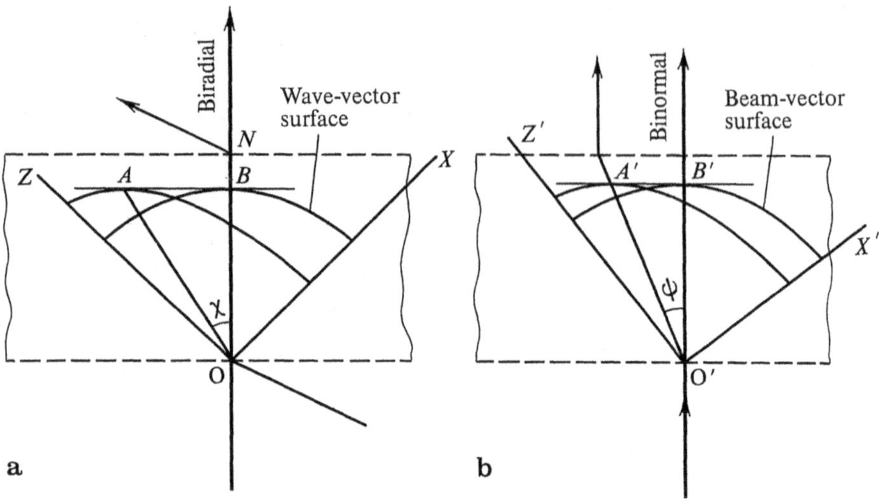

Fig. 7.8. External (**a**) and internal (**b**) conic refraction in a crystal

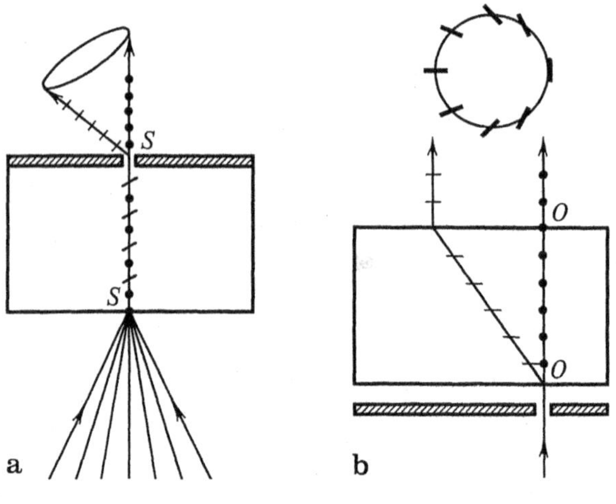

Fig. 7.9. Observation of the external (**a**) and internal (**b**) conic refraction: *SS* is the direction of the crystal biradial, 00 the direction of the optic axis of the crystal. The dashes and dots indicate the direction of the oscillations

Therefore, direction ON will be taken by the beam conjugate to the wave vector directed along OA. At the exit from the crystal we shall have a hollow cone of beams (Fig. 7.9a).

When observing the internal conical refraction (Fig. 7.9b), a plate is cut out of a biaxial crystal perpendicular to one of the optic axes. The plate surface is covered by a diaphragm with a small aperture. By directing a narrow beam of light towards the crystal one obtains, inside the crystal, a hollow cone of rays, which are distributed along a circle on emergence from the plate.

Both conical refractions were predicted theoretically by Hamilton in 1832 and found experimentally by Lloyd in 1833 [7.7], [7.2].

7.3 Birefringence in Crystals

7.3.1 Birefringence of Plane Waves at the Interface of Two Media

Having established the nature of the polarization of waves and the dependences of their refractive indices on the direction of propagation in crystals (wave surfaces), we should turn our attention to the laws of refraction of plane waves at an interface. To do this, we consider a plane wave falling from the air on the surface of an anisotropic medium (Fig. 7.10). We choose the coordinate axis so that the Z axis coincides with the direction of the normal to the interface, and the X axis lies in the plane of incidence, i.e., in the plane containing the normal to the interface and the wave vector of the incident wave $\boldsymbol{k}\,(n_y = 0)$. The incident wave produces one reflected and, in distinction to isotropic media, generally speaking, two linearly polarized refracted waves with vectors $\boldsymbol{n}'$ and $\boldsymbol{n}''$ making angles ψ' and ψ'' with the Z axis, these waves propagate with different velocities in the crystal.

The expressions for the fields of the incident wave and the refracted waves contain factors $\exp[\mathrm{i}\,\omega(t - \boldsymbol{nr}/c)]$. From the condition of field continuity at the interface, it follows that for any point $\boldsymbol{r}$ lying at the interface $Z = 0$ the relations $\boldsymbol{rn} = \boldsymbol{rn}' = \boldsymbol{rn}''$ hold true. Hence, it follows at once that $n_y = n_y' = n_y'' = 0$ and also $n_x = n_x' = n_x''$, i.e.,

$$\sin\varphi = n' \sin\psi' = n'' \sin\psi'' \ , \tag{7.19}$$

where φ is the angle of incidence. Thus, we can see that vectors $\boldsymbol{n}'$ and $\boldsymbol{n}''$ of the refracted waves lie in the plane of incidence, and each of them obeys the

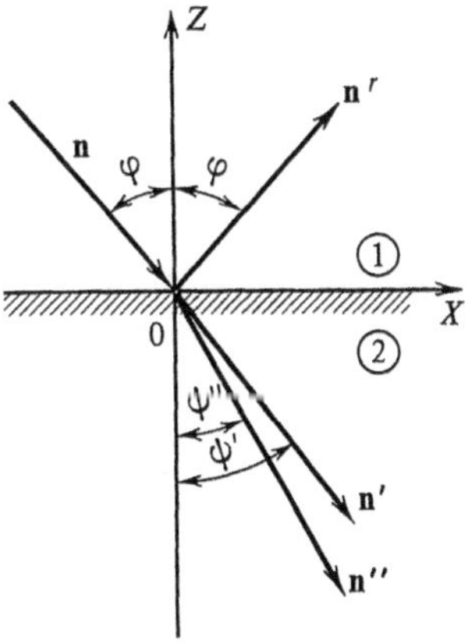

Fig. 7.10. Birefringence in a crystal

ordinary refraction law requiring the continuity of tangential component n_t of vector $\boldsymbol{n}$. It is also correct for reflection. The main difference from the case of isotropic media lies in the appearance of two refracted waves, which can therefore be termed *birefringence;* it was discovered on calcite crystals by Bartolini in 1669 [7.8] and gave rise to crystal optics. It should be remembered that to each wave vector in a crystal corresponds a beam vector of its own, coinciding with the direction of the energy flux vector; hence, there is also *birefringence* in the crystal. The direction of the beam vector does not coincide with the wave vector, and the former may generally lie outside the plane of incidence.

In uniaxial crystals, refraction produces an ordinary wave and an extraordinary wave (Sect. 7.2). The behavior of an ordinary wave is identical to that of a refracted wave in an isotropic medium, and its beam vector coincides with the wave vector. An extraordinary beam, however, does not obey the ordinary refraction law. Let us find, in particular, the direction of the extraordinary beam when the light falls obliquely onto the surface of a uniaxial crystal and when the interface is perpendicular to the optic axis. For this we write $\sin\varphi = n\sin\theta$, where θ is the refraction angle, and $1/n^2 = \cos^2\theta/\varepsilon_\perp + \sin^2\theta/\varepsilon_\parallel$ (7.13). Hence,

$$\tan\theta = \frac{\sqrt{\varepsilon_\parallel}}{\sqrt{\varepsilon_\perp}} \frac{\sin\varphi}{\sqrt{\varepsilon_\parallel - \sin^2\varphi}},$$

and since $\tan\theta' = (\varepsilon_\perp/\varepsilon_\parallel)\tan\theta$ (7.16), where θ' is the angle between the beam and the optic axis,

$$\tan\theta' = \frac{\sqrt{\varepsilon_\perp}}{\sqrt{\varepsilon_\parallel}} \frac{\sin\varphi}{\sqrt{\varepsilon_\parallel - \sin^2\varphi}}.$$

In biaxial crystals both beams may emerge from the plane of incidence on refraction.

7.3.2 Light Reflection from Transparent Crystals

As is well known, the solution of the boundary problem in the isotropic media optics is given by Fresnel equations [7.9]. Similar equations, expressing the amplitudes of the reflected and refracted wave in terms of the amplitude of the incident wave should be obtained for anisotropic media. Here, however, the situation is considerably more complicated as compared with isotropic media, since if we choose, for simplicity, the interface as one of the coordinate planes, it may not coincidence, generally speaking, with any one of the main planes of the dielectric constant tensor. We therefore restrict ourselves to uniaxial transparent crystals and a particular disposition of the interface [7.10].

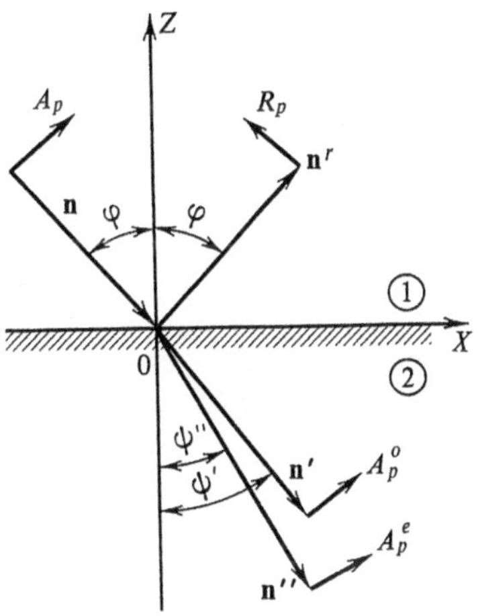

Fig. 7.11. Light reflection from a crystal. The optic axis of the crystal coincides with the Z axis

Suppose the interface is the main plane XY of tensor ε_{ik} and hence the optic axis of the crystal, Z, coincides with the normal to this plane (Fig. 7.11).

For the boundary conditions, we write, as usual, the continuity of the tangential components of field intensity vectors E_y, H_y and of the normal components of induction vectors D_Z, $B_Z=H_Z$:

$$\begin{aligned}&A_s+R_s=A_s^o/\varepsilon_\perp\ , \quad A_p+R_p=A_p^e v/c\ ,\\&A_p\sin\varphi+R_p\sin\varphi=A_p^e\sin\psi''\ ,\\&A_s\sin\varphi+R_s\sin\varphi=A_s^o\sin\psi' v_o/c\ .\end{aligned}\tag{7.20}$$

We assume all the vectors to be resolved into two components: the one normal to the plane of incidence (s) and the one lying in it (p). The amplitudes of the induction vectors $\boldsymbol{D}$ of the incident wave will be called A, those of the reflected wave R, those of the refracted waves, A^e and A^o; ψ' and ψ'' are the angles of refraction, and v and v_o are the velocities of the extraordinary and ordinary waves. Using the refraction law $\sin\varphi/\sin\psi'=c/v_o=\sqrt{\varepsilon_\perp}$, $\sin\varphi/\sin\psi''=c/v$, we obtain for the sought-for amplitudes:

$$R_p=\frac{\sin\psi''\cos\varphi-\dfrac{1}{\varepsilon_\perp}\cos\psi''\sin\varphi}{\dfrac{1}{\varepsilon_\perp}\cos\psi''\sin\varphi+\sin\psi''\cos\varphi}A_p\ ,$$

$$A_p^e=\frac{\sin 2\varphi}{\dfrac{1}{\varepsilon_\perp}\cos\psi''\sin\varphi+\sin\psi''\cos\varphi}A_p\ ,$$

$$R_s = \frac{\sin 2\varphi - \varepsilon_\perp \sin 2\psi'}{\varepsilon_\perp \sin 2\psi' + \sin 2\varphi} A_s \ ,$$

$$A_s^o = \frac{2\varepsilon_\perp \sin 2\varphi}{\varepsilon_\perp \sin 2\psi' + \sin 2\varphi} A_s \ . \tag{7.21}$$

By analogy with isotropic media, we find such an angle of incidence at which the reflected wave will be fully polarized, i.e., $R_p = 0$, so that $\tan\psi'' = 1/\varepsilon_\perp \tan\varphi$. With the aid of the refraction law for uniaxial crystals for the full polarization angle (the *Brewster angle*), we obtain the equation

$$\sin^2\varphi = \frac{\varepsilon_\parallel - \varepsilon_\perp \varepsilon_\parallel}{1 - \varepsilon_\perp \varepsilon_\parallel} \ , \tag{7.22}$$

which generalizes for this case the usual relation $\tan\varphi = n$ for isotropic media.

Such a result, where $n = n_0$, will be obtained for the full polarization angle only if a plate from a uniaxial crystal is cut out parallel to the optic axis.

From (7.22) it is seen that, in distinction to isotropic media, the light reflected from the crystal at the Brewster angle is not at all perpendicular either to the ordinary or extraordinary waves or to the beams in the crystal.

A thorough consideration of the problem of light reflection from transparent crystals was carried out by *Fedorov* [7.11, 12]. It was shown, in particular, that, in distinction to isotropic media, the angle of rotation of the polarization plane of the linearly polarized wave in reflection depends on the direction of vibrations of the incident wave in a more complicated manner. All the necessary relations describing complete internal reflection in crystals have been obtained.

7.4 Light Interference in Crystals

7.4.1 Elliptical Polarization

When a light is transmitted through a crystal plate, we have at the exit two coherent light oscillations with a certain phase difference $\delta = 2\pi d\mu/\mu$ (d is the plate thickness, μ is the difference between the refractive indices, and λ is the wavelength), polarized in two mutually perpendicular directions X and Y (Sect. 7.3), which can be represented as

$$E_x = \sin\tau \ , \quad E_y = k\sin(\tau - \delta) \ , \tag{7.23}$$

where $\tau = \omega t$, ω is the circular frequency of light oscillations, and k^2 is the relative wave intensity.

To determine the polarization of the resultant wave we eliminate parameter τ from these equations:

$$E_y^2-2kE_xE_y\cos\delta+k^2E_x^2=k^2\sin^2\delta\ .$$

This last equation shows that the terminus of the intensity vector of the resultant wave describes an ellipse; therefore, we say such a wave *elliptically polarized.*

Let us now reduce this equation to the conventional form by choosing a new main system of coordinates X' and Y' rotated through a certain angle β relative to the old one X and Y; then we can find the directions and semiaxes a and b of the oscillation ellipse. The parametric representation of the ellipse in the main axes will have the form:

$$E_{x'}=a\cos(\tau-\delta')=\cos\beta\sin\tau+k\sin\beta\sin(\tau-\delta)\ ,$$

$$E_{y'}=b\sin(\tau-\delta')=\sin\beta\sin\tau+k\cos\beta\sin(\tau-\delta)\ .$$

Then $E_{x'}^2/a^2+E_{y'}^2/b^2=1$.

Assuming $\tau=0$ and $\tau=\pi/2$ and denoting, for convenience, $\tan\alpha=k$ and $\tan\gamma=b/a$, we find

$$\tan2\beta=\tan2\alpha\cos\delta\ ,\quad \sin2\gamma=\sin2\alpha\sin\delta\ . \tag{7.24}$$

These equations give rise to two more (reciprocal) equations:

$$\cos2\alpha=\cos2\beta\cos2\gamma\ ,\quad \tan\delta=\tan2\gamma/\sin2\beta\ . \tag{7.25}$$

A particularly pictorial method for representing elliptical oscillations is their representation by a point on a *Poincaré sphere* with coordinates 2γ (latitude) and 2β (longitude) (Fig. 7.12). Now to all the points on the equator ($\gamma=0$) of the Poincaré sphere correspond linearly polarized waves ($b=0$), and to the pole points ($\gamma=\pm\pi/4$), circularly polarized waves ($a=b$) with

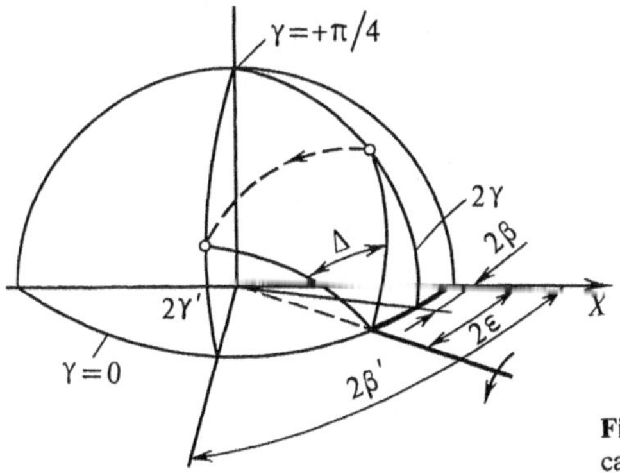

Fig. 7.12. Elliptical oscillations on the Poincaré sphere

oppositely directed scanning. Using the Poincaré sphere we can easily solve one of the principal problems of crystal optics concerning transmission of an elliptically polarized light (with parameters 2β and 2γ) through a crystal plate setting up a phase difference Δ, when one of the directions of oscillations in the plate makes an angle ε with the X axis. The new point on the sphere – $2\beta', 2\gamma'$ – which defines the parameters of the emerging elliptically polarized light, is obtained from the initial one by rotating the sphere through angle Δ about an axis lying in the equatorial plane at longitude 2ε.

Elliptical polarization can be described by Jones's vector $\boldsymbol{E}_0$ [7.13, 14] if the field intensity of the light wave is represented as $\boldsymbol{E} = \boldsymbol{E}_0 e^{i\tau} = (\boldsymbol{E}_1 + i\boldsymbol{E}_2) e^{i\tau}$, $\boldsymbol{E}_0 = \boldsymbol{E}_1 + i\boldsymbol{E}_2$. The equation defining the character of the polarization oscillations

$$\mathrm{Re}\{\boldsymbol{E}\} = \boldsymbol{R} = \boldsymbol{E}_1 \cos\tau - \boldsymbol{E}_2 \sin\tau$$

with the aid of relations $[\boldsymbol{R} \times \boldsymbol{E}_1] = [\boldsymbol{E}_1 \times \boldsymbol{E}_2] \sin\tau$ and $[\boldsymbol{R} \times \boldsymbol{E}_2] = [\boldsymbol{E}_1 \times \boldsymbol{E}_2] \cos\tau$ can easily be reduced to the form

$$[\boldsymbol{R} \times \boldsymbol{E}_1]^2 + [\boldsymbol{R} \times \boldsymbol{E}_2]^2 = [\boldsymbol{E}_1 \times \boldsymbol{E}_2]^2 \ .$$

Assuming $\boldsymbol{R} = \alpha\boldsymbol{E}_1 + \beta\boldsymbol{E}_2$ $(\alpha^2 + \beta^2 = 1)$, we obtain, using the Lagrange method, the extremal values (semiaxes):

$$R^2_{\mathrm{extr}} = \tfrac{1}{2}\left[E_1^2 + E_2^2 \pm \sqrt{(E_1^2 + E_2^2)^2 - 4(\boldsymbol{E}_1\boldsymbol{E}_2)^2}\right] \ .$$

The orientation of the oscillation ellipse is defined by the fact that vectors $\boldsymbol{E}_1$ and $\boldsymbol{E}_2$ are (as can be seen from equality $\boldsymbol{R} = \alpha\boldsymbol{E}_1 \pm \sqrt{1-\alpha^2}\boldsymbol{E}_2$) conjugate half diameters of the oscillation ellipse. In particular, if $\boldsymbol{E}_1 \perp \boldsymbol{E}_2$, then $|\boldsymbol{E}_1|$ and $|\boldsymbol{E}_2|$ are the main semiaxes of the ellipse. If the wave is linearly polarized, then $[\boldsymbol{E}_1 \times \boldsymbol{E}_2] = 0$. The conditions for the wave to be circularly polarized are equalities $\boldsymbol{E}_1\boldsymbol{E}_2 = 0$ and $E_1^2 = E_2^2$. If we assume, by tradition, that the light is left-polarized when during observation vector $\boldsymbol{R}$ rotates anticlockwise opposite to the propagation (τ increases), to this corresponds left-hand triad $\boldsymbol{E}_1, \boldsymbol{E}_2, \boldsymbol{n}$ and vice versa. All the necessary relations can be written by using a single vector $\boldsymbol{E}_0$ (or $\boldsymbol{E}$), thus obtaining [7.12] a completely invariant description of elliptical polarization. Such a representation is also convenient when solving problems of addition of coherent elliptical oscillations $\boldsymbol{E}_0'$ and $\boldsymbol{E}_0''$, because then $\boldsymbol{E}_0 = \boldsymbol{E}_0' + \boldsymbol{E}_0''$.

The character of polarization can also be judged by the ratio of the projection of field intensity vector $\boldsymbol{E}$ onto coordinate axes X and Y perpendicular to wave vector $E_y/E_x = \varkappa$. *Fedorov* [7.12] obtained equations giving angle β formed by the major axis of the ellipse with the X axis, and the ratio of the semiaxes b/a of the oscillation ellipse:

$$\tan 2\beta = \frac{\varkappa + \varkappa^*}{1 - |\varkappa|^2} \ , \quad \frac{b^2}{a^2} = \frac{1 + |\varkappa|^2 - \sqrt{(1 + |\varkappa|^2)^2 + (\varkappa - \varkappa^*)^2}}{1 + |\varkappa|^2 + \sqrt{(1 + |\varkappa|^2)^2 + (\varkappa - \varkappa^*)^2}} \ . \tag{7.26}$$

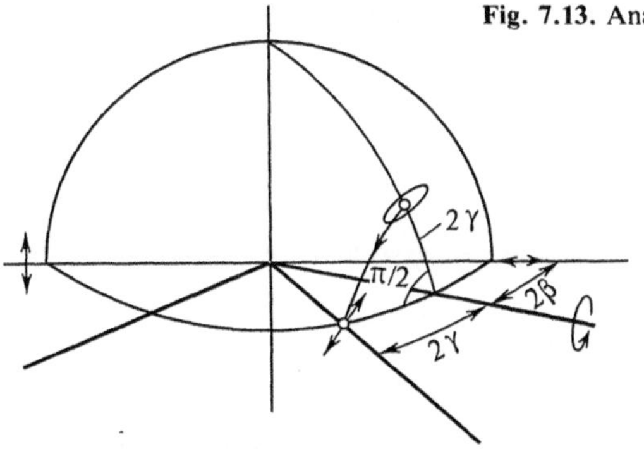

Fig. 7.13. Analysis of elliptical oscillations

It is easy to see that at $\varkappa = k\exp(-\mathrm{i}\delta)$ (7.23) this yields the same equations (7.24, 25). If we assume that X, Y, Z is a right-handed coordinate system and that the light propagates along the Z axis, then $\mathrm{Im}\{\varkappa\} > 0$ corresponds to a right-polarized light.

For analysis of an elliptically polarized light, i.e., for experimental determination of the ratio of the semiaxes $b/a = \tan\gamma$ and the direction of the main axis of the oscillation ellipse, it is customary to use a polarizer and a crystal plate producing a phase difference $\delta = \frac{\pi}{2}$ $(-\mu d = \lambda/4$, μ being the birefringence, and d the thickness), the so-called "quarter-wave" plate. The direction of the major axis of the oscillation ellipse is found readily from the position of the polarizer at the maximum darkening of the transmitted light. Setting the quarter-wave plate so that one of the directions of oscillations in it coincides with the axis of the oscillation ellipse, we turn the elliptically polarized light into a linearly polarized one, as can be seen from the construction of the Poincaré sphere (Fig. 7.13). It is also clear that rotation angle φ, through which the polarizer must be turned additionally to completely extinguish the light emerging from the "$\lambda/4$" plate, is related to the axes ratio $\tan\gamma$ by a simple dependence $\gamma = \varphi$.

7.4.2 Light Interference in Parallel Beams

Suppose a linearly polarized light beam falls along a normal to a crystal plate of thickness d. Owing to birefringence, two coherent waves with mutually perpendicular directions of oscillations and different velocities will propagate inside the crystal; on emergence from the crystal these waves will acquire a phase difference

$$\delta = \frac{2\pi d}{\lambda}(n'' - n') , \tag{7.27}$$

where n'' and n' are the refractive indices of the waves, and λ is the light wavelength in the air (Sect. 7.3). If we separate the components of the

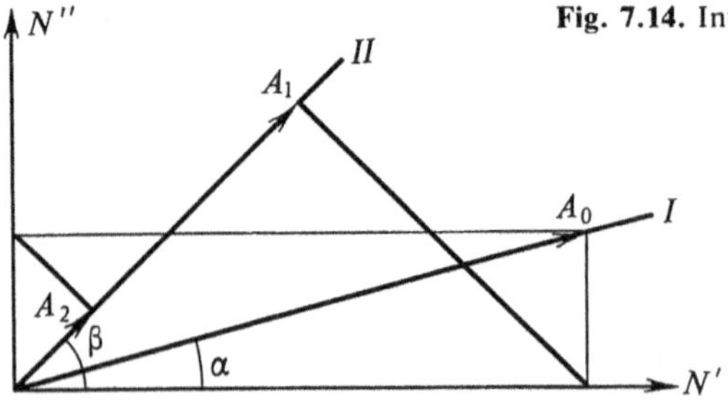

Fig. 7.14. Interference of polarized waves in a crystal

oscillations of the two waves in one direction with the aid of the output polarizer (analyzer), they will interfere. Let us calculate the intensity of the resultant wave at different orientations of the polarizer, crystal plate, and analyzer. In Fig. 7.14, I and II denote the directions of the oscillations passed by the polarizer and analyzer, N' and N'' being the directions of the oscillations of the light waves in the crystal.

The intensity of the light after passing the analyzer is determined by the time-average value of the square of the field intensity of the light wave

$$[A_1 \sin\omega t + A_2 \sin(\omega t - \delta)]^2$$
$$= [A_0 \cos\beta \cos\alpha \sin\omega t + A_0 \sin\beta \sin\alpha \sin(\omega t - \delta)]^2 \ , \qquad (7.28)$$

which yields

$$I = I_0 [\cos^2(\alpha - \beta) - \sin 2\alpha \sin 2\beta \sin^2(\delta/2)] \ , \qquad (7.29)$$

where $I_0 = A_0^2/2$ is the intensity of light incident upon the plate.

Proceeding from this equation [7.15] one can carefully trace the role of each of the variables α, β, and δ. For instance, on rotation of the crystal plate by a complete revolution, i.e., on change of angle α from 0 to 360°, with crossed polarizer and analyzer ($\beta - \alpha = \frac{\pi}{2}$) the intensity of the transmitted monochromatic light reduces to zero four times *(crystal extinction).*

Under a white light, component $\cos^2(\alpha - \beta)$ (white component), which is independent of the wavelength, goes to zero twice and takes the value of unity also twice as the analyzer rotates through 360°. The color component $-\sin 2\alpha \sin 2\beta \sin^2(\delta/2)$, which depends on phase difference δ, will take the extremal values four times; owing to this, two additional interference colors will alternate. Observations of the extinction of crystals make it possible to distinguish between crystals of the triclinic and monoclinic syngony and those of a higher symmetry. Since the orientation of the main axes of the Fresnel ellipsoid in crystals of orthorhombic and higher symmetry is completely defined by their symmetry (Sect. 7.1), the extinction of such crystals must be "direct," i.e., it must coincide in direction with the projection of the symmetry axes of the crystal onto the plane of the plate. For triclinic and monoclinic crystals the direction of extinction generally does not coincide with any crystallographic directions. Interference in crystal plates is used for developing many crystallo-optical devices: compensators for birefringence measurements, interference-polarization filters, etc.

7.4.3 Conoscopic Patterns in Uniaxial Crystals

Let us calculate the phase difference which is acquired by two beams on emergence from a crystal plate of thickness d with oblique incidence of light at an angle φ to its normal N (Fig. 7.15a). The sought-for phase difference is equal to

$$\delta = 2\pi(AC/\lambda'' + CD/\lambda - AB/\lambda') \ ;$$

here λ is the light wavelength in the air, and λ' and λ'' are the wavelengths in a crystal ($\lambda' = \lambda n'$, $\lambda'' = \lambda n''$). From the drawing it is clear that

$$AC = d/\cos\psi'' \ , \quad AB = d/\cos\psi' \ ,$$

$$CD = CB\sin\varphi = (EB - EC)\sin\varphi = d\sin\varphi(\tan\psi' - \tan\psi'') \ .$$

Using the refraction law (7.19) $\sin\varphi/\lambda = \sin\psi'/\lambda' = \sin\psi''/\lambda''$, we first find the exact expression

$$\delta = \frac{2\pi d}{\lambda}(n''\cos\psi'' - n'\cos\psi') \ .$$

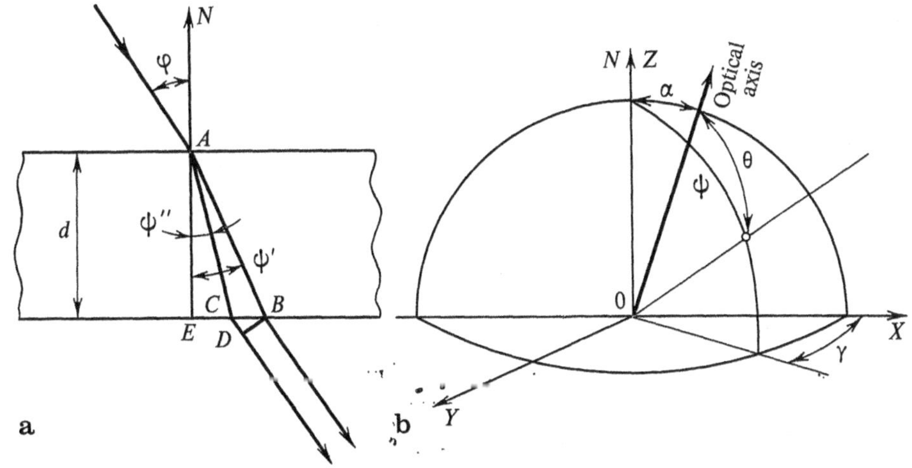

Fig. 7.15. Analysis of conoscopic patterns in a uniaxial crystal: **(a)** beam path in the plate; **(b)** the same pattern in spherical coordinates

We now assume that the birefringence in the crystal is low. Then for the difference $n''\cos\psi'' - n'\cos\psi'$, we can obtain the approximate value $(n''-n')/\cos\psi$, where ψ is some average refraction angle lying between ψ' and ψ''; hence,

$$\delta = \frac{2\pi d}{\lambda\cos\psi}(n''-n') \ .$$

As indicated above (7.15), for uniaxial crystals the difference of the refractive indices of an extraordinary wave and an ordinary wave is equal to $(n_e - n_0)\sin^2\theta$, where θ is the average angle made by the refracted waves with the optic axis of the crystal. Finally we have

$$\delta = \frac{2\pi d}{\lambda\cos\psi}(n_e - n_o)\sin^2\theta \ .$$

We now pass over to a spherical system of coordinates, as depicted in Fig. 7.15b, where the optic axis of the crystal is assumed to lie in plane ZX, the angle between the normal to the plate and the optic axis being denoted by α, so that

$$\cos\theta = \cos\alpha\cos\psi + \sin\alpha\sin\psi\cos\gamma \ . \tag{7.30}$$

For a phase difference in the approximation of the smallness of angle ψ, when we can assume that $1/\cos\psi \approx 1 + \frac{1}{2}\sin^2\psi$, we now have

$$\delta = \frac{2\pi d}{\lambda}[\sin^2\alpha + \sin^2\psi\,(\tfrac{1}{2} + \tfrac{1}{2}\cos^2\alpha - \sin^2\alpha\cos^2\gamma)$$
$$- \sin 2\alpha\sin\psi\cos\gamma](n_e - n_o) \ . \tag{7.31}$$

Using the Cartesian system of coordinates $x = \varrho\sin\psi\cos\gamma$, $y = \varrho\sin\psi\sin\gamma$, $\varrho = \sqrt{x^2+y^2+z^2}$, $z = d$, on the exit plane of the plate, we obtain, assuming angles ψ to be small,

$$\delta = \frac{2\pi}{d\lambda}(n_e - n_o)[x^2(\cos^2\alpha - \tfrac{1}{2}\sin^2\alpha)$$
$$+ y^2(\cos^2\alpha + \tfrac{1}{2}\sin^2\alpha) - xd\sin 2\alpha + d^2\sin^2\alpha] \ . \tag{7.32}$$

Proceeding to the analysis of this expression, we see that for curves of identical phase difference *(isochromes)*, i.e., at $\delta = \text{const}$, we obtain either ellipses ($\tan\alpha < \sqrt{2}$) or hyperbolas ($\tan\alpha > \sqrt{2}$). If $\alpha = 0$, when the optic axis coincides with the normal to the plate, the isochromic lines will be circles with center at the point of emergence of the optic axis. When observing

Fig. 7.16. Conoscopic pattern in a uniaxial crystal [7.2]

interference patterns in a converging white light, the crystal plate is usually placed on the stage of a polarization microscope between crossed polarizers. Then the interference picture *(conoscopic pattern)* shows, in addition to the colored isochromes, the uncolored dark cross with beams *(isogyres)* parallel to the directions of the oscillations passed by the polarizer and the analyzer. Their origin is not related to the phase difference, but is determined only by the direction of the oscillations. The projections of the directions of the oscillations of the considered light waves in the crystal on to the exit plane coincide with two straight-lines which pass through the point of emergence of the optic axis (i.e., the point of intersection of the binormal with the exit plane of the plate) and are parallel to the directions of the polarizer and analyzer oscillations. Therefore, one of the waves with such directions of the wave vector is not excited by the incident polarized light, and the other is extinguished by the analyzer.

The conoscopic pattern in a uniaxial crystal is shown in Fig. 7.16. In the general case (i.e., when the isogyres cross outside the center of the field of vision), as the crystal plate rotates about its normal, the beams of the dark cross will shift in the field of vision parallel to themselves. This manner of displacement of the isogyres makes it possible to distinguish uniaxial crystals from biaxial ones when the point of emergence of the optic axis lies beyond the field of vision.

7.4.4 Conoscopic Patterns in Biaxial Crystals

We have previously found that in biaxial crystals the difference in the refractive indices of the waves propagating in a crystal is defined by the expression [see (7.18)]

$$\frac{1}{(n')^2}-\frac{1}{(n'')^2}=\left(\frac{1}{\varepsilon_x}-\frac{1}{\varepsilon_z}\right)\sin\varphi_1\sin\varphi_2 \ ,$$

where φ_1 and φ_2 are the angles formed by the wave vector of the waves with optic axes. The phase difference at a low birefringence will be

$$\delta=\frac{2\pi\varrho}{\lambda}(n_x-n_z)\sin\varphi_1\sin\varphi_2 \ , \qquad (7.33)$$

where ϱ is the path of the wave in the crystal.

Let us determine the shape of the equal phase difference curves for a plate oriented perpendicular to the acute bisector of the angle between the optic axes. Denote by a_1 and a_2 the distances between the points of emergence of axes N_1 and N_2 and the exit of the direction of vector N of the light waves on the upper plane of the plate, and by α_1 and α_2 the angles between ON_1, N_1N, and ON_2, N_2N (Fig. 7.17a).

We have

$$\sin\varphi_1=\frac{a_1}{\varrho}\sin\alpha_1 \ , \quad \sin\varphi_2=\frac{a_2}{\varrho}\sin\alpha_2 \ ,$$

and we obtain for the equal phase difference curves

$$(a_1a_2/\varrho)\sin\alpha_1\sin\alpha_2=\text{const} \ . \qquad (7.34)$$

With a small angle between the binormals and a small field of vision, $\alpha_1\approx\sin\alpha_2\approx 1$, while ϱ can be assumed constant. Then we approximately obtain for the isochromes the ovals equation of Cassini:

$$a_1a_2=\text{const} \ . \qquad (7.35)$$

We now determine the shape of the isogyres of a biaxial crystal placed between crossed polarizes when the plate is oriented so that the points of emergence of the axes lie on the bisectors of the angles between the directions of the polarizer and analyzer (diagonal position), while the directions of oscillations coincide with those of the X and Y axes.

We shall show that here an equilateral hyperbola will serve as the isogyres. Indeed, for such a hyperbola the bisectors of the angles between the directions from the vertices to an arbitrary point of the hyperbola are parallel to their asymptotes ($\alpha_1=\alpha_2$, see Fig. 7.17b). Our statement about isogyres is therefore a corollary of the Fresnel theorem (Sect. 7.2).

The conoscopic pattern of a biaxial crystal cut out perpendicular to the bisector of the angle between the optic axes and disposed diagonally will thus have Cassini ovals as the isochromes and an equilateral hyperbola with vertices at the points of emergence of the optic axes at the isogyres (Fig. 7.18).

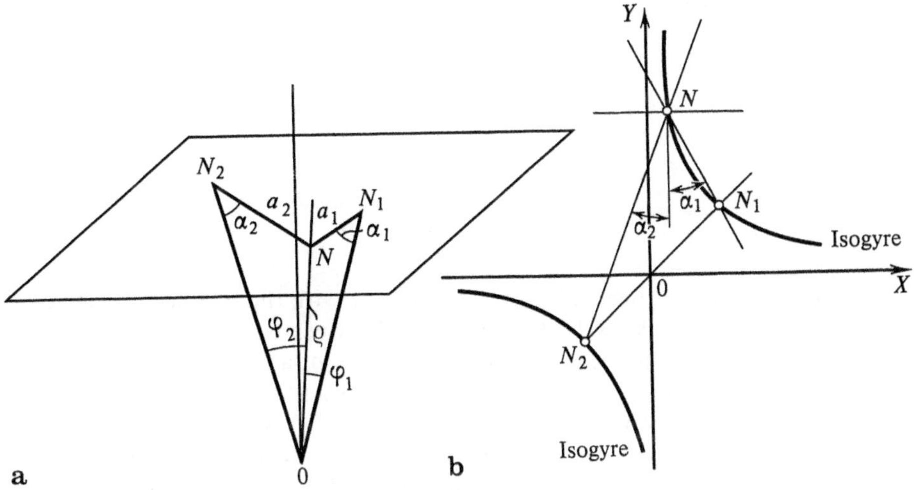

Fig. 7.17 a, b. Analysis of conoscopic patterns in a biaxial crystal

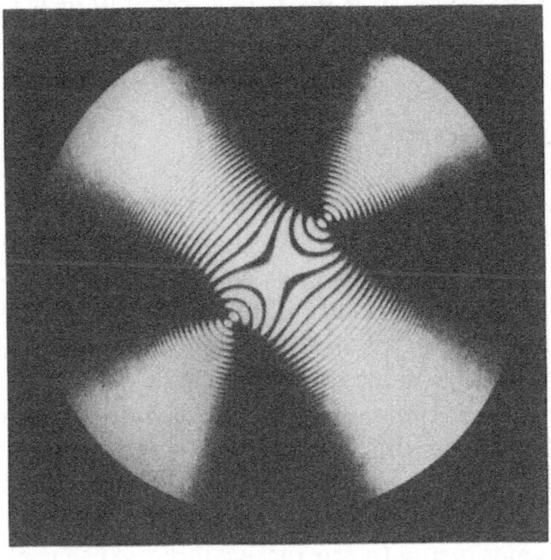

Fig. 7.18. Conoscopic pattern in a biaxial crystal [7.2]

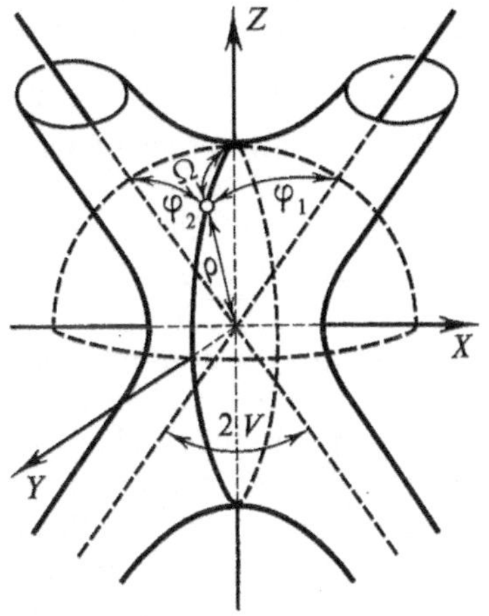

Fig. 7.19. General view of the surface of the constant phase difference in a crystal

To get an idea of the shape of isochromes for other orientations of the crystal plate, it is convenient to use so-called *Bertin surfaces,* whose radius vectors (ϱ) satisfy the equation [see (7.33)]

$$\varrho \sin \varphi_1 \sin \varphi_2 = \text{const} \ , \tag{7.36}$$

where φ_1 and φ_2 are the angles between the radius vector and the optic axis.

Let us consider a section of the Bertin surface by plane $x = 0$ (Fig. 7.19). Then

$$\cos\varphi_1 = \cos\varphi_2 = \cos V \cos\Omega\,, \quad \sin\varphi_1 \sin\varphi_2 = 1 - \cos^2 V \cos^2\Omega$$

and the equation of the curve of the section (isochrome) will have the form

$$\varrho = \varrho_0/(1 - \cos^2 V \cos^2\Omega)\ , \tag{7.37}$$

where Ω is the angle formed by the wave vector with the bisector of the angle between the optic axes, and $2V$ is the angle between the axes. As a result we obtain a closed oval-shaped curve with semiaxes ϱ_0 and $\varrho_0/\sin^2 V$. The other sections of the Bertin surface of biaxial crystals are found similarly.

Observations of conoscopic patterns of biaxial crystals, which are usually conducted in a polarizing microscope, make it possible to measure directly the angle between the optic axes $2V$, which is important for crystal diagnostics. It should be remembered that this angle is distorted by refraction at the air-crystal interface, and therefore a correction is made in accordance with the obvious equation $\sin V = \sin V_{\text{meas}}/n_2$, where V is the true angle, and n_2 is the average refractive index of the crystal.

7.4.5 Measurements of Refractive Indices of Crystals

The refractive indices are the principal optical constants of crystals and often serve as their diagnostic feature. One of the most accurate ($\Delta n \approx 1 \times 10^{-4} - 10^{-5}$) methods for determining refractive indices is the prism method (or the goniometric method). A prism is cut out of a crystal (Fig. 7.20) and is oriented (for instance, by the Federov method) so that one of the main axes of the indicatrix is parallel to the edge of the refraction angle of the prism [7.16]. Using a goniometer, one measures the angle φ of the least deviation of the beams transmitted through the prism, which is related to the refraction angle of the prism, α, and to its refractive index, n, by the dependence $n = \sin[(\alpha + \varphi)/2]/\sin(\alpha/2)$. The measurements are naturally made in a monochromatic polarized light. In measuring the refractive indices, use is also made of the refractometric method, based on the measurement of the limiting angle of total internal refraction of light, φ_0, from a crystal plate placed on a glass hemisphere with a known refractive index n_s larger than that of the crystal n_c (Fig. 7.21). Refractive index n_c is found from the equation

$$n_c = n_s \sin\varphi_0\ . \tag{7.38}$$

The simplest and most universal (although less accurate – $\Delta n \approx 10^{-3}$) is the immersion method, in which a finely crushed crystal is dipped into an immersion liquid with a known refractive index. Observing such specimens

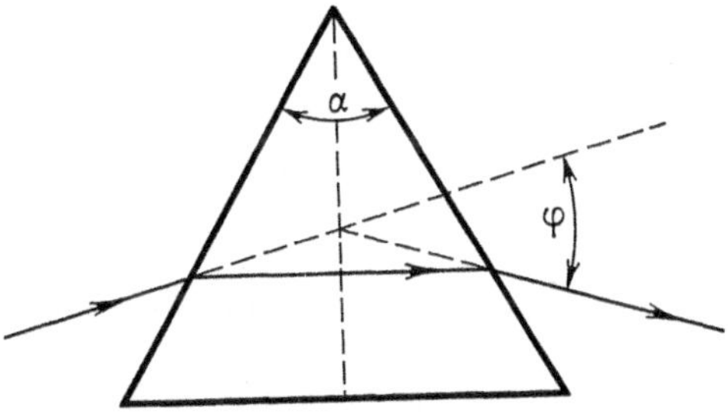

Fig. 7.20. Determination of refractive indices by the prism method

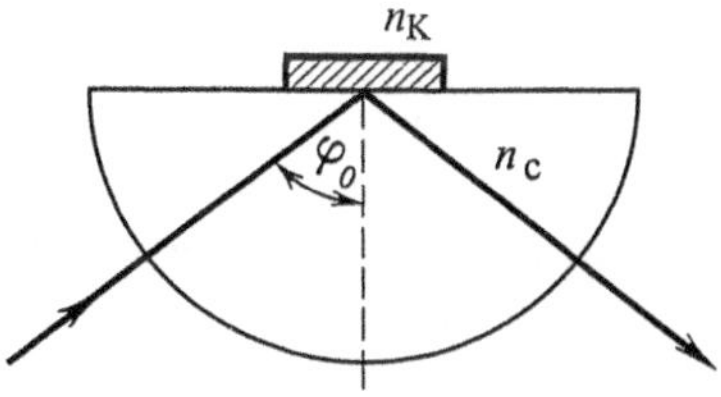

Fig. 7.21. Determination of refractive indices by complete internal reflection

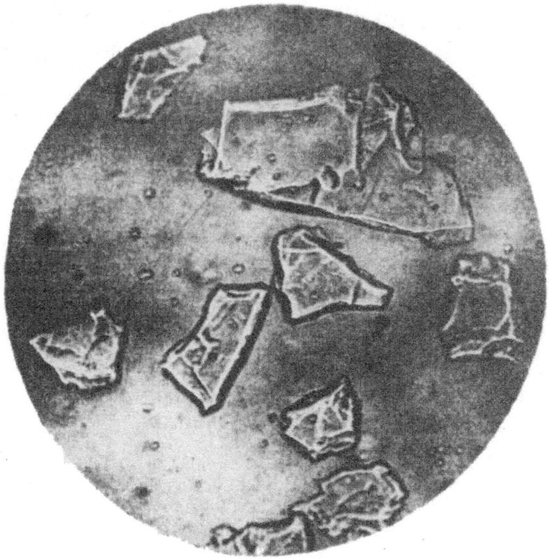

Fig. 7.22. Becke lines in an immersion preparation [7.19]

under the microscope, one finds a liquid most similar, as regards its refractive index, to the crystal under investigation. When comparing the refractive indices of the liquid and the crystal one observes a bright fringe (Becke strip) framing the crystal fragments (Fig. 7.22). The origin of this fringe becomes quite evident from the path of the beams at the edge of the crystal if we assume that the edge has the shape of a prism. When the microscope tube is raised, the fringe shifts toward the more refracting medium. If the refractive indices of the crystal and liquid in a monochromatic light are equal, the Becke strip disappears completely. The refractive indices are measured in a polarized light separately for the two waves in the position of crystal extinction [7.17 – 19].

The birefringence of crystals is characterized by the differences between the main refractive indices n_1, n_2, n_3. The measurement of these differences is based on the determination of the phase difference $\delta = 2\pi(n_i - n_k)d/\lambda$ (i, $k = 1,2,3$) for beams transmitted through a crystal plate perpendicular to the main axes of tensor ε_{ik}. The phase difference is usually measured by compensation methods, when an optical device (compensator) is interposed in the path of the light beam, thus producing an additional well-known phase difference with a sign opposite to that of the crystal. Changing the variable path difference in the compensator makes it equal in absolute value to the path difference in the crystal; then the total path difference reduces to zero, and the crystal loses its interference color. A quartz wedge is the simplest compensator. In rotary compensators of the Berek type the variable path difference is caused by a tilt of a crystal plate (usually calcite) cut out perpendicular to the optic axis. The path difference can also be determined by analyzing the parameters of the elliptical oscillation at the exit of the beams from the crystal plate (Sect. 7.4).

It has been indicated above (Sect. 7.2) that uniaxial crystals are usually divided into positive, for which $\varepsilon_\perp < \varepsilon_\parallel$ ($n_o < n_e$), and negative $\varepsilon_\perp > \varepsilon_\parallel$ ($n_o > n_e$). The sign of a uniaxial crystal can easily be defined for crystal sections perpendicular to the optic axis during observations in a polarizing microscope in a converging light. To do this, we use a thin quartz wedge (positive crystal) cut out parallel to the optic axis so that the direction of the oscillations of the wave with the higher refractive index in the quartz coincides with the short side of the wedge. Pushing the wedge, its thin end forward, from the fourth quadrant into the second, we observe the motion of the interference rings, which is different depending on the sign of the crystal. Suppose that we deal with a positive crystal; then for waves with directions of

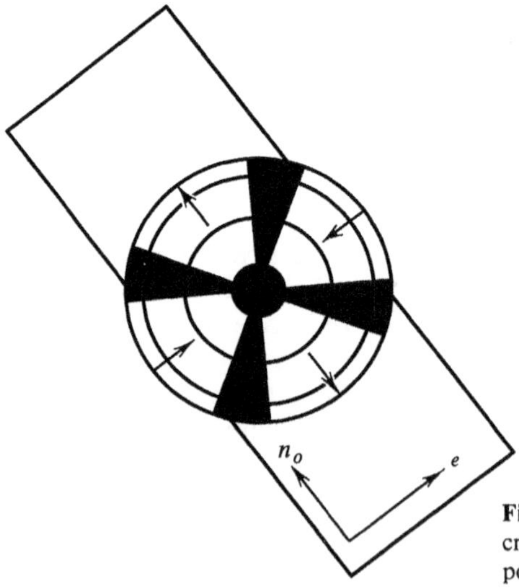

Fig. 7.23. Determination of the optical sign of a crystal with the aid of a quartz wedge (for a positive crystal)

the normal lying in the fourth and second quadrant the quartz edge will reduce the phase difference, and hence the radii of the isochrome circles will increase. It will seem to us that in these quadrants the interference rings move from the center of the field of vision to the periphery as the wedge progresses (Fig. 7.23). For a negative crystal, the motion of the isochromes will be reversed.

7.5 Light Absorption in Crystals

To describe the properties of absorbing crystals we introduce a symmetric conductivity tensor σ_{ik} relating the vector of current density $\boldsymbol{j}$ to electric field intensity $\boldsymbol{E}$: $j_i = \sigma_{ik} E_k$. Consequently only homogeneous damping waves with intensities proportional to $[\mathrm{i}(\omega t - \boldsymbol{k}\boldsymbol{r})]$, with a composite index $\hat{n} = n(1 - \mathrm{i}\varkappa)$ ($\varkappa$ being the absorption index), $\boldsymbol{k} = \dfrac{\omega}{c}\hat{n}\boldsymbol{n}^0$, and a real $\boldsymbol{n}^0$, we have, from the Maxwell equations,

$$\boldsymbol{H} = [\hat{\boldsymbol{n}} \times \boldsymbol{E}] \ , \quad -[\hat{\boldsymbol{n}} \times \boldsymbol{H}] = \boldsymbol{D} + \frac{4\pi}{\mathrm{i}\omega}\boldsymbol{j} = \hat{\boldsymbol{D}} \ , \quad \hat{\boldsymbol{n}} = \hat{n}\boldsymbol{n}^0 \ .$$

Now,

$$\hat{D}_{\mathrm{i}} = \left(\varepsilon_{ik} - \frac{4\pi\mathrm{i}}{\omega}\sigma_{ik}\right)E_k = \hat{\varepsilon}_{ik} E_k$$

with a complex symmetric tensor of the dielectric constant

$$\hat{\varepsilon}_{ik} = \varepsilon_{ik} - \frac{4\pi\mathrm{i}}{\omega}\sigma_{ik} \ ,$$

and the Maxwell equations will have exactly the same form as for transparent crystals.

Homogeneous waves can be excited in absorbing crystals only at a normal incidence of the exciting light. At an oblique incidence, inhomogeneous waves, characterized by complex vector $\boldsymbol{n}^0$, are excited in the crystal. The character of polarization of such waves is different for vectors $\boldsymbol{D}$ and $\boldsymbol{B}$.

The main axes of tensors ε_{ik} and σ_{ik} in crystals of the monoclinic and triclinic syngonies do not coincide; therefore, tensor $\hat{\varepsilon}_{ik}$ may not be reducible to the diagonal form even by means of a complex linear transformation. Then the use of the conventional coordinate representation of tensors involves fundamental difficulties, and the invariant method should be applied [7.12], [7.20]. For crystals with a symmetry not below orthorhombic, the introduction of a real main system of coordinates, in which complex symmetric tensor $\hat{\varepsilon}_{ik}$ is diagonal, is always possible. If we restrict ourselves to such crystals, then, acting as in the case of transparent crystals, [7.21] we again obtain, in the main system of coordinates, the Fresnel equation [see (7.9)]

$$\frac{(n_x^0)^2}{1/\hat{n}^2-1/\hat{\varepsilon}_x}+\frac{(n_y^0)^2}{1/\hat{n}^2-1/\hat{\varepsilon}_y}+\frac{(n_z^0)^2}{1/\hat{n}^2-1/\hat{\varepsilon}_z}=0 \ . \tag{7.39}$$

This means that here, too, to each direction of wave vector $\boldsymbol{n}^0$ correspond two (now complex) refractive indices $\hat{n}$.

To establish the nature of polarization of the proper waves, we choose, as before, a new system of coordinates, whose Z' axis coincides with the direction of the wave vector. Then from the equations $\boldsymbol{E}=\hat{\varepsilon}_{ik}^{-1}\hat{\boldsymbol{D}}$, $\hat{\boldsymbol{D}}=\hat{n}^2\boldsymbol{E}$ we have

$$\left(\frac{1}{\hat{n}^2}-\gamma_{11}\right)D_{x'}=\gamma_{12}D_{y'} \ , \quad \left(\frac{1}{\hat{n}^2}-\gamma_{22}\right)D_{y'}=\gamma_{12}D_{x'} \ . \tag{7.40}$$

Here $\gamma_{\alpha\beta}$ are the components of two-dimensional complex tensor $\hat{\varepsilon}_{\alpha\beta}^{-1}$ $(\alpha,\beta=x',y')$, whence we find $1/\hat{n}_{\pm}^2=(\gamma_{11}+\gamma_{22})/2\pm\sqrt{1/4(\gamma_{22}-\gamma_{11})^2+\gamma_{12}^2}$ and the ratio of the components of the induction wave vector

$$\begin{gathered}(D_{y'}/D_{x'})_+=k \ , \quad (D_{y'}/D_{x'})_-=-\frac{1}{k} \ , \\ k=\frac{1}{\gamma_{12}}\left[\tfrac{1}{2}(\gamma_{22}-\gamma_{11})+\sqrt{\tfrac{1}{4}(\gamma_{22}-\gamma_{11})^2+\gamma_{12}^2}\right] \ ,\end{gathered} \tag{7.41}$$

where k is the ellipticity of the waves propagating in the crystal. From these relations we conclude (Sect. 7.4) that the damping waves in absorbing crystals are polarized elliptically. The oscillation ellipses are similar, crossed, and are scanned in the same directions.

It can be shown that in crystals of intermediate syngonies $\gamma_{12}=0$, and hence in uniaxial crystals homogeneous waves are polarized linearly. In crystals of lower syngonies, however, homogeneous waves are polarized elliptically.

For weakly absorbing crystals we can neglect the square of absorption coefficient $\varkappa$, as compared to unity, and write $\hat{n}^2=n^2(1-\mathrm{i}\varkappa)^2\simeq n^2(1-2\mathrm{i}\varkappa)$; similarly, for the velocities we have $\hat{c}_n=\dfrac{c}{\hat{n}}\approx\dfrac{c}{n}(1+\mathrm{i}\varkappa)=c_n(1+\mathrm{i}\varkappa)$. The real part of the Fresnel equation yields the former expression (7.9) (as in transparent crystals), and the imaginary part is reduced to the form

$$\begin{aligned}&\varkappa c_n^2\left[\frac{(n_x^0)^2}{(c_n^2-c_x^2)^2}+\frac{(n_y^0)^2}{(c_n^2-c_y^2)^2}+\frac{(n_z^0)^2}{(c_n^2-c_z^2)^2}\right] \\ &=\frac{\varkappa_x c_x^2(n_x^0)^2}{(c_n^2-c_x^2)^2}+\frac{\varkappa_y c_y^2(n_y^0)^2}{(c_n^2-c_y^2)^2}+\frac{\varkappa_z c_z^2(n_z^0)^2}{(c_n^2-c_z^2)^2}\end{aligned} \tag{7.42}$$

and determines the value of absorption coefficent $\varkappa$ in relation to the direction of wave vector $\boldsymbol{n}^0$ and the state of polarization of the relevant wave. For a uniaxial crystal, when $\varepsilon_x = \varepsilon_y = \varepsilon_\perp$, $\varepsilon_z = \varepsilon_\parallel$, the Fresnel equation is divided into two: $\hat{c}_n' = \hat{c}_o$, $(\hat{c}_n'')^2 = \hat{c}_o^2 \cos^2\theta + \hat{c}_e^2 \sin^2\theta$ (θ being the angle between the wave vector and the optic axis of the crystal). Hence, for an ordinary wave $c_n' = c_o$, $\varkappa_n' = \varkappa_o$, and for an extraordinary wave $(c_n'')^2 = c_o^2 \cos^2\theta + c_e^2 \sin^2\theta$, $\varkappa_n''(c_n'')^2 = \varkappa_o c_o^2 \cos^2\theta + \varkappa_e c_e^2 \sin^2\theta$ ($c_o = c/n_o$, $c_e = c/n_e$; see Sect. 7.2).

To simplify the law of variations of $\varkappa$ in the general case of a biaxial crystal, we note that with no absorption we would have

$$D_i = -\frac{c^2 n_i^0 (\boldsymbol{E}\boldsymbol{n}^0)}{c_n^2 - c_i^2} ,$$

for the components of the induction vector and then we could write, as an approximation linear in $\varkappa$,

$$\varkappa c_n^2 = \varkappa_x d_x^2 c_x^2 + \varkappa_y d_y^2 c_y^2 + \varkappa_z d_z^2 c_z^2 ,$$

where $\boldsymbol{d}$ (the ort of induction vector $\boldsymbol{D}$) is determined for a crystal without absorption with the same values or real quantities ε_x, ε_y, and ε_z. Now, to determine $\varkappa$ it will suffice to know the directions of the oscillations, which, for a biaxial crystal, are defined by the Fresnel theorem (Sect. 7.2).

Thus, in crystals the coefficient of absorption of waves, as distinct from isotropic media, depends on the state of polarization of the waves. This phenomenon is called *dichroism*. If we assume that the birefringence in the crystal is low as well, then $\varkappa = \varkappa_x d_x^2 + \varkappa_y d_y^2 + \varkappa_z d_z^2$, so that an ovaloid with axes $\varkappa_x$, $\varkappa_y$, and $\varkappa_z$ can be called the absorption ovaloid.

As with transparent crystals, of special interest are those directions in crystals for which the complex refractive indices now coincide. These directions may be called optic axes, as before. The number of such axes, the refractive indices, and the nature of polarization of the waves propagating in these directions may be different. This difference is a useful feature for

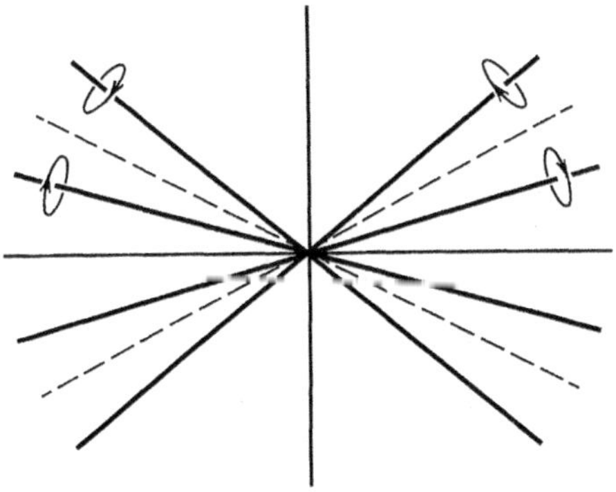

Fig. 7.24. Arrangement of circular axes in absorbing crystals of orthorhombic syngony (dashed lines denote the optic axes in the absence of absorption)

complete classification of absorbing crystals. We shall only note that crystals of the orthorhombic syngony may generally have four circular axes along which circularly polarized waves with the same refractive index may propagate (Fig. 7.24). Indeed, at $(\gamma_{22}-\gamma_{11})^2+4\gamma_{12}^2=0$ $n_+=n_-$ and from (7.41) follows $(D_{y'}/D_{x'})_+=\mathrm{i}$ and $(D_{y'}/D_{x'})_-=-\mathrm{i}$. The light reflection from absorbing crystals was considered in detail by *Fedorov* [7.12]. He also developed a method for determining the optical constants and the orientation of absorbing crystals from the analysis of the elliptical polarization of a reflected light.

7.6 Optical Activity of Crystals

The coupling equation $D_i=\varepsilon_{ik}^0 E_k$, which we have been considering so far, is only the first term of a more general linear relationship

$$D_i=\varepsilon_{ik}^0 E_k+\gamma_{ikl}\frac{\partial E_k}{\partial x_l}\ , \tag{7.43}$$

which takes into account the spatial-nonuniformity dependence of the field intensity in a light wave, i.e., terms of the order of a/λ, where a are atomic sizes, and λ is the wavelength. It can be shown that the principle of symmetry of the kinetic coefficients [7.1] requires the fulfillment of equalities $\gamma_{ikl}=-\gamma_{kil}$.

Suppose there is no absorption in the medium, so that for a monochromatic wave the time average of the energy-dissipation integral

$$\int\left(\boldsymbol{E}\frac{\partial \boldsymbol{D}^*}{\partial t}+\boldsymbol{E}^*\frac{\partial \boldsymbol{D}}{\partial t}\right)dv=\mathrm{i}\omega\int(\boldsymbol{E}\boldsymbol{D}^*-\boldsymbol{E}^*\boldsymbol{D})\,dv$$

must vanish. Considering that ε_{ik}^0 are real in a transparent medium and taking advantage of the relationship between $\boldsymbol{D}$ and $\boldsymbol{E}$, we get

$$\int\left(\gamma_{ikl}^*E_i\frac{\partial E_k^*}{\partial x_l}-\gamma_{ikl}E_i^*\frac{\partial E_k}{\partial x_l}\right)dv$$

$$=\int\left(\gamma_{ikl}^*E_i\frac{\partial E_k^*}{\partial x_l}+\gamma_{ikl}E_k\frac{\partial E_i^*}{\partial x_l}\right)dv=\int(\gamma_{ikl}^*+\gamma_{kil})E_i\frac{\partial E_k^*}{\partial x_l}\,dv\ . \tag{7.44}$$

Thus, the no-absorption condition requires that tensor γ_{ikl}, $\gamma_{ikl}^*=\gamma_{ikl}=-\gamma_{kil}$ also be real. For a plane wave with a wave vector $\boldsymbol{k}=\omega/c\boldsymbol{n}$, we have $D_i=\varepsilon_{ik}E_k$, where $\varepsilon_{ik}=\varepsilon_{ik}^0+\dfrac{\mathrm{i}\omega}{c}\gamma_{ikl}n_l$ (n_l being the components of vector $\boldsymbol{n}$).

Instead of a third-rank tensor γ_{ikl}, antisymmetric with respect to the first two subscripts, one can introduce gyration pseudotensor g_{ik}:

$$\frac{\omega}{c}\gamma_{ikl} = e_{ikm} g_{ml} \;,$$

where e_{ikm} is a completely antisymmetric unit tensor, and the axial gyration vector

$$g_i = g_{ik} n_k \;.$$

Crystals for which pseudotensor g_{ik} is different from zero are said to be *optically active*.

The symmetry of crystals imposes limitations on the components of pseudotensor g_{ik}. Thus, the presence of a center of symmetry reduces all the components g_{ik} to zero, i.e., completely eliminates the optical activity. According to the symmetry conditions, among the remaining acentric classes optical activity is nonexistent in another three classes: $\bar{6}$, $\bar{6}m2$, and $\bar{4}3m$. Therefore, only 18 classes remain in which optical activity can be observed. For these, there are 11 distinct types of gyration pseudotensor in the crystallophysical system of coordinates (Table 7.4).

To analyze the character of polarization in optically active crystals we write in the same approximation linear with respect to gyration vector g_i

$$\boldsymbol{E} = (\varepsilon^0)^{-1}\boldsymbol{D} + \mathrm{i}[\boldsymbol{D}\boldsymbol{G}] \;, \tag{7.45}$$

where $G_i = \frac{-1}{|\varepsilon^0|}\varepsilon^0_{ik} g_k$, $|\varepsilon^0|$ is the determinant of tensor ε^0_{ik}. We now choose a new system of coordinates in which the z' axis coincides with the direction of the wave vector, and the x' and y' axes are directed along the main axes of two-dimensional tensor $(\varepsilon^0_{\alpha\beta})^{-1}$. Now, from the coupling equation (7.45) and the relation $D_{x',y'} = n^2 E_{x',y'}$ (since $\boldsymbol{D} \perp \boldsymbol{n}$), we obtain a set of equations

$$\left(\frac{1}{n_{01}^2} - \frac{1}{n^2}\right) D_{x'} + \mathrm{i}G_{z'} D_{y'} = 0 \;, \quad \left(\frac{1}{n_{02}^2} - \frac{1}{n^2}\right) D_{y'} - \mathrm{i}G_{z'} D_{x'} = 0 \;, \tag{7.46}$$

where $1/n_{01}^2$ and $1/n_{02}^2$ are the main values of tensor $(\varepsilon^0_{\alpha\beta})^{-1}$. The condition for the existence of a solution will be the vanishing of the determinant of this system, whence we find, for the refractive indices,

$$\frac{1}{n^2} = \frac{1}{2}\left(\frac{1}{n_{01}^2} + \frac{1}{n_{02}^2}\right) \pm \left[\frac{1}{4}\left(\frac{1}{n_{01}^2} - \frac{1}{n_{02}^2}\right)^2 + G_{z'}^2\right]^{1/2} \;. \tag{7.47}$$

As follows from (7.46, 47), the ellipticity of the waves propagating in the crystal is equal to $(D_{y'}/D_{x'})_+ = \mathrm{i}k$ and $(D_{y'}/D_{x'})_- = -\mathrm{i}/k$, where

Table 7.4. Gyration tensors g_{ik} for crystals of different classes of symmetry [7.22]

Triclinic syngony
Class 1

$$\begin{Vmatrix} g_{11} & g_{12} & g_{13} \\ g_{21} & g_{22} & g_{23} \\ g_{31} & g_{32} & g_{33} \end{Vmatrix}$$

Monoclinic syngony

Class 2 ($2 \parallel X_2$)

$$\begin{Vmatrix} g_{11} & 0 & g_{13} \\ 0 & g_{22} & 0 \\ g_{31} & 0 & g_{33} \end{Vmatrix}$$

Class 2 ($2 \parallel X_3$)

$$\begin{Vmatrix} g_{11} & g_{12} & 0 \\ g_{21} & g_{22} & 0 \\ 0 & 0 & g_{33} \end{Vmatrix}$$

Class m ($m \perp X_2$)

$$\begin{Vmatrix} 0 & g_{12} & 0 \\ g_{21} & 0 & g_{23} \\ 0 & g_{32} & 0 \end{Vmatrix}$$

Class m ($m \perp X_3$)

$$\begin{Vmatrix} 0 & 0 & g_{13} \\ 0 & 0 & g_{23} \\ g_{31} & g_{32} & 0 \end{Vmatrix}$$

Orthorhombic syngony and class $\bar{4}2m$ of tetragonal syngony

Class 222

$$\begin{Vmatrix} g_{11} & 0 & 0 \\ 0 & g_{22} & 0 \\ 0 & 0 & g_{33} \end{Vmatrix}$$

Classes $mm2$, $\bar{4}2m$ ($m \perp X_1, X_2$)

$$\begin{Vmatrix} 0 & g_{12} & 0 \\ g_{21} & 0 & 0 \\ 0 & 0 & 0 \end{Vmatrix}$$

Trigonal, hexagonal, and tetragonal syngonies

Class $\bar{4}$

$$\begin{Vmatrix} g_{11} & g_{12} & 0 \\ g_{12} & -g_{11} & 0 \\ 0 & 0 & 0 \end{Vmatrix}$$

Class $\bar{4}2m$

$$\begin{Vmatrix} g_{11} & 0 & 0 \\ 0 & -g_{11} & 0 \\ 0 & 0 & 0 \end{Vmatrix}$$

Classes 3, 4, 6

$$\begin{Vmatrix} g_{11} & g_{12} & 0 \\ -g_{12} & g_{11} & 0 \\ 0 & 0 & g_{33} \end{Vmatrix}$$

Classes 32, 422, 622

$$\begin{Vmatrix} g_{11} & 0 & 0 \\ 0 & g_{11} & 0 \\ 0 & 0 & g_{33} \end{Vmatrix}$$

Classes $3m$, $4mm$, $6mm$

$$\begin{Vmatrix} 0 & g_{12} & 0 \\ -g_{12} & 0 & 0 \\ 0 & 0 & 0 \end{Vmatrix}$$

Table 7.4 (continued)

Cubic syngony
Classes 23, 432
$\begin{Vmatrix} g_{11} & 0 & 0 \\ 0 & g_{11} & 0 \\ 0 & 0 & g_{11} \end{Vmatrix}$

$$k = \frac{1}{G_{z'}}\left\{\frac{1}{2}\left(\frac{1}{n_{01}^2}-\frac{1}{n_{02}^2}\right)-\left[\frac{1}{4}\left(\frac{1}{n_{01}^2}-\frac{1}{n_{02}^2}\right)^2+G_{z'}^2\right]^{1/2}\right\}. \tag{7.48}$$

Hence, we conclude that in optically active crystals two elliptically polarized waves with ellipticity $k = b/a$, crossed ellipse axes, and opposite scanning directions can propagate in each direction with different velocities. The main ellipse axes coincide with the direction of the oscillations in an identical inactive crystal. On determining the value of k, pseudotensor g_{ik} may be considered symmetric.

The solution of the boundary problem on light transmission through a plate of an active crystal gives the following expression for the angle of rotation of the major axis of the resultant elliptical oscillation:

$$\tan 2\beta = \frac{\tan 2\alpha\cos\delta - \dfrac{2k}{1+k^2}\sin\delta}{\left(\dfrac{1-k^2}{1+k^2}\right)^2+\left(\dfrac{2k}{1+k^2}\right)^2\cos\delta+\dfrac{2k}{1+k^2}\tan 2\alpha\sin\delta}, \tag{7.49}$$

where α is the azimuth of the incident light, and δ is the phase difference. This expression differs from (7.24), obtained for an inactive crystal, by the terms proportional to k.

In the direction of the optic axis (when $n_{01} = n_{02} = n_o$), $k = -1$ (7.45, 46). Circularly polarized waves with opposite scanning directions and different velocities (if $G_{z'} \neq 0$) propagate along it. This leads to a rotation of the polarization plane of the initially linearly polarized wave, i.e., the optical activity is then manifested in the rotation of the polarization plane. Rotation angle θ is proportional to the difference between the indices $n' - n'' \approx n_o^3 G_{z'}$ of the circularly polarized waves and to the length of the light path in the crystal, l: $\theta = \pi G_{z'} n_o^3 l/\lambda$. Through rotation of the polarization plane, the optical activity was first discovered on quartz crystals in 1811 by Arago. The optical activity of crystals is usually characterized by *specific rotation* $\varrho = \theta/l$, i.e., the rotation of the polarization plane by a unit length. The specific rotation of some crystals is listed in Table 7.5.

Table 7.5. Specific rotation ϱ of the polarization plane of some crystals [7.24]

Crystal	Symmetry class	ϱ, deg/mm
Sodium uranylacetate $NaUO_2(C_2H_3O_2)_3$	23	1.48
Sodium chlorate $NaClO_3$	23	3.16
Lithium iodate $LiIO_3$	6	102.9
Paratellurite TeO_2	422	104.9
Quartz SiO_2	32	21.3
Benzyl $C_{14}H_{10}O_2$	32	25.0
Cinnabar HgS	32	560
Rochelle salt $KNaC_4H_4O_6 \cdot 4\ H_2O$	222	1.35
Iodous acid α-HIO_3	222	47.4
Tartaric acid $C_4H_4O_6$	2	10.7
Sugar $C_{12}H_{22}O_{11}$	2	$\varrho_1 = 22.0$[a] $\varrho_2 = -6.4$

[a] The values ϱ_1 and ϱ_2 correspond to the two optic axes.

During the transmission of a light through a plate cut out parallel to the optic axis, when the incident light is linearly polarized parallel or perpendicular to the plane containing the optic axis ($\alpha = 0$), the emerging light will be polarized elliptically, and the direction of the main axis of the oscillation ellipse, β, will oscillate about the initial direction of the oscillations, depending on the path difference of the two waves in the crystal, δ [7.23]. Figure 7.25 presents the result of calculations for a quartz crystal cut out parallel to the optic axis, when the path difference changes owing to the change in the wavelength of the incident light. The established dependence is described by the equation $\tan 2\beta = -2k \sin\delta$ (for small values of k).

We shall now give the values of the components of the gyration tensor in the main system of coordinates for right- and left-handed quartz (class 32) referring to the wavelength of 510 μm: $g_{11} = \pm 5.82 \times 10^{-5}$, and $g_{33} = \mp 12.96 \times 10^{-5}$. From these values we find, for specific rotation of the polarization plane along the optic axis, $\varrho = g_{33}\pi/n_o\lambda \approx 30$ deg/mm ($n_o \approx 1.5$). If the wave vector is perpendicular to the optic axis, we obtain for the wave ellipticity (k) an approximate value $k \approx g_{11}/2n_o(n_e - n_o) \approx 0.002$ ($n_e - n_o \approx 0.01$), which means that waves with small ellipticity propagate in the direction perpendicular to the axis.

Thus, the most conspicuous manifestation of the optical activity in the form of rotation of the polarization plane is observed in light propagation along the optic axis. In all the other directions of light propagation the optical activity is manifested in additional ellipticity of the transmitted light as compared with that due to linear birefringence, and its observation is rather difficult.

Crystals of enantiomorphous classes of symmetry (1, 2, 222, 3, 4, 6, 32, 422, 622, 23, 432) can exist in two modifications (right and left handed), and thus the optical activity in such crystals may have opposite signs.

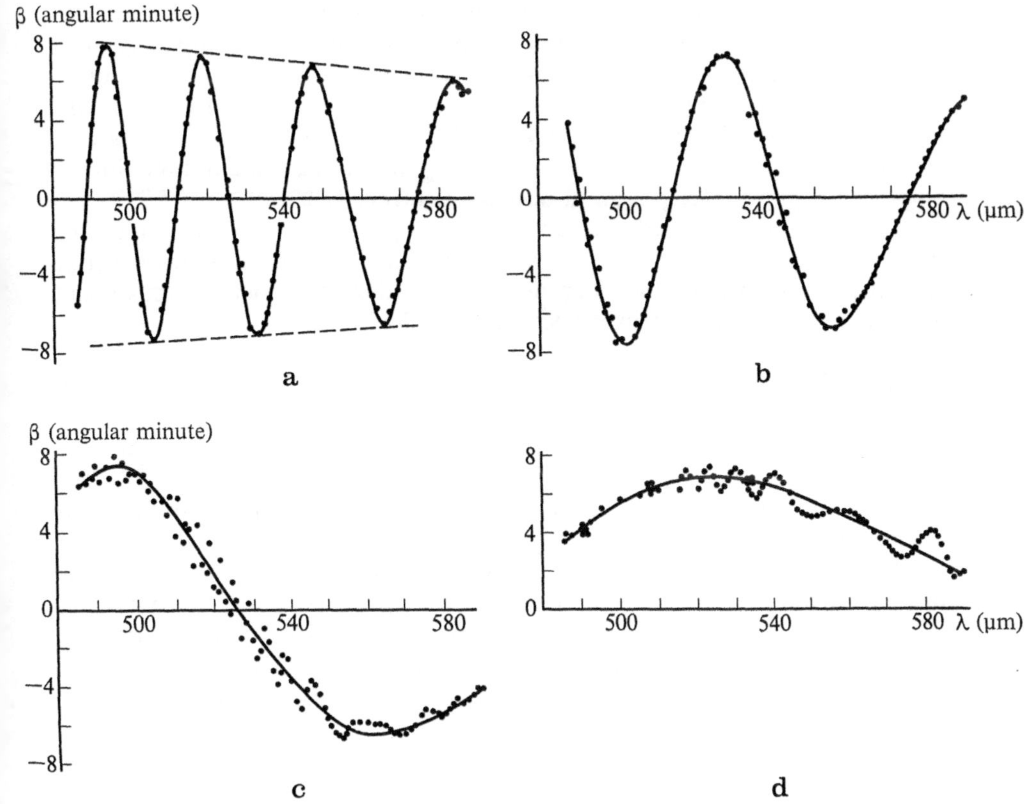

Fig. 7.25. Dependence of angle β, defining the orientation of the major axis of the polarization ellipse of the transmitted light, on wavelength λ for plates of right-handed quartz of various thickness d: **(a)** $d = 1$ mm, **(b)** $d = 0.5$ mm, **(c)** $d = 0.2$ mm, **(d)** $d = 0.1$ mm [7.25]

In crystals of intermediate syngonies, rotation is possible only along the optic axis. In biaxial crystals, rotation along different axes may differ both in sign and magnitude, depending on the symmetry and disposition of the plane of the optic axes.

Crystals with symmetries m, $mm2$, $\bar{4}$, and $\bar{4}2m$ are of interest. In crystals of classes $\bar{4}$ and $\bar{4}2m$, rotation along the axis is forbidden by the symmetry, and the optical activity can be observed only in directions differing from the optic axis. For instance, in crystals of cadmium thiogallate $CdGa_2S_4(\bar{4})$ at $\lambda = 0.487$ μm, rotation of the polarization plane was observed (at this wavelength the crystal becomes isotropic), and the components of the gyration tensor were calculated: $g_{11} = 11.7 \times 10^{-5}$ and $g_{12} = 7.8 \times 10^{-5}$ [7.26].

In crystals of classes m and $mm2$, rotation along optic axes is possible when the optic axes do not lie in the plane of symmetry. The rotation along both axes is then similar in magnitude and opposite in sign.

Crystals with symmetries $3m$, $4mm$ and $6mm$ (planar classes of intermediate syngonies), for which gyration pseudotensor g_{ik} is completely antisymmetric, require special consideration. They are called weakly gyrotropic. In these crystals, along any direction of the normal, vector $\boldsymbol{E}$ of the ordinary wave is polarized linearly. Vector $\boldsymbol{E}$ of the extraordinary wave is polarized elliptically. The optical activity in such crystals is also manifested in the ellipticity of the reflected light at oblique incidence [7.27].

The foregoing referred exclusively to transparent crystals, for which the dielectric constant and gyration tensors were real (7.44). If an optically active crystal is an absorbing one, then not only the dielectric constant tensor but also the gyration pseudotensor becomes complex:

$$g_{ik} = g'_{ik} + \mathrm{i} g''_{ik} \ . \tag{7.50}$$

The real part describes the circular birefringence, and the complex part, the circular dichroism, i.e, the different absorption of the right- and left-circularly polarized waves. The simultaneous presence of linear and circular dichroism results in elliptical dichroism.

The exact theory of optical activity of crystals, the inclusion of spatial dispersion $\varepsilon(\boldsymbol{k}, \omega)$, the laws of conservation of energy, and the boundary conditions are discussed in detail in the papers by *Federov* [7.28]. *Bokut'* et al. [7.29], *Agranovich* and *Ginzburg* [7.30]. In his survey, *Kizel'* demonstrated the possibility of utilizing the optical activity and circular dichroism for structure investigations [7.31].

7.7 Electro-optical Properties of Crystals

In a constant external electric field $\boldsymbol{E}$, an isotropic solid is known to become optically anisotropic [7.7]. To describe this phenomenon we assume the dielectric constant tensor to be dependent on electric field $\varepsilon_{ik}(\boldsymbol{E})$ and expand it into a series in powers of $\boldsymbol{E}$. For an isotropic body we can compose from the components of vector $\boldsymbol{E}$ two symmetric second-rank tensors $\alpha\delta_{ik}E^2$ and $\beta E_i E_k$ (α and β being scalars). If we add them to initial tensor $\varepsilon^0\delta_{ik}$, we get

$$\varepsilon_{ik} = (\varepsilon^0 + \alpha E^2)\delta_{ik} + \beta E_i E_k \ . \tag{7.51}$$

It is easy to see that one of the main axes of this tensor coincides with the direction of field $\boldsymbol{E}$, the respective main value of the tensor is equal to $\varepsilon_{\parallel} = \varepsilon^0 + (\alpha + \beta)E^2$, and the remaining two main values are $\varepsilon_{\perp} = \varepsilon^0 + \alpha E^2$. From this we conclude that an isotropic body in an electric field behaves as an optically uniaxial crystal.

In contrast to isotropic bodies, an electro-optical effect, linear in field $\boldsymbol{E}$, is possible in crystals. It is then more convenient to use reciprocal tensor

$\varepsilon_{ik}^{-1} = a_{ik}$ (polarization tensor). Changes in its components under the effect of an electric field will be written

$$\Delta a_{ik} = r_{ikl} E_l \ . \tag{7.52}$$

Here r_{ikl} is a third-rank tensor symmetric with respect to the first two subscripts, which is called the tensor of the electro-optical coefficients. The symmetry properties of this tensor coincide with the symmetry of the pseudo-electric coefficient tensors, and therefore the *linear electro-optical effect* (Pockels effect) may be found in crystals of the same 20 noncentrosymmetric classes which permit the piezoelectric effect (Sect. 3.3, [7.32]).

The type of these tensors for different classes in matrix form has been described above. Let us consider, as an example, the change in the surface of tensor a_{ik}, i.e., in optic indicatrix (Sect. 7.1), in the case of cubic crystals of class $\bar{4}3m$, when the acting electric field $\boldsymbol{E}$ is applied along the threefold axis of symmetry: $E_x = E_y = E_z = E/\sqrt{3}$. The optic indicatrix equation will then be [by (7.52)]

$$\frac{x^2+y^2+z^2}{n_0^2} + \frac{2r_{41}E}{\sqrt{3}}(yz+xz+xy) = 1 \ .$$

Reducing it to the main axes, we find the main refractive indices $n_1 = n_0 + n_0^3 r_{41} E/2\sqrt{3} = n_2$ and $n_3 = n_0 - n_0^3 r_{41} E/\sqrt{3}$, so that now the optic indicatrix is an ellipsoid with a rotation axis along the [111] direction, and the cubic crystal becomes optically uniaxial under the effect of the field thus directed. If the light propagates perpendicular to the direction of the field, we can obtain the expression $\delta = \sqrt{3}\,\pi l n_0^3 r_{41} E/\lambda$ for the field-induced phase difference of two waves in a crystal (l stands for the beam path in the crystal). If the electric field is applied in an arbitrary direction, the cubic crystal becomes optically biaxial.

Manifestations of the linear electro-optical effect at different orientations of the applied field and the direction of light propagation in less symmetric crystals can be considered similarly. It is found then that in optically uniaxial crystals (with the exception of $\bar{4}$ and $\bar{4}2m$) the field applied along the optic axis leaves the crystal uniaxial, and the field of arbitrary direction changes the shape of the indicatrix to a triaxial ellipsoid [7.33, 34].

All the crystals of dielectrics, as well as isotropic media, may also have a quadratic electro-optical effect when changes in polarization constants Δa_{ik} are proportional to the square of the electric field intensity (Kerr electro-optical effect):

$$\Delta a_{ij} = R_{ijkl} E_k E_l \ . \tag{7.53}$$

Fourth-rank tensor R_{ijkl}, i.e., the tensor of quadratic electro-optical coefficients, remains symmetric on permutation of individual subscripts in pairs

Table 7.6. Linear electro-optical coefficients of some crystals [7.33]

Crystal	Symmetry class	Electro-optical coefficients, 10^{-8} CGSE units	Wavelength λ, μm
Sphalerite ZnS	$\bar{4}3m$	$r_{41} = 6$	0.589
Copper chloride CuCl	$\bar{4}3m$	$r_{41} = 18.3$	0.56
Sodalite $Na_8\ [AlSiO_4]_6Cl_2$	$\bar{4}3m$	$r_{41} = 5$	0.535
Sodium chlorate $NaClO_3$	23	$r_{41} = 1.2$	–
Sodium uranylacetate $NaUO_2\ (CH_3COO)_3$	23	$r_{41} = 2.6$	–
Lithium niobate $LiNbO_3$	$3m$	$r_{13} = 25.8$ $r_{33} = 92.4$ $r_{42} = 84$ $r_{22} = 21$	0.633
Quartz SiO_2	32	$r_{11} = 0.59$ $r_{41} = 1.4$	0.59
Potassium dihydrophosphate KH_2PO_4 (KDP)	$\bar{4}2m$	$r_{63} = -30$ $r_{41} = 26$	0.54
Ammonium dihydrophosphate $NH_4H_2PO_4$ (ADP)	$\bar{4}2m$	$r_{63} = -25$ $r_{41} = 62$	0.54
Barium titanate $BaTiO_3$	$4mm$	$r_{13} = 24$ $r_{33} = 84$	0.633

(ij) and (kl), and the crystal symmetry, as usual, imposes limitations on the number of independent tensor components. For instance, only three distinct components of tensor R_{ijkl} exist for cubic crystals of classes $\bar{4}3m$, 432, and $m3m$ [7.35].

Measurements of the electro-optical coefficients are made by analyzing changes in conoscopic patterns and by the refractometric and interferometric techniques [7.36]. The photometric method is usually employed to measure the changes in the intensity of the light transmitted through a plate carved in a special way from a crystal placed between linear or circular polarizers, with an electric field applied to the crystal. Electro-optical coefficients can sometimes be calculated simply from the voltage necessary for obtaining a path difference of $\lambda/2$.

Among the cubic linear electro-optical crystals the most numerous are crystals with the zinc blende structure (ZnS, ZnSe, and ZnTe). All of them have high refractive indices ($n>2$) and are transparent in the visible and infrared regions of the spectrum (up to 12 μm for bulky specimens). Quantity r_{41} for ZnS has a value of $\sim 6 \times 10^{-8}$ CGSE units at $\lambda = 0.589$ μm (sodium radiation line) (Table 7.6).

Electro-optical crystals of the group of potassium dihydrophoshate KH_2PO_4 (KDP) of symmetry class $\bar{4}2m$ find the most extensive practical application. The region of transparency of a KDP crystal is 0.3 μm $<\lambda<1.15$ μm, refractive index ~ 1.5, birefringence $n_o - n_e \approx 0.04$, and electro-optical coefficient $r_{63} \approx -30 \times 10^{-8}$ CGSE units.

The electro-optical effect in ferroelectric crystals has specific features associated with spontaneous polarization and the domain structure, and therefore it is used in the corresponding procedures of experiments on phase transitions and other specific properties of ferroelectric crystals.

The extremely high response of the electro-optical effect explains the interest in the practical utilization of electro-optical crystals in devices for controlling certain light beam parameters, for instance, in light modulators with respect to intensity [7.37] and polarization parameters: azimuth and ellipticity, the angle of refraction or of internal reflection at the crystal boundary. Changes in the indicated parameters are due to changes in the magnitudes and directions of the main axes of the ellipsoid of the optic indicatrix under the effect of a constant or a variable electric field.

Along with the electro-optical effect, also possible in crystals is the electrogyration effect, i.e., a change or appearance of an optical activity under the effect of an electric field [7.38]. This effect is described by an expansion of the gyration pseudotensor in powers of the electric field

$$g_{ij}(E) = g_{ij}^0 + A_{kij} E_k + B_{ijkl} E_k E_l + \dots , \tag{7.54}$$

where g_{ij}^0 is the gyration pseudotensor in the absence of a field. For the symmetric part of tensor g_{ik}, A_{ijk} is a third-rank pseudotensor symmetric with respect to two subscripts ($A_{ijk} = A_{ikj} = A_{j\mu}$), whose form is given in Table 7,7, and B_{ijkl} is a fourth-rank pseudotensor symmetric with respect to the first and second pairs of subscripts.

Linear electrogyration ($A_{ijk} \neq 0$) can be exhibited by crystals of all classes with the exception of three: $\bar{4}3m$, 432, and $m3m$, in which $A_{ijk} = 0$.

Experimental observation of electrogyration is rather difficult because of the attending electro-optical effect. Linear electrogyration in the direction of the optic axis is not complicated by linear or quadratic electro-optical effects only in crystals of classes 3, $\bar{3}$, 4, $\bar{4}$, $4/m$, 6, $\bar{6}$, and $6/m$, when the field acts along the optic axis; in cubic crystals of classes 23 and $m\bar{3}$, when the field acts in the ⟨111⟩ direction; and in crystals of class $3m$, when the field acts in the ⟨011⟩, ⟨110⟩, and ⟨101⟩ directions. In experiments, uniaxial inactive crystals of classes $\bar{4}$, $4/m$, $\bar{6}$, and $6/m$ must be most convenient when the field is applied and the light propagates along the optic axis. Precisely in this manner, the appearance of electrogyration was discovered in a crystal of $PbMoO_4$ (class $4/m$): when a field of 10 kV was applied, the light polarization plane rotated through an angle of 1° at $\lambda = 0.633$ μm in a 2 mm plate [7.40].

Quadratic electrogyration is described by even-rank pseudotensor B_{ijkl}, and therefore it cannot occur in centrosymmetric crystals. Experimental

Table 7.7. Pseudotensor of linear electrogyration A_{ikl} for crystals of different classes of symmetry [7.41] ($A_{ilk} = A_{ikl} = A_{i\mu}$)

Triclinic syngony

Classes 1 and $\bar{1}$

$$\left\| \begin{matrix} A_{11} & A_{12} & A_{13} & A_{14} & A_{15} & A_{16} \\ A_{21} & A_{22} & A_{23} & A_{24} & A_{25} & A_{26} \\ A_{31} & A_{32} & A_{33} & A_{34} & A_{35} & A_{36} \end{matrix} \right\|$$

Monoclinic syngony

Classes 2, m, and $2/m$ ($2 \parallel X_2$, $m \perp X_2$)

$$\left\| \begin{matrix} 0 & 0 & 0 & A_{14} & 0 & A_{16} \\ A_{21} & A_{22} & A_{23} & 0 & A_{25} & 0 \\ 0 & 0 & 0 & A_{34} & 0 & A_{36} \end{matrix} \right\|$$

Classes 2, m, and $2/m$ ($2 \parallel X_3$, $m \perp X_3$)

$$\left\| \begin{matrix} 0 & 0 & 0 & A_{14} & A_{15} & 0 \\ 0 & 0 & 0 & A_{24} & A_{25} & 0 \\ A_{31} & A_{32} & A_{33} & 0 & 0 & A_{36} \end{matrix} \right\|$$

Orthorhombic syngony

Classes 222, $mm2$, and mmm

$$\left\| \begin{matrix} 0 & 0 & 0 & A_{14} & 0 & 0 \\ 0 & 0 & 0 & 0 & A_{25} & 0 \\ 0 & 0 & 0 & 0 & 0 & A_{36} \end{matrix} \right\|$$

Trigonal syngony

Classes 3 and $\bar{3}$

$$\left\| \begin{matrix} A_{11} & -A_{11} & 0 & A_{14} & A_{15} & -2A_{22} \\ -A_{22} & A_{22} & 0 & A_{15} & -A_{14} & -2A_{11} \\ A_{31} & A_{31} & A_{33} & 0 & 0 & 0 \end{matrix} \right\|$$

Classes 32, $3m$, and $\bar{3}m$ ($2 \parallel X_1$, $m \perp X_1$)

$$\left\| \begin{matrix} A_{11} & -A_{11} & 0 & A_{14} & 0 & 0 \\ 0 & 0 & 0 & 0 & -A_{14} & -2A_{11} \\ 0 & 0 & 0 & 0 & 0 & 0 \end{matrix} \right\|$$

Tetragonal and hexagonal syngonies

Classes 4, $\bar{4}$, $4/m$, 6, $\bar{6}$, and $6m$

$$\left\| \begin{matrix} 0 & 0 & 0 & A_{14} & A_{15} & 0 \\ 0 & 0 & 0 & A_{15} & -A_{14} & 0 \\ A_{31} & A_{31} & A_{33} & 0 & 0 & 0 \end{matrix} \right\|$$

Classes 422, $4mm$, $\bar{4}2m$, $4/mmm$, 622, $6mm$, $\bar{6}m2$, and $6/mmm$

$$\left\| \begin{matrix} 0 & 0 & 0 & A_{14} & 0 & 0 \\ 0 & 0 & 0 & 0 & -A_{14} & 0 \\ 0 & 0 & 0 & 0 & 0 & 0 \end{matrix} \right\|$$

Cubic syngony

Classes 23 and $m3$

$$\left\| \begin{matrix} 0 & 0 & 0 & A_{14} & 0 & 0 \\ 0 & 0 & 0 & 0 & A_{14} & 0 \\ 0 & 0 & 0 & 0 & 0 & A_{14} \end{matrix} \right\|$$

investigations of quadratic electrogyration is complicated by linear electrogyration and the electro-optical effect. Quadratic electrogyration can be detected in those directions in which it can be observed without these effects or by resolving a complex effect into its components. It is thus, for instance, that the quadratic electrogyration effect was discovered in quartz, when the light propagated along the optic Z axis and the field along the X axis. This effect proved to be commensurate with the electro-optical one.

7.8 Magneto-optical Properties of Crystals

If an external magnetic field $\boldsymbol{H}$ is applied to a crystal, dielectric constant tensor ε_{ik} must depend on the magnetic field: $\varepsilon_{ik} = \varepsilon_{ik}(\boldsymbol{H})$. The principle of symmetry of the kinetic coefficients in the presence of a magnetic field [7.1] requires that

$$\varepsilon_{ik}(\boldsymbol{H}) = \varepsilon_{ki}(-\boldsymbol{H}) \; . \tag{7.55}$$

At the same time, in the absence of absorption, tensor ε_{ik} remains Hermitian, as before, i.e., $\varepsilon_{ik} = \varepsilon^*_{ki}$ ($*$ is a complex conjugation); this means that its real and imaginary parts must be symmetric and antisymmetric tensors $\varepsilon_{ik} = \varepsilon'_{ik} + \mathrm{i}\varepsilon''_{ik}$, $\varepsilon'_{ik} = \varepsilon'_{ki}$, $\varepsilon''_{ik} = -\varepsilon''_{ki}$. Taking into account (7.55), it is easy to establish that ε'_{ik} is an even function and ε''_{ik} an odd function of $\boldsymbol{H}$. Reciprocal tensor $\varepsilon^{-1}_{ik} = a'_{ik} + \mathrm{i}a''_{ik}$ possesses the same properties.

The coupling equations relating the vectors of electric field intensity with electrical induction $\boldsymbol{E}$ and $\boldsymbol{D}$ take the form

$$E_i = (a'_{ik} + \mathrm{i}a''_{ik})D_k = a'_{ik}D_k + \mathrm{i}[\boldsymbol{D}\boldsymbol{G}]_i \; ; \tag{7.56}$$

here $\boldsymbol{G}(\boldsymbol{H})$ is an axial vector related to $a''_{ik}(\boldsymbol{H})$ by completely antisymmetric unit tensor e_{ikl}, $a''_{ik} = e_{ikl}G_l$.

Let us consider the basic optical properties of a medium with such coupling equations. As before (Sect. 7.5), we choose the direction of vector $\boldsymbol{n}$ for the direction of the Z' axis, and then $D_{z'} = 0$; $n'_x = n'_y = 0$, and for the transverse components of vector $\boldsymbol{D}$, we obtain $\left(a'_{\alpha\beta} + \mathrm{i}a''_{\alpha\beta} - \frac{1}{n^2}\delta_{\alpha\beta}\right)D_\beta = 0$ ($\alpha, \beta = x', y'$, summation over β). We direct the X' and Y' axis along the main axes of tensor $a'_{\alpha\beta}$, whose main values will be denoted by n_{01}^{-2} and n_{02}^{-2}. The condition of vanishing of the determinant of this system yields the equation

$$\left(\frac{1}{n^2} - \frac{1}{n_{01}^2}\right)\left(\frac{1}{n^2} - \frac{1}{n_{02}^2}\right) = G_{z'}^2 \; , \tag{7.57}$$

whose roots determine the refractive indices of the waves:

$$\frac{1}{n^2}=\frac{1}{2}\left(\frac{1}{n_{01}^2}+\frac{1}{n_{02}^2}\right)\pm\sqrt{\frac{1}{4}\left(\frac{1}{n_{01}^2}-\frac{1}{n_{02}^2}\right)^2+G_{z'}^2}\,. \tag{7.58}$$

The wave polarization is established by the fact that in one wave $(D_{y'}/D_{x'})_+ = \mathrm{i}k$, and in the other $(D_{y'}/D_{x'})_- = -\mathrm{i}/k$. Thus, both waves are (as in optically active crystals) elliptically polarized, with the ellipse axes ratio b/a equal to k. The polarization ellipses of the two waves are crossed and are scanned in opposite directions; here

$$k=\frac{1}{G_{z'}}\left[\frac{1}{2}\left(\frac{1}{n_{01}^2}-\frac{1}{n_{02}^2}\right)-\sqrt{\frac{1}{4}\left(\frac{1}{n_{01}^2}-\frac{1}{n_{02}^2}\right)^2+G_{z'}^2}\,\right]. \tag{7.59}$$

Because the real part a'_{ik} is even, the first correction terms in the expansion of components a'_{ik} into a series in the powers of weak field H will be of the second order, while for odd function a''_{ik} they will be of the first order. Therefore, it can be approximately assumed that the dependence on the magnetic field is contained in the imaginary part of tensor a''_{ik}.

If the light propagates in the direction of the optic axis of the crystals (judging by tensor a'_{ik}), then $n_{01}=n_{02}$; for the refractive indices of the waves we get $n_{\pm}^2 \approx n_0^2 \mp n_0^4 G_{z'}$, and for the ratio $(D_{y'}/D_{x'})_{\pm} = \pm\mathrm{i}$. Thus, two circularly polarized waves propagate at two different velocities along the optic axis of the crystal in the magnetic field, as well as in an isotropic medium. As a result, after a plane-polarized wave with a frequency $\nu = c/\lambda$ passes through a parallel-sided plate of a crystal with thickness l, the wave remains linearly polarized, with the polarization direction rotated through an angle

$$\theta = \pi G_{z'} n_0^3 l/\lambda\ ,$$

which is the so-called *Faraday effect*. The same expression can be represented as $(G_{z'} \sim H)$

$$\theta = VHl\ ,$$

where V is the Verdet constant and H is the magnetic field intensity. The values of the Verdet constant at room temperature for some materials are listed below [7.42].

Material:	H_2O	Quartz SiO_2	Cs_2	EuF_2	$Y_3Fe_5O_{12}$
V [ang. min/Oe · cm]	0.019	0.025	0.067	−1.0	34.5
λ [μm]	0.486	0.486	0.486	0.6	0.7

Note that if a light beam passes along the optic axis of the crystal twice, back and forth, the complete rotation of the polarization plane will be twice as large as with a single passage, since $\boldsymbol{G}$ is an axial vector.

In directions differing from the optic axis, the effect of the magnetic field is weak as compared with that of the birefringence and is manifested in the small ellipticity of the waves.

In diamagnetic substances such as quartz, magnetic fields on the order of $10^4 - 10^5$ Oe are required to rotate the polarization plane through several tens of degrees. Paramagnetics (for instance, EuF_2) have much larger Verdet constants, and V increases considerably with a decrease in temperature. The Verdet constant is rather high in ferromagnetics because these have spontaneous magnetization regions, and the magnetic moments of these regions orient themselves in the direction of the field under the effect of the external magnetic field. Therefore, in ferromagnetics the rotation of the polarization plane is proportional to the magnetization. In distinction to other substances, the ferromagnetics exhibit the *equatorial Kerr effect*, namely, the intensity of the light reflected from a ferromagnetic also depends on its magnetization. The magnetization vector must be perpendicular to the plane of incidence. The change in the intensity of the reflected light depends on the polarization of the incident wave. The greatest change in intensity (by 30%) has been obtained on the ferromagnetic EuO for $\lambda = 0.8$ μm as a result of its remagnetization at 10 K.

In diamagnetic and paramagnetic crystals, if axial vector $\boldsymbol{G}(\boldsymbol{H})$ and the direction of light propagation are mutually perpendicular, as a first approximation the magnetic field has no effect on the light. Here one must take into account the additions (quadratic in $\boldsymbol{H}$) in the expansion of the components of tensor a_{ik}, which describes artificial birefringence. If the light propagates along the optic axis, a birefringence due exclusively to the magnetic field arises. If the light propagates in an arbitrary direction, an additional birefringence, as compared with the usual, takes place. This is known as the *Cotton-Mouton effect*, which is similar to Kerr quadratic electro-optical effect.

Effects occurring during the interaction of a magnetic field with matter are widely used for scientific and technical purposes. They are employed in investigating the structure and properties of various compounds. The Faraday effect lies at the basis of devices used for light modulation and in laser techniques, such as optical valves and optical decoupling circuits.

7.9 Piezo-optical Properties of Crystals

An isotropic solid subjected to elastic deformations becomes optically anisotropic, as was first discovered by Brewster in 1818. For a phenomenological description of the elasto-optic effect, the dielectric constant tensor must be supplemented with terms proportional to the components of deformation tensor u_{ik}. For symmetry considerations, in an isotropic medium such terms additional to initial tensor $\varepsilon^0 \delta_{ik}$ can be only $c_1 u_{ik}$ and $c_2 u_{ll} \delta_{ik}$ (the summation is made over index l), so that

$$\varepsilon_{ik} = \varepsilon^0 \delta_{ik} + c_1 u_{ik} + c_2 u_{ll} \delta_{ik} \ . \tag{7.60}$$

Factors $c_1(\omega)$ and $c_2(\omega)$ are called the elasto-optic coefficients. If the main axes of the deformation tensor are chosen as the coordinate axes, then

$$\begin{aligned} \varepsilon_{xx} &= \varepsilon^0 + c_1 u_{xx} + c_2 u_{ll} \ , \\ \varepsilon_{yy} &= \varepsilon^0 + c_1 u_{yy} + c_2 u_{ll} \ , \\ \varepsilon_{zz} &= \varepsilon^0 + c_1 u_{zz} + c_2 u_{ll} \ . \end{aligned} \tag{7.61}$$

We can see that under the effect of a general-type deformation an isotropic medium becomes optically biaxial with three distinct main values of tensor ε_{ik}, and the main axes of the dielectric constant tensor coincide with those of the deformation tensor [7.43].

For crystals, we shall use reciprocal tensor $\varepsilon_{ik}^{-1} = a_{ik}$ (polarization tensor), and for the change in values a_{ik} under the effect of the deformation we write, in the general form,

$$\Delta a_{ik} = p_{ikrs} u_{rs} \ . \tag{7.62}$$

Here p_{ikrs} is a fourth-rank tensor symmetric with respect to the first two and the last two subscripts, which is called the tensor of *elasto-optic coefficients*. The crystal symmetry imposes limitations on the number of independent components of tensor p_{ikrs}; its type in matrix form for various classes of crystals is given in Table 7.8

One can similarly describe the changes in values a_{ik} under stresses

$$\Delta a_{ik} = \pi_{ikrs} \sigma_{rs} \ ,$$

where σ_{rs} are the components of the stress tensor; tensor π_{ikrs} is called the tensor of *piezo-optical coefficients*. Naturally, it possesses the same symmetry properties as tensor p_{ikrs}.

Let us consider, as an example, the piezo-optical properties of cubic crystals. The cubic crystals are divided into two groups in accordance with these properties: classes 23 and $m\bar{3}$ have four independent components, while the others have only three (Table 7.8). Suppose uniaxial tension σ acts on a first-group cubic crystal along axis 2. The equation of tensor ellipsoid ε_{ik}^{-1} (optic indicatrix) will be, instead of $x^2+y^2+z^2 = n_0^2$,

$$(x^2+y^2+z^2)/n_0^2 + \pi_{11}\sigma x^2 + \pi_{13}\sigma y^2 + \pi_{12}\sigma z^2 = 1 \ . \tag{7.63}$$

Table 7.8. Tensors of elasto-optic coefficients $p_{\lambda\mu}$ for crystals of various classes of symmetry [7.22], [7.35] ($p_{ijkl} = p_{jikl} = p_{ijlk} = p_{\lambda\mu}$)

Triclinic syngony

Classes 1 and $\bar{1}$

$$\begin{Vmatrix} p_{11} & p_{12} & p_{13} & p_{14} & p_{15} & p_{16} \\ p_{21} & p_{22} & p_{23} & p_{24} & p_{25} & p_{26} \\ p_{31} & p_{32} & p_{33} & p_{34} & p_{35} & p_{36} \\ p_{41} & p_{42} & p_{43} & p_{44} & p_{45} & p_{46} \\ p_{51} & p_{52} & p_{53} & p_{54} & p_{55} & p_{56} \\ p_{61} & p_{62} & p_{63} & p_{64} & p_{65} & p_{66} \end{Vmatrix}$$

Monoclinic syngony

Classes 2, m, and $2/m$

($2 \parallel X_2$, $m \perp X_2$)

$$\begin{Vmatrix} p_{11} & p_{12} & p_{13} & 0 & p_{15} & 0 \\ p_{21} & p_{22} & p_{23} & 0 & p_{25} & 0 \\ p_{31} & p_{32} & p_{33} & 0 & p_{35} & 0 \\ 0 & 0 & 0 & p_{44} & 0 & p_{46} \\ p_{51} & p_{52} & p_{53} & 0 & p_{55} & 0 \\ p_{61} & p_{62} & p_{63} & p_{64} & 0 & p_{66} \end{Vmatrix}$$

Classes 2, m, and $2/m$

($2 \parallel X_3$, $m \perp X_3$)

$$\begin{Vmatrix} p_{11} & p_{12} & p_{13} & 0 & 0 & p_{16} \\ p_{21} & p_{22} & p_{23} & 0 & 0 & p_{26} \\ p_{31} & p_{32} & p_{33} & 0 & 0 & p_{36} \\ 0 & 0 & 0 & p_{44} & p_{45} & 0 \\ 0 & 0 & 0 & p_{54} & p_{55} & 0 \\ p_{61} & p_{62} & p_{63} & 0 & 0 & p_{66} \end{Vmatrix}$$

Orthorhombic syngony

Classes 222, $mm2$, and mmm

$$\begin{Vmatrix} p_{11} & p_{12} & p_{13} & 0 & 0 & 0 \\ p_{21} & p_{22} & p_{23} & 0 & 0 & 0 \\ p_{31} & p_{32} & p_{33} & 0 & 0 & 0 \\ 0 & 0 & 0 & p_{44} & 0 & 0 \\ 0 & 0 & 0 & 0 & p_{55} & 0 \\ 0 & 0 & 0 & 0 & 0 & p_{66} \end{Vmatrix}$$

Trigonal syngony

Classes 3 and $\bar{3}$

$$\begin{Vmatrix} p_{11} & p_{12} & p_{13} & p_{14} & -p_{25} & p_{16} \\ p_{12} & p_{11} & p_{13} & -p_{14} & p_{25} & -p_{16} \\ p_{31} & p_{31} & p_{33} & 0 & 0 & 0 \\ p_{41} & -p_{41} & 0 & p_{44} & p_{45} & p_{52} \\ -p_{52} & p_{52} & 0 & -p_{45} & p_{44} & p_{41} \\ -p_{16} & p_{16} & 0 & p_{25} & p_{14} & p_{66} \end{Vmatrix}$$

Classes 32, $3m$, and $\bar{3}m$

$$\begin{Vmatrix} p_{11} & p_{12} & p_{13} & p_{14} & 0 & 0 \\ p_{12} & p_{11} & p_{13} & -p_{14} & 0 & 0 \\ p_{31} & p_{31} & p_{33} & 0 & 0 & 0 \\ p_{41} & -p_{41} & 0 & p_{44} & 0 & 0 \\ 0 & 0 & 0 & 0 & p_{44} & p_{41} \\ 0 & 0 & 0 & 0 & p_{14} & p_{66} \end{Vmatrix}$$

(Continued)

Table 7.8 (continued)

Tetragonal syngony

Classes 4, $\bar{4}$, and $4/m$

$$\left\|\begin{matrix} p_{11} & p_{12} & p_{13} & 0 & 0 & p_{16} \\ p_{12} & p_{11} & p_{13} & 0 & 0 & -p_{16} \\ p_{31} & p_{31} & p_{33} & 0 & 0 & 0 \\ 0 & 0 & 0 & p_{44} & p_{45} & 0 \\ 0 & 0 & 0 & -p_{45} & p_{44} & 0 \\ p_{61} & -p_{61} & 0 & 0 & 0 & p_{66} \end{matrix}\right\|$$

Classes 422, $4mm$, $\bar{4}2m$, and $4/mmm$

$$\left\|\begin{matrix} p_{11} & p_{12} & p_{13} & 0 & 0 & 0 \\ p_{12} & p_{11} & p_{13} & 0 & 0 & 0 \\ p_{31} & p_{31} & p_{33} & 0 & 0 & 0 \\ 0 & 0 & 0 & p_{44} & 0 & 0 \\ 0 & 0 & 0 & 0 & p_{44} & 0 \\ 0 & 0 & 0 & 0 & 0 & p_{66} \end{matrix}\right\|$$

Hexagonal syngony

Classes 6, $\bar{6}$, and $6/m$

$$\left\|\begin{matrix} p_{11} & p_{12} & p_{13} & 0 & 0 & p_{16} \\ p_{12} & p_{11} & p_{13} & 0 & 0 & -p_{16} \\ p_{31} & p_{31} & p_{33} & 0 & 0 & 0 \\ 0 & 0 & 0 & p_{44} & p_{45} & 0 \\ 0 & 0 & 0 & -p_{45} & p_{44} & 0 \\ -p_{16} & p_{16} & 0 & 0 & 0 & p_{66} \end{matrix}\right\|$$

Classes 622, $6mm$, $6m2$, and $6/mmm$

$$\left\|\begin{matrix} p_{11} & p_{12} & p_{13} & 0 & 0 & 0 \\ p_{12} & p_{11} & p_{13} & 0 & 0 & 0 \\ p_{31} & p_{31} & p_{33} & 0 & 0 & 0 \\ 0 & 0 & 0 & p_{44} & 0 & 0 \\ 0 & 0 & 0 & 0 & p_{44} & 0 \\ 0 & 0 & 0 & 0 & 0 & p_{66} \end{matrix}\right\|$$

Cubic syngony

Classes 23 and $m3$

$$\left\|\begin{matrix} p_{11} & p_{12} & p_{21} & 0 & 0 & 0 \\ p_{21} & p_{11} & p_{12} & 0 & 0 & 0 \\ p_{12} & p_{21} & p_{11} & 0 & 0 & 0 \\ 0 & 0 & 0 & p_{44} & 0 & 0 \\ 0 & 0 & 0 & 0 & p_{44} & 0 \\ 0 & 0 & 0 & 0 & 0 & p_{44} \end{matrix}\right\|$$

Classes 432, $\bar{4}3m$, and $m3m$

$$\left\|\begin{matrix} p_{11} & p_{12} & p_{12} & 0 & 0 & 0 \\ p_{12} & p_{11} & p_{12} & 0 & 0 & 0 \\ p_{12} & p_{12} & p_{11} & 0 & 0 & 0 \\ 0 & 0 & 0 & p_{44} & 0 & 0 \\ 0 & 0 & 0 & 0 & p_{44} & 0 \\ 0 & 0 & 0 & 0 & 0 & p_{44} \end{matrix}\right\|$$

The new main refractive indices can therefore be written

$$n_1 = n_0 - \frac{\pi_{11}}{2} n_0^3 \sigma \,, \quad n_2 = n_0 - \frac{\pi_{13}}{2} n_0^3 \sigma \,, \quad n_3 = n_0 - \frac{\pi_{12}}{2} n_0^3 \sigma \,,$$

and hence the crystal becomes biaxial. In second-group cubic crystals a uniaxial extension along the twofold axis turns them into optically uniaxial crystals (because for them $\pi_{12} = \pi_{13}$).

If second-group crystals are subjected to extension σ along the threefold axis (in the [111] direction), then, using the matrix of the piezo-optical coefficients, we find for the components of tensor $\varepsilon_{ik}^{-1} = a_{ik}$:

$$\begin{Vmatrix} a & b & b \\ b & a & b \\ b & b & a \end{Vmatrix}, \tag{7.64}$$

where $a = 1/n_0^2 + (\pi_{11} + 2\pi_{12})\sigma/3$, $b = \pi_{44}\sigma/3$.
Reducing it to the main axes, we obtain the principal values of the indices

$$\begin{aligned} n_1 = n_2 &= n_0 - \frac{(\pi_{11} + 2\pi_{12})\sigma n_0^3}{6} + \frac{\pi_{44}\sigma n_0^3}{6}, \\ n_3 &= n_0 - \frac{(\pi_{11} + 2\pi_{12})\sigma n_0^3}{6} - \frac{\pi_{44}\sigma n_0^3}{3}, \end{aligned} \tag{7.65}$$

and we see that under these stresses the crystal turns from cubic to uniaxial with the optic axes directed along [111].

If, during the observation of the piezo-optical effect in crystals of classes $\bar{4}3m$, 432, and $m3m$ under extension along [100], we direct the light along [001], the phase difference of the two waves in the crystal will be

$$\delta = \pi n_0^3 (\pi_{11} - \pi_{12})\sigma l/\lambda . \tag{7.66}$$

Under extension along [111] the phase difference for light waves propagating perpendicular to this direction is equal to

$$\delta = \pi n_0^3 \sigma l \pi_{44}/\lambda . \tag{7.67}$$

By determining these values experimentally we can find the difference $(\pi_{11} - \pi_{12})$ and the value of piezo-optical coefficient π_{44}. To determine the values π_{11} and π_{12} separately, we must make direct measurements of changes in index n_1 or n_2, for instance, by interferometric methods. The piezo-optical constants of some cubic crystals are given in Table 7.9. They are on the order of 10^{-11} cm^2/dyn; the elasto-optic coefficients are dimensionless and are on the order of 10^{-1}.

When considering the linear electro-optical effect in crystals we assumed the absence of crystal deformation (the crystal is clamped) and dealt with purely electrical changes in indicatrix, i.e., with the primary electro-optical effect. If the crystal is free, then, owing to the reciprocal piezo-electric effect, the electric field causes deformations, which, in turn, lead to a change in refractive index (*secondary electro-optical effect*). The effect caused by the

Table 7.9. Piezo-optical coefficients $\pi_{\lambda\mu}$ (10^{-11} cm^2/dyn) and elasto-optic coefficients $P_{\lambda\mu}$ of cubic crystals for $\lambda = 0.589$ [7.35]

Crystal	Class	π_{11}	π_{12}	π_{13}	π_{44}	P_{11}	P_{12}	P_{13}	P_{44}
Potash alum	$m\bar{3}$	3.7	9.1	8.5	−0.65	0.27	0.35	0.34	−0.0056
Lead nitrate	$m\bar{3}$	70.21	89.34	82.05	−1.39	8.5	8.78	8.67	−0.0191
Sodium chloride	$m3m$	0.25		1.46	−0.85	0.137		0.178	−0.0108
Fluorite	$m3m$	−0.29		1.16	0.698	0.0558		0.228	0.0236
Diamond	$m3m$	−0.43		0.37	−0.27	−0.125		0.325	−0.11

electric field in a free crystal is therefore the sum of the primary and secondary effects:

$$\Delta a_{ij} = r_{ijk} E_k + P_{ijkl} u_{kl} \ .$$

The order of magnitude of the secondary effect can be estimated from the coefficients of the reciprocal piezoelectric effect $u \approx 0.5 \times 10^{-12}$, and also from the values $p \approx 10^{-1}$. We can see that the secondary effect produces changes in the indices comparable with the primary effect. This can be used to intensify the action of the electric field, especially under mechanical resonance. The primary effect, on the other hand, can be measured by applying to the crystal a variable field of high frequency at which the deformations will be very small.

The elasto-optic effect is also observed during the propagation of sound waves in crystals [7.42]. If, for instance, we place a crystal plate of thickness l between crossed polarizers and excite a longitudinal standing acoustic wave of deformation

$$u = u_0 \cos Kz \cos \Omega t \ ,$$

where Ω is the frequency of sound oscillations, and $K = 2\pi/\Lambda$ is the wave number, then the following phase difference will arise in the crystal:

$$\Gamma = (2\pi l/\lambda) n_0^3 p u \ ,$$

where p is the effective elasto-optic coefficient, and the light intensity at the analyzer output will be

$$I = \frac{I_0}{D} \int_{-D/2}^{D/2} \sin^2 \frac{\Gamma(z)}{2} dz \ . \tag{7.68}$$

Here D is the with of the light beam, and I_0 is the light intensity at the input. Moreover, $2D/\Lambda$ is usually chosen as an integer; then

$$I = \frac{I_0}{2}\left[1 - J_0\left(\frac{2\pi n_0^3}{\lambda} p l u_0 \cos\Omega t\right)\right] , \quad (7.69)$$

where J_0 is a zero-order Bessel function. Thus, we find that the light at the exit will be modulated by even harmonics of frequency Ω. The refractive index of the medium in which the elastic standing wave has been excited varies by the law $n = n_0 + \Delta n \cos Kz \cos \Omega t$ because of the elasto-optic effect.

If the light beam propagates in the medium in the direction perpendicular to the refractive index gradient, it deviates by an angle equal to $\theta = l \operatorname{grad} n$. For narrow light beams near the nodes of the standing wave, $\operatorname{grad} n \approx (2\pi \Delta n/\Lambda) \cos \Omega t$, and the maximum deviation angle will be $\theta_{max} = (4\pi l/\Lambda)\Delta n$.

If the diameter of the light beam greatly exceeds acoustic wavelength Λ, the medium in which the acoustic wave has been excited represents for it a phase diffraction grating, which is fixed in the case of a standing wave or is moving with the speed of sound in the case of a running wave. The light beam will therefore be diffracted on it. If the acoustic column width l (Fig. 7.26) in the direction of incident light wave vector $\boldsymbol{k}$ is small, it can be assumed approximately that the light beams pass through it rectilinearly. With a standing acoustic wave its front at the exit from the crystal at $x = l$, which serves as the source of secondary waves in diffraction, will be described by the equation

$$E = E_0 \exp[\mathrm{i} k l (n_0 + \Delta n \cos Kz \cos \Omega t)] .$$

From diffraction theory we find that the light field away from $x = l$ is given by the expression

$$E = \frac{b}{\lambda} \int_{-L/2}^{L/2} \exp(\mathrm{i} k z \sin\theta) E \, dz ,$$

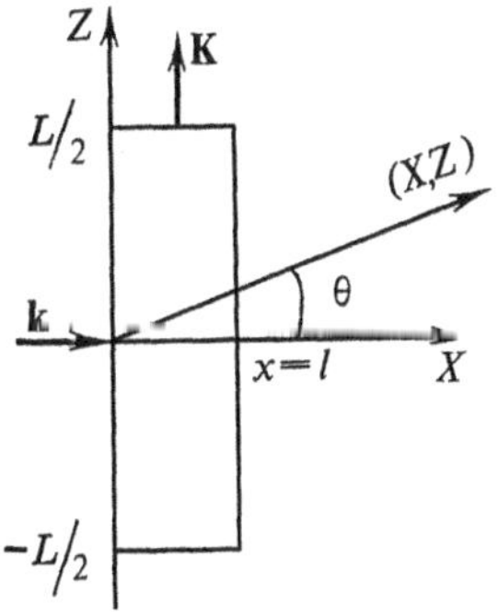

Fig. 7.26. Light diffraction on an ultrasonic wave [7.42]

where b and L are the transverse dimensions of the diffraction grating, and θ is the observation angle. Calculating this integral, we get

$$E = E_0 \frac{bL}{\lambda} e^{ikl} \sum_{m=-\infty}^{\infty} (-i)^m J_m (kl\Delta n \cos \Omega t)$$

$$\times \sin\left[\frac{\pi L}{\lambda}\left(\sin\theta - \frac{m\lambda}{\Lambda}\right)\right] \Big/ \frac{\pi L}{\lambda}\left(\sin\theta - \frac{m\lambda}{\Lambda}\right), \qquad (7.70)$$

where J_m is the an mth-order Bessel function. The direction of the main diffraction maxima is found for the condition $\theta_m = m\lambda/\Lambda$. The intensity at the mth-order diffraction maximum is proportional to the expression $J_m^2(\Gamma_0 \cos \Omega t)$, $\Gamma_0 = 2\pi l \Delta n/\lambda$.

For a running wave

$$E = E_0 \exp\{ikl[n_0 + \Delta n \cos(Kz - \Omega t)]\} .$$

A similar calculation yields the same directions of the maxima with time-independent intensities $J_m^2(\Gamma_0)$. The frequency of the maximum is shifted by $m\Omega$ owing to the Doppler effect. In a running wave $u^2 = 2P/S\varrho v^3$, where S is the cross-sectional area of the acoustic column, ϱ is the density of the medium, v is the sound velocity, and P is the acoustic power. At small Γ_0 the intensity at the first diffraction maximum will be

$$I_0 \frac{n^6 p^2}{\varrho v^3} \frac{2\pi^2 l^2}{\lambda^2 S} P .$$

The efficiency of the acousto-optical material is thus characterized by quantity $M = n^6 p^2/\varrho v^3$. The values of this parameter for some materials are listed in Table 7.10.

Generally, the problem of light diffraction on a sound wave reduces to the solution of the wave equation

$$\nabla^2 E - \frac{1}{c^2} \frac{\partial^2}{\partial t^2} (\varepsilon E) = 0 , \qquad (7.71)$$

which takes into account the dependence of dielectric constant ε on the coordinates and time in the sound wave. The solution essentially depends on the value of $Q = K^2 L/k$. When $Q \ll 1$ (the Raman-Nath region), we obtain the result already considered above. In the opposite case, i.e., at $Q \gg 1$ (the Bragg region) the intensities of all the diffraction maxima are low, with the exception of the zero maximum and the one observed at the angle

Table 7.10. Characteristics of some acousto-optical crystals [7.42]

Material	Elasto-optic coefficients, p_{ij}	n	ϱ, g/cm^3	v, 10^5 cm/s	$M = n^6 p^2/\varrho v^3$, 10^{18} s^3/g
1	2	3	4	5	6
Molten quartz	$p_{11} = 0.121$ $(p_{11} - p_{12})/2$ $p_{12} = 0.270$ $= -0.075$	1.46	2.2	5.95	1.51
Gallium phosphide GaP	$p_{11} = -0.151$ $p_{44} = -0.074$ $p_{12} = -0.082$	3.31	4.13	6.32	44.6
Rutile TiO_2	$p_{11} = 0.011$ $p_{13} = 0.168$ $p_{12} = 0.172$ $p_{33} = 0.058$ $p_{31} = 0.0965$	2.6	4.24	7.86	3.93
Lithium niobate $LiNbO_3$	$p_{31} = 0.178$ $p_{41} = 0.155$ $p_{13} = 0.092$ $p_{11} = 0.036$ $p_{33} = 0.088$ $p_{12} = 0.072$	2.20	4.7	6.57	6.99
Yttrium aluminium garnet $Y_3Al_5O_{12}$	$p_{11} = -0.029$ $p_{12} = 0.009$ $p_{44} = -0.0615$	1.83	4.2	8.53	0.012
Lithium tantalate $LiTaO_3$	$p_{11} = p_{12} = 0.0804$ $p_{44} = 0.022$ $p_{31} = 0.086$ $p_{33} = 0.150$ $p_{13} = 0.094$ $p_{41} = 0.024$ $p_{14} = 0.031$	2.18	7.45	6.19	1.37
Cadmium sulphide CdS	$p_{11} = 0.142$ $p_{12} = -0.066$ $p_{44} = 0.084$ $p_{31} = 0.041$ $p_{11} = 0.104$	2.44	4.82	4.17	12.1
Potassium dihydrophosphate KDP	$p_{11} = 0.251$ $p_{12} = 0.249$ $p_{31} = 0.225$ $p_{13} = 0.246$ $p_{33} = 0.221$ $p_{66} = 0.058$	1.51	2.34	5.50	3.83
Paratellurite TeO_2	$p_{11} = 0.0074$ $p_{33} = 0.240$ $p_{12} = 0.187$ $p_{44} = -0.17$ $p_{13} = 0.340$ $p_{66} = -0.0463$ $p_{31} = 0.090$	2.41	5.99	0.62	1200

θ_{-1} ($\sin\theta_{-1} = \lambda/2\Lambda$). The intensity of this maximum in the direction of the Bragg angle is equal to

$$I_{-1} = I_0 \frac{\pi l^2 \Delta n^2}{\lambda^2} \sin^2\left[\frac{\pi l \Delta n}{\lambda}\right] .$$

Apart from such ordinary Bragg diffraction, crystals may also cause diffraction with a change in the polarization plane of the incident wave (anisotropic diffraction). To find the direction of the diffraction maximum it should be borne in mind that light diffraction on an ultrasonic wave can be represented as the scattering of a photon with a wave vector $\boldsymbol{k}$ on a phonon with a wave vector $\boldsymbol{K}$ to produce a new light quantum with a momentum $\boldsymbol{k}'$. Considering that light oscillation frequency ω greatly exceeds the sound frequency Ω and proceeding from the laws of conservation of energy and momentum we can write

$$-k\sin\vartheta + K = k'\sin\vartheta' , \quad k\cos\vartheta = k'\cos\vartheta' .$$

Here $k = nk_0$; $k' = n'k_0$; n, n' are the refractive indices of the proper waves in a crystal with different polarization, whence we find

$$\begin{aligned} \sin\vartheta &= \frac{\lambda}{2n\Lambda}\left[1 + \frac{\Lambda^2}{\lambda^2}(n^2 - n'^2)\right] , \\ \sin\vartheta' &= \frac{\lambda}{2n'\Lambda}\left[1 - \frac{\Lambda^2}{\lambda^2}(n^2 - n'^2)\right] . \end{aligned} \tag{7.72}$$

At $n > n'$ and $\Lambda = \lambda/\sqrt{n^2 - n'^2} = V/\Omega^*$ angle of incidence ϑ is minimal, and diffraction angle $\vartheta' = 0$ (Fig. 7.27). The maximum and minimum sound frequencies at which anisotropic diffraction is possible ($\vartheta = 90°$) correspond to $\Lambda = \lambda/(n \mp n')$. For information an light scattering on elastic waves of biaxial crystals, see [7.44].

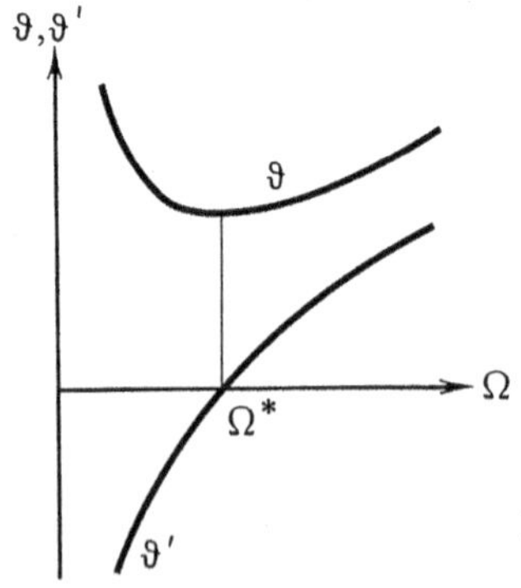

Fig. 7.27. Dependence of angle of incidence ϑ and angle of diffraction ϑ' on the sound frequency Ω [7.42]

Elasto-optic phenomena find numerous applications in devices for modulation and control of an optical, especially a laser, beam. The operating principle of such modulators is the same as for the electro-optical ones, except that the phase difference arises under the effect of mechanical stresses, and not of an electric field. When elasto-optic modulators are used, the frequencies on which a modulator can operate efficiently narrow down. Therefore, such modulators find application in narrow-band systems, for instance, in range finders.

Diffraction modulators are also widely used; the light diffracts on a grating formed by periodic changes in refractive index in the acoustic wave field.

7.10 Light Scattering in Crystals

Light scattering on deformation fluctuations caused by thermal elastic waves (Mandel'shtam-Brillouin scattering) in crystals is complicated by both elastic and optical anisotropy as compared with thermal scattering in liquids and gases [7.46–49].

Let us represent a displacement vector $\boldsymbol{e}$ in a plane elastic wave in a crystal by the expression

$$\boldsymbol{e} = (A/2)\{\exp i(\Omega t - \boldsymbol{K}\boldsymbol{r}) + \exp[-i(\Omega t - \boldsymbol{K}\boldsymbol{r})]\} ,$$

where Ω is the frequency, $\boldsymbol{K}$ is the wave vector of the elastic wave, and $\boldsymbol{A}$ is the wave amplitude. In a system of coordinates with orts ξ_k for vectors $\boldsymbol{K}$ and $\boldsymbol{A}$, we write $\boldsymbol{K} = K\varkappa_k\xi_k$, $\boldsymbol{A} = A\gamma_k\xi_k$. For the components of the tensor of deformations caused by such a wave, we find, in accordance with the theory of elasticity,

$$\begin{aligned}
u_{xx} &= \frac{\partial e_x}{\partial x} = -\tfrac{1}{2} iKA\varkappa_x\gamma_x\{\exp i(\Omega t - \boldsymbol{K}\boldsymbol{r}) - \exp[-i(\Omega t - \boldsymbol{K}\boldsymbol{r})]\} , \\
u_{yy} &= \frac{\partial e_y}{\partial y} = -\tfrac{1}{2} iKA\varkappa_y\gamma_y\{\exp i(\Omega t - \boldsymbol{K}\boldsymbol{r}) - \exp[-i(\Omega t - \boldsymbol{K}\boldsymbol{r})]\} , \qquad (7.73) \\
&\cdots\cdots\cdots \\
u_{xy} &= \frac{1}{2}\left(\frac{\partial e_x}{\partial y} + \frac{\partial e_y}{\partial x}\right) = -\tfrac{1}{4} iKA(\varkappa_y\gamma_x + \varkappa_x\gamma_y)\{\exp i(\Omega t - \boldsymbol{K}\boldsymbol{r}) \\
&\quad \exp[-i(\Omega t - \boldsymbol{K}\boldsymbol{r})]\} , \\
&\cdots\cdots\cdots
\end{aligned}$$

These deformations in a crystal are associated with a change in dielectric constant tensor, which is (as is known from the theory of the elasto-optic effect) equal to

$$\varepsilon_{xx}-\varepsilon_{xx}^0 = n^4(p_{11}u_{xx}+p_{12}u_{yy}+p_{13}u_{zz}+2p_{14}u_{yz}+2p_{15}u_{zx}+2p_{16}u_{xy}) ,$$
$$\dots\dots\dots\dots\dots\dots\dots\dots\dots\dots\dots\dots\dots\dots\dots$$
$$\varepsilon_{xy}-\varepsilon_{xy}^0 = n^4(p_{61}u_{xx}+p_{62}u_{yy}+p_{63}u_{zz}+2p_{64}y_{yz}+2p_{65}u_{zx}+2p_{66}u_{xy}) ,$$
$$\dots\dots\dots\dots\dots\dots\dots\dots\dots\dots\dots\dots\dots\dots\dots \tag{7.74}$$

where p_{rj} are the elasto-optic coefficients. Substituting the values of the components of the deformation tensor in an elastic wave, we obtain

$$\varepsilon_{ij}-\varepsilon_{ij}^0 = -\tfrac{1}{2}\,\mathrm{i}kAn^4\varphi_{ij}\{\exp\mathrm{i}(\Omega t-\boldsymbol{K}\boldsymbol{r})-\exp[-\mathrm{i}(\Omega t-\boldsymbol{K}\boldsymbol{r})]\} , \tag{7.75}$$

where $\varphi_{ij}=p_{ijkl}\varkappa_l\gamma_k$.

Let us now turn to the solution of the problem of light propagation in a medium whose dielectric constant varies in accordance with (7.75).

Suppose that in the absence of thermal waves the solutions of the Maxwell equations were $\boldsymbol{E}^0$, $\boldsymbol{D}^0$, $\boldsymbol{H}^0$, where $\boldsymbol{D}^0=\{\varepsilon^0\}\boldsymbol{E}^0$. Assuming the field of the scattered wave $\boldsymbol{E}'$, $\boldsymbol{H}'$ to be small as compared with the field of the incident wave $\boldsymbol{E}^0$, $\boldsymbol{H}^0$, we seek the solutions of the Maxwell equations in the form

$$\boldsymbol{E}=\boldsymbol{E}^0+\boldsymbol{E}' , \quad \boldsymbol{H}=\boldsymbol{H}^0+\boldsymbol{H}' .$$

Then the induction vector

$$\begin{aligned} D_i &= \varepsilon_{ik}E_k = (\varepsilon_{ik}^0+\Delta\varepsilon_{ik})(E_k^0+E_k') \\ &\approx \varepsilon_{ik}^0E_k^0+\varepsilon_{ik}^0E_k'+\Delta\varepsilon_{ik}E^0 \\ &= D_i^0+D_i'+\Delta\varepsilon_{ik}E_k^0 . \end{aligned} \tag{7.76}$$

The equations for fields $\boldsymbol{E}'$, $\boldsymbol{H}'$ will now be written

$$\begin{aligned} &c\,\mathrm{curl}\boldsymbol{E}'=-\frac{\partial\boldsymbol{H}'}{\partial t} , \quad c\,\mathrm{curl}\boldsymbol{H}'=\frac{\partial\boldsymbol{D}'}{\partial t}+4\pi\boldsymbol{j} , \\ &\mathrm{div}\boldsymbol{D}'=4\pi\varrho , \qquad \mathrm{div}\boldsymbol{H}'=0 ; \end{aligned} \tag{7.77}$$

here

$$j_i=\frac{1}{4\pi}\frac{\partial}{\partial t}\Delta\varepsilon_{ik}E_k^0 , \quad \varrho=-\frac{1}{4\pi}\mathrm{div}(\{\Delta\varepsilon\}\boldsymbol{E}^0) , \quad \boldsymbol{D}'=\{\varepsilon^0\}\boldsymbol{E}' . \tag{7.78}$$

Thus, the problem reduces to the solution of the Maxwell equations for a medium with currents and charges defined by (7.78). We shall give only the final solution abtained by *Ginzburg* [7.50]. It turns out, first of all, that here, as in an isotropic medium, the wave vector and frequency of a scattered wave satisfy the conditions

$$\boldsymbol{k}-\boldsymbol{k}_0 = \pm \boldsymbol{K} \; , \tag{7.79a}$$

$$\omega-\omega_0 = \pm \Omega \; , \tag{7.79b}$$

where $\boldsymbol{k}_0$, ω_0 and $\boldsymbol{k}, \omega$ are the wave vectors and frequencies of the incident and scattered wave, respectively.

In an anisotropic crystal, there are generally two light waves with different polarization for a fixed direction; accordingly, for incident waves we write

$$\boldsymbol{E}^0 = B\boldsymbol{\beta}^j \exp[\mathrm{i}(\omega t - \boldsymbol{k}_0^j \boldsymbol{r})] \; , \tag{7.80}$$

where $\boldsymbol{\beta}^j$ is the unit vector of the direction of oscillations of vector $\boldsymbol{E}^0$; in the scattered wave, the corresponding ort will be denoted by $\boldsymbol{\alpha}^i$. For the normal component of the vector of the Umov-Poynting energy flux of the scattered wave we get the following expression (assuming $\Omega \ll \omega$):

$$S_n \approx \frac{V}{R^2}\,\frac{\omega^4}{64\pi^3 c^3}\, n^8 \,\frac{n^i}{\cos^2\theta^i}\,\frac{K^2 A^2}{2}\, B^2 (\varphi_{lm}\,\alpha_l^i\,\beta_m^j)^2 \; ; \tag{7.81}$$

here V is the scattering volume, R is the distance to the observation point, and θ^i is the angle between vectors $\boldsymbol{E}$ and $\boldsymbol{D}$ of the scattered wave. It is obvious that $K^2/\varrho\Omega^2$ depends only on the elastic constants and density ϱ of the crystal, and $\varrho\Omega^2 A^2/2$ is the kinetic energy of the elastic thermal wave, which can be equated to kT. Since for the incident wave

$$S_{n_0} = \frac{c}{4\pi}\, n_0^j \cos^2\theta_0^j B^2 \; ,$$

we finally have, for the relative intensity of the scattered waves [7.49],

$$S_n = \frac{V}{R^2}\,\frac{\omega^4}{c^4}\,\frac{n^8}{16\pi^2}\,\frac{n^i}{n_0^j}\,\frac{1}{\cos^2\theta^i\cos^2\theta_0^j}\,\frac{K^2}{\varrho\Omega^2}\, kT (\varphi_{lm}\,\alpha_l^i\,\beta_m^j)^2 S_{n_0} \; . \tag{7.82}$$

In an optically anisotropic crystal, for each given scattering geometry, i.e., for fixed directions $\boldsymbol{k}_0^j$ and $\boldsymbol{k}^i$, (7.79a) is generally satisfied by four distinct vectors $\boldsymbol{K}^{ij}$, each of which corresponds to one of the four possible combinations of two incident and two scattered linearly polarized waves. In Fig. 7.28 are possibilities for the case where the scattering is observed at an angle of 90° to the direction of the incident waves. Further, to each direction of the wave vector of the elastic wave generally correspond three waves in a crystal, differing in polarization and velocity: one longitudinal and two transverse. In accordance with (7.79b), each of these waves will change the frequency in the scattered light so that we shall observe in the crystal a fine structure of the Rayleigh scattering line consisting of 12 long-wave and 12 short-wave satel-

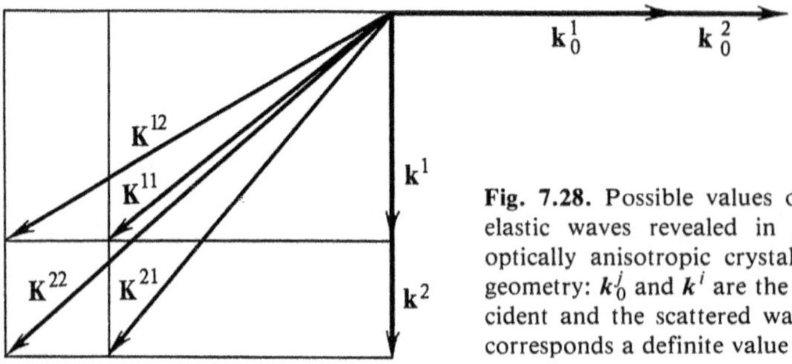

Fig. 7.28. Possible values of wave vectors $\boldsymbol{K}^{ij}$ of elastic waves revealed in light scattering in an optically anisotropic crystal in the 90° scattering geometry: $\boldsymbol{k}_0^j$ and $\boldsymbol{k}^i$ are the wave vectors of the incident and the scattered wave. To each pair $\boldsymbol{k}_0^j \boldsymbol{k}^i$ corresponds a definite value of $\boldsymbol{K}^{ij}$

lites, and one undisplaced line [7.51]. The displacements of the lines for the scattering geometry given in Fig. 7.28 are defined by

$$|\boldsymbol{K}| = \frac{2\pi}{\lambda}\sqrt{(n_0^j)^2 + (n^i)^2} \ , \tag{7.83}$$

and thus the frequency changes will be equal to

$$\Omega = \frac{2\pi v}{\Lambda} = \omega\frac{v}{c}\sqrt{(n_0^j)^2 + (n^i)^2} \ , \tag{7.84}$$

where v are the respective sound wave velocities, and Λ is the sound wave length. The splittings are usually on the order of fractions of a cm^{-1} and can be detected experimentally by high-resolution spectroscopy (for instance, using the Fabry-Perot reference scale).

To calculate the intensity of the Mandel'shtam-Brillouin components by (7.82) it is necessary to determine the velocity v and polarization γ_k of the respective elastic waves and also to know the elasto-optic coefficients. As an example we shall perform the following calculation for cubic crystals of classes $\bar{4}3m$, 432, and $m3$.

We set up equations defining the velocities and polarization of elastic waves:

$$(\Gamma_{ik} - \varrho v^2 \delta_{ik})\gamma_k = 0 \ , \tag{7.85}$$

where coefficients Δ_{ik} are determined by the following expressions:

$$\begin{aligned} \Gamma_{11} &= c_{11}\varkappa_2^2 + c_{44}\varkappa_2^2 + c_{44}\varkappa_3^2 \ , \quad & \Gamma_{23} &= (c_{12} + c_{44})\varkappa_2\varkappa_3 \ , \\ \Gamma_{22} &= c_{44}\varkappa_1^2 + c_{11}\varkappa_2^2 + c_{44}\varkappa_3^2 \ , \quad & \Gamma_{31} &= (c_{12} + c_{44})\varkappa_3\varkappa_1 \ , \\ \Gamma_{33} &= c_{44}\varkappa_1^2 + c_{44}\varkappa_2^2 + c_{11}\varkappa_3^2 \ , \quad & \Gamma_{12} &= (c_{12} + c_{44})\varkappa_1\varkappa_2 \ . \end{aligned} \tag{7.86}$$

The components of tensor φ_{ik} have the form

$$\varphi_{11} = p_{11}\varkappa_1\gamma_1 + p_{12}\varkappa_2\gamma_2 + p_{12}\varkappa_3\gamma_3 , \quad \varphi_{23} = p_{44}(\varkappa_3\gamma_2 + \varkappa_2\gamma_3) ,$$

$$\varphi_{22} = p_{12}\varkappa_1\gamma_1 + p_{11}\varkappa_2\gamma_2 + p_{12}\varkappa_3\gamma_3 , \quad \varphi_{31} = p_{44}(\varkappa_1\gamma_3 + \varkappa_3\gamma_1) ,$$

$$\varphi_{33} = p_{12}\varkappa_1\gamma_1 + p_{12}\varkappa_2\gamma_2 + p_{11}\varkappa_3\gamma_3 , \quad \varphi_{12} = p_{44}(\varkappa_2\gamma_1 + \varkappa_1\gamma_2) .$$

We shall now consider the following scattering geometry. Let the light fall along the edge of a cube in the X direction. We choose the direction of scattering along the Y axis. Then the components of the wave vectors of the light and elastic waves will be equal to $\boldsymbol{k}_0\,(k_0, 0, 0)$, $\boldsymbol{k}\,(0, k_0, 0)$, $\boldsymbol{K}\,(-k_0, k_0, 0)$ and $\varkappa_1 = -\sqrt{2}/2$, $\varkappa_2 = \sqrt{2}/2$, $\varkappa_3 = 0$. Coefficients Γ_{ik} in (7.85) have the form $\Gamma_{11} = \Gamma_{22} = \frac{1}{2}(c_{11} + c_{44})$, $\Gamma_{33} = c_{44}$, $\Gamma_{23} = \Gamma_{31} = 0$, $\Gamma_{12} = -\frac{1}{2}(c_{12} + c_{44})$. Using the set of equations (7.85), we find

1) $\varrho v^2 = \Gamma_{33} = c_{44}, \quad \gamma_1 = \gamma_2 = 0, \quad \gamma_3 = 1$,
2) $\varrho v^2 = \Gamma_{11} + \Gamma_{12} = \frac{1}{2}(c_{11} - c_{12}), \quad \gamma_1 = \gamma_2 = \sqrt{2}/2, \quad \gamma_3 = 0$,

for the transverse waves, and

3) $\varrho v^2 = \Gamma_{11} - \Gamma_{12} = \frac{1}{2}(c_{11} + c_{12} + 2c_{44}), \quad \gamma_1 = -\gamma_2 = \sqrt{2}/2, \quad \gamma_3 = 0$,

for the longitudinal wave. For further calculation it is necessary to assign the polarization of the incident and scattered waves. As an example, let us consider the following case:

$$\beta_1 = \beta_2 = 0 , \quad \beta_3 = 1 , \quad \alpha_1 = \alpha_2 = 0 , \quad \alpha_3 = 1 .$$

Then, calculating the components of tensor φ_{ik} and substituting them into (7.82), we find that the scattered light intensity is nonzero only for the longitudinal wave and is defined by

$$S_n = \frac{V}{R^2}\frac{\omega^4}{c^4}\frac{n^8}{16\pi^2}kT\frac{p_{12}^2}{\frac{1}{2}(c_{11} + c_{12} + 2c_{44})}S_{n_0} . \tag{7.87}$$

In the other cases, calculations are performed in the same way.

The results are presented in Table 7.11. As we see, the intensity of the scattered components essentially depends on the scattering geometry.

In deriving the expressions for the components of the tensor of change in dielectric constant due to the sound wave (7.75), we took into account only the symmetric part of the deformation tensor, which describes the contraction-extension and shear deformations, so that the components of the elasto-optic tensor satisfy condition $p_{ij(kl)} = p_{ij(lk)}$; that is, $p_{ij(kl)}$ is a tensor symmetric with respect to acoustic subscripts (kl). But an elastic wave in a crystal also produces another type of deformation, which is manifested as a rotation of a volume element of the crystal (i.e., the deformation is essentially nonuniform,

Table 7.11. Intensities of the Mandel'shtam-Brillouin components in cubic crystals ($\bar{4}3m$, 432, $m3$) in relation to the direction of polarization of the incident and the scattered light wave relative to the scattering plane [7.49]

Characteristic of elastic wave	$E_{\perp}^{inc}$		$E_{\parallel}^{inc}$	
	$E_{\perp}^{sc}$	$E_{\parallel}^{sc}$	$E_{\perp}^{sc}$	$E_{\parallel}^{sc}$
Transverse wave with direction of oscillations				
perpendicular to scattering plane, $\Omega_1 = K(c_{44}/\varrho)^{1/2}$	0	$\frac{1}{2}\frac{p_{44}^2}{c_{44}}$	$\frac{1}{2}\frac{p_{44}^2}{c_{44}}$	0
parallel to scattering plane $\Omega_2 = K\left(\frac{c_{11}-c_{12}}{2\varrho}\right)^{1/2}$	0	0	0	0
Longitudinal wave, $\Omega_3 = K\left(\frac{c_{11}+c_{12}+2c_{44}}{2\varrho}\right)^{1/2}$	$\frac{2p_{12}^2}{c_{11}+c_{12}+2c_{44}}$	0	0	$\frac{2p_{44}^2}{c_{11}+c_{12}+2c_{44}}$

the nonuniformity parameter being Λ) and is described by the antisymmetric part of the deformation tensor.[1] In an optically anisotropic crystal this type of deformation causes additional changes in the components of the dielectric constant tensor, and the corresponding part of the elasto-optic coefficient tensor is antisymmetric with respect to the acoustic subscripts, i.e., $p_{ij[kl]} = -p_{ij[lk]}$, its components being completely defined by the optical anisotropy of the crystal. Thus, a complete change in dielectric constant tensor due to the elastic wave is determined by the sum of contributions corresponding to the symmetric and antisymmetric parts of the deformation tensor. This was first pointed out by *Nelson* and *Lax* [7.52]. Taking this into account changes the scattering coefficients and the selection rules. Such changes have actually been detected experimentally [7.52 – 54].

7.11 Nonlinear Optical Properties of Crystals

In linear optics we assumed dielectric constant ε and susceptibility $\varkappa$ to be independent of the intensity of the electromagnetic waves propagating through the medium, and hence we considered the relationship between polarization vectors $\boldsymbol{P}$ and field $\boldsymbol{E}$ to be linear. This assumption was quite justified in experiments with low-power light beams, for which the intensity of the electric field in the wave is small as compared with the intra-atomic electric field. Indeed, light fields of 0.1 – 10 V/cm are attainable with the aid of conventional light sources producing intensities of 1 – 10 W/cm^2, while intra-atomic fields are on the order of $e/r_0^2 \approx 10^9$ V/cm (r_0 is the characteristic radius of the electron orbit $\sim 10^{-8}$ cm). In the fields of laser sources with an intensity of 10^{10} W/cm^2, the field intensity $E \approx 10^7$ V/cm cannot be considered negligibly small compared with the intra-atomic. Here one must introduce into the equation of relationship between $\boldsymbol{P}$ and $\boldsymbol{E}$ additional terms proportional to the powers of E:

$$P = \varkappa E + \chi E^2 + \theta E^3 \text{ (isotropic medium case)} \ . \tag{7.88}$$

Transition from linear relationship $P = \varkappa E$ to a nonlinear relationship leads to new qualitative phenomena, which are studied in nonlinear optics [7.55]. Thus, for instance, under the effect of an incident harmonic plane wave

$$E = A_1 \sin(\omega t - k_1 z) \ ,$$

there arises, in addition to the ordinary wave with a frequency ω, a polarization wave with a double frequency

---

1 As is well known, in the case of a uniform deformation the antisymmetric part of the deformation tensor describes the rotation of the crystal as a whole.

$$P^{(2\omega)} = A_1^2 \chi^{(2\omega)} \sin(2\omega t - 2k_1 z) \ ,$$

which excites an electromagnetic wave E_2 with a frequency 2ω in the medium.
From the Maxwell equations

$$\mathrm{curl}\,\boldsymbol{H} = \frac{1}{c}\frac{\partial \boldsymbol{D}}{\partial t} \ , \quad \mathrm{curl}\,\boldsymbol{E} = -\frac{1}{c}\frac{\partial \boldsymbol{H}}{\partial t}$$

and the relation $\varepsilon(2\omega) = 1 + 4\pi\varkappa(2\omega)$ we have a wave equation

$$\frac{\partial^2 E_2}{\partial z^2} + \frac{\varepsilon(2\omega)}{c^2}\frac{\partial^2 E_2}{\partial t^2} = -\frac{4\pi}{c^2}\frac{\partial^2 P^{(2\omega)}}{\partial t^2} \ . \tag{7.89}$$

For a forced wave we therefore obtain

$$E_2 = \frac{4\pi\chi^{(2\omega)} A_1^2 4\omega^2/c^2}{4k_1^2 - k_2^2} \sin(2\omega t - 2k_1 z) \ , \tag{7.90}$$

where $k_2 = (2\omega/c)\sqrt{\varepsilon(2\omega)}$.

Thus, in a nonlinear medium a *second harmonic* can be generated, which obtains energy from the main radiation via the polarization wave. The energy is transmitted efficiently and is accumulated over distances l, at which the phase difference between these waves $\Delta\varphi = l(k_2 - 2k_1)$ remains less than $\frac{\pi}{2}$, i.e., over distances of coherent length $l_{\mathrm{coh}} = \lambda_0/4(n_2 - n_1)$, where n_2 is the refractive index of the wave of the second harmonic, and n_1 is that of the main frequency. Because of the strong dispersion in the optical range $(n_2 - n_1) \approx 10^{-2} - 10^{-3})$, the coherent length is usually small, $l_{\mathrm{coh}} \approx 10^{-3}$ cm.

Other characteristic effects are also known in nonlinear optics, namely, optical detection, generation of the third harmonic, a change in the refractive index of the medium, and nonlinear light reflection [7.56].

In the nonlinear optical crystals we must generalize the coupling equation and write for the quadratic polarization term

$$P_i^{(\mathrm{nl})} = \chi_{ikl} E_k E_l \ , \tag{7.91}$$

where χ_{ikl} is the *nonlinear susceptibility tensor*, symmetric with respect to the last two subscripts and similar in symmetry properties to the tensors of the piezoelectric coefficients. The form of such a tensor for crystals of different classes has been given before (Chap. 3). (The values of the components of tensor χ_{ikl} for some crystals are indicated in Table 7.12.) The classical theory of nonlinear polarization dispersion, based on the consideration of the oscillations of electrons as in an anharmonic oscillator, shows that in the weak dispersion range tensor χ_{ikl} can be assumed symmetric with respect to all the three subscripts [7.57].

In distinction to isotropic media, where $n(2\omega) > n(\omega)$ in the normal dispersion region, synchronous nonlinear effects associated with the prop-

Table 7.12. Coefficients of nonlinear susceptibility tensor χ_{ikl} for some crystals[a] [7.22]

Crystal	Symmetry	Coefficients χ_{ikl} at	
		$\lambda = 0.69\ \mu m$	$\lambda = 1.06\ \mu m$
Potassium dihydrophosphate KH_2PO_4	$\bar{4}2m$	$\chi_{36} = 1.00$ $\chi_{14} = 0.95$	$\chi_{36} = 1.00$ $\chi_{14} = 1.01$
Ammonium dihydrophosphate $NH_4H_2PO_4$	$\bar{4}2m$	$\chi_{36} = 0.93$	$\chi_{14} = 0.89$
Gallium arsenide GaAs	$\bar{4}3m$	$\chi_{14} = 490$	$\chi_{14} = 580$
Lithium niobate $LiNbO_3$	$3m$	$\chi_{22} = 6.3$ $\chi_{31} = 11.9$ (for $\lambda = 1.15\ \mu m$)	$\chi_{33} = 83$
Barium titanate $BaTiO_3$	$4mm$	$\chi_{15} = 35$ $\chi_{33} = 14$	$\chi_{14} = 37$

[a] The adopted unit is the susceptibility χ_{36} of potassium dihydrophosphate equal to 6×10^{-9} CGSE units.

agation of waves of different polarization are possible in crystals. Thus, in the propagation of two plane electromagnetic waves with a frequency ω in a nonlinear medium, a wave due to the quadratic polarization terms is generated, which has a frequency of 2ω and a definite polarization. The condition for the formation of the second harmonic or the condition for vector synchronism during the interactions of an ordinary wave and an extraordinary wave of basic frequency (oe→e interaction) can be written as

$$n^{e}(\omega)+n^{0}(\omega)=2n^{e}(2\omega).$$

A similar synchronism is possible during the interaction of two ordinary waves [oo→e interaction: $n^{o}(\omega)=n^{e}(2\omega)$]. The direction of propagation of the second-harmonic wave makes an angle θ_0 with the optic axis and is determined by the sum of the wave vectors of the initial waves (Fig. 7.29). Angle θ_0 is called the *synchronism angle*.

In uniaxial negative crystals, when the dispersion of the main refractive and birefringence indices is such that along with $n_1^o<n_2^o$, $n_1^e<n_2^e$ one finds $n_1^o>n_2^e$, the synchronism condition $n_1^o=n_2^e(\theta_0)$ is fulfilled for the cone of the directions making with the optic axis in the crystal a synchronism angle θ_0, which is defined by the expression

$$\arctan\theta_0=\left\{\frac{(n_2^e)^2\,[(n_2^o)^2-(n_1^o)^2]}{(n_2^o)^2\,[(n_1^o)^2-(n_2^e)^2]}\right\}^{1/2}.$$

Subscript 1 corresponds to frequency ω, and subscript 2 to 2ω. Let us calculate, say, the amplitude of the second-harmonic wave.

Suppose an ordinary wave with a wave vector $\boldsymbol{k}_1$ propagates in a nonlinear crystal so that the azimuth of the main plane (containing the optic axis of the

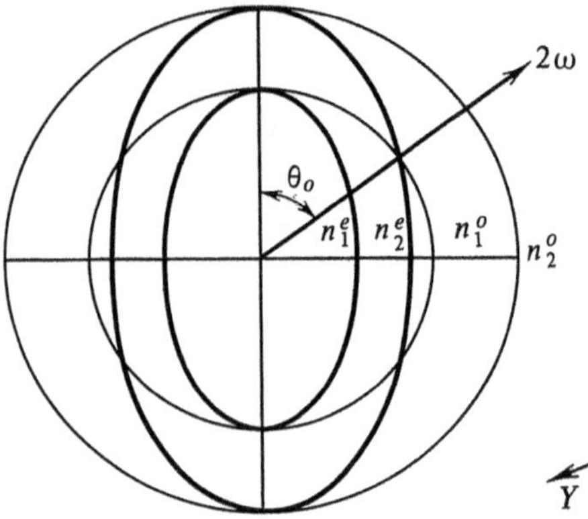

Fig. 7.29

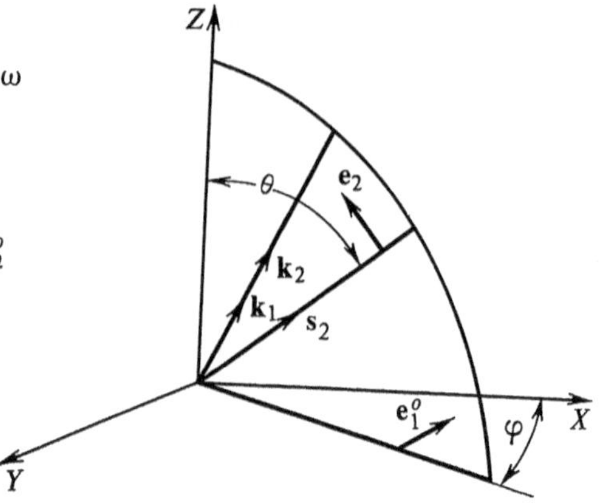

Fig. 7.30

Fig. 7.29. Synchronism condition in a uniaxial negative crystal

Fig. 7.30. Generation of the second harmonic in an $oo\rightarrow e$ interaction

crystal Z and vector $\boldsymbol{k}_1$) is φ (Fig. 7.30). Then the ort of the intensity vector of the electric field of the wave will have components $e_{1x}^o=\sin\varphi$, $e_{1y}^o=-\cos\varphi$, $e_{1z}^o=0$. The components of the vector of the polarization wave of the second harmonic are equal to

$$P_x^{(2\omega)}=\chi_{xyz}^{(2\omega)}e_{1z}^o e_{1z}^o=0\ ,$$

$$P_y^{(2\omega)}=\chi_{yxz}^{(2\omega)}e_{1x}^o e_{1z}^o=0\ ,$$

$$P_z^{(2\omega)}=\chi_{zyx}^{(2\omega)}e_{1y}^o e_{1x}^o=(-\chi_{zyx}^{(2\omega)}/2)\sin 2\varphi\ .$$

This means that an extraordinary wave of the second harmonic will propagate in the crystal, the ort of whose vector is $\boldsymbol{e}_2$, and whose amplitude E_2 is proportional to $\frac{1}{2}\chi_{zyx}^{(2\omega)}\sin 2\varphi\sin\theta$.

The fulfillment of the synchronism condition ensures a large coherent length and hence a high efficiency in frequency transformation. For positive crystals the synchronism condition can be fulfilled for ee→o, provided that $n_1^e>n_2^o$.

Additional possibilities for obtaining accumulating effects during the generation of the second harmonic in crystals arise when the radiation is not a single wave, but a set of plain waves whose wave vectors form a diverging beam. Then the syncronism condition can be satisfied for two-dimensional interaction, and not only in one direction. Indeed, suppose that we deal with a uniaxial negative crystal, whose sections of the wave-vector surface are presented in Fig. 7.31. It can be seen that the appropriately chosen pairs of waves of the basic radiation with wave vectors $\boldsymbol{k}^{1o}$ and $\boldsymbol{k}^{2o}$ yield radiation of

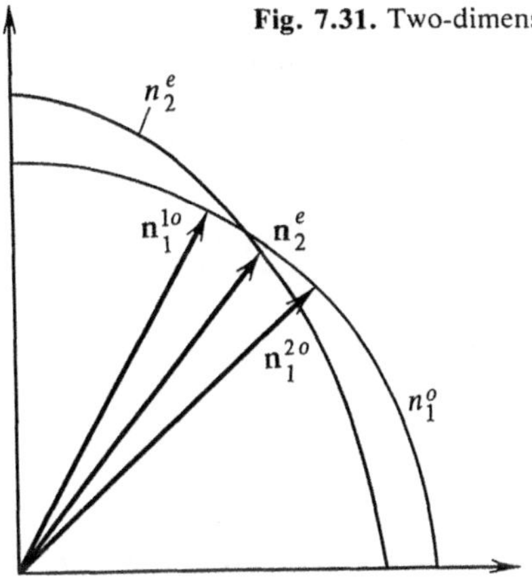

Fig. 7.31. Two-dimensional synchronism in a crystal

the second harmonic $\boldsymbol{k}_2^{\mathrm{e}}$, which is in synchonism with these waves, provided that $\boldsymbol{k}_1^{1\mathrm{o}}+\boldsymbol{k}_1^{2\mathrm{o}}=\boldsymbol{k}_2^{\mathrm{e}}$ [7.58].

By utilizing the nonlinear optical properties of crystals, one cannot only transform laser radiation into that of other fixed frequencies (harmonics), but also achieve smooth variation of the radiation frequency.

Suppose a powerful pumping wave of frequency ω_H (wave vector $\boldsymbol{k}_H$) falls on a nonlinear quadratic medium simultaneously with two weak waves with wave vectors $\boldsymbol{k}_1$ and $\boldsymbol{k}_2$, and frequencies ω_1 and ω_2 satisfying the condition $\omega_1+\omega_2=\omega_H$. Nonlinear polarization leads then to interaction of waves at frequencies ω_H and ω_2, to reradiation at a frequency ω_1, and also, owing to the interaction of waves with frequencies ω_H and ω_1, to radiation at a frequency ω_2. The synchronism condition here will be $k_H-k_2=k_1$. If this condition is fulfilled, the energy of the pumping wave is transmitted to the waves with frequencies ω_1 and ω_2. Therefore, if we place a nonlinear crystal into an optical resonator, i.e., between mirrors reflecting light at frequencies ω_1 and ω_2, generation at frequencies ω_1 and ω_2 arises in it at a sufficiently high pumping power.

The frequencies at which the synchronism condition is fulfilled depend on the direction of the interacting waves in the crystal and can be varied smoothly by changing the crystal orientation. Such light generators are called parametric.

Along with the quadratic terms in the nonlinear polarization expansion it is also possible to consider cubic ones:

$$P_i^{(\mathrm{nl})}=\theta_{iklm}E_kE_lE_m\ , \tag{7.92}$$

where θ_{iklm} is a fourth-rank tensor symmetric with respect to the last three subscripts. As in the quadratic term (7.91), for low dispersion it can be considered symmetric with respect to all the four subscripts. Thus, in acentric crystals, intricate nonlinear interactions of electromagnetic waves are possible. Of interest are papers by *Kurtz* and *Perry* [7.59] on the possibility of synchronous interaction and determination of the value of nonlinearity on crystal powders. This method provides information on the symmetry of the crystal and enables one to choose appropriate materials for use in nonlinear spectroscopy, and also in nonlinearly optical devices and instruments [7.60].

7.12 Essentials of the Crystal-Field Theory

7.12.1 Crystal-Field Model

So far we have been considering the manifestations of crystal anisotropy in optical phenomena, when a crystal behaves as an anisotropic continuous medium, all points of which have the same point symmetry. The discrete anisotropic structure of crystals with different symmetry of the positions in the lattice, which is described by its space symmetry, also finds its reflection in the optical properties of crystals. Among them are the spectroscopic features of absorption and luminescence of crystals with admixtures of ions of transition elements of the iron and rare-earth groups. Note that, in distinction to isotropic media, crystals have predominantly a single type of optically active center.

In interpreting the spectra of impurity crystals it is necessary to describe the interaction of an interstitial ion with the ions of the crystal lattice. With ionic crystals containing elements with an incomplete d or f shell as an impurity, this interaction can be considered the effect of some equivalent electrostatic field, induced by the surrounding ions, on the electron levels of the interstitial atom. This model lies at the basis of the crystal-field theory.

The main premises of present-day crystal-field theory are as follows:

1. In an activated crystal it is possible to single out an ion + medium complex (ligands), whose properties define the basic optical and magnetic characteristics of the crystal.
2. In considering the interaction of the activating ion with the medium, the effects of the overlapping of their wave functions can be neglected. The role of the surrounding ions consists in inducing an effective electrostatic field of a given symmetry, perturbing the states of the impurity ion.
3. The perturbing effect of the environment can be assumed equal for the excited and unexcited states of the impurity ion. For this we have to neglect the effect of the interstitial ion on the environment.

According to the crystal-field model [7.61], the effect of the crystal medium on the electron levels of the interstitial ion is regarded as the intracrystalline Stark effect due to the electrostatic field whose symmetry depends on the position of the ion in the crystal lattice [7.62]. Within the framework of this model the Hamiltonian of an impurity ion containing N electrons can be written as follows with due regard for the basic interactions:

$$H = H_0 + V_{ee} + V_{er} + V_{so} \ , \qquad (7.93)$$

where $H_0 = \sum_j [(p_j^2/2m) + U(r_j)]$ is the ion energy operator in the central field approximation, V_{ee} is the energy of the coulomb electrostatic interaction of the electrons of the ion, V_{cr} is the potential energy of the electron in the crystal field, and V_{so} is the spin-orbital interaction energy.

If we restrict our consideration to transitions inside a basic configuration, we can neglect the effect of the closed ion shells, which can be included in H_0, and consider only the system of d electrons of the unfilled shell. The task is now to find the splittings of the level of the given configuration under the effect of perturbations $V_{ee} + V_{cr} + V_{so}$. Three cases are distinguished in applied perturbation theory, depending on the ratio of the averages of these energies: the weak crystal-field approximation $V_{ee} \gg V_{so} \gg V_{cr}$, the average-field approximation $V_{ee} \gg V_{cr} \gg V_{so}$, and the strong-field approximation $V_{cr} \gg V_{ee} \gg V_{so}$. In particular, for iron group ions $V_{ee} \approx 10^5 - 10^6$, $V_{cr} \approx 10^2 - 10^5$ and $V_{so} \approx 10^2 \ \mathrm{cm}^{-1}$, so that here it is convenient to use the average- and strong-crystal-field approximations, whereas for rare earths the weak-field approximation is realized.

For calculating the potential of the intracrystalline field [7.63], it is often possible to use the point model, in which the ligands are replaced by charges at the vertices of the respective polyhedra of the structure. Such a model correctly reflects the main distinguishing feature of the problem, i.e., the crystal-field symmetry, and takes into account the strongest effect on the electrostatic ligand field on the impurity ion. As an example, we consider the calculation of the electric field potential for an octahedral environment.

We choose the origin at the center of the impurity ion. Then the coordinates of six ions of the environment will be $(a, 0, 0)$, $(-a, 0, 0)$, $(0, a, 0)$, $(0, -a, 0)$, $(0, 0, a)$, $(0, 0, -a)$.

The coulomb potential of the electric field at the point with radius vector $\boldsymbol{r}(x, y, z)$ is written as $\sum \dfrac{q}{|\boldsymbol{R}_j - \boldsymbol{r}|} = V_x + V_y + V_z$; here, $\boldsymbol{R}_j$ are radius vectors of ligands with a charge q, $V_x = q\,[(r^2 + a^2 - 2ax)^{-1/2} + (r^2 + a^2 + 2ax)^{-1/2}]$ and similarly for V_y and V_z. Since we are interested in the potential near the center of the octahedron, it can be assumed that $r < a$; then, restricting ourselves to the sixth-order terms in the expansion of the potential into a Taylor series with respect to the powers of the ratios $X = 2ax/A$, $Y = 2ay/A$, $Z = 2az/A$ [where $A = (r^2 + a^2)^{1/2}$], we have

$$V(xyz) = \frac{q}{A^{1/2}} \left[6 + \frac{3}{4}(X^2 + Y^2 + Z^2) + \frac{35}{64}(X^4 + Y^4 + Z^4) \right.$$
$$\left. + \frac{693}{64 \cdot 24}(X^6 + Y^6 + Z^6) \right] .$$

Table 7.13. Values of P_n and A_n in the expansion of the crystal-field potential [7.64]

$$V = \sum_{n,m>0} A_n^m P_n^m(x, y, z)$$

$P_2^0 = 3z^2 - r^2$	$P_2^2 = x^2 - y^2$
$P_4^0 = 35z^4 - 30r^2z^2 + 3r^4$	$P_4^2 = (7z^2 - r^2)(x^2 - y^2)$
$P_4^3 = xz(x^2 - 3y^2)$	$P_4^4 = x^4 - 6x^2y^2 + y^4$
$P_6^0 = 231z^6 - 315r^2z^4 + 105r^4z^2 - 5r^6$	$P_6^3 = (11z^2 - 3r^2)(x^2 - 3y^2)xz$
$P_6^4 = (11z^2 - r^2)(x^4 - 6x^2y^2 + y^4)$	$P_6^6 = x^6 - 15x^4y^2 + 15x^2y^4 - y^6$

$$V = \sum_n \sum_{m=-n}^{n} r^n B_n^m Y_n^m(\theta, \varphi), \quad P_n^m \sim r^n (Y_n^m + Y_n^{m\,*})$$

$A_2^0 = \dfrac{1}{\sqrt{2\pi}} \sqrt{\dfrac{5}{8}} B_2^0$	$A_2^2 = \dfrac{1}{\sqrt{2\pi}} \sqrt{\dfrac{15}{4}} B_2^2$
$A_4^0 = \dfrac{1}{\sqrt{2\pi}} \dfrac{3\sqrt{2}}{16} B_4^0$	$A_4^2 = \dfrac{1}{\sqrt{2\pi}} \dfrac{3\sqrt{5}}{4} B_4^2$
$A_4^3 = \dfrac{1}{\sqrt{2\pi}} \dfrac{3\sqrt{70}}{4} B_4^3$	$A_4^4 = \dfrac{1}{\sqrt{2\pi}} \dfrac{3\sqrt{35}}{8} B_4^4$
$A_6^0 = \dfrac{1}{\sqrt{2\pi}} \dfrac{\sqrt{26}}{32} B_6^0$	$A_6^3 = \dfrac{1}{\sqrt{2\pi}} \dfrac{\sqrt{26 \cdot 105}}{16} B_6^3$
$A_6^4 = \dfrac{1}{\sqrt{2\pi}} \dfrac{3\sqrt{13 \cdot 28}}{32} B_6^4$	$A_6^6 = \dfrac{1}{\sqrt{2\pi}} \dfrac{\sqrt{13 \cdot 21 \cdot 22}}{32} B_6^6$

Reverting to variables x, y, z, we obtain for the potential of the octahedral environment in Cartesian coordinates

$$\begin{aligned} V(xyz) = {} & \frac{6q}{a} + \frac{35q}{4a^5}\left(x^4 + y^4 + z^4 - \frac{3}{5}r^4\right) \\ & - \frac{21q}{2a^7}\left[(x^6 + y^6 + z^6) + \frac{15}{4}(x^2y^4 + z^4x^2 + y^2z^4\right. \\ & \left. + z^2x^4 + z^2y^4) - \frac{15}{14}r^6\right] = \frac{6q}{a} + \frac{7q}{16a^5}(P_4^0 + 5P_4^4) \ , \end{aligned}$$

where P_n^m is a homogeneous polynomial (Table 7.13).

Calculations of the energy levels of the ion in such a field will be performed with the aid of perturbation theory, and the matrix elements of the perturbation operator $V(x, y, z)$ will be calculated using the wave functions of a free ion expressed in terms of spherical functions $Y_l^m(\theta, \varphi)$. With this in view, potential $V(x, y, z)$ can also be conveniently expressed via the same

Table 7.14. Spherical harmonics in rectangular coordinates (Condon-Shortley phases, [7.65], [7.61])

$$Y_0^0 = \sqrt{\frac{1}{4\pi}}$$

$$Y_1^{-1} = \sqrt{\frac{3}{8\pi}}\,\frac{x-iy}{r}; \quad Y_1^0 = \sqrt{\frac{3}{4\pi}}\,\frac{z}{r}; \quad Y_1^1 = -\sqrt{\frac{3}{8\pi}}\,\frac{x+iy}{r};$$

$$Y_2^{-2} = \sqrt{\frac{5}{4\pi}}\sqrt{\frac{3}{8}}\,\frac{(x-iy)^2}{r^2}; \quad Y_2^{-1} = \sqrt{\frac{5}{4\pi}}\sqrt{\frac{3}{2}}\,\frac{z(x-iy)}{r^2};$$

$$Y_2^0 = \sqrt{\frac{5}{4\pi}}\sqrt{\frac{1}{4}}\,\frac{3z^2-r^2}{r^2}; \quad Y_2^1 = -\sqrt{\frac{5}{4\pi}}\sqrt{\frac{3}{2}}\,\frac{z(x-iy)}{r^2};$$

$$Y_2^2 = \sqrt{\frac{5}{4\pi}}\sqrt{\frac{3}{8}}\,\frac{(x+iy)^2}{r^2};$$

$$Y_3^{-3} = \sqrt{\frac{7}{4\pi}}\sqrt{\frac{5}{16}}\,\frac{(x-iy)^3}{r^3}; \quad Y_3^{-2} = \sqrt{\frac{7}{4\pi}}\sqrt{\frac{15}{8}}\,\frac{z(x-iy)^2}{r^3};$$

$$Y_3^{-1} = \sqrt{\frac{7}{4\pi}}\sqrt{\frac{3}{16}}\,\frac{(x-iy)(5z^2-r^2)}{r^3}; \quad Y_3^0 = \sqrt{\frac{7}{4\pi}}\sqrt{\frac{1}{4}}\,\frac{z(5z^2-3r^2)}{r^3};$$

$$Y_3^1 = -\sqrt{\frac{7}{4\pi}}\sqrt{\frac{3}{16}}\,\frac{(x+iy)(5z^2-r^2)}{r^3}; \quad Y_3^2 = \sqrt{\frac{7}{4\pi}}\sqrt{\frac{15}{8}}\,\frac{z(x+iy)^2}{r^3};$$

$$Y_3^3 = -\sqrt{\frac{7}{4\pi}}\sqrt{\frac{5}{16}}\,\frac{(x+iy)^3}{r^3};$$

$$Y_4^{-4} = \sqrt{\frac{9}{4\pi}}\sqrt{\frac{35}{128}}\,\frac{(x-iy)^4}{r^4}; \quad Y_4^{-3} = \sqrt{\frac{9}{4\pi}}\sqrt{\frac{35}{16}}\,\frac{z(x-iy)^3}{r^4};$$

$$Y_4^{-2} = \sqrt{\frac{9}{4\pi}}\sqrt{\frac{5}{32}}\,\frac{(x-iy)^2(7z^2-r^2)}{r^4};$$

$$Y_4^{-1} = \sqrt{\frac{9}{4\pi}}\sqrt{\frac{5}{16}}\,\frac{(x-iy)(7z^3-3zr^2)}{r^4};$$

$$Y_4^0 = \sqrt{\frac{9}{4\pi}}\sqrt{\frac{1}{64}}\,\frac{35z^4-30z^2r^2+3r^4}{r^4};$$

$$Y_4^1 = -\sqrt{\frac{9}{4\pi}}\sqrt{\frac{5}{16}}\,\frac{(x+iy)(7z^3-3zr^2)}{r^4};$$

$$Y_4^2 = \sqrt{\frac{9}{4\pi}}\sqrt{\frac{5}{32}}\,\frac{(x+iy)^2(7z^2-r^2)}{r^4};$$

$$Y_4^3 = -\sqrt{\frac{9}{4\pi}}\sqrt{\frac{35}{16}}\,\frac{z(x+iy)^3}{r^4}; \quad Y_4^4 = \sqrt{\frac{9}{4\pi}}\sqrt{\frac{35}{128}}\,\frac{(x+iy)^4}{r^4};$$

$$Y_5^{-5} = \sqrt{\frac{11}{4\pi}}\sqrt{\frac{63}{256}}\,\frac{(x-iy)^5}{r^5}; \quad Y_5^{-4} = \sqrt{\frac{11}{4\pi}}\sqrt{\frac{315}{128}}\,\frac{z(x-iy)^4}{r^5};$$

$$Y_5^{-3} = \sqrt{\frac{11}{4\pi}}\sqrt{\frac{35}{256}}\,\frac{(x-iy)^3(9z^2-r^2)}{r^5};$$

(Continued)

Table 7.14 (continued)

$$Y_5^{-2} = \sqrt{\frac{11}{4\pi}} \sqrt{\frac{105}{32}} \frac{(x-iy)^2 (3z^3 - zr^2)}{r^5} ;$$

$$Y_5^{-1} = \sqrt{\frac{11}{4\pi}} \sqrt{\frac{15}{128}} \frac{(x-iy)(21z^4 - 14z^2r^2 + r^4)}{r^5} ;$$

$$Y_5^0 = \sqrt{\frac{11}{4\pi}} \sqrt{\frac{1}{8}} \frac{63z^5 - 70z^3r^2 + 15zr^4}{r^5} ;$$

$$Y_5^1 = -\sqrt{\frac{11}{4\pi}} \sqrt{\frac{15}{128}} \frac{(x+iy)(21z^4 - 14z^2r^2 + z^4)}{r^5} ;$$

$$Y_5^{2} = \sqrt{\frac{11}{4\pi}} \sqrt{\frac{105}{32}} \frac{(x+iy)^2 (3z^3 - zr^2)}{r^5} ;$$

$$Y_5^3 = -\sqrt{\frac{11}{4\pi}} \sqrt{\frac{35}{256}} \frac{(x+iy)^3 (9z^2 - r^2)}{r^5} ;$$

$$Y_5^4 = \sqrt{\frac{11}{4\pi}} \sqrt{\frac{315}{128}} \frac{z(x+iy)^4}{r^5} ; \qquad Y_5^5 = -\sqrt{\frac{11}{4\pi}} \sqrt{\frac{63}{256}} \frac{(x+iy)^5}{r^5} ;$$

spherical functions (Table 7.14). Taking advantage of expansion $\frac{1}{|\boldsymbol{R}_j - \boldsymbol{r}|} = \sum_{n=0}^{\infty} \frac{r^n}{R_j^{n+1}} P_n(\cos\omega_j)$ for $R > r$, where ω_j is the angle between vectors $\boldsymbol{R}_j$ and $\boldsymbol{r}$, and of the addition theorem for Legendre polynomials

$$P_n(\cos\omega_j) = \frac{4\pi}{2n+1} \sum_{m=-n}^{n} (-1)^m Y_n^{-m}(\theta_j, \varphi_j)\, Y_n^M(\theta, \varphi) ,$$

we find for potential $V(r, \theta, \varphi)$ in spherical coordinates

$$V(r, \theta, \varphi) = \sum_n \sum_{m=-n}^{n} r^n B_n^m Y_n^m(\theta, \varphi) . \tag{7.94}$$

The coefficients of the expansion are determined by the equation

$$B_n^m = \sum_j \frac{4\pi}{2n+1} \frac{q_j}{R_j^{n+1}} (-1)^n Y_n^{-m}(\theta_j, \varphi_j) . \tag{7.95}$$

In particular, for the octahedral environment under consideration, after straightforward calculations using the explicit form of the spherical functions, we obtain – restricting ourselves to spherical harmonics up to the sixth order, inclusive –

$$V(r,\theta,\varphi) = D_4\{Y_4^0+(\tfrac{5}{14})^{1/2}[Y_4^4+Y_4^{-4}]\} + D_6\{Y_6^0-(\tfrac{7}{2})^{1/2}[Y_6^4+Y_6^{-4}]\}\;, \tag{7.96}$$

where

$$D_4 = \frac{7\sqrt{\pi}}{3}\,\frac{q}{d^5}\,r^4\;; \qquad D_6 = \frac{3}{2}\left(\frac{\pi}{13}\right)^{1/2}\frac{q}{d^7}\,r^6\;,$$

d being the distance from the central ion to the ligand ions.

In arranging the expressions for the crystal-field potential in the form of expansions in spherical functions it should be remembered that in the presence of an inversion center at the location of the central ion the potential will not contain harmonics with odd numbers n. It should also be taken into consideration that harmonics Y_n^0 possess an axial symmetry, Y_n^2 an orthorhombic, Y_n^4 tetragonal, and Y_n^6 trigonal, as can clearly be seen from their expressions in Cartesian coordinates (Table 7.14). The number of terms which must be retained in the expansion of the potential is due to the fact that not all of them yield nonzero matrix elements. For iron-group ions with configuration d^n, nonzero matrix elements will be obtained only for harmonics with polynomials not higher than the fourth, and for the rare-earth group (f^n), not higher than the sixth order. This result is a simple consequence of the rule of the triangle known in Legendre's theory of polynomials for integrals containing a product of three polynomials.

In calculating the matrix elements for one and the same term we can use the *operator-equivalent method*, in which, in place of the given operator, we construct another operator so that their matrix elements differ only by a constant factor. The new operator is chosen so that calculation of its matrix elements is as simple as possible. The proportionality coefficients can be determined by comparing the matrix elements calculated for these two operators. For instance, the operator equivalent for an axially symmetric potential $3\cos^2\theta-1$ is $3\hat{L}_z^2-L(L+1)=\mathrm{O}_2^0$ (Table 7.15). The calculation of

Table 7.15. Operators equivalents of polynomials (P_n^m ($\{AB\}_s = 1/2\,(AB+BA)$)) [7.64]

$O_2^0 = 3J_z^2 - J(J+1)\;;$

$O_2^2 = \frac{1}{2}(J_+^2 + J_-^2)\;;$

$O_4^0 = 35J_z^4 - 30J(J+1)J_z^2 + 25J_z^2 - 6J(J+1) + 3J^2(J+1)^2\;;$

$O_4^2 = \frac{1}{2}\{[7J_z^2 - J(J+1) - 5](J_+^2 + J_-^2)\}_s\;;$

$O_4^3 = \frac{1}{2}\{J_z(J_+^3 + J_-^3)\}_s\;;$

$O_4^4 = \frac{1}{2}(J_+^4 + J_-^4)\;;$

$O_6^0 = 231J_z^6 - 315J(J+1)J_z^4 + 735J_z^4 + 105J^2(J+1)^2J_z^2 - 525J(J+1)J_z^2 + 294J_z^2 - 5J^3(J+1)^3$
$\quad + 40J^2(J+1)^2 - 60J(J+1)\;;$

$O_6^3 = \frac{1}{2}\{[11J_z^3 - 3J(J+1)J_z - 59J_z](J_+^3 + J_-^3)\}_s\;;$

$O_6^4 = \frac{1}{2}\{[11J_z^2 - J(J+1) - 38](J_+^4 + J_-^4)\}_s\;;$

$O_6^6 = \frac{1}{2}(J_+^6 + J_-^6)\,.$

its matrix elements involves no particular difficulties if we take into account the rule for the transformation of wave function ψ_m, with the projection of orbital momentum L equal to m, under the effect of operator $\hat{L}_z : \hat{L}_z \psi_m = m\psi_m$.
In this direction, the calculation techniques have been completely simplified up to utilization of ready-made tables of matrix elements and proportionality coefficients.

7.12.2 A Single *d* Electron in a Cubic Field

(The following text is presented accordance with *Ballhausen* [7.61])

We begin the consideration of crystal-field theory with the simplest case of an ion with a single d electron. The ground state of the ion is term ^2D with $L = 2$, $S = \frac{1}{2}$, which is fivefold degenerate with respect to the orbit.

We have to find the splitting of this term, at first in the cubic symmetry field, say symmetry T_d for definiteness. We recall the table of characters of group T_d:

	E	$8C_3$	$3C_2$	$6\sigma_d$	$6S_4$
A_1	1	1	1	1	1
A_2	1	1	1	−1	−1
E	2	−1	2	0	0
T_1	3	0	−1	−1	1
T_2	3	0	−1	1	−1 .

The wave functions of a free ion (omitting the common multipliers)

$$\begin{aligned} d_{\pm 2} &= Y_2^{\pm 2} = \sqrt{\tfrac{3}{8}}\,(x \pm \mathrm{i}y)^2\;, \qquad & d_1 &= Y_2^1 = -\sqrt{\tfrac{3}{2}}\,(x+\mathrm{i}y)z\;, \\ d_{-1} &= Y_2^{-1} = \sqrt{\tfrac{3}{2}}\,(x-\mathrm{i}y)z\;, & d_0 &= Y_2^0 = \tfrac{1}{2}(3z^2 - r^2)\;, \end{aligned} \tag{7.97}$$

form the basis for the reducible representation of group T_d of dimension 5. Calculating the characters of this representation, we find

E	C_3	C_2	σ_d	S_4
5	−1	1	+1	−1 .

Character $\chi(C(\varphi)) = \dfrac{\sin(L+\frac{1}{2})\varphi}{\sin\varphi/2}$. To calculate the characters of reflections in a plane and of mirror rotations, we note that $S(\varphi) = IC(\pi + \varphi)$, $\sigma = IC_2$; here I is an inversion operation; $\varphi = 2\pi/n$ is the angle of rotation about symmetry axis, C_n; and C_2 is the operation of rotation through π about the twofold axis perpendicular to the σ plane.

Expanding this representation into irreducible representations of group T_d, we see that it decomposes into E and T_2, so that group theory predicts the

Table 7.16. Splitting of terms in fields of different symmetry for $J = n$ (n being an integer) [7.66]

J	$2J+1$	Symmetry		
		Cubic	Trigonal	Tetragonal
0	1	1(1)[a]	1(1)	1(1)
1	3	1(3)	1(1)+1(2)	1(1)+1(2)
2	5	1(2)+1(3)	1(1)+2(2)	3(1)+1(2)
3	7	1(1)+2(3)	3(1)+2(2)	3(1)+2(2)
4	9	1(1)+1(2)+2(3)	3(1)+3(2)	5(1)+2(2)
5	11	1(2)+3(3)	3(1)+4(2)	5(1)+3(2)
6	13	2(1)+1(2)+3(3)	5(1)+4(2)	7(1)+3(2)
7	15	1(1)+1(2)+4(3)	5(1)+5(2)	7(1)+4(2)
8	17	1(1)+2(2)+4(3)	5(1)+6(2)	9(1)+4(2)

[a] Orthorhombic symmetry fields show complete splitting. The figures in parentheses indicate the degree of degeneracy.

Table 7.17. Splitting of terms in fields of different symmetry for $J = 1/2\ (2n+1)$ [7.66]

J	$2J+1$	Cubic symmetry	Lower symmetry
1/2	2	1(2)	1(2)
3/2	4	1(4)	2(2)
5/2	6	1(2)+1(4)	3(2)
7/2	8	2(2)+1(4)	4(2)
9/2	10	1(2)+2(4)	5(2)
11/2	12	2(2)+2(4)	6(2)
13/2	14	3(2)+2(4)	7(2)
15/2	16	2(2)+3(4)	8(2)

splitting of term ^2D in the cubic symmetry field into one doubly and one triply degenerate (with respect to the orbit) energy levels (Tables 7.16, 17).

To find the proper functions transforming according to irreducible representations of the cubic group, we must consider thoroughly the transformation properties of the d orbits. This is done best by having the functions expressed via Cartesian coordinates and representing the operations of group T_d themselves also as the operators for replacement of the Cartesian coordinates. Let us say that rotation operation C_3 means the result of replacement

$$C_3 \begin{pmatrix} x \to y \\ y \to z \\ z \to x \end{pmatrix}, \text{ and further } C_2 \begin{pmatrix} z \to z \\ x \to -x \\ y \to -y \end{pmatrix}, \quad C_4 \begin{pmatrix} z \to z \\ x \to y \\ y \to -x \end{pmatrix}.$$

Operations E and C_n transform the wave functions as follows:

	E	c_3	c_2	c_4
d_2	d_2	$-\frac{1}{2}(d_1+d_{-1})-\frac{1}{4}(d_2+d_{-2})-\sqrt{\frac{3}{8}}\,d_0$	d_2	$-d_{-2}$
d_1	d_1	$(\mathrm{i}/2)(d_2-d_{-2})+\mathrm{i}(d_1-d_{-1})$	$-d_1$	$-\mathrm{i}d_1$
d_0	d_0	$-d_0/2+\sqrt{\frac{3}{8}}(d_2+d_{-2})$	d_0	d_0
d_{-1}	d_{-1}	$-(\mathrm{i}/2)(d_2-d_{-2})+(\mathrm{i}/2)(d_1-d_{-1})$	$-d_{-1}$	$\mathrm{i}d_{-1}$
d_{-2}	d_{-2}	$\frac{1}{2}(d_1+d_{-1})-\frac{1}{4}(d_2+d_{-2})-\sqrt{\frac{3}{8}}\,d_0$	d_{-2}	$-d_2$.

The matrix for the transformation of the wave functions of the d orbits, say for operation C_3, thus takes the form:

	d_2	d_1	d_0	d_{-1}	d_{-2}
d_2	$-\frac{1}{4}$	$-\frac{1}{2}$	$-\sqrt{\frac{3}{8}}$	$-\frac{1}{2}$	$-\frac{1}{4}$
d_1	$\mathrm{i}/2$	$\mathrm{i}/2$	0	$-\mathrm{i}/2$	$-\mathrm{i}/2$
d_0	$\sqrt{\frac{3}{8}}$	0	$-\frac{1}{2}$	0	$\sqrt{\frac{3}{8}}$
d_{-1}	$-\mathrm{i}/2$	$\mathrm{i}/2$	0	$-\mathrm{i}/2$	$\mathrm{i}/2$
d_{-2}	$-\frac{1}{4}$	$\frac{1}{2}$	$-\sqrt{\frac{3}{8}}$	$\frac{1}{2}$	$-\frac{1}{4}$.

As would be expected, character $\chi(C_3)$ in the reducible representation is equal to -1. It is also easy to check that $\chi(C_2)=1$.

From the above transformation properties it follows that functions d_0, d_1, and d_{-1} do not mix under transformation and are therefore pure orbits, so that we can write at once three regular wave functions of an electron in a cubic field: $e_g^a=d_0$, $t_{2g}^+=d_1$, $t_{2g}^-=d_{-1}$. Since functions d_2 mix with d_{-2}, the remaining two wave functions (we denote them by ψ_+ and ψ_-) will be sought in the form of linear combinations $\psi_\pm=ad_2\pm bd_{-2}$. Acting on them with operator C_2, we obtain $a=b$, and on normalizing

$$\psi_\pm=\frac{1}{\sqrt{2}}(d_2\pm d_{-2}) .$$

As a result, after a transformation corresponding to operator C_3, we get the matrix

	t_{2g}^0	t_{2g}^-	t_{2g}^+	e_g^a	e_g^b
t_{2g}^0	0	$-1/\sqrt{2}$	$-1/\sqrt{2}$	0	0
t_{2g}^-	$-\dfrac{\mathrm{i}\sqrt{2}}{2}$	$-\mathrm{i}/2$	$\mathrm{i}/2$	0	0
t_{2g}^+	$\mathrm{i}\sqrt{2}/2$	$-\mathrm{i}/2$	$\mathrm{i}/2$	0	0
e_g^a	0	0	0	$-\frac{1}{2}$	$\sqrt{3}/4$
e_g^b	0	0	0	$-\sqrt{3}/4$	$-\frac{1}{2}$,

Table 7.18. Linear combinations of spherical harmonics transforming according to representations of the cubic group [7.61]

	Linear combinations	Representation
S	Y_0^0	A_1
P	Y_1^{-1} Y_1^0 Y_1^1	T_1
D	$\frac{1}{\sqrt{2}}(Y_2^2 - Y_2^{-2})$ Y_2^1 Y_2^{-1}	T_2
	$\frac{1}{\sqrt{2}}(Y_2^2 + Y_2^{-2})$ Y_2^0	E
F	$\sqrt{3/8}\, Y_3^1 + \sqrt{5/8}\, Y_3^{-3}$ $\sqrt{3/8}\, Y_3^{-1} + \sqrt{5/8}\, Y_3^3$ Y_3^0	T_1
	$\sqrt{5/8}\, Y_3^1 - \sqrt{3/8}\, Y_3^{-3}$ $\sqrt{5/8}\, Y_3^{-1} - \sqrt{3/8}\, Y_3^3$ $\frac{1}{\sqrt{2}}(Y_3^2 + Y_3^{-2})$	T_2
	$\frac{1}{\sqrt{2}}(Y_3^2 - Y_3^{-2})$	A_2
G	$\frac{1}{\sqrt{2}}(Y_4^2 - Y_4^{-2})$ $\sqrt{1/8}\, Y_4^{-1} - \sqrt{7/8}\, Y_4^3$ $\frac{1}{\sqrt{8}}\, Y_4^1 - \sqrt{7/8}\, Y_4^{-3}$	T_2
	$\sqrt{7/12}\, Y_4^0 + \sqrt{5/12}\,(Y_4^4 + Y_4^{-4})\frac{1}{\sqrt{2}}$	A_1
	$\frac{1}{\sqrt{2}}(Y_4^2 + Y_4^{-2})$ $\sqrt{5/12}\, Y_4^0 - \sqrt{7/12}\,(Y_4^4 + Y_4^{-4})$	E
	$\frac{1}{\sqrt{2}}(Y_4^4 - Y_4^{-4})$ $\sqrt{7/8}\, Y_4^{-1} + \sqrt{1/8}\, Y_4^3$ $\sqrt{7/8}\, Y_4^1 + \sqrt{1/8}\, Y_4^{-3}$	T_1
H	$\frac{1}{\sqrt{2}}(Y_5^2 - Y_5^{-2})$ $\frac{1}{\sqrt{2}}(Y_5^4 - Y_5^{-4})$	E
	$\frac{1}{\sqrt{2}}(Y_5^2 + Y_5^{-2})$ $-\frac{\sqrt{120}}{16} Y_5^{-5} + \frac{\sqrt{112}}{16} Y_5^{-1} + \frac{\sqrt{24}}{16} Y_5^3$ $-\frac{\sqrt{120}}{16} Y_5^5 + \frac{\sqrt{112}}{16} Y_5^1 + \frac{\sqrt{24}}{16} Y_5^3$	T_2
	Y_5^0 $\frac{\sqrt{126}}{16} Y_5^{-5} + \frac{\sqrt{60}}{16} Y_5^{-1} + \frac{\sqrt{70}}{16} Y_5^3$ $\frac{\sqrt{126}}{16} Y_5^5 + \frac{\sqrt{60}}{16} Y_5^1 + \frac{\sqrt{70}}{16} Y_5^{-3}$	T_1
	$\frac{1}{\sqrt{2}}(Y_5^4 + Y_5^{-4})$ $\frac{\sqrt{10}}{16} Y_5^{-5} + \frac{\sqrt{84}}{16} Y_5^{-1} - \frac{\sqrt{162}}{16} Y_5^3$ $\frac{\sqrt{10}}{16} Y_5^5 + \frac{\sqrt{84}}{16} Y_5^1 - \frac{\sqrt{162}}{16} Y_5^{-3}$	T_1

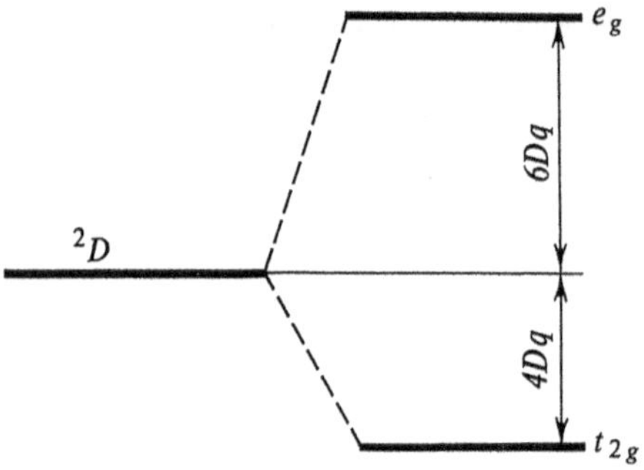

Fig. 7.32. Splitting of term ^2D in a cubic field

whence we see $\chi_{e_g}(C_3) = -1$, $\chi_{t_{2g}}(C_3) = 0$, as should be the case for irreducible representations of group T_d.

It can be verified that the transformation matrices of the constructed wave functions split into irreducible representations E and T_{2g} for the other operators of this group as well. Finally, for proper functions we have (Table 7.18):

$$t_{2g}^0 = \frac{1}{\sqrt{2}}(d_2 - d_{-2}) \ , \quad t_{2g}^- = d_{-1} \ , \quad t_{2g}^+ = d_1 \ ,$$
$$e_g^a = d_0 \ , \quad e_g^b = \frac{1}{\sqrt{2}}(d_2 + d_{-2}) \ . \tag{7.98}$$

Now we calculate the splitting between the two sublevels e_g and t_{2g}, into which term ^2D splits in a cubic field (Fig. 7.32). We denote the energy operator of the cubic field by V_0; we have already seen that $V_0 = D_4[Y_4^0 + \sqrt{\frac{5}{14}}(Y_4^4 + Y_4^{-4})]$, whose equivalent operator has the form $O_4^0 + 5O_4^4$. The value of matrix element $\langle d_2 | V_0 | d_2 \rangle$ will be denoted by Dq. Magnitude $10\,Dq$ is called the *crystal-field strength*. For the other matrix elements we get (Table 7.15):

$$\langle d_2 | V_0 | d_{-2} \rangle = 5\,Dq \ , \quad \langle d_1 | V_0 | d_1 \rangle = -4\,Dq \ ,$$
$$\langle d_0 | V_0 | d_0 \rangle = 6\,Dq \ , \quad \langle d_{-2} | V_0 | d_{-2} \rangle = Dq \tag{7.99}$$
$$\langle d_1 | V_0 | d_{-1} \rangle = -4\,Dq \ .$$

The corrections to the energy levels have the form

$$\langle \varphi_{e_g^a} | V_0 | \varphi_{e_g^a} \rangle = 6\,Dq \ , \quad \langle \varphi_{t_{2g}^0} | V_0 | \varphi_{t_{2g}^0} \rangle = -4\,Dq \ . \tag{7.100}$$

Thus, under the effect of a cubic field, term ^2D splits into level e_g, removed by a distance $6\,Dq$, and level t_{2g}, removed by $-4\,Dq$ (Table 7.19). The total splitting is found to equal $10\,Dq$, which serves as a pictorial explanation of the

Table 7.19. Splitting of terms in a weak octahedral field [7.61]

Main term of the ion	State and energy (in Dq units)	Main term of the ion	State and energy (in Dq units)
$d^1\ {}^2D$	${}^2T_{2g}(-4)$; ${}^2E_g(6)$	$d^6\ {}^5D$	${}^5T_{2g}(-4)$; ${}^5E_g(6)$
$d^2\ {}^3F$	${}^3T_{1g}(-6)$; ${}^3T_{2g}(2)$; ${}^3A_{2g}(12)$	$d^7\ {}^4F$	${}^4T_{1g}(-6)$; ${}^4T_{2g}(2)$; ${}^4A_{2g}(12)$
$d^3\ {}^4F$	${}^4A_{2g}(-12)$; ${}^4T_{2g}(-2)$; ${}^4T_{1g}(6)$	$d^8\ {}^3F$	${}^3A_{2g}(-12)$; ${}^3T_{2g}(-2)$; ${}^3T_{1g}(6)$
$d^4\ {}^5D$	${}^5E_g(-6)$; ${}^5T_{2g}(4)$	$d^9\ {}^2D$	${}^2E_g(-6)$; ${}^2T_{2g}(4)$
$d^5\ {}^6S$	${}^6A_{1g}(0)$		

term "cubic field strength." The same result can be obtained simply from the secular equation for d orbitals; we have specially emphasized the use of the symmetry properties in setting up regular wave functions of the zero approximation in crystal-field theory.

7.12.3 Ion with Configuration $3d^2$ in a Weak Cubic Field

In considering the terms of a free ion with configuration $3d^2$, as is done in atomic spectroscopy, their energies are expressed as follows via Slater integrals (F_0, F_2, F_4) or Racah parameters ($A = F_0 - 49F_4$, $B = F_2 - 5F_4$, $C = 35F_4$):

$$
\begin{aligned}
&E({}^1G) = F_0 + 4F_2 + F_4 \ , && E({}^3P) = F_0 + 7F_2 - 84F_4 \ , \\
&E({}^3F) = F_0 - 8F_2 - 9F_4 \ , && E({}^1S) = F_0 + 14F_2 + 126F_4 \ . \\
&E({}^1D) = F_0 - 3F_2 + 36F_4 \ , &&
\end{aligned}
\tag{7.101}
$$

We now pass on to calculation of the energy levels of these terms in a weak crystal cubic field.

The cubic field operator will be written as usual, in the form of an equivalent operator:

$$V_0 = O_4^0 + 5O_4^4 \ ,$$

where $O_4^0 = 35L_z^4 - [30(L+1)L - 25]L_z^2 - 6L(L+1) + 3L^2(L+1)^2$ and $O_4^4 = \frac{1}{2}(L_+^4 + L_-^4)$.

We have already seen that term D splits into terms whose wave functions are transformed according to irreducible representations E and T_2; we now similarly find the splittings of the other terms:

$$S \to A_1 \,, \qquad F \to A_2 + T_1 + T_2 \,,$$
$$P \to T_1 \,, \qquad G \to A_1 + E + T_1 + T_2 \,.$$
$$D \to E + T_2 \,,$$

Let us turn to the construction of the proper functions of the terms. We denote by ψ_M the wave functions of term $F(L=3)$, $M_L = M$. We have to construct sets of linear combinations of these functions being transformed as A_2, T_1, and T_2 under the effect of symmetry operations of group T_d. It can be shown that the wave functions which are transformed as T_1 will be (Table 7.18)

$$\Psi_1 = \sqrt{\tfrac{3}{8}}\psi_1 + \sqrt{\tfrac{5}{8}}\psi_{-3} \,, \quad \Psi_{-1} = \sqrt{\tfrac{3}{8}}\psi_{-1} + \sqrt{\tfrac{5}{8}}\psi_3 \,, \quad \Psi_0 = \psi_0 \,. \tag{7.102}$$

To this end, we write out wave functions ψ_M in Cartesian coordinates:

$$\begin{aligned} \psi_0 &= \tfrac{1}{2} z(5z^2 - 3r^2) \,, & \psi_1 &= -\sqrt{\tfrac{3}{16}}\,(x+\mathrm{i}y)(5z^2 - r^2) \,, \\ \psi_{-3} &= \sqrt{\tfrac{5}{16}}\,(x-\mathrm{i}y)^3 \,, & \psi_{-1} &= \sqrt{\tfrac{3}{16}}\,(x-\mathrm{i}y)(5z^2 - r^2) \,, \end{aligned} \tag{7.103}$$

etc. (the common multipliers are omitted), and perform symmetry transformations, for instance, C_3:

$$C_3\psi_0 = \sqrt{\tfrac{1}{4}}(5x^2 - 3r^2)x = \frac{1}{\sqrt{2}}\left[\sqrt{\tfrac{3}{8}}\,\psi_1 + \sqrt{\tfrac{5}{8}}\,\psi_{-3}\right]$$
$$- \frac{1}{\sqrt{2}}\left[\sqrt{\tfrac{3}{8}}\,\psi_{-1} + \sqrt{\tfrac{5}{8}}\,\psi_3\right] .$$

Finally, we find the following matrix for transformation C_3:

	Ψ_1	Ψ_0	Ψ_{-1}
Ψ_1	$-\mathrm{i}/2$	$1/\sqrt{2}$	$-\mathrm{i}/2$
Ψ_0	$1/\sqrt{2}$	0	$-1/\sqrt{2}$
Ψ_{-1}	$\mathrm{i}/2$	$1/\sqrt{2}$	$\mathrm{i}/2$

$\chi(C_3) = 0$, which should actually be the case in T_1. In precisely the same way we find the wave functions transformed (Table 7.18) according to T_2 and A_2:

$$\begin{aligned} &T_2 \begin{cases} \sqrt{\tfrac{5}{8}}\,\psi_1 - \sqrt{\tfrac{3}{8}}\,\psi_{-3} \\ \sqrt{\tfrac{5}{8}}\,\psi_{-1} - \sqrt{\tfrac{3}{8}}\,\psi_{-3} \,, \\ \dfrac{\psi_2 + \psi_{-2}}{\sqrt{2}} \end{cases} \\ &A_2 \left\{ \dfrac{\psi_2 - \psi_{-2}}{\sqrt{2}} \right. . \end{aligned} \tag{7.104}$$

Table 7.20. Energy levels of configuration d^2 in the cubic symmetry field [7.67]

Level type	Energy: Neglecting term interaction	Energy: Taking into account term interaction
$^3A_2$	$^3F+12Dq$	$^3F+12Dq$
$^3T_2$	^3F+2Dq	^3F+2Dq
$^3T_1$	$\begin{cases} ^3P \\ ^3F-6Dq \end{cases}$	$\frac{1}{2}\{^3P+{}^2F-6Dq\pm[(^3P-{}^3F+6Dq)^2+64(Dq)^2]^{1/2}\}$
$^1T_2$	$\begin{cases} ^1D-\frac{16}{7}Dq \\ ^1G-\frac{26}{7}Dq \end{cases}$	$\frac{1}{2}\{^1D+{}^1G-6Dq\pm[(^1D-{}^1G+\frac{10}{7}Dq)^2+\frac{4800}{49}(Dq)^2]^{1/2}\}$
^1E	$\begin{cases} ^1D+\frac{24}{7}Dq \\ ^1G+\frac{4}{7}Dq \end{cases}$	$\frac{1}{2}\{^1D+{}^1G+4Dq\pm[(^1D-{}^1G+\frac{20}{7}Dq)^2+\frac{19200}{49}(Dq)^2]^{1/2}\}$
$^1T_1$	^1G+2Dq	^1G+2Dq
$^1A_1$	$\begin{cases} ^1S \\ ^1G+4Dq \end{cases}$	$\frac{1}{2}\{^1S+{}^1G+4Dq\pm[(^1S-{}^1G-4Dq)^2+384(Dq)^2]^{1/2}\}$

We can now also calculate the splitting of these terms in a cubic field.

One of the wave functions belonging to term ^3F is

$$\psi(LM_LSM_S)=\psi(3211)=|\overset{+}{2}\overset{+}{0}|=\psi_2\ ,\quad \psi_{-2}=|\overset{+}{0}-\overset{+}{2}|\ ;$$

therefore,

$$E(^3A_2)=\int\frac{1}{\sqrt{2}}(\psi_2-\psi_{-2})V_0\left(\frac{\psi_2-\psi_{-2}}{\sqrt{2}}\right)d\tau$$

$$=\tfrac{1}{2}\int|\overset{+}{2}\overset{+}{0}|V_0|\overset{+}{2}\overset{+}{0}|d\tau-\tfrac{1}{2}\int|\overset{+}{2}\overset{+}{0}|V_0|\overset{+}{0}-\overset{+}{2}|d\tau$$

$$-\tfrac{1}{2}\int|\overset{+}{0}-\overset{+}{2}|V_0|\overset{+}{2}\overset{+}{0}|d\tau+\tfrac{1}{2}\int|\overset{+}{0}-\overset{+}{2}|V_0|\overset{+}{0}-\overset{+}{2}|d\tau\ .$$

Representing $|\overset{+}{2}\overset{+}{0}|=\begin{vmatrix} d_2(1) & d_0(1) \\ d_2(2) & d_0(2) \end{vmatrix}$ and calculating, we find

$$E(^3A_2)=12Dq\ . \tag{7.105}$$

Similarly,

$$E(^3T_2)=2Dq\ ,\quad E(^3T_1)=-6Dq\ . \tag{7.106}$$

For the energy levels originating from term ^1D we find analogously (Table 7.20):

$$E(^1E)=\tfrac{24}{7}Dq\ ,\quad E(^1T_2)=-\tfrac{16}{7}Dq\ . \tag{7.107}$$

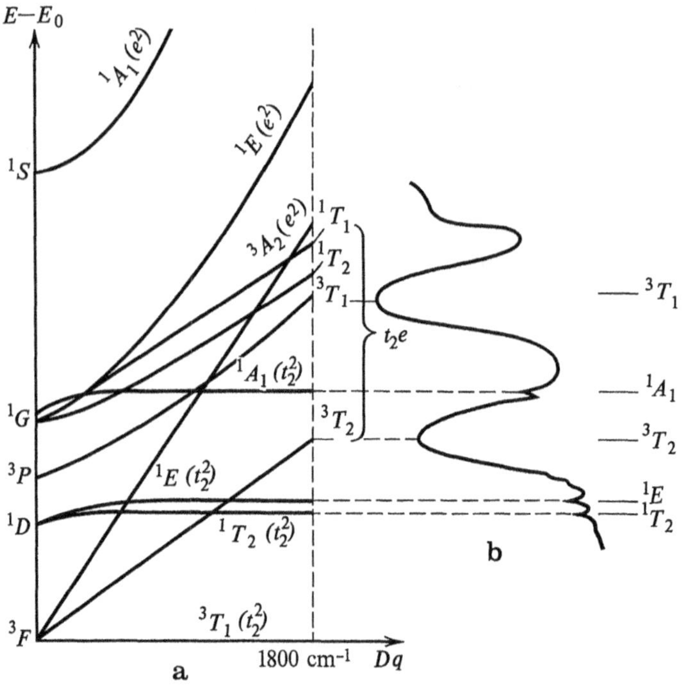

Fig. 7.33. Level scheme (**a**) and absorption spectrum (**b**) of the V^{3+} ion in corundum: $Dq = 1800$, $B = 680$, $C = 2500\ \mathrm{cm}^{-1}$ [7.67]

For term ^3P we have representation $^3T_1$, and therefore it will interact with the identical term $^3T_1$ originating from ^3F. Their interaction energy will be found by solving the secular equation

$$\begin{array}{cc} {}^3T_{1g}({}^3F) & {}^3T_{1g}({}^3P) \end{array}$$

$$\begin{vmatrix} -6Dq - \varepsilon + E & 4Dq \\ 4Dq & E - \varepsilon \end{vmatrix} = 0 \ ,$$

$$\varepsilon = \frac{E - 6Dq}{2} \pm \left[\frac{(E - 6Dq)^2}{4} + 6DqE + (4Dq)^2 \right]^{1/2} , \quad E = {}^3P - {}^3F \ .$$

In the same manner we can consider the splitting of the other terms of configuration $3d^2$ in the cubic field (Table 7.20 and Fig. 7.33).

7.12.4 Strong Crystal Field

So far we have been discussing the behavior of d electrons of ions in a weak crystal field. This meant that in the initial Hamiltonian the terms describing the interaction of electrons greatly exceeded the electron energy in the electric field of the crystal.

If the splitting parameter in a crystal field Dq is large enough, the state of orbital motion will primarily depend precisely on the effect of the crystal field rather than on the electron interaction. Quantum numbers L and S will no longer be "good" numbers. To determine the level scheme we must now consider, first, each electron in the crystal field separately, and only then take into account their interaction.

We know from crystal-field theory for a single d electron that state t_{2g} is the lower state. Therefore, low-energy group of states for a system of several electrons arises on filling of the whole t_{2g} shell at first. After all the six electrons appear in it, the filling of a higher energy level by electrons in the e_g state begins. The place of quantum numbers L and S is now occupied by the degrees of filling of orbits t_{2g} and e_g. After such a construction of electron configurations, one should take into account the coulomb repulsion and exchange interaction among the electrons. The estimate of the respective energies is obtained by calculating the number of electrons in orbits t_{2g} and e_g with $E(t_{2g}) = -4Dq$, $E(e_g) = 6Dq$, and we must add to the sum obtained the interaction energy among the electrons expressed by Slater integrals. This last value is found in the usual way as in atomic spectroscopy, when considering the free ion. This requires the knowledge of the coulomb, J, and the exchange, K, integrals for orbits t_{2g} and e_g. For instance $\left(\text{using the real wave functions of the orbits } \frac{1}{\mathrm{i}\sqrt{2}}(d_2 - d_{-2}) = \sqrt{3}\,(xy)\right)$,

$$J(xz, yz) = \int (xz)^*(1)\,(yz)^*(2)\,\frac{1}{r_{12}}\,(xz)(1)\,(yz)(2)\,d\tau = F_0 - 2F_2 - 4F_4\ ,$$

$$K(xz, yz) = \int (xz)^*(1)\,(yz)^*(2)\,\frac{1}{r_{12}}\,(yz)(1)\,(xz)(2)\,d\tau = 3F_2 + 20F_4\ .$$

Table 7.21. Coulomb (J) and exchange (K) integrals of set $t_{2g}(xz, xy, yz)$ and $e_g(x^2-y^2, z^2)$ [7.61]

$J(z^2, z^2)$	$= J(x^2-y^2, x^2-y^2) = J(xy, xy) = J(xz, xz) = J(yz, yz) = F_0 + 4F_2 + 36F_4$
$J(x^2-y^2, xz)$	$= J(x^2-y^2, yz) = J(xy, yz) = J(xy, xz) = J(xz, yz) = F_0 - 2F_2 - 4F_4$
$J(z^2, xz)$	$= J(z^2, yz) = F_0 + 2F_2 - 24F_2$
$J(z^2, xy)$	$= J(z^2, x^2-y^2) = F_0 - 4F_2 + 6F_4$
$J(x^2-y^2, xy)$	$= F_0 + 4F_2 - 34F_4$
$K(xy, yz)$	$= K(xy, xz) = K(xz, yz) = K(x^2-y^2, xz) = K(x^2-y^2, yz) = 3F_2 + 20F_4$
$K(z^2, x^2-y^2)$	$= K(z^2, xy) = 4F_2 + 15F_4$
$K(z^2, xz)$	$= K(z^2, yz) = F_2 + 30F_4$
$K(x^2-y^2, xy)$	$= 35F_4$

This method has been employed for compiling tables (Table 7.21), which are given in many books.

We now revert to the case of two d electrons and consider configuration $(t_{2g})^2$ in the strong-field approximation. For spin $S=1$ we have the following determinant functions taking into account Pauli's exclusion principle:

$$\psi_1=|(x^+z)(y^+z)|\,,\quad \psi_2=|(x^+z)(x^+y)|\,,\quad \psi_3=|(y^+z)(x^+y)|\,.$$

As usual, we find out the properties of these functions under symmetry transformations, say, of group T_d.

We write down the respective symmetry operations:

$$C_3\begin{pmatrix} x\to y \\ y\to z \\ z\to x \end{pmatrix},\quad C_2\begin{pmatrix} z\to z \\ x\to -x \\ y\to -y \end{pmatrix},\quad \sigma\begin{pmatrix} x\to y \\ y\to x \\ z\to z \end{pmatrix},\quad S_4\begin{pmatrix} x\to y \\ y\to -x \\ z\to -z \end{pmatrix}.$$

We now act on wave functions ψ_1, ψ_2, and ψ_3 by operator C_3:

$$C_3\psi_1=C_3|(xz)(yz)|=|(yx)(zx)|=-\psi_2\,,$$

$$C_3\psi_2=-\psi_3\,,\quad C_3\psi_3=\psi_1\,.$$

In the cubic-group representation generated by basis functions ψ_1, ψ_2, ψ_3, to element C_3 thus corresponds the matrix

$$C_3\to\begin{Vmatrix} 0 & -1 & 0 \\ 0 & 0 & -1 \\ 1 & 0 & 0 \end{Vmatrix}.$$

Character $\chi(C_3)$ in this representation is equal to zero. In a similar way we find the matrices and characters

$$C_2\to\begin{Vmatrix} 1 & 0 & 0 \\ 0 & -1 & 0 \\ 0 & 0 & -1 \end{Vmatrix},\quad \chi(C_2)=-1\,;\quad \sigma\to\begin{Vmatrix} -1 & 0 & 0 \\ 0 & -1 & 0 \\ 0 & 0 & 1 \end{Vmatrix},\quad \chi(\sigma)=-1\,;$$

$$S_4\to\begin{Vmatrix} 1 & 0 & 0 \\ 0 & 0 & 1 \\ 0 & -1 & 0 \end{Vmatrix},\quad \chi(S_4)=1\,.$$

Recall the table of characters of irreducible representations of group T_d or group O, which is isomorphous to it:

	E	C_3	C_2	σ	S_4
t_{2g}	3	0	-1	1	-1
t_{1g}	3	0	-1	-1	1
e_g	2	-1	2	0	0
A_{1g}	1	1	1	1	1
A_{2g}	1	1	1	-1	-1 .

Comparing the characters obtained, we see that ψ_1, ψ_2, ψ_3 are the basis functions of irreducible representation T_{1g}. Configuration $(t_{2g})^2$ includes nine wave functions generating the nine-dimensional reducible representation of the cubic group. To break it down into irreducible representations we set up characters in direct product $t_{2g} \times t_{2g}$; they will be:

E	C_3	C_2	σ	S_4
9	0	1	1	1 ,

which means that this reducible representation contains four irreducible representations of group T_d: $t_{2g} \times t_{2g} = A_{1g} + E_g + T_{1g} + T_{2g}$, in agreement with the result just obtained.

To find the other regular wave functions of configuration $(t_{2g})^2$ in the cubic field, we first write the wave functions for $S = 0$:

$$\Theta_1 = \frac{1}{\sqrt{2}}[|\overset{+}{(xz)}\overset{-}{(yz)}| - |\overset{-}{(xz)}\overset{+}{(yz)}|] \,, \quad \Theta_4 = |\overset{+}{(xz)}\overset{-}{(xz)}| \,,$$

$$\Theta_2 = \frac{1}{\sqrt{2}}[|\overset{+}{(xz)}\overset{-}{(xy)}| - |\overset{-}{(xz)}\overset{+}{(xy)}|] \,, \quad \Theta_5 = |\overset{+}{(yz)}\overset{-}{(yz)}| \,,$$

$$\Theta_3 = \frac{1}{\sqrt{2}}[|\overset{+}{(yz)}\overset{-}{(xy)}| - |\overset{-}{(yz)}\overset{+}{(xy)}|] \,, \quad \Theta_6 = |\overset{+}{(xy)}\overset{-}{(xy)}| \,.$$

Using linear combinations of these functions, we must compile three more sets of functions, which are the bases of irreducible representations ${}^1A_{1g}$, ${}^1T_{2g}$, and ${}^1E_g$. We consider the transformation properties of functions $\Theta_1 \ldots \Theta_6$ and choose these combinations. We then find that Θ_1, Θ_2, Θ_3 is the basis of T_{2g}; the basis of two-dimensional representation ${}^1E_g$ is

$$\frac{1}{\sqrt{2}}(\Theta_4 - \Theta_5) \,, \quad \frac{1}{\sqrt{6}}\Theta_4 + \frac{1}{\sqrt{6}}\Theta_5 - \sqrt{\tfrac{4}{6}}\,\Theta_6 \,,$$

and the basis of representation A_{1g}, the linear combination

Table 7.22. Values of nonzero matrix elements $(ab|1/r_{12}|cd)$ [7.61]

a	b	c	d	$(ab\|1/r_{12}\|cd)$	a	b	c	d	$(ab\|1/r_{12}\|cd)$
(xz)	(z^2)	(xz)	(x^2-y^2)	$-2\sqrt{3}F_2+10\sqrt{3}F_4$	(z^2)	(xy)	(yz)	(xz)	$\sqrt{3}F_2-5\sqrt{3}F_4$
(yz)	(z^2)	(yz)	(x^2-y^2)	$2\sqrt{3}F_2-10\sqrt{3}F_4$	(z^2)	(xz)	(xy)	(yz)	$2\sqrt{3}F_2-10\sqrt{3}F_4$
(xz)	(xz)	(z^2)	(x^2-y^2)	$\sqrt{3}F_2-\sqrt{3}F_4$	(x^2-y^2)	(xy)	(xz)	(yz)	$3F_2-15F_4$
(yz)	(yz)	(z^2)	(x^2-y^2)	$-\sqrt{3}F_2+5\sqrt{3}F_4$	(x^2-y^2)	(xy)	(yz)	(xz)	$-3F_2+15F_4$
(z^2)	(xy)	(xz)	(yz)	$\sqrt{3}F_2-5\sqrt{3}F_4$					

Table 7.23. Splitting of terms in a strong cubic field [7.61]

d^1	e	^2E
	t_2	$^2T_2$
d^2	e^2	$^1A_1+{}^3A_2+{}^1E$
	et_2	$^1T_1+{}^3T_1+{}^1T_2+{}^3T_2$
	$(t_2)^2$	$^1A_1+{}^1E+{}^3T_1+{}^1T_2$
d^3	e^3	^2E
	e^2t^2	$^2T_1+{}^4T_1+2{}^2T_1$
	$e(t_2)^2$	$^2A_1+{}^2A_2+2{}^2E+2{}^2T_1+{}^4T_1+2{}^2T_2+{}^4T_2$
	$(t_2)^3$	$^4A_2+{}^2E+{}^2T_1+{}^2T_2$
d^4	e^4	$^1A_1$
	e^3t_2	$^1T_1+{}^3T_1+{}^1T_2+{}^3T_2$
	$e^2(t_2)^2$	$2{}^1A_1+{}^1A_2+{}^3A_2+3{}^1E+{}^3E+{}^1T_1+3{}^3T_1+3{}^1T_2+2{}^3T_2+{}^5T_2$
	$e^2(t_2^3)$	$^1A_1+{}^3A_1+{}^1A_2+{}^3A_2+{}^1E+2{}^3E+{}^5E+2{}^1T_1+2{}^3T_1+2{}^1T_2+2{}^3T_2$
	$(t_2)^4$	$^1A_1+{}^1E+{}^3T_1+{}^1T_2$
d^5	e^4t_2	$^2T_2$
	$e^3(t_2)^2$	$^2A_1+{}^2A_2+2{}^2E+2{}^2T_1+{}^4T_1+2{}^2T_2+{}^4T_2$
	$e^2(t_2)^3$	$2{}^2A_1+{}^4A_1+{}^6A_1+{}^2A_2+{}^4A_2+3{}^2E+2{}^4E+4{}^2T_1+{}^4T_1+4{}^2T_2+{}^4T_2$
	$e(t_2)^4$	$^2A_1+{}^2A_2+2{}^2E+2{}^2T_1+{}^4T_1+2{}^2T_2+{}^4T_2$
	$(t_2)^5$	$^2T_2$

$$\frac{1}{\sqrt{3}}\Theta_4+\frac{1}{\sqrt{3}}\Theta_5+\frac{1}{\sqrt{3}}\Theta_6 \ .$$

When the regular wave functions of the zero approximation have been found, one should take into account (as mentioned above) the mutual repulsion and exchange interaction of electrons (Table 7.22) and finally calculate the energies of the terms of configuration $(t_{2g})^2$:

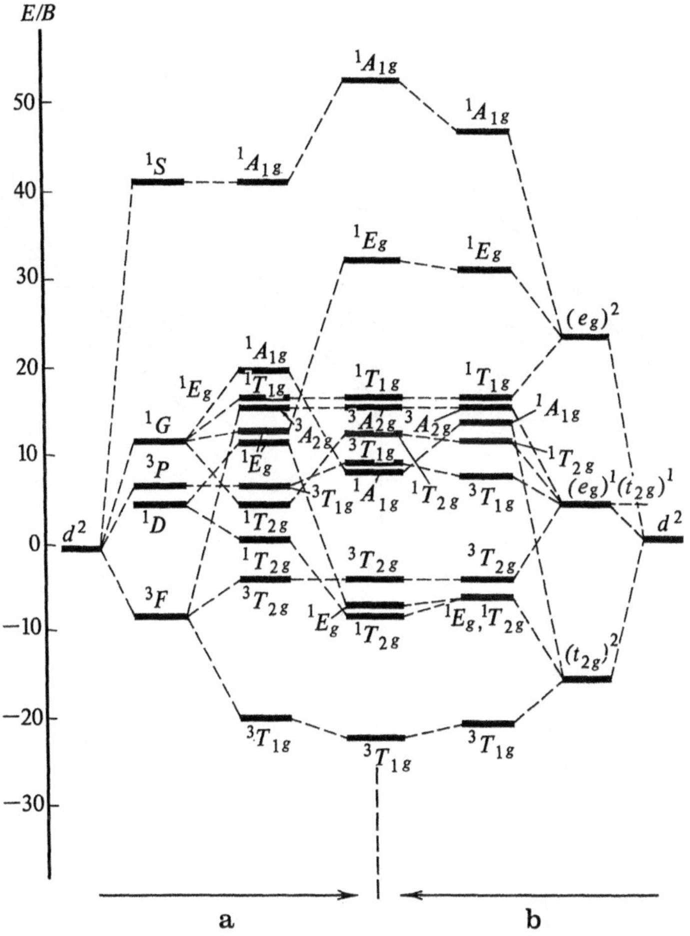

Fig. 7.34. Level scheme of the $3d^2$ configuration in the **(a)** weak- and **(b)** strong-field approximations [7.73]

$$
\begin{aligned}
&E(^3T_{1g}) = E(\psi_1) = J(xz, yz) - K(xz, yz) = F_0 - 5F_2 - 24F_4 \ ,\\
&E(^1T_{2g}) = F_0 + F_2 + 16F_4 \ , \qquad\qquad (7.108)\\
&E(^1E_g) \ = F_0 + F_2 + 16F_4 \ , \quad E(^1A_{1g}) = F_0 + 10F_2 + 76F_4 \ .
\end{aligned}
$$

To all these energies one should add $2E(t_{2g}) = -8Dq$. The terms of excited configurations $e_g t_{2g}$ and e_g^2 should be considered in the same manner.

In the strong-field approximation we ultimately have (Table 7.23):

$$
\begin{aligned}
&(t_{2g})^2 \ \rightarrow {}^1A_{1g} + {}^1E_{1g} + {}^3T_{1g} + {}^1T_{2g} \ ,\\
&(e_g t_{2g}) \rightarrow {}^1T_{1g} + {}^3T_{1g} + {}^1T_{2g} + {}^3T_{2g} \ ,\\
&(e_g)^2 \ \rightarrow {}^1A_{1g} + {}^3A_{2g} + {}^1E_{1g} \ .
\end{aligned}
$$

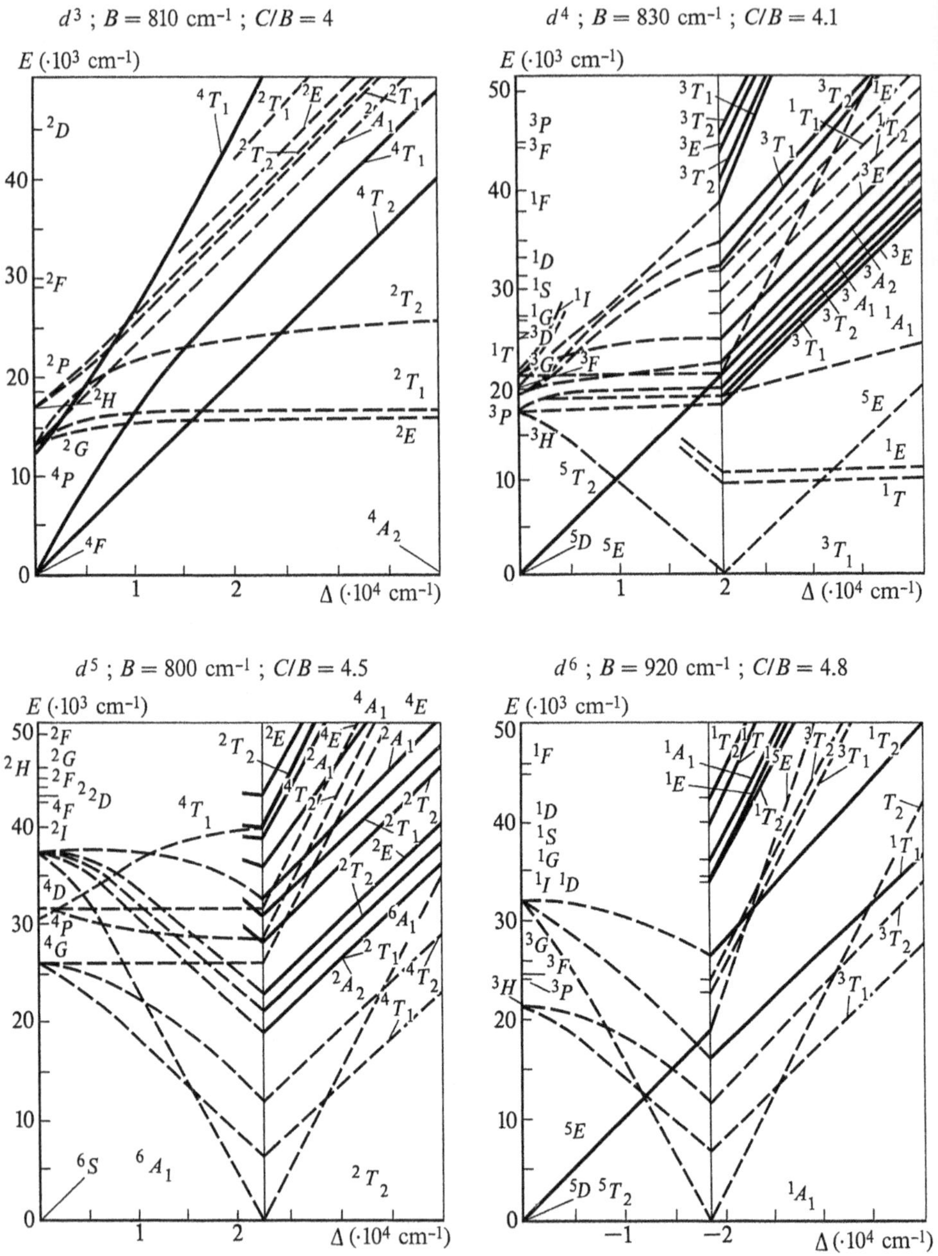

Fig. 7.35. Dependence of the level energies of d^n configurations on the strength of a cubic crystal field ($\Delta = 10Dq$) [7.68]. The verticals correspond to Dq values at which the change of the ground state occurs (translation from the weak field to the strong field)

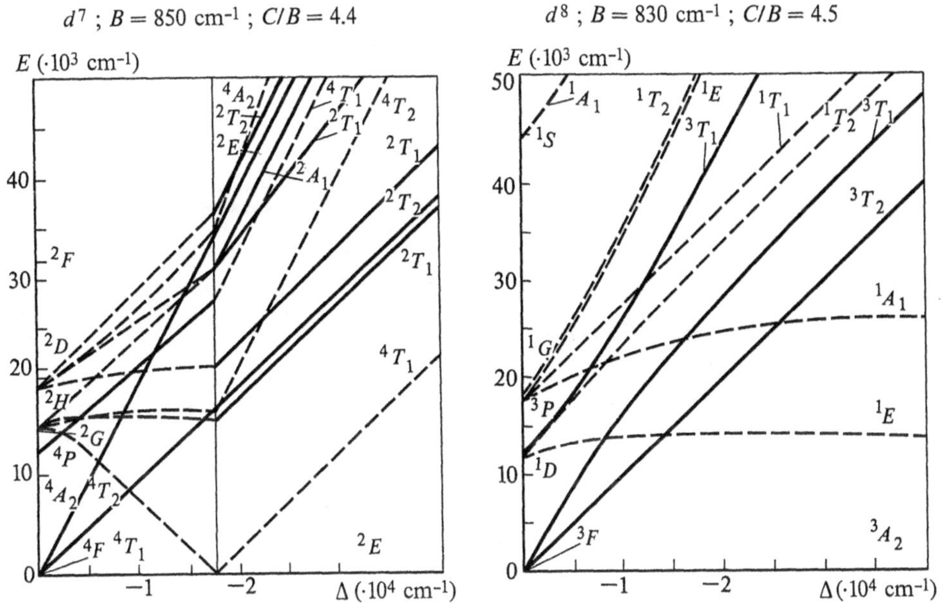

Fig. 7.35 (continued)

These are the same terms as those obtained in the weak-field approximation as originating from the terms of free ion ${}^3F$, ${}^1D$, ${}^3P$, ${}^1G$, and ${}^1S$. The entire pattern of splitting of the terms of configuration $3d^2$ in a cubic field and the relationship of the limiting cases of a weak field and a strong field is presented in Fig. 7.34.

The diagrams of the energy levels of all the d^n configurations in the strong-field scheme have been constructed [7.68-71] (Figs. 7.33,35).

7.12.5 Calculating the Spin-Orbital Interaction of Configuration d^2 in a Cubic Field

First let us use the *weak-field approximation* to consider the calculation of the elements of the energy matrix with due regard for the spin-orbital interaction in a weak cubic field. A group-theory consideration similar to the ones carried out previously shows that the energy levels of configuration d^2 belong to various irreducible representations of a double cubic group. In particular, representation Γ_1 (Bete's notation) includes terms ${}^1S_0$, ${}^1G_4$, ${}^3P_0$, and ${}^3F_4$. The matrix elements of block Γ_1 of the energy matrix have already been calculated, neglecting the spin-orbital interaction; they are given in Table 7.23. When this interaction is taken into account, the energy of the terms of block Γ_1 is calculated by the matrix [7.72]

$$\begin{array}{cccc} {}^1S_0 & {}^1G_4 & {}^3P_0 & {}^3F_4 \end{array}$$

$$\left\| \begin{array}{cccc} 14F_2+126F_4 & 4\sqrt{6}\,Dq & -2\sqrt{6}\,\lambda & 0 \\ & 4F_2+4F_4+4Dq & 0 & 2\lambda \\ & & 7F_2-84F_4-2\lambda & -4Dq \\ & & & -8F_2-9F_4-6Dq+3\lambda \end{array} \right\|$$

We shall show now, as an example, how the following matrix element of the spin-orbital interaction operator is calculated:

$$\langle {}^1S_0(\Gamma_1)|\sum \xi \boldsymbol{ls}|{}^3P_0(\Gamma_1)\rangle \; .$$

To do this, we must know wave functions ${}^1S_0$ and ${}^3P_0$. Wave function ${}^3P_0$ must be a linear combination $\psi(LM_LSM_S)$: $\psi(1\ \ -1\ \ 1\ \ 1)$, $\psi(1\ \ 0\ \ 1\ \ 0)$, $\psi(1\ \ 1\ \ 1\ \ -1)$. The coefficients of linear combinations are Wigner coefficents $(j_1\,1\,m_1 m_2\,|\,j_1\,1\,j\,m)$ [7.65]:

j	$m_2=1$	$m_2=0$
j_1+1	$\sqrt{\frac{(j_1+m)(j_1+m+1)}{(2j_1+1)(2j_1+2)}}$	$\sqrt{\frac{(j_1-m+1)(j_1+m+1)}{(2j_1+1)(j_1+1)}}$
j_1	$-\sqrt{\frac{(j_1+m)(j_1-m+1)}{2j_1(j_1+1)}}$	$\frac{m}{\sqrt{j_1(j_1+1)}}$
j_1-1	$\sqrt{\frac{(j_1-m)(j_1-m+1)}{2j_1(2j_1+1)}}$	$-\sqrt{\frac{(j_1-m)(j_1+m)}{j_1(2j_1+1)}}$

j	$m_2=-1$
j_1+1	$\sqrt{\frac{(j_1-m)(j_1-m+1)}{(2j_1+1)(2j_1+2)}}$
j_1	$\sqrt{\frac{(j_1-m)(j_1+m+1)}{2j_1(j_1+1)}}$
j_1-1	$\sqrt{\frac{(j_1+m+1)(j_1+m)}{2j_1(2j_1+1)}}$

For the case at hand, $j_1=1$, $j=0$ and $m=0$. From the last line of this formula we find:

$$\psi({}^3P_0)=\frac{1}{\sqrt{3}}\psi(1\ \ -1\ \ 1\ \ 1)-\frac{1}{\sqrt{3}}\psi(1\ \ 0\ \ 1\ \ 0)+\frac{1}{\sqrt{3}}\psi(1\ \ 1\ \ 1\ \ -1)\;.$$

The equations for the proper functions expressed via the determinant function can be obtained with the aid of stepping-up and depressing operators

$$\begin{aligned}
\psi(1\ \ -1\ \ 1\ \ 1) &= \sqrt{\tfrac{2}{5}}\,|\overset{+}{1}-\overset{+}{2}\,| - \sqrt{\tfrac{3}{5}}\,|\,\overset{+}{0}-\overset{+}{1}\,|\,,\\
\psi(1\ \ 0\ \ 1\ \ 0) &= \sqrt{\tfrac{2}{5}}\,|\overset{+}{2}-\overset{-}{2}\,| + \sqrt{\tfrac{2}{5}}\,|\overset{-}{2}-\overset{+}{2}\,| - \frac{1}{\sqrt{10}}\,|\overset{+}{1}-\overset{-}{1}\,|\\
&\quad - \frac{1}{\sqrt{10}}\,|\overset{-}{1}-\overset{+}{1}\,|\,,\\
\psi(1\ \ 1\ \ 1\ \ -1) &= \sqrt{\tfrac{2}{5}}\,|\overset{-}{2}-\overset{-}{1}\,| - \sqrt{\tfrac{3}{5}}\,|\overset{-}{1}-\overset{-}{0}\,|\,.
\end{aligned} \tag{7.109}$$

In the same way it can be shown that the wave function for energy level $^1S_0$ is written as

$$\psi(^1S_0) = \frac{1}{\sqrt{5}}\,|\overset{+}{0}\overset{-}{0}\,| - \frac{1}{\sqrt{5}}\,|\overset{+}{1}-\overset{-}{1}\,| - \frac{1}{\sqrt{5}}\,|-\overset{+}{1}\overset{-}{1}\,| + \frac{1}{\sqrt{5}}\,|\overset{+}{2}-\overset{-}{2}\,| + \frac{1}{\sqrt{5}}\,|-\overset{+}{2}\overset{-}{2}\,|\,. \tag{7.110}$$

We must now calculate the result of operator action $\sum \boldsymbol{ls} = \sum (\hat{l}_z\hat{s}_z + \frac{1}{2}\hat{l}_+\hat{s}_- + \frac{1}{2}\hat{l}_-\hat{s}_+)$ on the determinant wave functions by the known formulas. We thus find

$$\begin{aligned}
\sum \hat{\boldsymbol{l}}\hat{\boldsymbol{s}}\,|\overset{+}{0}\,\overset{-}{0}\,| &= \sqrt{6}\,|\overset{+}{0}-\overset{+}{1}\,| + \sqrt{6}\,|\overset{-}{1}\ \overset{-}{0}\,|\,,\\
\sum \hat{\boldsymbol{l}}\hat{\boldsymbol{s}}\,|\overset{+}{1}-\overset{-}{1}\,| &= |\overset{-}{2}-\overset{-}{1}\,| + |\overset{+}{1}-\overset{+}{2}\,| + |\overset{+}{1}-\overset{-}{1}\,|\,,
\end{aligned} \tag{7.111}$$

etc., for all the necessary functions. After that it is not difficult to obtain the matrix element. The ultimate result is

$$\lambda\langle ^1S_0(\Gamma_1)\,|\sum \boldsymbol{ls}\,|\,^3P_0(\Gamma_1)\rangle = -2\sqrt{6}\lambda\,. \tag{7.112}$$

Table 7.24 indicates all the other matrix elements of the spin-orbital interaction operator. We proceed in exactly the same way to calculate the matrix elements of the other blocks of the energy matrix. A complete calculation for this configuration was performed by *Liehz* and *Ballhausen* [7.72]. The calculation result is presented in Fig. 7.36 in the form of a plot of energy levels in relation to the strength of crystal field Dq.

Now we use the *strong-field approximation*, in which the spin-orbital interaction can be taken into account when using the wave functions already found.

For the orbital part of the ground state we have

$$\psi(A_{2g}) = |(x^2-y^2)(z^2)\,|\,. \tag{7.113}$$

At $S=1$ this state refers, in double cubic group O', to the representation of Γ_5. The three states corresponding to level Γ_5 are

Table 7.24. Effect of operator ls on a set of orbits in a cubic field [7.61]

$$ls(\overset{+}{z^2}) = -\frac{\sqrt{3}}{2}(\overset{-}{xz}) - \mathrm{i}\frac{\sqrt{3}}{2}(\overset{-}{yz})$$

$$ls(\overset{-}{z^2}) = -\frac{\sqrt{3}}{2}(\overset{+}{xz}) - \mathrm{i}\frac{\sqrt{3}}{2}(\overset{+}{yz})$$

$$ls(x^2\overset{+}{-}y^2) = \mathrm{i}(\overset{+}{xy}) - \mathrm{i}\tfrac{1}{2}(\overset{-}{yz}) + \tfrac{1}{2}(\overset{-}{xz})$$

$$ls(x^2\overset{-}{-}y^2) = -\mathrm{i}(\overset{-}{xy}) - \mathrm{i}\tfrac{1}{2}(\overset{+}{yz}) - \tfrac{1}{2}(\overset{+}{xz})$$

$$ls(\overset{+}{xy}) = -\mathrm{i}(x^2\overset{+}{-}y^2) + \tfrac{1}{2}(\overset{-}{yz}) + \mathrm{i}\tfrac{1}{2}(\overset{-}{xz})$$

$$ls(\overset{-}{xy}) = \mathrm{i}(x^2\overset{-}{-}y^2) - \tfrac{1}{2}(\overset{+}{yz}) + \mathrm{i}\tfrac{1}{2}(\overset{+}{xz})$$

$$ls(\overset{+}{xz}) = \mathrm{i}\tfrac{1}{2}(\overset{+}{yz}) - \tfrac{1}{2}(x^2\overset{-}{-}y^2) - \mathrm{i}\tfrac{1}{2}(\overset{-}{xy}) + \frac{\sqrt{3}}{2}(\overset{-}{z^2})$$

$$ls(\overset{-}{xz}) = -\mathrm{i}\tfrac{1}{2}(\overset{-}{yz}) - \tfrac{1}{2}(x^2\overset{+}{-}y^2) - \mathrm{i}\tfrac{1}{2}(\overset{+}{xy}) - \frac{\sqrt{3}}{2}(\overset{+}{z^2})$$

$$ls(\overset{+}{yz}) = -\mathrm{i}\tfrac{1}{2}(\overset{+}{xz}) + \mathrm{i}\tfrac{1}{2}(x^2\overset{-}{-}y^2) - \tfrac{1}{2}(\overset{-}{xy}) + \mathrm{i}\frac{\sqrt{3}}{2}(\overset{-}{z^2})$$

$$ls(\overset{-}{yz}) = \mathrm{i}\tfrac{1}{2}(\overset{-}{xz}) + \mathrm{i}\tfrac{1}{2}(x^2\overset{+}{-}y^2) + \tfrac{1}{2}(\overset{+}{xy}) + \mathrm{i}\frac{\sqrt{3}}{2}(\overset{+}{z^2})$$

$$^3A_{2g}(\Gamma_5)(e_g)^2 \begin{cases} |(x^2\overset{+}{-}y^2)(\overset{+}{z^2})|\,, \\ \frac{1}{\sqrt{2}}|(x^2\overset{+}{-}y^2)(\overset{-}{z^2})| + \frac{1}{\sqrt{2}}|(x^2\overset{-}{-}y^2)(\overset{+}{z^2})|\,, \\ |(x^2\overset{-}{-}y^2)(\overset{-}{z^2})|\,. \end{cases} \tag{7.114}$$

The first excited triplet state is $^3T_{2g}(t_{2g}e_g)$. In this case the orbital parts have the form

$$\begin{aligned} \varphi_1 &= |(xy)(z^2)|\,, \\ \varphi_2 &= \tfrac{1}{2}|(xz)(z^2)| + \sqrt{\tfrac{3}{2}}|(xz)(x^2-y^2)|\,, \\ \varphi_3 &= \tfrac{1}{2}|(yz)(z^2)| - \sqrt{\tfrac{3}{2}}|(yz)(x^2-y^2)|\,. \end{aligned} \tag{7.115}$$

In the double group the ninefold degeneration of the states produces states Γ_2, Γ_3, Γ_4, and Γ_5. To obtain the three wave functions being transformed according to Γ_5, we must choose – from among the nine possible combinations of orbital wave functions φ_1, φ_2, φ_3 with three spin functions $\psi_1 = \alpha\alpha$, $\psi_0 = \frac{1}{\sqrt{2}}(\alpha\beta + \beta\alpha)$, $\psi_{-1} = \beta\beta$ – such three linear combinations

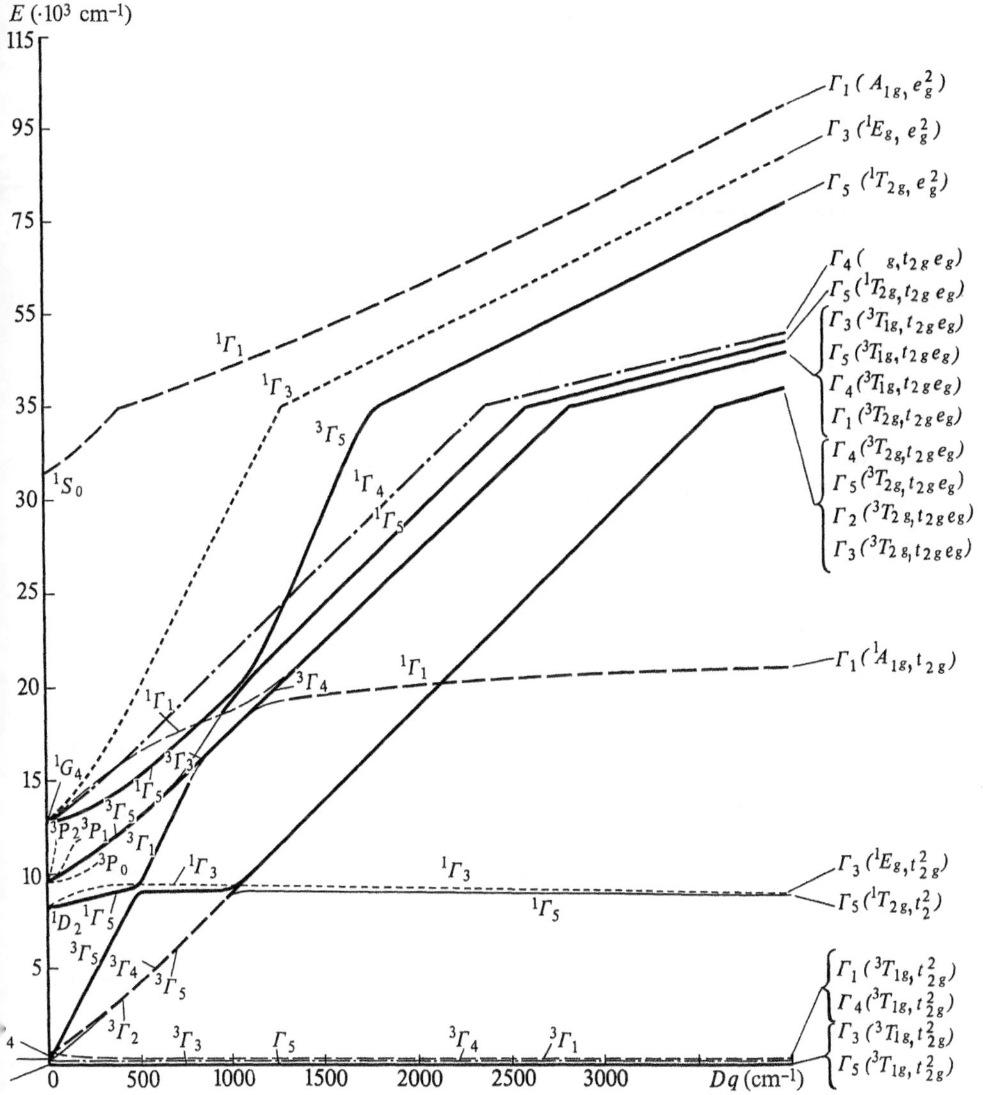

Fig. 7.36. Energy levels of the d^2 configuration in an octahedral field with due regard for spin-orbital interaction: $\lambda = 65\ \mathrm{cm}^{-1}$, $F_4 = 70\ \mathrm{cm}^{-1}$, $F_2 = 14F_4$ [7.72]

which are transformed according to this irreducible representation of double cubic group O′. It can be shown that such functions are

$$\frac{1}{\sqrt{2}}[\Psi_2\psi_{-1}+\mathrm{i}\Psi_3\psi_1]\ ,\quad \frac{1}{\sqrt{2}}[\Psi_2\psi_0+\mathrm{i}\Psi_1\psi_1]\ ,$$
$$\frac{1}{\sqrt{2}}[\Psi_3\psi_0-\mathrm{i}\Psi_1\psi_{-1}]\ , \tag{7.116}$$

where

$$\Psi_1=\varphi_1\ ,\quad \Psi_2=\frac{1}{\sqrt{2}}(\varphi_2+\mathrm{i}\varphi_3)\ ,\quad \Psi_3=\frac{1}{\sqrt{2}}(\mathrm{i}\varphi_2+\varphi_3)\ .$$

We now have to calculate in what way operator ls affects the set of orbits in the cubic field. For instance (Table 7.24),

$$\begin{aligned}\hat{l}\hat{s}(\overset{+}{z^2}) &= -\sqrt{\tfrac{3}{2}}(\overset{-}{xz})-\mathrm{i}\sqrt{\tfrac{3}{2}}(\overset{-}{yz})\ ,\\ \hat{l}\hat{s}(x^2\overset{-}{-}y^2) &= -\mathrm{i}(\overset{-}{xy})-\mathrm{i}\tfrac{1}{2}(\overset{+}{yz})-\tfrac{1}{2}(\overset{+}{xz})\ .\end{aligned} \tag{7.117}$$

Ultimately, one of the results will be

$$\lambda\langle{}^3T_{2g}(e_g t_{2g})\mid\Sigma\,\hat{l}\hat{s}\mid{}^3A_{2g}(e_g^2)\rangle=2\sqrt{2}\lambda\ . \tag{7.118}$$

In the same way we can calculate all the necessary matrix elements for each of the blocks of the energy matrix.

The wave functions of the ground state, taking into account mixing with level $\Gamma_5\,({}^3T_{2g})$ in the first order of perturbation theory, will be

$$\Psi(\Gamma_5^n)=\Gamma_5^n({}^3A_{2g})-\frac{2\sqrt{2}\lambda}{10Dq}\Gamma_5^n({}^3T_{2g})\ . \tag{7.119}$$

7.13 Laser Crystals

As has already been indicated, crystals with impurities of transition group elements are of special interest in connection with their application as active media in quantum light generators (lasers).

Laser operation is based on the effect of amplification of electromagnetic radiation on induced transitions. Let us consider the interaction of an atomic system consisting of two levels, 1 and 2, with a radiation of frequency $\nu_{12}=(E_2-E_1)/h$, where E_1 and E_2 are the energies of these states, and $\hbar$ is

Planck's constant. The increment in intensity of radiation I over thickness dx due to induced radiation will be

$$dI = (\sigma_{21} n_2 - \sigma_{12} n_1) I\,dx = (n_2 - n_1 g_2/g_1)\sigma_{21} I\,dx \; ; \tag{7.120}$$

here n_1 and n_2 are the populations and g_1 and g_2 are the multiplicities of degeneracy of levels 1 and 2; σ_{21} is the cross section of induced transitions $2 \rightarrow 1$ related to the probability of induced transitions W_{21} by the expression $\sigma_{21} = W_{21}/I$. Quantity $\alpha = \sigma_{21}(n_2 - n_1 g_2/g_1) = \Delta \cdot \sigma_{21}$ is called the amplification factor. For this quantity to exceed zero, it is necessary that $n_2 g_1 > n_1 g_2$, which is the condition for inverse population in the medium. Quantity $\Delta = n_2 - n_1 g_2/g_1$ is called the inverse population density,

The solution to differential equation (7.120) for a homogeneous rod of length l will be

$$I = I_0 \exp(\sigma_{21} \cdot \Delta \cdot l) \; ,$$

where I_0 is the intensity of the incident and I, of the intensified radiation. From the expression obtained it is seen that I can be increased by increasing l. This is achieved by repeated passage of the beam through the active medium, with the rod being placed in a Fabry-Perot interferometer (resonator) formed by two parallel-side mirrors.

To achieve generation in an optical resonator with an active substance, the inverse population density must exceed some threshold value Δ_{thr}. The threshold inverse population density can be estimated from the condition for compensation of the losses in the resonator by induced radiation. If one of the resonator mirrors has a reflection coefficient r and the other reflects completely, the radiation losses per unit length of the resonator are equal to $(1-r)/2l$, where l is the length of the active element (we assume that the resonator mirrors are deposited directly onto the end faces of the crystal). If the other losses are insignificant, we obtain the following expressions for the threshold value of inverse population:

$$\Delta_{\text{thr}} = n_2 - g_2 n_1/g_1 = (1-r)/2\sigma_{21} l \; . \tag{7.121}$$

The state of inverse population in activated crystals is achieved with the aid of radiation of exciting lamps (optical pumping).

The working circuits of lasers with optical pumping are divided into three- and four-level ones. Figure 7.37a exhibits a diagram of a three-level laser. Optical pumping is directed to the band corresponding to transition $1 \rightarrow 3$ with probability W_{13}. Subsequent relaxation $3 \rightarrow 2$ (radiationless, as a rule) populates level 2. Generation is effected at transition $2 \rightarrow 1$.

The four-level pattern of laser operation (Fig. 7.37b) is energetically more advantageous. The circuit differs from the preceding one in that the generation at transition $3 \rightarrow 2$ is completed not at the ground state, but at a higher

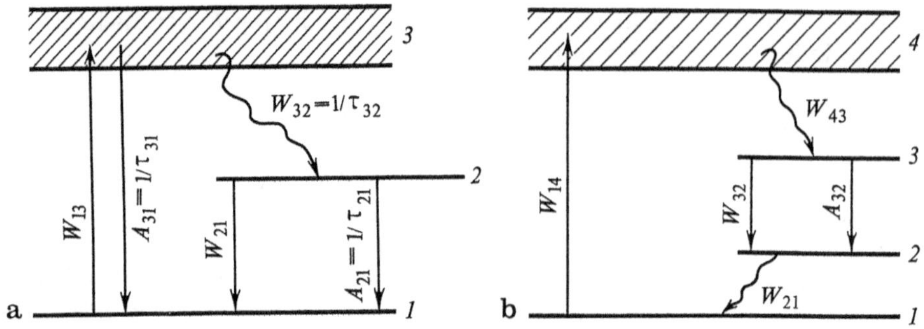

Fig. 7.37. Three-level (**a**) and four-level (**b**) laser circuits: W_{ij}, A_{ij} are the probabilities of induced and spontaneous $i \rightarrow j$ transitions

level, whose thermal population can be neglected. Therefore, to obtain inverse population in four-level active media, it is sufficient to transfer only an insignificant fraction of the atoms from the ground state to state 3. Therefore, four-level lasers require much less energy for their excitation than three-level ones, all things being equal.

Thus, for realization of the generation conditions the impurity ions in the crystal must have a metastable energy level at which the maximum possible number of excited ions can be accumulated. The lifetime at this level must depend mainly on radiative transitions with intensive and narrow luminescence lines. For effective pumping by a nonmonochromatic light source the absorption spectrum of the impurity ions must contain broad absorption bands from whose levels most of the ions transfer to the metastable state. Besides, the active substance must show high optical properties to ensure the high quality of the optical resonator.

Owing to the diversity of structures, ways of ingress, and state of impurities, crystals offer enormous opportunities for producing lasers for different spectral regions. Besides, in distinction to isotropic media (glasses, for example), the radiating centers in crystals are arranged in a regular manner, so that the pumping power is not expended, generally speaking, on the excitation of nongenerating centers. Advances in crystal growing make it possible to obtain laser materials of sufficiently high quality and required sizes.

The first laser with the use of crystals was one on a ruby [7.75], a crystal of aluminium oxide Al_2O_3 with an impurity of chromium ions (about 0.05%), which replace isomorphously the aluminium ions in the crystal lattice. The crystals belong to the rhombohedral symmetry; they are optically uniaxial and have refractive indices of about 1.76. It is worth mentioning the high mechanical strength and thermal conductivity (especially at low temperatures) of ruby crystals.

The diagram of the lower energy levels of Cr^{3+} ions ($3d^3$ configuration) in a trigonal crystal field of the Al_2O_3 lattice is presented in Fig. 7.38. The functioning transition is the one between metastable level ^2E (lifetime ~3 ms

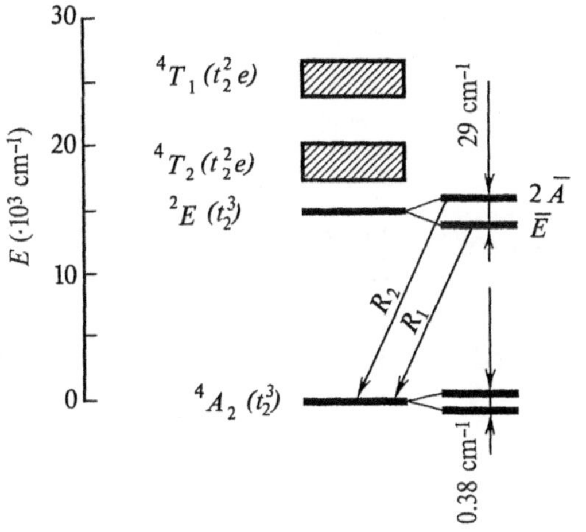

Fig. 7.38. Schemes of lower levels of the Cr^{3+} ion in Al_2O_3

at 300 K) and ground state $^4A_2$. Under the effect of the trigonal field and the spin-orbital interaction, level ^2E splits into two twice degenerate levels $\bar{E}$ and $2\bar{A}$, which are spaced 29 cm^{-1} apart. Transitions between these sublevels and the ground state correspond to radiation lines R_1 and R_2 with wavelengths of 6943 and 6929 Å. Each of these lines is a doublet, since the ground four-times degenerate level $^4A_2$ also consists of two sublevels with a distance of 0.38 cm^{-1}. As the temperature reduces, the lines narrow down from 11 cm^{-1} for 300 K to 0.3 cm^{-1} at the liquid nitrogen temperature and shift into the shortwave range of the spectrum from 6934 to 6919 Å.

The pumping radiation is absorbed by ruby in two broad dichroic absorption bands corresponding to transitions $^4A_2\,(t_2^3) \rightarrow {}^4T_2\,(t_2^2 e)$ and $^4A_2 \rightarrow {}^4T_1\,(t_2^2 e)$, which lie in the visible range of the spectrum, as shown in Fig. 7.38. Most of the excited ions transfer from these levels to the metastable radiationless one ^2E with a time $\tau_{32} \approx 5 \times 10^{-8}$ s. The energy scheme of a ruby laser thus has a characteristic form of a three-level system (Fig. 7.37a).

For estimating the cutoff population of levels $\bar{E}$ and $2\bar{A}$ in ruby during the generation of the R_1 line we assume $r = 0.5$, $l = 10$ cm and, taking into account that at room temperature $\sigma_{21} = 2.5 \times 10^{-20}$ cm^2, we find from (7.121) that $\Delta_{\mathrm{thr}} = 10^{18}$ cm^{-3}. According to the Boltzmann distribution, the ratio of the populations of levels $2\bar{A}$ and $\bar{E}$ at $T = 290$ K is equal to 0.87.

At the usual concentration of chromium ions ($\sim 0.05\%$) the number of ions per 1 cm^3 is $n_0 = 1.6 \times 10^{19}$ cm^{-3}. From the threshold condition

$$n(\bar{E}) - 1/2\,n(^4A_2) = 1 \times 10^{10^{18}}\ \mathrm{cm}^{-3}$$

and the balance equation

$$1.87\,n(\bar{E}) + n(^4A_2) = 1.6 \times 10^{19}\ \mathrm{cm}^{-3}\ ,$$

we obtain $n(\bar{E}) = 0.465 \times 10^{19}\,\mathrm{cm}^{-3}$, $n(2\bar{A}) = 0.405 \times 10^{19}\,\mathrm{cm}^{-3}$, and $n(^4A_2) = 0.73 \times 10^{19}\,\mathrm{cm}^{-3}$.

The kinetic equations describing the three-level system in symbols indicated in Fig. 7.37 a can be written

$$\begin{aligned} \frac{dn_1}{dt} &= W_{21}(n_2 - n_1) - W_{13}(n_1 - n_3) + \frac{n_2}{\tau_{21}} + \frac{n_3}{\tau_{31}}\,, \\ \frac{dn_2}{dt} &= -W_{21}(n_2 - n_1) + \frac{n_3}{\tau_{32}} - \frac{n_2}{\tau_{21}}\,, \\ \frac{dn_3}{dt} &= W_{13}(n_1 - n_3) + \frac{n_3}{\tau_{32}} - \frac{n_3}{\tau_{31}}\,, \end{aligned} \tag{7.122}$$

One of the equations can be eliminated by using the relation $n_1 + n_2 + n_3 = n_0$. Besides, if no accumulation of particles takes place at level 3 ($n_3 \ll n_1$, $dn_3/dt = 0$), we obtain $W_{13}n_1 = \frac{n_3}{\tau_{32}}\left(1 + \frac{\tau_{32}}{\tau_{31}}\right)$ from the last equation of the set, and then the change in inverse population (with due regard for the fact that in ruby $\tau_{31} \gg \tau_{32}$) will be

$$\frac{d\Delta}{dt} = -2\Delta W_{21} + n_0\left(W_{13} - \frac{1}{\tau_{21}}\right) - \Delta\left(W_{13} + \frac{1}{\tau_{21}}\right).$$

Under stationary laser operation $d\Delta/dt = 0$ and $\Delta = \frac{n_0(W_{13} - 1/\tau_{21})}{W_{13} + 1/\tau_{21} + 2W_{21}}$.

The generation condition $\Delta > 0$ or $W_{13} > 1/\tau_{21}$ is fulfilled if the pumping energy density averaged over the absorption band $\varrho_{13} > h\nu_{13}/V\sigma_{13}\tau_{21}$, since $W_{13} = V\sigma_{13}\varrho_{13}/h\nu_{13}$, where V is the volume of the active specimen. For a thin cylindrical specimen, estimates show that the threshold pumping power per unit specimen surface for an adsorption band with a maximum at 5600 Å is of the order of 200 W/cm^2.

The laser radiation of crystals may be polarized because of the anisotropy of the polarization characteristics of induced transitions. Thus, for ruby the probability of transition between levels $\bar{E}$ and $^4A_2$ is maximal for an ordinary wave, when the direction of the vector of the electric field of laser radiation is perpendicular to that of the optic axis. Optical generation of Cr^{3+} ions in yttrium-aluminium garnet has been discovered by *Sevast'yanov* et al. [7.76].

The laser effect has also been obtained on many other crystals containing active divalent rare-earth ions. An example is fluorite CaF_2 activated by Dy^{2+} [7.77]. Its absorption spectrum has a strong and broad band in the range of 2300–4900 Å and three weaker bands centered on 5800, 7150, and 9100 Å. Thus, almost the whole radiation energy in the range from 0.9 μm to the ultraviolet portion of the spectrum is used at pumping by xenon lamps, which ensures the low threshold energy level in the pulsed regime (1 J at 77 K). Induced radiation passes at a wavelength of 2.36 μm in the form of a very narrow line (less than 0.08 cm^{-1} at 77 K and 0.03 cm^{-1} at 27 K), which corresponds to transition between states $^5I_7 \rightarrow {}^5I_8$. The final level is located 35 cm^{-1} above the ground level. The lifetime at the metastable level is 10^{-2} s. Thus, generation proceeds according to the four-level scheme depicted in Fig. 7.37b.

Generation on trivalent ions of rare-earth elements in other single crystals proceeds along the same lines. But in distinction to the divalent ions of rare earths, the transitions between shells $4f \rightarrow 5d$ are in the ultraviolet range of the spectrum and are inconvenient for optical pumping. Excitation is effected by comparatively weak $4f \rightarrow 4f$ transitions. Of greatest practical importance is the Nd^{3+} ion, for which the laser effect is achieved at room temperature in many crystals. In some crystals continuous generation of Nd^{3+} ions has been realized e.g. in yttrium-aluminium garnet crystals [7.78].

The most intensive line in the luminescence spectrum of yttrium-aluminium garnet $Y_3Al_5O_{12}$ activated by Nd^{3+} ions ($\lambda = 1.0648$ μm) is associated with transition $^4F_{3/2} \rightarrow {}^4I_{11/2}$. Its width at room temperature is equal to 6.5 cm^{-1}. The lifetime at the metastable level of $^4F_{3/2}$ is about 200 μs. At this transition the final level of $^4I_{11/2}$ is removed from the ground level by 2000 cm^{-1}. Therefore, it is not populated even at room temperature, greatly facilitating the realization of inverse population. This situation is generally typical of the four-level scheme.

To increase the efficiency of optical pumping, an additional dopant (sensitizer) is introduced into the crystal, which transmits the absorbed energy to the generating ion. The sensitizer is selected so that its absorption spectrum overlaps as little as possible that of the main ion; thus, the energy is transmitted most efficiently. In a crystal of $Y_3Al_5O_{12}$ with Nd^{3+} ions, Cr^{3+} ions can serve as such a sensitizer. Energy transmission to Nd^{3+} ions is effected by these ions, which are in metastable state ^2E [7.79].

Laser crystals are widely used in science and technology owing to the exceptional features of laser radiation: light power, high monochromaticity, coherence, and small divergence of the light beam. The use of laser sources has lead to the development of new branches of crystal optics: nonlinear optics, excited-state spectroscopy, stimulated Raman scattering spectroscopy, holography, optical amplification, etc. Lasers are used in technology for machining hard and refractory materials, in medicine for performing complicated operations, in communications techniques, in data processing, and in other areas.

7.14 Polarized Luminescence in Crystals

According to the selection rules, electron transitions between energy levels of absorbing and luminescent centers in a crystal may be of different natures and may correspond to dipole (electric or magnetic) or quadrupole absorption and radiation. In the classical model, these transitions can be compared to the absorption and radiation of an electric or a magnetic dipole and a quadrupole with a characteristic space distribution of radiation and with polarization.

For an electric dipole, the absorption intensity is known to be proportional to $\cos^2\theta$, where θ is the angle between the direction of the oscillator axis and the electric field tension in the wave absorbed, and where the flux density of the energy radiated in the given direction at an angle φ to the dipole axis is proportional to $\sin^2\varphi$. The radiation is completely linearly polarized in the plane containing the dipole axis and the direction of observation. The space distribution of the radiation field of the magnetic dipole is the same as in the electric field, but the polarization plane is turned through angle $\pi/2$ along the direction of observation. The quadrupole radiation field can be represented as the summary field of two dipoles running in the same direction and oscillating with opposite phases.

The absorbing and luminescent centers are arranged in an orderly manner in crystals, and therefore the oscillators corresponding to the optical transitions are oriented in a definite way. For this reason the absorption and radiation of waves with different polarization propagating in crystals depend on the mutual orientation of the oscillators and on the direction of wave polarization. In absorption spectra this is manifested in dichroism (Sect. 7.5), and in luminescence spectra, in partial polarization of radiation.

The luminescence polarization may manifest itself in cubic, optically isotropic (as a continuum) crystals when absorption and luminescence are due to anisotropic optical centers oriented along the symmetry axes of the crystal. Owing to the equiprobable distribution of the centers among several symmetry axes of definite order, the microanisotropy of the centers is averaged out and concealed in the macroscopic properties of the cubic crystal. To reveal the anisotropy of the centers it is necessary to exert some anisotropic effect on the crystal, which differently affects the various groups of the centers. One of such methods is polarized luminescence of *Feofilov* [7.80], which enables one to determine the nature of elementary oscillators corresponding to the spectral band of the spectrum under study. The degree of polarization of luminescence is observed according to the following scheme: a crystal plate oriented in a definite way is excited by a polarized light, after which the dependence of the degree of polarization of luminescence on the angle of rotation of the plate about the observation axis is measured. Such azimuthal dependences have a characteristic shape and often make it possible to establish unambiguously the type of the oscillators and the character of their orientation in the crystal.

Let us consider the case where the absorbing and radiating oscillators are electric dipoles arranged along the fourfold symmetry axes of a cubic crystal. Suppose the plate is cut out parallel to the (110) plane, and the exciting light is polarized in the vertical plane, as shown in Fig. 7.39. The observed degree of polarization $P=(I_v-I_h)/(I_v+I_h)$, where I_v and I_h are the intensities of the light transmitted throug the polarizer in the vertical and horizontal

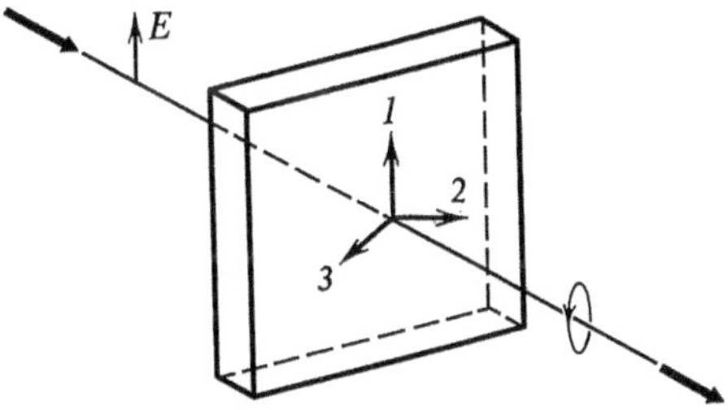

Fig. 7.39

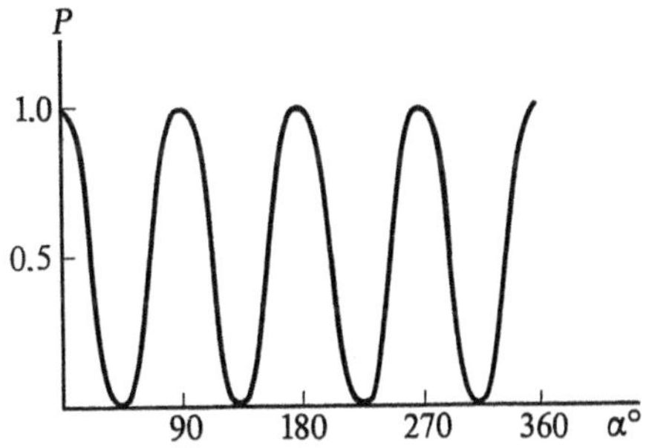

Fig. 7.40

Fig. 7.39. Observation of the azimuthal dependence of the degree of polarization in a cubic crystal: (*1, 2, 3*) directions of oscillator orientation, (*E*) polarization of exciting radiation

Fig. 7.40. Polarization diagram of luminescence calculated for the case depicted in Fig. 7.39 [7.80]. α-angle of rotation of the crystal relative to the observation axis

orientation, is equal for n radiators to $\sum P_i I_i / \sum I_i$, where P_i is the degree of polarization of the ith radiator, and I_i is the radiation intensity.

For three oscillators oriented along the fourfold axes we find

$$I_1 = \cos^2\alpha\ , \qquad P_1 = \cos 2\alpha\ ;$$
$$I_2 = \tfrac{1}{4}\sin^2\alpha\ , \qquad P_2 = -\cos 2\alpha\ ;$$
$$I_3 = \tfrac{1}{4}\sin^2\alpha\ , \qquad P_3 = -\cos 2\alpha\ ;$$

where α is the angle between the direction of the orientation of oscillator 1 and the polarization of the exciting radiation, E (Fig. 7.39). The degree of polarization of the total radiation

$$P = \frac{\cos 2\alpha\,(3\cos^2\alpha - 1)}{1 + \cos^2\alpha}\ .$$

The plot of this dependence is shown in Fig. 7.40. The luminescence intensity also depends on the azimuth α:

$$I = \sum I_i = (1 + \cos^2\alpha)/2\ .$$

On excitation by natural light, the observed degree of polarization (spontaneous polarization) $P_{sp} = \frac{1}{3}\cos 2\alpha$. Hence, in this case, even on excitation by an unpolarized light, the degree of polarization shows an azimuthal dependence. For oscillators oriented along the threefold axes, $P = \sin^2\alpha \dfrac{7 - 6\sin^2\alpha}{2 + \sin^2\alpha}$.

For oscillators oriented along the twofold axes, $P = \dfrac{2 + 7\sin^2\alpha - 6\sin^4\alpha}{6 + \sin^2\alpha}$. One

Table 7.25. Deformation splitting of lines in a cubic crystal with orientation of centers along different axes [7.81]

Axis along which contraction F is directed	Number of components of splitting		
	⟨100⟩	⟨111⟩	⟨110⟩
4	2	1	2
3	1	2	2
2	2	2	3

can consider other cases of oscillator arrangement and orientation of the crystal plate in the same way and see that the interpretation of such experiments is unambiguous.

Another important means for revealing the anisotropy of optical centers and their orientation in the crystal lattice is the study of the patterns of splitting of the spectral absorption and luminescence lines under the effect of uniform elastic deformation of contraction or extension of crystals (piezospectroscopy, see [7.81]). The position of the energy centers in a crystal depends on the intracrystalline field, and therefore such an effect causes a slight shift of these levels, which is different for specimens differently oriented along the direction of contraction of the centers. Single spectral bands in the absorption and radiation spectra will suffer splitting in a stressed crystal.

In cubic crystals the anisotropic centers are oriented along the symmetry axes, and therefore the number of components of splitting depends on the number of the different positions of the symmetry axes along which the centers are oriented, relative to the direction of contraction. This is true if the elastic deformation causes only a line shift. When the lines are due to transitions to degenerate levels and elastic deformation affects the crystal, this degeneracy may be lifted, and the picture becomes much more complicated. For the simplest cases, when contraction is effected along one of the symmetry axes, the number of components of splitting is as indicated in Table 7.25.

Thus, the picture of line splitting on application of deformation along symmetry axes 2, 3, and 4 determines unambiguously the orientation of the centers in cubic crystals.

The components of splitting are, generally speaking, polarized, depending on the nature of the elementary oscillators, the polarization of the exciting light, and the direction of observation. It is important to know whether the absorption and excitation occur at the same center, or whether they proceed separately and independently (the excitation is transferred from another center) and hence the excitation can be considered isotropic. If the absorption and radiation occur within the same anisotropic center, then even when the luminescence is excited by a beam of an unpolarized light, groups of centers differently oriented with respect to this beam have different probabilities of luminescence excitation. This is established by calculating the probabilities of

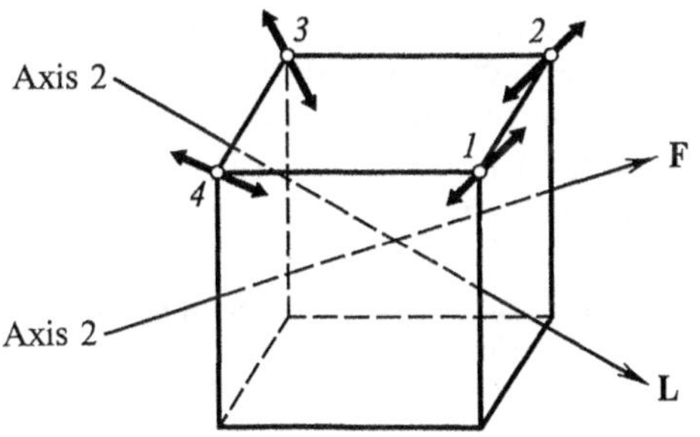

Fig. 7.41. Observation of deformation splitting of luminescence lines in a cubic crystal: *1, 2, 3* and *4* are oscillators oriented along threefold axes [7.81]

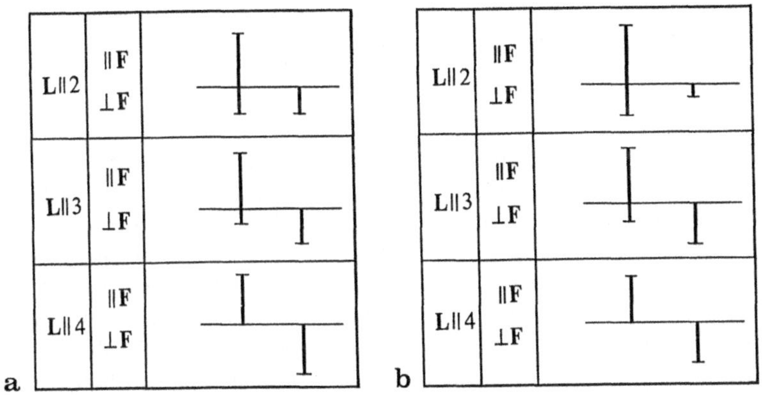

Fig. 7.42. Polarized splitting of the lines of the luminescence spectra of electrical dipoles oriented along the threefold axes in the case of isotropic (**a**) and anisotropic (**b**) excitation: the direction of deformation $F \parallel 2$ and $\perp$ to the direction of observation L, the exciting beam $\parallel L$ [7.81]

light absorption by groups of centers disposed along different axes of the same order.

Consider, for example, isotropic excitation while the electric dipoles of the absorbing and luminescent centers are arranged along the threefold axes. Suppose the direction of excitation and observation L is parallel to axis 2, and the direction of contraction F is also parallel to axis 2, but perpendicular to L (Fig. 7.41). Oscillators 1 and 3 give off radiation which is in one line of luminescence with a relative intensity of $\frac{2}{3}$ and is polarized perpendicular to F. Oscillators 2 and 4 radiate in another line (Table 7.25). The relative intensity of radiation in this line of luminescence, when it passes through a polarizer oriented $\perp$ or $\parallel$ to F, is equal to $\frac{2}{3}$ or $\frac{4}{3}$, respectively.

The overall pattern of splitting is shown in Fig. 7.42a, where the vertical segments correspond to the separate components of the splitting of the spectral band, while the lengths of these segments above and below the horizontal line are proportional to the intensity of the components polarized $\parallel$ and $\perp$ to F, respectively. Other orientations of the observation axis $L-L \parallel 3$, $L \perp F$ and $L \parallel 4$, $L \perp F$ – are calculated in a similar manner.

A different result is obtained if the radiation and absorption proceed at the same center. For the same case $L \parallel 2$, when the excitation light is not polarized, oscillators 1 and 3 radiate in one line with relative intensities of $O(\parallel F)$ and $\frac{1}{9}$ $(\perp F)$, and oscillators 2 and 4 radiate in another line with intensities of $\frac{2}{3}$ $(\parallel F)$ and $\frac{1}{3}$ $(\perp F)$. The distribution of luminescence among the lines and polarizations is shown in Fig. 7.42b.

The total set of such patterns for various types of oscillators suggests that piezospectroscopy is a tool for revealing the nature of luminescence oscillators and their disposition in cubic crystals.

7.15 Electron Paramagnetic Resonance of Impurity Ions in Crystals

A knowledge of the electron paramagnetic resonance (EPR) is essential in studying the valency state, the symmetry of the closest environment, and the interaction of the ions of the impurity of transition-group elements in the crystal lattice. The EPR phenomenon, discovered by *Zavoisky* in 1944, is known to consist in resonance absorption of the energy of quanta of a high-frequency electromagnetic field under Zeeman transitions between energy levels of a paramagnetic placed in an external magnetic field [7.64], [7.66], [7.82]. The effect is caused by ions of transition-group elements, which retain, in the crystal, a nonzero magnetic moment of an unfilled electron shell. The continuous absorption of the energy of the high-frequency field in a paramagnetic is associated with the mechanisms which return the system, after its excitation under the magnetic transition, to the initial, thermodynamically equilibrium state by means of interaction with the lattice ions. Such mechanisms determine the processes and times of paramagnetic relaxation. During spin-lattice relaxation the oscillations of crystal ions modulate the intracrystalline electric field, which changes the orbital motion of the electrons and affects the state of the spins owing to spin-orbital interaction.

The Hamiltonian of an ion in a crystal, which we considered in crystal-field theory $H = H_0 + V_{\rm cr} + V_{\rm so}$, must now be supplemented with the Zeemann energy of the electrons in the external magnetic field $\boldsymbol{H}$:

$$\mathcal{H}_z = \beta(\boldsymbol{L} + 2\boldsymbol{S})\boldsymbol{H} \ , \tag{7.123}$$

where $\boldsymbol{L}$ is the orbital moment operator, $\boldsymbol{S}$ is the spin moment operator, $\boldsymbol{H}$ is the magnetic field vector, and β is the Bohr magneton. Besides, we should also take into account the energy of the weak spin-spin interaction $V_{\rm ss}$, which is of the same order of magnitude ($\sim 1\,\rm cm^{-1}$) as the Zeeman term. *Abragam* and *Pryce* [7.83] developed a very important method, which enables one to calculate the splitting of the ground energy level by the perturbation-theory method (the *spin-Hamiltonian method*). Calculation of the matrix elements of the perturbation operator $V_{\mathrm{so}}+V_{\mathrm{ss}}+\mathscr{H}_z$ will first be made with the aid of coordinate wave functions of the unperturbed problem. As a result the perturbation energy will be the function of spin operator $\hat{S}$. This function is called the *spin Hamiltonian*.

The form of the spin operators appearing in the Hamiltonian is established from symmetry considerations, since they define the wave functions of the ion in the crystal field of the nearest neighbors [7.66]; their dependence on spin matrices S_x, S_y, and S_z is the same as the dependence of equivalent operators O_n^m on orbital momentum $\boldsymbol{L}$ in crystal-field theory. As a result, for different types of symmetry of the electric field of point charges we have (in the crystallophysical coordinate system):

a) cubic symmetry:

$$\mathscr{H}=\mathrm{B}_4^0(\mathrm{O}_4^0+5\,\mathrm{O}_4^4)+\mathrm{B}_6^0(\mathrm{O}_6^0-21\,\mathrm{O}_6^4)+g\beta \boldsymbol{HS}\;;$$

b) hexagonal symmetry:

$$\mathscr{H}=\mathrm{B}_2^0\mathrm{O}_2^0+\mathrm{B}_4^0\mathrm{O}_4^0+\mathrm{B}_6^0\mathrm{O}_6^0+g_{\parallel}\beta H_zS_z+g_{\perp}\beta(H_xS_x+H_yS_y)\;;$$

c) trigonal symmetry:

$$\mathscr{H}=\mathrm{B}_2^0\mathrm{O}_2^0+\mathrm{B}_4^0\mathrm{O}_4^0+\mathrm{B}_6^0\mathrm{O}_6^0+\mathrm{B}_4^3\mathrm{O}_4^3+\mathrm{B}_6^3\mathrm{O}_6^3+\mathrm{B}_6^6\mathrm{O}_6^6$$
$$+g_{\parallel}\beta H_zS_z+g_{\perp}\beta(H_xS_x+H_yS_y)\;;$$

d) tetragonal symmetry:

$$\mathscr{H}=\mathrm{B}_2^0\mathrm{O}_2^0+\mathrm{B}_4^0\mathrm{O}_4^0+\mathrm{B}_6^0\mathrm{O}_6^0+\mathrm{B}_4^4\mathrm{O}_4^4+\mathrm{B}_6^4\mathrm{O}_6^4+g_{\parallel}\beta H_zS_z+g_{\perp}\beta(H_xS_x+H_yS_y)\;;$$

e) orthorhombic and lower symmetries:

$$\mathscr{H}=\mathrm{B}_2^0\mathrm{O}_2^0+\mathrm{B}_2^2\mathrm{O}_2^2+\mathrm{B}_4^0\mathrm{O}_4^0+\mathrm{B}_4^2\mathrm{O}_4^2+\mathrm{B}_4^4\mathrm{O}_4^4+\mathrm{B}_6^0\mathrm{O}_6^0+\mathrm{B}_6^2\mathrm{O}_6^2$$
$$+\mathrm{B}_6^4\mathrm{O}_6^4+\mathrm{B}_6^6\mathrm{O}_6^6+g_x\beta H_xS_x+g_y\beta H_yS_y+g_z\beta H_zS_z\;. \qquad (7.124)$$

The Zeeman energy is now expressed with the aid of a g factor with main values g_x, g_y, and g_z, which generalizes the usual value of the g factor for a free ion

$$g(S,L,J)=1+\frac{J(J+1)-L(L+1)+S(S+1)}{2J(J+1)}\;,$$

equal to 2 for a pure spin state ($L=0$). Deviations from this value are caused by the contribution from the orbital motion.

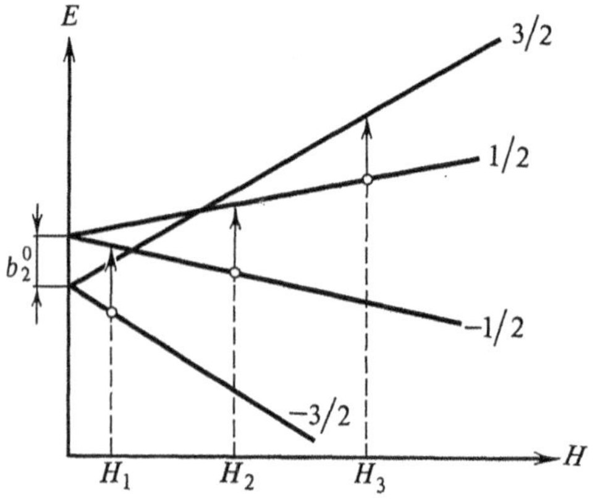

Fig. 7.43. Scheme of energy levels and allowed transitions (H_1, H_2, and H_3) for the Cr^{3+} ion (d^3, $S = \frac{3}{2}$) in a magnetic field H_0 parallel to the optic axis of the crystal

The number of spin operators O_n^m defining the initial splittings of the ground level (these must be retained in the spin Hamiltonian) depends on the total spin of ground state S. Thus, for iron-group elements the nonzero matrix elements will yield only operators O_n^m with $m = 0$; 2; 3; and 4 with respect to the spin functions. For rare earths, the other above-indicated operators will also be present.

The total number of energy levels into which the level with the spin quantum number S splits in the external magnetic field is equal to $2S+1$, so that one may observe $2S$ absorption lines satisfying the selection rule $\Delta M_s = \pm 1$ for strong magnetic fields, when the Zeeman term becomes the determining one. These lines form the *fine structure* of the EPR spectrum. For instance, for Cr^{3+} ions ($3d^3$ configuration) the ground state, according to Hund's rule, will be term ^4F with a spin $S = \frac{3}{2}$, and the fine structure of the spectrum contains three absorption lines (Fig. 7.43).

The spin-lattice relaxation time also strongly depends on the valency state. Whereas Cr^{3+}, V^{2+} and other ions usually produce (in an octahedral environment) narrow lines already at room temperature, rare-earth group ions almost always (with the exception of Eu^{2+} and Cd^{3+}) require helium temperatures for observation of EPR spectra. Note that, on the strength of the well-known Kramers theorem for ions with an odd number of electrons, an electric field cannot fully eliminate the degeneracy of the ground level; it remains twice degenerate (Kramers doublet). This degeneracy is lifted by applying a magnetic field, and this creates favorable conditions for the observation of EPR.

Thus, the number of fine structure lines, which is equal to $2S$, is determined by the valency state of the ion of the paramagnetic impurity and serves as a reliable basis for its identification.

The position of the lines in the spectrum depends on the orientation of the crystal in the external magnetic field (angular dependence of the fine structure) and reflects the local symmetry of the crystal lattice around the

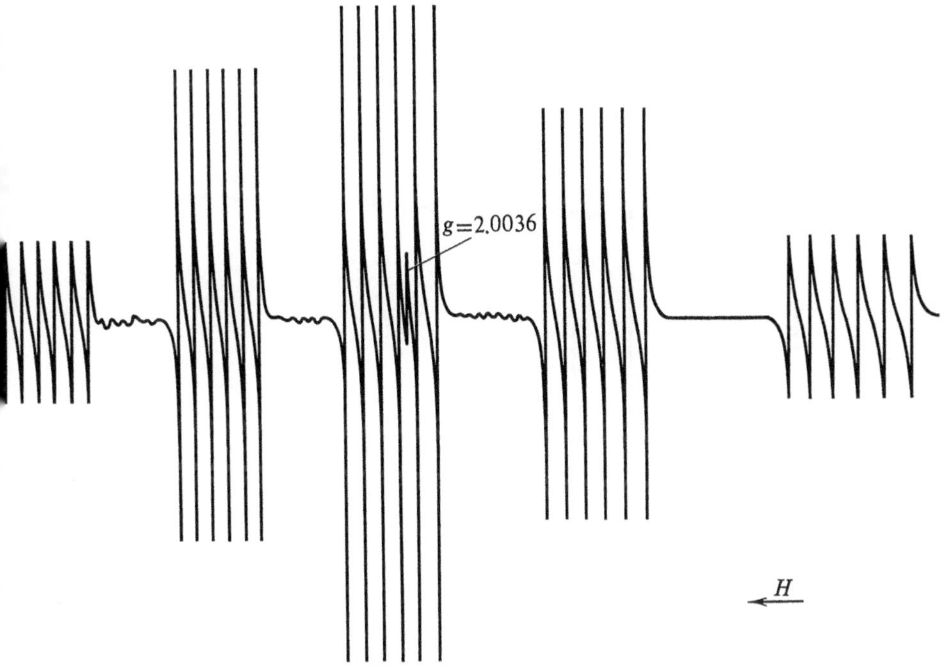

Fig. 7.44. The EPR spectrum of the Mn^{2+} ion ($S = \frac{5}{2}$, $I = \frac{5}{2}$) in a crystal of Zn, Na metasilicate in the SHF range ($\lambda = 3$ cm) ($H_0 \parallel$ to axis 2, the lines of the spectra of two equivalent Mn ions coincide) [7.85]

paramagnetic ion (site symmetry). Investigation of the angular dependences reveals a latent local anisotropy and establishes the symmetry of the crystal field and the number of equivalent (differing only in orientation) lattice sites occupied by paramagnetic ions, their population [7.84] density, and the way of compensation of the charge on replacement of crystal lattice ions.

If the nucleus of a paramagnetic ion contains an isotope with a nonzero magnetic moment, then, owing to the interactions of the electronic and nuclear magnetic moments, each line of the fine structure reveals an additional *hyperfine structure* (HFS) of $(2I+1)$ – (I being the spin of the nucleus) components, nearly equidistant (in strong fields) and of equal intensity (Fig. 7.44). When there are several such isotopes, the intensities of the HFS components are in line with the distribution of the isotopes of the natural mixture or the degree of enrichment.

If a covalent bond with ligand ions is formed the electron spin can interact with the magnetic moments of the nuclei of the nearest neighbors and lead to the formation of a *superhyperfine structure* (SHFS) of the EPR spectrum. The nature of this structure depends on the environment symmetry. If n identical ligands have nuclear spins equal to I, each line of an ordinary hyperfine structure will split into $2nI+1$ components, whose intensities obey

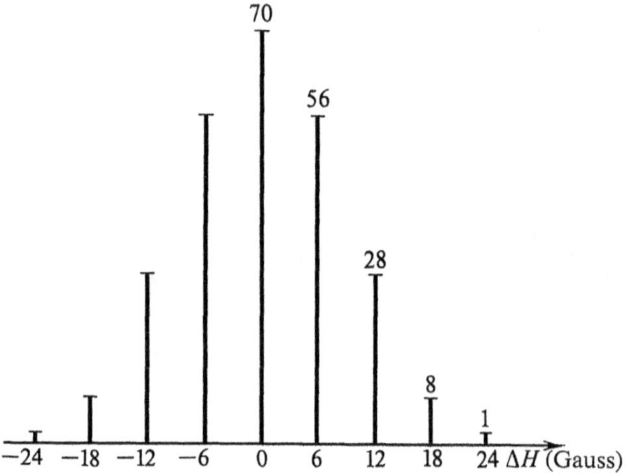

Fig. 7.45. Theoretical ratio of the intensities of the components of the superhyperfine structure of the EPR spectrum of the Mn^{2+} ion in a BaF_2 crystal due to the interaction with the neighboring equivalent nuclei of F^{19} ($n = 8, I = \frac{1}{2}$): ΔH indicates the distance, in gauss, from the center of the group

the binomial law (Fig. 7.45). Observations of the HFS may supply additional information on the way of ingress of an impurity ion and its electron-density distribution among the ligands.

In describing the EPR spectrum the spin-Hamiltonian constants are usually determined from experiment by the position of the resonance values of the field for the absorption lines. To do this, one has to set up a secular equation and find its eigenvalues by perturbation-theory methods or by some computerized numerical method (for high orders of the equation) [7.66]. The application of the spin-Hamiltonian technique will be considered on the same example. The ground state of an ion with such a configuration will be spin triplet ^3F. We assume that of all the constants B_n^m only B_2^0 is different from zero (this corresponds to the axial symmetry of the intracrystalline field) and write the spin Hamiltonian of the problem:

$$\mathscr{H} = \mathrm{B}_2^0 \mathrm{O}_2^0 + g\beta HS \ , \tag{7.125}$$

where $\mathrm{O}_2^0 = 3S_z^2 - S(S+1)$ is the spin operator.

Suppose that external magnetic field H is at first directed along the Z axis coinciding with the tetragonal axis of the crystal ($H \parallel Z$). Then

$$\mathscr{H} = b_2^0 [\hat{S}_z^2 - \tfrac{1}{3} S(S+1)] + g_\parallel \beta H S_z \ . \tag{7.126}$$

Since we are interested only in the distances between the energy levels, we add a value $\frac{2}{3}\, b_2^0$ to the diagonal terms and pass on to dimensionless values

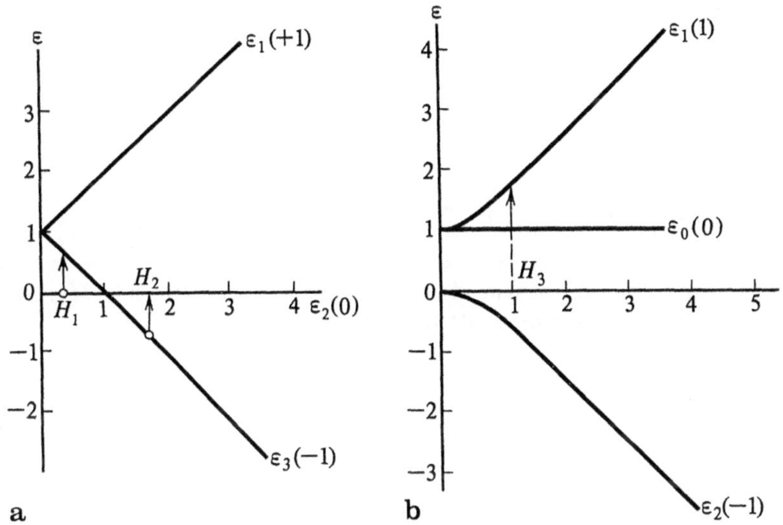

Fig. 7.46. Schemes of energy levels for the d^2 configuration ($S=1$) in a crystal of nickel fluorosilicate: **(a)** $H \| Z$, **(b)** $H \perp Z$. The Z axis coincides with axis 4 of the crystal; H_1, H_2, and H_3 are the magnetic fields of allowed transitions for $h\nu < b_2^0$ [7.66]

$x = g_{\|}\beta H/b_2^0$, $\varepsilon = E/b_2^0$, where $b_2^0 = 3\,\mathrm{B}_2^0$ is a parameter defining the initial splitting of the levels at $H=0$ (Fig. 7.43). Using wave functions ψ_{MS} (ψ_1, ψ_0, ψ_{-1}) as the basis ones, we write the secular equation

$$\begin{vmatrix} 1+x-\varepsilon & 0 & 0 \\ 0 & -\varepsilon & 0 \\ 0 & 0 & 1-x-\varepsilon \end{vmatrix} = 0 \;,$$

whence we obtain the energy levels

$$\varepsilon_1 = 1+x \;, \quad \varepsilon_2 = 0 \;, \quad \varepsilon_3 = 1-x \;.$$

If one observes, in the EPR spectrum, two resonance transitions in fields H_1 and H_2 for a quantum of high-frequency field $h\nu < b_2^0$, as indicated in Fig. 7.46a, then evidently

$$b_2^0 = \frac{h\nu(H_2+H_1)}{(H_2-H_1)} \;, \quad g_{\|} = \frac{2h\nu}{\beta(H_2-H_1)} \;. \tag{7.127}$$

Similarly, when $H \perp Z$,

$$\mathscr{H} = g_{\perp}\beta H S_x + \mathrm{B}_2^0 \mathrm{O}_2^0 \;,$$

and from the solution of the secular equation

$$\begin{vmatrix} 1-\varepsilon & x/\sqrt{2} & 0 \\ x/\sqrt{2} & -\varepsilon & x/\sqrt{2} \\ 0 & x/\sqrt{2} & 1-\varepsilon \end{vmatrix} = 0 \ ,$$

where $x = g_\perp \beta H/b_2^0$, $\varepsilon = E/b_2^0$, we easily find $\varepsilon_2 = 1$, $\varepsilon_3 = \frac{1}{2}\ (1-\xi)$, $\varepsilon_1 = \frac{1}{2}\ (1+\xi)$, $\xi = \sqrt{1+4x^2}$. The level scheme in this case is shown in Fig. 7.46 b. If the transition observed corresponds to magnetic field H_3, then for determining the main value $g_\perp$ of tensor g we get the equation

$$g_\perp = \sqrt{h\nu(h\nu + b_2^0)}/\beta H_3 \ . \qquad (7.128)$$

8. Liquid Crystals

8.1 Basic Characteristics of Liquid Crystals

The *liquid-crystal* (mesomorphous) state inherent in some organic substances consisting of elongated molecules (Fig. 8.1) is characterized by parallel stacking of such molecules, i.e., by a long-range orientation order [8.1, 8.2]. The corresponding thermodynamic phase has a definite region of its own in the phase diagram. Some substances can form not one, but several liquid-crystal phases differing in structure [2.5]. As regards their macroscopic rheological properties, liquid crystals occupy an intermediate position between true solid crystals and true (isotropic) liquids: they are fluid, but anisotropic. The viscosity of liquid crystals ranges from that of water to that of a thick oil. "Vitrified" liquid crystals resulting from supercooling are also known. The anisotropy of liquid crystals manifests itself in a number of their properties: optical, electrical, magnetic, etc. [8.1 – 8.15].

Three basic types of liquid crystals are distinguished: *smectic, nematic,* and *cholesteric.* Both nematic and smectic liquid crystals are characterized by a parallel arrangement of the molecules. In nematic liquid crystals the centers of gravity of the molecules exhibit no positional ordering, whereas in smectic phases periodic distribution is observed parallel to the main Z axis, along which the long axes of the molecules are disposed (Fig. 8.2a, b). Smectic liquid crystals may also show ordering in the layer perpendicular to the Z axis

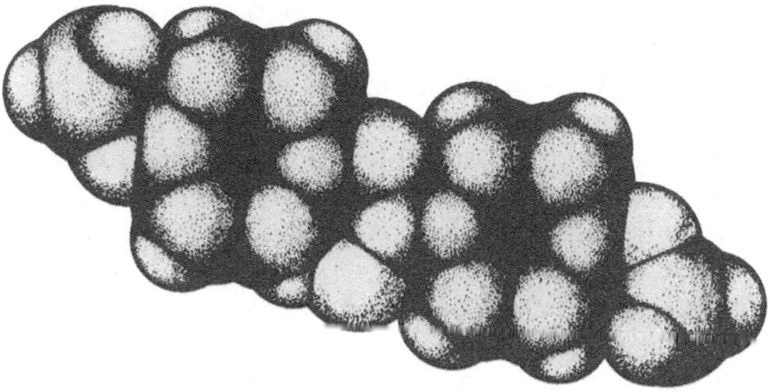

Fig. 8.1. Shape of a molecule of para-azoxyanisole forming a liquid crystal phase

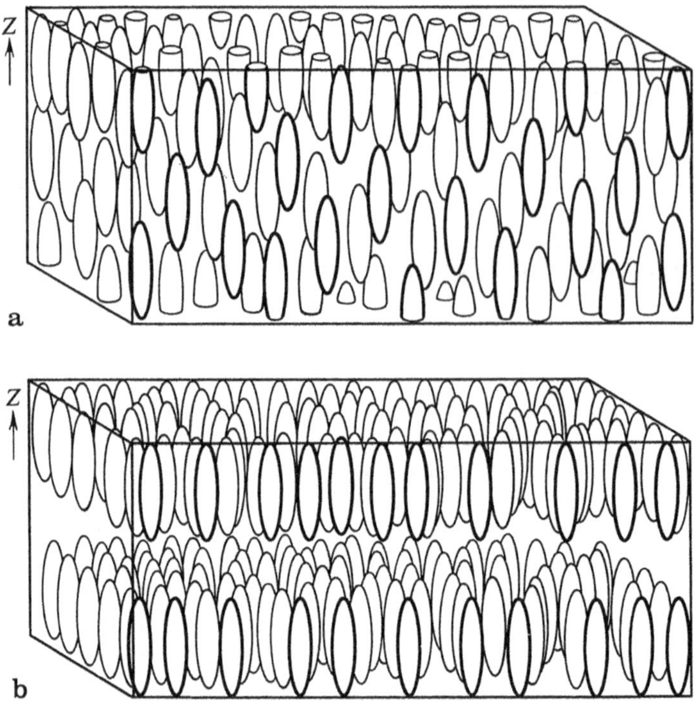

Fig. 8.2. Structures of nematic (**a**) and smectic (**b**) liquid crystals [8.1]

(Fig. 8.2b). In the cholesteric phase, at comparatively short distances, the molecules are arranged as in the nematic phase, except that at long distances the orientational structure has a helical shape (Fig. 8.3).

Some types of ordering intermediate between the smectic and nematic types are also known. In recent years, liquid crystals have been discovered which consist of disclike molecules stacked in liquid columns forming a two-dimensional crystal lattice [8.15].

Thermotropic and *lyotropic* liquid crystals are distinguished in accordance with the method of crystallization. Thermotropic liquid crystals are obtained by heating a solid crystal or cooling an isotropic liquid; they exist within a definite temperature range. This range may vary widely from one substance to another. Compounds with a liquid-crystal phase of negative temperatures up to 300 – 400 °C are known at present.

Lyotropic liquid crystals form on dissolution of solid crystals in certain solvents. They also have a closed region of their own in the phase diagram.

Speaking of the types of liquid crystals – smectic, nematic, and cholesteric – one implies their structure at the molecular level. In macroscopic specimens, regions of 10^{-5} to 10^{-2} cm always arise, with an ordering corresponding to a given liquid crystal, and the specimen represents a set of

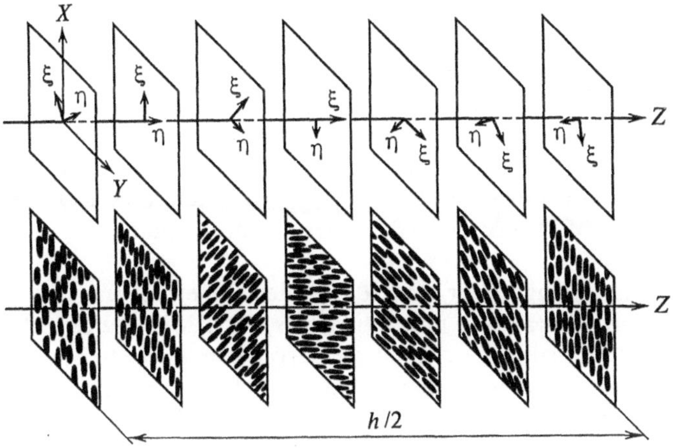

Fig. 8.3. Helical structure of cholesteric liquid crystals: h is the pitch of the spiral structure

regions with a uniform molecular orientation (domains), which are oriented randomly or regularly (texture). Under certain conditions a single-domain specimen may form.

A thin layer (10 – 50 μm) of a nematic liquid crystal is transparent, and its transparency slightly increases on disappearance of disinclinations (an analog of dislocations in solid crystals). Uniformly oriented "single-crystal" cholesteric and smectic layers are also transparent. The boundaries between domains and their mutual ordering are clearly observed in an optical microscope.

Large volumes of liquid crystals scatter light intensively and have a turbid appearance. This is caused by light scattering on inhomogeneities – orientational fluctuations and also domain boundaries and disinclinations. Even if strong orienting factors are used, for instance, a magnetic field of 20000 Oe, the substance still remains turbid, which indicates that although the molecules have largely lined up along the field, the specimen nevertheless has orientational inhomogeneities.

If a thin layer of a nematic liquid crystal is subjected to an orienting effect, the specimen may turn into a large single domain. If the optic axis of such a domain is normal to the glass plates enclosing the thin layer of liquid crystal, it is quenched between the crossed nicols of a polarizing microscope. One can, however, see a multitude of flickering light dots against the dark background of the domain. This is due to the local deviation of the optic axis of the crystal from the normal. The phenomenon can be observed if around 10^5 molecules take part simultaneously in thermal fluctuation causing local changes in the direction of the optic axis. This suggested previously that the liquid-crystal medium consists of more or less rigid "swarms" containing about 10^5 molecules parallel to each other. But it is clear at present that there is no sufficient ground for speaking of rigid "swarms" with sharply defined boundaries, although the "response" of a liquid crystal to a thermal or some other effect is actually of a cooperative nature: it involes whole ensembles of molecules. At the same time, the concept of a "swarms" is useful when one deals with heterophase fluctuations, i.e., the nucleation of a new phase near the phase transition point. In the initial phase, local and time heterophase fluctuations arise, which are molecular "swarms" of the new phase. Thus, "swarms" are dynamic formations; they disintegrate, are replaced by new ones, etc.

Liquid crystals readily change their structure under the effect of various external factors (radiation, electric and magnetic fields, etc.), which alter their optical, electrical, and other properties. Hence, it becomes possible to control the properties of a liquid-crystal specimen by weak external effects, or to detect these effects. In distinction to solid crystals, in which voltages of hundreds and thousands of volts are used to control, for instance, their optical properties, in liquid crystals voltages of 2 to 20 V are sufficient. The wide technical application of these unique properties, the importance of liquid-crystal structures in molecular biology, and the extensive material obtained by studying liquid crystals for the development and perfection of the general theory of the structure of condensed media have resulted in a vigorous growth of investigations in this branch of science during the past decade.

8.2 Texture and Optical Properties of Liquid Crystals

8.2.1 Smectic Liquid Crystals

The term "smectic" derives from the Greek "smegma" (σμηγμα-soap), because liquid crystals of this type were first discovered in soaps. An example of a substance forming smectic-type liquid crystals is ethyl ether of para-azoxybensoic acid:

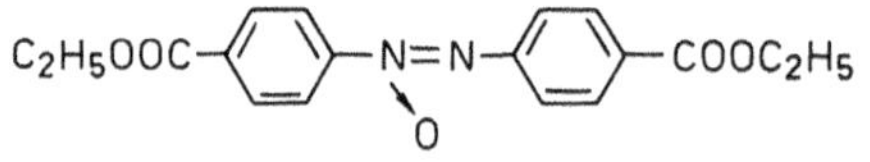

Solid crystals of this substance melt at 114 °C and form smectic liquid crystals. These melt into an ordinary isotropic liquid at 120 °C. If an isotropic melt is cooled, individual nuclei of the liquid crystal phase appear at 120 °C, which have the shape of elongated rods (Fig. 8.4). Then the rods fuse together, forming a so-called confocal texture, which can be studied conveniently in a polarizing microscope with crossed polaroids (Fig. 8.5). Recall that an ordinary isotropic liquid is not visible between crossed polaroids, i.e., the field of vision of the microscope remains dark. Considering that the birefringent texture shown in Fig. 8.5 has a regular structure, it becomes evident that the smectic liquid crystals, although being fluid, differ radically from ordinary liquids.

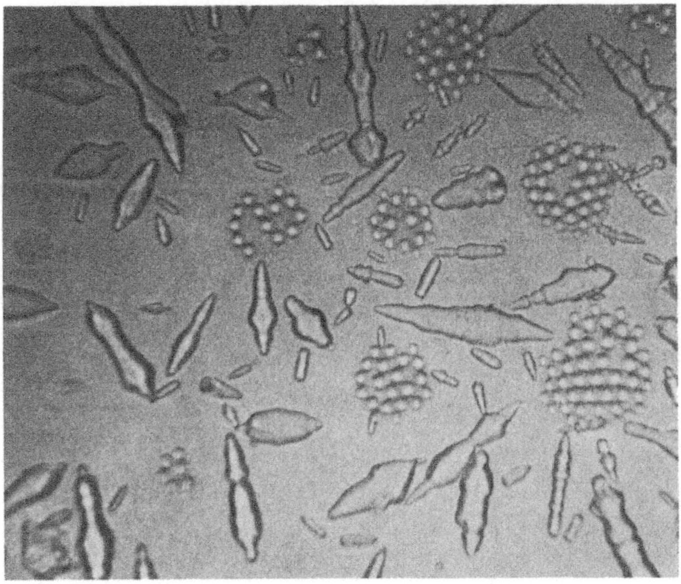

Fig. 8.4. Nuclei of a smectic liquid crystal in the form of rods floating in an isotropic melt (magnification 200×, natural light) [8.2]

Fig. 8.5. Smectic confocal texture in crossed nicols (magnification 400×) [8.2]

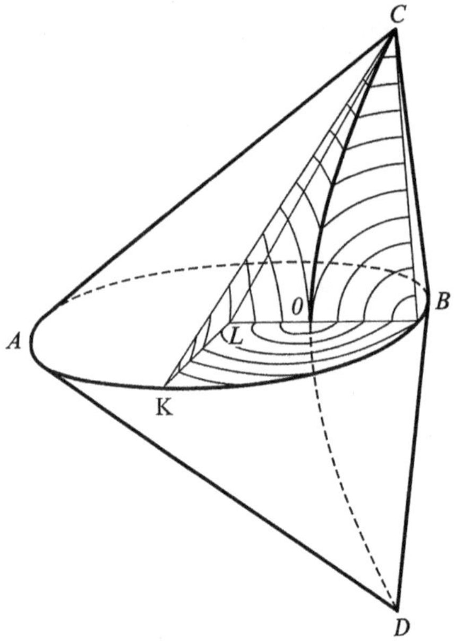

Fig. 8.6. Structure of a confocal domain

The texture shown in Fig. 8.5 consists of separate free regions – confocal domains. The geometry of each domain is rather complicated. The smectic layers in them are wound up so that they form a family of surfaces, which are called Dupin cyclides in topology. The arrangement of smectic layers in an individual domain is shown in Fig. 8.6. The basis for the formation of a family of cyclides is a confocal pair: ellipse AB and hyperbola COD. At any place of the domain the direction of the long molecular axes coincides with that of a straight line drawn through three points: the center of a given molecule, a point of the ellipse, and a point of the hyperbola. For instance, BD, BC, or KC in Fig. 8.6 may be such lines. These lines also represent the directions indicating the arrangement of the optic axes in the domain. Such a structure of confocal domains explains the optical pattern observed under the microscope (Fig. 8.5).

Sets of domains are interrelated more or less regularly, forming characteristic textures (polygonal, fanlike, etc.), which help one to identify clearly the smectic phase. One can also observe consistent changes in domain structure on transition from one smectic polymorphous modification to another.

Among the many structural features of smectic liquid crystals we note one more, most prominently displaying their peculiarity, namely, the formation of stepped drops. If we place a little smectogenic substance on a very clean and even surface, say on a mica flake freshly split off along the cleavage plane, the after melting of solid crystals the smectic phase appears in the form of stepped flat drops. Their structure is presented schematically in Fig. 8.7. Stepped drops result from the layer structure of a smectic liquid crystal. The thickness

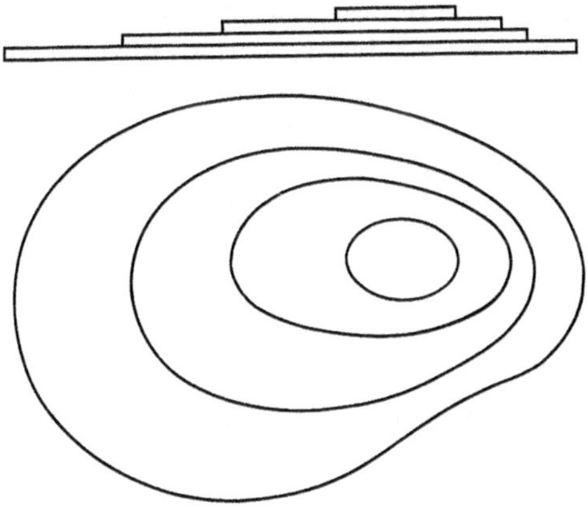

Fig. 8.7. Structure of stepped smectic drops in two projections

of each step is a multiple of the thickness of a separate smectic layer. When a drop is subjected to a mechanical effect, one can see that the separate layers slip freely relative to each other.

It is possible to obtain thin, uniformly oriented layers of a liquid crystal – single-domain specimens – between glass plates. Investigation of such specimens in a converging light showed that the conoscopic picture is similar to the pattern obtained from ordinary single-crystal plates.

Several smectic polymorphous modifications are distinguished [Ref. 2.5, Chap. 2]. Modifications *A, B,* and *C* have been studied most thoroughly. In smectic *A*, the long molecular axes are perpendicular to the smectic layers. Inside the layers there is only a short-range positional molecular order. In distinction to smectic *A*, smectic *B* has an apparently positional long-range (hexagonal) order in the arrangement of the molecules inside the layer. Phases *A* and *B* are optically uniaxial, their optical symmetry being ∞/mm.

Liquid crystals whose long molecular axes are consistently inclined towards the smectic planes (phase *C*) and form some angle θ with the normal to the layers are optically biaxial. Uniformly oriented smectic specimens can then be assigned the point symmetry $2/m$, the same as in the family of indicatrices of biaxial crystals. Smectic phase *C* can be assigned a local (at short-range molecular distances) monoclinic symmetry with a point group containing axis 2. The axis is parallel to the smectic layers. When molecular compounds are chiral (i.e., are in one of the two possible enantiomorphous forms), there are no other symmetry elements. The molecular dipoles are directed along axis 2. Such phases ($\tilde{C}$) possess specific ferroelectric properties. If a phase is a racemate (i.e., if it consists of both types of chiral molecules), the point group of symmetry is $2/m$

In distinction to smectics *C*, in smectics $\tilde{C}$ the molecules deviating from the normal to the layer by angle θ rotate from layer to layer relative to the normal, forming a helical structure (a helicoid) with a pitch

$$h = 2\pi\tilde{t}/\alpha \ ,$$

where $\tilde{t}$ is the thickness of the smectic layer, and α is the azimuthal angle characterizing the rotation of the molecules in two neighboring layers. Such chiral $\tilde{C}$ structures have a polarization $\boldsymbol{P}$ in the layer plane. Vector $\boldsymbol{P}$ is perpendicular to the axes of the molecules and rotates about the normal with a pitch of several micrometers. An example of a compound with a ferroelectric $\tilde{C}$ phase is *n*-decyloxybenzylidene-*n'*-amino-2-methyl-butyl-cinnamate:

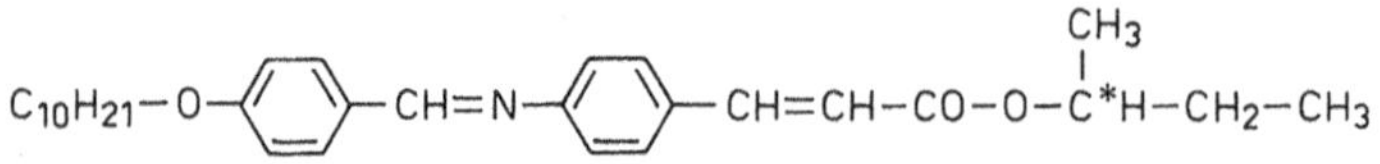

The asymmetric atom of carbon C* causes the chirality of the $\tilde{C}$ phase and point symmetry 2.

In addition to the *A, B, C,* and $\tilde{C}$ phases, several more varieties of smectic polymorphous structures are known, and their number will evidently increase as experimental data accumulate.

8.2.2 Nematic Liquid Crystals

A typical example of a substance yielding nematic liquid crystals is para-azoxyanisole:

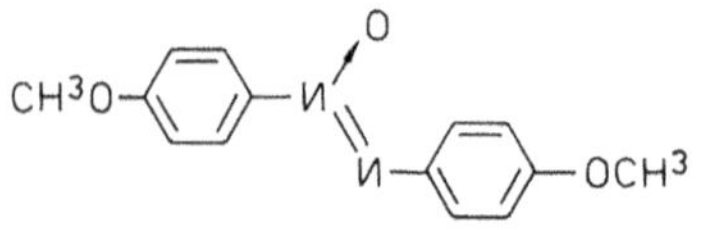

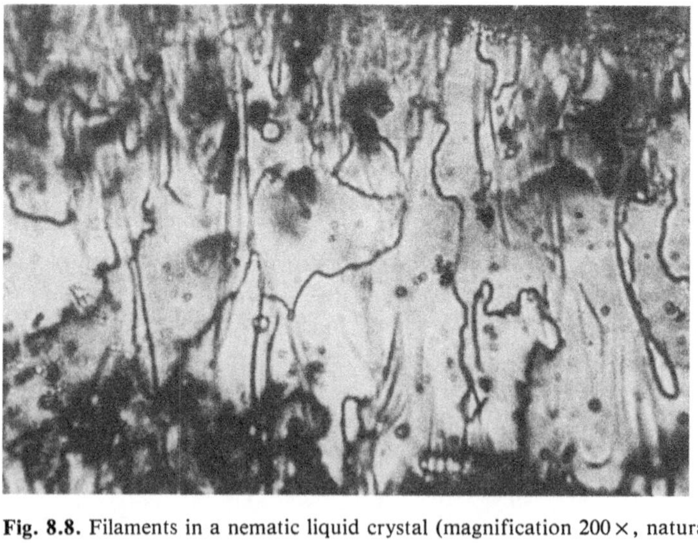

Fig. 8.8. Filaments in a nematic liquid crystal (magnification 200×, natural light)

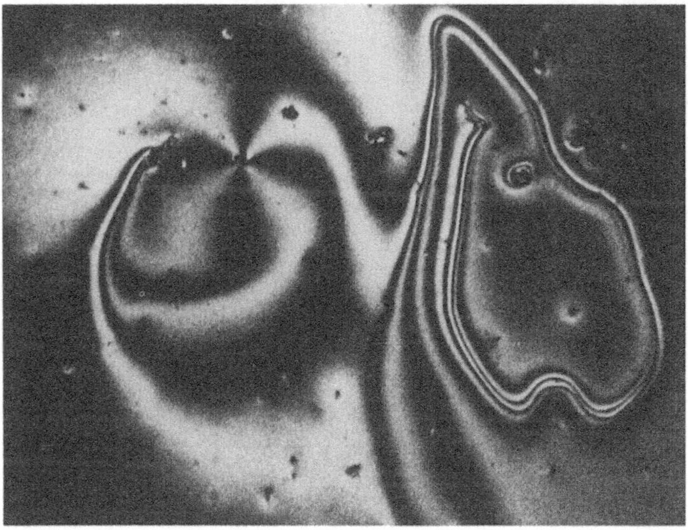

Fig. 8.9. Liquid-crystal nuclear texture of para-azoxyanisole (magnification 120×)

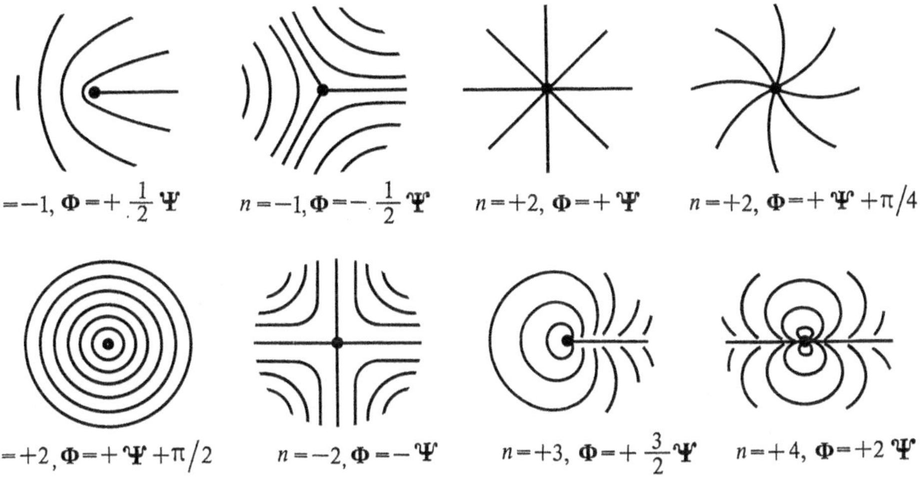

Fig. 8.10. Arrangement of molecules near nuclei [8.3]: the lines denote the direction of the long axes of the liquid crystal

Liquid crystals of para-azoxyanisole exist in the temperature range from 116 to 136 °C. The specimen shows numerous dark filaments. The term "nematic" derives from the Greek "nema" ($\nu\eta\mu\alpha$), i.e., filament. The filaments are mobile and clearly visible in a natural light (Fig. 8.8). They are sites of rupture of optical continuity of the medium; here the orientation of the elongated molecules changes sharply. By analogy with dislocations in

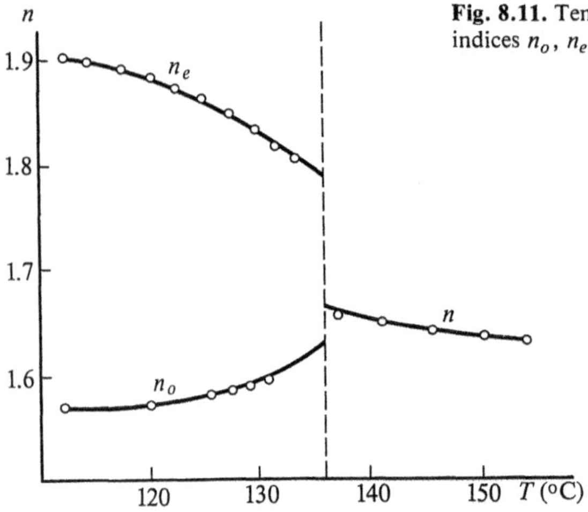

Fig. 8.11. Temperature dependence of the refractive indices n_o, n_e, and n for para-azoxyanisole [8.2]

ordinary crystals, the filaments are called *disinclinations.* Disinclinations are often perpendicular to the surface of the specimen, and their emergences at the surface are clearly seen. They have the appearance of dark spots, "nuclei." Dark branches extending from the nuclei are noticeable between the crossed polaroids (Fig. 8.9). These are the parts of the crystal where the directions of the long axes of the molecules coincide with the vibrations of the electrical vector of the light wave in the polarizer and the analyzer and which are therefore invisible. The molecular pattern near disinclinations may vary. The possible mutual arrangements of molecules near disinclinations are shown in Fig. 8.10. Here n, ψ, and φ are the parameters of (8.35), which describes the molecular ordering of a liquid crystal.

Using special techniques for treating the glass plates enclosing the substance, one can obtain "single-crystal," disinclination-free nematic layers. Investigations in a converging light beam show that nematic liquid crystals are uniaxial and optically positive. Their point symmetry is ∞/mm. Note that external effects on a liquid crystal may change the spatial symmetry. Thus, mechanical rotation of the glass plates enclosing the specimen leads to the winding up of its structure. Superposition of an electric field on a specimen may result in optical biaxiality and transition to symmetry *mmm*.

The magnitude of birefringence of nematic phases is rather high. Thus, for para-azoxyanisole at 117 °C the difference between the refractive indices of an ordinary and an extraordinary beam is $\Delta n = (n_e - n_o) = 0.268$. The magnitude of birefringence changes with the temperature (Fig. 8.11), and reduces to zero at the temperature of transition of the liquid crystal to an isotropic liquid.

8.2.3 Cholesteric Liquid Crystals

As follows from the name, cholesterine derivatives cholesteric-type liquid crystals produce. For instance, cholesterylcinnamate, whose structure formula is

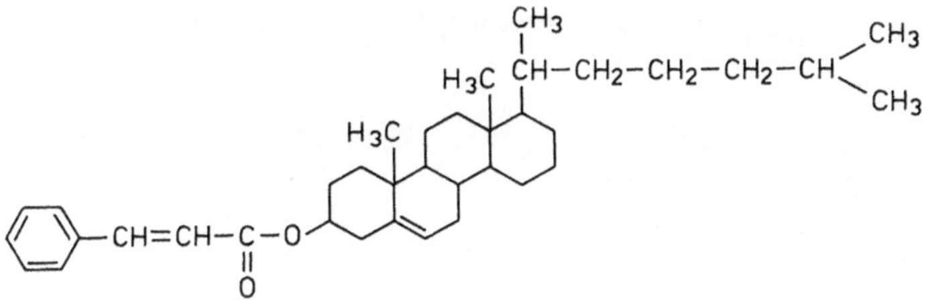

forms a cholesteric liquid crystal phase in the temperature range from 156 to 197 °C.

An example of a cholesteric liquid crystal which is not a cholesterine compound is active amyl-p-(4-cyanobenzylidene-amino)-cinnamate:

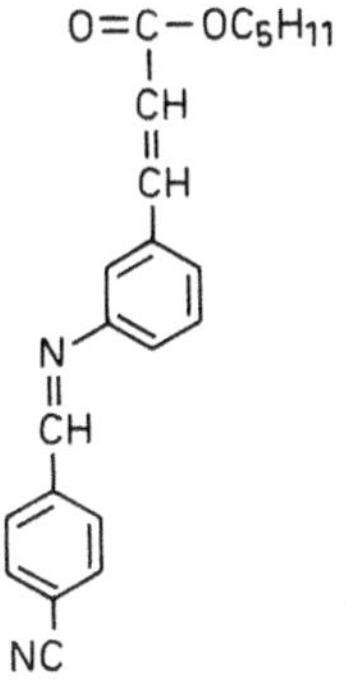

We also recall that the point symmetry group of cholesteric liquid crystals ($\infty/2$) causes to the winding up of their structure. We have previously shown how mechanical winding up of a nematic specimen transforms it into a spatially inhomogeneous (screw-type) liquid crystal. Such a transition can also be caused by adding a small amount of optically active impurity to a nematic substance. The helical structure and the ability of rotating the light polarization plane are what distinguishes a cholesteric liquid crystal from a nematic one. In contrast to the nematic substances, the cholesteric liquid crystals are weakly biaxial and optically negative. This means that the light velocity in the direction perpendicular to the molecular layers is maximal.

The texture of the cholesteric liquid crystals resembles the smectic one as regards is layer arrangement; it may contain different types of confocal domains (Fig. 8.12).

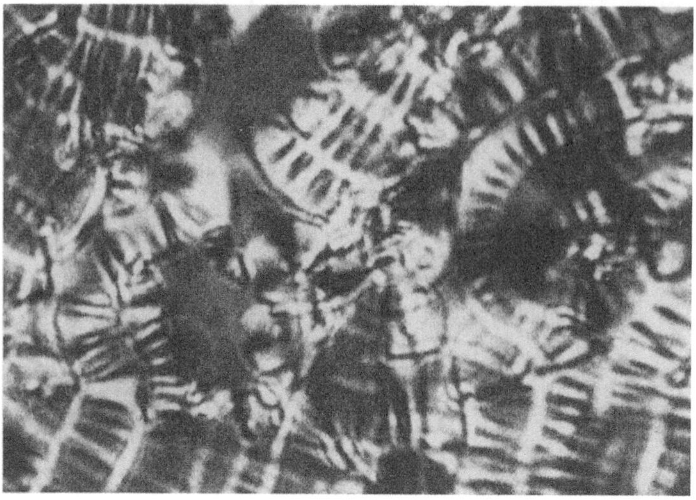

Fig. 8.12. Cholesteric confocal texture of cholesterylcinnamate (magnification 270×) [8.2]

The molecules of the cholesteric mesophase are enantiomorphous and are arranged in quasi-nematic layers, in which the direction of the long molecular axes is approximately the same and corresponds to "two-dimensional" nematic ordering (Fig. 8.3). But this direction changes gradually from layer to layer, by a constant angle α, the result being helical superposition of layers on each other. As angle α between the molecules in neighboring layers is small (only a few angular minutes), the pitch of the helical structure, h, is very large and may reach from 10^3 to 10^4 Å. Hence, a single domain of a cholesteric liquid crystal (plane Grandjean texture) (Fig. 8.13) has a periodic structure with period h (Fig. 8.3).

The specific molecular structure of the cholesteric liquid crystals explains their unique optical properties. If a linearly polarized light passes in the direction of the axis of the helical structure, the direction of the electrical vector of the light wave rotates successively along the turns of the spiral (to the left or to the right, depending on the substance and the wavelength). Therefore, the light polarization plane also rotates through an angle proportional to the thickness of the layer of the liquid crystal. This is similar to the appearance of optical activity on passage of a polarized light through a stack of helically disposed birefringent thin crystal plates. The optical activity of a cholesteric liquid crystal may be as high as 10^5-10^6 deg/mm. This is a very large value, considering that ordinary crystals or liquids have an optical activity of the order of $10-10^2$ deg/mm (for instance about 20 deg/mm for quartz).

The magnitude of optical activity Ω is defined by the equation

$$\Omega = a h n^2/\lambda^2 \ ,$$

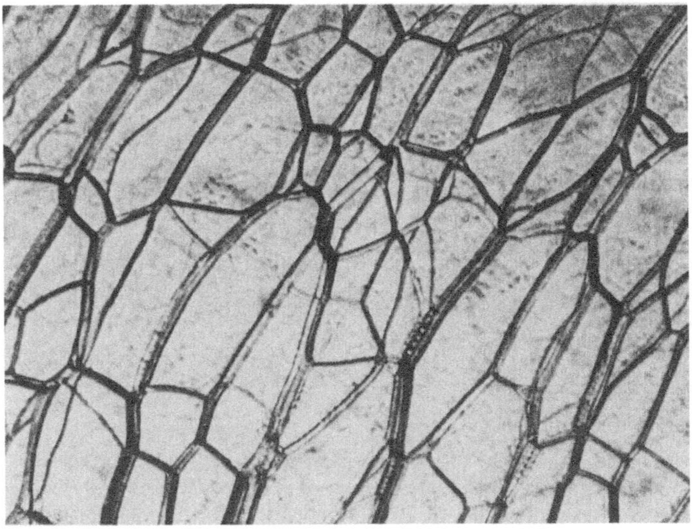

Fig. 8.13. Single domain of a cholesteric crystal containing a disinclination network (magnification 300×) [8.2]

where a is a constant, h is the structure period, λ is the light wavelength, and n is the average refractive index.

Another important property of a cholesteric plane texture is selective light reflection. If a beam of white light is directed at a liquid crystal, the reflection law fails: with one and the same angle of incidence the angles of reflection will be different for different wavelengths. As a result, a plate of liquid crystal will seem to be colored differently when observed at different angles. This may be attributed to the fact that the beam meets a periodic structure with period h along the spiral.

Indeed, according to the Bragg-Wulff law, in the direction of observation the maximum intensity will be shown by a light with a wavelength λ_0 at which the beams reflected by equivalent planes arrive in phase:

$$n h \sin(\theta/2) = k \lambda_0 \ ,$$

where k is the order of magnitude of diffraction.

Since period h is $10^3 - 10^4$ Å, the wavelengths, which obey the Bragg-Wulff condition, are often within the range of the visible spectrum, causing selective reflection.

The optical properties of cholesteric liquid crystals strongly depend on the temperature, mechanical effects, electric fields, etc., these being widely used in practice.

The temperature sensitivity of cholesteric liquid crystals enables one to visualize the temperature distribution on the surface; this property is utilized

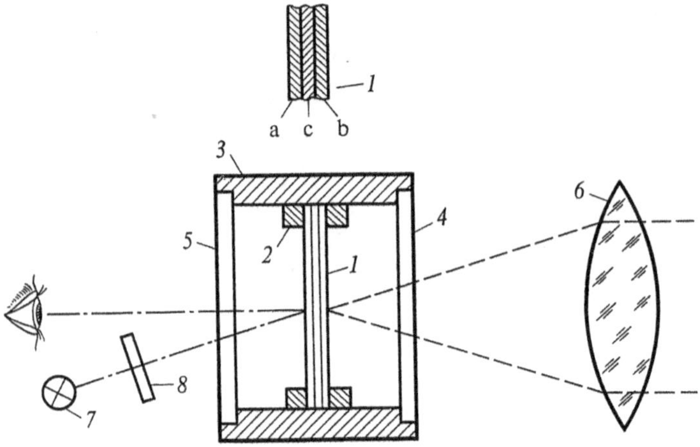

Fig. 8.14. Infrared-to-visible image converter: (*1*) liquid-crystal membrane [(*a*) supporting film, (*b*) blacking layer, (*c*) liquid crystal], (*2*) membrane-holding ring, (*3*) converter body, (*4*) IR-transparent window, (*5*) chamber window, (*6*) IR-image-forming lens, (*7*) white-light source, (*8*) thermal filter

in introscopy (nondestructive testing), in medicine for diagnosing a number of diseases, and in various temperature pickups.

When used for thermographical purposes, a liquid crystals is placed on a thin black substrate, which is previously deposited on the surface area under investigation. The change of color makes it possible to study the temperature distribution visually: even a negligible temperature difference of a few hundredths or even thousandths of a degree becomes visible.

Discoloration on irradiation makes it possible to visualize IR, UV, and VHF radiations (the last-mentioned type of radiation can be used for VHF holography).

The sensitivity of selective reflection to shear can be utilized for visualizing the oscillation distribution on the surface (e.g., of piezoquartz) and designing mechanical-oscillation transmitters.

Finally, the effect of electric and magnetic fields changes the helical pitch h, enabling one to control the color of flat cholesteric films. This can also be used in various technological devices (modulators, regulated monochromators, etc.). Similar techniques are employed in investigating the temperature distribution in radio circuits (multilayer printed circuits, diode and resistance matrices, etc.), thus detecting overheatings, breakdowns, and short circuits.

Cholesteric liquid crystals have found an interesting application in infrared techniques. They served as a basis for an infrared-to-visible image converter (Fig. 8.14). A liquid-metal film is deposited on a thin black membrane, which absorbs the infrared radiation and transfers the heat to the liquid-crystal layer. Since its color is temperature dependent, white light illumination produces a visible image of the heat radiation.

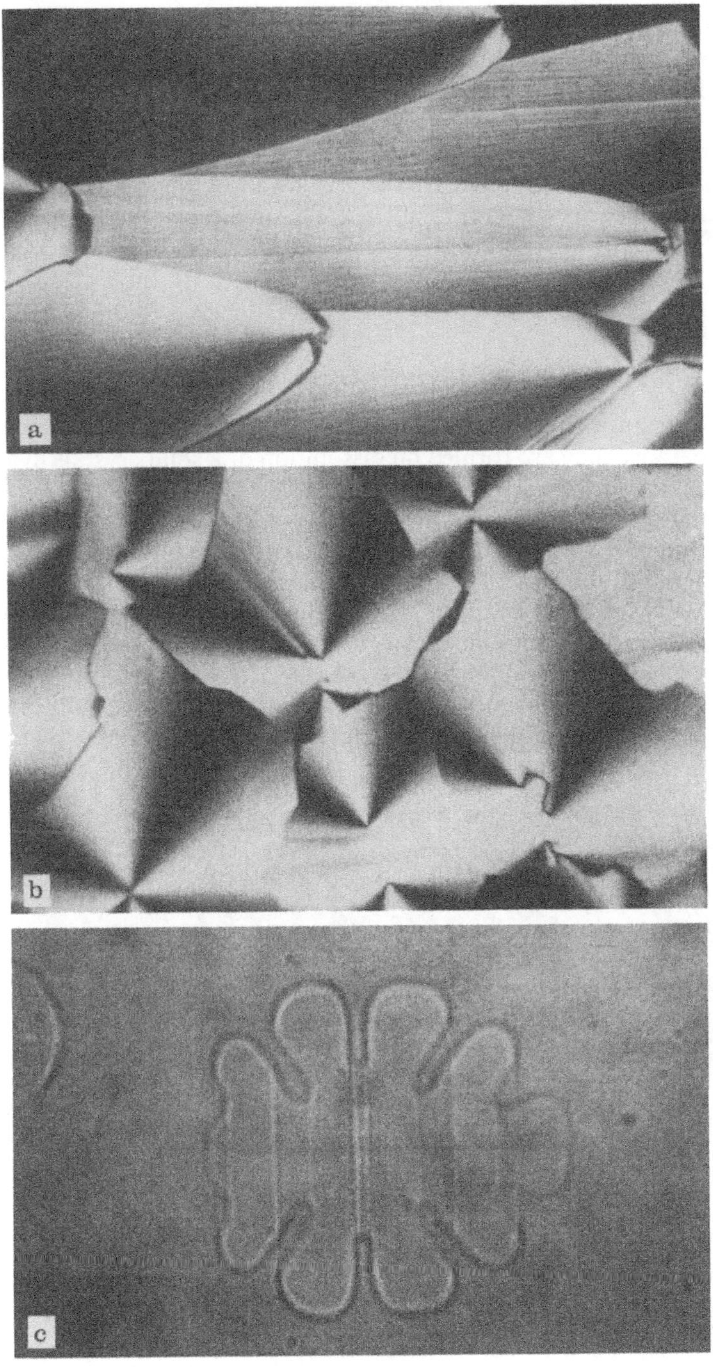

Fig. 8.15. Lyotropic smectic liquid crystals of potassium oleate: (**a**) texture at specimen edges (magnification 300×), (**b**) texture at the specimen center, (**c**) single domain in the form of a body of revolution (magnification 800×, the nicols are crossed) [8.2]

Films with a thickness on the order of 10 – 20 μm are usually employed in practice. The time sensitivity (the time of appearance of the color after the excitation of such films) is 20 – 30 ms. The power required for the excitation and appearance of a selective reflection is on the order of 10^{-3} W/cm^2. The temperature resolution may be as good as thousandths of a degree.

Films of liquid crystals with such characteristics allow one to visualize thermal fields on the surface of micromoduli with the smallest circuit elements on the order of 2 μm. This means that the method makes it possible to record information of up to 10^9 bits/cm^2.

8.2.4 Lyotropic Liquid Crystals

Many colloid systems are in the liquid-crystal state. Such mesophases are called lyotropic, because they are formed on dissolution of a solid crystals in a solvent.

A substance highly convenient for investigating the texture of lyotropic crystals is a water-alcohol solution of potassium oleate. If we place a drop of such a solution between the slide and the cover glasses, after several hours smectic lyotropic liquid crystals begin to grow at the edges of the cover glass (Fig. 8.15a). The texture show in Fig. 8.15b forms closer to the center of the preparation. If a single domain of such a texture grows freely (without interference from neighboring domains) it has the shape of a body of revolution (Fig. 8.15c). After the domain increases in diameter in the course of growth

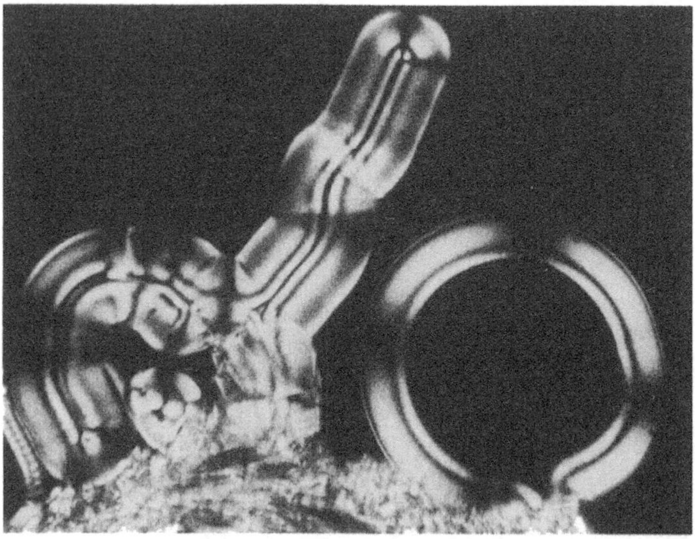

Fig. 8.16. Myelin forms of liquid crystals of a cholesterine-glycerin mixture (magnification 800×, the nicols are crossed) [8.2]

Fig. 8.17. Lamellar (**a**), cylindrical hexagonal (**b**), and complex hexagonal (**c**) lyotropic textures

and comes in contact with the glasses, the smectic layers perpendicular to the domain axis wind up into Dupin cyclides, and a complicated texture is formed (Fig. 8.15b).

There are many types of lyotropic liquid-crystal textures, of which we shall mention just one. The system solvent-smectic liquid crystal often gives rise to a peculiar texture consisting of so-called myelin shapes (Fig. 8.16). The appearance of myelin shapes is best observed in the cholesterine-glycerin system. If this system is heated, at a definite temperature the cholesterol combines with the glycerin to produce smectic-type liquid crystals. The latter absorb the excess glycerin, this leading to the formation of myelin shapes, which move in the course of their growth, meshing together into a "tangle of snakes." Myelin tubes of diameter 0.01 – 0.02 mm have a bright interference color.

The great diversity of lyotropic textures is due to the different intramolecular structure, which is more complicated than in thermotropic liquid crystals. The structural units here are not molecules, but molecular complexes, micelles. Micelles may be lamellar (platelike), cylindrical, spherical, or rectangular. The complexes are distributed in the solvent medium, forming different structures. Figure 8.17 shows some of them.

Lyotropic liquid crystals are particularly widespread among bio-organic substances. They form polypeptides, cholesterol esters, cerebrosides, viruses, etc. Many structural formations of a living organism are characterized by a liquid-crystal state (the myelin shell of the nerves, etc.).

8.3 Theory of Liquid-Crystal State

A fundamental property of a liquid crystal, which distinguishes it from an isotropic liquid and likens it to a solid, is its having an orientational degree of freedom, which characterizes the macroscopic ordering of the identified molecular axes in space. This degree of freedom of an anisotropic liquid explains the unique properties of the medium, which are associated with the high sensitivity of the orientational ordering distribution to changes in temperature and mixture concentration, and also to the effects of electrical and magnetic fields, elastic stresses, and viscous flow. The continuum theory of structural transformations provides a qualitative and quantitative explanation of the different types of instability in liquid crystals [8.14].

The theory of instabilities in a liquid-crystal medium reflects simultaneously the great diversity and complexity of phenomena in dissipative anisotropic systems which include liquid crystals in the general case. Indeed, a liquid crystal is, as a rule, a complex system, which has, along with the above degree of freedom, other degrees of freedom: ionic and neutral impurities, uniform and nonuniform elastic deformations, partial positional ordering of molecular centers of masses, and nonuniform flow velocity distributions are being developed. These degrees of freedom interact, and this complicates consideration of many phenomena, particularly those accompanied by irreversible processes of electric charge, mass, and heat transfer.

Thus, the problem of structural instabilities in liquid crystals includes various aspects of continuum theory, among them the thermodynamics and dynamics of the medium investigated. By considering these aspects as an entity, one can explain theoretically a wide range of the phenomena observed. The current objects of studies are systematization of thermodynamic states of liquid crystals based on the theory of groups and irreducible concepts, determination of the corresponding phase transition parameters, classification of the types of thermodynamic instability, and description of the properties of modulated structures resulting from various types of instability. Theories of formation of weakly nonuniform structures in a dissipative anisotropic medium are elaborated, investigation of nonlinear models of instability of liquid crystals is made, and description of orientational turbulence in a liquid-crystal medium is undertaken.

8.3.1 Thermodynamic States of Liquid Crystals

The mesophases are usually classified by the type of density function $\varrho(\boldsymbol{r})$ and the local orientation of molecules $\boldsymbol{n}(\boldsymbol{r})$. The unit vector $\boldsymbol{n}(\boldsymbol{r})$ ("director") indicates the direction along which the identified molecular axes (for instance, long ones) are oriented, on the average, at a point with coordinate $\boldsymbol{r}$. A phase with $\varrho = \text{const}$ and $\boldsymbol{n} = \text{const}$ is a nematic liquid crystal (NLC). Mesophases in which the function $\varrho(\boldsymbol{r})$ is periodic along the identified Z axis and constant along the XY planes are smectic liquid crystals (SLC). As to fluidity properties, NLC are closest to an ordinary isotropic liquid with $\varrho = \text{const}$. Smectic liquid crystals possess fluidity along the layers and behave almost like solids with respect to uniaxial loading perpendicular to the layers. Cholesteric liquid crystals (ChLC) are characterized by the value $\varrho = \text{const}$ and by a macroscopically modulated structure $\boldsymbol{n}(z)$.

Description of the diversity of liquid-crystal structures and of the possible structural transformations calls for a comprehensive consideration of the symmetry properties of liquid crystals by studying mutual correlations betwen the positions of different atoms [8.16]. It is often sufficient to investigate the point symmetry of correlation functions; for instance, as regards crystals, it will suffice to consider the relevant crystal classes. Deeper insight into the modulated structures in systems with chiral molecules is provided by Lifshitz invariants, which play the same role here as in ordinary crystals.

It is occasionally sufficient to use the pairwise correlation function $\varrho_{12}(\boldsymbol{r}_{12})$, where $\boldsymbol{r}_{12}$ is the spacing between two atoms. It cannot, however, be employed to describe structures without a center of symmetry, since the function $\varrho_{12}(\boldsymbol{r}_{12})$ possesses the property of parity. Description of chiral structures requires more complex correlation functions, for instance, four-particle correlations between the atomic positions. The presence or absence of a center of symmetry in a molecular system may be characterized by the pairwise correlation function $\varrho_{12}^{m}(\boldsymbol{r}_{12}^{m}, \boldsymbol{e}_1, \boldsymbol{e}_2)$ depending on the distance between the centers of masses of the molecules, $\boldsymbol{r}_{12}^{m}$, and the orientations $\boldsymbol{e}_1$ and $\boldsymbol{e}_2$ of the long axes of the molecules 1 and 2. In particular, in a system of chiral molecules whose axes $\boldsymbol{e}$ are nonpolar, ϱ_{12}^{m} contains a pseudoscalar:

$$(\boldsymbol{e}_1 \boldsymbol{e}_2)(\boldsymbol{r}_{12}^{m}[\boldsymbol{e}_1 \boldsymbol{e}_2]) \ . \tag{8.1}$$

Liquid crystals whose molecules also possess isolated short axes require a still more sophisticated description. In the group-theory approach the structure of a liquid crystal can be described uniformly in terms of the corresponding symmetry groups, and this permits, in principle, taking into account all the possible changes in the degrees of freedom under phase transformations.

The most symmetric phase is an isotropic liquid whose point symmetry group may be either a complete orthogonal group $\infty\infty/m$ or a complete rotational group $\infty\infty$ in the case of chiral molecules. The emergence of uniaxial nematic, cholesteric, and smectic phases from an isotropic phase is interpreted as the lowering of symmetry $\infty\infty/m \to \infty/mm$ (NLC and type A SLC), $\infty\infty \to \infty/2$ (ChLC and type A* SLC), which is induced by the five-dimensional representations G_g and G, respectively. The bases of these representations are linear combinations of products $n_i n_j$ of the base functions n_i of three-dimensional vector representations F and F_u of groups $\infty\infty$ and $\infty\infty/m$. (Here the subscripts g und u correspond to the evenness and oddness of the bases G and F relative to the inversion operation.) Representations F and F_u must induce spontaneous polarization, i.e., uniaxial phases with axial symmetry groups ∞ and ∞m, which have not yet been detected experimentally. The same is true for a phase with a symmetry ∞/m (the presence of spontaneous magnetization) induced by the representation F_g of the group $\infty\infty/m$.

The representations G and G_g correspond to ordering of the long axes of the molecules. Since the symmetric cubes $[G^3]$ and $[G_g^3]$ of the representations G and G_g contain a unit representation, these transformations can only be first-order phase transitions. The antisymmetric square of the representation G contains a vector representation $\{G^2\} = F$; therefore, there is an invariant

$$e_{ijk}\, n_i\, n_m \frac{\partial}{\partial x_j} (n_k\, n_m) \ , \tag{8.2}$$

where e_{ijk} is a totally antisymmetric unit tensor. This invariant determines the modulated orientational structure, namely, the ChLC structure, and corresponds to the corrections (8.1) in the correlation function ϱ_{12}^m.

In such cases, a possible orientational order parameter is the tensor $\hat{S}$ with components

$$S_{ik}(\boldsymbol{r}) = \langle e^i(\boldsymbol{r})\, e^k(\boldsymbol{r})\rangle - \tfrac{1}{3}\langle e^j(\boldsymbol{r})\, e^j(\boldsymbol{r})\rangle\, \delta_{ik} = S(\boldsymbol{r})\, [n_i(\boldsymbol{r})\, n_k(\boldsymbol{r}) - \tfrac{1}{3}\delta_{ik}] \ , \tag{8.3}$$

which are the local averages of quadratic combinations formed by projections of the unit vector of the isolated molecular axis $\boldsymbol{e}$, for instance, the long axis in an elongated molecule or the short axis in a disklike molecule. The value of $S(\boldsymbol{r})$ determines the share of molecular axes parallel to the director $\boldsymbol{n}$ at a given point. In the isotropic phase, $S=0$.

The transformation from isotropic liquid (IL) to nematic phase (N) is described by expansion of the density of free energy $F_N(S)$, which has the following form in accordance with symmetry properties:

$$F_N = a_N S^2 + b_N S^3 + c_N S^4 + \dots \ , \tag{8.4}$$

where $a_N = a'(T - T_0)$, $a'_N > 0$, $b_N < 0$, $c_N > 0$. The coefficients of such an expansion are parameters of the substance and depend on the molecular structure. The point T_{IN} of a first-order phase transition $IL \leftrightarrow N$ and the corresponding jump of the S-order parameter are given by relations

$$a_N(T_{IN}) = b_N^2/4c_N \ , \qquad S(T_{IN}) = \tilde{S} = -b_N/2c_N \ .$$

The theoretical dependence $S(T)$ is in good agreement with experiment (Fig. 8.18).

In type A and A^* SLC with point symmetry groups ∞/mm and $\infty/2$, the following density wave arises along with orientational ordering:

$$\varrho(z) = \varrho_0 + |\Psi_1| \cos(\boldsymbol{k}\boldsymbol{r} + \beta_1) \ , \tag{8.5}$$

where $|\Psi_1|$ and β_1 are the wave amplitude and phase, $\boldsymbol{k}\boldsymbol{r} = kz$, $k = 2\pi/\tilde{t}$, $\tilde{t} \sim l$, l is the length of the molecule, and $\tilde{t}$ is the thickness of one smectic layer. Strictly speaking, the periodic function $\varrho(z)$ in a three-dimensional body is "washed out" by thermal fluctuations, i.e., unidimensional lattices in unbounded A and A^* phases are unstable. Real SLC layers of small thickness d are stable, however, since the mean square of lattice displacements $\langle u^2\rangle$ is very small as compared with l^2 if $d < 10^{-4}$ m.

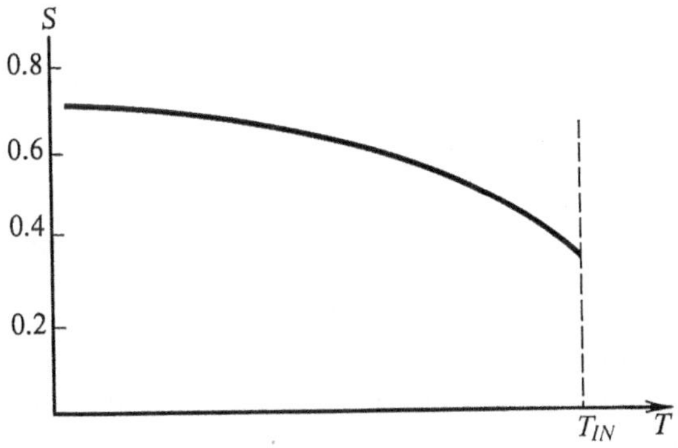

Fig. 8.18. Change in degree of ordering S in a nematic phase in relation to the temperature (T_{IN} is the temperature of transition of LC into an isotropic liquid)

In the general case, description of phase transitions in SLC calls for analysis of space symmetry groups. Sometimes, however, just as for ordinary crystals, it will suffice to restrict oneself to consideration of the point symmetry, for instance, if phase transitions in SLC leave the smectic layers intact. The same is true for cases where the space group contains no screw axes or glide planes.

Transformations from phase N to phase A and cholesteric phase Ch to phase A^* leave the point symmetry group intact, but the translation group changes. The appearance of $\tilde{t}$ translations, i.e., the formation of a unidimensional crystal with a two-component order parameter $\Psi_1 = |\Psi_1| \exp[\mathrm{i}(kz+\beta_1)]$, is described by expansion of the density of free energy $F(\psi_1)$, which has the following form in accordance with the translation symmetry:

$$F_A = a_A|\Psi_1|^2 + c_A|\Psi_1|^4 - f|\Psi_1|^2 S \ , \tag{8.6}$$

where $a_A > 0$, $c_A > 0$, $f > 0$. The positive sign of the coefficients a_A *and* c_A reflects the experimental fact of the absence of SLC with the value of the parameter of orientational order $S = 0$. The positive sign of f means physically that the additional orientational ordering due to the density wave corresponds to an increase in attraction forces operating between molecules, i.e., reduces the system energy.

Investigation of the phase diagram using (8.4, 6) shows that transitions to the A phase are possible subject to sufficiently large orientational ordering S in the N phase, and at $fS > a_A$ the A phase arises directly from the IL phase. At $fS < a_A$, successive phase transitions $IL \leftrightarrow N \leftrightarrow A$ occur. The point T_{NA} of the $N \leftrightarrow A$ phase transition is found from the equation $fS(T_{NA}) - a_A$, this transformation being a second-order phase transition if $f^2 > 8c_A|a_N(T_{NA})|$, and a first-order phase transition if the inequality has the opposite sign. The equality $f^2 = 8c_A|a_N(T_{\mathrm{tc}})|$ defines the tricritical point T_{tc}, for instance, on the line of phase transitions $T_{NA}(l)$. All these situations and the dependence of the temperature and phase transition heat on the molecular length l are observed experimentally.

Similar results follow from the consideration of the transition of isotropic phase IL^* to Ch phase and that of the phase diagram $IL \leftrightarrow Ch \leftrightarrow A^*$, $IL^* \leftrightarrow A^*$. Strictly speaking, these results are valid if the correlation length r_c, characterizing the orientational order in a system, is small as compared with the pitch of the cholesteric spiral $h = 2\pi/q$, $r_c \ll h$, i.e., the $IL^* \leftrightarrow Ch$ transformation is not too close to a second-order phase transformation. If $r_c \gg h$, this transformation is very close to the second-order phase transition and the $ChLC$ structure on h scales may differ substantially from the usual nematic one. In particular, orientational ordering becomes biaxial, which significantly affects the transition character. A situation arises here in which the presence of the third-order invariant in the expression for F does not lead to a first-order phase transition, since the phase transformation is described by a multicomponent order parameter. Here a second-order phase transition is possible provided the main components responsible for the transformation enter into the corresponding invariants only to an even power, while the remaining components, such as perturbation, form odd-order invariants. At $r_c \gg h$, quantities $2S'n'_x n'_y$ and $S'(n'^2_x - n'^2_y)$ (where S' is the degree of ordering of the transverse axes of the molecules along the direction $\boldsymbol{n}'(z) \perp \boldsymbol{q}$, $\boldsymbol{n}' \perp \boldsymbol{n}$, and $\boldsymbol{q}$ is the wave vector of the cholesteric spiral with the Z axis) become such main components. Now the tensor $\hat{S}$ does not have the form (8.3), and the invariant $S_{ij} S_{jm} S_{mi}$ does not forbid a second-order phase transition, because the indicated components may appear in the invariant expressions only to an even power and type (8.3) ordering arises as perturbation.

Phase transitions $A \leftrightarrow C$ and $A^* \leftrightarrow \tilde{C}$ in SLC are induced by irreducible vector representations E_{1g} of group ∞/mm and E_1 of group $\infty/2$, respectively. The optical biaxiality of the C and $\tilde{C}$ phases is associated with space symmetry groups $\tilde{t}2/m$ and $\tilde{s}_{k/q}/2$ of these phases. This symmetry is similar to that of chain molecules (here $\tilde{t}$ is a translational displacement, $s_{k/q}$ is a screw displacement consisting of rotation through an angle q/k with a simultaneous shift along the crystal axis Z, $\boldsymbol{k}$ is the wave vector of the crystal lattice, and $\boldsymbol{q}$ is the wave vector of the modulated orientational structure). The $A(A^*)$ and $C(\tilde{C})$ phases differ in the value of the angle $\theta = |[\boldsymbol{k}\boldsymbol{n}]|/k$, formed by the vector $\boldsymbol{k}$ and the director $\boldsymbol{n}$, θ being zero in the $A(A^*)$ phase. Therefore, the $C(\tilde{C})$ phase can be defined by an order parameter $S(\boldsymbol{n}\boldsymbol{k})\ [\boldsymbol{n}\boldsymbol{k}]\ \Psi_1$ satisfying the symmetry requirements. By writing the free energy density F as an invariant expression containing the order parameters of the A and C phases and taking into account the interaction of these parameters, one can obtain a phase diagram of $N \leftrightarrow A \leftrightarrow C$ and $N \leftrightarrow C$ transformations and find the corresponding tricritical points.

If the degree of the orientational order S and the crystalline structure $(|\Psi_1|, k, \beta_1)$ are fixed, while only the orientation of the director $\boldsymbol{n}$ changes, the $C(\tilde{C})$ phase is actually described by a two-component order parameter [8.16]

$$(\xi_1, \xi_2) = (n_z n_x, n_z n_y) \ , \tag{8.7}$$

which is transformed according to the representation $E_{1g}(E_1)$. Since the symmetric cubes $[E_1^3]$, $[E_{1g}^3]$ of the representations E_1 and E_{1g} do not contain unit representation, the $A \leftrightarrow C$ and $A^* \leftrightarrow \tilde{C}$ phase transformations may be second-order phase transitions. Since the antisymmetric square $\{E_1^2\}$ of the representation E_1 contains a unidimensional vector representation, the symmetry group $\infty/2$ admits of the existence of the Lifshitz invariant

$$\lambda\left(\xi_1\frac{\partial\xi_2}{\partial z}-\xi_2\frac{\partial\xi_1}{\partial z}\right), \tag{8.8}$$

which explains the appearance of the screw axis in the $\tilde{C}$ phase.

Under $A \leftrightarrow C$ and $A^* \leftrightarrow \tilde{C}$ phase transitions, ordering must arise in orientation of the transverse molecular axes as an improper phenomenon induced by representations E_{1g} and E_1, since the symmetric squares $[E_{1g}^2]$ and $[E_1^2]$ contain vector representations E_{2g} and E_2. Accordingly, the degree of such ordering S' is proportional to the square of the angle θ, $S' \sim \theta^2$, this being corroborated by experiment.

Since $N \leftrightarrow A$ and $A(A^*) \leftrightarrow C(\tilde{C})$ phase transitions are described by two-component order parameters (Ψ_1, Ψ_1^*) and (ξ_1, ξ_2), it is possible, by using a symmetrically formal analogy with transformations to the superfluid state, to ascribe the corresponding critical behavior to the former. For instance, to the order parameters $|\Psi_1|$ and θ, the correlation radius r_c, the susceptibility χ, and the constant-pressure heat capacity c_p correspond critical indices β, ν, γ, and α, e.g.,

$$\theta \sim (T_{AC}-T)^{\beta}, \quad |\Psi_1| \sim (T_{NA}-T)^{\beta}, \quad r_c \sim |T-T_c|^{-\nu},$$
$$\chi \sim |T-T_c|^{-\gamma}, \quad c_p \sim |T-T_c|^{-\alpha},$$

with $\beta \approx 0.33$, $\nu \approx 0.66$, $\alpha \approx 0.02$, and $\gamma \approx 1.3$. It should be noted, however, that the measurements of these indices are still too scant to definitely ascribe the transition in liquid helium to these transformations. Besides, one cannot rule out the effect of dissolved impurities, which damp out, at a given concentration, large-scale fluctuations in a system and, in particular, may weaken the singularity of some thermodynamic characteristics. Therefore, the qualitative and quantitative description of such transformations in terms of phenomenological theory is reasonable.

In the general case, the SLC density function $\varrho(z)$ can be represented by the Fourier expansion:

$$\varrho(z) = \mathrm{Re}\left\{\sum_m |\Psi_m| \exp(\mathrm{i}mkz+\beta_m)\right\},$$

in which the set $\{|\psi_m|, \beta_m\}$ is determined from the thermodynamic equilibrium conditions. Here the system's thermodynamic potential is written as an expansion in the powers of the values $|\Psi_m|$, which contains invariants of (8.6) formed by such harmonics, and also takes into account the interaction of the latter, for instance, in the form

$$-g|\Psi_1^2||\Psi_2|\cos(2\beta_1-\beta_2) \ ,$$

where $g>0$. The last term is due to the fact that the product of three unidimensional density waves with wave numbers k, k and $-2k$ forms an invariant. The positive sign of the constant g here means that condensation of one of the waves provides conditions for condensation of another. Existence, in experiment, of a type A SLC with one or two density waves (A and A_{tw} phases) is associated with the specific structure of the molecules. Molecules whose polar parts are located in the central region of the molecular contour form, as a rule, SLC with a single density wave. Smectic liquid crystals with two density waves consist of molecules with a clearly defined asymmetry in the arrangement of the polar and nonpolar parts. The like parts of the molecules are predominantly located in separate planes of the smectic layer; distortions of one of the these planes destabilize the other plane and, vice versa, stabilization of one of them (condensation of one of the density waves) promotes stabilization of the other (condensation of the other wave).

With varying external parameters, such as pressure or concentration of the mixture, the indicated form of the thermodynamic potential gives rise to a phase diagram of $N\leftrightarrow A\leftrightarrow A_{tw}$ and $N\leftrightarrow A_{tw}$ transformations, and to the existence of tricritical points on the lines of $A\leftrightarrow A_{tw}$ and $N\leftrightarrow A_{tw}$ phase transitions. Here the $N\leftrightarrow A$ and $N\leftrightarrow A_{tw}$ transitions resulting in density waves with arbitrary phases β_1 and $\beta_2=2\beta_1$ are similar to transitions to the superfluid state, while the $A\leftrightarrow A_{tw}$ transition with the phase β_1 fixed is similar to the phase transformation in the Ising model.

As in ordinary crystals, tricritical phenomena in liquid crystals are due to several interacting order parameters. In fact, third-order invariants being absent, the factor c_{ef} at the fourth-order invariant is renormalized: $c_{ef}\leqslant 0$ if the interaction is large enough. First-order phase transitions to which correspond values $c_{ef}<0$ are accompanied by hysteresis phenomena. In particular, supercooling of the substance may take place.

8.3.2 Supercooled Liquid-Crystal Phases

In the general case of first-order phase transformations the transition temperatures are defined as points of intersection of thermodynamic potentials $\mathcal{T}(T, P, \mathcal{N})$ at given P and mixture concentration $\mathcal{N}$. By ruling out one of the phases in experiment, one can, in principle, investigate the potential of the other phase at T, P, and $\mathcal{N}$ corresponding to the region of metastable existence of this phase. Harder modifications cannot, as a rule, be perceptibly overheated above the transition point owing to the relative ease of nucleation of a disordered liquid phase at the interface. Formation of the solid phase requires overcoming considerable energy barriers by the molecules so that the molecular complexes residing in the liquid phase could disintegrate and the molecules could form a crystal lattice. Therefore, whenever the high-temperature phase contains, besides the crystallization nuclei, molecular groupings differing substantially from those of the low-temperature phase in their structure, the substance admits of large supercooling and can be transformed to the glassy state.

Liquid crystals can have several structural modifications with characteristic molecular groupings reflecting the short-range positional order in mesophases. These modifications differ as to the temperature ranges of metastable states, and this affects the positions of intersections of the surfaces of thermodynamic potentials of the $\mathscr{T}(T, P, \mathscr{N})$ mesophases. In some cases one observes only the metastable existence of some low-temperature phases which do not form on heating of the substance, but arise as a result of supercooling of the high-temperature phase. Such properties are exhibited particularly often by low-symmetry smectic modifications, but they are also inherent in the A phase under certain conditions. For instance, SLC with two density waves is in the metastable state at high pressures, which is ascribed to strong repulsion of the polar portions of molecules when they draw closer together; there also exists a supercooled nematic phase over a wide range of temperatures, which may be due to the fact that the N phase shows smectic fluctuations, so-called cybotactic clusters.

In the last case the potentials of the nematic and smectic phases $\mathscr{T}_N$ and $\mathscr{T}_A$ as a function of temperature in the supercooling region of the A phase touch at a single point $T = T_0$ at $P = P_0$, and at $P > P_0$ the region of metastable existence of the A phase disappears, while at $P < P_0$ and when the temperature falls off, one observes successive transitions between supercooled phases: $N_{\text{s.c.}} \to A_{\text{s.c.}} \to N_{\text{s.c.}}$. Since at the point of tangency $\mathscr{T}_N = \mathscr{T}_A$ and $\partial \mathscr{T}_N / \partial T = \partial \mathscr{T}_A / \partial T$, the expansion of the potentials in the powers of deviations $(T - T_0)$ and $(P - P_0)$ and the conditions of equality $\mathscr{N}_N = \mathscr{T}_A$ at the points of $N_{\text{s.c.}} \to A_{\text{s.c.}}$ and $A_{\text{s.c.}} \to N_{\text{s.c.}}$ transformations lead to the equation

$$P = P_0 - \zeta (T - T_0)^2 ,$$

where $\zeta > 0$, which corresponds to the phase diagram observed.

If, under certain conditions, such phase transformations are close to second-order phase transitions, they can be described proceeding from the Landau expansion

$$\mathscr{T} = a |\Psi_1|^2 + c |\Psi_1|^4 + e |\Psi_1|^6 ,$$

where $c < 0$, $e > 0$, $a = a'[P - P^* + \zeta (T - T_0)^2]$ at $a' > 0$. This expression corresponds to the absolute instability of the A phase at $P > P_0 = P^* + (c^2/4ea')$ and the presence of two transitions $N \leftrightarrow A$ at $P < P_0$ at points

$$T_{NA,AN} = T_0 \pm \sqrt{(P_0 - P)/\zeta}\ .$$

Note that with an increase in parameter $c^2/4ea'$, characterizing hysteresis phenomena in the system, including the entropy jump, the critical pressure value P_0 must also increase, as it does in experiment.

8.3.3 Two-Dimensional Ordering and Plastic SLC

It is common knowledge that a two-dimensional body cannot have a density function $\varrho(x, y)$ periodic in two dimensions, but such a density function is possible in a three-dimensional body. Liquid crystals were the first to demonstrate this possibility: systems with disklike molecules have revealed phases described by the function $\varrho(x, y)$ in which the molecular centers of mass are staggered randomly along lines ("liquid strands") forming regular two-dimensional lattices [8.15].

A two-dimensional molecular system may, in the absence of a long-range crystalline order, have a finite value of the shear modulus, as in an amorphous body, provided the temperature is low enough. At a higher temperature, "melting" takes place, i.e., the shear modulus reduces to zero. Below this melting point, the correlations ϱ_{12} of the positions of the two particles diminish at large distances according to the power law, and in the pulse space q we have

$$\varrho_{12}(q) \sim |q-k|^{-2+2\Delta_\psi}\ , \tag{8.9}$$

where k is the reciprocal lattice vector, and Δ_ψ is the critical index. In a system (stack) of flat molecular layers, a three-dimensional crystal order arises in the presence of an indefinitely weak interaction. The degree of the three-dimensional positional order ψ depends on the interplanar interaction constant w according to the law

$$\psi \sim w^{\Delta_\psi/(2-2\Delta_\psi)}\ , \tag{8.10}$$

while the structural factor $\varrho_{12}(q)$ takes the form of the δ function.

It follows from the foregoing that the so-called smectic phase B cannot exist as a system of independent ($w=0$) parallel-sided solid ($\psi \neq 0$) layers. A short-range positional, long-range orientational order is, however, possible, under which the axes of molecules or molecular groupings (clusters) should be parallel or the order in the orientations of bond angles can take place in the stack of layers, while in a two-dimensional (monomolecular) layer the orientational correlations must diminish by the power law. In this phase (hexatic), to the short-range positional order in a two-dimensional layer corresponds the correlation function

$$\varrho_{12}(q) \sim \frac{r_0^2}{1+r_0^2(q-k)^2}\ , \tag{8.11}$$

where r_0 is the correlation length at which positional correlations decrease exponentially. To such a *B* phase model corresponds an asymmetry of the correlation function ϱ_{12}, described, e.g., by the point group $6/mmm$, a subgroup of the symmetry group ∞/mm, which means that the $A \leftrightarrow B$ transformation can be realized as an orientational phase transition.

As the temperature increases, a three-dimensional plastic (with a weak interplanar interaction) type-*B* crystal must lose its positional ordering described by (8.10), but retain the orientational order, which then disappears on transition to the *A* phase. The first of these phase transitions may occur by the dislocation melting mechanism, i.e., with the formation of individual dislocations. Experiment indicates the observance of the law (8.9) in thin films and the existence of a plastic crystalline state in thick type-*B* layers of SLC. The three-dimensional hexatic *B*-phase is also observed. Explanation of some experimental data within the framework of the short-range positional, long-range orientational order model requires assuming a scale of $r_0 \approx 10^{-5}$ cm for the molecular cluster sizes in (8.11).

A further lowering in the symmetry of smectic phases, such as the *B* phase with a point symmetry group $6/mmm$, is considered by analogy with transformations in crystals and provides a set of various orientational phase transitions. Irrespective of the model, these transitions can be interpreted as an increase in orientational ordering of molecular groupings (clusters, unit cells) and band orientations, because the point symmetry is revealed at short distances. For instance, one can have transformations $6/mmm \rightarrow 2/m$ of two types: the first-order phase transition, according to the representation E_{2g}, to SLC of the kind of the *E* phase observed, and the second-order phase transition, according to the representation E_{1g}, to SLC of the kind of the *G* phase observed.

First-order phase transitions are also possible with conservation of the point symmetry $2/m$. They result in molecular groupings of different types. A further lowering of symmetry $2/m \rightarrow \bar{1}$ corresponds to the formation of the *H* phase under $G \rightarrow H$ and $E \rightarrow H$ transformations. Experiment also indicates the existence of chiral phases $\tilde{G}$ and $\tilde{E}$, in which the above molecular groupings acquire an orientational curl in space. It is possible that smectics *E*, *G* and *H* are three-dimensional solids.

Experimental data show that $A \leftrightarrow B$, $E \leftrightarrow H$, $B \leftrightarrow G \leftrightarrow H$, and other transformations are first-order phase transitions. Contrary to the conclusions of phenomenological theory, possible physical causes for the actual existence of first-order phase transitions are homogeneous and inhomogeneous elastic deformations in plastic SLC, which possess such properties of solids as anisotropy of elastic vibrations and shear moduli. Such SLC obey the general theorem on first-order orientational phase transitions, which are close to second-order phase transitions because the shear moduli are small [8.14].

8.3.4 Orientational Distortions of Liquid Crystals

Away from the phase transition point, the parameter of orientational order S changes only slightly with temperature and is not subject to strong thermal fluctuations; therefore, S can be considered here a certain parameter. The orientation of the director $\boldsymbol{n}$ undergoes perceptible thermal fluctuations away from the transition temperature as well, and is also rather readily affected by external fields. This is the most important cause for practically all instabilities of the orientational structure. Description of orientational deformations is given by the expression for the total free energy $\mathscr{F}$ of distorted NLC, which satisfies their symmetry properties.

If function $\boldsymbol{n}(\boldsymbol{r})$ varies slowly along the body, i.e., the derivatives of this function along the coordinates are small, the corresponding density of free energy $F_0(\mathscr{F}_0 = \int F_0\,dV)$ contains three independent scalars [8.1]:

$$F_0 = \tfrac{1}{2}\{K_1(\operatorname{div}\boldsymbol{n})^2 + K_2(\boldsymbol{n}\operatorname{curl}\boldsymbol{n})^2 + K_3[\boldsymbol{n}\operatorname{curl}\boldsymbol{n}]^2\}\ , \tag{8.12}$$

where K_j are the Frank coefficients characterizing the orientational elasticity of NLC (Fig. 8.19). In *ChLC*, the macroscopic nonuniformity $\boldsymbol{n}(\boldsymbol{r})$ at distances q_0^{-1}, which are larger than molecular sizes $l(q_0 l \ll 1)$, is due to the above invariant (8.2), which can be written as

$$\tfrac{1}{2}K_2(q_0 + \boldsymbol{n}\operatorname{curl}\boldsymbol{n})^2\ . \tag{8.13}$$

The values of K_j are proportional to the square of the order parameter S.

In an anisotropic medium its characteristics, such as the dielectric constant ε and diamagnetic susceptibility χ_m, are tensor quantities. The presence here of a vector field $\boldsymbol{n}(\boldsymbol{r})$ explains the form of the components of the second-rank tensors ε_{ij} and χ_{ij}:

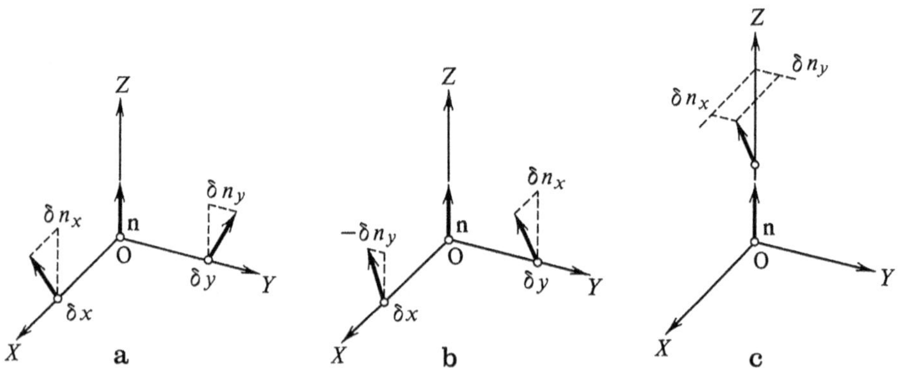

Fig. 8.19. Molecular orientation vector components describing transverse bending **(a)**, torsion **(b)**, and longitudinal bending **(c)** deformations

$$\begin{aligned} \varepsilon_{ij} &= \varepsilon_\perp \delta_{ij} + \Delta\varepsilon n_i n_j \ , \quad \Delta\varepsilon = \varepsilon_\parallel - \varepsilon_\perp \ , \\ \chi_{ij} &= \chi_\perp \delta_{ij} + \Delta\chi n_i n_j \ , \quad \Delta\chi = \chi_\parallel - \chi_\perp \ , \end{aligned} \tag{8.14}$$

where $\varepsilon_\parallel$, $\varepsilon_\perp$ and $\chi_\parallel$, $\chi_\perp$ are the main values of these tensors as measured along and across the orientation of the director $\boldsymbol{n}$. In the presence of an external magnetic field $\boldsymbol{H}$ (with given currents, i.e., field sources) or an electric field $\boldsymbol{E}$ (at fixed potentials across the external conductors), the following terms appear, respectively, in the expression for the free energy density:

$$F_e = -\frac{1}{8\pi} D_i E_i \ , \qquad F_m = -\frac{1}{2} M_i H_i \ ,$$

where $D_i = \varepsilon_{ij} E_j$ are the components of the electrical induction vector, and $M_i = \chi_{ij} H_j$ are the components of the magnetic-field-induced magnetization. Taking into account (8.14), F_e and F_m are written

$$F_e = -\frac{1}{8\pi} [\varepsilon_\perp E^2 + \Delta\varepsilon (\boldsymbol{n}\boldsymbol{E})^2] \ , \qquad F_m = -\frac{1}{2} [\chi_\perp^2 H^2 + \Delta\chi (\boldsymbol{n}\boldsymbol{H})^2] \ , \tag{8.15}$$

where $\boldsymbol{E}$ is the field already altered by the dielectric, while $\boldsymbol{H}$ corresponds to the field which would have been induced by the given sources (currents) in the absence of a medium magnetized.

Since the director orientation in a type-A SLC and a type-A* SLC is rigidly connected with the crystal axis, in such layered systems the moduli K_2 and K_3 are anomalously high, while the corresponding orientational deformations are forbidden. Therefore, only the first summand is retained here in expression (8.12) for F_0. Accordingly, the values of moduli $K_2 \sim K_3 \sim r_c \sim (T - T_{NA})^{-\nu}$ increase anomalously in NLC and *ChLC* as the temperature decreases and as we approach the points of transition to the A and A* phases. Another result of the decrease in *ChLC* temperature and of approach to the point of transition to SLC is an anomalous increase in the pitch of the cholesteric spiral. This phenomenon is due to the greater fluctuations of the smectic order $\langle |\Psi_1|^2 \rangle$ along the axis of a unidimensional crystal, which badly distort the spiral, because as a result of such fluctuations the molecules in the cholesteric planes have to rotate through large angles about the spiral's axis and orient themselves along the crystal axis.

In SLC and *ChLC*, as in layered systems, it is expedient to consider uniaxial extension and contraction along the crystal Z axis. The contribution of the elastic displacement $u(r)$ of the smectic and cholesteric planes to the free energy density is given by the following expression:

$$\begin{aligned} F_d &= \frac{1}{2} c_{33} \left(\frac{\partial u}{\partial z}\right)^2 + \frac{1}{8} c_{11} \left[\left(\frac{\partial u}{\partial x}\right)^2 + \left(\frac{\partial u}{\partial y}\right)^2\right]^2 \\ &\quad + \frac{1}{2} c_{13} \frac{\partial u}{\partial z} \left[\left(\frac{\partial u}{\partial x}\right)^2 + \left(\frac{\partial u}{\partial y}\right)^2\right] \ , \end{aligned} \tag{8.16}$$

where c_{ij} are the moduli of elasticity. Within the "coarse-grain distortions" model [8.1], the modulus $c_{33} \to \tilde{c}_{33} = K_2 q_0^2$, $\tilde{c}_{11} \sim \tilde{c}_{13} \sim \tilde{c}_{33}$ in the *ChLC*. The displacements u along the Z axis are related to the orientational distortions of the director $\boldsymbol{n}$ and the unit vector $\boldsymbol{m} = q_0/|q_0|$ by

$$n_x = \frac{\partial u}{\partial x}\,, \qquad n_y = \frac{\partial u}{\partial y}\,, \qquad m_x = \frac{\partial u}{\partial x}\,, \qquad m_y = \frac{\partial u}{\partial y}\,.$$

Averaging of F_0 on large scales in *ChLC* makes it possible to write it here as

$$\tilde{F}_0 = \tfrac{1}{2}\tilde{K}_1(\operatorname{div}\boldsymbol{m})^2\,, \qquad \tilde{K}_1 = \tfrac{3}{8}K_3\,. \tag{8.17}$$

The set of (8.12 – 17) describes all the orientational rearrangements caused in liquid crystals by external fields in the absence of dissipative processes [8.1, 8.14]. These transformations can be regarded as field-induced phase transitions, and the corresponding orientational distortions, as the parameters of such transitions. The critical-field value and the dependence of the transition parameter on the field are calculated by minimizing the concrete functionals $\mathscr{F}$ under given conditions at the boundaries of a liquid-crytals layer of thickness d. The results for a number of orientational instabilities are given below.

A small finite homogeneous extension $(du/dz)_c$ of the sample of a type-A SLC along the crystal axis leads to a wavelike deformation of smectic layers (undulation mode), which has the form of a square lattice with a period s_c, and

$$\left(\frac{du}{dz}\right)_c = \frac{2\pi}{d}\,\frac{(K_1 c_{33})^{1/2}}{c_{13}}\,, \qquad s_c = 2(2\pi d)^{1/2}\left(\frac{K_1}{c_{33}}\right)^{1/4}\,. \tag{8.18}$$

The same situation takes place in *ChLC*, except that one must make replacements $K_1 - \tilde{K}_1$, $c_{ij} - \tilde{c}_{ij}$ in (8.18). Note that the threshold deformation in SLC has an order of magnitude l/d, and in *ChLC*, h/d, i.e., the structure of *ChLC* is much more stable than that of SLC.

When the electric field $\boldsymbol{E}$ acts on the layer of the A phase with a negative dielectric anisotropy $\Delta\varepsilon$ (the field is oriented along the crystal axis), a structural instability arises similar to the effect of homogeneous extension, with threshold characteristics

$$U_c = E_c d = 2\pi\left|\frac{2d}{\Delta\varepsilon}\right|^{1/2}(K_1 c_{33})^{1/4}\,, \qquad s_c = 2(2\pi d)^{1/2}\left(\frac{K_1}{c_{33}}\right)^{1/4}\,, \tag{8.19}$$

if $|\Delta\varepsilon| \ll \varepsilon_\parallel$. At $|\Delta\varepsilon| \gtrsim \varepsilon_\parallel$, one should replace $\Delta\varepsilon \to \Delta\varepsilon \cdot \varepsilon_\parallel/\varepsilon_\perp$ in (8.19). A similar phenomenon is observed for *ChLC* with a positive dielectric anisotropy ($\Delta\varepsilon > 0$). Here

$$E_c \approx \frac{(2\pi)^{3/2}(6K_2K_3)^{1/4}}{(\Delta\varepsilon h d)^{1/2}}, \qquad s_c \approx (2hd)^{1/2}\left(\frac{3K_3}{2K_2}\right)^{1/4}, \tag{8.20}$$

if $d \gg h$ and $\Delta\varepsilon \ll \varepsilon_\perp$. At $\Delta\varepsilon \gtrsim \varepsilon_\perp$, one should replace $\Delta\varepsilon \to 2\Delta\varepsilon \cdot \varepsilon_\perp/(\varepsilon_\perp + \varepsilon_\parallel)$ in (8.20). At $d \lesssim h$, the order of magnitudes of (8.20) is retained, but the distorted orientational structure of *ChLC* becomes striplike, the orientation of the strips depending on the boundary conditions. In SLC, the modulated structure becomes striplike if $\Delta\varepsilon > 0$ and if the field $\boldsymbol{E}$ is parallel to the smectic plane, the strips being perpendicular to the direction of $\boldsymbol{E}$.

Unlike the enumerated second-order phase transitions, SLC may exhibit a first-order phase transition under which an orientation of smectic layers advantageous to the given field direction arises jumpwise in the specimen bulk. Such a gain in bulk energy is achieved at the expense of a loss in surface energy, which increases jumpwise owing to the formation of a system of defects, such as dislocations arising on contact of smectic layers near the solid surface with those in the specimen interior. From the energy-balance condition if follows that

$$E_c^2 \sim \frac{8\pi W_d}{|\Delta\varepsilon| dl}, \tag{8.21}$$

where $W_d \sim c_{33} l^2$ is the energy of a dislocation nucleus of diameter l. As seen from (8.19, 21), the threshold values of the field are comparable far from the point of phase transition T_{NA}, but at $T \to T_{NA}$ defect formation due to the relationship $c_{33} \sim |\Psi_1|^2 \sim (T_{NA} - T)$ becomes more probable. The situations listed above are observed experimentally.

External electric and magnetic fields also affect ordinary phase transitions by shifting the points of transformations and changing their character. For instance, an external field changes the character of transition from the uniaxial (N_1) to the biaxial (N_2) state of NLC. In this case, the curve $T_{N_1N_2}(E)$ contains a tricritical point (T_{tc}, E_{tc}), and at $E > E_{tc} \sim 10^8$ V/m this transformation becomes a second-order phase transition. Another example is the shift of the phase transition temperature T_{AC} under the effect of the external field. Thus, a magnetic field parallel to the crystal axis increases T_{AC} at a negative diamagnetic anisotropy ($\Delta\chi < 0$) and reduces T_{AC} at $\Delta\chi > 0$, in agreement with experiment. Finally, the same role can be played by boundary conditions if the liquid-crystal layer is thin enough. In particular, no $IL \leftrightarrow N$ phase transition occurs in a layer of thickness $\lesssim 10^{-6}$ cm, since the solid walls effectively order the molecular orientation at this depth. At a certain critical thickness $d = d_c$, a critical point T_c exists which is similar to that in the liquid-gas system.

We have considered the effects of orientational instability which are quadratic with respect to the external field. These effects also include unwinding of the cholesteric spiral by the external field and the Frederiks effect in

NLC, which are discussed in more detail in subsequent sections. Here we will give the critical field values obtained by the above method. The field E, perpendicular to the spiral axis of $ChLC$ with a positive dielectric anisotropy ($\Delta\varepsilon > 0$), leads to a $ChLC \rightarrow$ NLC transformation at the critical value

$$E_c = \frac{\pi^2}{h_0}\left(\frac{4\pi K_2}{\Delta\varepsilon}\right)^{1/2} .$$

A uniformly oriented layer of NLC with $\Delta\chi > 0$ and rigid boundary conditions ($\boldsymbol{n} = \boldsymbol{n}_0$) is unstable at threshold magnetic field values ($\boldsymbol{H} \perp \boldsymbol{n}_0$)

$$H_{tj} = \frac{\pi}{d}\left(\frac{K_j}{\Delta\chi}\right)^{1/2} ,$$

where $K_j = K_3$ if the director $\boldsymbol{n}_0$ is parallel to the normal $\boldsymbol{v}$ to the solid surface; $K_j = K_1$ if $\boldsymbol{n}_0 \perp \boldsymbol{v}$ and $\boldsymbol{H} \| \boldsymbol{v}$, $K_j = K_2$ if $\boldsymbol{n}_0 \perp \boldsymbol{v}$ and $\boldsymbol{H} \perp \boldsymbol{v}$. In the above expressions, the electrical and magnetic analogs of the indicated effects correspond to replacement

$$\Delta\chi H^2 \rightleftarrows \Delta\varepsilon E^2/4\pi .$$

With a mutually perpendicular orientation of the director ($\boldsymbol{n}_0 \perp \boldsymbol{v}$), the so-called twist effect takes place at the threshold voltage

$$U_{th} = \pi\left[\frac{4\pi}{\Delta\varepsilon}\left(K_1 - \frac{1}{2}K_2 + \frac{1}{4}K_3\right)\right]^{1/2} .$$

With an inclined orientation ($\boldsymbol{n}_0\, \boldsymbol{v} \neq 1$ and 0) at the NLC layer boundaries the Frederiks effect occurs without a finite threshold, i.e., the director orientation changes at arbitrarily small values of E.

8.3.5 Polarized States of Liquid Crystals

Liquid-crystal systems do not exhibit proper ordering of constant electrical dipole moments of the molecules, and macroscopic polarization is associated here with definite molecular ordering. In particular, polarization $\boldsymbol{P}$ may arise in NLC as a result of bending and twisting of the director $\boldsymbol{n}$ in space

$$\boldsymbol{P} = f_1 \boldsymbol{n} \operatorname{div} \boldsymbol{n} + f_2 [\boldsymbol{n} \operatorname{curl} \boldsymbol{n}] , \quad (8.22)$$

where f_1 and f_2 are the so-called flexoelectric coefficients. Supplementing the free-energy density by a summand $\boldsymbol{EP}$,

$$F = F_o + F_e - \boldsymbol{EP} , \quad (8.23)$$

and minimizing the functional $\mathscr{F}=\int F\,dV$, taking into account (8.12, 15, 22) and concrete boundary conditions, one can obtain all the consequences of the flexoelectric effect in NLC caused by a uniform electric field.

In a nonuniform field, expression (8.23) is supplemented by a summand

$$-f_3 \boldsymbol{n}(\boldsymbol{n}\nabla)\boldsymbol{E}\ , \tag{8.24}$$

describing the interaction of the macroscopic quadrupole moment with a nonuniform field. Note that $f_1 \sim f_2 \sim S^2$ and $f_3 \sim S$. The summand (8.24) above the transition point T_{IN} leads to an analog of the Kerr effect; the induced birefringence is here determined by the proportionality $S \sim f_3\,|\partial E_i/\partial x_j|$.

In a type-A SLC with molecular orientation rigidly perpendicular to the smectic planes, the quantity F includes the following invariants with appropriate coefficients

$$E_z\left(\frac{\partial^2 u}{\partial x^2}+\frac{\partial^2 u}{\partial y^2}\right),\qquad E_x\frac{\partial^2 u}{\partial x\,\partial z}+E_y\frac{\partial^2 u}{\partial y\,\partial z}\ ,\qquad E_z\frac{\partial^2 u}{\partial z^2}\ ,$$

where u is the displacement of the smectic plane along the crystal Z axis.

In an electric field, NLC may have both unidimensional flexodeformations, depending on a single coordinate, and two-dimensional distortions, depending on two coordinates. The former reduce to a surface effect and are realized only under nonrigid boundary conditions, while the latter are volumetric and exist under any boundary conditions. If $\boldsymbol{n}_0 \parallel \boldsymbol{v}$ in the absence of a field, the general form of boundary conditions is

$$w_{1,2}\,\theta \pm K_3\frac{d\theta}{dz}=(p_{1,2}\pm f_2)\,E\ , \tag{8.25}$$

where w_1 and w_2 are the energies of anchoring with the upper and lower surfaces, p_1 and p_2 are the corresponding surface polarizations, and θ is the angle between the director $\boldsymbol{n}$ and the normal $\boldsymbol{v}$.

Unidimensional flexodeformation exists in an arbitrarily small field $\boldsymbol{E} \perp \boldsymbol{v}$ if $\Delta\varepsilon = 0$, and it is the smaller, the larger the energies $w_{1,2}$. At $\Delta\varepsilon < 0$ and $w_1 = w_2 = 0$, the maximum flexodeformation arises near the walls, while inside the layer the distortion is exponentially small in strong fields. At $\Delta\varepsilon > 0$, when $\boldsymbol{n}_0 \parallel \boldsymbol{v}$ and $\boldsymbol{E} \perp \boldsymbol{v}$, the flexodeformation arises irrespective of the Frederiks effect if $w_1 = w_2$ and $E < E_{t3}$, while under nonsymmetric boundary conditions ($w_1 \neq w_2$) an increase in field leads to an increase in perturbation θ at the center of the layer, which causes a Frederiks effect without a definite threshold.

If $\boldsymbol{n}_0 \parallel \boldsymbol{v}$ and $\boldsymbol{E} \parallel \boldsymbol{v}$, the appearance of flexodeformation acquires a threshold character. For instance, at $\Delta\varepsilon = 0$, $w_2 = 0$, $w_1 \to \infty$, and $p_1 = -p_2 = p$, the threshold voltage is

$$U_f^0 = E_f^0 d = \frac{K_3}{f_1+f_2+p} \;;$$

at $w_1 = w_2 = 0$, there is no threshold, and at $w_1 = w_2 = w \to \infty$ we have

$$U_f \approx \frac{1}{2} U_f^0 \left(\frac{wd}{K_3}\right) \to \infty \; .$$

With a finite dielectric anisotropy $\Delta\varepsilon > 0$ the solution to the problem (8.23, 25) shows that no flexodeformation arises in this situation if $\Delta\varepsilon \gtrsim 1$, while at $\Delta\varepsilon < 0$ the solution describes the Frederiks effect with allowance for the flexoeffect. The latter reduces the threshold value $E_{\rm th}$ for one polarity sign and increases it for the opposite sign, provided asymmetry $w_1 \neq w_2$ takes place.

It should be emphasized that the above surface effects arise provided the anchoring energy is small at least on one of the two surfaces of the NLC layer, and are absent under rigid conditions at both interfaces. In the latter case, at $\boldsymbol{E} \parallel \boldsymbol{v}$ and $\boldsymbol{n}_0 \perp \boldsymbol{v}$, a threshold flexoelectric effect exists which is volumetric and leads to a two-dimensional modulated orientational structure with threshold characteristics

$$U_{\rm th} = \frac{2\pi\bar{K}}{|f_1-f_2|(1+\eta)} \, , \qquad s_{\rm th} = 2{\rm d}\left(\frac{1+\eta}{1-\eta}\right)^{1/2} , \tag{8.26}$$

where η must satisfy the inequality $-1 \leqslant \eta = \Delta\varepsilon K/4\pi(f_1-f_2)^2 \leqslant 1$, and $\bar{K}$ is the average of K_j, $\boldsymbol{s} \perp \boldsymbol{n}_0$, $\boldsymbol{s} \perp \boldsymbol{v}$. At $U \gg U_{\rm th}$, the structure period $s \sim dU_{\rm th}/U$. The indicated dependences are observed in experiment. Expressions (8.26) admit of generalization with allowance for the nonuniformity of the electric field and of the corresponding contribution of (8.24) to the free energy. The field nonuniformity in the NLC layer may be attributed to the distribution of the electric space charge in the electrolyte.

According to (8.22), the induced medium polarization resulting from the flexoelectric effect is a quadratic function of the perturbations of orientation $\boldsymbol{n}$ – the polar angle $\theta \sim \sin(2\pi y/s)$ and the azimuthal angle $\varphi \sim \cos(2\pi y/s)$. The spontaneous polarization may be due to the spontaneous modulation of the director orientation, for instance, in the $\tilde{C}$ phase, where

$$n_x = \sin\theta\cos\varphi \, , \qquad n_y = \sin\theta\sin\varphi \, , \qquad n_z = \cos\theta \, , \qquad \varphi = \varphi(z) \; . \tag{8.27}$$

The induced polarization $\boldsymbol{P}$ in the $\tilde{C}$ phase need not coincide with axis 2 in direction and, generally speaking, contains three independent contributions:

$$\boldsymbol{P} = \boldsymbol{P}_0 + \boldsymbol{P}_\parallel + \boldsymbol{P}_\perp \; ,$$

where the vector $\boldsymbol{P}_0$ is parallel to axis 2, and its components, as well as those of the transition component (8.7), are transformed according to the irreducible representation E_1 of the symmetry group $\infty/2$; the vector $\boldsymbol{P}_{\|}$ is parallel to the crystal Z axis; and the vector $\boldsymbol{P}_{\perp}$ is perpendicular to the vectors $\boldsymbol{P}_0$ and $\boldsymbol{P}_{\|}$. Accordingly, the $\tilde{C}$ phase has three flexoelectric coefficients μ_0, $\mu_{\|}$ and $\mu_{\perp}$, which enter into the invariants

$$-\mu_0\left(P_x\frac{\partial\xi_1}{\partial z}+P_y\frac{\partial\xi_2}{\partial z}\right)\,,\qquad -\mu_{\|}P_z\left(\frac{\partial\xi_1}{\partial x}+\frac{\partial\xi_2}{\partial y}\right)\,, \tag{8.28}$$

etc. The coefficients μ_0 and $\mu_{\perp}$ correspond to different polarization components proportional to $\theta\partial\varphi/\partial z$ and to $\partial\theta/\partial z$. At small angles θ the difference $(\mu_0-\mu_{\perp})\sim\theta^2$, i.e., it is small near the phase transition point.

At the same time, the $\tilde{C}$ phase has a spontaneous polarization $\boldsymbol{P}_0$, which is associated with the symmetry properties of the chiral system [8.16]. When the chiral molecules possess transverse dipole moments, the $\tilde{C}$ phase with a point symmetry 2 must be polarized owing to the piezoeffect

$$(P_{0x},P_{0y})\sim(\xi_2,-\xi_1)\,, \tag{8.29}$$

while F contains the invariant

$$-\mu_{\rm p}(P_{0x}\xi_2-P_{0y}\xi_1)\,, \tag{8.30}$$

where $\mu_{\rm p}$ is the piezomodulus.

The expression (8.29) reflects the pseudoproper character of spontaneous polarization, which is spiral modulated because of the Lifshitz invariant.

Variation, according to polarization $\boldsymbol{P}$, of the quantity F, which includes the even powers of the order parameters (ξ_1,ξ_2), (P_x,P_y) and the invariants (8.8, 28, 30), leads [with allowance for (8.27)] to the expressions

$$P_x=P_0\sin\varphi\,,\qquad P_y=-P_0\cos\varphi\,,\qquad P_0=\chi_{\rm e}\left(\mu_{\rm p}-\mu_0\frac{\partial\varphi}{\partial z}\right)\theta\,, \tag{8.31}$$

$$F=a\theta^2+c\theta^4+\frac{1}{2}K\theta^2\left(\frac{\partial\varphi}{\partial z}-q_0\right)^2+g_{\|}\left(\frac{\partial\theta}{\partial z}\right)^2+\dots\,, \tag{8.32}$$

where $a=a'(T-T_{\rm c})$, $q_0=-(\lambda+\chi_{\rm e}\mu_{\rm p}\mu_0)/K$; $T_{\rm c}$ is the phase transition point. Here K, $g_{\|}$ are the moduli of elasticity; $\chi_{\rm e}$ is the dielectric susceptibility; and λ is the chirality parameter, which is generally also given by the expansion

$$\lambda=\lambda_0+\lambda'\theta^2+\lambda''\theta^4+\dots\,.$$

Therefore,

$$q_0=q_{\rm c}+q'\theta^2+q''\theta^4+\dots\,,\qquad q_{\rm c}=-(\lambda_0+\chi_{\rm e}\mu_{\rm p}\mu_0)/K\,.$$

The parameter λ_0 is relatively small because few atoms per molecule (~1) occupy an asymmetric position, while the magnitude of the K modulus is determined by the interaction of many atoms ($\sim 10^2$). Therefore, the wave number $q_0 \sim \lambda/K$ is usually on the order of $10^{-2} l^{-1}$.

Expressions (8.31, 32) show that below the point T_c the system acquires a slope of the molecular axes $\theta = \theta_0 = (-a/2c)^{1/2}$, an azimuthal distribution $\varphi = \varphi_0 = q_0 z$, and a helical pseudoproper polarization with an amplitude $|P_0|$. When $\lambda + \chi_e \mu_p \mu_0 = 0$, a macroscopically homogeneous ferroelectric state is possible.

Macroscopic polarization $\bar{\boldsymbol{P}} = \int \boldsymbol{P} dV$ can be induced by an external field $\boldsymbol{E}$, which contributes $-\boldsymbol{PE}$ to F. The field $\boldsymbol{E}$, perpendicular to the Z axis, distorts the helical structure of the $\tilde{C}$ phase, and the perturbations $[\theta(z) - \theta_0]$ and $[\varphi(z) - \varphi_0(z)]$ are slight and mutually independent. These distortions introduce a correction $\delta\chi_e$ to the dielectric susceptibility, i.e., to the linear response of the system to the external field [8.14]. In strong fields, the helix unwinds completely, as in *ChLC*, provided $E \gg E_c$,

$$E_c = \frac{\pi^2}{16} \frac{K q_0^2 \theta_0}{\chi_e \mu_p} . \tag{8.33}$$

The macroscopic mean of polarization $\bar{\boldsymbol{P}}$ in the $\tilde{C}$ (at $E \ll E_c$) and A^* phases is equal to

$$\bar{\boldsymbol{P}} = (\chi_e + \delta\chi_e)\boldsymbol{E} \ ,$$
$$\delta\chi_e = \begin{cases} \dfrac{1}{2}\chi_e^2 \left\{ \dfrac{\mu_p^2}{K q_0^2} + \dfrac{[\mu_p - (\mu_0 - \mu_\perp) q_0]^2}{g_\parallel q_0^2 - 4a} \right\} , & T < T_c \ , \\ \chi_e^2 \mu_p^2 / (2a + K q_c^2) \ , & T \geq T_c \ . \end{cases} \tag{8.34}$$

It should be emphasized that $\delta\chi_e$ does not go to infinity at the point T_c, i.e., at $a = 0$, since the $\tilde{C}$ phase is modulated. At $a \gg K q_c^2$, the expression (8.34) is similar to the Curie-Weiss Law, since in this case the correlation radius $r_c \sim (K/a)^{1/2}$ is much smaller than the spiral pitch $h_c = 2\pi/q_c$.

The frequency dispersion of dielectric susceptibility exhibits at least two relaxation times τ_θ and τ_φ in the $\tilde{C}$ phase, and one time τ_θ in the A^* phase. The time $\tau_\varphi \sim \gamma_1 / K q_0^2$, where γ_1 is the SLC viscosity, characterizes the slow attenuation of azimuthal perturbations $(\varphi - \varphi_0)$ at a constant angle $\theta = \theta_0$, i.e., the process of helix reconstruction. The time $\tau_\theta \sim \gamma_1 / (g_\parallel q_0^2 - 4a)$ characterizes the relatively rapid relaxation of the perturbations of the polar angle $\theta - \theta_0$ with a constant spiral pitch $h = h_0$. Experiment points to such a dispersion dependence of the dielectric susceptibility and also to the temperature dependence of the susceptibility (8.34) and of the critical field value (8.33).

The above results refer to behavior of thick layers of the $\tilde{C}$ phase. In thin layers of thickness of several molecular lengths, the ferroelectric properties of the $\tilde{C}$ phase are specific. Here the spontaneous polarization $\bar{P}$ may have a finite value owing to the dipole-dipole interaction $\mathscr{W}$, which suppresses the orientational fluctuations in the two-dimensional system, at least in the vicinity of the point T_c. The quantity $\bar{P}$ is small, while the correlation radius r_c and susceptibility χ_e are large because $\mathscr{W}$ is small:

$$\bar{P} \sim \mu_p \int \xi_1 \, dV \sim \mu_p \, \mathscr{W}^{\Delta_\xi/(1-2\Delta_\xi)} \; , \qquad r_c \sim \mathscr{W}^{-1/(1-2\Delta_\xi)} \; ,$$
$$\chi_e \sim \mathscr{W}^{-(2-2\Delta_\xi)/(1-2\Delta_\xi)} \; ,$$

where Δ_ξ (a critical index depending on temperature) has the same meaning as the index Δ_ψ in (8.10). The parameter $\mathscr{W}$, as well as the parameters λ and μ_p, is strongly dependent on the difference of the concentration $\mathscr{N}$ of the "right"- and "left"-handed molecules in the SLC mixture: $\mathscr{W} = \lambda = \mu_p = 0$ at $\mathscr{N} = 0$.

In a sufficiently strong field

$$\bar{P} \sim E^{\Delta_\xi/(2-\Delta_\xi)} \; .$$

Experiment points to a high susceptibility χ_e in thin films of the $\tilde{C}$ phase. The index Δ_ξ is calculated approximately; the value of $\Delta_\xi(T_c)$ is within the range $1/8 \lesssim \Delta_\xi(T_c) \lesssim 1/\pi$, depending on the phase transition model. One of such models is disinclination melting, which is similar to dislocation melting. As a result, an individual disinclination ("vortex") appears in the high-temperature phase, while in the low-temperature phase unlike disinclinations are paired up.

The appearance of disinclinations in the smectic plane of the $\tilde{C}$ phase is similar to the situation in NLC (Fig. 8.10) and is described by singular solutions to an equation obtained by varying the orientational free energy $\mathscr{F}_0$. In both cases this equation has the following form in the XY plane:

$$\frac{\partial^2 \varphi}{\partial x^2} + \frac{\partial^2 \varphi}{\partial y^2} = 0 \; , \tag{8.35}$$

where $\varphi(x, y)$ is the azimuthal angle of director orientation. The solutions to this equation correspond to disinclinations of different types:

$$\Phi(r) = \tfrac{1}{2} n \, \Psi(r) + \Phi_0 \; ,$$

where $\Psi(r)$ is the angular coordinate of the point r (the origin at the singular point $r = 0$); $n = \pm 1, \pm 2, \ldots$ in NLC; $n = \pm 2, \pm 4, \ldots$ in SLC.

The energy of the individual disinclination is $\mathscr{W} = \pi l K \ln(R/l)$ or (by definition of scale dimensionality) Δ_ξ, $\mathscr{W} = (T/4\Delta_\xi) \ln(R/l)$, where R is the "vortex" radius. The "vortex" entropy $\mathscr{E}$ is the logarithm of its area, i.e., $\mathscr{E} = 2 \ln(R/l)$. Thus, the change in the thermodynamic potential $\delta\mathscr{T}$ due to the birth of the disinclination is

$$\delta \mathscr{T} = \mathscr{W} - T\mathscr{E} = T\left(\frac{1}{4\Delta_\xi} - 2\right) \ln \frac{R}{l} \; .$$

It becomes negative at $\Delta_\xi < \frac{1}{8}$, i.e., $\Delta_\xi(T_c) = \frac{1}{8}$ in the disinclination melting model.

8.3.6 Electrohydrodynamic Phenomena in Liquid Crystals

The dynamics of an anisotropic liquid, which is regarded as a continuum, is concerned with macroscopic phenomena associated with the movement of such a liquid. The set of equations for this medium is as follows [8.1, 8.6, 8.14, 8.15]:

1. The *continuum equation*

$$\partial \varrho/\partial t + \mathrm{div}(\varrho \boldsymbol{v}) = 0 \ , \tag{8.36}$$

where ϱ is the density of the substance, and $\boldsymbol{v}$ is the movement velocity.

2. The *Navier-Stokes equation*

$$\frac{\partial}{\partial t}(\varrho v_i) = f_i - \frac{\partial \Pi_{ki}}{\partial x_k} \ , \qquad \Pi_{ki} = \varrho v_k v_i - \sigma^*_{ki} - \tilde{\sigma}_{ki} \ , \tag{8.37}$$

where $\boldsymbol{f}$ is the volumetric force acting on the medium in an arbitrary electric field $\boldsymbol{E}$ and σ^*_{ki} is the stress tensor, which determines, in the absence of external fields, the part of the pulse flow unconnected with the pulse transfer directly by the mass of the moving liquid. Usually, $\sigma^*_{ki} = -P\delta_{ki}$, where P is the pressure; however, with large deviations of the director $\boldsymbol{n}$ from the equilibrium orientation $\boldsymbol{n}_0$, the value of σ^*_{ki} includes the "distortion stress tensor"

$$\left\{ -\frac{\delta F_0}{\delta(\partial n_j/\partial x_k)} \frac{\partial n_j}{\partial x_i} \right\} .$$

The "viscous" stress tensor $\tilde{\sigma}_{ik}$ in incompressible NLC is

$$\tilde{\sigma}_{ij} = \alpha_1 n_i n_j A_{km} n_k n_m + \alpha_2 n_i N_j + \alpha_3 n_j N_i + \alpha_4 A_{ij} + \alpha_5 n_i n_k A_{kj} + \alpha_6 A_{ik} n_k n_j \ , \tag{8.38}$$

where

$$\boldsymbol{N} = \frac{d\boldsymbol{n}}{dt} + \frac{1}{2}[\boldsymbol{n}\ \mathrm{curl}\,\boldsymbol{v}] \ ; \qquad \hat{A} = \langle A_{ij} \rangle = \left\{ \frac{1}{2}\left(\frac{\partial v_i}{\partial x_j} + \frac{\partial v_j}{\partial x_i} \right) \right\} \ ;$$

$\alpha_1, \ldots, \alpha_6$ are coefficients depending on the order parameter S; $\alpha_1 \sim S^2$; $\alpha_2 \sim \alpha_3 \sim \alpha_5 \sim \alpha_6 \sim S$; $\alpha_4 = \mathrm{const}$ at $S \to 0$. The coefficients α_n are related by

$$\alpha_6 - \alpha_5 = \alpha_2 + \alpha_3 \ , \qquad \alpha_4 > 0, \qquad \alpha_3 > \alpha_2 \ ; \qquad \alpha_1 + \alpha_4 + \alpha_5 + \alpha_6 > 0 \ ,$$
$$(\alpha_3 - \alpha_2)(2\alpha_4 + \alpha_5 + \alpha_6) > (\alpha_6 - \alpha_5)^2 \ .$$

Allowing for strange charges with a bulk density

$$\varrho_s = \frac{1}{4\pi}\frac{\partial}{\partial x_i}(\varepsilon_{ij}E_j) \;, \tag{8.39}$$

the bulk force f has components

$$f_i = -\frac{\partial}{\partial x_i}\left[P - \frac{1}{8\pi}\varrho\left(\frac{\partial \varepsilon_{jm}}{\partial \varrho}\right)_T E_j E_m\right] + \varrho_s E_i - \frac{1}{8\pi}\frac{\partial \varepsilon_{jm}}{\partial x_i} E_j E_m \tag{8.40}$$

in an anisotropic liquid medium.

3. The *law of conservation of strange charges in a liquid medium*

$$\partial\varrho_s/\partial t + v\nabla\varrho_s + \mathrm{div}\,\boldsymbol{j} = 0 \;, \quad j_i = \sigma_{ik}(E_k - \beta_{km}\,\partial\vartheta/\partial x_m) \;, \tag{8.41}$$

where σ_{ik} is a symmetric electrical conductivity tensor, β_{ik} is the electrokinetic coefficient tensor, and ϑ is the chemical potential of the electrolyte solution.

4. The *law of conservation of the electrolyte mass*

$$\varrho\left(v\nabla\mathcal{N} + \frac{\partial\mathcal{N}}{\partial t}\right) + \mathrm{div}\,\boldsymbol{i} = 0 \;, \quad i_k = -\frac{\varrho}{(\partial\vartheta/\partial\mathcal{N})_{P,T}}\mathcal{D}_{km}\frac{\partial\vartheta}{\partial x_m} + \beta_{km}j_m \;, \tag{8.42}$$

where $\mathcal{N}$ is the concentration of the electrolyte solution and $\mathcal{D}_{ik}$ is the symmetric diffusion tensor.

5. The *director movement equation*

$$I\frac{d}{dt}\left[n\frac{dn}{dt}\right] = [\boldsymbol{n}\boldsymbol{g}] - [\boldsymbol{n}(\gamma_1 N + \gamma_2 \boldsymbol{n}\hat{A})] \;, \tag{8.43}$$

where $\gamma_1 = \alpha_3 - \alpha_2$, $\gamma_2 = \alpha_3 + \alpha_2$, I is the inertia moment density, and $\boldsymbol{g}$ is the bulk force compelling the director to take an equilibrium orientation. The force $\boldsymbol{g}$ is determined by means of variation $\delta\mathcal{F}$ of the total free energy $\mathcal{F}$:

$$\delta\mathcal{F} = -\int \boldsymbol{g}\,\delta\boldsymbol{n}\,dV \;. \tag{8.44}$$

6. The *general equation for heat transfer* in incompressible *NLC*

$$\varrho T\left(\frac{\partial\tilde{\mathscr{E}}}{\partial t} + v\nabla\tilde{\mathscr{E}}\right) = \tilde{\sigma}_{ji}\frac{\partial v_i}{\partial x_j} + \frac{dn}{dt}(\gamma_1 N + \gamma_2 \boldsymbol{n}\hat{A}) + \frac{\partial}{\partial x_j}\left(\varkappa_{ji}\frac{\partial T}{\partial x_i}\right) + \boldsymbol{jE} - \boldsymbol{i}\nabla\vartheta \;, \tag{8.45}$$

where $\varkappa_{ij}$ is the symmetric heat-conduction tensor of the substance, and $\tilde{\mathscr{E}}$ is the entropy of the mass unit of the substance.

Equations (8.36–45) describe the effects of electrohydrodynamic (EHD) instability of NLC. The thresholds of EHD effects are usually found as a linear approximation from perturbations $(\boldsymbol{n}-\boldsymbol{n}_0)$, $\boldsymbol{v}$, and ϱ_s. For instance, when $\boldsymbol{E}\perp\boldsymbol{n}_0$ ($\boldsymbol{E}\parallel Z$ axis, $\boldsymbol{n}_0\parallel X$ axis) the linearized equations take the form

$$
\begin{aligned}
&\frac{\partial v_z}{\partial t}+\frac{1}{\tau_v}v_z-\Gamma_1\Theta-\Gamma_2 E\varrho_s=0\ ,\\
&\frac{\partial\Theta}{\partial t}+\frac{1}{\tau_o}\Theta+\Gamma_3 v_z=0\ ,\qquad \frac{\partial\varrho_s}{\partial t}+\frac{1}{\tau_e}\varrho_s+\sigma_a E\Theta=0\ ,
\end{aligned}
\tag{8.46}
$$

where $\Theta=\partial n_z/\partial x$, v_z is the projection of the flow velocity onto the Z axis, $v_z\sim\Theta\sim\varrho_s\sim\exp[\mathrm{i}(kx+k_0 z)]$, $k_0=\pi/d$, and d is the thickness of the NLC layer. Here τ_v, τ_o, and τ_e are the effective relaxation times with orders of magnitude

$$
\tau_v\approx\frac{\varrho k^2}{\bar{\alpha}(k^2+k_0^2)^2}\ ,\qquad \tau_o\approx\frac{\bar{\alpha}}{\bar{K}(k^2+k_0^2)}\ ,\qquad \tau_e\approx\frac{\bar{\varepsilon}}{4\pi\bar{\sigma}}\ ;
$$

the parameters Γ_1, Γ_2, Γ_3, and σ_a have the following orders of magnitude:

$$
\Gamma_1\approx\frac{\bar{K}k^2}{\varrho}\ ,\qquad \Gamma_2\approx\frac{k^2}{\varrho(k^2+k_0^2)}\ ,\qquad \Gamma_3\approx k^2\ ,\qquad \sigma_a\approx\sigma_\parallel-\sigma_\perp\ .
$$

Here $\bar{\alpha}$, $\bar{K}$, $\bar{\sigma}$, and $\bar{\varepsilon}$ are the averages of viscosity, elasticity, electrical conductivity, and dielectric constant, respectively.

It follows from (8.46) that out of three possible EHD regimes, two exist if we neglect the inertia of the system (formally $\varrho\to 0$), while the third implies its presence [8.14]. Among the former two, distinction is made between the conductance regime, which is achieved at frequencies ω of the external field $E\sim\sin(\omega t)$, much less than the inverted relaxation time of the bulk electric charge ($\omega\tau_e\ll 1$), and the dielectric regime, at $\omega\tau_e\gg 1$. Low-frequency EHD instability of NLC is accompanied by modulated perturbations $(\boldsymbol{n}-\boldsymbol{n}_0)$ and $\boldsymbol{v}$, which hardly change in time, and by perturbation ϱ_s, which oscillates almost in phase with field variation $E(t)$. The dielectric regime exhibits a different situation: the volume charge ϱ_s oscillates weakly about some mean value, while the director orientation and velocity vary strongly with the external field frequency. These instabilities differ substantially in the frequency dependence of the threshold characteristics – voltage $U_{th}=E_{th}d$ and the wave number of the arising modulated structure k_{th}. The threshold characteristics are determined from the minimum of the dispersion dependence $U(k)$, which is found provided the existence of nontrivial solutions of the system (8.46). At $\omega\tau_e\ll 1$, we have

$$U_{\text{th}} \approx 2\pi(\pi\bar{\sigma}\bar{K}/\sigma_{\text{a}}\bar{\varepsilon})^{1/2}\ , \qquad k_{\text{th}} \approx k_0\ , \tag{8.47}$$

and at $\omega\tau_{\text{e}} \gg 1$ $U_{\text{th}}(\omega) \sim d\omega^{1/2}$, $k_{\text{th}}(\omega) \sim \omega^{1/2}$. Note that in the dielectric regime the threshold voltage strongly depends on the thickness of the NLC layer, while the wave number does not.

The third regime of EHD instability must arise at $\omega\tau_{\text{e}} \gg 1$ and $\varrho\bar{\sigma}d^2/\bar{\alpha} \gg 1$ owing to the inertial effect. This regime differs fundamentally from the above regimes in that here the volume charge oscillates almost in counterphase with field variation, whereas the director orientation and the flow velocity vary only slightly about the mean values. The physical consequence of this is a qualitatively different behavior of the threshold characteristics depending on the field frequency, the material parameters, and the thickness of the NLC layer:

$$U_{\text{th}}^{\text{in}} \approx \omega\left(\frac{\pi\bar{\varepsilon}\bar{K}}{\bar{\sigma}\sigma_{\text{a}}}\right)^{1/2}\ , \qquad k_{\text{th}}^{\text{in}} \approx k_0\ . \tag{8.48}$$

To achieve this EHD effect one needs large values of electrical conductivity of the substance and the external field frequency. Since this effect exists at high frequencies (usually in the region of dielectric relaxation and frequency dispersion of electrical conductivity), the real dependence $U_{\text{th}}^{\text{in}}(\omega)$ may differ appreciably from linear; among other things, it may weaken.

For the inclined director at the layer boundaries (in the XZ plane, at an angle θ_0 to the X axis; $\boldsymbol{E}\|Z$ axis), another type of EHD instability is possible caused by azimuthal perturbations $(\boldsymbol{n}-\boldsymbol{n}_0)$ in the XY plane. In the conductance regime, this EHD effect has threshold characteristics

$$U_{\text{th}}^{\text{az}} \approx \frac{4\pi}{\sin\theta_0}\left(-\frac{\pi\bar{\alpha}\bar{\sigma}\bar{K}}{\alpha_3\sigma_{\text{a}}}\right)^{1/2}\ , \qquad k_{\text{th}}^{\text{az}} \approx \left|\frac{\alpha_2}{\alpha_3}\right|^{1/2} k_0\ . \tag{8.49}$$

Thus, this instability is possible if the combinations of inequalities $\sigma_{\text{a}} > 0$, $\alpha_3 < 0$ or $\sigma_{\text{a}} < 0$, $\alpha_3 > 0$ are fulfilled. The former combination is realized in NLC far away from the point of transition to SLC, and the latter, in NLC in the vicinity of the point of transition to SLC, where these phenomena are actually observed; in the former case, as a rule, $U_{\text{th}}^{\text{az}} \gg U_{\text{th}}$ and $k_{\text{th}}^{\text{az}} \gg k_{\text{th}}$ by virtue of the inequality $|\alpha_2/\alpha_3| \gg 1$.

Nonlinear effects arise in EHD-instability phenomena, usually above the indicated thresholds. Prior to transition to the turbulent state the system passes through several quasi-stationary regimes having their own excitation thresholds [8.14]. Intermediate orientational instabilities are due to a growth of flow velocity $\mathscr{V}$ and of deviations of the director $(\boldsymbol{n}-\boldsymbol{n}_0)$ in the regions of primary perturbations upon increase in electric field. Estimates of amplitudes $\mathscr{V}_0$ and θ_0 of such deviations –

$$\theta_0 \sim \frac{\bar{\alpha}d}{K}\,\mathscr{V}_0 \sim \left(\frac{U^2 - U_{\text{th}}^2}{U_{\text{th}}^2}\right)^{1/2} -$$

and of the conditions for the development of secondary perturbations $\mathscr{V}_0 \sim \bar{w}/\bar{\alpha}\theta_0$, if $d\bar{w} < \bar{K}$, or $\mathscr{V}_0 \sim \bar{K}/\bar{\alpha}d\theta_0$, if $d\bar{w} > K$, where $\bar{w}$ is the surface anchoring energy, show that transition to orientational turbulence must occur at voltages U'_{th} which exceed the values of U_{th}, $(U'_{th} - U_{th}) \sim U_{th}$, by just a few times.

A developed orientational turbulence can be described qualitatively by means of a dimensionless parameter, the Ericksen number $\mathrm{Er} = \bar{\alpha}\mathscr{V}d/\bar{K} \gg 1$. The usual Reynolds numbers $\mathrm{Re} = \varrho\mathscr{V}d/\bar{\alpha}$ are on the order of $\mathrm{Re} \sim \varrho^{1/2} \times U/\bar{\alpha} \lesssim 1$, and therefore one cannot neglect the viscosity of the substance in liquid crystals. At large scales $(\sim d)$ of the pulsational movement we can ignore the orientational elasticity $\bar{K}$ by virtue of the inequality, $\mathrm{Er_d} \sim U^2/\bar{K} \gg 1$; the change in velocity here has a scale $\mathscr{V} < E^2 d/\bar{\alpha}$, and the change in angles, $\theta \sim 1$. At the smallest scales of x_0, the number $\mathrm{Er}_{x_0} \approx 1$, and here the elasticity $\bar{K}$ plays an essential role, $x_0 \sim \bar{K}^{1/2} E^{-1}$, $v_{x_0} \sim \bar{K}^{1/2} E/\bar{\alpha}$. At intermediate scales of x, the pulsational velocity variation $v_x \sim (E^2/\bar{\alpha})x$, and the number of degrees of freedom of such a motion is of the order of $U^2/\bar{K}$. Experiment shows that a developed orientational turbulence (or the "dynamic scattering" regime) is characterized by a system of mild isotropic scatterers with effective radii inversely proportional to the electric field intensity.

In EHD effects (8.47–49) the main physical cause for orientational instability is anisotropy of liquid-crystal electrical conductivity, at finite values of which the relevant thresholds may be on the order of 10 V. At a low electrical conductivity anisotropy, these threshold values are anomalously large, and then the liquid crystals exhibit electroconvective instability, which is inherent in an ordinary isotropic liquid as well. With increasing rate of isotropic electroconvection, this instability becomes observable because of orientational instability as a secondary phenomenon, which is manifested as a change in the optical characteristics of the liquid-crystal medium.

8.4 Magnetic Properties of Liquid Crystals

Liquid crystals are diamagnetic materials, as are most of the organic compounds. In a magnetic field $\boldsymbol{H}$ they acquire a magnetic moment $\boldsymbol{M}$ opposite to $\boldsymbol{H}$. In nematic and smectic liquid crystals, which are mostly aromatic compounds, this effect manifests itself particularly prominently. Two, three, or more benzene rings are arranged along the axis of the molecule. Owing to this the magnetic properties of the liquid crystals are anisotropic [8.17] (Fig. 8.20). The direction of the highest magnetic susceptibility χ tends to coincide with that of external field $\boldsymbol{H}$. Since this direction lies in the plane of the benzene ring, the long axes of liquid-crystal molecules are directed parallel to the magnetic field.

However, the interaction energy of an individual molecule with field $\boldsymbol{H}$ is low. It is about 5×10^{-21} erg at $H = 10$ kOe, which is 10^6 times lower than the

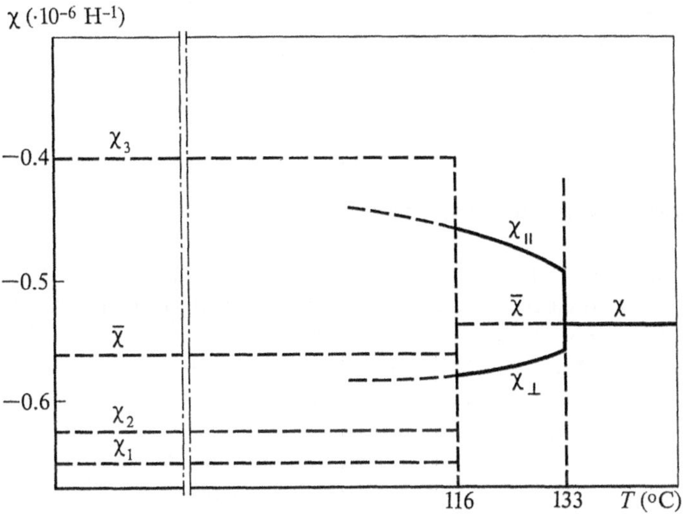

Fig. 8.20. Temperature dependence of diamagnetic susceptibility χ of para-azoxyanisole: $\chi_\perp$, $\chi_\parallel$ are the diamagnetic susceptibilities in the direction perpendicular and parallel to the long molecular axes and to the external magnetic field; χ_1, χ_2, χ_3 are the main susceptibilites of a solid crystal; χ is the susceptibility of an isotropic liquid; $\bar{\chi}$ is the mean diamagnetic susceptibility [8.4]

thermal energy kT. Since a liquid crystal is still well oriented at $H = 10$ kOe, it should be concluded that the magnetic field interacts with molecular ensembles, which rotate and orient themselves simultaneously when a field is superimposed.

The orienting effect of the magnetic field on the layers of the substance adjoining the walls of the flat container, in which the liquid crystal is investigated, is hindered by the action of the forces of adhesion between the molecules and the walls. In the case of a nematic sustance this effect extends to a distance from the wall $d(H) = c/H$, where $c \approx 1$ Oe · cm. Thus, at $H \approx 10^4$ Oe the perturbations introduced by the walls extend only through a thickness of $d \approx 1$ μm. In practice, by using sufficiently strong fields it is possible to obtain a single-crystal nematic specimen well oriented in the direction of $\boldsymbol{H}$.

Smectic liquid crystals cannot be oriented by magnetic fields under 30000 Oe because of their high viscosity. But they can be oriented by cooling an isotropic melt in a magnetic field. The long axes of the molecules are parallel to the field, i.e., in the same way as in a nematic substance. If the field is then switched off, the orientation persists, which is not the case in nematic specimens.

The effect of a magnetic field on cholesteric liquid crystals is peculiar. If, for instance, a magnetic field is applied perpendicular to the helical axis of the plane texture and $\Delta\chi = \chi_\parallel - \chi_\perp > 0$, the molecules tend to line up parallel to the field. As $\boldsymbol{H}$ grows, the pitch of the helical structure will increase. Finally, at a certain critical value H_{cr} all the molecules orient themselves in the same direction, and a nematic structure is formed.

8.5 Electrical Properties of Liquid Crystals

8.5.1 Dielectric Properties of Liquid Crystals

The liquid crystals are dielectrically anisotropic. Substances with a positive and a negative dielectric anisotropy are distinguished. Mesophases in which the dipole molecular moment is arranged along the long axis of the molecules or makes a small angle with it, are dielectrically positive. Then the dielectric constant along the optic axis (the direction of the long axes of the molecules) $\varepsilon_{\parallel}$ is higher than in the perpendicular direction $\varepsilon_{\perp}$, and $\Delta\varepsilon = \varepsilon_{\parallel} - \varepsilon_{\perp} > 0$. Mesophases in which the dipoles are disposed across the long axis of the molecules and $\Delta\varepsilon = \varepsilon_{\parallel} - \varepsilon_{\perp} < 0$ are said to be dielectrically negative. The temperature dependence of the dielectric anisotropy for para-azoxyanisole ($\Delta\varepsilon < 0$) is shown in Fig. 8.21. It is seen that the dielectric anisotropy decreases with an increase in the temperature of the nematic phase, which is due to the decrease of the ordering in the arrangement of the molecules.

The behavior of liquid crystals in an electric field depends on many factors: the sign of the dielectric anisotropy, the initial orientation of the specimen, the presence of impurity ions, etc.

The electrical and electro-optical properties of liquid crystals are usually studied on thin oriented "single-crystal" layers. The orientation is achieved by special treatment of the cell surfaces between which the crystal is located. To obtain layers with a "planar" orientation (the optic axis of the layer is parallel to the cell surface) the "rubbing" method is often used, in which both cell surfaces are covered with microscratches (in the direction of rubbing). The long

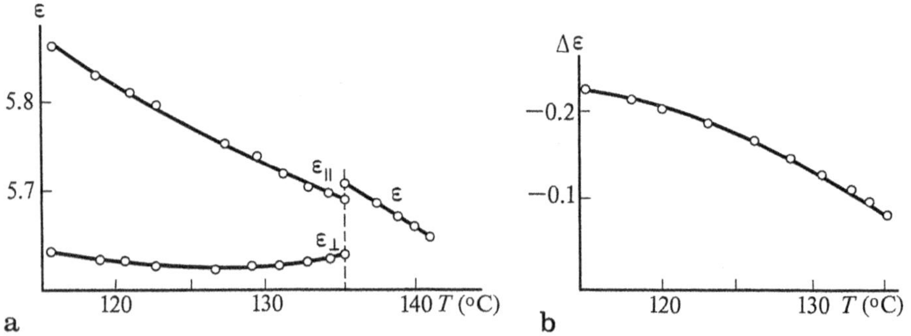

Fig. 8.21. Temperature dependence of the dielectric constant **(a)** and the dielectric anisotropy **(b)** of para-azoxyanisole: $\varepsilon_{\parallel}$ and $\varepsilon_{\perp}$ are the main dielectric constants of the liquid crystal; ε is the dielectric constant of the isotropic melt

axes of the molecules are oriented parallel to the scratches, because this reduces the crystal energy. To obtain "homeotropic" layers, whose optic axis is perpendicular to the cell surfaces, these surfaces are thoroughly cleaned and treated with certain surface-active substances.

To investigate the field effects, transparent electrodes of tin oxide SnO_2 or indium oxide InO_2 are deposited on the internal surfaces of the cell. The effect of field $\boldsymbol{E}$ on a liquid-crystal specimen generally amounts to the following:

1) a gradual change in the direction of the optic axis (layer deformation effect, Sect. 8.3.4),
2) the appearance of a domain structure (Sect. 8.3.6),
3) dynamic light scattering (Sect. 8.3.6).

Let us consider some other best-studied effects arising in thin layers of a nematic liquid crystal in an electric field.

8.5.2 Transverse Domains in Nematics with a Negative Dielectric Anisotropy

Consider the behavior of a uniformly oriented layer of a nematic liquid crystal of thickness 15 – 30 μm with a negative dielectric anisotropy ($\varepsilon_{\parallel} - \varepsilon_{\perp} < 0$). The molecules of such substances possess a dipole moment arranged at a large angle to the long axis of the molecule. An example is *p*-azoxyanisole, in which dipole moment $\mu = 2.48 \times 0^{-18}$ CGSE units and forms an angle of 57.5° with the long axis of the molecule.

Suppose we have prepared a planar specimen, the long axes of whose molecules are parallel to the X axis of the cell (Fig. 8.22). Field $\boldsymbol{E}$ is directed along the Z axis. When threshold tension $U_{th} \approx 5 - 7$ V is achieved (8.47), a system of parallel long light and dark bands appears in the microscope field of vision; this is a domain pattern, which is visible both in an ordinary and a polarized light (Fig. 8.23 a). The long domains are directed across the X axis and are parallel to the Y axis. These domains are called transverse, because they are perpendicular to the initial direction of the long axes of the molecules.

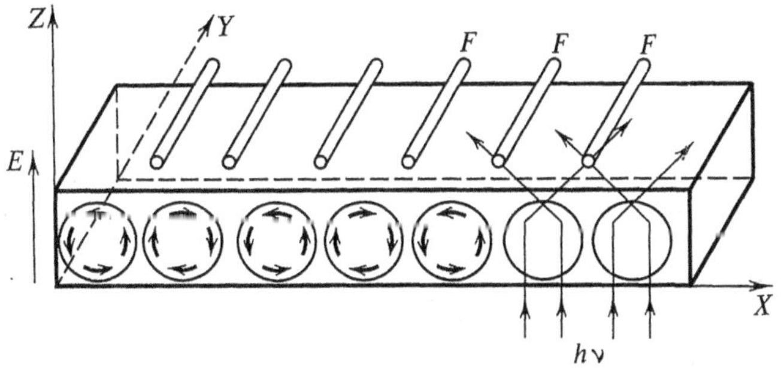

Fig. 8.22. Scheme explaining the origin of transverse domains

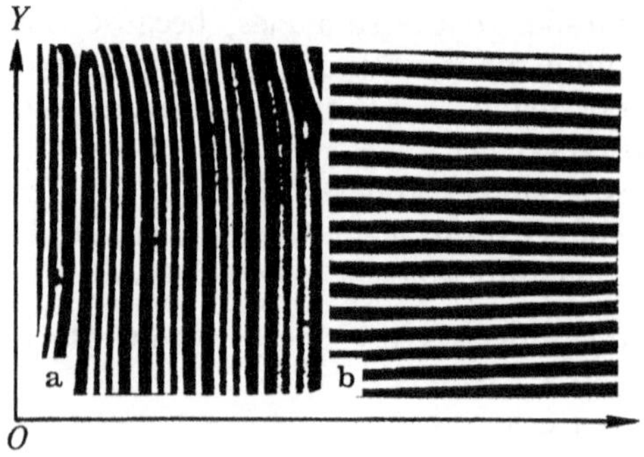

Fig. 8.23. Transverse (**a**) and longitudinal (**b**) domains arising in a nematic liquid crystal under the effect of an electric field. The long axes of the molecules are parallel to the x axis (magnification 120×, polarized light)

The formation of transverse domains is due to the fact that when field $\boldsymbol{E}$ is superimposed along the Z axis in a liquid crystal, a system of parallel cylindrical eddies arises (the bases of these cylinders are denoted by circles in Fig. 8.22). The eddies, like cylindrical lenses, focus the light beams so that a system of parallel focal lines, visualizing the domain pattern of the crystal, is observed in the microscope.

A liquid crystal shows an electrical-conductivity anisotropy $\sigma_{\|} > \sigma_{\perp}$, because the ion mobility is higher in the direction of the long axes of the molecules. This induces space charges in the liquid-crystal layer. When constant electric fields are superimposed, the space charges are supplemented by injection of electrons from the cathode.

The moving ions cause the rotation of molecules and a mass flow. This, coupled with the orienting effect of field $\boldsymbol{E}$ on the molecules, brings about electrohydrodynamic instability. Eddy tubes (domains) arise, in which the flow reorients the molecules. A system of "cylindrical lenses" is formed, which focuses the light beams and produces a regular system of focal lines parallel to the Y axis. A change in field tension and frequency results in the formation of complex systems of domains (herringbone domains, hexagonal systems, etc.).

8.5.3 Effect of Dynamic Light Scattering

In nematics, when the voltage increases and the second critical threshold $U'_{\text{th}} \approx 12-15$ V (Sect. 8.3.6) is reached, the ordered mass flow is disturbed, and a turbulence arises, which causes dynamic light scattering. The scattering centers are intensively moving regions of liquid crystals, $1-5$ μm in size.

Dynamic light scattering and domains appear both in constant and alternating electric fields. Threshold voltages U_{th} and U'_{th} increase with the frequency of the exciting field. Dynamic scattering due to turbulence may not arise in constant fields in low-conductivity compounds. Two instability regimes, separated by critical frequency ω_{cr}, are observed, as a rule. Below ω_{cr} the space charge manages to follow the field, and a stationary domain

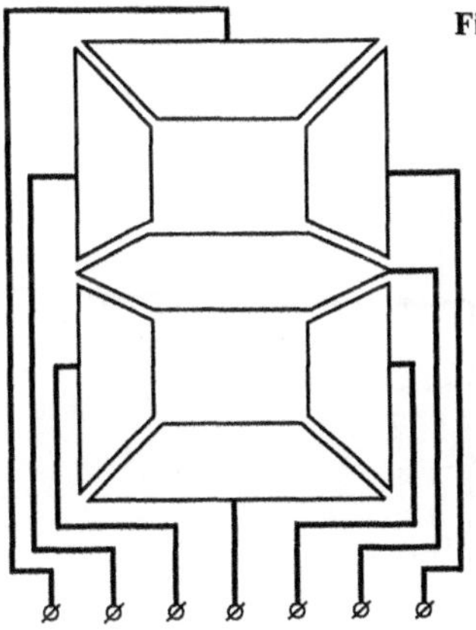

Fig. 8.24. Design of a digital indicator on a liquid crystal

structure is observed (electrical-conductivity regime). Above ω_{cr}, the vector characterizing the local molecular orientation experiences angular oscillations (dielectric regime). Then a strong electro-optical effect is observed, which is similar to dynamic light scattering, but has a shorter relaxation time.

The dynamic scattering effect can be utilized for producing information reflection systems in letter-digital indicators, matrix systems, etc. The cells can be made according to the "in the clear" or "reflection" scheme (one of the electrodes is a mirror electrode). The dynamic scattering changes either the transparency or the intensity of the reflection. One of the electrodes is made solid, and the image-forming electrode, shaped. For instance, the digital indicator consists of seven segments (Fig. 8.24). By commuting the voltage on the segments it is possible to obtain the image of the respective digit.

The maximum image contrast coefficient obtained on this effect may reach $20-30$ at a voltage of $20-50$ V. The time required for the effect to appear is $1-15$ ms, depending on the current density. The disappearance of the image depends on the temperature and the substance, and equals $20-1000$ ms. The use of circuits reducing the switching-off time shortens it to $1-5$ ms.

To obtain a more complicated image, one can use a matrix system with deposited crossed XY electrodes with a liquid-crystal layer placed between them (Fig. 8.25). Application of a voltage to definite XY electrodes induces a field at the respective point of the matrix cross hairs, which makes the point glow.

The electrodichroism phenomenon makes it possible to construct color pickups. The phenomenon consists in a change of light under the effect of an electric field in nematic substances with a small addition of dichroic dyes. usually, $0.5-10$ wt % of azo dye (for instance, methylene red, indophenone blue) is added. In the absence of an electric field, the liquid-crystal molecules are parallel to the electrodes and serve as a matrix which makes the elongated dye molecules orient themselves similarly. If the incident white light has an

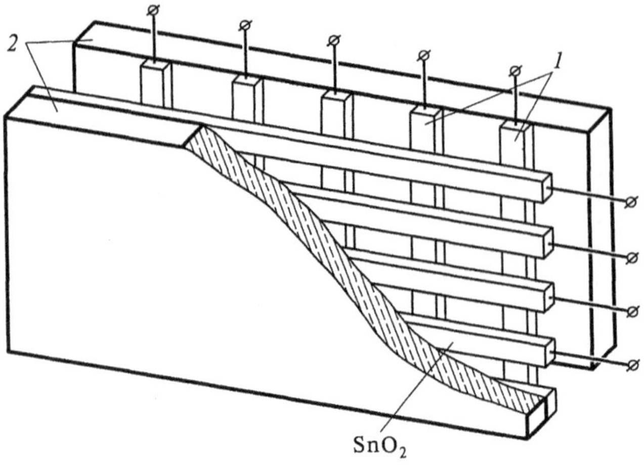

Fig. 8.25. Matrix system with crossed electrodes (*1*) system of *XY*-electrodes, (*2*) glasses enclosing the liquid-crystal layer [8.5]

electric field tension vector parallel to the long axes of the dye molecules, it is absorbed by them, and the mixture acquires the color of the dye. Switching on a field of about 1 V/μm changes the molecular orientation by 90°. The light absorption decreases, and the mixture becomes colorless.

By adding a nondichroic dye together with the dichroic one to a nematic liquid crystal it is possible to produce a two-color pickup. The ordered arrangement of the molecules of the dichroic dye does not affect the characteristic color of the nondichroic one. The color of the system, which is characteristic of a combination of two dyes, will change on application of a field to the color inherent in a nondichroic dye. Electrodichroism can be used for color digital indicators, adjustable filters, laser modulators, etc.

The dynamic scattering effect is also used for other purposes: amplitude modulators, optrons, optical attenuators, etc. We shall quote a well-known example of utilization of dynamic scattering in an electro-optical amplifier.

An electro-optical amplifier (Fig. 8.26) is a liquid crystal (*1*) and semiconductor (*2*) couple enclosed between two transparent electrodes (*4, 4'*). The conductor surfaces are coated with thin layers of a dielectric (*5, 5'*); layer *5'* is coated with a light-reflecting metal layer (*3*). A low-frequency alternating voltage is applied to electrodes *4, 4'*. The voltage is chosen so that, in the absence of an image on the semiconductor (its left-hand surface is photosensitive), most of the voltage drops on it, and the image on the liquid crystal does not exceed the threshold of dynamic scattering. When an image arises on photoconductor *2*, its resistance in the light-affected regions decreases. As a result, the voltage drop on liquid crystal *1* increases and it transforms to the state of dynamic light scattering.

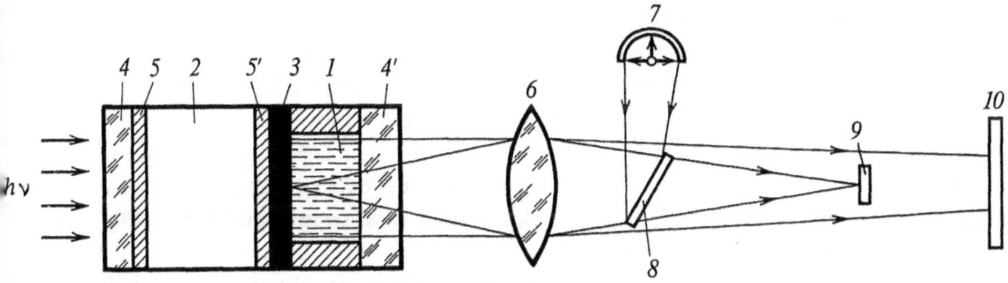

Fig. 8.26. Electro-optical amplifier on a liquid crystal: (*1*) liquid-crystal layer, (*2*) semiconductor, (*3*) metal mirror, (*4* and *4′*) transparent electrodes, (*5* and *5′*) dielectric films, (*6*) lens, (*7*) projection lamp, (*8*) translucent mirror, (*9*) aperture, (*10*) screen

On the opposite side of the electro-optical amplifier the radiation of a projection lamp (*7*) is focused on reflecting layer *3* with the aid of a translucent mirror (*8*) and lens (*6*). In the absence of an image on the photoconductor, when the liquid-crystal layer is transparent, the narrow light beam of lamp *7* reflected from layer *3* is stopped by a diaphragm (*9*). In the presence of an image and hence of dynamic scattering in the liquid crystal, the light scattered by layer *1*, bypassing diaphragm *9*, reaches a screen (*10*), producing on it an image formed on the light-sensitive surface of the semiconductor. Since the light of the projection lamp can be made many times more intensive than the image, the amplification can be very large ($\sim 10^3$).

Let us now dwell briefly on the "memory" effect. The time to dynamic scattering (1 – 15 ms) is, as a rule, much less than the switching-off time (20 – 1000 ms) when the field is lifted. This effect is often undesirable and is eliminated by using a high-frequency (2 – 20 kHz) switching-off signal. Its use shortens the time of light scattering reduction ten- to fifteenfold.

Sometimes the "memory" effect has to be enhanced. To do this, for instance, a mixture of a nematic and a cholesteric (5 – 10 wt %) compound is often used. Then the dynamic scattering "acts" for a long time, which depends on the mixture composition, even after the excitation field is switched off. The "memory" can be deleted by applying a low-frequency field to the cell.

8.5.4 Effect of Deformation of a Nematic Liquid-Crystal Layer

If a liquid crystal ($\Delta\varepsilon < 0$) is initially oriented so that the long axes of the molecules are perpendicular to the electrodes and their optic axis is parallel to the Z axis, after field $\boldsymbol{E}$ is applied the molecules begin to rotate because the dipoles tend to orient themselves parallel to $\boldsymbol{E}$. This results in uniform deformation of the liquid-crystal layer and tilting of the optic axis relative to the Z axis. After that the above-described mechanism of electrohydrodynamic instability again comes into play, domains are formed, and then dynamic light scattering begins.

In addition to the dynamic scattering effect, one can observe the effect of deformation of the liquid-crystal layer (Sect. 8.3.4). If the normal (the long axes of the molecules are perpendicular to the electrodes) liquid-crystal layer with a negative dielectric anisotropy ($\Delta\varepsilon < 0$) is placed between crossed polaroids, the field of vision will be dark, because the optic axis is parallel to the light beam. On application of a voltage the molecules experience a torque and rotate through a definite angle. As a result, and interference color appears, which depends on the voltage. This system can be used for color indicators. The contrast coefficient of such a cell is on the order of 1000 at voltage of about 5 V. For liquid crystals with a positive dielectric anisotropy ($\Delta\varepsilon > 0$) one should obtain a layer with an optic axis parallel to the plane of the cell, and quench it between the crossed polaroids by rotating the cell. On application of a field the long axes of the molecules rotate, tending to line up parallel to the field, which results in birefringence.

Also of interest is a system with a liquid-crystal layer wound up by 90°, in which $\Delta\varepsilon > 0$. The winding up is due to the mutually perpendicular direction of rubbing of the opposite electrodes enclosing the liquid-crystal layer. Here, at both electrodes, the molecules orient themselves parallel to the rubbing. Thus, if these directions are perpendicular, the molecular orientation gradually rotates by 90° in passing from one electrode to another, forming a wound up liquid-crystal layer. In crossed polaroids such a layer is not quenched, since it rotates the light polarization plane through 90°. Quenching occurs with parallel nicols. When field $\boldsymbol{E}$ is superimposed, the molecules rotate, birefringence occurs, and the system begins to transmit light (twist effect).

8.5.5 Longitudinal Domains in Substances with a Negative Dielectric Anisotropy

Of special interest is the behavior, in an electric field, of thinner (less than 10 μm) layers of a liquid crystal with a negative dielectric anisotropy. Here, too, a domain system arises when the threshold voltage is achieved. But, in distinction to the above-considered case (Fig. 8.23 a), domains line up not perpendicular, but parallel to the initial direction of the optic axis of the crystal: longitudinal domains are formed [8.18] (Fig. 8.23 b). This effect evidently cannot be attributed to the electrohydrodynamic flow. This is also attested by the fact that there is no dynamic scattering effect up to and including breakdown voltages and that longitudinal domains form better in specially pure substances with a minimum number of impurity ions.

The domain width for this case depends on tension E, decreasing with the field growth. The change in domain width is accompanied by the appearance of dielectric hysteresis loops. We evidently have here the manifestation of the polar symmetry of the liquid crystal under the effect of the electric field (flexoelectricity, Sect. 8.3.5).

The regularities observed during the formation of domains can be utilized for a number of devices. For instance, the effect of formation of longitudinal domains parallel to the initial direction of the optic axis of a nematic single crystal can be used for light beam control. The width of longitudinal domain bands, s, is inversely proportional to the tension of field E, i.e., $s = \alpha/E$, where $\alpha = \text{const}$. The domain structure possesses the properties of a phase diffraction grating, and the angle of deviation of the diffraction maximum changes in accordance with the relation

$$\sin\varphi = m\lambda E/\alpha \ ,$$

where m is the order of magnitude of the diffraction maximum, and λ is the wavelength. Thus, the period of the diffraction grating can be changed by means of the electric field. The same principle can naturally be used to control the deviation of the light beam by large angles ($\sim 80°$), using small voltages (15 – 70 V). True enough, the frequency of deviations is still restricted to several tenths of hertz. The same principle can be used for amplitude and phase modulation of light.

8.5.6 Domains in Liquid Crystals with a Positive Dielectric Anisotropy

In a substance with a positive dielectric anisotropy the parallel domain pattern usually does not appear at any arbitrary initial orientation of the molecules. The reason is that both the motion of the impurity ions and field $\boldsymbol{E}$ make the molecules orient themselves in one and the same direction, parallel to $\boldsymbol{E}$. The appearance of domains and of the dynamic scattering effect can be ensured by applying a magnetic field $\boldsymbol{H}$ perpendicular to $\boldsymbol{E}$. If the magnetic field prevails over the electric one and makes the long axes of the molecules orient themselves perpendicular to $\boldsymbol{E}$, a situation favoring electrohydrodynamic instability arises.

8.5.7 Behavior of Cholesteric and Smectic Liquid Crystals in an Electric Field

The effect of the electric field changes the texture of cholesteric liquid crystals. The direction of the axis of the helical structure and its pitch undergo changes.

The effect of constant electric fields $\boldsymbol{E}$ on cholesteric liquid crystals depends on the sign of dielectric anisotropy $\Delta\varepsilon$. In ordinary cholesterics (aliphatic ethers of cholesterol) $\Delta\varepsilon < 0$. When field $\boldsymbol{E}$ is parallel to the axis of the *helical* structure the molecules do not reorient themselves, because the molecular dipoles are already oriented along $\boldsymbol{E}$. If, however, $\boldsymbol{E}$ is applied perpendicular to this axis, the molecules tend to orient their dipoles along $\boldsymbol{E}$, and molecular reorientation occurs. For instance, if a transverse field $\boldsymbol{E}$ is superimposed on the confocal domain structure, the latter transforms into a uniformly oriented Grandjean texture.

With holesterylhalides ($\Delta\varepsilon > 0$) the application of $\boldsymbol{E}$ increases the pitch of the helix, and after a certain critical value E_c (Sect. 8.3.4) is achieved the cholesteric completely unwinds and transforms to the nematic state [8.1]. A change in the pitch of the helical structure changes the maximum wavelength of selective light reflection, λ_m. If λ_m lies in the visible region of the spectrum, its change affects the color of the flat cholesteric layer. This effect has found practical application in the production of changeable-color letter-digital indicators. If field $\boldsymbol{E}$ acts on the cholesteric structure of a mixture of a nematic with a small addition of some chiral substance, on attainment of $\boldsymbol{E}_c$ (8.20) a periodic domain structure arises in the form of a rectangular grating.

Because of their high viscosity, smectic liquid crystals orient themselves in electric fields much more reluctantly than nematics. Still, it is possible to orient smectics, provided the smectic phase appears on cooling of an isotropic or nematic phase. The nature of this orientation depends on the sign of the dielectric anisotropy and electrical conductivity of the liquid crystal. Orientation in electric and magnetic fields is observed only in smectic modifications A, B, C, and $\tilde{C}$.

In an A-type smectic, one can observe an electrohydrodynamic instability and dynamic light scattering in substances with a high electrical conductivity ϱ. If ϱ is low, only molecule reorientation and transition of the single-domains state is observed.

In smectic liquid crystals of the type C, as well as in nematics, a domain system forms under the effect of constant and variable fields. But in the compounds which also have a nematic phase (for instance, heptyloxybenzoic acid) the appearance of domains in smectic phase C requires a voltage about three times as high as in the nematic phase. The width of domains in a smectic is one order of magnitude smaller than in the nematic phase.

If an electric field acts on chiral smectic $\tilde{C}$, then at some critical value of the field, $\boldsymbol{E}_c$, a polydomain specimen changes into a single domain one because the helical structure winds off (Sect. 8.3.3).

8.6 Thermal Properties of Liquid Crystals

Recall that a substance is in the liquid-crystal phase within a definite temperature range. In some substances this range may be rather extensive. Thus, n-propyl-p-azoxycinnamate is in the liquid-crystal state between 123 and 243 °C. Substances are also known whose liquid-crystal range is only a few degrees. Thus, in methylbenzal-p-aminobenzene-p-oxybenzoate this range is 3° (from 174 to 177 °C).

Clarification temperature T_{cl} (point of phase transition of a liquid crystal to an isotropic liquid) characterizes the thermal stability of a liquid crystal: the higher T_{cl}, the higher the thermal stability. To overcome the molecular interaction forces ensuring the liquid-crystal ordering and to achieve melting into an isotropic liquid, a high thermal energy kT is required.

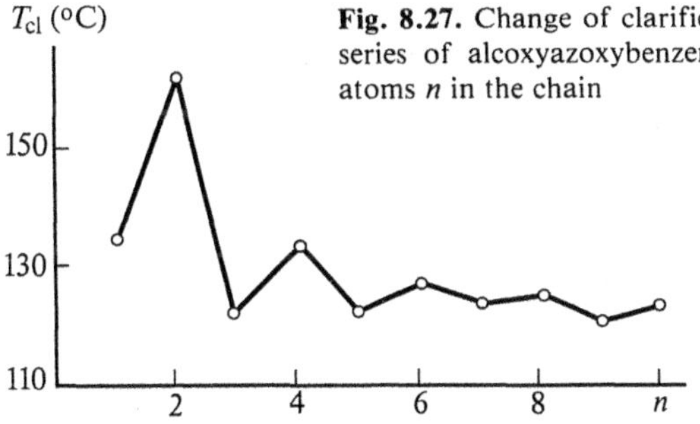

Fig. 8.27. Change of clarification temperature T_{cl} in the homologous series of alcoxyazoxybenzene in relation to the number of carbon atoms n in the chain

Clarification temperature T_{cl} is extremely sensitive to a change in molecular structure. This can be demonstrated on the change of T_{cl} in the homologous series of alcoxyazoxybenzene (the first term of the series being para-azoxyanisole):

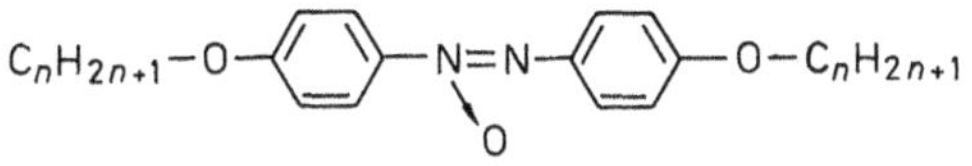

Figure 8.27 shows how T_{cl} changes for homologs of this series in relation to the number n of carbon atoms of aliphatic chains. The jumpwise change of T_{cl} on transition from an odd to an even homolog is attributed to the fact that the change in the number of homologs is accompanied by an addition of a C–C bond either parallel to the long axis of the molecule (even-numbered homologs) or at an angle of about 70° (odd homologs). Since the polarizability of the C–C bond parallel to the axis of the molecule is higher than in the perpendicular direction, a polarization anisotropy arises, which leads to a jumpwise change of T_{cl}.

The practical utilization of liquid crystals had long been delayed by the fact that only substances forming liquid crystals at temperatures considerably exceeding room temperature were known. At present these difficulties have been overcome. A large number of compounds which are in the liquid-crystal state at room temperatures are known. An example is p-methoxybenzylidene-butylaniline,

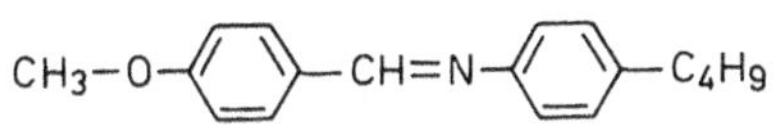

which forms a nematic mesophase in the temperature range from 20 to 43 °C. Other lower-temperature substances are also known, so that the temperature range of existence of the liquid-crystal phase (−50 to +100 °C) required for the operation of electronic equipment is no obstacle, in principle, to the practical utilization of liquid crystals.

To extend the temperature range of existence of the liquid-crystal phase and to shift it one way or the other, one can use, in addition to the synthesis of

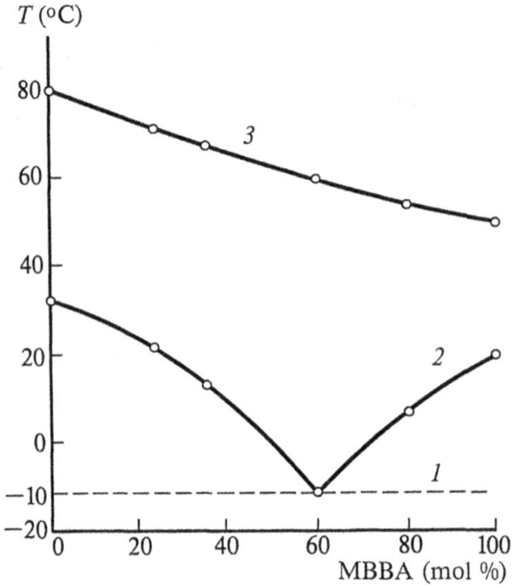

Fig. 8.28. Phase diagram of a mixture of p-meotxybenzylidenebutylaniline (MBBA) with p-ethoxybenzylidenebutylaniline: (*1*) solid phase, (*2*) liquid crystal, (*3*) isotropic liquid

new compounds, the method for obtaining multicomponent systems. Thus, the phase diagram of the mixture of p-methoxybenzylidenebutylaniline with p-ethoxybenzylidenebutylaniline (Fig. 8.28) shows that by changing the percentage ratio of the components one can obtain systems with different ranges of existence of the liquid-crystal phase.

The transition of a liquid crystal to an isotropic liquid, as well as the transition of a solid crystal to a liquid one, is characterized by the latent heat of crystallization. Thus, for p-azoxyanysole the specific heat of fusion of a solid crystal into a nematic liquid crystal is 29570 J/mol, and the specific heat of fusion of a nematic liquid crystal into an isotropic liquid is 547 J/mol. The latter figure is very small, which is understandable, because during the transition to an isotropic melt only the long-range orientation order in the arrangement of the molecules due to weak dispersion forces is destroyed.

The transition to an isotropic liquid is accompanied by a jumpwise change in specific volume, which, together with clearly defined transition temperature and magnitude of the latent heat of crystallization, indicates that this transition is always a first-order phase transition (Sect. 8.3.1). Note that transitions between different polymorphous smectic modifications may be either of the first or of the second order.

If liquid crystals appear both on cooling and on heating of a substance, the liquid-crystal phase is called enantiotropic. The diagram of phase transitions in this case is as follows:

$$\mathrm{SC} \rightleftarrows \mathrm{LC} \rightleftarrows \mathrm{IL} \ ;$$

here SC stands for the solid-crystal phase, LC for the liquid-crystal phase, and IL for the isotropic liquid phase. The arrows indicate that the liquid-crystal phase appears both in the course of heating and cooling of the substance (the phase transitions are reversible). Examples of such substances are para-azoxy-anisole, anizaldazine, p-methoxybenzylidenebutylaniline, and cholesterylcinnamate.

Substances are also known in which liquid crystals appear only on super-cooling (Sect. 8.3.2). Here the liquid-crystal phase is said to be monotropic in respect to the solid state. The diagram of phase transformations in this case is as follows:

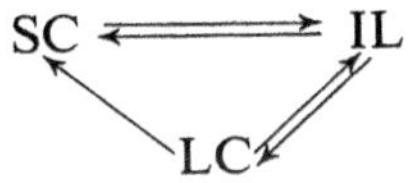

An example of such a substance is cholesterylacetate. Solid crystals of this substance melt directly into an isotropic liquid at 114 °C. If the isotropic melt is cooled very slowly, liquid crystals may not appear at all, since solid crystals being to grow immediately. On rapid cooling, at a temperature of about 90 °C, the specimen transforms to the liquid-crystal phase, and if cooling continues, the specimen crystallizes out with time.

Some substances have several liquid-crystal phases. Thus, in the temperature range of 70 – 73 °C p-nonoxybenzaltoluidine produces smectic-type liquid crystals, and in the range of 73 – 76 °C, nematic-type liquid crystals. The scheme of phase transformations is as follows:

$$\mathrm{SC} \rightleftarrows \mathrm{SLC} \rightleftarrows \mathrm{NLC} \rightleftarrows \mathrm{NL} \; .$$

An ever-increasing amount of information on substances with still more liquid-crystal phases is constantly becoming available (Sect. 8.3.3). Thus,

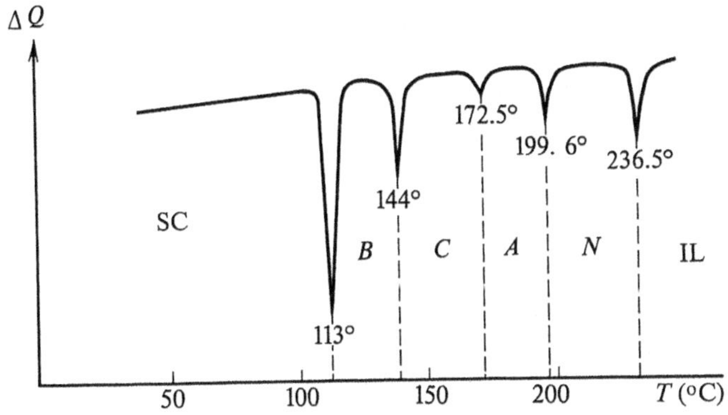

Fig. 8.29. Thermogramm of terephthal bis-(4-n-butylaniline)

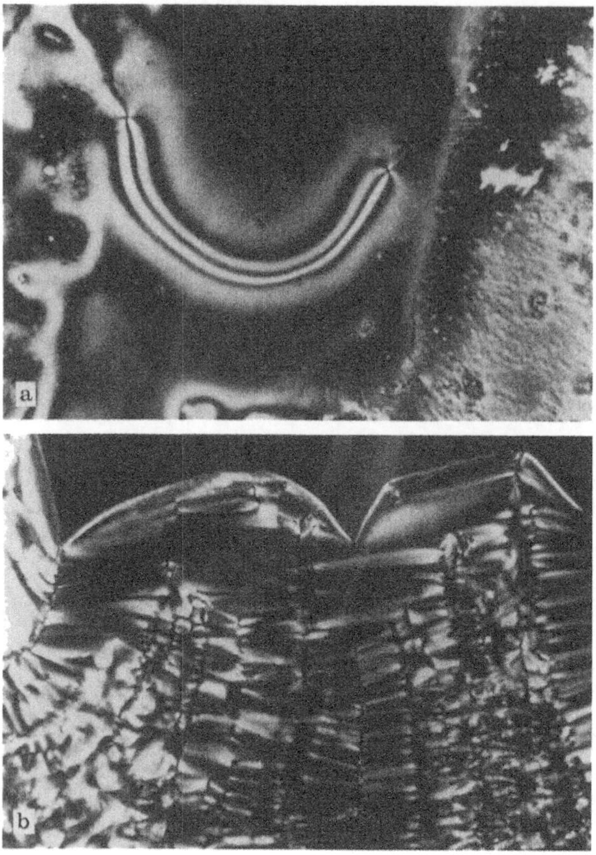

Fig. 8.30. Examples of typical textures for nematic (**a**), smectic (**b**) liquid crystals and a nemato-cholesteric mixture (**c**, **d**)

terephthalbis-(4-n-butylaniline) has three smectic phases, *A*, *B*, and *C*, and one nematic, *N*. Phase transitions are clearly recorded on a thermogram (Fig. 8.29). The peaks on the curve are due to the release of latent heat on corresponding phase transitions. The scheme of phase transformations for this substance is written thus:

$$\mathrm{SC} \xrightleftharpoons{113\,^{\circ}\mathrm{C}} \mathrm{SLC}\ (B) \xrightleftharpoons{144\,^{\circ}\mathrm{C}} \mathrm{SLC}(C) \xrightleftharpoons{172.5\,^{\circ}\mathrm{C}} \mathrm{SLC}(A)$$

$$\xrightleftharpoons{199.6\,^{\circ}\mathrm{C}} \mathrm{NLC} \xrightleftharpoons{236.1\,^{\circ}\mathrm{C}} \mathrm{IL}\ .$$

Phase transitions in liquid crystals are usually investigated by an integrated method (x-ray and differential-thermal analysis, transparency measurement,

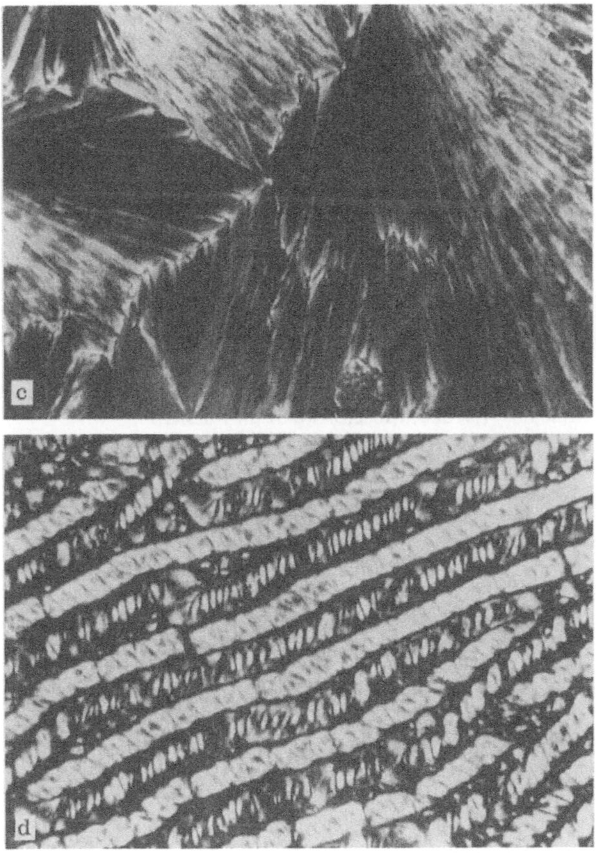

Fig. 8.30 c, d

etc.). Of great importance for identification of a phase is a polarizing-microscope study of the liquid-crystal texture. A phase transition is generally accompanied by a clearly defined change in texture, and in most cases the growth front of the new phase can be observed visually. The textures of liquid crystals are varied, but they lend themselves to classification. Figure 8.30 shows some typical textures for nematics, smectics, cholesterics, and also for complicated liquid-crystal systems.

References

Chapter 1

1.1 B.K. Vainshtein: *Modern Crystallography I.* Springer Ser. Solid-State Sci., Vol. 15 (Springer, Berlin, Heidelberg, New York 1981)
1.2 N.E. Kochin: *Vektornoye ischisleniye i nachala tenzornogo ischisleniya* (Vector Calculus and Elements of Tensor Calculus) (Nauka, Moscow 1965) (in Russian)
1.3 H.A. Jahn: Acta Crystallogr. **2**, 30 (1949)
1.4 Y.I. Sirotin, M.P. Shaskol'skaya: *Osnovy Kristallifiziki.* (Nauka, Moscow 1979)/English transl.: *Fundamentals of Crystal Physics* (Mir, Moscow 1982)
1.5 J.F. Nye: *Physical Properties of Crystals* (Clarendon Press, Oxford 1964)
1.6 A.V. Shubnikov: Izv. AN SSSR, Ser. fiz. **13**, 347 (1949)
1.7 I.S. Zheludev: *Simmetriya i eyo prilozheniya* (Symmetry and Its Applications) (Atomizdat, Moscow 1976) (in Russian)
1.8 B.A. Tavger: Sov. Phys.-Crystallogr. **3**, 341 (1958)
1.9 A.V. Shubnikow, E.E. Flint, G.B. Bokii: *Osnovy kristallografii* (Fundamentals of Crystallography) (Izd-vo An SSSR, Moscow-Leningrad 1940) (in Russian)
1.10 A.V. Shubnikov: *Izbrannyie trudy po kristallografii* (Selected Works on Crystallography) (Nauka, Moscow 1975) (in Russian)
1.11 S. Bhagavantam, T. Venkatarayudu: *Theory of Groups and Its Application to Physical Problems* (Andhra Univ., Waltair 1951)
1.12 A.V. Shubnikow: Usp. Fiz. Nauk **59**, 591 (1956) (in Russian)
1.13 I.S. Zheludev, L.A. Shuvalov: Kristallografiya **1**, 681 (1956) (in Russian)
1.14 L.A. Shuvalov: Sov. Phys.-Crystallogr. **4**, 371 (1959)
1.15 L.A. Shuvalov: J. Phys. Soc. Japan **28**, Suppl., 38 (1970)
1.16 K. Aizu: J. Phys. Soc. Japan **27**, 387 (1969)

Chapter 2

2.1 J.F. Nye: *Physical Properties of Crystals* (Clarendon Press, Oxford 1957)
2.2 U. Wooster: *Text Book of Crystal Physics* (University Press Cambridge 1949)
2.3 L.D. Landau, E.M. Lifshitz: *Electrodynamics of Continuous Media* (Pergamon Press, London 1960)
2.4 R. Keyes: Solid State Physics **20**, 37 (1967)
2.5 B.K. Veinshtein, V.M. Fridkin, V.L. Indenbom: *Modern Crystallography II,* Springer Ser. Solid-State Sci., vol. 21 (Springer, Berlin, Heidelberg 1983)
2.6 A. Schaefer, L. Bergman: Naturwiss. **22**, 685 (1934)
2.7 I.M. Sil'vestrova, Yu.V. Pisarevsky: private comm.
2.8 M.J. Musgrave: *Crystal Acoustics* (San-Francisco, Holden-Day 1970)
2.9 W.C. Dash: Bull. Amer. Phys. Soc. **4**, 47 (1959)
2.10 J.P. Hirth, J. Lothe: *Theory of Dislocations* (McGraw Hill, London 1968)
2.11 G.J. Taylor: Proc. Roy. Soc. A. **116**, 16 (1927)

2.12 D.K. Bowen, V. Christian: Philos. Mag. **12**, 36 (1965)
2.13 H.J. Van Bueren: *Imperfections in Crystals* (North-Holland, Amsterdam 1960)
2.14 V.L. Indenbom, A.A. Urusovskaya: Sov. Phys.-Crystallogr. **4**, 84 (1960)
2.15 G.V. Berezhkova: *Nitevidnye kristally* (Whiskers) (Nauka, Moscow 1969) (in Russian)
2.16 R. Thyagaradzhan, A.A. Urusovskaya: Sov. Phys.-Solid State **7**, 64 (1965)
2.17 M. Classen-Nekludowa: Z. Phys. **55**, 555 (1929)
2.18 M.V. Klassen-Neklyudova: *Plasticheskiye svoistva i prochnost' kristallov* (Plastic Properties and Strength of Crystals) (Gostekhizdat, Leningrad-Moscow 1933) (in Russian)
2.19 A.L. Roitburd: "Sovremennoye sostoyanie teorii martensitnykh prevrashchenii" (Modern State of the Theory of the Martensitic Transformations), in: *Nesovershenstva kristallicheskogo stroeniya i martensitnye prevrashcheniya* (Imperfections in Crystal Structure and Martensitic Transformations) ed. by Yu.A. Osip'yan, R.I. Entin (Nauka, Moscow 1972, p. 7) (in Russian)
2.20 V.I. Al'shits, V.L. Indenbom: Sov. Phys.-Uspekhi **18**, 1 (1975)
2.21 M.I. Kaganov, V.Ya. Kravchenko, V.D. Natsik: Sov. Phys.-Uspekhi **16**, 878 (1974)
2.22 Ya.I. Frenkel', T.A. Kontorova: Zhurn. Eksper. i Teor. Fiziki **8**, 1340 (1938) (in Russian)
2.23 F.R.N. Nabarro: Adv. Phys. **1**, 269 (1952)
2.24 F.C. Frank, W.T. Read: Phys. Rev. **79**, 722 (1950)
2.25 W.C. Dash: J. Appl. Phys. **27**, 138 (1956)
2.26 J.J. Gilman: Trans. AIME **212**, 783 (1958)
2.27 A.A. Predvoditelev: Sov. Phys.-Crystallogr. **7**, 759 (1963)
2.28 A.A. Urusovskaya, G.F. Dobrzhanski, N.L. Sizova, V.G. Govorkov, Yu.N. Martyshev: Sov. Phys.-Crystallogr. **13**, 899 (1969)
2.29 A.A. Urusovskaya: Sov. Phys.-Uspekhi **11**, 631 (1969)
2.30 E.M. Savitski, G.S. Burkhanov: *Monokristally tugoplavkikh i redkikh metallov i splavov* (Single Crystals of Refractory and Rare Metals and Alloys) (Nauka, Moscow 1972) (in Russian)
2.31 a) J.W. Christian: *Plastic Deformation of BCC Crystals,* in: Proc. 2nd Intern. Conf. on Strength of Metals and Alloys (ASM, Cleveland, Ohio, 1970)
b) K.R. Evans: Treatise Mater, Sci. Techn. **4**, 113 (1974)
2.32 G.T. Horne, R.B. Roy, H.W. Paxton: J. Iron and Steel Inst. **201**, 161 (1963)
2.33 A.A. Urusovskaya, N.L. Sizova, I.A. Rachkov, Ya.A. Zakharin, V.M. Dobryak: Phys. Status Solidi (a) **41**, 443 (1977)
2.34 V.N. Rozhanski, A. Predvoditelev: Dokl. AN SSSR **158**, 835 (1964) (in Russian)
2.35 J. Friedel: *Les Dislocations* (Gauthier-Villar, Paris 1956)
2.36 R.W. Whitworth: Adv. Phys. **24**, 203 (1975)
2.37 A.W. Stepanow: Z. Phys. **81**, 560 (1933)
2.38 P. Haasen: Nachr. Akad. Wiss. Göttingen, II. Math.-Phys. Klasse, 1970 N 6, S. 1
2.39 M.F.Ashbee: Acta Metallurg. **14**, 679 (1966)
2.40 A.A. Shtolberg: Phys. Status Solidi (b) **43**, 523 (1971)
2.41 E.K. Naimi: Izvestiya Vuzov, Ser. Fizika, N 3, 94 (1975) (in Russian)
2.42 A. Granato, K. Lücke: J. Appl. Phys. **27**, 583; 798 (1956)
2.43 V.L. Indenbom, V.M. Chernov: Phys. Status Solidi (a) **14**, 347 (1972)
2.44 M.M. Savin, V.M. Chernov: Sov. Phys.-Crystallogr. **21**, 253 (1976)
2.45 V.M. Chernov, M.M. Savin: Phys. Status Solidi (a) **47**, 45 (1978)
2.46 P.G. Bordoni: J. Acoust. Soc. Amer. **26**, 495 (1954)
2.47 N.D. Zakharov, V.N. Rozhanski, P.L. Korchazhkina: Sov. Phys.-Solid State **16**, 926 (1975)
2.48 D.M. Barnett, W.D. Nix: Acta Metallurg. **21**, 557 (1973)
2.49 Yu.Z. Eotrin, A.A. Urusovskaya, G.G. Knab: Sov. Phys.-Crystallogr. **17**, 141 (1972)
2.50 J.T. Michalak: Acta Metallurg. **13**, 213 (1965)
2.51 A. Evans, R. Rawlings: Phys. Status Solidi **34**, 9 (1969)
2.52 A.H. Kottrell: *Dislocations and Plastic Flow in Crystals* (Oxford University Press, Fair Lawn, New York 1953)
2.53 W. Frank: Phys. Status Solidi **29**, 391; 767 (1968)
2.54 W.S. Gorsky: Phys. Z. Sow. (Sov. Physics) **6**, 77 (1934)

2.55 A.M. Kosevich: Sov. Phys.-Uspekhi **17**, 920 (1975)
2.56 V.L. Indenbom, A.N. Orlov: Fiz. Metallov i Metallovedenie **43**, 469 (1977) (in Russian)
2.57 H. Tertsch: *Die Festigkeitserscheinungen der Kristalle* (Springer-Verlag, Wien 1949)
2.58 E. Schmid, B. Boas: *Plasticity of Crystals* (Hughes, London 1950)
2.59 A. Pabst: Bull. Geol. Soc. Amer. **66**, 89 (1955)
2.60 A. Kelly, G.W. Groves: *Crystallography and Crystal Defects* (Longman, London 1970)
2.61 I.V. Obreimov, V.I. Startsev: Sov. Phys.-JETP **35**, 743 (1958)
2.62 V.I. Startsev, V.M. Kosevich, Yu.S. Tomenko: Kristallografiya **1**, 429 (1956) (in Russian)
2.63 L.A. Thomas, W.A. Wooster: Proc. Roy. Soc. A **208**, 43 (1951)
2.64 E.V. Tsinzerling: *Iskusstvennoye dvoinikovaniye kvartsa* (Artificial Twinning of Quartz) (Izd. AN SSSR, Moscow 1961) (in Russian)
2.65 R.D. Baëta, K.H.G. Ashbee: Philos. Mag. **22**, 601; 625 (1970)
2.66 M.V. Klassen-Neklyudova, M.A. Chernysheva, A.A. Shternberg: Dokl. AN SSSR **63**, 527 (1948) (in Russian)
2.67 M.A. Chernysheva: Dokl. AN SSSR **74**, 117 (1950) (in Russian)
2.68 M.A. Chernysheva: Dokl. AN SSSR **81**, 1065 (1965) (in Russian)
2.69 M.V. Burkart, T.A. Read: J. Metals **197**, 1516 (1953)
2.70 C.L. Pearson, W.T. Read, W.L. Feldman: Acta Metallurg. **5**, 181 (1957)
2.71 A.F. Joffe, M.V. Kirpicheva, M.A. Levitskaya: Z. Phys. **22**, 286 (1924)
2.72 J.E. Gordon: "Current Research on the Strength of Whiskers and Thin Films at Tube Investment Laboratory", in: *Growth and Perfection of Crystals,* ed. by R.H. Doremus, B.W. Roberts, D. Turnbull (Wiley, New York 1958), p. 219
2.73 S.S. Brenner: Science **128**, 569 (1958)
2.74 Z. Gyulai: Z. Phys. **138**, 317 (1954)
2.75 M.V. Klassen-Neklyudova: Techn. Phys. USSR **5**, 827 (1938)
2.76 S.N. Zhurkov: Phys. Z. Sow. **1**, 123 (1932)
2.77 A.P. Aleksandrov, S.N. Zhurkov: *Yavleniye khrupkogo razryva* (Brittle Rupture Phenomenon) (Gostekhizdat, Moscow 1933) (in Russian)
2.78 W.B. Hillig: J. Appl. Phys. **32**, 741 (1961)
2.79 N.V. Belov, M.V. Klassen- Neklyudova: Zhurn. Eksper. i Teoret. Fiz. **18**, 265 (1948) (in Russian)
2.80 V.L. Indenbom, M.A. Chernysheva: Izv. AN SSSR, Ser. Fiz. **22**, 1469 (1958) (in Russian)
2.81 R.L. Lyles, Jr., H.G.F. Wilsdorf: Acta Metallurg. **23**, 269 (1975)
2.82 P.L. Raffo: J. Less-Common Metals **17**, 133 (1969)
2.83 V.L. Indenbom, A.N. Orlov: "Sovremennye predstavleniya o podvizhnosti dislokatsii" (Modern Conceptions of Dislocation Mobility) in: *Dinamika dislokatsii* (Dislocation Dynamics) (FTINT AN Ukr. SSR, Khar'kov 1968) (in Russian)
2.84 V.R. Regel, A.M. Leksovski, S.N. Sakiev: Fiz. Metallov i Metallovedeniye **40**, 812 (1975) (in Russian)
2.85 V.L. Indenbom, V.I. Nikitenko, B.M. Strunin: Priroda **4**, 74 (1974) (in Russian)
2.86 V.L. Indenbom, A.N. Orlov: Sov. Phys.-Uspekhi **5**, 272 (1962)
2.87 C. Zener: "The Micromecanism of Fricture", in: *Fricturing of Metals* (Amer. Inst. of Metals, Cleveland, Ohio 1948), p. 3
2.88 N.F. Mott: J. Phys. Soc. Japan **13**, 650 (1955)
2.89 A. A. Griffith: Trans. Roy. Soc. A **221**, 163 (1920)
2.90 M.Kh. Blekherman, V.L. Indenbom: "Kriterii Griffitsa v mikroskopicheskoi teorii treshchin" (Griffith Criterion in Microscopic Theory of Cracks), in: *Uspekhi mekhaniki deformiruemykh sred* (Progress in Mechanics of Deformed Media), ed. by G.S. Shapiro (Nauka, Moscow 1975), p. 74 (in Russian)
2.91 E. Orowan: *Internal Stresses* (Inst. Metals, London 1948)
2.92 V.I. Likhtman, E.D. Shchukin, P.A. Rebinder: *Fiziko-khimicheskaya mekhanika metallov* (Physico-Chemical Mechanics of Metals) (Izd, AN SSSR, Moscow 1962) (in Russian)
2.93 J.J. Gilman: Austr. J. Phys. **13A**, 327 (1969)
2.94 Yu.S. Boyarskaya: *Deformirovaniye kristalov pri ispytanii na mikrotverdost'* (Deformation of Crystals in Microhardness Testing) (Shtiintsa, Kishinev 1972) (in Russian)

2.95 N.P. Yushkin: *Mekhanicheskiye svoistva mineralov* (Mechanical Properties of Minerals) (Nauka, Leningrad 1971) (in Russian)
2.96 L.L. Beranek: *Acoustic Measurements* (Wiley, New York; Chapman and Hall, London, 1949, VIII)
2.97 B.P. Belikov, K.S. Aleksandrov, T.V. Ryzheva: *Uprugiye svoistva mineralov i gornykh porod* (Elastic Properties of Minerals and Rocks) (Nauka, Moscow 1970) (in Russian)
2.98 V.S. Postnikov: *Vnutrennee treniye v metallakh* (Internal Friction in Metals) (Metallurgiya, Moscow 1969) (in Russian)
2.99 G.A. Dubov, V.R. Regel: Zhurn. Tekhn. Fiz. **25**, 2542 (1955) (in Russian)
2.100 N.Yu. Ikornikova: "Metodika issledovaniya mikrotverdosti korunda" (Procedure for Investigation of Corundum Microhardness), in: *Mikrotverdost'* (Microhardness), ed. by M.M. Khrushchev (Izd. AN SSSR, Moscow 1951), p. 100 (in Russian)
2.101 S.N. Zhurkov: Vestnik AN SSSR, N 3, 6 (1968) (in Russian)
2.102 V.R. Regel, A.I. Slutsker, E.E. Tomashevski: Sov. Phys.-Uspekhi **15**, 45 (1972)
2.103 V.R. Regel, A.I. Slutsker, E.E. Tomashevski: *Kinetichneskaya priroda prochnosti tverdykh tel* (Kinetic Nature of Strength of Solids) (Nauka, Moscow 1974) (in Russian)
2.104 V.L. Indenbom: Pis'ma v Zhurn. Eksper. i Teor. Fiz. **12**, 526 (1970) (in Russian)

Chapter 3

3.1 V.V. Gladkii, I.S. Zheludev: Sov. Phys.-Crystallogr. **10**, 50 (1965)
3.2 V.V. Gladkii, I.S. Zheludev: Sov. Phys.-Crystallogr. **12**, 788 (1967)
3.3 A.V. Shubnikov, I.S. Zheludev, V.P. Konstantinova, I.M. Sil'vestrova: *Issledovaniye piezoelectricheskikh tekstur* (Investigation of Piezoelectric Textures) (Isd-vo AN SSSR, Moscow 1955) (in Russian)
3.4 H.F. Kay, P. Vousden: Phyl. Mag. **40**, 1019 (1949)
3.5 I.S. Zheludev: *Fizika Kristallicheskikh dielectrikov* (Physics of Crystalline Dielectrics) (Nauka, Moscow 1968) (in Russian)
3.6 W.J. Merz: Phys. Rev. **76**, 1221 (1949)
3.7 W.J. Merz: Phys. Rev. **91**, 513 (1953)
3.8 W.J. Merz: J. Appl. Phys. **27**, 938 (1956)
3.9 G.E. Bacon, R.S. Pease: Proc. Roy. Soc. A **220**, 397 (1953)
3.10 B. A. Strukov, S.D. Toshev: Sov. Phys.-Crystallogr. **9**, 349 (1964)
3.11 A. von Arx, W. Bantle: Helv. Phys. Acta **17**, 298 (1944)
3.12 G. Busch: Helv. Phys. Acta **11**, 269 (1938)
3.13 J. Hablutzel: Helv. Phys. Acta **12**, 489 (1939)
3.14 F. Iona, G. Shirane: *Ferroelectric Crystals* (Pergamon, Oxford 1962)
3.15 S. Hoshino, T. Mitsui, F. Iona, R. Pepinsky: Phys. Rev. **107**, 1255 (1957)
3.16 E. Sawaguchi, I. Kittaka: J. Phys. Soc. Japan, **7**, 336 (1957)
3.17 W. Cochren: Adv. Phys. **10**, 401 (1961)
3.18 G. Shirane, R. Nathans, V.F. Minkiewicz: Phys. Rev. **157**, 396 (1967)
3.19 E.M. Lifshitz: Zhur. Eksper. Teor. Fiz. **11**, 255; 269 (1941) (in Russian)
3.20 V.L. Ginzburg: Usp. Fiz. Nauk **38**, 490 (1949) (in Russian)
3.21 I.E. Dzyaloshinskii: Zhur. Eksper. Teor. Fiz. **19**, 960 (1964); Ibid. **20**, 223; 665 (1965)
3.22 A.P. Levanyuk, D.G. Sannikov: Sov. Phys.-Solid State **18**, 235; 1122 (1976)
3.23 V.L. Indenbom: Izv. AN SSSR, Ser. Fiz. **24**, 1180 (1960)

Chapter 4

4.1 S.V. Vonsovski: *Magnetism* (Nauka, Moscow 1971) (in Russian)
4.2 J.S. Smart: *Effective Field Theories of Magnetism* (W.B. Saunders company, Philadelphia-London 1966)
4.3 L.D. Landau, E.M. Lifshitz: Sow. Phys. **8**, 157 (1935)

4.4 H.J. Williams, R.C. Sherwood: "Structures of Ferromagnetic Domains", in: *Magnetic Properties of Metals and Alloys* (Amer. Soc. for Metals, Cleveland, Ohio 1958)
4.5 K. Honda, H. Masumoto: Sci. Rep. Tohoku Imp. Univ. **20**, 323 (1931) (in Japanese)
4.6 R.M. Bozorth: Phys. Rev. **96**, 311 (1954)
4.7 R.M. Bozorth: *Ferromagnetism* (Van Nostrand, Toronto, New York, London 1951)
4.8 W.R. Mason: Phys. Rev. **82**, 715 (1951)
4.9 W.R. Mason: Phys. Rev. **96**, 302 (1954)
4.10 R. Becker, W. Döring: *Ferromagnetismus* (Springer-Verlag, Berlin 1939)
4.11 N.S. Akulov. *Ferromagnetism* (ONTI, Moscow 1939) (in Russian)
4.12 J. Smit, H.P.J. Wijn: *Ferrites* (Philips' Technical Library 1959)
4.13 S. Krupička: *Physik der Ferrite und der verwandten magnetischen Oxide* (Prag 1973)
4.14 I.E. Dzyalosninsky: Soviet Physics-JETP **5**, 1259 (1957)
4.15 R. Pauthenet, Y. Barnier, G. Rimet: J. Phys. Soc. Japan **17**, suppl. B-1, 309 (1962)
4.16 A.S. Borovik-Romanov: Sov. Physics-JETP **9**, 1390 (1959)
4.17 D.N. Astrov: Sov. Physics-JETP, **11**, 708 (1960)

Chapter 5

5.1 J.C. Slater: *Quantum Theory of Molecules and Solids, Vol. 3. Insulators, Semiconductors, and Metals* (McGraw Hill, New York 1967)
5.2 C.S.M. Rooymans: *Structural Investigations on Some Oxides and Other Chalcogenides at Normal and Very High Pressures* (Academic Proetschrift, Amsterdam 1968)
5.3 S.I. Pekar: *Issledovaniya po elektronnoi teorii kristallov* (On Electronic Theory of Crystals) (Gostekhizdat, Moscow, Leningrad 1951) (in Russian)
5.4 E.A. Ukshe, N.G. Bukun: *Tverdye elektrolity* (Solid electrolytes) (Nauka, Moscow 1977) (in Russian)
5.5 R.A. Smith: *Semiconductors* (Academic Press, New York, London 1963)
5.6 V.L. Bonch-Bruevich, S. G. Kalashnikov: *Fizika poluprovodnikov* (Physics of semiconductors) (Nauka, Moscow 1977) (in Russian)
5.7 V.N. Bogomolov: *Ustroistva s datchikami Kholla i datchikami magnitosoprotivleniya* (Devices with Hall and magnetoresistance transducers) (Gosenergoizdat, Moscow, Leningrad 1961) (in Russian)
5.8 A.G. Milnes, D.L. Feucht: *Heterojunctions and Metal-Semiconductor Junctions* (Academic Press, New York, London 1972)
5.9 S.A. Semiletov: *Elektronograficheskoe issledovanie struktury plenok germaniya* (Electron-diffraction study of germanium thin films). Kristallografiya **1**, 542 (1956) (in Russian)
5.10 J.W. Mayer, L. Eriksson, J.A. Davis: *Ion Implantation in Semiconductors* (Academic Press, New York, London 1970)
5.11 T.S. Moss, G.S. Burrell, B.L. Ellis: Semiconductor Optoelectronics (Butterworths, London 1973)
5.12 O.V. Bogdankevich, S.A. Darznek, P.G. Eliseev: *Poluprovodnikovye lazery* (Semiconductor lasers) (Nauka, Moscow 1976) (in Russian)

Chapter 6

6.1 A.C. Beer: *Galvanomagnetic Effects in Semiconductors* (Academic Press, New York 1963)
6.2 E.H. Putley: *The Hall Effect and Related Phenomena* (Butterworth, London 1960)
6.3 I.M. Tsidilkovskii: *Thermomagnetic Effects in Semiconductors* (Academic Press, New York, London 1963)
6.4 R.T. Delves: Thermomagnetic Effects in Semiconductors and Semimetals. Rept. Progr. Phys. **28**, 249 (1965)
6.5 A.A. Abrikosov: *An Introduction to the Theory of Normal Metals* (Academic Press, New York 1972)

6.6 L.D. Landau, E.M. Lifshitz: *Electrodynamics of Continuous Media* (Pergamon Press, London 1960)
6.7 C.S. Smith: Macroscopic Symmetry and Properties of Crystals. Solid State Phys. **6**, 175 (1958)
6.8 L.N. Bulaevski: Superconductivity and Electronic Properties of Layered Compounds. Sov. Phys.-Uspekhi **18**, 514 (1976)
6.9 A.F. Ioffe: *Poluprovodnikovye termoelementy* (Semiconductor Thermoelements) (Izd-vo AN SSSR, Moscow, Leningrad 1960) (in Russian)

Chapter 7

7.1 L.D. Landau, E.M. Lifshitz: *Electrodynamics of Continuous Media* (Pergamon Press, London 1960)
7.2 F. Pockels: *Lehrbuch der Kristalloptik* (Leipzig, Berlin 1906)
7.3 S. Szivessy: *Kristalloptik.* Handbuch der Physik. Bd 20 (Springer, Berlin 1928)
7.4 R.W. Ditchburn: *Light* (Blackie & Son, London 1963)
7.5 R.W. Wood: *Physical Optics* (Macmillan, New York 1934)
7.6 E.S. Larsen, H. Berman: *The Microscopic Determination of the Nonopaque Minerals* 2nd ed. (Washington Gov. Print Off. 1934)
7.7 M. Born, W. Wolff: *Principles of Optics. Electromagnetic Theory of Propagation, Interference and Diffraction of Light* 2nd ed. (Pergamon Press, Oxford, London, Edinburgh, New York, Paris, Frankfurt 1964)
7.8 I.I. Shafranovsky: *Istoriya kristallografii* (The History of Crystallography) (Nauka, Moscow 1978) (in Russian)
7.9 V.A. Kizel': *Otrazheniye sveta* (Light Reflection) (Nauka, Moscow 1973) (in Russian)
7.10 Y.V. Dashevsky: (Zh. Eksp. Teor. Fiz. **8**, 1007 (1938) (in Russian)
7.11 F.I. Fedorov, V.V. Filippov: *Otrazheniye i prelomleniye sveta prozrachnymi kristallami* (Reflection and Refraction of Light by Transparent Crystalls) (Nauka i Tekhnika, Minsk 1976) (in Russian)
7.12 F.I. Fedorov: *Optika anizotropnykh sred* (Optics of Anisotropic Media) (Nauka, Moscow 1958) (in Russian)
7.13 W.A. Shurkliff: *Polarized Light Production and Use* (Harvard Univ. Press, Cambridge, Mass. 1962)
7.14 A. Gerrard, J.M. Burch: *Introduction to Matrix Methods in Optics* (John Wiley & Sons, London, New York, Sydney, Toronto 1975)
7.15 A.V. Shubnikov: *Opticheskaya kristallografiya* (Optical Crystallography) (Izd-vo AN SSSR, Moscow 1950) (in Russian)
7.16 V.S. Sobolev: *Fedorovskii metod* (The Fedorov Method) (Gos. nauch.-tekh. izd-vo lit-ry po geologii i okhrane nedr, Moscow 1954) (in Russian)
7.17 V.N. Lodochnikov: *Osnovy kristallooptiki* (Fundamentals of Crystal Optics) (Gosgeolizdat, Moscow, Leningrad 1947) (in Russian)
7.18 D.S. Beljankin, W.P. Petrov: *Kristalloptik* (Übers. H. Werner, "Technik", Berlin 1954)
7.19 N.M. Melankholin, S.V. Grum-Grzimailo: *Metody issledovaniya opticheskikh sboistv kristallov* (Methods for Investigation of Optical Properties of Crystals) (Izd-vo AN SSSR, Moscow 1954) (in Russian)
7.20 A.M. Goncharenko: Sov. Phys.-Crystallogr. **4**, 688 (1959)
7.21 M. Born: Optik. Ein Lehrbuch der elektromagnetischen Lichttheorie von Dr. Max Born (Springer, Berlin 1933)
7.22 Yu. I. Sirotin, M.P. Shaskol'skaya: *Fundamentals of Crystal Physics* (Mir, Moscow 1982)
7.23 F.I. Fedorov, A.F. Konstantinova: Optics & Spectrosc. **12**, 223 (1962)
7.24 *Tekhnicheskaya entsiklopediya* (Technical Encyclopedia) (Sov. entsiklopediya, OGIZ, Moscow 1932) (in Russian)
7.25 A.F. Konstantinova, N.R. Ivanov, B.N. Grechushnikov: Sov. Phys.-Crystallogr. **14**, 222 (1969)
7.26 M.V. Hobden: Acta Crystallogr. **A24**, 676 (1968)

7.27 F.I. Fedorov, B.V. Bokut', A.F. Konstantinova: Sov. Phys.-Crystallogr. **7**, 738 (1962)
7.28 F.I. Fedorov: *Theoriya girotropii* (Gyrotropy Theory) (Nauka, Moscow 1976) (in Russian)
7.29 B.V. Bokut', A.N. Serdyukov, F.I. Fedorov: Sov. Phys.-Crystallogr. **15**, 871 (1970)
7.30 V.M. Agranovich, V.L. Ginzburg: *Spatial Dispersion in Crystal Optics and Theory of Excitons* (John Wiley & Sons, London, New York, Sydney 1966)
7.31 V.A. Kizel', Y.J. Krasilov, V.I. Burkov: Uspekhi fiz. nauk **114**, 295 (1974) (in Russian)
7.32 I.S. Zheludev: *Elektricheskiye kristally* (Electrical Crystals) (Nauka, Moscow 1969) (in Russian)
7.33 A.S. Sonin, A.S. Vasilevskaya: *Elektroopticheskiye kristally* (Electrooptical Crystals) (Atomizdat, Moscow 1971) (in Russian)
7.34 T.R. Sliker: JOSA, **54**, 1348 (1964)
7.35 J.F. Nye: Physical Properties of Crystals (Clarendon Press, Oxford 1964)
7.36 V.A. Shamburov: Sov. Phys.-Crystallogr. **7**, 476 (1962)
7.37 I.P. Kaminow, E.N. Turner: App. Optics **5**, 1612 (1966)
7.38 I.S. Zheludev, O.G. Vlokh: Sov. Phys.-Crystallogr. **5**, 368 (1960)
7.39 O.G. Vlokh, T.D. Krushel'nitskaya: Sov. Phys.-Crystallogr. **15**, 504 (1970)
7.40 O.G. Vlokh, I.S. Zheludev, I.M. Klimov: Dokl. AN SSSR **223**, 1391 (1975) (in Russian)
7.41 I.S. Zheludev: Soc. Phys.-Crystallogr. **9**, 418 (1964)
7.42 E.R. Mustel', V.N. Parygin: *Metody modulyatsii i skanirovaniya sveta* (Methods for Modulation and Scanning of Light) (Nauka, Moscow 1970) (in Russian)
7.43 M.M. Frocht: *Photoelasticity, Vols. 1–2* (John Wiley & Sons, New York 1941, 1948)
7.44 Yu.V. Pisarevskii, I.M. Sil'vestrova: Sov. Phys.-Crystallogr. **18**, 630 (1974)
7.45 A. Korpel: "Acusto-optics", in: *Applied Solid State Science,* Vol. 3, ed. by R. Wolfe (Academic Press, New York 1972), p. 72
7.46 V. Chandrasekhar: Proc. Ind. Acad. Sci. **A33**, 183 (1951)
7.47 M.V. Vol'kenshtein: *Molekularnaya optika* (Molecular Optics) (Gostekhizdat, Moscow, Leningrad 1951) (in Russian)
7.48 I.R. Fabelinskii: *Molekulyarnoye rasseivaniye sveta* (Molecular Light Scattering) (Nauka, Moscow 1965) (in Russian)
7.49 G.P. Motulevich: "Molekulyarnoye rasseyaniye v kristallakh" (Molecular Scattering in Crystals), in: *Trudy FIAN* (Works of the Physical Institute of the USSR Acad. Sci.) Vol. 5, (1950) p. 9 (in Russian)
7.50 V.L. Ginzburg: Zhurn. Eksper. i Teor. Fiz. **10**, 601 (1940) (in Russian)
7.51 V.V. Vladimirsky: Dokl. AN SSSR **31**, 866 (1941) (in Russian)
7.52 D.F. Nelson, M. Lax: Phys. Rev. Letters **24**, 379 (1970)
7.53 D.F. Nelson, P.D. Lazay: Phys. Rev. Letters **25**, 1187 (1970)
7.54 O.V. Kachalov: Zhurn. Eksper. i Teor. Fiz. **61**, 1352 (1971) (English transl.: Sov. Phys.-JETP **34**, 719 (1972))
7.55 N. Blombergen: *Nonlinear Optics. A Lecture Note* (Benjamin, New York, Amsterdam 1965)
7.56 S.A. Akhmanov, R.V. Khokhlov: *Problemy nelineinoi optiki* (Problems of Nonlinear Optics) (All-Union Inst. of Scientific & Technical Information, Moscow 1964) (in Russian)
7.57 D.A. Kleinman: Phys. Rev. **128**, 1761 (1962)
7.58 M.P. Golovei, I.N. Kalinkina, G.I. Kosourov: Optics Spectrosc. **28**, 535 (1970)
7.59 S.K. Kurtz, T.T. Perry: J. Appl. Phys. **39**, 3798 (1968)
7.60 F. Zernike, J. Midwinter: *Applied Nonlinear Optics* (Academic Press, New York 1973)
7.61 C.I. Ballhausen: *Introduction to Ligand Field Theory* (McGraw Hill, New York 1962)
7.62 H. Bethe: Ann. Phys. **3**, 133 (1929)
7.63 M. T. Hutchings: Solid State Phys. **16**, 227 (1964)
7.64 A. Abragam, B. Bleaney: *Electron Paramagnetic Resonance of Transition Ions* (Clarendon Press, Oxford 1970)
7.65 E.U. Condon, G.H. Shortley: *The Theory of Atomic Spectra* (Reprint. J. Cambridge Univ. Press, Cambridge 18957)
7.66 S.A. Altshuler, B.M. Kozyrev: *Elektronnyi paramagnitnyi rezonans soyedinenii elementov promezhutochnykh grupp* (Electron Paramagnetic Resonance of Compounds of Intermediate-Group Elements) (Nauka, Moscow 1972) (in Russian)

7.67 S.V. Vonsovsky, S.V. Grum-Grzimailo, V.I. Cherepanov et al.: *Teoriya kristallicheskogo polya i opticheskiye spektry primesnykh ionov s nezapolnennoi d-obolochkoi* (Nauka, Moscow 1969) (in Russian)

7.68 S. Sugano, Y. Tanabe, H. Kamimura: *Multiplets of Transition Metal Ions in Crystals* (Academic Press, New York, London 1970)

7.69 D.T. Sviridov, R.K. Sviridova, Yu.F. Smirnov: *Opticheskiye spektry ionov perekhodnykh metallov v kristallakh* (Optical Spectra of Ions of Transition Metals in Crystals) (Nauka, Moscow 1976) (in Russian)

7.70 D.T. Sviridov, Yu.F. Smirnov: *Teoriya opticheskikh spektrov ionov perekhodnykh metallov* (Theory of Optical Spectra of Ions of Transition Metals) (Nauka, Moscow 1977) (in Russian)

7.71 T.F. Veremeichik, B.N. Grechushnikov, I.N. Kalinkina, D.T. Sviridov: Zh. Prikl. Spektrosk. **26**, 131 (1977) (in Russian)

7.72 A.D. Liehr, C.I. Ballhausen: Ann. Physics **6**, 134 (1959)

7.73 H.L. Schläfer, G. Glieman: *Einführung in die Ligandenfeldtheorie* (Akad. Verlag, Frankfurt am Main 1967)

7.74 A.L. Mikaelyan, M.L. Ter-Mikaelyan, J.G. Turkov: *Opticheskiye generatory na tvyordom tele* (Optical Generators on a Solid) (Sov. Radio, Moscow 1967) (in Russian)

7.75 T.H. Maiman: Nature **187**, 493 (1960)

7.76 B.K. Sevast'yanov, Kh. S. Bagdasarov, L.B. Pasternak, S.Yu. Volkov, V.P. Orekhova: JETP Letters **17**, 47 (1973)

7.77 Z.J. Kiss, R.C. Duncan: Proc. IRE **50**, 1531 (1962)

7.78 A.A. Kaminskii: *Laser Crystals, Their Physics and Properties* (Springer, Berlin, Heidelberg, New York 1981)

7.79 Z.J. Kiss, R.C. Duncan: Appl. Phys. Letters **5**, 200 (1964)

7.80 P.P. Feofilov: *Polyarizovannaya luminestsentsiya atomov, molekul i kristallov* (Polarised Luminescence of Atoms, Molecules and Crystals) (Fizmatgiz, Moscow 1959) (in Russian)

7.81 A.A. Kaplyanskii: Optika i spektrosk. **7**, 677 (1959) (in Russian)

7.82 W. Low: *Paramagnetic Resonance in Solids.* Solid State Physics, Supplement 2 (Academic Press, New York, London 1960)

7.83 A. Abragam, M.H.I. Pryce: Proc. Roy. Soc. **A205**, 135 (1951)

7.84 S. Geschwind, J.R. Remeika: Phys. Rev. **122**, 757 (1961)

7.85 V.F. Koryagin, B.N. Grechushnikov: "Spektr EPR iona Mn^{2+} v kristallakh Zn, Na-metasilikata" (EPR Spectrum of Mn^{2+} Ion in Crystals of Zn, Na-metalsilicate), in: *Spektroskopiya kristallov* (Crystal Spectroscopy), ed. by S.V. Grum-Grzimailo (Nauka, Moscow 1970), p. 323 (in Russian)

Chapter 8

8.1 P.G. de Gennes: *The Physics of Liquid Crystals* (Clarendon Press, Oxford 1974)

8.2 L.D. Landau, E.M. Lifshitz: *Statistical Physics,* Vol. 1, 3rd ed. (Pergamon Press, Oxford 1980)

8.3 G.W. Gray: *Molecular Structure and the Properties of Liquid Crystals* (Academic Press, London, New York 1962)

8.4 I.G. Chistyakov: *Zhidkiye kristally* (Liquid Crystals) (Nauka, Moscow 1966) (in Russian)

8.5 I.G. Chistyakov: *Zhidkiye kristally* (Liquid Crystals) Usp. Fiz. Nauk **98**, 563 (1966) (in Russian)

8.6 M.J. Stephen, J.P. Straley: Physics of Liquid Crystals, Rev. Modern Phys. **46**, 617 (1974)

8.7 G.W. Gray, P.A. Winston: *Liquid Crystals and Plastic Crystals* (Oxford Univ. Press, Oxford 1974)

8.8 B.K. Vainshtein, I.G. Chistyakov: "Simmetriya, struktura i svoistva zhidkikh kristallov" (Symmetry, Structure and Properties of Liquid Crystals), in: *Problemy sovremennoi kristallografii* (Problems of Modern Crystallography), ed. by B.K. Vainshtein, A.A. Chernov (Nauka, Moscow 1975), pp. 12–26 (in Russian)

8.9 G.H. Brown (ed.): *Advances in Liquid Crystals* (Academic Press, London, New York) Vol. 1 (1975), Vol. 2 (1976), Vol. 3 (1978)
8.10 J.F. Johnson, R.S. Porter (eds.): *Liquid Crystals and Ordered Fluids,* Vol. 1 (1976), Vol. 2 (1977), Vol. 3 (1978) (Plenum, New York, London)
8.11 A.P. Kapustin: *Eksperimental'nyie issledovaniya zhidkikh kristallov* (Experimental Investigations of Liquid Crystals) (Nauka, Moscow 1978) (in Russian)
8.12 N.A. Plate, V.P. Shibayev: Grebneobraznyie polimery i zhidkiye kristally (Comblike Polymers and Liquid Crystals) (Khimiya, Moscow 1980) (in Russian)
8.13 L.M. Blinov: *Elektro- i magnitooptika zhidkikh kristallov* (Electro- and Magnetooptics of Liquid Crystals) (Nauka, Moscow 1978) (in Russian)
8.14 S.A. Pikin: *Strukturnyie prevrashcheniya v zhidkikh kristallakh* (Structural Transformations in Liquid Crystals) (Nauka, Moscow 1981) (in Russian)
8.15 S. Chandrasekhar: *Liquid Crystals* (Cambridge Monographs on Physics, Cambridge 1977)
8.16 S.A. Pikin, V.L. Indenbom: Thermodynamic States and Symmetry of Liquid Crystals, Usp. Fiz. Nauk **125**, 251 (1978) English transl.: Sov. Phys.-Uspekhi **21**, 487 (1978)
8.17 V.N. Tsvetkov: Magnitnaya vospriimchivost' zhidkikh kristallov (Magnetic Susceptibility of Liquid Crystals) Zhurn. Eksper. i Teor. Fiz. **9**, 602 (1939) (in Russian)
8.18 L.K. Vistin': Novoye elektrostrukturnoye yavleniye v zhidkikh kristallakh nematicheskogo tipa (New Electrostructural Phenomenon in Nematic – Type Liquid Crystals), Dokl. AN SSSR **194**, 1318 – 1325 (1970) (in Russian)

Subject Index

GPSR Compliance

The European Union's (EU) General Product Safety Regulation (GPSR) is a set of rules that requires consumer products to be safe and our obligations to ensure this.

If you have any concerns about our products, you can contact us on ProductSafety@springernature.com

In case Publisher is established outside the EU, the EU authorized representative is:

Springer Nature Customer Service Center GmbH
Europaplatz 3
69115 Heidelberg, Germany

Batch number: 10761599

Printed by Printforce, the Netherlands